Green Energy and Technology

More information about this series at http://www.springer.com/series/8059

Roland Schmehl

Editor

Airborne Wind Energy

Advances in Technology Development
and Research

 Springer

Editor
Roland Schmehl
Faculty of Aerospace Engineering
Delft University of Technology
Delft
The Netherlands

ISSN 1865-3529   ISSN 1865-3537   (electronic)
Green Energy and Technology
ISBN 978-981-13-3824-3   ISBN 978-981-10-1947-0   (eBook)
https://doi.org/10.1007/978-981-10-1947-0

Printed on acid-free paper

This Springer imprint is published by the registered company Springer Nature Singapore Pte Ltd. part of Springer Nature
The registered company address is: 152 Beach Road, #21-01/04 Gateway East, Singapore 189721, Singapore

To Wubbo Ockels, 1946-2014
Pioneer of airborne wind energy

Foreword

We live in a strange world.

On the one hand, we all want to save energy: the average world temperature is rising, with one record year following the other; almost all scientists and politicians agree that greenhouse gases, notably CO_2, are the main cause of this change; they agree that this change will continue to cause significant damage to the livelihoods of hundreds of millions of people, and that the only way to limit climate change is to reduce the world's greenhouse gas emissions; most of the world's governments just committed themselves in Paris to the ambitious aim to limit global warming to less than two degrees Celsius; in summary, nearly everyone agrees that the use of fossil fuels needs to be significantly reduced.

On the other hand, we consume more fossil fuels than ever: the world has the lowest prices for fossil fuels since long; oil costs only 50 dollar per barrel, about 30 cents per liter; and western politicians and most of their electorates are happy about these low prices; the US-led shale gas revolution helps to limit the costs of fossil energy for decades to come; the world oil consumption and the world coal consumption continue to grow; in summary, the greenhouse gas emissions have never been higher than today, and continue to grow year after year.

Most of the authors of this volume, and many other people concerned with the world's climate, including myself, have a dream: we want to live in a world where fossil fuels are mostly replaced by renewable energy sources.

I believe that this dream can become reality. We experience already that solar and wind power installations grow year after year. They start to become competitive at many places, thanks to government support and technological advances. Already today, the yearly electricity output of wind and solar power installations exceeds the output of about one hundred nuclear power plants worldwide, providing the equivalent total energy needs of about 20 million people with European lifestyle. Unfortunately, much more investment in renewable energy sources is needed to replace most of fossil energy production. This investment into renewable energy sources will only happen under two conditions:

- First, renewable energy sources must become available in even larger quantities and at a reasonable price for both the economy and the environment. Here, research on new and even more competitive renewable energy technologies can play an important role, including storage and distribution technologies, and research non new and possibly disruptive new concepts. Airborne wind energy, that could tap into resources not commercially available with conventional wind turbine technology, could become a game changer, if successful. The present book is both, a motivation for airborne wind energy, as well as a testimony of the advancements that have been achieved in this young field in the last years. It is a reason for optimism.
- Second, fossil fuels must become more expensive (or forbidden). This point is often forgotten, probably because it is difficult to achieve and thus a reason for pessimism. But without higher prices for fossil fuels, the transition to a carbon free economy can unfortunately not become reality. The low price for oil, notably for ship diesel, was arguably one of the main reasons why one of the most developed airborne wind energy technologies, traction kites for cargo vessels, did not sell well on the market, forcing the company SkySails GmbH to file for bankruptcy at the start of this year. One might hope that this bankruptcy was an exception and will not become the rule, but as a matter of fact, any new energy technology has to compete with fossil fuels, and is bound to loose on a free market without carbon pricing, as I will argue below.

One might hope that OPEC and the other oil producers might one day decide to reduce oil production, such that the oil price will rise again. This would give another short time boost for renewable energies, as in the 1970's and in the previous decade. But unfortunately, one cannot reasonably hope that the world prices for fossil fuels would ever become high enough to render large amounts of renewable energy sources competitive. The reason is simple: the production costs of oil, gas and coal are extremely low at the best locations, and will stay low for decades to come. For example, the production costs for a barrel of oil in Saudi Arabia, with its vast reserves, are below 10 Euro, that is 6 cents per liter. Combined with the ease of storing, transporting and converting them, fossil fuels are simply too attractive to be driven out of the market without external intervention.

What would happen if the world relies more and more on carbon free technologies? Would fossil fuel producers reduce their production accordingly? Unfortunately not. If significantly fewer people than now would need oil, its price would sink close to its technical lower limit, about 6 cents per liter. It is impossible to imagine that this low cost, together with the ease of use, can be beaten by any other energy source. Thus, exactly if renewable energy sources become successful and start to replace fossil fuels, which we all hope, fossil fuels would become so cheap that new renewable energy installations would no longer be economical. The world would consume more and more of the cheap energy, and some of it would be renewable, but the total greenhouse gas emissions would remain high.

There is only one way to solve this dilemma: to put a realistic price on greenhouse gas emissions, in order to internalize the costs that they cause. The cap-and-trade scheme that is tried in the European Union should not serve as an example. By construction, such a scheme does not lead to a fixed and predictable extra price on energy usage, which is needed in order to make long-term investments in energy savings and renewable energy sources attractive.

A better way would be via a "climate protection" or "carbon tax", that puts a fixed price on all equivalent greenhouse gas emissions. Ideally, this tax would be raised at the same rate in most countries of the world, and the raised money could be redistributed equally to each country's population (or, in utopia, equally to the world's population). When countries that raise the tax trade goods with countries that do not raise it, tariffs on imported goods can take their carbon footprint into account and thus correct for undesired market distortions. Air and sea traffic needs special attention, but should be included into the tax scheme. Because production of fossil fuels is more centralized than their consumption, the tax would more easily be levied at the production side, acknowledging the fact that all the carbon contained in fossil fuels will ultimately end in the atmosphere.

The desired result would be that the price of fossil fuels, of all carbon intensive technologies, and of all the goods that they produce, would rise in the zone where the climate protection tax is raised. The tax would automatically ensure that the higher the carbon footprint of a good, the higher would be its price. One could start with a tax level that implicitly amounts to e.g. 30 cents per liter of oil. Important is that no exceptions are made for large consumers of coal, oil or gas, and that the tax is guaranteed to remain in place for at least one decade or more, to ensure predictability.

The higher prices of all carbon intensive technologies would have two major effects: first, they would serve as an incentive to reduce fossil energy consumption where it hurts least and can be done most efficiently. Second, and equally important, it would help making the best carbon free energy technologies competitive. Airborne wind energy could become one of these. A climate protection tax would thus be one of the few taxes that distort the market in a desirable way. Large economic zones with significant carbon emissions such as China, the US, India or the European Union would be ideally suited to start, in the hope that other regions would follow.

In summary, we need to work on two sides in order to make renewable energies and in particular airborne wind energy successful: on technology development as well as on carbon pricing. None of the two sides can be successful without the other.

The present book with its 30 interesting and well-written chapters, for which I want to congratulate authors and editor, is not only a pleasure to read; it is also a testimony that technology development of airborne wind energy advances well, and that many smart people work successfully on topics ranging from system modeling and optimization, the many practical issues related to design and real-world implementation, to the socio-economic implications of airborne wind energy. As a veteran in the field, who did his first tether drag and (unsuccessful) crosswind flight control experiments on motorways and football fields in Hamburg in the early 1990s, who experienced the excitement and secrecy that accompanied the new, patentable ideas in the early 2000s, and who witnessed and enjoyed the emergence of a small, but open research community on airborne wind energy in the early 2010s, I am absolutely delighted to see this research community being as large and productive as it is now, in particular the many protoptypes that are now in successful operation. Many of the authors have met at the Airborne Wind Energy Conference AWEC 2015 in Delft, which created, in addition to this book, a highly recommendable video resource (http://www.awec2015.com) of public presentations on many of the topics covered in the book. I am curious about the further developments in the field and look forward to the upcoming meeting of the airborne wind energy community on 5-6 October 2017, in Freiburg (http://www.awec2017.com). I sincerely hope that one or more of the companies which are now active in the field will have created commercial products soon. Most important, I do hope that renewable energy sources such as airborne wind energy will not continue to be suffocated by the low price of fossil fuels – and that scientists do not forget, and politicians start to implement, the one important ingredient that is missing in today's climate policy, without which the transition to a carbon neutral economy will not happen: a carbon tax.

Keeping this in mind, I wish all readers of this volume pleasure while learning about the recent advances of airborne wind energy!

Freiburg, Germany, June 2016   *Moritz Diehl*

Preface

Dear readers,

This book is about the use of kites or, more generally, tethered aerodynamic lift devices for wind energy generation. Not much more than a decade ago this subject was pursued by only a few visionary pioneers, but it has since become a rapidly evolving field of activity of a global community of scientists, researchers, developers and investors. While this development is clearly motivated by the urge to explore innovative and cost-effective technologies to reduce our dependency on fossil fuels and to aid the transition towards renewable energies, it is the conceptual simplicity and potential of airborne wind energy that exerts a certain fascination.

For conventional wind turbines the tower and foundation transfer the bending moment of the resultant aerodynamic load to the ground. Airborne wind energy systems, on the other hand, are designed as tensile structures and thus require far less material to transfer forces of similar magnitude. As consequence, the system costs can be lower and the environmental footprint can be reduced substantially. Furthermore, the operational altitude can be selected by design and with far less impact on the costs or technical feasibility. This makes it possible to not only adjust the operation dynamically to the available wind resource, but also to access an unexploited large source of energy: wind at higher altitudes.

Clearly, the use of tethered flying devices also entails a number of technical challenges and this has lead to controversial discussions about the economic viability of the technology at large. Because the motion of the devices is only constrained by one or more flexible tethers and, in general, is also inherently unstable, a reliable and robust control is crucial for the commercial use of the technology. It has also become clear that automatic launching and landing will be an enabling technology component, as will be durable and lightweight flexible materials that can sustain a large number of load cycles.

The current research and development activities address these challenges in different ways. Since the publication of the first textbook on Airborne Wind Energy in October 2013, the key industry players have advanced rapidly with building next-generation prototypes. Following the acquisition by Google in 2013, the team of

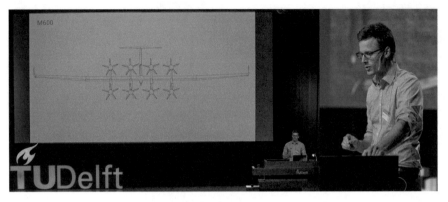

Fig. 1 Damon Vander Lind presenting the M600 energy kite at the AWEC 2015
(http://www.awec2015.com/presentations)

Makani Power has developed a 600 kW energy kite (see Fig. 1) and is testing this flying machine with an impressive 30 m wingspan and eight onboard wind turbines in the vicinity of San Francisco. Having grown to 40 employees, Ampyx Power has developed two rigid wing aircraft prototypes and registered these with the aviation authorities as aircraft. EnerKíte, TwingTec and Kitemill are testing rigid wing prototypes, Kitepower a soft wing prototype with automatic launching and landing capability while Kite Power Systems is testing an implementation which uses two separate soft kites that operate on the same generator. The "IG Flugwind" was founded in Germany as an airborne wind energy interest group and now includes most European system developers. The group has started to systematically approach the regulation and certification of airborne wind energy systems, with the aim to define safety standards for the operation of the system and the interaction with other users of the air space or ground surface.

These activities indicate that the investment climate for the commercial development of innovative wind energy solutions is improving steadily. Notable recent governmental funding of commercial activities has come from the ARPA-E (Makani Power) and SBIR (eWind Solutions, WindLift and Altaeros Energies) programs of the US government, in Europe from the SME Instrument (Ampyx Power) and the Fast Track to Innovation pilot (industrial/academic consortium with Kitepower as central partner) and from several national governments. In Germany, for example, the funding program ZIM for small and medium-sized enterprises (High Altitude Wind Network HWN500) and the projects OnKites I and II (Fraunhofer Institute). In my opinion, the increasing maturity level of the technology has a distinctly positive effect on the success rate of grant applications. The resulting global map of academic and industrial contributors is shown in Fig. 2.

Recent academic key contributions are the ERC project Highwind (Moritz Diehl) and the Swiss network project A2WE, both focusing on the control aspects of airborne wind energy, as well as the European Training Network AWESCO. The latter combines 14 PhD projects at eight universities and four industry partners into a truly multidisciplinary approach which covers four central themes: (1) Modeling

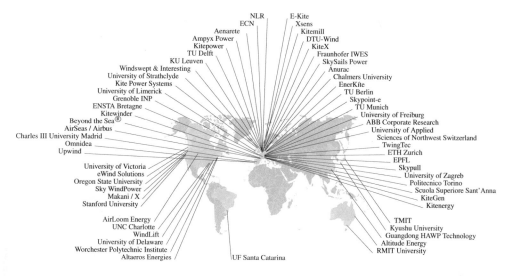

Fig. 2 Airborne wind energy research and development activities in 2017

and Simulation, (2) System Design and Optimization, (3) Sensors and Estimation and (4) Control Systems. Airborne wind energy is also an increasingly present topic at conferences and wind energy conventions. To name some examples: the major international control conferences regularly feature invited tracks on control aspects of the new technology and, on the other hand, the key industry players provide now frequent updates on their development status at the conferences of the European Wind Energy Association (EWEA) and corresponding national events. The increasing recognition of the new technology is also reflected in the fact that the TU Delft Wind Energy Institute (DUWIND) includes airborne wind energy in the R&D program 2015-2020 as one of its five research lines on conversion system level. This is in my opinion a notable achievement because airborne wind energy has so far been developed quite independently of conventional wind energy.

But lastly, it is the annual Airborne Wind Energy Conference (AWEC) which connects the global community of scientists, researchers, developers and investors and which makes the field of activity a very special one. This conference and exhibit was held in 2009 for the first time and in June 2015 I had the honor and opportunity to organize a very successful event with more than 200 international participants in Delft (see Fig. 3). The resulting 54 oral presentations, 17 poster presentations, the screening of a unique documentary movie and the many discussions were not only inspiring and motivating but also an indicator for a systematic advancement of the technology development. It was this particular event and the success of the first textbook (almost 50,000 chapter downloads in two years) which triggered the idea

Fig. 3 Group photo at the AWEC 2015 in Delft (16 June 2015)

to produce the present, second textbook on airborne wind energy. Following the call for chapter contributions in August 2015, 33 manuscripts were received in total, of which 28 were based on conference presentations and 5 were new contributions. Because of this dominant proportion of conference material the present book can factually be regarded as scientifically expanded conference proceedings, however of a selected subset of the presented material. Each manuscript was peer-reviewed by at least two and up to five anonymous reviewers and improved within two to three consecutive review iterations. The names of the 96 reviewers are listed on the following pages xv–xvii and I would like to express my appreciation for their fast, competent and constructive feedback. Based on the recommendations I accepted 30 chapters for publication which are divided into five parts.

Following an introductory chapter on "Emergence and Economic Dimension of Airborne Wind Energy" is Part I on "Fundamentals, Modeling & Simulation" which contains 7 contributions that describe fundamental aspects of the technology, quasi-steady as well as dynamic models and simulations of airborne wind energy systems or individual components. Part II on "Control, Optimization & Flight State Measurement" combines 5 chapters on control of kite and ground station, 3 chapters on system optimization and 1 chapter on flight state measurement. Part III on "Concept Design & Analysis" comprises 5 chapters presenting and analyzing novel launching and landing concepts as well as novel energy harvesting concepts. Part IV on "Implemented Concepts" contains 4 chapters about established system concepts. The final Part V on "Technology Deployment" comprises 4 chapters on various aspects that are relevant for the commercial deployment of the technology.

I hope that the present book can contribute to the discussion by providing scientific evidence for the technical feasibility of the innovative technology and its economic potential.

Delft, Netherlands, July 2017   *Roland Schmehl*

Acknowledgements

This book would not have been possible without the support and engagement of reviewers selected from the international scientific community. The quality of any peer-reviewed scientific book largely depends on the willingness of the reviewers to share their expertise and knowledge with colleagues from all over the world. As a minor token of the editors' appreciation for their diligence and work, the names of all reviewers for this book are listed hereafter:

Eva Ahbe, Swiss Federal Institute of Technology Zurich, Switzerland
Hammad Ahmad, GE Power & Water, U.S.A.
William Anderson, Updraft Technologies, Netherlands
Damian Aregger, University of Applied Sciences Northwestern Switzerland
Florian Bauer, Technical University of Munich, Germany
Wim Bierbooms, Delft University of Technology, Netherlands
Nedeleg Bigi, ENSTA Bretagne, France
Alexander Bormann, Enerkite GmbH, Germany
Jaap Bosch, Ampyx Power B.V., Netherlands
Joep Breuer, Airborne Technology Centre B.V., Netherlands
Merlin Bungard, Anurac, Germany
Lode Carnel, KiteMill AS, Norway
Antonello Cherubini, Sant'Anna School of Advanced Studies, Italy
Roger Coenen, Delft University of Technology, Netherlands
Joseph Coleman, University of Limerick, Ireland
Mirko Čorić, University of Zagreb, Croatia
Dino Costa, TwingTec AG, Switzerland
Sean Costello, Swiss Federal Institute of Technology Lausanne, Switzerland
Piyush Jadhav D, Indian Institute of Technology Madras, India
Marcelo De Lellis, Federal University of Santa Catarina, Brazil
Bertrand Delprat, Calidris SARL, France
Moritz Diehl, University of Freiburg, Germany
Sjoerd Dirksen, Sjoerd Dirksen Ecology, Netherlands
Matthew Doe, European Space Agency (ESA), Netherlands

Storm Dunker, A-Z Chuteworks LLC, U.S.A.
Hisham Eldeeb, Technical University of Munich, Germany
Michael Erhard, SkySails GmbH, Germany
Pietro Faggiani, Delft University of Technology, Netherlands
Lorenzo Fagiano, ABB Switzerland Ltd, Switzerland
Uwe Fechner, Delft University of Technology, Netherlands
Carlos Simão Ferreira, Delft University of Technology, Netherlands
Mikko Folkersma, Delft University of Technology, Netherlands
Marco Fontana, Sant'Anna School of Advanced Studies, Italy
Falko Fritz, SkySails GmbH, Germany
Allister Furey, Kite Power Systems Ltd, Great Britain
Cédric Galliot, TwingTec AG, Switzerland
Pieter Gebraad, National Renewable Energy Laboratory, U.S.A.
Fabian Girrbach, Xsens Technologies B.V., Netherlands
Ben Glass, Altaeros Energies, U.S.A.
Flavio Gohl, TwingTec AG, Switzerland
Leo Goldstein, Airborne Wind Energy Labs, U.S.A.
Thomas Haas, University of Leuven (KU Leuven), Belgium
Ahmad Hably, Grenoble Institute of Technology, France
Christoph M. Hackl, Technical University of Munich, Germany
Thomas Hårklau, KiteMill AS, Norway
Stefan Haug, Mayser GmbH & Co. KG, Germany
Henrik Hesse, Swiss Federal Institute of Technology Zurich, Switzerland
Jeroen Hol, Xsens Technologies B.V., Netherlands
Max ter Horst, E-Kite B.V., Netherlands
Corey Houle, University of Applied Sciences Northwestern Switzerland
Jan Hummel, Technical University of Berlin, Germany
Thomas Jann, German Aerospace Center (DLR), Germany
Claudius Jehle, Fraunhofer Institute for Transportation & Infrastructure Systems, Germany
Colin Jones, Swiss Federal Institute of Technology Lausanne, Switzerland
Linda Kamp, Delft University of Technology, Netherlands
Axel Kilian, Princeton University, USA
Jonas Koenemann, Ampyx Power B.V., Netherlands
Michiel Kruijff, Ampyx Power B.V., Netherlands
Gijs van Kuik, Delft University of Technology, Netherlands
Patrick Lauffs, Technical University of Munich, Germany
Richard Leloup, Océa S.A./beyond the sea, France
Jean-Baptiste Leroux, ENSTA Bretagne, France
Rachel Leuthold, University of Freiburg, Germany
Haocheng Li, Worcester Polytechnic Institute, U.S.A.
Rolf Luchsinger, TwingTec AG, Switzerland
Elena C. Malz, Chalmers University of Technology, Sweden
Prabu Sai Manoj Mandru, Delft University of Technology, Netherlands
Johan Meyers, University of Leuven, Belgium
Milan Milutinovic, University of Zagreb, Croatia

Mark D. Moore, National Aeronautics and Space Administration (NASA), U.S.A.
Alain Nême, ENSTA Bretagne, France
René van Paassen, Delft University of Technolgy, Netherlands
Navi Rajan, Delft University of Technolgy, Netherlands
Maximilian Ranneberg, Enerkite GmbH, Germany
Volkan Salma, Delft University of Technolgy, Netherlands
Ramiro Saraiva, Federal University of Santa Catarina, Brazil
Stephan Schnez, ABB Switzerland Ltd, Switzerland
Jochem De Schutter, University of Freiburg, Germany
Coert Smeenk, E-Kite B.V., Netherlands
Garrett Smith, Wind Fisher S.A.S., France
Alain de Solminihac, ENSTA Bretagne, France
Bernd Specht, SkySails GmbH, Germany
Andy Stough, Windlift, U.S.A.
Matthias Stripf, Karlsruhe University of Applied Sciences, Germany
Paul Thedens, SkySails GmbH, Germany
Paolo Tiso, Swiss Federal Institute of Technology Zurich, Switzerland
Alexandre Trofino, Federal University of Santa Catarina, Brazil
Jos Vankan, National Aerospace Laboratory (NLR), Netherlands
Christopher Vermillion, University of North Carolina, U.S.A.
Axelle Viré, Delft University of Technolgy, Netherlands
Stefan Wilhelm, Airborne Wind Energy Network HWN 500, Germany
Paul Williams, Ampyx Power B.V., Australia
Andrea Zanelli, University of Freiburg, Germany
Mario Zanon, Chalmers University of Technology, Sweden
Dimitrios Zarouchas, Delft University of Technolgy, Netherlands
Aldo Zgraggen, Swiss Federal Institute of Technology Zurich, Switzerland

Also appreciated is the contribution of the editorial assistants, Roger Coenen,
Tommy Smits and Prabu Sai Manoj M, all from Delft University of Technology.
The financial support of the European Commission through the doctoral training
network AWESCO (H2020-ITN-642682) and the "Fast Track to Innovation" project
REACH (H2020-FTIPilot-691173) is gratefully acknowledged.

Contents

Part II Control, Optimization & Flight State Measurement

Part III Concept Design & Analysis

Editor Biography

Roland Schmehl is Associate Professor in the Section of Wind Energy at the Faculty of Aerospace Engineering of Delft University of Technology. He is head of the Kitepower Research Group and co-founder of the startup company Kitepower. He is coordinator of the Marie Skłodowska-Curie Initial Training Network AWESCO (Airborne Wind Energy System Modelling, Control and Optimisation) which addresses key challenges of airborne wind energy technologies. The multidisciplinary training network includes 14 PhD researchers at 12 European consortium partners and is funded by the European Union's Horizon 2020 Framework Programme and the Swiss federal government. He further coordinates the Horizon 2020 "Fast Track to Innovation" project REACH (Resource Efficient Automatic Conversion of High-Altitude Wind) which aims at commercially developing a 100 kW mobile kite power system based on softwing technology.

He teaches the MSc course "Airborne Wind Energy" and has supervised more than 50 MSc graduation projects on this topic. He has been co-promotor, defense committee member or external assessor of 7 completed PhD dissertations. He regularly tutors student teams in a Design Synthesis Exercise which is the final highlight of the Aerospace Engineering BSc curriculum. He is author of more than 80 scientific publications in various fields of science and engineering. He co-edited the pioneering book "Airborne Wind Energy", published in 2013 with Springer, comprising 36 contributed peer-reviewed chapters. He was speaker at the TEDxDelft in 2012, organiser of the 6th International Airborne Wind Energy Conference (AWEC) 2015 in Delft and a co-organizer of the AWEC 2017 in Freiburg, Germany. The two conferences attracted each more than 200 international participants from industry and academia. Video recordings of the presentations, the posters and the illustrated books of abstracts are available online through open access.

He graduated in 1994 with an MSc degree in Mechanical Engineering from Karlsruhe Institute of Technology, where he also completed his research on the computational modeling of liquid droplet dynamics and liquid fuel preparation for jet engines with a PhD degree cum laude. In 2002 he was awarded a post-doctoral research fellowship at the European Space Research and Technology Centre (ESTEC) in the Netherlands. In this function and later also as consultant, he was involved in the accident analysis of the Ariane 5 upper stage propulsion system (AESTUS), the reentry analysis of the Automated Transfer Vehicle (ATV) and in other studies on propulsion and life support systems. In 2005 he started as Software Architect at TNO Automotive Safety Solutions in Delft, developing a Computational Fluid Dynamics module for the simulation of airbag deployment. He returned to academia in 2009, combining his multidisciplinary industry experience with applied research and education to develop and commercialize a challenging innovative wind energy technology.

Nomenclature

a	acceleration [m/s^2]
A	surface area [m^2]
$A\!R$	aspect ratio
b	wing span [m]
c_a	availability factor
c_f	capacity factor
B	magnetic field [mGauss]
C_D	aerodynamic drag coefficient
C_L	aerodynamic lift coefficient
C_M	aerodynamic moment coefficient
C_p	power coefficient
CF	crest factor
d	diameter [m]
D	duty cycle
D or F_D	aerodynamic drag force [N]
E	energy [J]
E	elastic modulus [N/m^2]
f	frequency [1/s]
f	reeling factor
$\mathbf{F}_a$	resultant aerodynamic force [N]
F_D or D	aerodynamic drag force [N]
F_L or L	aerodynamic lift force [N]
$\mathbf{g}$	gravitational acceleration [m/s^2]
h	altitude above ground [m]
I	electrical current [A]
I	moment of inertia [kg m^2]
L	power losses [W]
L or F_L	aerodynamic lift force [N]
l	length [m]
$\mathbf{M}_a$	aerodynamic moment [Nm]
m	mass [kg]

n	normal vector
p	static pressure [N/m^2]
P	power [W]
r	radius [m]
r	position [m]
S	surface area [m^2]
S	safety factor
t	time [s]
T	temperature [K]
T or F_t	tether force [N]
u	control vector
U	electrical voltage [V]
v	velocity [m/s]
v_a	apparent wind velocity [m/s]
v_w	wind velocity [m/s]
v_∞	freestream or upstream velocity [m/s]
x	state vector
α	angle of attack [rad]
β	elevation angle [rad]
β_s	sideslip angle [rad]
χ	course angle [rad]
γ	flight path angle [rad]
ζ	power factor
η	efficiency
κ	camber
λ	tangential velocity factor, crosswind factor, tip speed ratio
μ	coefficient of viscous friction [Nms]
μ	dynamic viscosity [Ns/m^2]
ν	kinematic viscosity [m^2/s]
ρ	air density [kg/m^3]
τ	torque [Nm]
τ_μ	friction torque [Nm]
ω	angular velocity [rad/s]

Subscripts

a	apparent
c	cycle
e	electrical
f	force
g	ground
i	reel-in
k	kite

m	mechanical
n	normal
p	pumping
o	reel-out
r	radial
t	tether
v	velocity
w	wind
τ	tangential

Coordinates and rotation sets

P, Q, R	roll, pitch, yaw angular velocities [1/s]
r, θ, ϕ	radial distance, polar/elevation angle, azimuthal angle [rad]
x, y, z	Cartesian coordinates [m]
ϕ, θ, ψ	roll, pitch, yaw angles [rad]

Chapter 1
Emergence and Economic Dimension of Airborne Wind Energy

Udo Zillmann and Philip Bechtle

Abstract Airborne wind energy has the potential to evolve into a fundamental cornerstone of sustainable electricity generation. In this contextual analysis we discuss why this technology is emerging at this very moment in time and why it has the potential to disrupt the wind energy economy in the short term and the global energy markets in the longer term. We provide an order-of-magnitude estimate of the economic dimension of this scenario. Following this introductory chapter, the current technology status, principles and challenges of designing, building and flying an airborne wind energy device will be discussed.

1.1 A Digital Product Conquers the Air: Drones

Drones will eventually be "as ubiquitous as pigeons", futurist Liam Young recently predicted [10]. Even now, drones are omnipresent. They first belonged to the realm of the military and were unaffordable for anyone else. Today, they have become so inexpensive that hobbyists and even children can afford them. Drones have conquered the extreme ends of the market for technical goods: multimillion dollar military drones as well as low-cost consumer products. Between these two extremes lies the market for commercial drones, which is still largely untapped. It is our aim to change this dramatically. We have presented the subject at the 6th International Airborne Wind Energy Conference (AWEC) 2015 [79] and discussed it in [78].

Udo Zillmann (✉) · Philip Bechtle
Airborne Wind Europe, Avenue de la Renaissance 1, 1000 Brussels, Belgium
e-mail: zillmann@airbornewindeurope.org

Philip Bechtle
Physikalisches Institut, Rheinische Friedrich-Wilhelms-Universität Bonn, Nussallee 12, 53115 Bonn, Germany
e-mail: bechtle@physik.uni-bonn.de

© Springer Nature Singapore Pte Ltd. 2018
R. Schmehl (ed.), *Airborne Wind Energy*, Green Energy
and Technology, https://doi.org/10.1007/978-981-10-1947-0_1

While much has been written and speculated about a huge commercial market for drones, it is quite uncertain to which uses commercial drones might be put. A drone in this sense is every autonomous flying object, and such broadly defined drones can be used for a surprising variety of tasks. Much media attention was paid to Amazons', Google's and DHL's announcement of using delivery drones. The use of drones in surveillance e.g. for detecting fires, cracks in pipelines or illegal wood logging can already be considered a classic use of drones [5, 36]. Drones can also monitor farmland in detail for precision farming. Autonomous solar powered drones can be used to hover at high altitude over an area for months to provide wireless communication similar to a satellite. Facebook and Google have invested in startup companies in this field. Also there is the potential to replace pilot-controlled aircraft for the transportation of humans or goods with drones, as investigated e.g. by NASA [8] or Joby Aviation [48].

But there are other disruptive uses for autonomous flying objects—or drones—of which the current debate is largely unaware.

One example is Elon Musk and his company SpaceX. He is successfully working on routinely landing and later reusing Falcon rockets after they have delivered their payload into space. It is impossible for a pilot to control a precision upright landing of a rocket that literally falls out of the sky with several times the speed of sound. Only cutting-edge drone technology can do this. If the rocket was to be recycled it would lower the flight costs from the cost of building a rocket to the cost of refueling it. That is US\$ 200,000 instead of US\$ 55 million [68].

The business potential for such a "rocket drone" is enormous. And it is clear that once such reusable rockets exist, other rocket manufacturers cannot compete any more on the basis of their non-reusable rockets. Digital drone technology completely disrupts the rocket market. In fact it already has, even before SpaceX managed to reuse a single rocket. In a reaction to a failed landing attempt of SpaceX both western rocket manufacturers, United Launch Alliance and Airbus Defense and Space, announced that they will develop new rocket systems with the capability to land and reuse at least the most costly parts of the rocket [18]. After the first successful landings of Falcon rockets [77] (and others [11]) their commitment to this decision was surely reinforced.

1.2 The Airborne Wind Energy Device: A "Wind Drone"

A similar disruption is to be expected in the wind energy market by AWE devices. It was Miles Loyd who first worked on the idea of airborne wind energy [51] during the energy crises of the late 1970s at Lawrence Livermore National Laboratory. He had the radical idea of building a wind generator without a tower, only using a flying wing connected to the ground by a tether, much like a kite. A sketch of such a device is shown in Fig. 1.1. His basic idea was very simple: the aerodynamically most efficient part of a wind turbine is the outermost area of the blades, since it harvests the wind from the largest effective area compared to its own size.

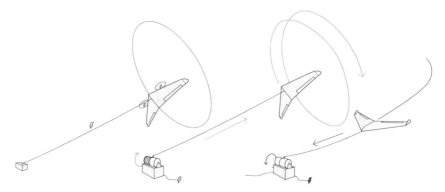

Fig. 1.1 The fundamental idea of airborne wind energy, here exemplified using the most common form of a crosswind device, as imagined by Miles Loyd [51]: use a tethered wing on a tether to replace the blade tips of a wind turbine, generating electricity either in drag mode (left) or in lift mode. For the latter the wing is mostly operated in a pumping cycle, alternating between reel-out of the tether (center) and reel-in (right)

Would it be possible to just keep this area, and do away with all the inefficient rest of the blades, and the huge tower? The simple answer is a crosswind AWE device. If an automatically—that is autonomously—flown kite or tethered aircraft flies perpendicularly to the wind while pulling on the tether, it can either harvest the wind energy via repellers and generators on board (the "drag mode") or via pulling the tether out against the force of a generator on ground (the "lift mode"). For a more detailed introduction into the physics of current implementations of AWE see [21] and other chapters of the 2013 AWE book [3]. This fundamental idea, once it works, opens entirely new realms of wind energy: there is no direct constructive limitation on the area swept by the wind energy device anymore. Instead of optimizing it for efficiency in a given area, as the rotor disk of a wind turbine, it can be optimized for total energy output. Thus it sweeps larger areas than wind turbines, using only a tiny fraction of the material, and with the theoretical potential to go to higher altitudes with less intermittent wind than the area of about 100 to 200 meters above ground accessible to wind turbines.

Miles Loyd calculated the expected energy output of his "flying wind generator". Based on the formula he first established—today known as Loyd's Formula—he found that a tethered aircraft with the size, weight and aerodynamics of one of the largest aircraft of the 1970s could produce 6.7 MW of power. Even larger wings with an output of 45 MW seemed feasible. Loyd filed a patent [53] and published the now famous article [51] on this new technology.

But here the story of Miles Loyd ends. He also wanted to build a flying wind generator but could not find funding. Not because the physics would have been wrong, but because—amongst many challenges—he had no answer to one question: how to control such a flying wing without a pilot?

Today, we have a technology that lets us control flying objects without a pilot. It is called: drones.

If we apply this new technology to Loyd's old formula we can build a new type of drone: we call this the "wind drone".

Miles Loyd explained in his foreword to the 2013 AWE Book that his computer at the time had 64 kByte of RAM [52]. Today, a personal computer in everybody's home usually has 8 to 16 GByte of RAM, so computing power has increased by about one million times, if roughly estimated by RAM size. It was impossible for Miles Loyd to build an airborne wind energy device that could fly the required patterns autonomously, adapt to the wind gusts and changes in direction and to autonomously start and land with the computing power and sensor technology of the 1970s. Only after the turn of the millennium did drone technology become powerful enough that turning Miles Loyd's invention into reality started to become feasible. It is no coincidence that we saw the establishment of the first research teams at Universities at this time and shortly thereafter the establishment of the first generation of AWE companies like Makani Power, Ampyx Power, Kitegen, Enerkite and so many others.

This is also the reason why all AWE devices can be rightfully called "wind drones": No matter whether they are controlled from the ground or from the flying object, or whether they are rigid or soft, or whether they are almost stationary or crosswind devices: all of them are or ultimately will be autonomously controlled flying objects, hence they are drones. Drones on a leash, that is. And their advent is made possible by the same technological revolution which also made autonomous free-flying drones and autonomously landing rocket drones possible: the tremendous increase in the precision of smaller and lighter sensors, and of computing power over the last decades.

But even in the last decade, since the first practical AWE projects started, Moore's law continued to do its work and computing power has since doubled roughly every 2.7 years. Today, a lack of drone and control technology is no longer the main obstacle to AWE that it once was. Today, drone and control technology is the enabler for AWE.

1.3 Airborne Wind Energy in 2017

A lot has happened in the last decade in the area of wind drones [17]. Small-scale wind drone prototypes were built by many research teams and companies and have been flying for many years. Makani, which had been acquired in 2013 by Google and is currently one of the "moonshot" projects of the Alphabet subsidiary X, is the first company to take the next step towards the approximate scale of present commercial wind turbines. It has already built a fully functional demonstrator with 600 kW output, 26 m wingspan and a mass of 2 tons. Illustrated in Fig. 1 and detailed in [56], this demonstrator is currently being tested and will be installed in a wind park in Hawaii [7, 58].

Makani will be first to show a wind drone with power outputs comparable to today's wind turbines. But they are not the only ones who have realized that drone

Fig. 1.2 Overview of major companies and institutions who are stakeholders in airborne wind energy in 2017[a]

---

[a] Investments into startups and/or research conducted by stakeholders companies and institutions:
Rabobank, KLM, Schiphol, WWF, Statkraft: Ampyx Power
Google/Alphabet: Makani
Honeywell: own research, patent, presentation at AWEC 2010
Alstom, Festo, Zürcher Kantonalbank: Twingtec
Sabic: KiteGen
3M: NTS [4]
DSM: Skysails
Softbank, Mitsubishi Heavy Industries: Altaeros Energies [70]
GL Garrad Hassan (now merged with DNV GL): own research [34]
E.ON, Schlumberger, Shell, Scottish Investment Bank: Kite Power Systems [39]
E.ON: own research, poster at EWEA Offshore 2015 [59], cooperation with Ampyx Power [40]
Fraunhofer IWES: own research, workshop 2012 [63], presentations at AWEC 2013 & 2015
NASA: own research [64, 65]
ABB: own research [30, 31]

technology is now ripe to take on Loyd's formula. Several companies, including for example 3M, ABB, Alstom, E.ON, Honeywell, Mitsubishi Heavy Industries, Sabic, Shell, Softbank and Statkraft, have conducted research on wind drones and/or financed one of the dozens of AWE startups worldwide. An overview of these and more is shown in Fig. 1.2, a global listing of airborne wind energy research and development activities is displayed in Fig. 2 of the preface to this book. GE sent a research team to the Airborne Wind Energy Conference 2015 and Siemens is contemplating to partner up with Google Makani [20]. Bill Gates has called AWE the potential "magic solution" for the energy problem and named it one of his best bets for a game-changing energy breakthrough [2, 50].

How can one be confident that wind drones become a success? It is a logical combination of two factors: one is the understanding that drone technology is mature enough to build wind drones. The other is the understanding that the economic

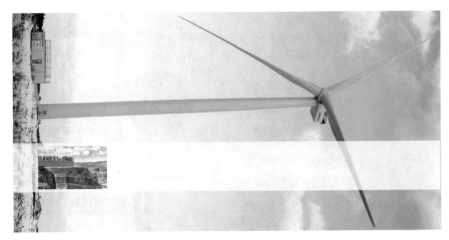

Fig. 1.3 Crazy architecture versus routine in wind turbine design: cantilevering structures. The large photo shows the horizontal axis wind turbine Vestas V164-8.0 MW® installed at the Østerild wind turbine test field, Denmark, with a tip height of 187 m, a hub height of 118 m and a blade length of 80 m [62] (photo credits: Frank Boutrup Schmidt). The photo is rotated by 90° (and mirrored) to illustrate the structural challenge of withstanding the aerodynamic loading of the turbine rotor. The small inset photo, set at approximately the same scale, shows the Skywalk Grand Canyon cantilevering 21 m from the canyon wall (photo credits: Fabrizio Marchese)

advantages of wind drones over wind turbines are so big that once operational, wind drones will outperform wind turbines by a large margin [21].

1.4 The Success Formula of Wind Drones: Replacing Concrete and Steel with Cleverness

It is drone technology that makes building wind drones possible, but it is a mechanical reason that makes them so competitive with wind turbines: wind drones can be built without large lever arms.

If horizontal axis wind turbines are analyzed from a structural perspective they can be seen as a combination of four large lever arms. The tower is a cantilever beam that needs to support the horizontal aerodynamic loading of the entire turbine. Similarly, the three rotor blades need to support the generated aerodynamic forces.[1] Wind energy currently holds the world record in building the longest cantilevering structure, at about 220 m from base of the tower to the tip of the blade [61]. This principle works well due to a tremendous amount of good engineering, but if there is a chance to avoid it, a lot can be gained by omitting all the aerodynamically unnecessary material. A graphical sketch of the lever arm problem of conventional wind is shown in Fig. 1.3.

---

[1] The rotational motion contributes an apparent centrifugal stiffening effect

Wind drones on the other hand, can be built in a way that the tether is almost parallel to the generated aerodynamic force. In addition, in some wind drone designs the tethered wing or kite can be supported by a bridle line system to even reduce lever arms on the wing itself, allowing for very lightweight structures. The mechanical fact that wind drones can be built as tensile structure without such huge lever arms has two important economic implications: first, wind drones can be built in a substantially less massive way than wind turbines and second, wind drones can be built to reach the winds at higher altitudes.

Therefore, the comparison between wind turbines and wind drones is easy: wind turbines achieve mechanical stability with the help of very unfortunate lever arms. Wind drones replace mechanical stability with autonomous cleverness to stay up in the air.

Of course, the drastic change of the operating model of wind energy, from a fixed structure to a dynamically controlled tethered aircraft, comes with a set of challenges and possible downsides. Amongst them is also the question of aviation regulations. Together with the European Aviation Safety Agency (EASA), companies like Ampyx Power are working on addressing these issues. The use of wind drones for wind energy generation is already mentioned in EASA's risk assessments [26].

1.5 An Order-Of-Magnitude Comparison of Capital Costs

The omission of a large lever arm makes it possible to build very lightweight wind drones. This means that wind drones can be expected to be cheaper to build than wind turbines, once large-scale production begins. Half of the total capital costs of wind turbines, which make up the bulk of the total costs of wind energy, are the costs for the massive structural elements, the tower, the blades, the foundation and the rotor hub [44].[2] Within the costs for these components the material costs rather than production costs dominate. The material effort is extremely high: up to 700 tons of steel are used for the tower, another 100 tons of steel for the rotor hub [72], up to 100 tons of glass fiber reinforced plastic for the blades [62], and up to 4000 tons of concrete for the foundation of a single wind turbine.[3] An overview over the cost components of these components is given in Fig. 1.4.

Thanks to the efficient geometry of wind drones, they lack these massive structures. The tower is replaced by a thin tether. A wind drone with the power of the largest currently existing wind turbine (8 MW) requires a tether that is 6 cm thick [21]. Without bending moments only minimal foundations are needed and the wings can be much lighter, requiring only 1 to 10% of the material of the blades of a wind turbine.[4] The Google Makani 600 kW wing weighs 2 tons including the tether and on-board generators [38]. A 600 kW wind turbine weighs between 50 and

---

[2] Additional Operations & Maintenance costs are 20% of the total costs

[3] All data is for the MHI Vestas V164-8.0 MW®, currently the worldwide largest wind turbine.

[4] A detailed explanation of the higher efficiency of the wind drone wings is beyond the scope of this article. See [21] for details.

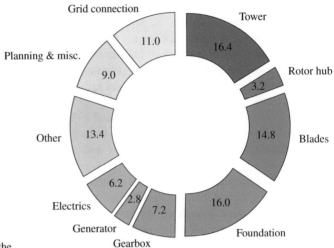

Fig. 1.4 Breakdown of the capital costs of existing wind turbines in percent [45]

100 tons without foundations, while sporting a smaller capacity factor and higher intermittency.

In comparison, the required components for power generation are cheap: the costs for the electricity producing generator amounts to less than 3% of total costs. Certainly, wind drones will need more and better sensors, processors and other control components, but these cost much less than the materials saved. As a result, and in the absence of public prices for grid-scale energy production with a commercial AWE system, this order-of-magnitude comparison suggests that a wind drone can be built for about half the costs of a wind turbine with the same rated power.

1.6 High-Altitude Wind Resource

Since wind drones are not restricted by lever arms, they can fly higher than the hub height of a wind turbine. Wind drones could easily reach altitudes twice to four times as high as normal wind turbine towers, so 300 to 600 m instead of 150 m. On average the wind speed increases with altitude. And higher wind speed means more wind power. Since wind power increases with the cube of the wind speed, doubling the wind speed therefore means that wind power is multiplied by a factor of eight.

The difference the altitude makes can be clearly seen from the wind data measured above central London shown in Fig. 1.5. Even at 120 m altitude, London city is a reasonably good wind resource with an average wind speed of 7.0 m/s. But at 250 m altitude, this figure increases to 9.3 m/s. Due to the cubic relationship, wind

Fig. 1.5 Mean wind speed and wind power density profiles above central London, a city with a large energy demand. The wind speed has been measured by [22] over a period of 4578 hours using a Doppler lidar. The wind power density is calculated using this wind speed data in conjunction with the barometric altitude formula for constant temperature [73] using a reference density of $\rho_0 = 1.225$ and a value $H_\rho = 8550$

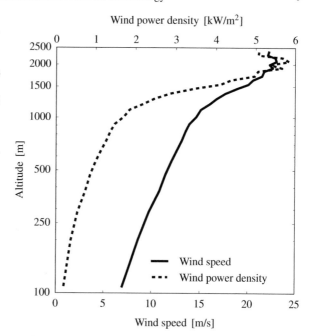

power more than doubles, from 210 to 480 W/m^2. The world's largest offshore wind park "London Array" has a comparable average wind speed of 9.2 m/s at 100 m hub height [71]. The reason for this similarity between offshore wind sites and high-altitude wind is simple: obstacles on land like forests, hills and buildings slow the wind down. Offshore winds partly owe their strength to the lack of obstacles. The same applies to high-altitude winds: no obstacles exist to slow the winds down. At a height of 500 m, the average wind speed is 11.6 m/s and wind power again approximately doubles to 924 W/m^2 with respect to a height of 250 m. This is a higher average wind speed than any operational offshore wind park in the world can offer.[5]

Flying wind drones *directly* above the city center of London might not be very realistic, but the ability of wind drones to harvest high-altitude winds leads to three important economic implications that will become relevant.

1.7 Higher Energy Production means Lower Energy Price

The cost of wind energy depends not only on the cost to install a wind turbine with a certain capacity but also on the amount of energy this wind turbine produces. This is why wind energy is affordable at good wind sites but expensive at less windy

---

[5] The highest wind speed of any operational offshore wind park worldwide is 10.46 m/s [1], and it was only achieved by building a floating wind turbine (Hywind Project) that is anchored in 200 m deep waters.

Fig. 1.6 Utilization factor
for a simulated wind drone
design based on the Enerkite
concept [27]

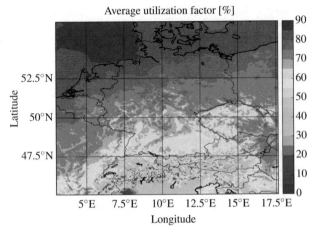

sites. Due to strong winds, offshore wind turbines often run at full capacity and do
not stay idle much. The average energy production of offshore turbines is therefore
twice as high as that of onshore turbines with the same rated capacity. [44]. The so-
called capacity factor of offshore turbines, which is the percentage of a power plant's
maximum potential that is actually achieved over time is therefore twice as high as
for onshore turbines. But since offshore turbines cost two to three times as much
as onshore turbines, the advantage of higher production with the same nameplate
capacity does not result in lower costs per kWh. Offshore wind energy is still more
costly than onshore wind [44].

According to research conducted by E.ON, Germany's largest utility, offshore
wind drones can even boost the offshore wind turbines' high yields by another 50%.
Their capacity factor can be as high as 69% per annum [59]. In many onshore locations the advantage for wind drones is even higher and capacity factors can more
than double compared to wind turbines. This was also shown in a recent study by
EWC Weather Consult, in which the expected capacity factor of the Enerkite wind
drone EK200 [24] was analyzed for a deployment in Germany, taking into account
the local and temporal variation of the wind speed at different altitudes over the duration of two years. The result is shown in Fig. 1.6 [27]. It can be seen that for most
of Germany the capacity factor is above 60%, and for *all* of the Netherlands it is
even above 75%.[6] This compares to a current average capacity factor of 18.3% for
wind turbines in Germany [12].

So, wind drones could be cheaper to install while producing more. Under the
assumption that installation costs per kW would be reduced by half and that output
per installed kW would be doubled at the same time, the price per kWh of electricity

---

[6] It should be noted that the Enerkite EK200 is optimized towards high capacity factor. A comparison between a wind turbine and a wind drone optimized towards high power output can be found
in Fig. 1.9.

Fig. 1.7 Average wind power density in Germany at altitudes of 100 m (left) and 200 m (right) above ground [46]. IRENA: Global Atlas, Map data: DTU 2015, OpenStreetMap contributors

produced by wind drones would be a quarter of the price of today's wind energy.[7] This would make wind drones highly competitive with wind turbines. It would also make wind drone energy cheaper than energy from fossil fuels.

This gain in competitiveness does not only mean a big difference for the individual investor or purchaser of energy, but also for the global world economy. The International Energy Agency (IEA) estimates that in order to implement the climate investment pledges made to the UN by world leaders, the global energy industry must invest \$ 13.5 trillion through 2030 [41]. Savings in the order of trillions of dollars could be made if airborne wind energy was used to provide the energy at least partially.

1.8 Offshore-Quality Wind—Almost Everywhere!

The higher average wind speed and the cubic dependency of wind power on the wind speed means that many sites open up for wind energy production that, at the hub height of wind turbines, do not have a wind resource that is suitable for economic wind energy generation. The wind maps of Germany at 100 and 200 m altitude depicted in Fig. 1.7 show the dramatic improvement of the wind resource with altitude. The maps only show areas with a good wind resource in excess of 350 W/m^2. At 100 m only the north of Germany can offer such good wind sites, while at 200 m the majority of the south also becomes suitable for economic wind energy generation.

---

[7] This estimate is also confirmed by recent studies on the cost of a wind farm based on lift mode wind drones [19].

But it is not only Germany: Google has calculated that less than 16% of all the onshore US sites are suitable for economic wind energy production with wind turbines. For wind drones this figure more than quadruples to 66% of the United States becoming viable [6, 57]. Vast areas of the world that do not have strong enough winds for conventional turbines will open up for wind energy production.

The fact that many new locations become economically viable energy producers by using wind drones has important economic implications for the total cost of the energy system, even for countries that do have good wind sites. Currently, for example Germany builds various high-voltage direct current (HVDC) power lines to transport wind energy from the north of the country, where the vast majority of wind energy is produced, to the south of the country, which has large demand for energy. The German government estimates that this and other grid enforcements to integrate wind energy will cost 21 Billion Euro for Germany alone [67]. In general, the electric grid is a large cost factor in the total cost of electricity. Grid costs for private households were for example in Germany 6.76 Euro ct per kWh while the costs for power production was 7.12 Euro ct per kWh [14].[8]

Airborne wind energy would turn most regions into economically viable energy producers, which could make local or regional energy grids possible with less need to build large, strong and costly (supra)national grids. This would especially be a chance for developing countries in which the costly large grid systems do not yet exist. But building wind drones close to the demand centers—if the regulatory framework allows—will also lower the costs that developed countries have to invest to make their grid capable of dealing with a large percentage of renewable energy.

1.9 The Reliable Renewable

We have seen that wind drones should be able to produce cheaper energy and that they should be able to produce wind energy much more independently from the quality of the low-altitude wind resource. In the mid to long term a third characteristic of wind drones might turn out to be the economically most relevant: the high capacity factor does not only decrease the cost of energy, it also means that wind energy is available for the most time of the year and that wind energy is less fluctuating. An energy source with higher quality: the reliable renewable.

Both, solar photovoltaics (solar PV) and conventional wind energy are highly intermittent. The average capacity factor in the USA is 37% for wind and only 20% for solar [69]. This causes considerable concerns. Electricity grid operators face the challenge of matching the fluctuating production of renewables with the demand.[9] Current scenarios foresee the necessity to invest billions in stronger grids

---

[8] On the basis of a consumption of 3500 kWh per year. The various surcharges for renewable feed-in tariffs was another 6.62 Euro ct.

[9] In 2015, in Germany alone more than 1 billion Euro of running costs had to be spent to maintain the integrity of the grid and to avoid blackouts due to power fluctuations [49]. These costs already today amount to enormous sums and they are expected to grow considerably if a higher fraction

and energy storage to enable a 100% renewable future. Therefore, while solar PV and conventional wind have at some places already reached so-called grid-parity, making them as cheap or cheaper than other sources of power, they can still not simply replace fossil fuel plants due to intermittency. If wind drones can produce with a capacity factor of about 70%, as envisaged by E.ON, or as high as 90% with optimized devices in optimal locations, they could largely replace coal, nuclear and gas power plants with a significantly reduced necessity for investments in grid and storage. Grid and distribution costs already make up for the greater part of electricity bills. The high quality of wind drone power could become a decisive factor, even more important than its low cost which it can reach at the same time.[10]

In a study of the French Utility EDF a European Grid is simulated in which 40% of the energy is contributed by fluctuating wind and solar PV [15]. If an additional 100 MW of conventional wind power are installed in such a scenario, only 20 MW of base load power stations, mostly fossil or nuclear, can be retired. If 100 MW of solar PV are installed, no base load power plants can be retired at all due to the even higher intermittency of solar power. With only little renewable penetration of the grid, as was the case at the beginning of large-scale solar and wind deployment in the 90s and the 2000s, every bit of renewable power was useful, since peak power plants had to work less often. Today, in Germany, there are certain days in which wind and solar produce more power than the country consumes. Wind power plants have to be shut down on more and more days because the power is not needed. Currently, the feed in tariffs in Germany and in many countries provide rules that pay the wind park owners for the virtual amount of power that they were not allowed to produce and to feed into the grid. This cannot and will not continue in the future, if such situations occur more often due to increased penetration of the grid by renewables. This means that, at a certain point, renewables will have to face market reality in which power prices are mostly dependent on the availability of other renewable power sources. If the sun is shining and the wind is blowing power prices fall, often close to zero. Sometimes even negative power prices are the result of such oversupply [37]. This means that in the future, renewable power producers will only be able to sell their renewable power with profit at those times at which other renewable energy sources cannot produce.

For airborne wind energy this means that it will be able to receive the best prices for its power not in the very windy time when also conventional wind turbines produce, but in those less windy times when they do not. To become a competitive

---

of power is produced with highly intermittent power sources or sources far away from power consumption.

[10] It should be noted that the capacity factor of every wind energy device, hence also for conventional wind, can of course be increased by design, simply by optimizing it for lower wind speeds. But optimizing for higher capacity factors than the 37% reached in the USA would make conventional wind more expensive. When optimizing for capacity factor, AWE has two advantages over wind turbines: first, the possibility to reach higher altitudes with stronger and steadier winds allows to reduce intermittency and increase capacity factor without design changes, and second, the low material use of wind drones reduces the costs of designs with very large wings that are required to lower the rated wind speed and to further increase the capacity factor.

product wind drones will therefore have to be able to produce with a high capacity factor and low intermittency.

Furthermore, only in this case can they help to replace base load power plants on a large scale. Today, both politics and the energy industry have finally realized that in future power will have to be 100% renewable. But due to the low capacity factor and high intermittency of solar PV and conventional wind, the current realization is that ultra strong grids and vast amounts of storage capacity have to be built in addition to the renewables production capacity. However, a cheap and abundant storage technology does not exist yet. Storage technology is therefore often called the "holy grail" of the energy world [23, 29]. This perception may in part prove being incorrect once AWE enters the market with its high capacity factor that comes close to that of today's base load plants [69].

It is even possible to provide continuous electricity completely without storage, if various AWE parks that are connected by the grid are distributed over a larger area with different local winds. While no detailed studies on this presently exist for AWE, it is shown that even conventional wind can be distributed so that it covers local intermittency well [54]. Thus, given the much lower expected intermittency of AWE, this should be even less of a problem for AWE.

The true economic advantage of the technology is therefore not limited to the lower price of energy compared to conventional wind. The savings in storage and grid capacity must be added to assess the full economic value of AWE to the world economy. A recent study assessed that an investment of over US$ 4.5 trillion was needed just for storage solutions for a 100% renewable energy system based on current solar PV, solar thermal and conventional wind technology [66]. A significant part of this investment could be saved with the help of AWE.

1.10 Quantifying the Increase in Reliability: Measuring the Intermittency

As discussed above, high-altitude winds not only have a higher average wind speed, but are also less intermittent and therefore a more reliable source of electricity. This is illustrated in Fig. 1.8 comparing statistical data at altitudes 145 and 265 m at Dresden, Germany. In this context, less intermittent means that very low wind speeds occur less frequently. This can be seen in Fig. 1.8 (top).

The example demonstrates that the much higher average wind power density at higher altitudes can be used to design a wind drone with a very reliable power output. The wind drone can be designed for a lower *relative* rated wind speed compared to the average wind speed that would be economically feasible for a wind turbine at lower altitudes. This trade-off, which is economically only available for very high average wind power densities, can be of paramount importance for an efficient power grid operation and for the economic success of renewables.

But how much is the intermittency of the power output of a wind drone compared to a wind turbine at the same site? In this context, we can measure the intermittency

Fig. 1.8 An example for the much lower intermittency of the wind at higher altitudes, taken from measurements at an onshore location in Dresden, Germany [33]. At 265 m the wind is still highly variable, as it is at 145 m, the maximum hub height of current wind turbines. But it can be clearly seen that for a given design wind speed of the wind drone, which could be around $v_w = 10$ m/s, there is wind above that cut-off for at least 40% of the time for an altitude of 265 m, and only in 10% of the time for 145 m. This means that the much higher average wind power density at higher altitudes can be much more economically traded into a lower intermittency than at lower altitudes

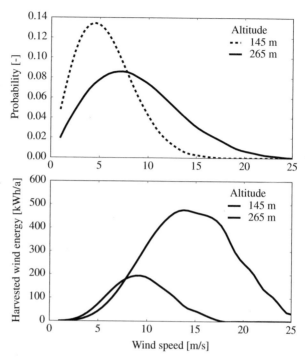

as the fraction of time in which the wind turbine produces less than a certain percentage of its nameplate power. Using measured wind data [33], a Vestas V90 wind turbine with 2 MW rated power and 105 m hub height is compared to a simulation of an Ampyx Power aircraft AP4 with the same rated power. The result is displayed in Fig. 1.9.

Fig. 1.9 Intermittency of the power output of a simulated Ampyx Power aircraft AP4 compared to a real Vestas V90 wind turbine [25] at an on-shore site, based on measured wind data from [33]. Both have a nameplate capacity of 2 MW. Due to limitations in the available wind data, no dynamic optimization of the flight altitude of the tethered aircraft was included. Thus, the observed intermittency can be seen as a conservative estimate

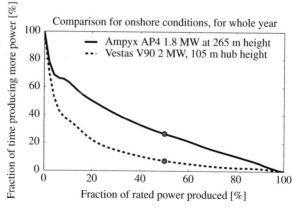

It can be seen that the wind drone dramatically improves the reliability of the power output. For a threshold of 50% of the rated power, the wind turbine produces power at this level or above for 8% of the time, averaged over a full year. For the wind drone this figure more than triples to 28%[11] of the time. This translates into a doubling of the capacity factor, from 16.5% for the wind turbine to 33.8% for the wind drone at this low wind location.[12] The data shows one significant advantage of wind drones compared to wind turbines: the lower intermittency of wind drones leads to a higher quality of renewable energy and will be an important factor on the way to a 100% renewable future.[13]

1.11 The Market for Airborne Wind Energy

It has been shown that wind drones will be competitive with wind turbines and can be considered a superior energy source amongst renewables, providing reliability benefits in addition to low price. The following is an attempt to assess the size of the market potential for this technology.

To begin with, the global wind turbine market is a large market. Its volume amounted to US$ 80 billion in 2013 [35]. Its growth rate averaged 25% per year over the last decade [43] and given the political environment and the need to abandon fossil fuels, it is safe to assume that the market will continue to grow[14] beyond the record year of 2015, when 63 GW of wind energy was installed [28, 55].

But wind energy is still only a small part of the total energy market and accounts for less than 1% of total global energy use.[15] This will change. And it is mostly a question of competitiveness, more than of politics. Onshore wind turbines are on the brink of becoming competitive with coal and natural gas. This so called grid-parity has been reached in some regions. It means that wind energy is already the cheapest source of electricity even without subsidies. Wind drones have the potential to produce at one quarter of the costs, can provide steadier production, and have the

---

[11] It should be noted, that the wind drone was not optimized for a deployment at relatively low wind speeds, and that the available wind data did not allow for a dynamic optimization of the flight altitude of the wind drone. Also, no wind data was available for the maximum flight altitude of the wind drone of up to 450 m. Thus, it can be expected that the real increase in reliability of the power output of AWE can be even higher.

[12] This increase by a factor of two corresponds to the ratio of the integrals of the two curves in Fig. 1.9.

[13] Optimizing the intermittency might involve flying at altitudes higher than the design value at times of low wind. This might not be their most efficient mode of operation, but if there is only wind at several hundred meters altitude, wind drones can use these high-altitude winds to produce at least some energy. This is in stark contrast to wind turbines, which cannot adjust their height to where the wind happens to be. Technological challenges like less efficient angles of attack and increased tether drag will be hurdles on the way to these altitudes, but nevertheless it remains a possibility and ideas for future improvements are numerous, for example using dancing kites [76].

[14] See the accelerated market forecast in [42].

[15] 0.3% in 2011: wind 434 TWh, total energy demand: 13,070 Mtoe (= 152,000 TWh) [43]

ability to be deployed almost anywhere. This means that wind drones do not only compete with wind turbines in their niche, but have a realistic chance of becoming the cheapest source of electricity: cheaper than coal, gas, nuclear and hydro power.

And since electric cars are on the rise, the electricity produced by wind drones will be able to compete with oil as a transportation fuel. Electrification of transport will end a monopoly for oil that has allowed oil to be one of the most costly sources of energy for a long time. In the fully electrified economy of the future all sources of energy can compete with each other [13]. And oil will have a hard time competing with wind drones, even at the current low oil prices. Taking into account the inefficiencies of the combustion engine, oil at US$ 60 per barrel is still a more expensive source of power for a car than the electricity produced by today's wind turbines.[16] Based on the analysis above, oil would have to sell at a quarter of that price, below US$ 15 per barrel to compete with wind drone energy on a pure cost of fuel basis.

So how large is the market potential for wind drones? This is hard to say exactly, but looking at Table 1.1, which lists the companies with the largest revenue worldwide, we can get a rough idea of the full market potential for AWE. In the next two decades, the market for fossil fuels has to be taken over by renewables in order to keep our planet from overheating. Since the UN Climate Change Conference 2015 in Paris it is the declared will of all governments in the world to make this happen [74]. If the current policies regarding CO_2 emissions are continued, by 2034 the burning of fossil fuels would have to be stopped completely to comply with the 2°C

Rank	Name	Country	Industry	Revenue [bn US$]
1	Walmart	US	Retail	485
2	Sinopec	China	Petroleum*	446
3	Shell	UK	Petroleum*	431
4	China Petrol	China	Petroleum*	428
5	ExxonMobil	US	Petroleum*	382
6	BP	UK	Petroleum*	358
7	State Grid	China	Power*	339
8	Volkswagen	Germany	Mobility	268
9	Toyota	Japan	Mobility	247
10	Glencore	Switzerland	Metal/Energy*	221
11	Total	France	Petroleum*	212
12	Chevron	US	Petroleum*	203

Table 1.1 The top of the Forbes Global 500 list of 2015 [32]. Nine of the top 12 companies with the largest revenue are in the energy market (marked with a *)

---

[16] One barrel of oil contains the chemical energy of 1628 kWh. At US$ 60 per barrel of oil this would translate to 3.6 cents per kWh. Taking into account the combustion engines' theoretical maximum efficiency of about 50% this means an oil price of 7.2 cents per kWh of usable mechanical energy. However, the average real life efficiency of the combustion engine of a car over the average driving cycle is closer to 25%, which would result in an oil price of 14.4 cents per kWh.

goal [16]. Given the technical and cost advantages of wind drones, not a small part of the total trillion dollar energy market might go to companies utilizing AWE.

1.12 Future Goals for Airborne Wind Energy

We have seen how the laws of mechanics in combination with sensors, chips and smart algorithms can replace the tons of steel and concrete that wind turbines are made of. This can make power produced by the first generation of wind drones cheaper than electricity from fossil fuels. Their ability to harvest stronger winds at higher altitudes gives wind drones the potential to provide power where it is needed irrespective of the existing wind resource. Cost-effective electricity made by wind drones could even provide the basis for the clean synthetic fuels of the future. And this fuel could be available at less than today's oil price—even taking the historically low prices of at the beginning of the year 2016 into account.

Once AWE matures, a lack of wind will no longer be a problem. We have seen how the wind resource dramatically improves by increasing the altitude to 250 m or even 500 m. But this is only the first hop of the very first generation of wind drones into the air. Once these altitudes are mastered, it will be tempting to gradually go higher, for example towards 1,000 to 2,000 meters.[17] The wind speed profile above the center of London, illustrated in Fig. 1.5, shows how much is to be gained by increasing the altitude even further, at least at certain locations. At 1,500 m the average wind speed above London is 20 m/s, corresponding to almost storm-like conditions, which is defined as a wind above 20.8 m/s. This more or less continuous storm leads to a four-fold increase of available wind power compared to the situation at 500 m and to a twenty-fold increase compared to 120 m, the altitude of today's wind turbines, that is 4143 vs. 924 vs. 210 W/m^2.

The ultimate dream of airborne wind energy would be reaching the jet stream at an altitude of about 10 km. Before this is possible, many technical and legal problems will have to be solved, however. And it might turn out in the end that it is not the most economical solution. But given the strong development the field has taken in the last 10 years, it is reasonable to expect that it will be earnestly attempted. The wind resources at this altitude are simply too enticing. A map of the winds at the altitude of present wind turbines and at jet stream altitudes is shown in Fig. 1.10. The median energy density over New York at this height is more than 10 kW/m^2 [6], of which about 5 kW/m^2 can theoretically be harvested.[18] The total average continuous energy consumption per person in the US amounts to 10.5 kW. This includes all electricity use, heating, car and aviation fuels, and even industrial energy consumption [75]. This means that harvesting wind in an area of 2 m^2 per person, the size of an open front door, could on average provide all our energy. If ten wind turbines

---

[17] To produce wind energy at these altitudes economically, technical solutions for lowering the tether drag have to be implemented, for example by using the dancing kite concept [76].

[18] The theoretical maximum is the Betz limit 16/27 or 59%. Modern wind turbines are very close to this with efficiencies of about 50%, including losses in generators, drivetrains etc.

Fig. 1.10 Global distribution of the wind power density on 1 December 2014 9:00 UTC at altitudes of 100 m (top) and 10,500 m (bottom) [9]. Dark blue, no stripes: negligible, dark blue = below $0.5\,kW/m^2$, brown = 0.5 to $2\,kW/m^2$, medium blue = 2 to $10\,kW/m^2$, turquoise = 10 to $40\,kW/m^2$, red = 40 to $70\,kW/m^2$, pink = over $70\,kW/m^2$. Data: GFS / NCEP / US National Weather Service

with today's dimensions[19] (but of course with vastly different specifications) were installed at that altitude above New York, they could have the same rated power as an average nuclear power plant—over 1 GW.

High-altitude wind energy is not only an extremely concentrated source of energy, it is also abundant. It is estimated that it could provide 100 times more energy than humanity consumes today [47, 60]. High-altitude wind energy could allow us to live a greener lifestyle without the need to reduce our use of energy. For the energy sector this could mean nothing less than finally solving the conflict between economy and ecology.

---

[19] The Vestas V164-8.0 MW® with a blade length of 82 m features a swept area of $21,124\,m^2$. With $10\,kW/m^2$ and 50% efficiency, this results in 105 MW per turbine or over 1 GW for ten turbines [72].

Of course it might seem unrealistic to reach these altitudes with the present day technology, with tethers, their weight and aerodynamics being the major hindrances. However, when Miles Loyd invented crosswind airborne wind energy in the second half of the 1970s, it seemed inconceivable to use AWE at all. Now, 40 years later, it is a reality, and thus those who made AWE a reality should not stop dreaming big.

1.13 Conclusion

The past decades have seen a revolution of digital technology, in the processing power of computers, in terms of robustness, and in the accompanying improvements of mostly digital sensors such as accelerometers and GPS. This revolution has brought us into the age of drones. Drones will have a very wide array of applications, and maybe the most important one will be in airborne wind energy, the "wind drone".

Wind drones can harvest stronger and steadier winds at higher and at varying altitudes. Wind drones require only a fraction of the material investment of wind turbines, since they replace the mechanical sturdiness of the wind turbine by a fragile flight, reliably controlled by clever autonomous control algorithms. Since they implement only the aerodynamically most efficient part of the wind turbine, the tip of the rotor blade, leaving out all the inefficient rest, they can fly higher in more reliable winds and cover larger areas. This leads to a higher capacity factor with lower intermittency and, very roughly estimated, a quarter of the cost of energy. Wind drones have the unique potential to represent the most reliable *abundant* renewable energy, thus potentially saving billions of dollars otherwise required for making the power grids more robust against variations in power production and consumption in space and time. Given the enormous technical and financial challenges the grid operators are facing even now, it is possible that the indirect cost savings of wind drones compared to conventional wind might be even bigger than the direct cost advantage.

The economic future of wind drones looks bright and the development of the first commercial prototypes is in full swing. At the same time, new concepts are being explored by a growing number of researchers and start-ups.

Burning fossil fuels started the industrial revolution. It enabled the advances of mankind in the last 200 years and the strong exponential growth the world economy experienced ever since. Without fossil fuels, feeding 7 billion people on this planet would be impossible. But fossil fuels also destroy and pollute nature, poison our cities and homes and cause an ever more dangerous climate change. These so called external effects lead not only to human suffering but also to large costs that are not paid for by those responsible for the pollution, but by the global community. Nonetheless, these costs have to be paid and cause slower growth and development.

Solving huge environmental and economic problems with existing technologies would require unimaginable resources, which is why Bill Gates and others have called for an energy miracle.

Wind drones could be such a miracle. They can tap the richest and most concentrated source of renewable energy that is otherwise literally out of reach for humanity: high-altitude winds. And wind drones are digital products that can be scaled up and brought to utility scale very fast. We believe that wind drones are the digital products that will revolutionize the energy market, the largest market in the world. The digital revolution quickly disrupts the new market it enters, leaving old technology no chance to compete. Wind drones can become for oil majors what Amazon.com was for bookstores.

When mankind started to burn fossil fuels it made a huge leap forward. The day it stops to burn fossil fuels, it will make another big step towards a better world. Wind drones might bring this day much closer than most of us believe today.

References

1. 4C Offshore Ltd: Global wind speed rankings of offshore wind farms. http://www.4coffshore.com/windfarms/windspeeds.aspx. Accessed 18 Jan 2016
2. Afams, C., Thornhill, J.: Gates to double investment in renewable energy projects. Financial Times, 25 June 2015. http://on.ft.com/1U2Btc2
3. Ahrens, U., Diehl, M., Schmehl, R. (eds.): Airborne Wind Energy. Green Energy and Technology. Springer, Berlin Heidelberg (2013). doi: 10.1007/978-3-642-39965-7
4. Ahrens, U., Pieper, B., Töpfer, C.: Combining Kites and Rail Technology into a Traction-Based Airborne Wind Energy Plant. In: Ahrens, U., Diehl, M., Schmehl, R. (eds.) Airborne Wind Energy, Green Energy and Technology, Chap. 25, pp. 437–441. Springer, Berlin Heidelberg (2013). doi: 10.1007/978-3-642-39965-7_25
5. Allen, J., Walsh, B.: Enhanced oil spill surveillance, detection and monitoring through the applied technology of unmanned air systems. International Oil Spill Conference Proceedings 2008(1), 113–120 (2008). doi: 10.7901/2169-3358-2008-1-113
6. Archer, C. L., Caldeira, K.: Global Assessment of High-Altitude Wind Power. Energies 2(2), 307–319 (2009). doi: 10.3390/en20200307
7. Bangert, R.: Google Energy Kite Nears Launch Date. Alameda Sun, 8 Oct 2015. http://alamedasun.com/news/google-energy-kite-nears-launch-date
8. Barnstorff, K.: Ten-Engine Electric Plane Completes Successful Flight Test. NASA Langley Research Center, 30 Apr 2015. https://www.nasa.gov/langley/ten-engine-electric-plane-completes-successful-flight-test Accessed 27 Jan 2016
9. Beccario, C.: Earth Wind Map. http://earth.nullschool.net. Accessed 18 Jan 2016
10. Bliss, L.: 'Ubiquitous As Pigeons': Imagining Life in the City of Drones. Citylab, 5 Aug 2014. http://www.citylab.com/tech/2014/08/ubiquitous-as-pigeons-imagining-life-in-the-city-of-drones/375568/ Accessed 18 Jan 2016
11. Blue Origin: The Technology of the Blue Origin System. https://www.blueorigin.com/technology. Accessed 27 Jan 2016
12. Boccard, N.: Capacity factor of wind power realized values vs. estimates. Energy Policy 37(7), 2679–2688 (2009). doi: 10.1016/j.enpol.2009.02.046
13. Brognaux, C., Ward, N.: When Fuels Compete: The Evolving Dynamic of Global Energy Markets. bcg.perspectives, 15 July 2015. https://www.bcgperspectives.com/content/articles/energy-environment-when-fuels-compete-evolving-dynamic-global-energy-markets/ Accessed 18 Jan 2016
14. Bundesverband der Energie- und Wasserwirtschaft e.V.: Strompreisanalyse März 2015. http://kitepower.tudelft.nl/AWEbook/bdew-strompreis-2015.pdf. Accessed 10 Oct 2017

15. Burtin, A., Silva, V.: Technical and Economic Analysis of the European Electricity Systems with 60% RES, EDF Research and Development Division, Paris, France, 17 June 2015. http://energypost.eu/edf-study-download-15/ Accessed 18 Jan 2016
16. Carbon Tracker: Carbon Budgets. http://www.carbontracker.org/wp-content/uploads/2014/08/Carbon-budget-checklist-FINAL-1.pdf (2013). Accessed 27 Jan 2016
17. Cherubini, A., Papini, A., Vertechy, R., Fontana, M.: Airborne Wind Energy Systems: A review of the technologies. Renewable and Sustainable Energy Reviews 51, 1461–1476 (2015). doi: 10.1016/j.rser.2015.07.053
18. Coppinger, R.: Airbus' Adeline Project Aims to Build Reusable Rockets and Space Tugs. Space.com, 10 June 2015. http://www.space.com/29620-airbus-adeline-reusable-rocket-space-tug.html Accessed 18 Jan 2016
19. De Lellis, M., Mendonça, A. K., Saraiva, R., Trofino, A., Lezana, Á.: Electric power generation in wind farms with pumping kites: An economical analysis. Renewable Energy 86, 163–172 (2016). doi: 10.1016/j.renene.2015.08.002
20. Deckstein, D., Hammerstein, K. von: Unter Geiern. Innovationen: Getrieben vom rasenden technologischen Wandel, muss Siemens-Chef Joe Kaeser Europas größten Hightech-Konzern zukunftsfähig machen. Der Spiegel 49, 86–91 (2015). http://magazin.spiegel.de/EpubDelivery/spiegel/pdf/140036942
21. Diehl, M.: Airborne Wind Energy: Basic Concepts and Physical Foundations. In: Ahrens, U., Diehl, M., Schmehl, R. (eds.) Airborne Wind Energy, Green Energy and Technology, Chap. 1, pp. 3–22. Springer, Berlin Heidelberg (2013). doi: 10.1007/978-3-642-39965-7_1
22. Drew, D. R., Barlow, J. F., Lane, S. E.: Observations of wind speed profiles over Greater London, UK, using a Doppler lidar. Journal of Wind Engineering and Industrial Aerodynamics 121, 98–105 (2013). doi: 10.1016/j.jweia.2013.07.019
23. Dunn, B., Kamath, H., Tarascon, J.-M.: Electrical Energy Storage for the Grid: A Battery of Choices. Science 334(6058), 928–935 (2011). doi: 10.1126/science.1212741
24. EnerKite GmbH: Technical Data – EK200. http://www.enerkite.de/downloads/EnerKite_200_Technical_Data_EN_SM.pdf. Accessed 18 Jan 2016
25. Engelen, S., Ruiterkamp, R.: Simulation of the intermittency of an AP4 PowerPlane compared to a Vestas V90 wind turbine of the same rated power. Private communication, 9 Feb 2016
26. European Aviation Safety Agency: Transposition of Amendment 43 to Annex 2 to the Chicago Convention on remotely piloted aircraft systems (RPAS) into common rules of the air, EASA NPA 2014-09, 3 Apr 2014. https://www.easa.europa.eu/system/files/dfu/NPA%202014-09.pdf
27. European Weather Consult. https://www.weather-consult.com. Accessed 18 Jan 2016
28. Evans, S.: Mapped: How China dominates the global wind energy market. Carbon Brief, 19 Apr 2016. http://www.carbonbrief.org/mapped-how-china-dominates-the-global-wind-energy-market Accessed 9 May 2016
29. Evans-Pritchard, A.: Holy Grail of energy policy in sight as battery technology smashes the old order. The Telegraph, 10 Aug 2016. http://www.telegraph.co.uk/business/2016/08/10/holy-grail-of-energy-policy-in-sight-as-battery-technology-smash/
30. Fagiano, L., Marks, T.: Design of a Small-Scale Prototype for Research in Airborne Wind Energy. IEEE/ASME Transactions on Mechatronics 20(1), 166–177 (2015). doi: 10.1109/TMECH.2014.2322761
31. Fagiano, L., Schnez, S.: On the Take-off of Airborne Wind Energy Systems Based on Rigid Wings. Renewable Energy 107, 473–488 (2017). doi: 10.1016/j.renene.2017.02.023
32. Fortune Magazine: Gobal 500 2015. http://fortune.com/global500/2015. Accessed 18 Jan 2016
33. Geiss, C.: Untersuchungen zum vertikalen Windprofil in Sachsen. Student Project Report, Chemnitz University of Technology, Jan 2012. https://www.tu-chemnitz.de/etit/eneho/lehre/studentischearbeiten.php
34. GL Garrad Hassan: Market Status Report High Altitude Wind Energy, now merged with DNV GL, Aug 2011

35. Global Wind Energy Council (GWEC): Global Wind Report – Annual Market Update 2013, Apr 2014. http://www.gwec.net/wp-content/uploads/2014/04/GWEC-Global-Wind-Report_9-April-2014.pdf
36. Grenzdörffer, G.: Investigations on the use of airborne remote sensing for variable rate treatments of fungicides, growth regulators and N-fertilisation. In: Stafford, J., Werner, A. (eds.). Precision Agriculture. Proceedings of the 4th European Conference on Precision Agriculture, pp. 241–246, Berlin, Germany, 15–19 June 2003. doi: 10.3920/978-90-8686-514-7
37. Gross, D.: The Night They Drove the Price of Electricity Down. Slate, 18 Sept 2015. http://www.slate.com/articles/business/the_juice/2015/09/texas_electricity_goes_negative_wind_power_was_so_plentiful_one_night_that.html Accessed 1 Feb 2016
38. Hardham, C.: Response to the Federal Aviation Authority. Docket No.: FAA-2011-1279; Notice No. 11-07; Notification for Airborne Wind Energy Systems (AWES), Makani Power, 7 Feb 2012. https://www.regulations.gov/#!documentDetail;D=FAA-2011-1279-0014
39. Hill, J. S.: Kite Power Systems Secures £2 Million Investment From Scottish Investment Bank., 22 June 2017. https://cleantechnica.com/2017/06/22/kite-power-systems-secures-2-million-investment-scottish-investment-bank/
40. Hirtenstein, A.: The Next Plan for Drones? Tethered Aircraft Generating Power. Bloomberg, 11 Apr 2017. https://www.bloomberg.com/news/articles/2017-04-11/flying-drones-that-generate-power-from-wind-get-backing-from-eon
41. International Energy Agency (IEA): Energy and Climate Change: World Energy Outlook Special Briefing for COP21, 21 Oct 2015. http://www.iea.org/media/news/WEO_INDC_Paper_Final_WEB.PDF
42. International Energy Agency (IEA): Medium-Term Renewable Energy Market Report 2015, OECD Publishing, Paris, 2 Oct 2015. doi: 10.1787/renewmar-2015-en
43. International Energy (IEA): World Energy Outlook 2013, OECD Publishing, Paris, 12 Nov 2013. doi: 10.1787/weo-2013-en
44. International Renewable Energy Agency: Global Atlas for Renewable Energy. http://irena.masdar.ac.ae. Accessed 18 Jan 2016
45. International Renewable Energy Agency: Renewable Energy Cost Analysis – Wind Power. IRENA Working Paper, June 2012. http://www.irena.org/DocumentDownloads/Publications/RE_Technologies_Cost_Analysis-WIND_POWER.pdf
46. International Renewable Energy Agency (IRENA): DTU Global Wind Atlas. http://irena.masdar.ac.ae/?map=103. Accessed 12 May 2016
47. Jacobson, M. Z., Archer, C. L.: Saturation wind power potential and its implications for wind energy. Proceedings of the National Academy of Sciences (PNAS) **109**(39), 15679–15684 (2012). doi: 10.1073/pnas.1208993109
48. Joby Aviation, Inc.: Lotus. http://www.jobyaviation.com/lotus/. Accessed 27 Jan 2016
49. Kwasniewski, N.: Blackout-Abwehr kostete 2015 eine Milliarde Euro. Spiegel Online, 17 Jan 2016. http://www.spiegel.de/wirtschaft/unternehmen/blackout-abwehr-kostete-2015-eine-milliarde-euro-a-1072438.html Accessed 18 Jan 2016
50. Lay, J., Price-Waldman, S.: Bill Gates and the Quest for Sustainable Energy. The Atlantic, 13 Oct 2015. http://www.theatlantic.com/video/index/410011/bill-gates-and-the-quest-for-sustainable-energy/
51. Loyd, M. L.: Crosswind kite power. Journal of Energy **4**(3), 106–111 (1980). doi: 10.2514/3.48021
52. Loyd, M. L.: Foreword. In: Ahrens, U., Diehl, M., Schmehl, R. (eds.) Airborne Wind Energy, Green Energy and Technology. Springer, Berlin Heidelberg (2013). doi: 10.1007/978-3-642-39965-7
53. Loyd, M. L.: Wind driven apparatus for power generation. US Patent 4,251,040, Dec 1978
54. MacDonald, A. E., Clack, C. T. M., Alexander, A., Dunbar, A., Wilczak, J., Xie, Y.: Future cost-competitive electricity systems and their impact on US CO_2 emissions. Nature Climate Change **6**, 526–531 (2016). doi: 10.1038/nclimate2921

55. Magill, B.: If a Power Plant Is Built in U.S., It's Likely to be Renewable. Climate Central, 25 Mar 2016. http://www.climatecentral.org/news/if-a-power-plant-is-built-in-us-chances-are-its-renewable-20175
56. Makani. https://x.company/makani/technology/. Accessed 10 Oct 2017
57. Makani Power: FAQ. http://www.google.com/makani/faq/. Accessed 18 Jan 2016
58. Makani Power: Makani Power Google+ Site. https://plus.google.com/+makani/posts/BWaiJZWfMsV. Accessed 18 Jan 2016
59. Mann, S., Gunn, K., Harrison, G., Beare, B., Lazakis, I.: Wind Yield Assessment for Airborne Wind Energy. Poster presented at the EWEA Offshore Conference, Copenhagen, Denmark, 10–12 Mar 2015. http://www.ewea.org/offshore2015/conference/allposters/PO090.pdf
60. Marvel, K., Kravitz, B., Caldeira, K.: Geophysical limits to global wind power. Nature Climate Change 3, 118–121 (2013). doi: 10.1038/nclimate1683
61. MHI Vestas Offshore Wind: V164-8.0 MW breaks world record for wind energy production. http://www.mhivestasoffshore.com/v164-8-0-mw-breaks-world-record-for-wind-energy-production/. Accessed 18 Jan 2016
62. MHI Vestas Offshore Wind: V164-8.0 MW testing programme to be ramped up with installation of two additional onshore turbines in Denmark. http://www.mhivestasoffshore.com/v164-8-0-mw-testing-programme-to-be-ramped-up-with-installation-of-two-additional-onshore-turbines-in-denmark/. Accessed 18 Jan 2016
63. Montnacher, J.: Kite-Steuerungsplattform – Bodeneinheit zur Höhenwindnutzung. In: Proceedings of the workshop "Flugwindenergie", pp. 38–51, Bremerhaven, Germany, 20 Nov 2012. http://publica.fraunhofer.de/dokumente/N-223493.html
64. Moore, M.: NASA Wind Energy Airborne Harvesting System Study. http://awtdata.webs.com. Accessed 10 May 2016
65. North, D. D., Aull, M. J.: Tethered vehicle control and tracking system. US Patent 8,922,041, 2014. https://technology.nasa.gov//t2media/tops/pdf/LAR-TOPS-40.pdf
66. Pleßmann, G., Erdmann, M., Hlusiak, M., Breyer, C.: Global Energy Storage Demand for a 100% Renewable Electricity Supply. Energy Procedia 46. Proceedings of the 8th International Renewable Energy Storage Conference and Exhibition (IRES 2013), 22–31 (2014). doi: 10.1016/j.egypro.2014.01.154
67. Presse- und Informationsamt der Bundesregierung: Wie teuer wird der Ausbau der Stromtrassen und wie lange wird er dauern? https://www.bundesregierung.de/Webs/Breg/DE/Themen/Energiewende/Fragen-Antworten/2_Netzausbau/2_netzausbau/_node.html#doc605896bodyText4. Accessed 18 Jan 2016
68. Puiu, T.: How SpaceX's Elon Musk wants to drop space launch prices 100 fold with reusable rockets. ZME Science, 21 Aug 2013. http://www.zmescience.com/space/spacex-reusable-rocket-100-times-cheaper-0432423/ Accessed 18 Jan 2016
69. Randall, T.: Solar and Wind Just Passed Another Big Turning Point: It has never made less sense to build fossil fuel power plants. Bloomberg, 6 Oct 2015. http://www.bloomberg.com/news/articles/2015-10-06/solar-wind-reach-a-big-renewables-turning-point-bnef Accessed 18 Jan 2016
70. Shu, C.: Aerostat startup Altaeros gets $7.5M from SoftBank to bring broadband wireless to rural areas. TechCrunch, 8 Aug 2017. http://tcrn.ch/2ulTEPv
71. Siemens AG: Siemens to provide 175 wind turbines for the world's largest offshore wind farm London Array. Press Release, 19 May 2009. http://www.siemens.com/press/pi/ERE200905050e Accessed 18 Jan 2016
72. Snieckus, D.: Vestas V164 tower on a roll. Recharge News, 27 Nov 2013. http://www.rechargenews.com/wind/europe_africa/article1344738.ece Accessed 18 Jan 2016
73. Stull, R. B.: Meteorology for Scientists and Engineers. 2nd ed. Brooks/Cole Publishing Company, Pacific Grove (2000)
74. United Nations Framework Convention on Climate Change (UNFCCC) Conference of the Parties (COP): Adoption of the Paris Agreement. Proposal by the President, Decision 1/CP.21, 12 Dec 2015. http://undocs.org/FCCC/CP/2015/L.9/Rev.1

75. US Energy Information Administration: How much energy does a person use in a year? http://www.eia.gov/tools/faqs/faq.cfm?id=85&t=1. Accessed 18 Jan 2016
76. Zanon, M., Gros, S., Andersson, J., Diehl, M.: Airborne Wind Energy Based on Dual Airfoils. IEEE Transactions on Control Systems Technology **21**(4), 1215–1222 (2013). doi: 10.1109/TCST.2013.2257781
77. Zhang, S.: SpaceX's Falcon Rocket Finally Sticks the Landing. Wired Science, 21 Dec 2015. http://www.wired.com/2015/12/spacex-just-landed-rocket-ground-first-time/ Accessed 18 Jan 2016
78. Zillmann, U.: The Trillion Dollar Drone. European Energy Review, 24 June 2015. http://www.europeanenergyreview.eu/the-trillion-dollar-drone/ Accessed 30 July 2016
79. Zillmann, U.: The Trillion Dollar Drone – A Change of Perspective. In: Schmehl, R. (ed.). Book of Abstracts of the International Airborne Wind Energy Conference 2015, p. 60, Delft, The Netherlands, 15–16 June 2015. doi: 10.4233/uuid:7df59b79-2c6b-4e30-bd58-8454f493bb09. Presentation video recording available from: https://collegerama.tudelft.nl/Mediasite/Play/aca3c3f29eb54dc3b6ca4cde1a68084c1d

Part I
Fundamentals, Modeling & Simulation

Chapter 2
Tether and Bridle Line Drag in Airborne Wind Energy Applications

Storm Dunker

Abstract This chapter discusses the physics of tether and bridle line drag based on literature, describes the typical flight regimes for airborne wind energy and identifies regimes of elevated drag caused by vortex-induced vibration and movement-induced excitation such as galloping. The presented laboratory tests show increases of aerodynamic drag due to vortex-induced vibration up to 300% and due to galloping up to 210%. Given that tether drag is a primary limitation to an airborne wind energy system's ability to fly faster and produce more energy, understanding the regimes of elevated drag as well as the mechanisms to suppress the causing phenomena are important. The chapter provides a basic overview of these phenomena as well as potential solutions for drag reduction. The information and material presented should provide an airborne wind energy developer a useful introduction to the considerations of tether and bridle line aerodynamic drag.

2.1 Introduction

As the name implies, airborne wind energy (AWE) is the conversion of wind energy by of one or more flying, buoyant or otherwise lifted devices into electrical energy. All conversion concepts employ one or more tethers to mechanically connect the lift devices to the ground. Many concepts use additional bridle lines to further distribute the load transfer from the lifting device to the tethers. AWE systems are either of the Ground-gen variant (generators located on the ground, operated by a tether wrapped around a coupled drum, the reeling out and in of the tether converting linear motion into shaft power) or of the Fly-gen variant (generators located on the lifting device are driven by impellers, the generated electrical power transmitted down a tether of fixed length). Ground-gen systems are, for example, developed by Enerkite and TU

Storm Dunker (✉)
A-Z Chuteworks LLC., Houston, TX, USA
e-mail: storm@azc-llc.com

Delft, while Fly-gen systems are developed by Makani and Altaeros Energies. More comprehensive overviews of implemented concepts are presented in [7, 26].

Tethers and bridle lines are textile components forming a tensile structure that is designed and optimized for the transfer of tensile loads but unsuitable for supporting compression loads. Bridle lines cascade out of the tether, are thinner than tethers due to distributed load, generally have a shorter length and connect to the lifting device in several attachment points. Tethers for Fly-gen systems are generally thicker and heavier because they incorporate additional conductive wires to transmit electric energy. These wires can be contained inside the core of a braided tether or can be braided among the other braiding carriers as part of the braid itself.

Loyd [20] derived analytical models for the achievable power output from simple kites that perform only a reel out motion and kites that are additionally flown in crosswind maneuvers. A key parameter in these models is the aerodynamic lift-to-drag ratio L/D of the kite. Power output in Loyd's crosswind model, in which systems principally operate at relative airspeeds up to L/D times higher than for simple, non-maneuvering kites, is especially sensitive to the aerodynamic drag contributions of tethers and bridle lines. In particular the drag of long tethers can represent a significant part of total drag of crosswind kites. Reduced tether drag directly increases L/D which according to Loyd's theory increases the flight velocity, the tether tension and consequently also the power output.

The aerodynamic properties of tethers and bridle lines depend on the local relative flow conditions, which, for a static setup includes the orientation with respect to the relative flow, the flow cross section and the surface characteristics of the exposed material. However, as a result of its inherent elasticity and inertia, the tensile structure can be exited by the relative flow to oscillate. Aero-structural coupling phenomena such as Vortex-Induced Vibrations (VIV) and galloping can increase drag significantly and cause other unwanted dynamic effects.

This chapter is structured as follows. Section 2.2 presents a mathematical framework for the description of tensioned tethers and bridle lines in a cross flow environment and discusses various assumptions to simplify the physical problem. Section 2.3 discusses the operating envelopes assumed for all AWE applications that could use tethers and bridle lines. Section 2.4 will then review basic background physics, specifically from early chapters of Blevins' Flow-Induced Vibrations [3], relevant to the domain of aerodynamic drag for tethers and bridle lines. Related VIV and galloping experiments of bridle lines, either from the author or from literature, are presented in Sect. 2.5. Finally, in Sect. 2.6, potential tether design solutions are introduced that could help control elevated drag regimes.

2.2 Mathematical Framework and Assumptions

Throughout this chapter, the following descriptions and assumptions apply unless stated otherwise. The tether and the bridle lines are evaluated as an elastic flexible structure that can stretch, twist and dynamically oscillate along and perpendicular to the relative flow. The movement of this structure in the wind field is described in the wind reference frame x_w, y_w, z_w, which has its origin **O** located at the ground attachment point of the tether. The x_w-axis of this reference frame is aligned with the wind velocity $\mathbf{v}_w$, which is assumed to be constant in time and uniform in space, while its z_w-axis is pointing upward. The tether is assumed to be straight and accordingly the radial coordinate r can be used to describe positions on the tether

$$\mathbf{r} = r\mathbf{e}_r. \tag{2.1}$$

This configuration is illustrated in Fig. 2.1. The value of the radial coordinate varies between 0 at the origin and the tether length l_t at the kite **K**. A corresponding non-dimensional tether coordinate can be defined as

$$R = \frac{r}{l_t}. \tag{2.2}$$

The tether length is generally not constant but varies as a result of the reeling motion and to a minor degree also the strain of the tether. It is important to note that the radial coordinate r is a geometric measure which does not describe material points on the tether. Because of the reeling motion, the radial velocity $\mathbf{v}_{t,r}$ of material points is constant along the tether when neglecting strain, and equal to the radial velocity $\mathbf{v}_{k,r}$ of the kite

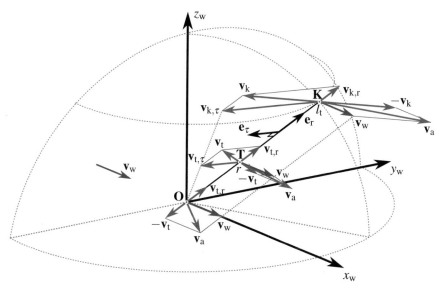

Fig. 2.1 Relative flow conditions at the origin **O**, an arbitrary point **T** along the tether and the kite **K**. A point on the tether moves with the material velocity $\mathbf{v}_t$, consisting of radial and tangential components $\mathbf{v}_{t,r}$ and $\mathbf{v}_{t,\tau}$, respectively. The tip of the tether at $r = l_t$ moves with the kite velocity, i.e. $\mathbf{v}_t = \mathbf{v}_k$. The radial unit vector $\mathbf{e}_r = \mathbf{v}_{k,r}/v_{k,r}$ is aligned with the tether, while the tangential unit vector $\mathbf{e}_\tau = \mathbf{v}_{k,\tau}/v_{k,\tau}$ is perpendicular and pointing in the flight direction of the kite

$$\mathbf{v}_{t,r} = \mathbf{v}_{k,r}, \qquad \text{for} \quad 0 \leq R \leq 1. \tag{2.3}$$

The tangential velocity $\mathbf{v}_{t,\tau}$ is constrained to zero at the ground attachment point $\mathbf{O}$ and is identical to the tangential velocity $\mathbf{v}_{k,\tau}$ of the kite at the kite attachment point $\mathbf{K}$. It can be formulated as a linear function of the radial coordinate

$$\mathbf{v}_{t,\tau} = \mathbf{v}_{k,\tau} R, \qquad \text{for} \quad 0 \leq R \leq 1. \tag{2.4}$$

The material velocity $\mathbf{v}_t = \mathbf{v}_{t,r} + \mathbf{v}_{t,\tau}$ of a point on the tether can thus be related to the radial and tangential velocity components of the kite by

$$\mathbf{v}_t = \mathbf{v}_{k,r} + \mathbf{v}_{k,\tau} R, \qquad \text{for} \quad 0 \leq R \leq 1. \tag{2.5}$$

The apparent wind velocity of a material point on the tether is defined as

$$\mathbf{v}_a = \mathbf{v}_w - \mathbf{v}_t, \tag{2.6}$$

$$= \mathbf{v}_w - \mathbf{v}_{k,r} - \mathbf{v}_{k,\tau} R, \qquad \text{for} \quad 0 \leq R \leq 1, \tag{2.7}$$

which is visualized in Fig. 2.1 for an arbitrary point $\mathbf{T}$ along the tether and for the end points $\mathbf{O}$ and $\mathbf{K}$, respectively. From Eq. (2.7) and Fig. 2.1 it is obvious that the apparent wind velocity along the moving tether varies in magnitude and direction.

The relative flow conditions are further detailed in Fig. 2.2. The angle of attack

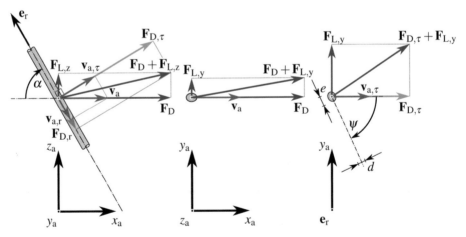

(a) Side view of a cylindrical tether segment, inclined towards the relative flow

(b) Top view of a cylindrical tether segment, inclined towards the relative flow

(c) Axial view of a twisted tether with deformed elliptical cross section

Fig. 2.2 (a) Definition of the local angle of attack α of a cylindrical tether segment, relative flow velocity $\mathbf{v}_a = \mathbf{v}_{a,\tau} + \mathbf{v}_{a,r}$ and total aerodynamic force $\mathbf{F}_a = \mathbf{F}_D + \mathbf{F}_{L,y} + \mathbf{F}_{L,z} = \mathbf{F}_{D,\tau} + \mathbf{F}_{L,y} + \mathbf{F}_{D,r}$ in a side view, (b) same configuration in top view with the resulting elliptical cylinder section, (c) definition of the local angle of incidence ψ of a twisted tether with deformed elliptical cross section in an axial view. The relative flow reference frame x_a, y_a, z_a is constructed from the local apparent wind velocity vector $\mathbf{v}_a$ and the radial unit vector $\mathbf{e}_r$. By definition $\mathbf{F}_{D,\tau}$ is aligned with $\mathbf{v}_{a,\tau}$

α of the tether or bridle line segment is measured from the local apparent wind velocity $\mathbf{v_a}$ to the tether axis, which coincides with the radial unit vector $\mathbf{e_r}$, and relates the radial and tangential velocity components to the magnitude as follows

$$v_{a,r} = v_a \cos \alpha \tag{2.8}$$
$$v_{a,\tau} = v_a \sin \alpha. \tag{2.9}$$

The radial and tangential components of the apparent wind velocity are defined as

$$\mathbf{v_{a,r}} = (\mathbf{v_a} \cdot \mathbf{e_r}) \mathbf{e_r}, \tag{2.10}$$
$$\mathbf{v_{a,\tau}} = \mathbf{v_a} - \mathbf{v_{a,r}}, \tag{2.11}$$

while the magnitudes of these components can be calculated from Eq. (2.7) as functions of the corresponding wind and kite velocity components

$$v_{a,r} = v_{w,r} - v_{k,r}, \tag{2.12}$$
$$v_{a,\tau} = v_{w,\tau} - v_{k,\tau} R. \tag{2.13}$$

As illustrated in Fig. 2.2 the aerodynamic force $\mathbf{F_a}$ acting on the tether segment can be represented in the relative flow reference frame x_a, y_a, z_a by a drag component $\mathbf{F_D}$ and two perpendicular lift components $\mathbf{F_{L,y}}$ and $\mathbf{F_{L,z}}$. The lift component $\mathbf{F_{L,z}}$ is caused by the inclination of the cylinder, while the lift component $\mathbf{F_{L,y}}$ is generally fluctuating as a result of unsteady flow separation from the cylinder. Alternatively, the aerodynamic force can be decomposed into a tangential drag force $\mathbf{F_{D,\tau}}$ acting perpendicularly to the tether and in line with $\mathbf{v_{a,\tau}}$, an axial drag force $\mathbf{F_{D,r}}$ acting in line with the tether and a transverse lift force $\mathbf{F_{L,y}}$ acting perpendicularly to the tether and to $\mathbf{v_{a,\tau}}$. This alternative representation will be used in Sect. 2.4.1 to theoretically construct the aerodynamic loading of a tether segment that is inclined with respect to the relative flow.

As a first approximation a tether or line segment can be represented by a circular cylinder. However, there are many practical situations where such approximation is not appropriate. For example, when tapes are used as part of the bridle line system [34] or when the originally cylindrical line is twisted under tension such that the cross section deforms significantly. To characterize deviations from the circular cross section the ellipse ratio e/d is introduced. The definition of the twist angle ψ of a tether with deformed elliptical cross section is shown in Fig. 2.2(c). Because this angle characterizes the orientation of the cross section with respect to the normal component of the relative flow, it can also be regarded as incidence angle. The inclination of the elliptical shape leads to a steady transverse lift component $\mathbf{F_{L,y}}$. A tether with circular cross section is characterized by $e/d = 1$, which, for simplicity, is the assumed shape unless stated otherwise. While yaw of the flying device will add twist to the tether, this chapter assumes no yaw of the flying device.

It is also assumed, unless due to VIV or plunge galloping defined later, that the nominal orientation of the tether section parameter d is perpendicular to the relative flow, as shown in Figs. 2.2(c). Reference test data is often only available for rigid circular cylinders. The angle of twist ψ varies along the tether and by that also the local lift and drag contributions. Depending on the torsional stiffness of the tether, the relative flow can induce an aero-structural coupling phenomenon which is denoted as torsional galloping.

A tether or line is also able to vibrate with a transverse motion such as seen during VIV and a phenomenon denoted as plunge galloping. The vibrations typically have a high frequency and a time scale that is much shorter than the flight dynamic time scale of the kite. The kinematics and the mechanism of the aero-elastic phenomenon are illustrated in Fig. 2.3. The amplitude of the vibration is A_y, the trans-

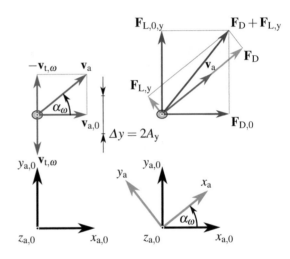

Fig. 2.3 Transverse oscillation of the tether with velocity $v_{t,\omega}$ and resulting vibration-induced angle of attack α_ω (left), decomposition of the resulting aerodynamic force $\mathbf{F}_a$ into drag and lift components $\mathbf{F}_D$, $\mathbf{F}_{L,y}$ and $\mathbf{F}_{L,z}$ which can be transformed back into force components $\mathbf{F}_{D,0}$, $\mathbf{F}_{L,0,y}$ and $\mathbf{F}_{L,0,z}$ in the mean relative flow reference frame $x_{a,0}, y_{a,0}, z_{a,0}$ (right)

verse velocity of the tether or line experiencing transverse vibration is $v_{t,\omega}$ and the resulting vibration-induced angle of attack is α_ω. Both $v_{t,\omega}$ and α_ω vary with time and are often assumed to follow a sine curve for steady vibrations. The transverse velocity $v_{t,\omega}$ requires the modification of the apparent wind velocity from v_a to the vibration-induced relative flow velocity

$$v_a = \sqrt{v_{a,0}^2 + v_{t,\omega}^2}, \tag{2.14}$$

where $v_{a,0}$ is the mean relative flow velocity. Since the transverse vibration mode is possible without twist of the tether or line, the flow-induced transverse vibrations can lead to a significant increase of aerodynamic lift and drag.

Throughout the chapter, the tethers and bridle lines are evaluated as discrete components rather than with system-level interactions. This chapter does not attempt to cover impacts of system-level dynamics, such as other than aerodynamic influences, e.g. wave transmission along the tether from the flying vehicle, varying tension forcing functions from the flying vehicle, elastic behavior of a textile tether, inertial resistance to vehicle motion, etc.

2.3 Operating Envelope for Airborne Wind Energy Applications

Prior to discussing the applicable physics of tethers and bridle lines in a cross flow environment, the ranges of potential operating and environmental conditions are defined. The variables involved are the apparent wind velocity, air density, temperature, tether diameters, tether angles of attack and derived nondimensional parameters such as Reynolds number.

For the purposes of the study, all of the pursued AWE concepts are considered and a set of generic ranges of these parameters is proposed. The considerations include systems that operate at low altitudes and not yet access high altitudes, static and dynamic AWE systems with tethers that are still or move, Ground-gen and Fly-gen systems, ranges of tether or line diameter and temperature ranges based on geography and operating altitude.

A simplified range of airspeeds would be from a low speed of 2.5 m/s low (static tether AWE system) up to a high speed of 80 m/s high (dynamic tether AWE system) [12]. The air density could theoretically range from below sea level, say 1.235 kg/m^3 standard day, towards the upper region of the conceived operating area, basically the jet stream or approximately 0.253 kg/m^3, which is not yet accessible due to existing airspace restrictions and possible technical challenges. Temperatures could range from $-60°$ C at upper altitudes to $+45°$ C in lower deserts. It is noted that the minimum temperature occurs at maximum altitude and vice versa.

Tether and line diameters scale with the power output of the AWE systems because the tensile force is the primary dependency. The anticipated range of diameters covering the smallest lines to the largest tethers is assumed to be 1 to 50 mm.

Assuming steady and uniform wind, a maximum tether angle of attack could theoretically be at or near $90°$ for a portion of a tether when a system overflies the wind window. This could occur due to reel in of the tether and due to the wing's inertia when on a flight trajectory with a continuously increasing inclination angle. Most lines used in bridling have individual angles of attack different from the tether (and each other) where some of these could very likely encounter a $90°$ angle of attack. A minimum tether angle of attack could also be very low, depending on non-nominal wind conditions and landing maneuvers. For the purpose of this study, a range from 0 to $90°$ is considered possible for AWE tethers and bridle lines.

2.4 Physics of Tensioned Cables in Cross Flows

Tethers and bridle lines are essential for the load transfer from the airborne lifting device to the ground. The movement of this tensile structure in a wind field creates an additional aerodynamic loading. While the fluid-dynamic pressure on the leading edge of the cylindrical components is higher than in free stream, the pressure on the sides and trailing edge is lower. The integral pressure and shear stress results in an aerodynamic force on the tensile structure. The trailing wake flow is often turbulent and organized by discrete swirling vortices that shed from the sides of the structure in an alternating phase. The resulting aerodynamic forces are unsteady and can interact with the structure producing movement or deformation, leading to a coupling of fluid and structural motion (fluid-structure interaction). This section details the relevant physical processes.

2.4.1 Aerodynamic Forces and Flow Regimes

As outlined in Sect. 2.2, the aerodynamic force on a cylindrical structure can be decomposed into a drag component, acting in x_a-direction which is aligned with the relative flow, and lift components, acting in y_a- and z_a-directions which are perpendicular to the relative flow. The alternating vortex shedding produces a cyclic loading which can initiate or propagate vibrations of the tensile structure. The vibrations are substantially amplified if the frequency of vortex shedding, i.e. the period of the load cycles, is near a specific harmonic resonance of the structure, at its natural frequency.

According to Hoerner [15] and Bootle [4] the aerodynamic force acting on an inclined circular cylinder in a low-speed flow can be approximated as a superposition of a normal drag contribution $F_{D,\tau}$, depending on the normal velocity component $v_{a,\tau}$, and an axial drag contribution $F_{D,r}$, depending on the axial velocity component $v_{a,r}$. The essence of this "cross flow principle" is that the two perpendicular drag components illustrated in Fig. 2.2(a) are evaluated independently

$$F_{D,\tau} = \frac{1}{2}\rho C_{D,\tau} l_t d v_{a,\tau}^2, \tag{2.15}$$

$$F_{D,r} = \frac{1}{2}\rho C_f l_t \pi d v_{a,r}^2, \tag{2.16}$$

where $C_{D,\tau}$ is the drag coefficient of a cylinder at $\alpha = 90°$, C_f is the skin friction drag coefficient, $l_t d$ is the flow cross section of a cylinder segment, $l_t \pi d$ is the wetted surface area of the segment and ρ is the fluid density.

The two force components can be transformed back to the mean relative flow reference frame and, using Eqs. (2.8) and (2.9), expressed as functions of the relative flow velocity v_a

$$F_D = \frac{1}{2}\rho \left(C_{D,\tau} \sin^3 \alpha + C_f \pi \cos^3 \alpha\right) l_t d v_a^2, \tag{2.17}$$

$$F_{L,z} = \frac{1}{2}\rho \left(C_{D,\tau} \sin^2 \alpha \cos \alpha - C_f \pi \cos^2 \alpha \sin \alpha\right) l_t d v_a^2, \tag{2.18}$$

which leads to the following aerodynamic drag and lift coefficients [4]

$$C_D = C_{D,\tau} \sin^3 \alpha + C_f \pi \cos^3 \alpha, \tag{2.19}$$

$$C_{L,z} = C_{D,\tau} \sin^2 \alpha \cos \alpha - C_f \pi \cos^2 \alpha \sin \alpha. \tag{2.20}$$

In extension of the "crosswind principle", the transverse aerodynamic lift force acting in y_a-direction on a cylinder with arbitrary cross section can be represented as

$$F_{L,y} = \frac{1}{2}\rho C_{L,\tau} l_t d v_{a,\tau}^2, \tag{2.21}$$

where $C_{L,\tau}$ is the lift coefficient at $\alpha = 90°$. This leads to the following transverse lift coefficient

$$C_{L,y} = C_{L,\tau} \sin^2 \alpha. \tag{2.22}$$

In summary, it can be stated that the coefficients defined by Eqs. (2.19), (2.20) and (2.22) are multiplied by the square of the relative flow velocity v_a and a term $1/2\rho l_t d$ to determine the drag and lift force components F_D, $F_{L,z}$ and $F_{L,y}$ acting on the inclined cylinder segment.

The aerodynamic forces generated by a transverse oscillation of the cylinder are illustrated in Fig. 2.3. The vibration-induced force components F_D and $F_{L,y}$ can be transformed into the mean flow reference frame as follows

$$F_{D,0} = F_D \cos \alpha_\omega - F_{L,y} \sin \alpha_\omega, \tag{2.23}$$

$$F_{L,0,y} = F_D \sin \alpha_\omega + F_{L,y} \cos \alpha_\omega. \tag{2.24}$$

To account for the varying relative flow conditions along the moving tether, higher order models generally discretize the tether into connected segments [6]. The aerodynamic forces are evaluated per segment, based on the local apparent wind velocity v_a and angle of attack α, and the equations of motion are solved by stepwise integration over the tether elements.

For the subsonic flows that are relevant within the scope of the chapter, the aerodynamic coefficients depend primarily on the Reynolds number

$$\mathrm{Re} = \frac{v_a d}{v}, \tag{2.25}$$

where v is the dynamic viscosity of the air. In essence, the non-dimensional number is a measure for the ratio of inertial forces and viscous forces in the fluid flow around the cable.

Considering the ranges of apparent wind velocity, tether diameter and kinematic viscosity of air discussed in Sect. 2.3, the expected range of the Reynolds number for AWE applications in general is $42 < \mathrm{Re} < 2.9 \times 10^5$. A specific AWE application will have a much narrower range than this. With the exception of the thickest diameter tether, the majority of tethers and bridle lines operate below the critical Reynolds number condition of $\mathrm{Re}_{crit} \approx 3.5 \times 10^5$ within the sub-critical range defined as $300 < \mathrm{Re} < 1.5 \times 10^5$ [3].

Using the stated operating regimes, the cross flow drag coefficient and the axial skin friction drag coefficient can be plotted as functions of the Reynolds number and together against the tether or line diameter. The result is illustrated in Fig. 2.4. The plotted data for the drag coefficient is time-averaged and does not resolve the fluctuations caused by the unsteady flow separation from the cylinder. The data is

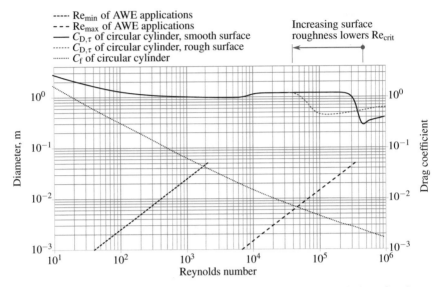

Fig. 2.4 Minimum and maximum Reynolds numbers Re_{min} and Re_{max}, respectively, as functions of cylinder diameter d, mean cross flow drag coefficients $C_{D,\tau}$ of circular cylinders with smooth and rough surface and skin friction drag coefficient C_f of a circular cylinder in axial flow [15, 25]

sourced from circular cylinders which can be assumed to be smooth and have low surface roughness when compared to braided AWE tethers and bridle lines. The critical Reynolds number trends toward lower values and shallower dips with increasing roughness [11, 15, 25]. For cylinders with a rough surface, the critical Reynolds number region can be as low as 3.0×10^4, which is within the range of the top end Reynolds numbers for larger diameter tethers. However, the reduction of drag at the critical Reynolds number diminishes as surface roughness increases. The roughness performance data is based on tests using sand grains of a specific size adhered to a cylinder surface, with surface roughness being between 0.005 to 0.02 (sand grain size to cylinder diameter). In the reference data, drag reduction at the critical Reynolds number appears to trend towards no or negligible drag reduction at a Reynolds number of about 3×10^4 for surface roughness greater than 0.02.

From Fig. 2.4, the approximate range of aerodynamic drag coefficients for a cylinder with $\alpha = 90°$ in the AWE Reynolds number range of interest is $0.98 < C_D < 1.8$. The skin friction coefficients vary to a much greater extent. The skin friction coefficient values are less than 10% of the drag coefficient for Reynolds numbers above about 500. While the friction coefficient is much lower than the aerodynamic drag coefficient, it should be noted that the wetted area is at least π times the section area used for aerodynamic drag calculation, depending on angle of attack, which increases the relative importance of skin friction.

2.4.2 Unsteady Vortex Shedding

The principles of unsteady vortex formation from a circular cylinder are well documented by Blevins [3]. The relevant flow phenomenon for AWE applications is that of a fully turbulent vortex street, occurring in the range from $300 < \text{Re} < 2.9 \times 10^5$ and illustrated in Fig. 2.5.

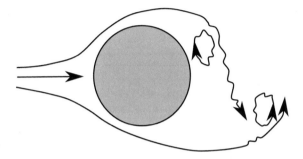

Fig. 2.5 Illustration of a fully turbulent vortex street relevant for AWE applications [19]

When evaluating in two dimensions, the vortex shedding from a rigid stationary cylinder generates a resultant force vector that oscillates in magnitude and direction, as shown in Fig. 2.6. In reference to Drescher [8], the direction of the resultant force vector, which is composed of lift and drag components, varies between $-45°$ and $45°$ at $\text{Re} = 1.12 \times 10^5$. The pressure oscillation occurs at a specific frequency and can be described in terms of the Strouhal number. This non-dimensional number is defined as

$$\text{St} = \frac{f_\text{s} d}{v_\text{a}} \tag{2.26}$$

and used to characterize oscillating flow mechanisms. It relates the shedding frequency f_s to the freestream velocity v_a and the characteristic length d of a subject

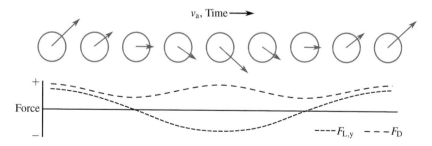

Fig. 2.6 Generic resultant pressure vector for one complete vortex shedding cycle compared against chronologically aligned plot of notional flow-aligned and -transverse force components

body or diameter for tethers and bridle lines. By inverting Eq. (2.26) the shedding frequency can be represented as a function of the Strouhal number.

An approximation of various Strouhal numbers for rough and smooth surfaces, taken from [3], is shown in Fig. 2.7 and combined with observations of Strouhal numbers from AWE relevant testing of kitesurfing line from Dunker [9].

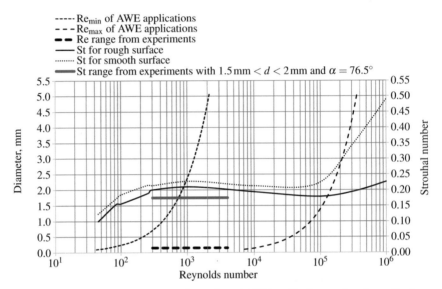

Fig. 2.7 Minimum and maximum Reynolds number of AWE applications as functions of tether diameter, Reynolds number range from experiments, Strouhal number as function of Reynolds number from literature sources for circular cylinders at $\alpha = 90°$ and from experiments with bridle lines at $\alpha = 76.5°$

Given the structural composition and design of a braided tether or bridle line, its surface can be assumed to be rough. Excluding Reynolds numbers below about 300, the Strouhal number for AWE applications is about 0.20. This corresponds to one vortex shedding cycle for every 5 body diameters of airflow past the tether, bridle line or cylinder. The performed tests with braided lines of diameters ranging from 15 to 20 mm, showed Strouhal numbers ranging from 0.17 to 0.18 [9]. The tests were performed with $\alpha = 76.5°$, which according to Eq. (2.17) results in a 10% decrease of drag.

In his original research, Strouhal noted that audible tones resulting from flow around a cylinder were not a function of tension or cylinder length, which in essence means that the natural frequency did not affect the production of the Strouhal frequency tones [30]. Rather, an increase in length of a cylinder produced a louder tone of the same frequency.

2.4.3 Effect of Inclination

An inclination of the cylinder with respect to the relative flow has a predictable effect on shedding frequencies for small deviations from the perpendicular orientation (indicated by subscript n). King [18] proposed a correlation

$$f_s = f_{s,\tau} \sin \alpha, \qquad \text{for} \quad \alpha > 60°, \tag{2.27}$$

where $f_{s,\tau}$ is the value for $\alpha = 90°$.

Naudascher [21] reported that the prediction of vibration frequencies is more complicated for $\alpha < 60$. The shedding phenomenon becomes increasingly three-dimensional and there can be a drastic decrease in vortex strength. Shedding frequencies can also depend on the design of the cylinder tips. The lift force component, which is perpendicular to the relative flow, oscillates at the shedding frequency, but the drag force component, aligned with the flow, oscillates at twice the shedding frequency. This difference in oscillation frequencies is seen in Fig. 2.6. Seemingly, the lift force oscillations would reduce as vortex strength decreases.

Dunker recorded the dominant vibration frequency on several occasions at two times the shedding frequency, where the Reynolds number was close to 300, for $\alpha = 76.5°$. Several secondary non-dominant frequencies were observed at two times the shedding frequency for Reynolds numbers up to about 1000.

2.4.4 Natural Frequency of the Tensile Structure

The free vibration characteristics of the tensile structure also determine how the structure responds to the fluctuating aerodynamic loading caused by unsteady vortex shedding. The effect on the aerodynamic drag can be substantial, for example, a string vibrating at its natural frequency can experience a higher than 300% increase compared to a non-vibrating string.

The natural frequencies f_n of an elastic string are the integer multiples of the fundamental frequency and can be formulated as

$$f_n = \frac{n}{2l_t} \sqrt{\frac{F_t}{\lambda}}, \qquad n = 1, 2, 3, \ldots, \tag{2.28}$$

where n is the vibration node number, l_t the length of string, F_t the tensile force and λ the mass per unit length, which is also denoted as linear density.

A common assumption for a vibrating string is that both ends of the string are fixed. For AWE applications the situation is different. The upper end of the tether is attached to a flying device which moves in space and exerts a traction force on the tether. However, compared to the tether the flying device has enough mass to consider it for the vibration dynamics as an end point with prescribed motion. The lower end of the tether is generally reeled from a winch at a fixed position. Many of the implemented AWE systems use the winch control algorithm to constantly adjust the reeling speed to maintain the tether force below a permitted maximum value. It is clear that this particular setup and the effect on the vibration dynamics requires further investigation.

To assess the range of natural frequencies relevant for AWE applications and to compare with related applications, the lengths and tensions occurring during nominal flight conditions were estimated. To eliminate variations due to different cable materials Dyneema® SK75, was chosen as the primary tether material, with approximately $\lambda = 6.5 \times 10^{-6}$ kg/m per unit strength $F_t = 9.81$N provided. The particular material is a common selection for AWE and related industries due to its superior strength to weight and size properties. Various types of Dyneema material and the competing Spectra® material exist, both based on High Molecular Weight Polyethylene High Modulus Polyethylene (HMPE). Additional information about HMPE tethers is presented in Bosman [5].

Some AWE applications are based on ram-air wings or leading edge inflatable tube kites, which are also used for skydiving, paragliding and kite boarding [10]. The comparison of natural frequencies also includes these applications, using the strength of bridle lines of common commercial products. To setup a generalized comparison matrix, the bridle lines used for skydiving are rated at 2256 N (230 kgf), the lines for paragliding at 1128 N (115 kgf) and the lines for kite boarding at 2256 N (230 kgf). Multiplying these force ratings by $6.5/9.81 \times 10^{-6}$ kg/(Nm) yields the mass per unit length λ for each case. The strength of the AWE tether was selected to provide a minimum of 4 times the tension occurring in crosswind flight operation, based on commercially available materials. Natural frequencies occurring in AWE and related application areas are compared in Table 2.1.

Line tension, N	Length of tether or bridle line, m								
	1	7	8	10	15	30	50	400	>400
5	40.9	5.8	5.1	4.1					
15	70.9	10.1	8.9	7.1	3.3	1.7			
20	57.9	8.3	7.2	5.8	3.9	1.9			
70	108.9	15.5	13.5	10.8	7.2	3.6			
500					19.3	9.7	8.2	1.0	<1.0
5000							6.5	0.8	<0.8
50000							7.2	0.9	<0.9

		λ, kg/m
·············	Parachute bridling (skydiving)	0.00149
---------	Parachute and paraglider bridling	0.00075
-----	Paraglider bridling	0.00149
-··-··-···	Kite boarding tether	0.00149
———	AWE tether	500 N	0.00075
		5000 N	0.01193
		50000 N	0.09542

Table 2.1 Comparison of natural frequencies occurring in AWE and related application areas

The lengths of tethers or bridle lines are based on personnel-sized commercially available products for the various industries listed. Generally speaking, ram-air sky-diving parachutes and paragliders have the shortest bridle lines (upper cascading of suspension lines). The lines below the cascades are much longer, more so on paragliders than skydiving parachutes, which represent the top end of the limits. For kite boarding a short cascaded bridling up at the kite is used with much longer lines below the cascades down to the kite boarder. An arbitrary minimum length of 50 m is used for AWE applications.

The line tensions used for skydiving, paragliding and kite boarding are based on average human weights (plus assumptions for relevant equipment) distributed into the common tethering or bridling structure of the wings for that industry. This takes into account the number of lines used, cascading of lines and generic distribution of load among the lines. For AWE applications, generic line tensions were used. Values for line lengths and tensions were rounded.

2.4.5 Mass Ratio, Damping Factor and Mass Damping

An important indicator for the susceptibility of a tether or line to vortex-induced vibration is the mass ratio

$$m^* = \frac{\lambda}{\rho d^2}, \tag{2.29}$$

where λ is the tether mass per unit length and ρd^2 is proportional to the displaced fluid mass per unit length. The tether mass generally includes an added mass term representing a contribution of the fluid. Because of the large density ratio this term can be considered negligible.

The mass ratio relates two primary driving factors for vibrations of a cylinder in a transverse fluid flow. The cylinder mass in the numerator is a measure for the acceleration that the cylinder experiences in response to an external force. The fluid mass in the denominator, on the other hand, is a measure for the force that the fluid flow exerts on the cylinder. A higher cylinder mass or a lower fluid density result in a higher mass ratio and decrease the susceptibility to VIV. Conversely, a lower mass or higher density lead to a lower mass ratio and increase the susceptibility to VIV. It is inferred therefore that lower mass ratios, being more susceptible to VIV, are also associated with higher vibration amplitudes. Empirical data from Dale, Feng and Scruton described in [3] support this conclusion.

Therefore, tether or bridle lines made of fabric materials are likely more susceptible to VIV than heavier materials, such as wire or cable alternatives. This has a generally negative effect on AWE, since the tethers are desired to be as light as possible to minimize the airborne mass. The AWE industry has currently converged on Dyneema® and equivalent Spectra® materials with a density of around 0.97 g/cm^3. In laboratory tests, Dyneema lines with mass ratios of about $m^* = 720$ have exhibited strong VIV effects with corresponding significant increases in drag [9].

Slightly larger mass ratios can be expected from recent improvements in braiding efficiency or even from using tethers with sheaths and unbraided parallel fiber cores.

It should be noted that increasing the strength and diameter of a tether does not significantly change the mass ratio. As tethers are made stronger by including more fiber content into the braid, the raw material fiber mass and volume remain unchanged. This suggests that any tether using HMPE fiber material is strongly susceptible to VIV and drag increase.

However, the mass ratio m^* is not the only parameter determining the vibration amplitude A_y. The damping factor quantifies how much energy is shed off and dissipated per cycle relative to the total oscillation energy

$$\zeta = \frac{E_c - E_{c+1}}{4\pi E_c}. \tag{2.30}$$

The oscillations of a string in a static fluid, such as in a pluck test, will decay over the oscillation cycles due to resistive surface pressures generated by the fluid during the oscillatory motion. According to Blevins [3], the natural logarithm of the amplitude ratio of any two successive cycles of a lightly damped structure in free decay equals to $2\pi\zeta$. For AWE applications, this would require making amplitude measurements or peak velocity measurements of a selected tether during a pluck test.

Assuming a logarithmic decrement of the oscillations, the product term $m^*\zeta$ is a non-dimensional parameter denoted as mass damping factor

$$\delta_r = \frac{2\lambda(2\pi\zeta)}{\rho d^2}, \tag{2.31}$$

which is identical to the Scruton number Sc. The mass damping factor describes the effect of the tether diameter d on the oscillation amplitude A_y [3, 24].

The above considerations imply that the tensile structure is linear and viscously damped, suspended between fixed points. However, structural damping can also be due to absorption of energy by the flying wing, kite, winch and other connected system components. Consequently, it may prove difficult to use existing theory from literature alone to determine the oscillation characteristics.

2.4.6 Vibration Amplitude and Effect on Drag

Sarpkaya [24] has derived a simplified expression for the maximum oscillation amplitude A_y of a taut string or cable as a function of the mass damping factor δ_r

$$A_y = \frac{0.369d}{\sqrt{0.06 + \left(2\pi St^2 \delta_r\right)^2}}. \tag{2.32}$$

From Blevins [3], Eq. (2.32) agrees well over the range $2 \times 10^2 < Re < 2 \times 10^5$. Understanding that other formulations of the maximum amplitude are proposed in this reference, they are considered to be within 15% of each other. Further, when the mass damping term is greater than 64, the peak amplitudes are normally less than $0.01d$.

It is well known that larger oscillation amplitudes result in a larger effective aerodynamic drag $C_{D,\mathrm{eff}}$. For rigid cylinder experiencing transverse oscillations, a near linear fit exists from data compiled from multiple sources [24, 31, 33]

$$\frac{C_{D,\mathrm{eff}}}{C_D} = 1 + 2.1\frac{A_y}{d}, \tag{2.33}$$

where C_D is the cylinder drag at $A_y = 0$.

It can be concluded that a string of lower mass in a higher density fluid shall experience larger oscillation amplitudes and thus drag coefficients well above nominal. Conversely, a string of higher mass in a lower density fluid is expected to experience lower amplitudes and drag coefficients closer to nominal. Pressure vectors associated with the small tether and large tether examples are illustrated in Fig. 2.8.

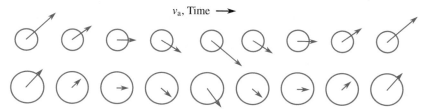

v_a, Time →

Fig. 2.8 Generic pressure vectors during one vortex shedding cycle of a string of lower mass in a higher density fluid (top), and one cycle of a string of higher mass in a lower density fluid (bottom)

The presented conventional framework of mass damping does not closely associate the magnitude of A_y to the natural frequency f_n. For the higher harmonic oscillations ($n = 2,3,\dots$) the tether body has less time to react to changes in the pressure vector direction and magnitude. The effect of shortened vortex shedding cycle times has been observed to reduce the oscillation amplitude and consequently the elevated drag associated with VIV [9].

For AWE applications, further analysis and likely testing will be required to determine the trend in VIV sensitivity due to mass damping as the diameter of tether increases and in the change of vibration amplitude A_y as the natural frequency f_n increases.

2.5 Elevated Drag Regimes

Vortex-Induced Vibrations (VIV) lock-in and galloping are two different aero-elastic coupling phenomena which can occur in AWE applications and which can substantially affect the aerodynamic drag of the tensile structure connecting the aerodynamic lifting device with the ground. Because increased drag leads to reduced flight speeds the occurrence of these phenomena can negatively impact the performance of the AWE system. This section further details the aero-elastic mechanisms and maps the relevant regimes on the basis of wind tunnel measurements of Dunker [9].

2.5.1 Vortex-Induced Vibrations and Lock-in

Vortex-Induced Vibrations (VIV) are caused by unsteady flow separation from an elastic structure and the resulting cyclic variation of fluid forces. In reaction to these, the structure deforms, changing its kinetic and potential energy. The deformation motion in turn changes the relative flow with corresponding changes in the fluid forces. This fluid-structure coupling mechanism exhibits all physical contributions that are required for forced oscillations: an exciting periodic force, an elastic restoring force and inertia as well as aerodynamic damping.

The main implication for AWE is the case when the flow shedding frequency is in harmonic resonance with a natural frequency f_n of the tether and bridle line system. This type of resonance produces vibration amplitudes and aerodynamic drag many times larger than sub- and super-harmonic resonance where the vibration of the structure is at a specific multiple or fraction of the shedding frequency, respectively.

In basic physical terms, lock-in is the alignment of vortex shedding frequencies with the natural frequency of the vibrating structure. The alignment can be up- or down-shifted from the shedding frequency of the stationary structure to match the natural frequency over a range of freestream velocities. Since vortices are commonly formed at maximum displacement of a transverse vibration, the vibration frequency exerts some influence over vortex wake position as well as its phasing. A common range provided in literature as prone to lock-in is when $0.7 f_s > f_n > 1.3 f_s$. During lock-in, the effective drag $C_{D,eff}$ remains elevated over a range of velocities. Across this range, the tether or bridle line experiences a near constant vibration at the nearest natural frequency.

Vortex shedding produces cyclic drag and lift force components that act on the tensile structure, as shown in Fig. 2.4. As consequence of the vibrations, the shedding along the axis of the structure becomes more correlated [32] organizing the wake in three dimensions, the vortices become stronger [13], the drag increases [2] and in the case of traverse vibration, the ability of the shedding frequency to lock-in to the vibration frequency is increased [13]. For certain conditions, the propensity for a given vibration to create stronger vortices, to in turn create stronger amplitude vibrations and so on, can be seen. For a stationary structure with cylindrical cross section, lock-in can occur with as much as $\pm 40\%$ deviation from the nominal shedding frequency.

A more detailed analysis of the physical phenomena governing lock-in, including the necessary illustrations to effectively communicate the vortex street formation

and phase, as well as the more complex principles of hysteresis and sub- and super-
harmonics, can be further sought in Blevins [3] and Naudascher [21].

Dunker [9] studied the lock-in characteristics of braided lines on the basis of wind
tunnel measurements. As illustrated in Fig. 2.9, a line with $d = 1.5$ mm is seen to
have a dominant vibration frequency matching a Strouhal number of St $= 0.172$. For

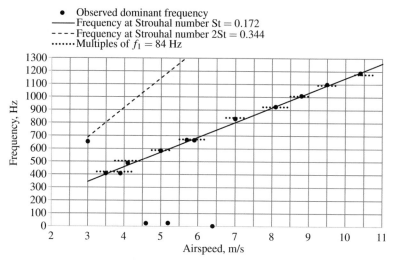

Fig. 2.9 Dominant vibration frequencies and lock-in for Dyneema® line with $d = 1.5$ mm and
$\alpha = 76.5°$ at Re $= 300$ to 1000. Adapted from Dunker [9]

instances where a single lock-in was documented across a velocity range, one each
vibration condition was observed on either side of the stationary shedding frequency
Strouhal curve. The lock-in ranges occurring at 408 and 665 Hz are examples of
this. The fundamental frequency for this line was obtained from a pluck test and
was observed to be $f_1 = 84$ Hz.

Here it can be seen that the dominant frequency observed at the lowest measurement is at twice the shedding frequency. This is unusual in that the tether angle of
attack is relatively small for this characteristic to occur, where normally this can
occur for $\alpha < 60°$.

Several instances of a major secondary vibration frequency were observed at
twice the shedding frequency, however, these were not always the dominant vibration frequency. As mentioned previously, the lift force oscillations occur at the
Strouhal frequency, but drag force oscillations can occur at twice the Strouhal shedding frequency. The vibration spectrum map in Fig. 2.10 displays both vibration
modes. The magnitude of the peaks represents the relative dominance of the vibration frequency.

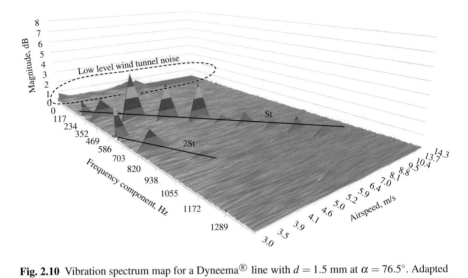

Fig. 2.10 Vibration spectrum map for a Dyneema® line with $d = 1.5$ mm at $\alpha = 76.5°$. Adapted from Dunker [9]

2.5.2 Impact on Aerodynamic Drag Characteristics

Because the vibration amplitude is maximum during lock-in, the resonance phenomenon negatively impacts the aerodynamic drag characteristics of the tether and bridle line system. Continuing with the example of a line with $d = 1.5$ mm, Fig. 2.11 compares the measured effect of lock-in on the line drag with the theoretical drag

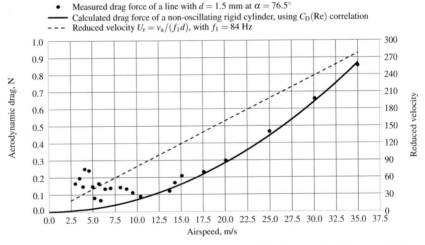

Fig. 2.11 Measured aerodynamic drag force and theoretical drag force of a rigid, non-vibrating cylinder. For the definition of the reduced velocity see Eq. (2.34). Adapted from Dunker [9]

of a rigid, non-vibrating cylinder. The most important observation is that several drag measurements exceed the theoretical baseline value by more than 300%. These instances of elevated drag mostly, but not exclusively, occur during observations of vibration and/or audible harmonics for the stated condition.

Here it is seen that higher velocities, those with expected smaller vibration amplitudes, match closer to the theoretical C_D-curve than lower velocities, even though the Strouhal frequency vibration mode is present. This closer correlation begins around airspeed 17.5 or 20 m/s, which is around the natural frequency f_{23}.

2.5.3 Galloping

Galloping is a movement-induced excitation of elastic tensile structures with non-circular cross sections which can result in very large amplitude vibrations at much lower frequencies than the shedding frequency. It is an aero-elastic phenomenon in which aerodynamic forces cause a distortion of the position or orientation of the structure. When the structure vibrates in a transverse uniform fluid flow its orientation with respect to the relative fluid flow is constantly changing, as depicted in Fig. 2.3. Consequently, the lift and drag components of the fluid force vary as the velocity of the structure under vibration varies. Above the critical freestream velocity of the structure, the result is a condition in which energy is fed into the vibration motion.

This is important for AWE applications as many tethers are braided, flexible and subject to distortion from braid style, repetitive winding and other handling operations and a tether with circular cross section can develop a non-circular cross section. A tether or line may also have a non-circular cross section upon manufacture. In power cables with twisted strand design, ice buildup of just 10% of the diameter is sufficient to produce plunge galloping [29].

Galloping generally comes in two forms, plunge galloping and torsional galloping, where the former is a transverse motion with one degree of freedom relative to the free stream flow v_a, as shown in Fig. 2.8, and the latter is a torsion motion along the structure axis, denoted as twist, as shown in Fig. 2.3. Mixed plunging modes are possible and involve inertial coupling, while the one degree of freedom models assume no inertial coupling [3]. Generally, the flow velocity v_a must be much higher than the vibration velocity $v_{t,\omega}$ of the structure, such that the fluid flow has time to react to the structural motion and to account for the varying vibration-induced angle of attack α_ω (see Fig. 2.3) or the angle of incidence ψ (see Fig. 2.2). For this reason, quasi-steady fluid dynamics can be assumed, which means that fluid forces are calculated using the instantaneous relative velocity and orientation of the structure.

Plunge galloping is sensitive to the aerodynamic characteristics of the tether of bridle line. For galloping to occur, a negative C_L at positive α_ω is required and vice versa [3]. Generally, plunge galloping is possible for cases where $f_s \gg f_1$, for a reduced velocity [3]

$$U_r = \frac{v_a}{f_1 d} > 20. \tag{2.34}$$

This non-dimensional parameter quantifies the distance covered by the mean flow during a single vibration cycle, measured in terms of the diameter, which is the characteristic length of the flow problem. Almost all AWE systems will nominally operate at $U_r > 20$, which is also a condition underlying the measurements described in Fig. 2.11.

One example of plunge galloping, or potentially also combined plunge and torsional galloping, of a tensioned line at Reynolds numbers relevant for AWE is available from Siefers [28]. The investigated braided line of rectangular cross section with rounded corners, effective diameter of $d = 1.8$ mm and airspeed $v_a = 27.5$ m/s experienced a vibration frequency of $f = 35.4$ Hz with an amplitude of about $A_y = 5d$. The Strouhal frequency under the same conditions is approximately $f_s = 2250$ Hz. The aspect ratio of the cross section of between $e/d = 2$ to 3 is consistent with the predictions of Naudascher [21] which indicate a possible galloping range of $e/d = 1$ to 3.

Fig. 2.12 Tensioned line with rectangular cross section and rounded corners experiencing plunge galloping in wind tunnel tests [28]. Stationary line indicated by blue lines, up/down vibration of the line in this photo is at an amplitude of about $A_y = 4d$ at a reduced velocity $U_r = 441$

The tests presented by Siefers [27] indicate only minimal increases of aerodynamic drag by plunge galloping. While it is important to note that very large amplitude oscillations are possible under plunge galloping, e.g. power lines have experienced amplitudes as high as 10 m [21], this may not impact the drag values substantially. However, although system level interactions are not discussed in this chapter, such amplitudes would provide a potential inertial-based forcing function from the tether into the flying apparatus which would need to be considered when a system level analysis is performed.

In torsional galloping, the tether is reoriented about its axis in the relative flow with velocity v_a via the twist angle of incidence ψ and the corresponding angular velocity $\dot{\psi}$ [23]. Torsional galloping leads to varying angles of attack along the entire length of the structure. The phenomenon is sensitive to the torsional spring constant k, a material-structure property resisting twist deformation, and the aerodynamic moment coefficient C_M.

Blevins [3] provides a torque galloping onset parameter, but as it is not based on a threshold value as in plunge galloping, rather instead on a number of parameters dependent on cross section shape other than circular, a guideline for AWE applications cannot be extracted. Torsional galloping will need to be evaluated on a case-by-case basis for tethers or bridle lines with non-circular cross sections.

Tethers or bridle lines with cross section aspect ratios of $e/d > 1$ can have cross sectional aerodynamic characteristics C_D, C_L and C_M which vary significantly with the twist angle ψ and should thus be considered strongly susceptible to torsion galloping. For torsional galloping to occur generally a negative moment coefficient C_M at positive ψ is required and vice versa [3].

Siefers [27] also observed torsional galloping with C_D up to 210% larger than the average static result. It is noted that with torsional galloping an increase in surface area is exposed to the freestream flow, which could potentially contribute to the changes in drag force, however, torsional amplitude resulted in ψ less than 19° for this testing.

The prediction of galloping for tethers and bridle lines is dependent on knowing the coefficients of lift, drag and moment for various angles of incidence ψ with respect to the relative flow, and the torsional spring constant, which will all be unique for each tether or line. The potential for galloping and elevated drag forces is present, however this result will remain unique to each tether or line.

Tethers, while not perfectly cylindrical, are normally axis-symmetric, of general round cross section and have cross section aspect ratios of around $e/d = 1$ and are thus not strongly prone to galloping, especially torsional galloping. However, as mentioned, caution should be taken when any deformation from this round cross section does occur, e.g. during manufacturing or from processes such as winding or other handling. Full scale testing of tethers should be attentive to vibration modes observed.

2.5.4 Evaluation of Tether Usages

AWE systems with a static tether that predominantly operates in a fixed location, such as for buoyant lifting devices, are generally exposed to more uniform freestream velocities than dynamic tether systems. Additionally, much lower airspeeds are expected since the tether is not moving. This results in a lower reduced velocity, as defined in Eq. (2.34), but still with a value $U_r > 20$, which is the general threshold for galloping to occur. For static tether systems both lock-in vibration and galloping modes can thus occur.

AWE systems with a tether that pivots about a fixed ground location, are subject to varying local conditions of airspeed, Reynolds number and angle of incidence as functions of the radial distance from the ground. The basic kinematics are illustrated in Fig. 2.1. Although not in all situations, such as in downwind flight, the relative velocity generally increases as a function of the radial distance along the tether. This affects the local shedding frequency along the tether or bridle line, which likewise increases with tether length.

Assuming that vibration lock-in is possible within the vortex shedding regime $0.7f_s < f_n < 1.3f_s$, the susceptible regions of a tether can be determined. Potentially many different localized lock-in regions could coexist along the tether. These regions would be separated by regions of tether that are not locked-in. An example of a tether with multiple lock-in regions is shown in Fig. 2.13. It is yet uncertain how multiple lock-in regions and corresponding dormant regions would interact and whether multiple localized modes could be achieved in one contiguous tether.

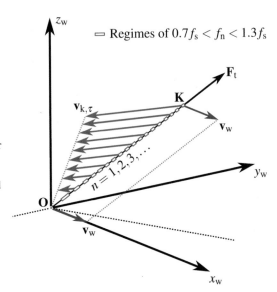

Fig. 2.13 Tether with lock-in regions determined on the basis of the local relative flow velocity at the tether. Compare also with Fig. 2.1 for more kinematic details. The tether regions outlined in red beyond this first region would increment by 1 and would dependent on wind velocity and local tether velocity. The lock-in regions become narrower with increasing distance from the ground attachment point

Both VIV and galloping are conventionally analyzed at a constant freestream velocity v_a. It can be assumed that the constantly changing relative flow conditions of dynamic tether systems impede the formation and sustainment of steady galloping cycles. Vortex shedding, however, will always occur. When vortex-induced vibrations are possible and lock-in conditions are met, the lock-in can occur much quicker than a gallop, appearing instantaneously. A challenge for AWE will be to understand the reaction of vortex forces that occur at fundamentally different frequencies along a line.

Audible tones or whistling of a tether in flight are generally strong indicators for vibration lock-in. While the induced vibrations substantially increase the aerodynamic drag, attention should also be given to eliminate the audible tones that emanate from tethers and bridle lines that are not obviously vibrating, i.e. Strouhal frequency tones. The literature on the topic is insufficient to determine whether elimination of sound has an impact on drag as a result of advantageously controlling the shedding frequency.

The need to investigate multiple vibration regimes in a single tether is a valid concern. Multiple vibration modes have at least occurred in subspan galloping, where, by definition, a localized subsection of a cable oscillates at a different frequency and amplitude than other sections of the same cable.

2.6 Mechanisms to Attenuate Vibration and Reduce Drag

Mechanisms to reduce aerodynamic drag are based mostly on the attenuation of vibration phenomena. A number of additive design concepts exist to reduce VIV for tensile structures with cylindrical cross section. These comprise helical strakes, shrouds, slats, fairings, splitters, ribbons, guide vanes and spoiler plates.

Blevins [3] provides illustrated examples and further references of these concepts. The challenge for AWE applications is to incorporate these into a flexible, elastic, braided textile that is typically wound onto a large drum under tension. Any concept that considers tethers or bridle lines with other than round cross section will need a serious solution to maintaining desired orientation or risk elevated nominal drag, galloping, tether controllability, winding ability onto a spool and more.

Although thicker in cross section, a kite boarding line with helical strake braiding yielded a lower drag over some velocities of interest than a thinner line with round cross section [9]. Examples are shown in Fig. 2.14. The strake disrupts synchro-

Fig. 2.14 Examples of helical strake braiding of kite boarding lines

nization of trailing vortices and modulates the local Strouhal number [22]. The drag force improvement was attributed to the reduced susceptibility to VIV. Although some vibration effects were still observable, the concept can potentially be used for AWE applications. Further development and understanding of the relevant vibration regimes in laboratory and full scale environments are required.

Another option is the addition of an aerodynamically shaped fairing to the tether [1, 14]. These concepts promise not only better flow control and less susceptibility to all forms of vibration-induced drag, but also consistently reduce the drag across a wide range of scales and airspeeds.

Jung [16] has shown that the addition of latex coating to a variety of fast rope braids for helicopters consistently reduced the drag coefficient by 15 to 50%, depending on size and braid construction. Several of the analyzed ropes are in the diameter range of large-scale AWE applications. However, braiding styles were generally much rougher and fuzzier, using different materials than common for AWE. Given the tested range of braid geometries, the direct carry-over of this technique to AWE applications is uncertain.

Laboratory experiments of selective air ejection along the cylinder axis have shown to decrease drag coefficients by approximately 20% [17]. Although a round cross section could be maintained on the basis of this concept, the complexity of the air duct and ejection system seems prohibitive for AWE applications.

2.7 Conclusion

This chapter has introduced the physics of tether and bridle line drag in airborne wind energy (AWE) applications. It has further provided a compilation of experimental data and correlations to analyze line drag and identify regimes of elevated drag caused by Vortex-Induced Vibration (VIV) and by galloping. AWE developers and researchers should find relevant ranges of Reynolds numbers, drag and friction coefficients, Strouhal numbers and natural frequencies or the information necessary to retrace these in the given literature sources.

Correlations for predicting the vibration amplitude and the corresponding effective drag coefficients are presented in the application context of AWE. Further references have been provided to supplement finer details and complexities beyond the scope of this introductory chapter. Lock-in and galloping phenomena were shown in laboratory experiments to have elevated drag regimes, where drag forces were observed to be over 300% for lock-in and up to 210% for torsional galloping.

Both lock-in and galloping have been shown to be relevant vibration modes for static tether systems but only lock-in as being potentially relevant for dynamic tether systems. Dynamic tether systems will experience shedding frequencies sufficient to align with a natural frequency of the tensile structure and to achieve lock-in. However, because of the constantly varying airspeed along the tensile structure operated in crosswind motion, the vibration response to multiple lock-in frequencies that occur along the tensile structure is not yet certain and subject to future research. Finally, a collection of conventional and more recent VIV suppression solutions were presented which could be utilized by AWE applications.

References

1. Bevirt, J.: Tether Sheaths and Aerodynamic Tether Assemblies. US Patent 0,266,395, 2011
2. Bishop, R. E. D., Hassan, A. Y.: The Lift and Drag Forces on a Circular Cylinder in a Flowing Fluid. Proceedings of the Royal Society of London A: Mathematical, Physical and Engineering Sciences **277**(1368), 51–57 (1964). doi: 10.1098/rspa.1964.0005
3. Blevins, R. D.: Flow Induced Vibration. 2nd ed. Krieger Publishing Company, Malabar, FL, USA (2001)

4. Bootle, W. J.: Forces on an inclined circular cylinder in supercritical flow. AIAA Journal **9**(3), 514–516 (1971). doi: 10.2514/3.6213

5. Bosman, R., Reid, V., Vlasblom, M., Smeets, P.: Airborne Wind Energy Tethers with High-Modulus Polyethylene Fibers. In: Ahrens, U., Diehl, M., Schmehl, R. (eds.) Airborne Wind Energy, Green Energy and Technology, Chap. 33, pp. 563–585. Springer, Berlin Heidelberg (2013). doi: 10.1007/978-3-642-39965-7_33

6. Breukels, J.: An Engineering Methodology for Kite Design. Ph.D. Thesis, Delft University of Technology, 2011. http://resolver.tudelft.nl/uuid:cdece38a-1f13-47cc-b277-ed64fdda7cdf

7. Cherubini, A., Papini, A., Vertechy, R., Fontana, M.: Airborne Wind Energy Systems: A review of the technologies. Renewable and Sustainable Energy Reviews **51**, 1461–1476 (2015). doi: 10.1016/j.rser.2015.07.053

8. Drescher, H.: Messung der auf querangeströmte Zylinder ausgeübten zeitlich veränderten Drücke. Zeitschrift für Flugwissenschaften und Weltraumforschung **4**, 17–21 (1956)

9. Dunker, S.: Experiments in Line Vibration and Associated Drag for Kites. AIAA Paper 2015-2154. In: Proceedings of the 23rd AIAA Aerodynamic Decelerator Systems Technology Conference, Daytona Beach, FL, USA, 30 Mar–2 Apr 2015. doi: 10.2514/6.2015-2154

10. Dunker, S.: Ram-Air Wing Design Considerations for Airborne Wind Energy. In: Ahrens, U., Diehl, M., Schmehl, R. (eds.) Airborne Wind Energy, Green Energy and Technology, Chap. 31, pp. 517–546. Springer, Berlin Heidelberg (2013). doi: 10.1007/978-3-642-39965-7_31

11. Fage, A., Warsap, J. H.: Effects of turbulence and surface roughness on drag of circular cylinders. Reports and memoranda 1283, Aeronautical Research Committee, Oct 1929. http://naca.central.cranfield.ac.uk/reports/arc/rm/1283.pdf

12. Goldstein, L.: Airborne wind energy conversion systems with ultra high speed mechanical power transfer. In: Ahrens, U., Diehl, M., Schmehl, R. (eds.) Airborne Wind Energy, Green Energy and Technology, Chap. 13, pp. 235–247. Springer, Berlin Heidelberg (2013). doi: 10.1007/978-3-642-39965-7_13

13. Griffin, O. M., Ramberg, S. E.: On vortex strength and drag in bluff-body wakes. Journal of Fluid Mechanics **69**(04), 721 (1975). doi: 10.1017/S0022112075001656

14. Griffith, S., Lynn, P., Montague, D., Hardham, C.: Faired tether for wind power generation systems. Patent WO 2009/142762, 26 Nov 2009

15. Hoerner, S. F.: Fluid-Dynamic Drag. Bricktown, Brick Town, NJ, USA (1965)

16. Jung, T. P.: Wind Tunnel Study of Drag of Various Rope Designs. AIAA Paper 2009-3608. In: Proceedings of the 27th AIAA Applied Aerodynamics Conference, San Antonio, TX, USA, 22–25 June 2009. doi: 10.2514/6.2009-3608

17. Kim, J., Choi, H.: Distributed forcing of flow over a circular cylinder. Physics of Fluids **17**, 033103 (2005). doi: 10.1063/1.1850151

18. King, R.: A review of vortex shedding research and its application. Ocean Engineering **4**(3), 141–171 (1977). doi: 10.1016/0029-8018(77)90002-6

19. Lienhard, J. H.: Synopsis of lift, drag, and vortex frequency data for rigid circular cylinders. Research Division Bulletin 300, Washington State University, Pullmann, WA, USA, 1966. http://www.uh.edu/engines/vortexcylinders.pdf

20. Loyd, M. L.: Crosswind kite power. Journal of Energy **4**(3), 106–111 (1980). doi: 10.2514/3.48021

21. Naudascher, E., Rockwell, D.: Flow-Induced Vibrations: An Engineering Guide, Chap. 2–3, 7, 9. A. A. Balkema Publishers, Rotterdam, The Netherlands (1994)

22. Nebres, J. V., Batill, S. M.: Flow around a cylinder with a spanwise large-scale surface perturbation. AIAA Paper 93-0657. In: Proceedings of the 31st Aerospace Sciences Meeting, Reno, NV, USA, 11–14 Jan 1993. doi: 10.2514/6.1993-657

23. Oudheusden, B. W. van: Investigations of an aeroelastic oscillator: Analysis of one-degree-of-freedom galloping with combined translational and torsional effects. LR-707, Delft University of Technology, Dec 1992. http://resolver.tudelft.nl/uuid:0d3c0eac-2bab-422a-9e0f-8c74da19dcd0

24. Sarpkaya, T.: Fluid Forces on Oscillating Cylinders. Journal of the Waterway, Port, Coastal and Ocean Division **104**, 275–290 (1978)
25. Schlichting, H.: Boundary-layer theory. McGraw-Hill (1979)
26. Schmehl, R. (ed.): Book of Abstracts of the International Airborne Wind Energy Conference 2015. Delft University of Technology, Delft, The Netherlands (2015). doi: 10.4233/uuid: 7df59b79-2c6b-4e30-bd58-8454f493bb09
27. Siefers, T. M., Campbell, J. P., Clark, D. K., McLaughlin, T. E., Bergeron, K.: Quantification of Drag from Flat Suspension Line for Parachutes and the Influence of Flow Induced Vibrations. AIAA Paper 2016-1777. In: Proceedings of the 54th AIAA Aerospace Sciences Meeting, San Diego, CA, USA, 4–8 Jan 2016. doi: 10.2514/6.2016-1777
28. Siefers, T., Greene, K., McLaughlin, T., Bergeron, K.: Wind and Water Tunnel Measurements of Parachute Suspension Line. AIAA Paper 2013-0064. In: Proceedings of 51st AIAA Aerospace Sciences Meeting, Grapevine (Dallas/Ft. Worth Region), TX, USA, 7–10 Jan 2013. doi: 10.2514/6.2013-64
29. Simpson, A.: Fluid-Dyamic Stability Aspects of Cables. In: Shaw, T. L. (ed.) Mechanics of Wave-Induced Forces on Cylinders, pp. 90–132. Pitman Publishing Ltd., London (1979)
30. Strouhal, V.: Ueber eine besondere Art der Tonerregung. Annalen der Physik und Chemie **5**(10), 216–251 (1878). http://www.deutschestextarchiv.de/strouhal_tonerregung_1878
31. Tanida, Y., Okajima, A., Watanabe, Y.: Stability of a circular cylinder oscillating in uniform flow or in a wake. Journal of Fluid Mechanics **61**(4), 769–784 (1973). doi: 10.1017/s0022112073000935
32. Toebes, G. H.: The Unsteady Flow and Wake Near an Oscillating Cylinder. ASME Journal of Basic Engineering **91**(3), 493–502 (1969). doi: 10.1115/1.3571165
33. Tørum, A., Anand, N. M.: Free Span Vibrations of Submarine Pipelines in Steady Flows-Effect of Free-Stream Turbulence on Mean Drag Coefficients. Journal of Energy Resources Technology **107**(4), 415–420 (1985). doi: 10.1115/1.3231212
34. Vlugt, R. van der, Peschel, J., Schmehl, R.: Design and Experimental Characterization of a Pumping Kite Power System. In: Ahrens, U., Diehl, M., Schmehl, R. (eds.) Airborne Wind Energy, Green Energy and Technology, Chap. 23, pp. 403–425. Springer, Berlin Heidelberg (2013). doi: 10.1007/978-3-642-39965-7_23

Chapter 3
Analytical Tether Model for Static Kite Flight

Nedeleg Bigi, Alain Nême, Kostia Roncin, Jean-Baptiste Leroux, Guilhem Bles,
Christian Jochum and Yves Parlier

Abstract The use of traction kites as auxiliary propulsion systems for ships appears
to be a high-potential alternative for fuel saving. To study such a system a tether
model based on the catenary curve has been developed. This model allows calculating static flight positions of the kite on the edge of the wind window. The effect
of the wind velocity gradient is taken into account for the evaluation of the aerodynamic forces acting on kite and tether. A closed-form expression is derived for the
minimum wind velocity required for static flight of the kite. Results are presented
for a kite with a surface area of 320 m^2 and a mass of 300 kg attached to a tether
with a diameter of 55 mm and a mass per unit length of 1.20 kg m^{-1}. The minimum
wind speed measured at 10 m altitude to launch the kite is found to be around 4.5
m/s. After the launching phase, we show that the optimal tether length for static
flight is 128.4 m with a minimum wind speed of 4.06 m/s. The presented approach
shows an error up to 9% for a zero-mass kite model with a straight massless tether
regarding the maximal propulsion force estimation.

3.1 Introduction

This study is part of the beyond the sea® research program led by the ENSTA Bretagne school of engineering. The project attempts to develop a kite system as an
auxiliary propulsion device for merchant ships. Such a system is a high-potential
alternative to conventional fossil fuel based propulsion systems, as indicated by

Nedeleg Bigi (✉) · Alain Nême · Kostia Roncin · Jean-Baptiste Leroux · Guilhem Bles · Christian
Jochum
ENSTA Bretagne, FRE CNRS 3744, IRDL, 29200 Brest, France
e-mail: nedeleg.bigi@ensta-bretagne.org

Y. Parlier
beyond the sea, 110 avenue de l'Europe, 33260 La Teste de Buch, France
e-mail: yves.parlier@beyond-the-sea.com

© Springer Nature Singapore Pte Ltd. 2018
R. Schmehl (ed.), *Airborne Wind Energy*, Green Energy
and Technology, https://doi.org/10.1007/978-981-10-1947-0_3

several authors [7, 17, 19, 23]. Depending on the maritime route and the seasons, weather conditions vary. Thus, to achieve a tether and a kite design according to the encountered weather condition, one of the main inputs of a functional specification is the minimum wind velocity enabling kite flight. The capacity of the system to tow a ship at a certain wind condition must be evaluated as well.

Moreover, wind speed increases with altitude, and as it has been highlighted by many authors, notably by Leloup et al. [17], this is a benefit for the kite to generate a propulsive force. This benefit is directly dependent on the tether length, the higher the kite is, the stronger the wind generally is. Therefore, during the early design stage, studies taking into account tether effects have to be performed. At this stage of the design a wide range of potential solutions must be investigated. This work aims therefore to provide an early design step accurate enough to be realistic.

Tethers are currently made of fiber materials such as Dyneema®(Ultra-high-molecular-weight polyethylene, UHMWPE) for example. This means that compression, transverse shear, bending and torsional stiffness of the tether can be neglected compared to its tensile stiffness. In addition, the tether shape is highly dependent on aerodynamic loading acting on the tether surface and tether gravity acting on the tether volume. This kind of structure has been studied for other industrial applications such as electrical power lines, anchored offshore structures, tethered underwater vehicles or sling loads. Tether models for airborne wind energy applications were inspired by these fields.

Williams et al. [24] developed a so-called lumped mass model for dynamic flight. The mass of each element is concentrated on each node and the distance between each node remains constant. Breukels and Ockels [4] used discrete element modeling with inelastic bar elements. Argatov et al. [1] accounted for tether sag due to wind load and gravity, assuming that the tension along the tether is constant. They proposed a method to calculate wind load by neglecting the tangential wind component relatively to the line. They showed how tether effects decrease the power production for a dynamic flight. A model considering the tether as a straight elastic spring to account for material stiffness has been used to study the stability of the kite during a dynamic flight by Terink et al. [21]. To identify the low wind limit for kite flight, the most restrictive flight case is assumed to be the static flight because the apparent wind velocity is, most of the time, lower compared to a dynamic flight.

Kite deployment and recovering phases can also reasonably be considered quasi-static. Indeed, with a constant reeling velocity and neglecting the dynamic effect on the tether, the kite follows a straight path at a constant speed, thus the kite flight can be considered as equivalent to a static flight. Moreover, Leloup et al. [17] have shown for upwind sailing that a static flight could be more efficient than dynamic flight for fuel saving. All these tether models have been developed for dynamic flight and are still valid for static flight. Nevertheless, for discrete model, artificial structural damping needs to be added to reach static equilibrium as reported by Breukels and Ockels [4]. Considering low wind velocities, tether sag could be important, therefore a single straight elastic spring modeling the tether [21] is not a realistic enough assumption. Varma and Goela [22] developed a soft kite tether model for static kite flights at zero azimuth angle. Their model is based on the catenary curve

[18]. Varma and Goela [22] consider a flexible tether of constant length and mass per unit length. Indeed, the average aerodynamic loading on the tether is not significantly modified by the increasing length of the tether due to its tensile stiffness.

Hobbs [10] studied the influence of the wind velocity gradient effect on the tether shape for static kite flights at zero azimuth angle. He concluded his study on the wind profile influence arguing that the main factor influencing the line shape is the mean wind velocity according to the altitude. Being analytical, the model presented by Varma and Goela [22] has the potential to sufficiently reduce computation times to perform tether analysis early in the design.

The study presented in this chapter provides an analytical formulation of the catenary curve [18, 22] to model a flexible tether of constant length for any static kite flight position, with an arbitrary attachment point altitude on the ship deck, and with a wind velocity gradient law for kite forces estimation. The preliminary content of the present chapter has been presented at the Airborne Wind Energy Conference 2015 [3]. The determination of tether's shape and tension only requires the solution of a one-dimensional transcendental equation with a fixed-point algorithm. This procedure improves the reliability and the convergence rate of 2D Newton's method suggested in [14]. A closed-form expression is determined to evaluate a mean aerodynamic loading on the tether according to the wind velocity gradient effect. These developments are then used to identify a new analytical low wind speed limit for kite flight. Then, results highlighting the capability of the model for an early design stage are presented.

3.2 Mathematical Model

3.2.1 Tether Model

The tether model is based on the well-known catenary curve [18]. As illustrated in Fig. 3.1, the points $\mathbf{S}$ and $\mathbf{K}$ mark the extremities of the tether namely the ship attachment point and the position of the kite. A constant load per unit length is applied on the tether, which is assumed to be flexible, of constant length and with no transverse shear and no bending stiffness. Consequently, the tether remains in a plane defined by $(\mathbf{S}, \mathbf{y}_t, \mathbf{z}_t)$ of the R_t coordinate system. $R_0 = (\mathbf{x}_0, \mathbf{y}_0, \mathbf{z}_0)$ denotes a coordinate system attached to the ship sailing at constant speed on a straight course, where $\mathbf{z}_0$ is opposed to the earth gravity. The unit vector $\mathbf{z}_t$ is defined by the load per unit length $\mathbf{q}$ as

$$\mathbf{q} = -\|\mathbf{q}\| \, \mathbf{z}_t = -q\mathbf{z}_t. \tag{3.1}$$

The unit vector $\mathbf{x}_t$ is defined as $\mathbf{x}_t = (\mathbf{SK} \times \mathbf{z}_k) / \|\mathbf{SK} \times \mathbf{z}_k\|$, where "$\times$" denotes the cross product operator. In order to obtain a direct orthonormal coordinate system, the unit vector $\mathbf{y}_t$ is given by $\mathbf{y}_t = \mathbf{z}_t \times \mathbf{x}_t$.

With tension $\mathbf{T}$ along the tether, s the curvilinear abscissa, T_{y_t} and T_{z_t} denoting the projections $\mathbf{T} \cdot \mathbf{y}_t$ and $\mathbf{T} \cdot \mathbf{z}_t$, the following equations define the static equilibrium

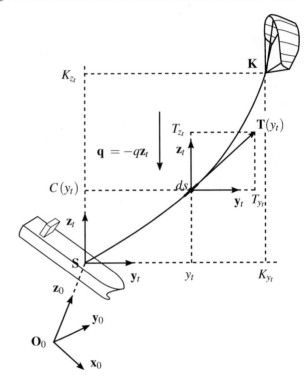

Fig. 3.1 Coordinate systems used for the development of the catenary equation

of an infinitesimal length of tether ds projected on $\mathbf{y}_t$ and $\mathbf{z}_t$

$$\frac{dT_{y_t}}{ds} = 0, \tag{3.2}$$

$$\frac{dT_{z_t}}{ds} - q = 0. \tag{3.3}$$

According to Eqs. (3.2) and (3.3), a catenary function C must fulfill the following equation

$$\frac{q}{T_{y_t}} = \frac{C''(y_t)}{\sqrt{1 + C'(y_t)^2}}, \tag{3.4}$$

where C' and C'' denote the first and second derivative of the function. Therefore, by integration of Eq. (3.4), C could be expressed as follows

$$C(y_t) = \frac{T_{y_t}}{q} \cosh\left(\frac{q}{T_{y_t}} y_t + K_1\right) + K_2, \tag{3.5}$$

where K_1 and K_2 are two constants of integration. They are determined with the boundary conditions

$$0 = \frac{T_{y_t}}{q} \cosh(K_1) + K_2, \tag{3.6}$$

$$K_{z_t} = \frac{T_{y_t}}{q} \cosh\left(\frac{q}{T_{y_t}} K_{y_t} + K_1\right) + K_2, \tag{3.7}$$

$$l_t = \int_0^{K_{y_t}} \sqrt{1 + C'^2(y_t)} \, dy_t, \tag{3.8}$$

$$= \frac{T_{y_t}}{q} \left[\sinh\left(\frac{q}{T_{y_t}} K_{y_t} + K_1\right) - \sinh(K_1) \right], \tag{3.9}$$

where Eq. (3.9) is derived from constant tether length l_t, Eq. (3.6) from the coordinates of $\mathbf{S} = [0,0,0]_{R_t}^\top$ and Eq. (3.7) from $\mathbf{K} = [0, K_{y_t}, K_{z_t}]_{R_t}^\top$. Using trigonometric identities, constants K_1 and K_2 can be expressed thanks to the boundary conditions in order to obtain the function C which can be expressed as

$$C(y_t) = \frac{K_{z_t} \sinh(\omega y_t) + \lambda \left\{ \sinh(\omega y_t) - \sinh(\omega K_{y_t}) + \sinh[\omega(K_{y_t} - y_t)] \right\}}{\sinh(\omega K_{y_t})}, \tag{3.10}$$

where λ and ω are defined by

$$\lambda = \frac{l_t \sinh(\omega K_{y_t}) - K_{z_t}[\cosh(\omega K_{y_t}) - 1]}{2[\cosh(\omega K_{y_t}) - 1]}, \tag{3.11}$$

$$\omega^2 (K_{z_t}^2 - l_t^2) = 2[1 - \cosh(\omega K_{y_t})]. \tag{3.12}$$

It can be noticed that the catenary function does not depend on the load per unit length, q. Equation (3.12) can be rearranged in order to compute the value of ω. With $u = \omega^2 K_{y_t}^2$, $\bar{l}_t = \frac{l_t}{K_{y_t}}$ and $\beta = \frac{K_{z_t}}{K_{y_t}}$ we arrive at

$$u = \left\{ \text{argcosh} \left[\frac{u\left(\bar{l}_t^2 - \beta^2\right)}{2} + 1 \right] \right\}^2. \tag{3.13}$$

The value of u is computed by applying the fixed-point algorithm to Eq. (3.13) achieving convergence for all positive values of u. Thus, for a given kite position $\mathbf{K}$ and a given ship attachment point position $\mathbf{S}$, tether tension is expressed by

$$\mathbf{T}(y_t) = \left[0, \frac{q}{\omega}, \frac{q}{\omega} C'(y_t)\right]_{R_t}^\top. \tag{3.14}$$

It can be noticed that the inverse of ω is directly proportional to the tension in the y_t direction with the factor q. Consequently, tether shape and tension along the tether are determined for any kite and ship attachment point positions.

By contrast to the previous approach, an expression giving the kite location $\mathbf{K}$, for a known tension at $\mathbf{K}$, is relevant in order to determine the minimal wind velocity permitting a static flight. This expression is then developed. The tension is tangential to the tether, which means at $\mathbf{K}$

$$C'(K_{y_t}) = \sinh\left(\frac{q}{T_{y_t}}K_{y_t} + K_1\right) = \frac{T_{z_t}}{T_{y_t}}. \tag{3.15}$$

Then, using Eqs. (3.6), (3.7) and (3.9), expressions for the kite location **K** with a given tether tension at **K** are

$$K_{y_t} = \frac{T_{y_t}}{q}\left[\text{argsinh}\left(\frac{T_{z_t}}{T_{y_t}}\right) - \text{argsinh}\left(\frac{T_{z_t} - ql_t}{T_{y_t}}\right)\right], \tag{3.16}$$

$$K_{z_t} = \frac{T_{y_t}}{q}\left(\sqrt{1 + \left(\frac{T_{z_t}}{T_{y_t}}\right)^2} - \sqrt{1 + \left(\frac{T_{z_t} - ql_t}{T_{y_t}}\right)^2}\right). \tag{3.17}$$

Equations (3.16) and (3.17) are similar to [14, Eqs. (1.27) and (1.28)] in case of an flexible tether of constant length with very large Young's modulus.

3.2.2 Wind Model

It has been observed that the wind above the sea increases with the altitude due to the friction stress on the free surface within the atmospheric boundary layer. This phenomenon, called wind velocity gradient effect, can be taken into account with a simple formula according to ITTC [12]. The true wind velocity $\mathbf{V}_{TW}$ at a given altitude z_0 is calculated as

$$\mathbf{V}_{TW} = \mathbf{U}_{ref}\left(\frac{z_0}{z_{ref}}\right)^n, \tag{3.18}$$

from the known reference wind velocity $\mathbf{U}_{ref}$ measured at an altitude z_{ref}. The coefficient n denotes the friction effect due to the free surface. A typical value of $n = 1/7$ is given by ITTC [12] for sea friction. Figure 3.2 illustrates the evolution of the wind

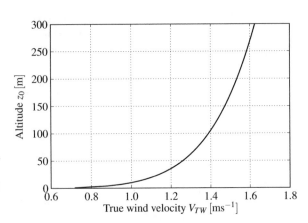

Fig. 3.2 Wind velocity gradient evolution against altitude according to Eq. (3.18) using $z_{ref} = 10$ m, $U_{ref} = 1$ m s^{-1} and $n = 1/7$

velocity with the altitude range from 1 m to 300 m with a wind of 1 m s^{-1} at the reference altitude $z_{ref} = 10$ m.

Consequently, since the relative wind velocity $\mathbf{V}_{RW}$ is given by the difference between the true wind velocity and the ship velocity $\mathbf{V}_s$, $\mathbf{V}_{RW}$ can be expressed as

$$\mathbf{V}_{RW} = \mathbf{U}_{ref} \left(\frac{z_0}{z_{ref}} \right)^n - \mathbf{V}_s. \tag{3.19}$$

3.2.3 Tether Load Model

The load per unit length on the tether is given by

$$\mathbf{q}(s) = \mathbf{q}_w(s) + \mathbf{q}_g = \mathbf{q}_w(s) - m_t g \mathbf{z}_0, \tag{3.20}$$

where $\mathbf{q}_w$ denotes the load per unit length due to wind and $\mathbf{q}_g$ denotes weight distribution, along the curvilinear abscissa, m_t the mass per unit length of tether and g the acceleration due to gravity ($g = 9.81 \text{ m s}^{-2}$).

Aerodynamic tether loading q_w is very sensitive since a tether can encounter a wide range of Reynolds number. The flow around circular cylinder has been widely studied in the past and is still a research topic as demonstrated by Sarpkaya in his literature review [20] and in Chap. 2 of this book. In addition, a textile rope has not exactly a circular section. Jung [15] performed wind tunnel experiments for various rope sections and various roughness surface at a Reynolds number $\text{Re} = 84.0 \times 10^3$. According to his measurements the drag coefficient can vary from 0.76 to 1.56 with orthogonal flow.

Nevertheless, since the Reynolds effect and the surface roughness are out of the scope of the paper, the Hoerner formulation [11] is used similar to many other authors involved in airborne wind energy. As mentioned in Sect. 3.2.2, $\mathbf{V}_{RW}$ depends on altitude, and therefore $\mathbf{q}_w$ as well. Since the catenary tether model requires only constant load per unit length, as can be seen in Sect.3.2.1, an approximation of constant wind tether load must be achieved. The determination of an equivalent altitude z_{q0} to evaluate $\mathbf{q}_w$ is proposed here. It is assumed that the tether is a straight line between $\mathbf{S}$ and $\widetilde{\mathbf{K}}$. $\widetilde{\mathbf{K}}$ is the kite position calculated with the static flight model described in [17] and summarized in Sect.3.2.5.

As illustrated in Fig. 3.3, the wind load $\mathbf{q}_w$ can be decomposed into drag force $\mathbf{q}_d$ and lift force $\mathbf{q}_l$

$$\mathbf{q}_w = \mathbf{q}_l + \mathbf{q}_d. \tag{3.21}$$

Both components are determined from the Hoerner formulas [11] as

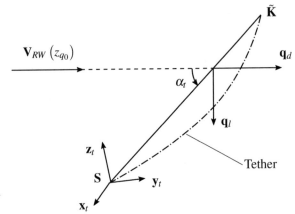

Fig. 3.3 Diagram of the tether
wind load model

$$\mathbf{q}_d = \frac{1}{2}\rho_a d_t \left[1.1 \sin^3(\alpha_t) + 0.02\right] \left\| \mathbf{V}_{RW}(z_{q0}) \right\| \mathbf{V}_{RW}(z_{q0}), \tag{3.22}$$

$$\mathbf{q}_l = \frac{1}{2}\rho_a d_t \left[1.1 \sin^2(\alpha_t) \cos(\alpha_t)\right] \left\| \mathbf{V}_{RW}(z_{q0}) \right\|$$

$$\frac{\mathbf{V}_{RW}(z_{q0}) \times \left[\mathbf{V}_{RW}(z_{q0}) \times \mathbf{S}\tilde{\mathbf{K}}\right]}{\left\| \mathbf{V}_{RW}(z_{q0}) \times \mathbf{S}\tilde{\mathbf{K}} \right\|} \tag{3.23}$$

where ρ_a is the air density, d_t is the tether diameter, α_t is the angle of attack between the wind and the tether as described in Fig. 3.3 and assuming a base drag coefficient of 1.1 for orthogonal flows ($\alpha_t = \pi/2$).

With respect to the ship velocity $\mathbf{V}_S$ and according to Eq. (3.18), the relative wind velocity at the altitude z_{q0} is given by

$$\mathbf{V}_{RW}(z_{q0}) = \mathbf{U}_{ref}\left(\frac{z_{q0}}{z_{ref}}\right)^n - \mathbf{V}_S. \tag{3.24}$$

In order to conserve approximately the total force acting on the tether, z_{q0} is defined such that the following equation must be fulfilled

$$\left\| \mathbf{V}_{RW}(z_{q0}) \right\|^2 = \frac{1}{(K_{z0} - S_{z0})} \int_{S_{z0}}^{K_{z0}} \left\| \mathbf{V}_{RW}(z) \right\|^2 dz, \tag{3.25}$$

which, by keeping only the largest root, leads to a second degree polynomial equation in z_{q0}^n

$$0 = \frac{\|\mathbf{U}_{ref}\|^2}{z_{ref}^{2n}} z_{q0}^{2n} - 2\frac{\mathbf{V}_s \cdot \mathbf{U}_{ref}}{z_{ref}^n} z_{q0}^n + \|\mathbf{V}_s\|^2 - \|\mathbf{U}_{ref}\|^2 \frac{K_{z0}^{2n+1} - S_{z0}^{2n+1}}{(2n+1)\left(K_{z0} - S_{z0}\right) z_{ref}^{2n}}$$

$$+ 2\mathbf{V}_s \cdot \mathbf{U}_{ref} \frac{K_{z0}^{n+1} - S_{z0}^{n+1}}{(n+1)\left(K_{z0} - S_{z0}\right) z_{ref}^n} - \|\mathbf{V}_s\|^2 \quad (3.26)$$

It must be noticed that the definition of the equivalent altitude z_{q0}, Eq. (3.25) is not correct to conserve the total force acting on the tether. Indeed, the load direction varies with the altitude which is not considered in Eq. (3.25). A better definition could have been

$$\mathbf{V}_{RW}\left(z_{q0}\right) = \frac{\mathbf{V}_2}{\sqrt{\|\mathbf{V}_2\|}}, \quad (3.27)$$

with,

$$\mathbf{V}_2 = \frac{1}{K_{z0} - S_{z0}} \int_{S_{z0}}^{K_{z0}} \|\mathbf{V}_{RW}\| \mathbf{V}_{RW} dz. \quad (3.28)$$

However, our proposition should be reasonable in order to achieve a closed-form formulation of the equivalent altitude z_{q0}.

3.2.4 Aerodynamic Kite Model

For a static flight, forces acting on the kite must be opposed to the tether tension and vary with altitude due to the wind velocity gradient. Applying the first Newton's law to the kite we obtain

$$0 = -\mathbf{T}\left(K_{y_t}\right) + \mathbf{L} + \mathbf{D} + \mathbf{W}, \quad (3.29)$$

where $\mathbf{T}\left(K_{y_t}\right)$ is the tether tension at kite location, $\mathbf{L}$ is the lift kite aerodynamic force, $\mathbf{D}$ is the drag kite aerodynamic force and $\mathbf{W} = -M_K g \mathbf{z}_0$ is the kite weight calculated from the kite mass M_K. For static flight, the lift-to-drag ratio angle ε is assumed to be constant. $\mathbf{D}$ is by definition in the direction of the relative wind and can be determined as follows

$$\mathbf{D} = \frac{1}{2} \rho_a A_K C_{L_K} \tan\left(\varepsilon\right) \|\mathbf{V}_{RW}\| \mathbf{V}_{RW}, \quad (3.30)$$

where ρ_a is the air density, A_K is the kite area and C_{L_K} is the kite lift coefficient. According to the assumption of a constant lift-to-drag ratio, the magnitude of the lift can be determined as

$$\|\mathbf{L}\| = \frac{\|\mathbf{D}\|}{\tan\left(\varepsilon\right)} \quad (3.31)$$

and the orthogonality of lift and drag components is formally expressed as

$$\mathbf{L} \cdot \mathbf{D} = 0. \quad (3.32)$$

One additional equation is needed to determine the lift. As a balance is expected between kite forces and tether tension, we know that at least they must stay in the plane $(\mathbf{S}, \mathbf{y}_t, \mathbf{z}_t)$. This is a consequence of the projection of Eq. (3.29) on axis $\mathbf{x}_t$, which can be expressed as

$$(\mathbf{L} + \mathbf{D} + \mathbf{W}) \cdot \mathbf{x}_t = 0. \tag{3.33}$$

The component L_{x_t} can be derived from Eq. (3.33) as

$$L_{x_t} = -\left(D_{x_t} + W_{x_t} \right). \tag{3.34}$$

Equations (3.31) and (3.32) lead to a second order polynomial equation in L_{z_t}

$$L_{z_t} = \frac{\sqrt{\Delta} - L_{x_t} D_{x_t} D_{z_t}}{\left(D_{z_t}^2 + D_{y_t}^2 \right)}, \qquad \text{with} \tag{3.35}$$

$$\Delta = \frac{D_{y_t}^2 \|\mathbf{D}\|}{\tan^2(\varepsilon)} \left[D_{y_t}^2 + D_{z_t}^2 - L_{x_t}^2 \tan^2(\varepsilon) \right]. \tag{3.36}$$

Equation (3.36) describes the discriminant Δ of the second order polynomial equation in L_{z_t}. Using Eq. (3.32) L_{y_t} can be derived as

$$L_{y_t} = -\frac{L_{x_t} D_{x_t} + L_{z_t} D_{z_t}}{D_{y_t}} \tag{3.37}$$

The condition $\Delta \geq 0$ is a necessary condition to allow a static kite flight.

3.2.5 Zero-Mass Model

In this study, the zero-mass model developed by Leloup et al. [17] is used as reference model. This model has been expressed for static and dynamic kite flight. For a commodity purpose the corresponding static flight formulation is recalled with the present coordinate system. Neglecting the kite mass and the tether mass, the first Newton's law applied to the kite can be expressed by the following equation

$$0 = -\mathbf{T} + \mathbf{L} + \mathbf{D}, \tag{3.38}$$

where $\mathbf{T}$ is given according to Leloup et al. [17] as

$$\mathbf{T} = \frac{\frac{1}{2} \rho_a A_k C_{L_K} \|\mathbf{V}_{RW}\|^2}{\cos(\varepsilon)} \frac{\mathbf{SK}}{\|\mathbf{SK}\|}. \tag{3.39}$$

We define the reference frame $R_{RW} = (\mathbf{S}, \mathbf{x}_{RW}, \mathbf{y}_{RW}, \mathbf{z}_0)$, with $\mathbf{x}_{RW} = \mathbf{V}_{RW} / \|\mathbf{V}_{RW}\|$ and $\mathbf{y}_{RW} = \mathbf{z}_0 \times \mathbf{x}_{RW}$, as illustrated in Fig. 3.4, to obtain

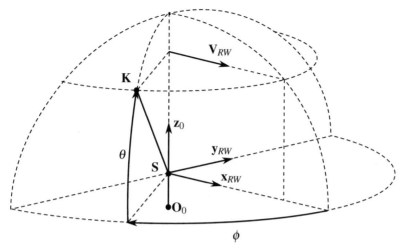

Fig. 3.4 Zero-mass model parametrization of Leloup et al. [17]. In this diagram, the azimuth angle ϕ is negative

$$\mathbf{SK} = [l_t \cos(\phi)\cos(\theta), l_t \sin(\phi)\cos(\theta), l_t \sin(\theta)]^\top_{R_{RW}} .\qquad(3.40)$$

The static flight condition is calculated by Leloup et al. [17] as

$$\phi = \pm \arccos\left[\frac{\sin(\varepsilon)}{\cos(\theta)}\right]\qquad(3.41)$$

3.2.6 Kite Static Equilibrium

The equilibrium equation of the kite, Eq. (3.29), can be solved by coupling these models. The initial kite position is determined according to the zero-mass formulation of the kite balanced equation, as shown in Sect. 3.2.5. From this position, the tether load and coordinate system R_t are calculated, as shown in Sect. 3.2.1, and kept constant until the kite equilibrium position is reached. A Newton-Raphson algorithm, in Eq. (3.42), is used to solve the kite static equilibrium in plane $(\mathbf{y}_t, \mathbf{z}_t)$

$$\begin{bmatrix} K_{y_t} \\ K_{z_t} \end{bmatrix}_{(k+1)} = \begin{bmatrix} K_{y_t} \\ K_{z_t} \end{bmatrix}_{(k)} - \begin{bmatrix} \dfrac{\partial F_{y_t}}{\partial K_{y_t}} & \dfrac{\partial F_{y_t}}{\partial K_{z_t}} \\ \dfrac{\partial F_{z_t}}{\partial K_{y_t}} & \dfrac{\partial F_{z_t}}{\partial K_{z_t}} \end{bmatrix}^{-1}_{(k)} \begin{bmatrix} F_{y_t} \\ F_{z_t} \end{bmatrix}_{(k)},\qquad(3.42)$$

where $\mathbf{F} = -\mathbf{T} + \mathbf{L} + \mathbf{D} + \mathbf{W}$ and k represents the iteration number.

3.2.7 Verification of the Implementation

The implementation of the presented model is verified on the basis of the experimental data of Irvine and Sinclair in [13]. In this experiment, the two extremities of a cable were horizontally attached. The cable length was 1.20 m, the cable cross sectional area was 1.58×10^{-6} m^2 and the Young's modulus of the cable was 1.00×10^{11} N m^{-2}. The horizontal distance between the attachment point was 1.00 m. A total of 20 weights of 2.45 N were added to the cable with ferrules in order to neglect the cable bending stiffness. From the attachment point, the weights were attached with a distance of 0.03 m and the weights were equally spaced each other by a distance of 0.06 m. The weight of the cable, ferrules and weights were 50 N. Figure 3.5 represents the cable corresponding to the experiment in [13] (dashed line) and the corresponding cable shape calculated with model, Eqs. (3.10) – (3.13). The experimentally measured shape of the cable has been extracted from [13].

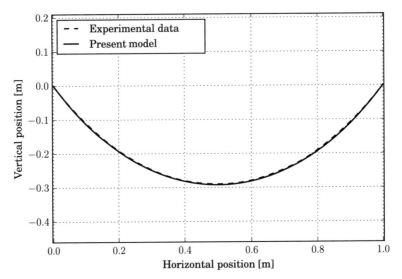

Fig. 3.5 Comparison of the tether shape computed with the present model and measured by [13]

The present model fits pretty well with the experimental data [13] and can be considered validated. Nevertheless, a comparison between the entire present model and static kite flight must be investigated as well.

3.3 Low Wind Limit for Kite Flight

Most kite launch step begins by quasi-static flight at zero azimuth angle. Therefore the low wind limit for static kite flight at zero azimuth angle is an important param-

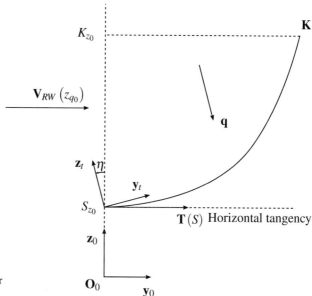

Fig. 3.6 Diagram of the lower limit static flight case

eter. Obviously, a tether should not touch the ground or the free surface. In that case, friction with the ground could have a dramatic effect on the material durability and kite control. This leads to the mathematical condition that the whole tether must be above the attachment point, as shown in Fig. 3.6. The mathematical expression of this limit is given by

$$\mathbf{T}(S) \cdot \mathbf{z}_0 = 0. \tag{3.43}$$

In the static kite flight case at zero azimuth angle, the first Newton's law applied to the tether and projected on axis $\mathbf{z}_0$, in accordance with the condition given by Eq. (3.43), leads to

$$L - W + l_t \mathbf{q} \cdot \mathbf{z}_0 = 0. \tag{3.44}$$

Therefore, the relative wind at the kite location is given by

$$V_{RW} = \sqrt{\frac{2(W - l_t \mathbf{q} \cdot \mathbf{z}_0)}{\rho_a A_k C_{L_K}}}. \tag{3.45}$$

In the static kite flight at zero azimuth angle, the kite position in reference frame R_0, compared to the position in reference frame R_t, is defined by the angle

$$\eta = \arctan\left(-\frac{\mathbf{q} \cdot \mathbf{y}_0}{\mathbf{q} \cdot \mathbf{z}_0}\right). \tag{3.46}$$

Kite altitude in R_0 is given by

$$K_{z_0} = S_{z_0} + K_{y_t}\sin(\eta) + K_{z_t}\cos(\eta). \tag{3.47}$$

Inserting Eqs. (3.45) and (3.47) into Eq. (3.24), assuming $\mathbf{V}_{RW}$ and $\mathbf{V}_S$ are collinear, and reorganizing leads to the wind velocity at the measurement altitude

$$U_{ref,min} = \frac{z_{ref}^n}{\left[S_{z_0} + K_{y_t} \sin(\eta) + K_{z_t} \cos(\eta)\right]^n} \left[\sqrt{\frac{2(W - l_t \mathbf{q} \cdot \mathbf{z}_0)}{\rho_a A_k C_{L_K}}} + V_s\right]. \quad (3.48)$$

As indicated by Eqs. (3.20) – (3.24), $\mathbf{q}$ depends on $U_{ref,min}$ which means that Eq. (3.48) needs to be solved. However, rather than solving the problem numerically, for example by iterative methods, a closed-form approximation of the minimal wind velocity required for a static flight is provided assuming that the load per unit length on the tether is only due to the gravity (this hypothesis is discussed at the end of the Sect. 3.5 and is illustrated in Fig. 3.10). Therefore, $\mathbf{z}_t$ is equal to $\mathbf{z}_0$ and the load per unit length is $\mathbf{q} = \mathbf{q}_g$. Then, the closed-form Eq. (3.45) becomes

$$V_{RW} = \sqrt{\frac{2(W + m_t l_t g)}{\rho_a A_k C_{L_K}}}, \quad (3.49)$$

where $g = 9.81 \text{ m s}^{-2}$ is the acceleration due to gravity. Using Eqs. (3.17) and (3.24), the lower limit is

$$U_{ref,min}^- = \frac{z_{ref}^n \left(\sqrt{\frac{2g(M_k + l_t m_t)}{\rho_a A_k C_{L_K}}} + V_s\right)}{\left\{S_{z_0} + \tan(\varepsilon)\left(l_t + \frac{M_k}{m_t}\right)\left[\sqrt{1 + \left(\frac{m_t l_t}{(m_t l_t + M_k)\tan(\varepsilon)}\right)^2} - 1\right]\right\}^n}. \quad (3.50)$$

Using the following dimensionless problem parameters

$$\tilde{U} = U_{ref,min}^- \sqrt{\frac{A_k \rho_a C_{L_K}}{2W}}, \qquad \tilde{l}_t = \frac{m_t l_t}{M_K}, \qquad \tilde{S} = \frac{S_{z_0}}{z_{ref}},$$

$$\tilde{V}_s = V_s \sqrt{\frac{A_k \rho_a C_{L_K}}{2W}}, \qquad \tilde{z} = \frac{z_{ref} m_t}{M_K},$$

Eq. (3.50) can be normalized to

$$\tilde{U} = \frac{\sqrt{1 + \tilde{l}_t + \tilde{V}_s}}{\left\{\tan(\varepsilon)\frac{(1 + \tilde{l}_t)}{\tilde{z}}\left[\sqrt{1 + \left[\frac{\tilde{l}_t}{\tan(\varepsilon)(1 + \tilde{l}_t)}\right]^2} - 1\right] + \tilde{S}\right\}^n}. \quad (3.51)$$

The parameter $\tilde{l}_t$ can be interpreted as the dimensionless tether length. The attachment point altitude is normalized by the wind measurement altitude. The parameter $\tilde{z}$ characterizes the tether mass per unit length compared to the kite mass. This last parameter provides information on the structural and material design priority be-

tween the tether and the kite, it increases when the ratio of safety factors between
the line and the kite increases.

3.4 Case Study

The following example calculations are based on the case study of Dadd [5] where
kite parameters have been extrapolated from data measured by Dadd et al. [6] for
a Flexifoil® Blade III kite with 3 m² surface area. Kite and tether characteristics
are summarized in Table 3.1. Tether diameter and mass per unit length have been
estimated using a dynamic flight load case calculated with the analytical zero-mass
model developed by Leloup et al. [17]. The flight trajectory is taken from Argatov
et al. [2] with a polar angle amplitude of $16°$ and an azimuth angle amplitude of
$66°$. For a true wind speed of $17\ \mathrm{m\,s^{-1}}$ and for a cruising ship speed of $7.5\ \mathrm{m\,s^{-1}}$,
according to the model of Leloup et al. [17], maximum tether tension is given for a
true wind angle[1] of $110°$. At this configuration the tether tension is 1.5×10^6 N.

Flexifoil® Blade III characteristics extrapolated by [5]		
Wing surface area	A_K	320 m²
Wing mass*	M_K	300 kg
Aerodynamic lift coefficient	C_L	0.776 -
Lift-to-drag ratio angle	ε	12.02 deg
Tether characteristics		
Length	l_t	300 m
Mass per unit length*	m_t	1.20 kg m⁻¹
Diameter*	d_t	55.0 mm

Table 3.1 Kite and tether characteristics for the study. Estimated values are marked by an asterisk (*)

The chosen material of the tether is Ultra-High Molecular Weight Polyethylene
(UHMWPE), which is also known under the brand name Dyneema®. According to
its ultimate specific stress [8, 9, 16] of $1.46 \times 10^3\ \mathrm{J\,g^{-1}}$ and a safety factor of 1.2, a
maximal load of 1.25×10^6 N is allowed. This leads to a tether with a mass per unit
length of $1.2\ \mathrm{kg\,m^{-1}}$ and a diameter of 55 mm.

For the results presented in Sects. 3.5 and 3.6, the ship attachment point altitude
S_{z0} is 10 m and the true wind speed is measured at an altitude of $z_{ref} = 10$ m.
According to the ITTC [12], the wind velocity gradient parameter is $n = 1/7$.

---

[1] The true wind angle is the angle between the ship path and the true wind velocity at the reference
altitude [17]

3.5 Minimal Wind Velocity

In order to identify a low wind limit for static kite flight, the criterion of Sect. 3.3 is applied with $\varepsilon = 12.02°$. Figure 3.7 represents three surface plots of the non-dimensional minimal wind velocity $\tilde{U}$ defined by Eq. (3.51) as a function of the non-dimensional parameters $\tilde{z}$ and $\tilde{l}_t$ for $\tilde{V}_s = 0$ and for three specific values of $\tilde{S}$.

Assuming a given value of $\tilde{l}_t$, Fig.3.7 shows that the non-dimensional minimal wind velocity $\tilde{U}$ increases when the ratio $m_t l_t / M_K$ increases or when the ratio $\tilde{S}$ increases. These results make sense in a natural way. The explanation of the variation of $\tilde{U}$ according to the tether length is less obvious as it can be observed in Fig.3.7. An optimal tether length can appear to minimize $\tilde{U}$. Nevertheless the main result is that the non-dimensional minimal wind velocity increases when the tether length increases beyond a finite value which can be zero. This result is important for keeping the kite airborne. Finally the effective minimal wind velocity $U^-_{ref,min}$ is obtained by dividing $\tilde{U}$ by the factor $\sqrt{A_k \rho_a C_{LK}/(2W)}$. The latter increases when the kite weight to lift coefficient ratio increases.

For the investigated case described in Sect. 3.4 and a zero ship velocity, the minimal wind velocity $U^-_{ref,min}$ given by Eq. (3.50) has been plotted in Fig. 3.8 for different tether lengths from 0 to 400 m. For a tether length $l_t = 0$, the minimal required wind speed is 4.44 m s^{-1}. Then, the minimal wind speed required increases to reach a maximum at 4.48 m s^{-1} for a tether length of 8 m. With longer tether, the minimal required wind speed decreases to 4.06 m s^{-1} for $l_t = 128$ m. The third part

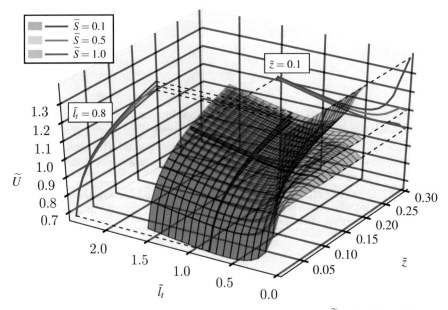

Fig. 3.7 Surface plots of the non-dimensional minimum wind velocity $\tilde{U}$ as function of the non-dimensional parameters $\tilde{z}$ and $\tilde{l}_t$ for $\tilde{S} = 0.1, 0.5$ and 1

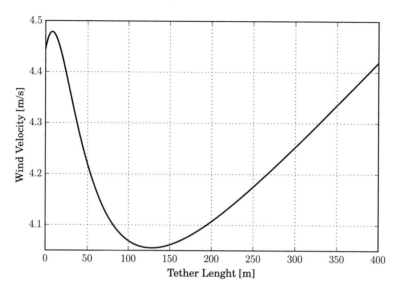

Fig. 3.8 Minimal wind velocity U_{ref} at $z_{ref} = 10$ m for a tether length l_t from 0 to 400 m

of the curve increases. A tether length of 128.4 m is optimal to allow a static flight for a minimal true wind speed. For this tether length the required true wind speed is $4.06 \, \text{m s}^{-1}$.

On one hand, the longer the tether, the higher the kite is. So, due to the wind velocity gradient effect, the relative wind speed at the kite increases with the tether length. On the other hand, tether weight increases with tether length. In Fig. 3.8 it can be noticed that for tether length such as $8 \leq l_t \leq 128$ m the increase of the relative wind speed is more significant than the increase of the tether weight $m_t l_t$. This has the effect of reducing the minimal wind speed required to allow a static flight. For tether length such as $0 \leq l_t \leq 8$ m and $l_t \geq 128$ m, the phenomenon is reversed. The increase of the wind speed is no longer sufficient to counteract the increase of the tether gravitational load.

Figure 3.9 shows the minimum wind required to allow a static flight for different lift-to-drag ratio angles $\varepsilon = \arctan(D/L)$. Here, for low range of ε, the wind velocity required increases linearly with ε.

In this section, the effect of gravity on the tether has been taken into account while the influence of the aerodynamic loading has been neglected. For a static flight case at zero azimuth angle and for a tether length of 300 m, a comparison between aerodynamic load and gravitational load on the tether is displayed in Fig. 3.10.

The diagram clearly shows that for the minimum wind speed of $4.5 \, \text{m s}^{-1}$, which is required for static flight according to Fig. 3.8, the corresponding wind load is less than 12.5% of the gravitational load. Thereby, in order to obtain a closed-form for-

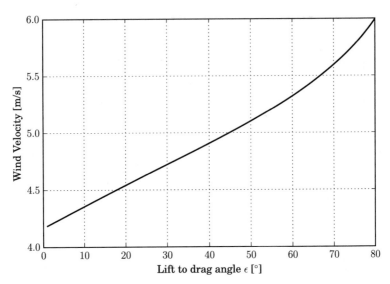

Fig. 3.9 Minimal wind velocity U_{ref} at $z_{ref} = 10$ m for a lift-to-drag angle ε range from $1°$ to $80°$

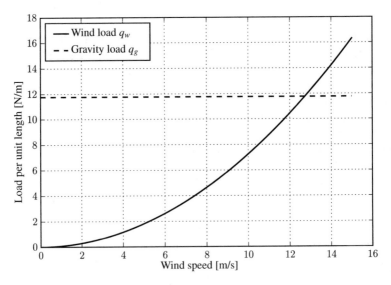

Fig. 3.10 Tether load per unit length due to wind q_w and tether load per unit length due to gravity q_g for a wind U_{ref} range from $0\,\mathrm{m\,s}^{-1}$ to $15\,\mathrm{m\,s}^{-1}$ at $z_{ref} = 10$ m

mula to determine the minimal wind speed to allow a static flight, the assumption of neglecting the aerodynamic force can be considered as being reasonable. However, it can be expected that within the framework of this hypothesis, the low wind speed limit is underestimated.

3.6 Tether Tension

Figure 3.11 represents tether tension for all azimuth angles enabling a static flight at wind condition $\mathbf{U}_{ref} = [0, 7.5, 0]^\top_{R_0} \, \mathrm{m\,s}^{-1}$ and ship velocity $\mathbf{V}_s = [7.5, 0, 0]^\top_{R_0} \, \mathrm{m\,s}^{-1}$. Two models are compared. Solid and dashed lines are respectively the tension calculated with the present model at ship position $\mathbf{S}$ and kite position $\mathbf{K}$. The dotted line represents static flight tension calculated with the model of [17]. Red, black and blue lines are the tension projected on the unit vector $\mathbf{x}_0$, $\mathbf{y}_0$ and $\mathbf{z}_0$, respectively.

For the presented model, a difference in tension between the ground and kite attachment points can be noticed. This difference is significant for the tensions projected on the axis $\mathbf{z}_0$ and it is caused by the tether weight and aerodynamic loads but as well by the tension direction differences between $\mathbf{S}$ and $\mathbf{K}$.

Leloup et al. [17] consider a straight tether and do not take into account tether loading and kite weight, leading therefore to no differences in tension between the kite and the attachment point. Moreover for a given azimuth, kite altitude is higher for the zero-mass model than for the present model. Combined with the wind veloc-

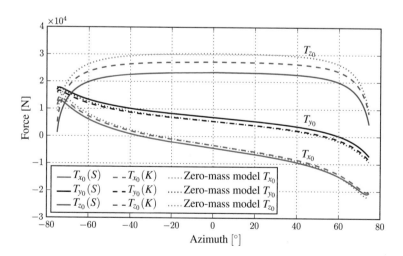

Fig. 3.11 Tether forces projected in frame R_0 for azimuth angle from $-75°$ to $75°$ calculated with the present model at ship attachment point $\mathbf{S}$ and kite position $\mathbf{K}$ and with the zero-mass model presented in Sect. 3.2.5

ity gradient effect, this leads to a higher kite aerodynamic force with the model of Leloup et al. [17].

The propulsive part of the tether tension is the tension projected along the ship path, $T_{x_0}(\mathbf{S})$. The optimal position of the kite to generate the maximum propulsive force is reached for an azimuth angle of $-73.67°$ for both models. We denote F_{P_0} as the propulsive force obtained with the zero-mass model of Leloup et al. [17]. At this azimuth angle the dimensionless difference between F_{P_0} and $T_{x_0}(\mathbf{S})$, defined by

$$\Delta_p = \frac{F_{P_0} - T_{x_0}(\mathbf{S})}{T_{x_0}(\mathbf{S})} \tag{3.52}$$

for a ship velocity $\mathbf{V}_s = [7.5, 0, 0]_{R_0}^\top \, \mathrm{m\,s^{-1}}$ and for a wind velocity range from $\mathbf{U}_{ref} = [0, 5.0, 0]_{R_0}^\top \, \mathrm{m\,s^{-1}}$ to $\mathbf{U}_{ref} = [0, 20.0, 0]_{R_0}^\top \, \mathrm{m\,s^{-1}}$, is plotted in Fig. 3.12. It can be

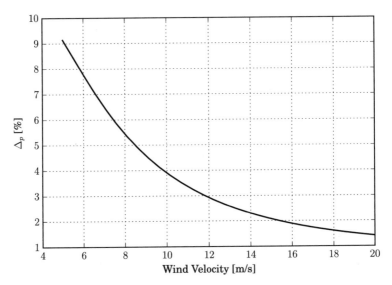

Fig. 3.12 Plot of Eq. (3.52), the dimensionless difference of propulsion forces calculated with the present model compared to propulsion forces calculated with the zero-mass model described in Sect. 3.2.5 for a wind range from $5 \, \mathrm{m\,s^{-1}}$ to $20 \, \mathrm{m\,s^{-1}}$ at $z_{ref} = 10 \, \mathrm{m}$

noticed from this diagram that the relative difference between F_{P_0} and $T_{x_0}(S)$ is up to 9% for a wind speed of $5 \, \mathrm{m\,s^{-1}}$ and decreases to almost 1.5% for a wind speed of $20 \, \mathrm{m\,s^{-1}}$. The error decreases with an increasing wind velocity because of tether loading. Kite weight and sag effects become smaller compared to the kite aerodynamic force. This shows that it is particularly important to take into account tether deformation due to tether loading at low wind speed. By contrast, at high-speed wind, the propulsive force error tends to 1%.

3.7 Conclusion

A three-dimensional analytical model for the tether deformation due to gravity and aerodynamic loading has been derived in this chapter. The effect of the wind velocity gradient has been taken into account both for the tether as well as for the kite aerodynamics. A method for determining a low wind speed limit for kite flight has been developed. This method allows for a comparative study between the influence of tether design, in terms of length and mass per unit length, and kite design, in terms of lift-to-drag ratio angle, lift coefficient and mass. For the present case study, an optimal tether length around 128 m has been identified in order to allow a static flight at a minimal wind velocity. A wind speed above 4.5 m s^{-1} is required to launch the kite. Finally, the presented model indicates an error up to 9% for the zero-mass kite model with a straight zero-mass tether on propulsion force estimation with a static kite flight case.

For this study a tether drag coefficient of 1.1 has been assumed. Jung [15] shows how coating can be used to effectively reduce the aerodynamic drag of the tether. Because a coated rope is heavier it should be analyzed whether the increased tether mass justifies the achievable drag reduction when considering the low wind limit for static kite flight. The presented model could be an interesting starting point to study these competing effects.

In addition, the solution for the static position of a kite in a wind field with velocity gradient is computed in less than 1 s on a common PC. Because of these short calculation times the model is suitable for coupled simulations of a ship towed by a kite for maneuvering and seakeeping assessments. The accuracy of the presented model will further be assessed by comparison with finite element simulations which is currently in progress.

References

1. Argatov, I., Rautakorpi, P., Silvennoinen, R.: Apparent wind load effects on the tether of a kite power generator. Journal of Wind Engineering and Industrial Aerodynamics **99**(5), 1079–1088 (2011). doi: 10.1016/j.jweia.2011.07.010
2. Argatov, I., Rautakorpi, P., Silvennoinen, R.: Estimation of the mechanical energy output of the kite wind generator. Renewable Energy **34**(6), 1525–1532 (2009). doi: 10.1016/j.renene.2008.11.001
3. Bigi, N., Nême, A., Roncin, K., Leroux, J.-B., Bles, G., Jochum, C., Parlier, Y.: A Quasi-Analytical 3D Kite Tether Model. In: Schmehl, R. (ed.). Book of Abstracts of the International Airborne Wind Energy Conference 2015, p. 43, Delft, The Netherlands, 15–16 June 2015. doi: 10.4233/uuid:7df59b79-2c6b-4e30-bd58-8454f493bb09. Presentation video recording available from: https://collegerama.tudelft.nl/Mediasite/Play/0ed3695b3d284dd190f5162cdc5b806a1d
4. Breukels, J.: Kite launch using an aerostat. Technical Report, Delft University of Technology, 21 Aug 2007. http://repository.tudelft.nl/view/ir/uuid%3A1a0c6dfd-6115-461f-ac04-bd8751efd6fb

5. Dadd, G. M.: Kite dynamics for ship propulsion. Ph.D. Thesis, University of Southampton, 2013. http://eprints.soton.ac.uk/id/eprint/351348
6. Dadd, G. M., Hudson, D. A., Shenoi, R. A.: Determination of kite forces using three-dimensional flight trajectories for ship propulsion. Renewable Energy 36(10), 2667–2678 (2011). doi: 10.1016/j.renene.2011.01.027
7. Duckworth, R.: The application of elevated sails (kites) for fuel saving auxiliary propulsion of commercial vessels. Journal of Wind Engineering and Industrial Aerodynamics 20(1–3), 297–315 (1985). doi: 10.1016/0167-6105(85)90023-6
8. Fer, F.: Thermodynamique macroscopique, vol. 2. Gordon & Breach (1971)
9. Fitzgerald, J. E.: A tensorial Hencky measure of strain and strain rate for finite deformations. Journal of Applied Physics 51(10), 5111–5115 (1980). doi: 10.1063/1.327428
10. Hobbs, S. E.: A Quantitative Study of Kite Performance in Natural Wind with Application to Kite Anemometry. Ph.D. Thesis, Cranfield University, 1986. https://dspace.lib.cranfield.ac. uk/bitstream/1826/918/2/sehphd2a.pdf
11. Hoerner, S. F.: Fluid-Dynamic Drag. Bricktown, Brick Town, NJ, USA (1965)
12. International Towing Tank Conference: ITTC Symbols and Terminology List Version 2014, Sept 2014. http://ittc.info/media/4004/structured-list2014.pdf
13. Irvine, H. M., Sinclair, G. B.: The suspended elastic cable under the action of concentrated vertical loads. International Journal of Solids and Structures 12(4), 309–317 (1976). doi: 10.1016/0020-7683(76)90080-9
14. Irvine, H. M.: Cable structures. MIT Press, London (1981)
15. Jung, T. P.: Wind Tunnel Study of Drag of Various Rope Designs. AIAA Paper 2009-3608. In: Proceedings of the 27th AIAA Applied Aerodynamics Conference, San Antonio, TX, USA, 22–25 June 2009. doi: 10.2514/6.2009-3608
16. Leclère, G., Nême, A., Cognard, J., Berger, F.: Rupture simulation of 3D elastoplastic structures under dynamic loading. Computers & Structures 82(23–26), 2049–2059 (2004). doi: 10.1016/j.compstruc.2004.03.073
17. Leloup, R., Roncin, K., Behrel, M., Bles, G., Leroux, J.-B., Jochum, C., Parlier, Y.: A continuous and analytical modeling for kites as auxiliary propulsion devoted to merchant ships, including fuel saving estimation. Renewable Energy 86, 483–496 (2016). doi: 10.1016/j. renene.2015.08.036
18. Lewis, W. J.: Tension Structures: Form and Behaviour. Thomas Telford, London (2003)
19. Naaijen, P., Koster, V.: Performance of auxiliary wind propulsion for merchant ships using a kite. In: Proceedings of the 2nd International Conference on Marine Research and Transportation, pp. 45–53, Naples, Italy, 28–30 June 2007. http://www.icmrt07.unina.it/Proceedings/ Papers/c/26.pdf
20. Sarpkaya, T.: A critical review of the intrinsic nature of vortex-induced vibrations. Journal of Fluids and Structures 19(4), 389–447 (2004). doi: 10.1016/j.jfluidstructs.2004.02.005
21. Terink, E. J., Breukels, J., Schmehl, R., Ockels, W. J.: Flight Dynamics and Stability of a Tethered Inflatable Kiteplane. AIAA Journal of Aircraft 48(2), 503–513 (2011). doi: 10.2514/ 1.C031108
22. Varma, S. K., Goela, J. S.: Effect of wind loading on the design of a kite tether. Journal of Energy 6(5), 342–343 (1982). doi: 10.2514/3.48051
23. Wellicome, J. F.: Some comments on the relative merits of various wind propulsion devices. Journal of Wind Engineering and Industrial Aerodynamics 20(1–3), 111–142 (1985). doi: 10.1016/0167-6105(85)90015-7
24. Williams, P., Lansdorp, B., Ockels, W. J.: Modeling and Control of a Kite on a Variable Length Flexible Inelastic Tether. AIAA Paper 2007-6705. In: Proceedings of the AIAA Modeling and Simulation Technologies Conference and Exhibit, Hilton Head, SC, USA, 20–23 Aug 2007. doi: 10.2514/6.2007-6705

Chapter 4
Kite as a Beam: A Fast Method to get the Flying Shape

Alain de Solminihac, Alain Nême, Chloé Duport, Jean-Baptiste Leroux, Kostia Roncin, Christian Jochum and Yves Parlier

Abstract Designing new large kite wings requires engineering tools that can account for flow-structure interaction. Although a fully coupled simulation of deformable membrane structures under aerodynamic load is already possible using Finite Element and Computational Fluid Dynamics methods this approach is computationally demanding. The core idea of the present study is to approximate a leading edge inflatable tube kite by an assembly of equivalent beam elements. In spanwise direction the wing is partitioned into several elementary cells, each consisting of a leading edge segment, two lateral inflatable battens, and the corresponding portion of canopy. The mechanical properties of an elementary cell—axial, transverse shear, bending, and torsion stiffness—and the chordwise centroid position are determined from the response to several imposed elementary displacements at its boundary, in the case of a cell under an uniform pressure loading. For this purpose the cells are supported at their four corners and different non-linear finite element analyses and linear perturbation computations are carried out. The complete kite is represented as an assembly of equivalent beams connected with rigid bodies. Coupled with a 3D non-linear lifting line method to determine the aerodynamics this structural model should allow predicting the flying shape and performance of new wing designs.

Alain de Solminihac · Alain Nême (✉) · Chloé Duport · Jean-Baptiste Leroux · Kostia Roncin ·
Christian Jochum
ENSTA Bretagne, FRE CNRS 3744, IRDL, 29200 Brest, France
e-mail: alain.neme@ensta-bretagne.fr

Yves Parlier
Beyond the sea, 1010 avenue de l'Europe, 33260 La Teste de Buch, France
e-mail: yves.parlier@beyond-the-sea.com

4.1 Introduction

The aim of this research project is the development of tethered kite systems as auxiliary devices for the propulsion of merchant ships. It is part of the *beyond the sea*® program which is lead by the Institut de Recherche Dupuy de Lôme (IRDL) of EN-STA Bretagne. The goal is to design leading edge inflatable tube kites with surface area larger than 300 m^2. This requires a significant upscaling of common sports kites which generally do not exceed a surface area of 30 m^2. This upscaling process raises several issues: What are the relevant physical effects to take into account? Is it possible to use the same materials as for sports kites? Which geometry should the bridle system have?

Point mass and rigid body models have been used for real-time or faster-than-real-time simulation of the kite dynamics [7, 9]. A typical application of these type of models is control engineering or flight path optimization. However, to get a deeper understanding of the steering behavior and aerodynamic performance of a highly flexible wing the shape deformation plays a crucial role. Breukels [3, 4] developed an engineering model of a deformable flying kite, discretizing the tubular frame by chains of rigid bodies connected by rotational springs and the canopy by arrays of elastic springs and damper elements. All mechanical properties were derived from basic experiments and the aerodynamic load distribution was prescribed by an empirically determined correlation framework. The approach allows modeling of aeroelastic effects.

Bosch [2] applied a geometrically non-linear finite element framework to the kite, discretizing the tubular frame by beam elements and the canopy by custom-made shell elements. This model was used to determine the quasi-static deformation resulting from changes in the boundary conditions, such as aerodynamic loading and steering line displacements. However, only macro-scale Fluid-Structure Interaction (FSI) effects of spanwise torsion and bending of the wings were taken into account. Gaunaa [8] developed a computationally efficient method for determining the aerodynamic performance of kites. The approach iteratively couples a Vortex Lattice Method (VLM) with 2D airfoil data to account for the effects of airfoil thickness and of viscosity. Deformation of the wing is not considered.

The aim of the present study is to develop an engineering tool which enables kite designers to efficiently determine the flying shape of new kites. Given a few design parameters such as the global wing shape, the material used and the wind conditions, it should be possible to predict the flying shape and aerodynamic performance. Structural non-linearity and macro-scale FSI calculations are conducted as major influence factors. Contrary to the point mass and rigid body approaches, the proposed method underlines the importance of considering the inflatable kite as a deformable membrane structure. The method can be used to identify critical aerodynamic peak loads and to take design measures to alleviate these.

In Sect. 4.2 and 4.3 the wing design and basic methods are first outlined, introducing the dicretization concept of the elementary cell, then followed by the identification of mechanical and inertial properties and the assembly of several elementary cells into a model of the complete wing. In Sect. 4.4 and 4.5 results are presented, discussed and interpreted, elaborating also on the fast potential flow-based method used to derive the instantaneous aerodynamic loading of the wing. The preliminary content of the present chapter has been presented at the Airborne Wind Energy Conference 2015 [17].

4.2 General Design Parameters

In this section the general problem definition is outlined. The kite design is developed in several steps starting from a 3D baseline. The required material properties are then discussed, followed by a specification how the aerodynamic loading is determined for different operational modes of the kite.

4.2.1 Design Geometry

The spanwise shape of the wing design is defined by a 3D curve. The chord, sweep, dihedral, and twist of the wing are specified by evolution laws along this baseline. The inflatable tubular frame is detailed by specifying the attachment points of the inflatable battens at the leading edge tube as well as all tube cross section geometries. Each pair of neighboring battens and the corresponding part of the leading edge tube spans a wing section. To complete the definition of the design geometry the design camber of each of these wing sections is defined. This property describes the maximum deviation of the canopy from the mean chord of the wing section. An example of the resulting wireframe representation of the wing is shown in Fig. 4.1.

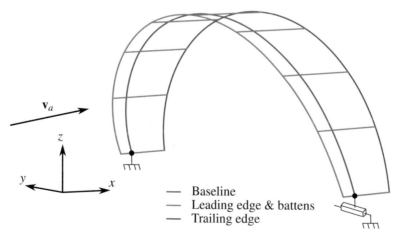

Fig. 4.1 Wireframe representation of a wing including the kite-fixed reference frame (x, y, z) and the apparent wind velocity vector $\mathbf{v}_a$

4.2.2 Material Properties

The fabric mechanical properties for the inflatable beams and the canopy are defined using some data per unit mass (J/kg) for specifying the specific Young's modulus E_m, and some data per unit area (kg/m^2) for specifying the fabric density μ. The Poisson's ratio ν of the fabric should also be specified. In this study only isotropic materials are considered, but the present model could be extended in the case of anisotropic materials.

4.2.3 Relative Wind Conditions

The relative flow conditions at the wing are defined by the apparent wind velocity

$$\mathbf{v}_a = \mathbf{v}_w - \mathbf{v}_p - \mathbf{v}_k, \tag{4.1}$$

with $\mathbf{v}_w$ denoting the true wind velocity and $\mathbf{v}_p$ the ship velocity, both known properties, and $\mathbf{v}_k$ denoting the kite velocity relative to the ship.

Kites can be used in two different flight modes to generate a traction force for towing ships. In static flight mode the kite has a fixed position with respect to the ship and the apparent wind velocity can be readily calculated from Eq. (4.1) by setting $\mathbf{v}_k = 0$. In dynamic flight mode the kite is operated perpendicularly to the tether and the kite velocity is a variable. It is possible to use a simple dynamic flight model, such as the zero mass model [6, 12–14], to calculate $\mathbf{v}_k$ as a function of time and to use this in Eq. (4.1) to derive the apparent wind velocity.

The model described further in the next section requires an a priori estimation of the pressure loading of the canopy, because its geometrical stiffness must be considered. From the apparent wind velocity, and given an estimate a priori of the aerodynamic lift coefficient, this can be achieved by calculating

$$P_m = \frac{1}{2}\rho C_L v_a^2, \tag{4.2}$$

with ρ denoting the air density and C_L the aerodynamic lift coefficient. Given the relatively high lift-to-drag ratio of the wings involved, and given an approximate pressure loading is only required, the effect of the aerodynamic drag coefficient is neglected here.

4.3 Structural Model of the Wing

In this section the structural model of the wing is built up in steps, starting from individual elementary cells which are assembled into a structural model of the entire flexible wing.

4.3.1 Elementary Cell Concept

The structural discretization of the kite is based on a spanwise division of the wing into sections. The proposed concept of the elementary cell accounts for the particular structural design of a membrane wing with inflatable tubular frame. As illustrated in Fig. 4.2 each cell is composed of a segment of the inflatable leading edge, two inflatable battens and the corresponding portion of the canopy. The mechanical be-

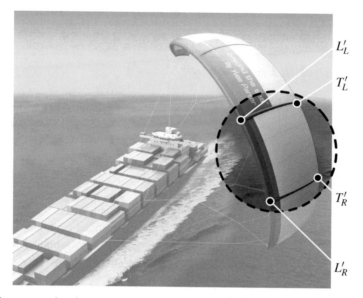

Fig. 4.2 A representative elementary cell with corner points L'_L, L'_R, T'_R and T'_L which are located at the extremities of the inflatable battens

havior of the elementary cell is approximated by an equivalent beam. The stiffness of this simplified structure is matched precisely with the stiffness of the elementary cell under nominal inflation pressure of the tubes.

Because the geometry of the wing is double-curved an elementary cell Q' : L'_L, L'_R, T'_R, T'_L is generally not planar. To further simplify the cell geometry the planar approximation $Q : L_L, L_R, T_R, T_L$ is introduced. Using the midpoints M_1 and M_2 on the left and right batten segments L'_L, T'_L and L'_R, T'_R the spanwise dimension L and the mean chord H of the planar approximation Q are defined as

$$L = \|M_2 - M_1\|, \tag{4.3}$$

$$H = \frac{1}{2}(\|L'_L - T'_L\| + \|L'_R - T'_R\|). \tag{4.4}$$

A local coordinate system $(\mathbf{e}_1, \mathbf{e}_2, \mathbf{e}_3)$ is defined by the unit vector along the spanwise direction

$$\mathbf{e}_1 = \frac{1}{L}(M_2 - M_1), \tag{4.5}$$

the unit vector perpendicular to the plane

$$\mathbf{e}_3 = \frac{(T'_R - L'_L) \times (T'_L - L'_R)}{\|(T'_R - L'_L) \times (T'_L - L'_R)\|}, \tag{4.6}$$

and a third unit vector defined as cross product

$$\mathbf{e}_2 = \mathbf{e}_3 \times \mathbf{e}_1. \tag{4.7}$$

4.3.2 Equivalent Beam Concept

The equivalent beam is introduced to describe the mechanical behavior of the elementary cell by means of an idealized structural object. The following beam properties are identified on the basis of finite element analysis of the elementary cell under various loads:

- Beam centroid distance from the leading edge,
- Tension/Compression stiffness,
- Bending stiffness,
- Torsion stiffness,
- Shear coefficients.

The structural analysis is performed with the finite element solver Abaqus[TM].

4.3.3 Finite Element Model of the Elementary Cell

As a conclusion of a convergence analysis the canopy of the elementary cell is discretized by 2000 rectangular linear membrane elements. The mechanical properties used for the canopy are the in-plane stiffness $E_C = \mu_C E_{m,C}$ and the Poisson ratio v_C. The subscript C indicates properties of the canopy. It is possible to adapt these mechanical properties for the different regions of the canopy, as for instance at the trailing edge if canopy reinforcement effects have to be investigated.

The canopy of the elementary cell is supported by the leading edge tube and two battens. These inflatable elements are modeled as straight beams and discretized by 200 linear beam elements in total for three tubes. Starting from the known beam radius R, fabric stiffness $E_B = \mu_B E_{m,B}$ and Poisson ratio v_B, where subscript B indicates properties of the beam, the section properties are estimated as:

- Elongation stiffness: $2\pi R E_B$,
- Bending stiffness: $\pi R^3 E_B$,
- Transverse shear stiffness [5] : $\dfrac{0.53}{1 + v_B} \pi R E_B$,

- Torsion stiffness: $\dfrac{\pi}{1+v_B}R^3 E_B$.

Because in the final wing model the elementary cells are connected the stiffness of the finite element beam representing a batten is only 50% of the stiffness of the full batten. Underlying is a linear superposition assumption. In this study, it is assumed that the kite design geometry allows to consider the tips of the wing as battens. For these, the stiffness of the corresponding finite element beam is 100% of the stiffness of the full batten.

4.3.4 Pressurization of the Elementary Cell

The geometrical stiffness of the canopy must be considered because it is comparable to the stiffness of the beam frame. The initial shape of the canopy before applying the pressure loading is expressed in the Cartesian frame (e_1, e_2, e_3) with origin at L_L

$$\mathbf{x} = x_1 \mathbf{e}_1 + x_2 \mathbf{e}_2 + x_3 \mathbf{e}_3, \tag{4.8}$$

with x_3 given by the following analytic expression

$$x_3 = \lambda H \sin\left(\pi \frac{x_1}{L}\right) \sin\left(\pi \frac{x_2}{H}\right) \tag{4.9}$$

and λ denoting the design camber of the canopy with a value of $\lambda \approx 5\%$.

The first computation step is a non-linear geometrical analysis. The four corners (T_L, L_L, L_R, T_R) are clamped in space and the elementary cell is loaded with the estimated homogeneous pressure as described in Sect. 4.2.3.

Since membrane elements have no bending stiffness, a damping factor of 5×10^6 is introduced in the Abaqus^{TM} simulation [16] to achieve convergence of the nodal force balance at the end of the time step (100 seconds). Then a second computation step is conducted without damping to check the validity of the obtained solution. A representative simulation result is shown in Fig. 4.3. As a last step the characteristics of the elementary cell under homogeneous pressure loading are determined.

4.3.5 Computation in Linear Perturbation Mode

Starting from this pressurized structure, five linear perturbation calculation cases are completed in order to evaluate the stiffnesses of the elementary cell with respect to the different global degrees of freedom. The cases are listed in Table 4.1 where (a) represents traction along e_1, (b) out-of-plane shear along e_3, (c) in-plane shear along e_2, (d) in-plane bending about e_3 and (e) torsion about e_1. The elementary displacement is given by a and ω is determined by

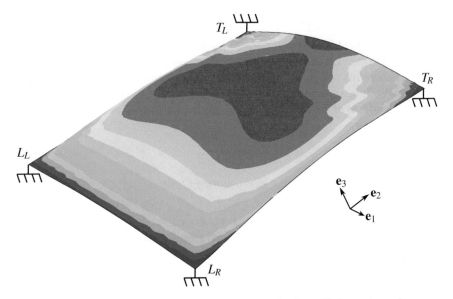

Fig. 4.3 Shape of the canopy under homogeneous pressure loading with the contour value representing the out-of-plane displacement x_3 (deformation scale factor = 1)

Case		(a)	(b)	(c)	(d)	(e)
T_R	U	[a,0,0]	[0,0,a]	[0,a,0]	[-a,0,0]	[0,0,a]
	UR	[0,0,0]	[0,0,0]	[0,0,0]	[0,0,ω]	[ω,0,0]
L_R	U	[a,0,0]	[0,0,a]	[0,a,0]	[a,0,0]	[0,0,-a]
	UR	[0,0,0]	[0,0,0]	[0,0,0]	[0,0,ω]	[ω,0,0]

Table 4.1 Boundary conditions in displacements (U) and rotations (UR) for the load cases (a)–(e), components expressed in the frame $(\mathbf{e}_1, \mathbf{e}_2, \mathbf{e}_3)$

$$\omega = \frac{2a}{H}. \tag{4.10}$$

Numerical results depend linearly on a since a linear perturbation mode is used. Reaction forces at the right corner points, T_R and L_R, are measured for each load case in the direction of the elementary displacement. $F_{TR,X}$ is the reaction force at the trailing edge and $F_{LR,X}$ is the reaction force at the leading edge for load case X.

The computed deformation of the elementary cell is shown for two representative load cases. Figure 4.4 shows the deformation for traction along $\mathbf{e}_1$ while Fig. 4.5 shows the deformation for torsion about $\mathbf{e}_1$.

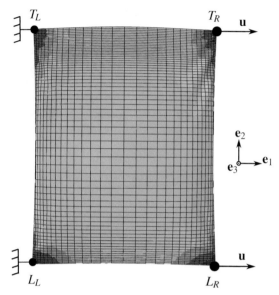

Fig. 4.4 Case (a): traction along $\mathbf{e}_1$. The contour value represents the displacement x_1

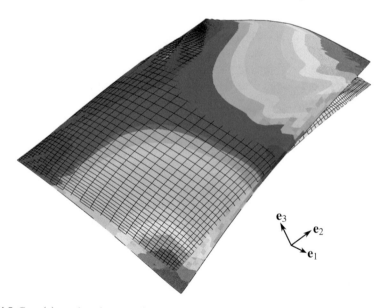

Fig. 4.5 Case (e): torsion along $\mathbf{e}_1$. The contour value represents the displacement x_3

4.3.6 Identification of Beam Properties

The objective of the equivalent beam is to model the mechanical behavior of the elementary cell. The specific load cases (a), (b), (c), (d) and (e) are studied. Static equilibrium on a Timoshenko beam, without warping effect, allows getting relationships between measured reaction forces, elementary displacements and equivalent beam properties. Solving this system of equations, the equivalent beam can be completely described. For better understanding, this approach is presented for the torsion load case

$$M_{(e),ElementaryCell} = (H - D)F_{TR,(e)} - DF_{LR,(e)} + M_{LR,(e)} + M_{TR,(e)}, \tag{4.11}$$

$$M_{(e),EquivalentBeam} = \frac{2aGJ}{HL}. \tag{4.12}$$

For the same torsion angle per unit length the torque is considered to be the same for the elementary cell and the equivalent beam

$$M_{(e),ElementaryCell} = M_{(e),EquivalentBeam}, \tag{4.13}$$

which leads to

$$\frac{2aGJ}{HL} = (H - D)F_{TR,(e)} - DF_{LR,(e)} + M_{LR,(e)} + M_{TR,(e)}, \tag{4.14}$$

from which GJ can be calculated.

The distance of the beam from the leading edge is computed as

$$D = \frac{H}{2} \left(1 + \frac{F_{LR,(e)} + F_{TR,(e)}}{F_{LR,(b)} + F_{TR,(b)}} \right) \tag{4.15}$$

and the equivalent beam extremities B_1 and B_2 are given by

$$B_1 = M_1 - \left(\frac{H}{2} - D \right) e_2, \tag{4.16}$$

$$B_2 = M_2 - \left(\frac{H}{2} - D \right) e_2. \tag{4.17}$$

The stretching stiffness is calculated as

$$EA_0 = L \frac{(F_{LR,(a)} + F_{TR,(a)})}{a} \tag{4.18}$$

and the torsional stiffness from Eq. (4.14) as

$$GJ = \frac{HL}{2a} \left[(H - D)F_{TR,(e)} - DF_{LR,(e)} + M_{LR,(e)} + M_{TR,(e)} \right]. \tag{4.19}$$

The in-plane bending stiffness about e_3 is determined as

$$EI_3 = -\frac{L}{4a}\left[L^2(F_{LR,(c)} + F_{TR,(c)}) - 2HD(F_{LR,(d)} + F_{TR,(d)})\right.$$
$$\left. +2H^2F_{TR,(d)} - 2H(M_{LR,(d)} + M_{TR,(d)})\right]. \qquad (4.20)$$

The strain energy ratio of transverse shear stiffness along e_3 and bending stiffness along e_2 is conventionally evaluated as $12EI_2/(GA_{03}L^2)$. For all the studied cases, this ratio is approximately 3, which is expected for standard leading edge inflatable tube kites. According to this property the transverse shear stiffness along e_3 can be estimated as

$$GA_{03} = \frac{L(F_{LR,(b)} + F_{TR,(b)})}{a}. \qquad (4.21)$$

The transverse shear stiffness along e_2 is given by

$$GA_{02} = \frac{12EI_3L(F_{LR,(c)} + F_{TR,(c)})}{12EI_3a - L^3(F_{LR,(c)} + F_{TR,(c)})} \qquad (4.22)$$

and the bending stiffness about e_2 is evaluated as

$$EI_2 = \frac{L^2GA_{03}}{6[aGA_{03} - 2(M_{LR,(b)} + M_{TR,(b)})]}(M_{LR,(b)} + M_{TR,(b)}). \qquad (4.23)$$

4.3.7 Wing Assembly

To build the kite structure, equivalent beams representing elementary cells are gathered and connected together with rigid bodies. This method is illustrated in Fig. 4.6. The equivalent beams edges are not at the same position for two successive elementary cells. It is, then, assumed here that two neighbor beams share similar displacements and rotations at their extremities similarly to a virtual rigid body connecting these two sections.

4.4 Case of Study

The data used for the case of study is summarized in Table 4.2. A kite with $35\,\text{m}^2$ surface area is considered having a chord measuring between 1.2 m and 2.4 m. The leading edge tube has a radius of 0.1 m whereas the batten tubes have radii of 0.05 m. These values have been applied in a numerical model with specific boundary conditions for estimation of the spanwise bending of the wing. In the following, the spanwise bending of the wing is characterized by the closing of the kite which is defined as the distance between tips.

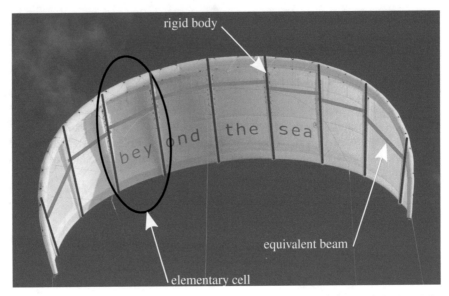

Fig. 4.6 Equivalent beams connected by rigid bodies modeling the mechanical behavior of the flexible wing

	Specific Young's Modulus	2300 J/g
Canopy	Surface Weight	52 g/m^2
	Poisson's Coefficient	0.15
Inflated tubes	Specific Young's Modulus	2200 J/g
	Surface Weight	146 g/m^2
	Poisson's Coefficient	0.15

Table 4.2 Material properties of the kite

4.4.1 Flow Model—Non-Linear 3D Lifting Line Method

The non-linear 3D lifting line is based on an extension of Prandtl's lifting line theory. This extension is intended for wings with variable dihedral and sweep angles. Leloup introduces in [14] a linear implementation while the present method is considering the non-linearity of the aerodynamic lift coefficient. The finite wing and its wake are represented by a set of horseshoe vortices of different circulation strengths Γ. The aim of the algorithm presented below is to calculate the circulation of each horseshoe vortex. Once obtained, the local effective flow for each wing section allows calculating the local aerodynamic forces and torques along the wing span. The numerical iterative solution is based on Anderson [1, Chap. 5, Sect. 5.4], the calculation of effective local incidence angles was adapted to the cases of wings which are non-straight and non-planar. The horseshoe vortices used for discretisation and calculation of their influences are derived from Katz and Plotkin [10, Chap. 12, Fig. 12.2 (a)].

The wing is divided in a finite number of parallel sections, each one represented by a horseshoe vortex. A horseshoe vortex consists of six vortex segments. The bound vortex is located at the quarter chord length, carefully perpendicular to the plane of the considered section. Each of the two trailing vortices are separated into two parts: the first one extends parallel to the chord over one chord length and the second one extends parallel to the local free stream over several chord lengths. Finally the starting vortex closes the horseshoe. It is important to note that even with a swept wing, the bound vortex along the lifting line is orthogonal to the two adjacent trailing vortices. This is illustrated in Fig. 4.7. This leads to a piecewise constant discretisation of the lifting line, but it is necessary to have a correct match between the local lift calculated from the Kutta formula or from the polar of the section.

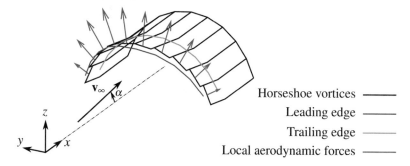

Horseshoe vortices ——————
Leading edge ——————
Trailing edge ——————
Local aerodynamic forces ——————

Fig. 4.7 Example of a coarsely discretized wing in translation at a flow incidence angle of $\alpha = 10°$ and with local aerodynamic resulting forces (local torques not represented to improve readability)

In order to calculate the aerodynamic forces, the circulation is initialized by an elliptic distribution along the wing span. With the Biot-Savart law, the induced velocities by each vortex segment can be calculated and summed at each point of the lifting line. Combined with the local free stream velocity, the effective wind and the effective incidence angle are obtained for each section. The current local bound vortex strength is then calculated using the polar of the section, which leads to the local lift force, and via the Kutta formula, which converts this force in local circulation. The circulation value is ultimately updated by weighting between current and previous values using a damping factor. This whole process is repeated until the circulation distribution converges. The lift, drag and torque of each section of the wing are then post processed with the converged local circulation value, which leads to integrated local loads, and after being carefully summed, to global loads of the wing.

4.4.2 Abaqus^{TM} Procedure

The structural analysis is conducted using the commercial solver Abaqus^{TM}. The computation of the equivalent beam deformation under aerodynamic loading is performed with a large-displacement formulation from the initial configuration of the equivalent beam which accounts for its stress-free geometry. The large-displacement formulation of Timoshenko beam elements used in Abaqus^{TM} [16] is based on a multiplicative decomposition of the deformation gradient into a *stretch* part ($\mathbf{F}^s$) and a *distorsion* part ($\mathbf{F}^d$). The strain tensor is obtained by addition of the logarithm of $\mathbf{F}^s$ and the Green-Lagrange formula applied to $\mathbf{F}^d$. No artificial damping forces were introduced into the finite element model. Since the geometrical location of the finite element beam lies on the lifting line, its local section direction $\mathbf{n}_1$ is determined with the orthogonal projection of the point M, defined as the geometric center of the beam element, on the equivalent beam which is located at the distance D (see Eq. (4.15)) from the leading edge. If P represents the projection of M, it can be determined from

$$P - B_1 = [(M - B_1) \cdot \mathbf{e}_1] \mathbf{e}_1. \tag{4.24}$$

If $\mathbf{t}$ stands for the unit vector along the beam element axis, the unit vector $\mathbf{n}_1$ is obtained from

$$\begin{cases} \mathbf{n}_1' = (P - M) - [(P - M) \cdot \mathbf{t}] \mathbf{t}, \\ \mathbf{n}_1 = \dfrac{\mathbf{n}_1'}{\|\mathbf{n}_1'\|}. \end{cases} \tag{4.25}$$

The second local section direction of the beam element $\mathbf{n}_2$ is such that

$$\mathbf{n}_2 = \mathbf{t} \times \mathbf{n}_1. \tag{4.26}$$

We assume that the location of the beam element section centroid is expressed in the local beam element frame $(\mathbf{t}, \mathbf{n}_1, \mathbf{n}_2)$ as

$$[0, \|\mathbf{n}_1'\|, 0]^\top. \tag{4.27}$$

The beam element section properties are the same as in the Eqs. (4.18) to (4.23) assuming the local beam element frame $(\mathbf{t}, \mathbf{n}_1, \mathbf{n}_2)$ is matching the equivalent beam frame $(\mathbf{e}_1, \mathbf{e}_2, \mathbf{e}_3)$.

4.4.3 Boundary and Wind Conditions

To model the closing and opening of the kite under load the specific boundary conditions listed in Table 4.3 are chosen.

By definition the apparent wind velocity is aligned with the x-axis as illustrated in Fig. 4.1. It has a value of 30 m/s at an air density of $1.2 \, kg/m^3$. No twist is considered for the stress-free geometry of the kite and the wind is parallel to its symmetry

Table 4.3 Boundary conditions in displacements (U)
and rotations (UR) for the kite
opening calculation case

	Left	Right
U_x	0	0
U_y	0	Free
U_z	0	0
UR_x	Free	Free
UR_y	0	0
UR_z	0	0

plan. So, the attack angle of $10°$ is directly the angle between the apparent wind velocity vector and the center kite chord. According to these assumptions, the initial aerodynamic load computed with the 3D lifting line method outlined in Sect. 4.4.1 leads to a C_L value of 0.707. Consequently, the mean pressure given by Eq. (4.2) and used for the identification of the equivalent beam properties is 382 Pa.

4.4.4 Fluid-Structure Coupling

An iterative algorithm [15] is used in a single artificial time increment corresponding to the kite evolution from its stress-free configuration up to its deformed configuration under aerodynamic loading. The first beam loading is computed with the 3D lifting line method considering the stress-free configuration. A similar procedure is used in [18]. The same line is used for both flow model and structure calculation and both lines have the same mesh. Fluid computations provide nodal aerodynamic forces and moments reduction whereas solid calculations determine nodal displacements and rotations. Note that the deformation of the kite does not change the 2D characteristics of the wing section used for fluid computations. The convergence of the procedure is observed through two physical values: lift and kite closing.

4.4.5 Results

The CPU times observed on a classical computer[1] are respectively 0.12 s and 1.3 s to obtain the non-linear lifting line and Abaqus[TM] converged solutions. Generally, six fluid-structure coupling loops are required to achieve the convergence, as can be seen in Fig. 4.8.

This is quite similar to the convergence observed on former study with a shell finite element modeling of the canopy [11]. In Fig. 4.9, the torsion of the canopy can be observed whereas in Fig. 4.10 the opening is shown. The presented design does not contain any bridle system hence such large displacements can be noticed.

---

[1] Intel® Xeon® CPU E31220 @ 3.10GHz / 4.00 GB RAM / 64 bits

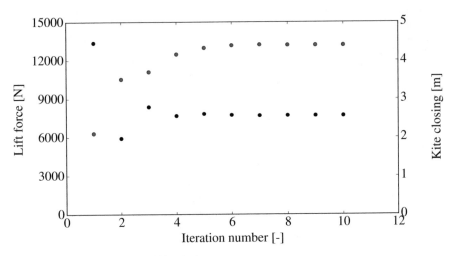

Fig. 4.8 Convergence of lift and kite closing

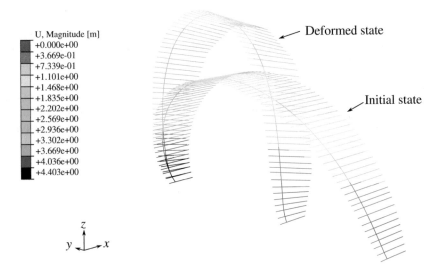

Fig. 4.9 Global deformation of the kite (deformation scale factor = 1)

4.5 Discussion

This method allows a first estimation of the flying shape, the drag and the lift of a deformable kite. As highlighted in Fig. 4.8 the global lift force is around 40% lower for a soft kite (converged point) than for a rigid kite (first point). This result tends to justify the soft kite approach for the simulation of kite performances. For realistic displacements, the bridle system and tethers must also be taken into account. It is

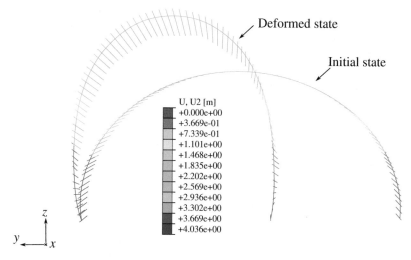

Fig. 4.10 Closing of the kite under load

important to notice that the structural behavior of the kite largely depends on its loading. As an example, torsional stiffness of an equivalent beam increases significantly with the pressure, as illustrated in Fig. 4.11.

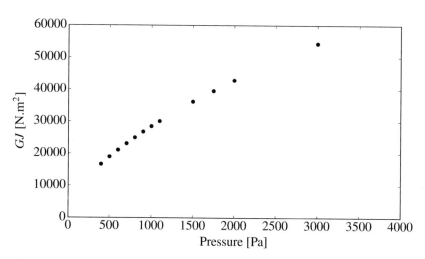

Fig. 4.11 Increase of torsional stiffness of an equivalent beam with applied pressure on the canopy

4.6 Conclusion

In this study the complex structural behavior of a soft kite was simplified to a simple arrangement of beams. The parameters of the beams were calculated from finite element analysis of so-called elementary cells, which model the canopy of a single kite cell, under homogeneous pressure. This pressure was derived from the global lift coefficient of the initial kite geometry.

Coupled with a fluid model, the simplified structure model approach presented in this study allows a prediction of the flying shape and helps obtaining a better understanding of the main phenomena which have to be considered. It is as well a quick (a couple of minutes) and convenient way to get a first estimation of the kite performance accounting for fluid-structure interaction.

However, the "kite as a beam" model has not been compared to more detailed structural models. This analysis is currently ongoing. Additionally, the "kite as a beam" approach presented here does not directly address local aspects like stresses and strains in the canopy and in the inflatable structure. These aspects have to be investigated with fully coupled FEA / CFD computations. Overall validation requires relevant experiments that are currently under progress at the institute.

The next step to extend the "kite as a beam" model would be the inclusion of bridles and tethers to improve their design and for better towing force estimations. In parallel, it will be necessary to develop a more realistic beam frame model for the kite structure. A parametric formulation of the influence of the geometry of the canopy on the kite stiffness will be developed. Hence, the stiffness of the elementary cells will depend on the aerodynamic pressure. The new model should also enable an improvement of the design of the inflatable leading edge and the battens according to stress limit and buckling condition.

References

1. Anderson, J. D.: Fundamentals of Aerodynamics. 5th ed. McGraw-Hill (2014)
2. Bosch, A., Schmehl, R., Tiso, P., Rixen, D.: Dynamic nonlinear aeroelastic model of a kite for power generation. AIAA Journal of Guidance, Control and Dynamics 37(5), 1426–1436 (2014). doi: 10.2514/1.G000545
3. Breukels, J.: An Engineering Methodology for Kite Design. Ph.D. Thesis, Delft University of Technology, 2011. http://resolver.tudelft.nl/uuid:cdece38a-1f13-47cc-b277-ed64fdda7cdf
4. Breukels, J., Schmehl, R., Ockels, W.: Aeroelastic Simulation of Flexible Membrane Wings based on Multibody System Dynamics. In: Ahrens, U., Diehl, M., Schmehl, R. (eds.) Airborne Wind Energy, Green Energy and Technology, Chap. 16, pp. 287–305. Springer, Berlin Heidelberg (2013). doi: 10.1007/978-3-642-39965-7_16
5. Cowper, R. G.: The Shear Coefficient in Timoshenko's Beam Theory. Journal of Applied Mechanics 33(2), 335–340 (1966). doi: 10.1115/1.3625046
6. Dadd, G. M., Hudson, D. A., Shenoi, R. A.: Determination of kite forces using three-dimensional flight trajectories for ship propulsion. Renewable Energy 36(10), 2667–2678 (2011). doi: 10.1016/j.renene.2011.01.027

7. Fechner, U., Vlugt, R. van der, Schreuder, E., Schmehl, R.: Dynamic Model of a Pumping Kite Power System. Renewable Energy (2015). doi: 10.1016/j.renene.2015.04.028. arXiv:1406. 6218 [cs.SY]
8. Gaunaa, M., Paralta Carqueija, P. F., Réthoré, P.-E. M., Sørensen, N. N.: A Computationally Efficient Method for Determining the Aerodynamic Performance of Kites for Wind Energy Applications. In: Proceedings of the European Wind Energy Association Conference, Brussels, Belgium, 14–17 Mar 2011. http://findit.dtu.dk/en/catalog/181771316
9. Gohl, F., Luchsinger, R. H.: Simulation Based Wing Design for Kite Power. In: Ahrens, U., Diehl, M., Schmehl, R. (eds.) Airborne Wind Energy, Green Energy and Technology, Chap. 18, pp. 325–338. Springer, Berlin Heidelberg (2013). doi: 10.1007/978-3-642-39965-7_18
10. Katz, J., Plotkin, A.: Low-speed aerodynamics. 2nd ed. Cambridge University Press (2001)
11. Leloup, R., Bles, G., Roncin, K., Leroux, J., Jochum, C., Parlier, Y.: Prediction of the stress distribution on a Leading Edge Inflatable kite under aerodynamic load. In: Proceedings 14èmes Journées de l'Hydrodynamique, Val de Reuil, France, 18–20 Nov 2014
12. Leloup, R., Roncin, K., Behrel, M., Bles, G., Leroux, J.-B., Jochum, C., Parlier, Y.: A continuous and analytical modeling for kites as auxiliary propulsion devoted to merchant ships, including fuel saving estimation. Renewable Energy **86**, 483–496 (2016). doi: 10.1016/j. renene.2015.08.036
13. Leloup, R.: Modelling approach and numerical tool development for kite performance assesment and mechanical design; application to vessels auxiliary propulsion. Ph.D. Thesis, ENSTA Bretagne/University of Western Brittany, 2014
14. Leloup, R., Roncin, K., Bles, G., Leroux, J.-B., Jochum, C., Parlier, Y.: Estimation of the Lift-to-Drag Ratio Using the Lifting Line Method: Application to a Leading Edge Inflatable Kite. In: Ahrens, U., Diehl, M., Schmehl, R. (eds.) Airborne Wind Energy, Green Energy and Technology, Chap. 19, pp. 339–355. Springer, Berlin Heidelberg (2013). doi: 10.1007/978-3-642-39965-7_19
15. Sigrist, J. F.: Fluid-Structure Interaction: An Introduction to Finite Element Coupling. Wiley (2015)
16. Simulia: Abaqus Analysis User's Guide. v6.14. (2014)
17. Solminihac, A. d., Nême, A., Roncin, K., Leroux, J.-B., Jochum, C., Parlier, Y.: Kite as Beam – An Analytical 3D Kite Tether Model. In: Schmehl, R. (ed.). Book of Abstracts of the International Airborne Wind Energy Conference 2015, p. 44, Delft, The Netherlands, 15–16 June 2015. doi: 10.4233/uuid: 7df59b79-2c6b-4e30-bd58-8454f493bb09. Presentation video recording available from: https://collegerama.tudelft.nl/Mediasite/Play/ 0551a13079294bc88f7b0e32e8944d121d
18. Trimarchi, D., Rizzo, C. M.: A FEM-Matlab code for Fluid-Structure interaction coupling with application to sail aerodynamics of yachts. In: Proceedings of the 13th Congress of the International Maritime Association of the Mediterranean, Istanbul, Turkey, 12–15 Oct 2009. http://eprints.soton.ac.uk/id/eprint/69831

Chapter 5
Dynamic Model of a C-shaped Bridled Kite Using a few Rigid Plates

Jelte van Til, Marcelo De Lellis, Ramiro Saraiva and Alexandre Trofino

Abstract This chapter presents a dynamic model of a flexible wing as the main component of an airborne wind energy system for crosswind operations. The basic components are rigid plates that are interconnected by gimbal joints and allow for rotational degrees of freedom which mimic the basic deformations of a C-shaped kite. Realistic steering is accomplished through length-varying bridle lines that are actuated by a kite control unit. This suspended cable robot is connected to the ground by a tether model which uses linked rigid line elements and allows for reel out at a constant speed. The simulation results show that the developed model is robust and that the steering behavior of a C-shaped kite can be reproduced. The main deformation modes are captured and the model has the potential to run real-time, making it suitable for control simulation purposes.

5.1 Introduction

The aim of this work is to develop a kite model that mimics the deformations observed in a physical bridled kite. Previous studies [3] have shown that deformation modes have a significant influence on the dynamic behavior of the kite. In principle, the shape of the kite is determined by the equilibrium of aerodynamic forces, bridle line forces, internal structural forces and acceleration forces, the latter being comparatively small for a lightweight membrane wing. An imbalance of these forces, for example induced by actuation of the bridle lines, results in a deformation of the

Jelte van Til (✉)
Delft University of Technology, Faculty of Mechanical, Maritime and Materials Engineering, PO Box 5, 2600 AA Delft, The Netherlands
e-mail: jeltevtil@gmail.com

Marcelo De Lellis · Ramiro Saraiva · Alexandre Trofino
Federal University of Santa Catarina, Department of Automation and Systems. Roberto Sampaio Gonzaga, 88040-900, Florianopolis, Brazil

© Springer Nature Singapore Pte Ltd. 2018
R. Schmehl (ed.), *Airborne Wind Energy*, Green Energy and Technology, https://doi.org/10.1007/978-981-10-1947-0_5

wing shape. The orientation of the aerodynamic lift forces change as a result of the deformation, hence affecting the motion of the kite. Therefore a good deformation model should be used in kite control to ensure a stable flight and optimal energy harvest. As the kite is led into turning maneuvers, one is interested in observing two main deformation modes: spanwise bending, and torsion [2, 4].

The kite pumping system considered in this chapter is shown in Fig. 5.1. The inflatable kite is connected to an airborne Kite Control Unit (KCU) through four bridle lines. Two of the lines, connected to the kite trailing edge, are variable in length to accomplish steering and de-powering. The kite transfers traction forces to a main tether which connects the KCU to the ground. Power is harvested while the tether is reeled out and the kite is steered into figure-eight patterns to maximize power generation.

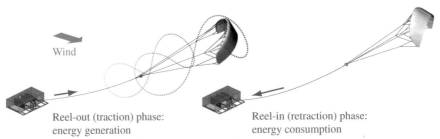

Reel-out (traction) phase: Reel-in (retraction) phase:
energy generation   energy consumption

Fig. 5.1 Kite power system using a bridled kite, a Kite Control Unit, a main tether and a ground-based winch containing a generator/motor module [13]

One of the current state-of-the-art mechanical models of a kite found in the literature and used for control purposes is the four-point mass model developed by Fechner et al. [7]. This model represents very basic deformations, through spring/damper elements connecting the masses, while still capable of real-time simulation. The translational displacements between the masses however do not accurately mimic the deformations observed in a physical C-shaped kite.

Several other models can be found that aim to reproduce the deformations in a kite. Bosch et al. [2] presented a dynamic nonlinear aeroelastic model of a kite. In [12] fluid and structural dynamics solvers have been coupled to study the deformation behavior of wind turbines. Although these models may produce accurate deformation results, they are at least two orders of magnitude slower than real-time and are thus not suitable for real-time simulation and control.

Aiming at studying the behavior of the airborne subsystem of a pumping kite system, in this chapter we present the development of a dynamic model comprising the C-shaped kite, four bridle lines, the KCU, and the main tether connecting the KCU to the ground winch. The kite itself is modeled with three rigid planes interconnected by gimbal joints, each of which introduces three rotational degrees of freedom. Rotational springs and dampers are associated with each of these six resulting degrees of freedom. This configuration allows for the basic torsion and bending deformation modes.

The remaining chapter is organized as follows. In Sect. 5.2 the computational modeling approach is explained. In Sect. 5.3 the model parameters are tuned to make the behavior match as closely as possible the behavior of the four-point mass model introduced earlier. The main results in relationship to the deformations are also presented: the displacement angle profile in the gimbal joints, the angle of attack on each rigid plate, and the steering behavior. In Sect. 5.4 the analysis is concluded by some final remarks. The preliminary content of the present chapter has been presented at the Airborne Wind Energy Conference 2015 [11].

5.2 Modeling Approach

The primary focus of this work is to show the effects of deformation of the kite in a simple way. It was reasoned that, if the model would prove accurate in representing these phenomena with reasonable computational speed in a non-optimal framework such as the SimMechanics™ toolbox of Matlab®[9], one could proceed in optimizing computational speed of the same model by implementing it in Python, for instance. SimMechanics™ offers several advantages:

- Rapid construction of mechanical systems is possible, without the need to implement the equations of motion. Only force definitions are needed.
- Integrated numerical fine-tuning is present in the software, which usually is very time-consuming to perform manually.

5.2.1 SimMechanics™ Modeling Environment

SimMechanics™ allows to interconnect *bodies* (which can be given coordinate systems, masses and inertias) with *joints*, which can be chosen according to the Degrees of Freedom (DoFs) wished for. Furthermore, *constraints* are used (e.g. distance drivers between bodies in order to control reel-out velocity), as well as *force elements*: linear/rotational springs and dampers.

5.2.2 Wind model

In order to determine the wind speed at the height of the kite and at the height of the tether, the power law was used [1],

$$v_{\text{w}} = v_{\text{w,ref}} \left(\frac{z}{z_{\text{ref}}} \right)^{\alpha}, \tag{5.1}$$

where $v_{w,ref}$ is the reference wind speed; a value of 9.39 m/s was used, as in Fechner's four-point mass model [7]. Likewise, a reference height z_{ref} of 6 m and an exponent $\alpha = 1/7$ have been used.

5.2.3 Tether Model

The tether was modeled as three line sections of equal length $r_0/3$, separated by Universal (2 DoF rotational) joints, with point masses at the location of the joints, as shown in Fig. 5.2.

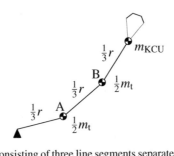

Fig. 5.2 Tether model consisting of three line segments separated by point masses at A and B. Each of the junctions A and B are represented by universal joints, allowing for two rotational degrees of freedom—all but rotation about the axis of the line segments

The drag force on the point masses A and B comprising the tether was determined by considering half of the total tether length, below and above the middle. The orientation vectors of the bottom and top segments, s_b and s_t, were taken as the vector from the ground to the position of the bottom mass $p_{t,b}$, and the vector connecting the position of the top mass $p_{t,t}$ to the position of the KCU p_{KCU}, respectively

$$s_b = p_{t,b}, \tag{5.2}$$

$$s_t = p_{KCU} - p_{t,t}. \tag{5.3}$$

The apparent velocities of each of the tether segments were calculated using the wind velocity at each of the point masses, together with their own velocities

$$v_{a,t,b/t} = v_w|_{p_{t,b/t}} - v_{t,b/t}. \tag{5.4}$$

The cross-sectional areas of the segments were projected to give the component perpendicular to the apparent wind velocity vectors of each of the point masses

$$A_{t,b/t} = \frac{3}{2} s_{b/t} d_t \sqrt{1 - \left(\frac{s_{b/t}}{s_{b/t}} \cdot \frac{v_{a,s,b/t}}{v_{a,s,b/t}} \right)^2}, \tag{5.5}$$

where d_t is the diameter of the tether. The factor 3/2 is to approximate the total cable area, since only the top and bottom segments are used. The aerodynamic drag force on each of the tether segments was finally computed using the air density ρ and the constant tether drag coefficient $C_{D,t}$

$$\mathbf{F}_{d,t,b/t} = \frac{1}{2} C_{D,t} \rho \, v_{a,t,b/t} A_{t,b/t} \mathbf{v}_{a,t,b/t}. \tag{5.6}$$

5.2.4 Implementation of Forces

Several forces act on the kite system: aerodynamic, gravitational and apparent forces (centripetal and Coriolis) [6]. The aerodynamic forces were implemented by hand, while the gravitational and apparent forces were automatically implemented in Simulink®, in which the SimMechanics™ toolbox is used, in a so-called *Machine Environment*.

Aerodynamic Forces The aerodynamic lift and drag forces acting on each of the three wings of the kite were computed using their respective velocities and positions, made available through *body sensors*. Once the forces were computed, they were applied to the center of mass of the respective objects by means of *body actuators*. In case of the kite wings, the position was monitored at several points on the wing in body-fixed coordinate systems to set up the coordinate systems for each of the wings. The definitions of these forces are explained further in Sects. 5.2.3 and 5.2.5.

Gravitational Forces Gravitational forces were applied to point masses A and B comprising the tether, the KCU, and each of the three kite wings. SimMechanics™ automatically implements these forces through the Machine Environment, and the general formula for a body i with mass m_i states

$$\mathbf{F}_{g,i} = -m_i g \hat{\mathbf{z}}. \tag{5.7}$$

Apparent Forces SimMechanics™ also automatically calculates the centripetal and centrifugal forces applied to bodies in a rotating system. The magnitude of these forces were found to be relatively small w. r. t. aerodynamic forces. The most significant force is the centrifugal force, which acts upon the kite along the tether. It can be approximated as

$$\mathbf{F}_{C,k}(t) = \frac{(m_k + m_{KCU})}{r^4} ||\mathbf{v}_{KCU} \times \mathbf{r}||^2 \mathbf{r}, \tag{5.8}$$

where $\mathbf{r}$ is the vector pointing from the ground winch to the kite, i. e. giving the position of the kite. The centrifugal forces applied to the bottom and top tether segments are approximately

$$\mathbf{F}_{\mathrm{C,t,b}}(t) = \frac{3\,m_t}{2\,r^4}||\mathbf{v}_{t,b} \times \mathbf{r}_b||^2 \mathbf{r}_b, \tag{5.9}$$

$$\mathbf{F}_{\mathrm{C,t,t}}(t) = \frac{3\,m_t}{2\,r^4}||\mathbf{v}_{t,t} \times \mathbf{r}_t||^2 \mathbf{r}_t. \tag{5.10}$$

Vectors $\mathbf{r}_b$ and $\mathbf{r}_t$ describe the center of mass of the bottom and top tether segments, respectively. The tether mass is taken as constant, since it varies very little with respect to the compound mass of kite and KCU.

5.2.5 Aerodynamic Model of the Three-Plate Kite

The lift and drag coefficients (C_L and C_D) as a function of the angle of attack α were taken from the results of previous models of respective coefficients of stalled and unstalled airfoils [10]. These curves were modified by [7], based on experience, to form a better match to a Leading Edge Inflatable (LEI) tube kite. The resulting curves, after interpolation with polynomials of degree 9, are shown in Fig. 5.3.

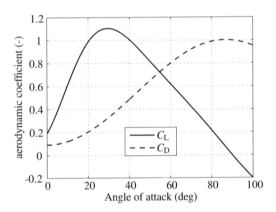

Fig. 5.3 Lift and drag coefficients as function of the angle of attack, obtained through interpolation of the curves from [7] with 9-degree polynomials

A concise way to represent the shape of a C-shaped bridled kite, including bending and torsion deformations, is represented in Fig. 5.4. The kite is modeled as three rigid plates (labeled A, B and C), which are connected by gimbal joints to allow for three rotational DoFs between each two adjacent plates. These degrees of freedom are decomposed in a spanwise bending δ and two torsions: around the local y-axes (τ_y) and the z-axes (τ_z) of the top plate at the respective joint locations, as shown in Fig. 5.5. Each rectangular plate has a constant density ρ_k and thickness t_k, thus having its center of mass in the middle. The kite's flexibility stems from rotational spring/damper combinations for each DoF, giving a total of six springs and six dampers, also shown in Fig. 5.5. The kite bridle lines are represented by distance constraints: the power lines (in front, leading edge) with fixed length l_p, and the steering lines (in back, trailing edge) with time-varying lengths, imposed by the kite controller. Each of the steering lines contains a translational spring/damper element to represent the response of the steering lines to the control input, as well as to relax the constraints to the system, i.e. to reduce its stiffness.

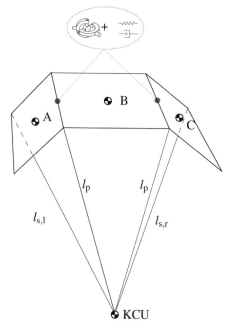

Fig. 5.4 Three-plane kite model, with red dots showing the gimbal joints with rotational springs and dampers [8]. For an illustration of the degrees of freedom governed by these joints, the reader is referred to Fig. 5.5

As the plates (foils) rotate about each other they may intersect, but this intersection is carried out as if it were physically absent. This behavior characterizes another approximation made in the model. The power lines are attached to the top of the side foils, rather than to the front corners of the kite. This allows the top foil to open up independently from the side foils. In this configuration a steering line control input causes a much larger change in the angles of attack of the side foils than if the steering lines were connected to the corners, making the kite more suitable for steering manoeuvres.

For each of the plates, all forces are applied to the center of mass. The aerodynamic forces are determined by the apparent wind velocity of each airfoil and the respective angle of attack $\alpha_{A/B/C}$. This apparent air velocity for a given airfoil is calculated as

$$\mathbf{v}_{a,A/B/C} = \mathbf{v}_w - \mathbf{v}_{A/B/C}, \tag{5.11}$$

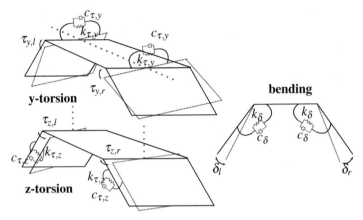

Fig. 5.5 Three rotational deformations in the kite, consisting of the spanwise bending (right) and two types of torsion: y-torsion (top) and z-torsion (bottom). The axes of torsional deformations are indicated by the red dotted line. The rotational springs and dampers associated to these rotations represent the material stiffness and damping. The left and right plates, as seen from the leading edge, are denoted with subscripts l and r

where $\mathbf{v}_w$ is taken to be the wind velocity at the KCU, which is always within a few meters of the kite foils, and $\mathbf{v}_{A/B/C}$ is the velocity of the foil A, B or C. The angle of attack is determined by the apparent air velocity at each airfoil and its orientation, which is determined by the following coordinate system

$$\hat{\mathbf{x}}_{A/B/C} = \frac{\mathbf{P}_{f,A/B/C} - \mathbf{P}_{CG,A/B/C}}{||\mathbf{P}_{f,A/B/C} - \mathbf{P}_{CG,A/B/C}||}, \tag{5.12}$$

$$\hat{\mathbf{y}}_{A/B/C} = \frac{\mathbf{P}_{r,A/B/C} - \mathbf{P}_{CG,A/B/C}}{||\mathbf{P}_{r,A/B/C} - \mathbf{P}_{CG,A/B/C}||}, \tag{5.13}$$

$$\hat{\mathbf{z}}_{A/B/C} = \hat{\mathbf{x}}_{A/B/C} \times \hat{\mathbf{y}}_{A/B/C}. \tag{5.14}$$

The setup of this coordinate system is illustrated in Fig. 5.6.

With the definition of the foil coordinate system, the deformation angles can be computed as

$$\tau_{y,l/r} = \arccos((\hat{\mathbf{x}}_B \times \hat{\mathbf{x}}_{A/C}) \cdot \hat{\mathbf{y}}_B), \tag{5.15}$$

$$\tau_{z,l/r} = \arccos((\hat{\mathbf{x}}_B \times \hat{\mathbf{x}}_{A/C}) \cdot \hat{\mathbf{z}}_B), \tag{5.16}$$

$$\delta_{l/r} = \pm(\arcsin(\hat{\mathbf{y}}_{A/C} \cdot \hat{\mathbf{y}}_B) + \pi) + \Theta_0, \tag{5.17}$$

where Θ_0 is the angle between the adjacent airfoils, as seen from the leading edge, when there is no deformation. In order to compute the angles of attack, the apparent velocity needs to be projected onto the x-z plane, resulting in

$$\mathbf{v}_{a,xz} = \mathbf{v}_a - (\mathbf{v}_a \cdot \hat{\mathbf{y}})\hat{\mathbf{y}}. \tag{5.18}$$

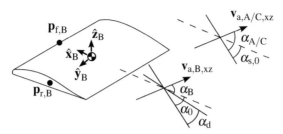

Fig. 5.6 Definition of the foil coordinate system and angles of attack, where the top foil is taken as example. The angles of attack of the side foils are shown in the top right. The chord line is represented by the dotted line, and the angles α_0, α_B and α_d lie in the x-z plane

The angle of attack of the top airfoil then becomes

$$\alpha_B = \arcsin\left(\frac{\mathbf{v}_{a,B,xz} \cdot \hat{\mathbf{z}}_B}{v_{a,B,xz}}\right) - \alpha_d + \alpha_0, \tag{5.19}$$

where α_d is the de-power angle, that results from reeling out an equal length of both steering lines, and α_0 is the angle between the chord line of the top foil and the x-axis of the kite reference frame (see [7]) when the kite is fully powered, i.e. $\alpha_d = 0$. We can refer to α_0 as the *base angle of attack*: it results from how the bridle lines are assembled (its geometry), and therefore α_0 is a constant quantity throughout the flight, whereas α_d can be used as a control input and thus may vary with time. Observe that, because α_0 is the only component which is constant in Eq. (5.19), α_B must be computed at each time step of the simulation.

The angles of attack of the side wings, which also must be computed at each time step, are calculated as

$$\alpha_A = \arcsin\left(\frac{\mathbf{v}_{a,A,xz} \cdot \hat{\mathbf{z}}_A}{v_{a,A,xz}}\right) + \alpha_{s,0}, \tag{5.20}$$

$$\alpha_C = \arcsin\left(\frac{\mathbf{v}_{a,C,xz} \cdot \hat{\mathbf{z}}_C}{v_{a,C,xz}}\right) + \alpha_{s,0}, \tag{5.21}$$

where $\alpha_{s,0}$ is the base angle of attack of each of the side foils. The drag force for each of the three foils can be readily calculated considering the projected kite area A and the side kite area A_{side} (the sum of the areas of foils A and C):

$$\mathbf{F}_{d,B} = \frac{1}{2}\rho K_D v_{a,B}^2 A C_D(\alpha_B)\frac{\mathbf{v}_{a,B}}{v_{a,B}}, \tag{5.22}$$

$$\mathbf{F}_{d,A} = \frac{1}{2}\rho K_D v_{a,A}^2 A \frac{A_{side}}{A} C_D(\alpha_A)\frac{\mathbf{v}_{a,A}}{v_{a,A}}, \tag{5.23}$$

$$\mathbf{F}_{d,C} = \frac{1}{2}\rho K_D v_{a,C}^2 A \frac{A_{side}}{A} C_D(\alpha_C)\frac{\mathbf{v}_{a,C}}{v_{a,C}}. \tag{5.24}$$

The factor K_D is taken to be the same as in the four-point mass model [7], $K_D = 1$ - $\frac{A_{side}}{A}$. Its purpose is to obtain the same L/D ratio as in a 1-point mass approach of the kite, also presented in [7]. The lift forces are computed as

$$\mathbf{F}_{l,B} = \frac{1}{2}\rho\, v_{a,B,xz}^2 A C_L(\alpha_B)\frac{\mathbf{v}_{a,B} \times \hat{\mathbf{y}}_B}{||\mathbf{v}_{a,B} \times \hat{\mathbf{y}}_B||}, \qquad (5.25)$$

$$\mathbf{F}_{l,A} = \frac{1}{2}\rho\, v_{a,A,xz}^2 A \frac{A_{side}}{A} C_L(\alpha_A)\frac{\mathbf{v}_{a,A} \times \hat{\mathbf{y}}_A}{||\mathbf{v}_{a,A} \times \hat{\mathbf{y}}_A||}, \qquad (5.26)$$

$$\mathbf{F}_{l,C} = \frac{1}{2}\rho\, v_{a,C,xz}^2 A \frac{A_{side}}{A} C_L(\alpha_C)\frac{\mathbf{v}_{a,C} \times \hat{\mathbf{y}}_C}{||\mathbf{v}_{a,C} \times \hat{\mathbf{y}}_C||}. \qquad (5.27)$$

5.2.6 Controller Design

For controlling the pumping kite system we considered a decentralized topology with independent control laws for the subsystems of ground winch (reeling-in/out) and kite flight. Regarding the latter subsystem, we implemented the cascade control approach with two loops, similarly to [5] and to Chap. 14. In short, the outer loop uses Bernoulli's lemniscate as a reference for the lying-eight trajectory, as depicted in Fig. 5.7, where $\mathbf{e}_\theta$ and $\mathbf{e}_\phi$ are unitary vectors in the direction of the kite coordinates θ (polar angle) and ϕ (azimuth angle), and $\mathbf{v}_{k,\tau}$ is the kite velocity vector projected onto the tangent plane $(\mathbf{e}_\theta, \mathbf{e}_\phi)$. Based on the distance between the kite

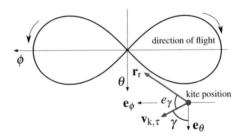

Fig. 5.7 Bernoulli's lemniscate as a reference for the lying-eight flight trajectory [5]. Vector $\mathbf{r}_r$ represents the course angle reference, whereas vector $\mathbf{v}_{k,\tau}$ defines the current course angle γ

position and the lemniscate, the outer loop generates a reference γ_{ref} for the course angle γ and feeds such reference to the inner loop. The task of the inner loop is to make the course angle tracking error $e_\gamma = \gamma_{ref} - \gamma$ converge to zero by manipulating the steering input Δl_s according to the control law

$$\Delta l_s = \frac{1}{2}(l_1 - l_r) = -K_{ctrl}\frac{e_\gamma^3}{||e_\gamma||}, \qquad (5.28)$$

where K_{ctrl} is the proportional gain. The controller signal was chosen to be quadratic in the error to speed up the kite re-alignment for relatively large tracking errors, without having to increase the gain, which would lead to noisiness and possibly control instability in facing wind turbulence.

Yaw damping A rotational damper with constant c_{yaw} was placed on the kite to counteract noisy yaw motions caused by aggressive steering. In order to implement this damper, a (virtual) massless rod was introduced connecting the ground winch position to the center of gravity of the top kite plate. The rod was connected to the ground winch through a universal joint, blocking rotation around its own axis. At the kite top plate the rod was connected with a gimbal joint, where a rotational damper was placed upon the torsional degree of freedom between the kite plate and the rod. Although this problem could also be tackled by fine-tuning the flight controller, the damper was found to make the overall flight behavior of the kite more smooth and robust, therefore it was left in the model.

5.3 Model Calibration and Results

The initial position of the kite was set to the center of the eight-figure trajectory, with the kite pointing with an initial velocity towards the top-right of the window, as seen from the ground winch. The first half period of the flight was removed from the results, allowing the kite system to reach a periodic flight regime—hence $t = 0$ corresponds to the kite flying toward the top-left. The parameters in the kite model were identified using reference values for the angles of attack from Fechner's four-point mass model [7], after running both models with the same controlled trajectories and wind profile. Once the parameters were determined, results related to the deformation behavior were gathered, to be presented in Sect. 5.3.2.

5.3.1 Choice of Parameters

Most of the design parameters were chosen according to Table 5.1. In a second step, the remaining parameters, related to the spring and damping constants, were fine-tuned, in order to make the behavior of the three-plate kite model here developed to more closely match the behavior of the four-point mass model. These optimized values can be found in Table 5.2.

The damping constants of the rotational springs in the kite structure were all set to relatively low values, as it was found that increasing these values led to a significant drop in computational speed. The length of the steering lines was reduced by an empirical value of 5 cm to compensate for the stretching of the steering lines, causing asymmetry with respect to the infinitely stiff front lines. This way h_b was

Parameter name	Symbol	Value	Unit
Kite mass*	m_{kite}	6.21	kg
Kite projected area*	A	10.18	m^2
Angle between airfoils	Θ_0	110	deg
Kite height*	h_{kite}	2.23	m
Relative side area*	A_{side}/A	30.6	%
Span of side airfoils	l_s	2.37	m
Span of top airfoil	l_t	3.34	m
Chord of top airfoil	d_t	1.5	m
Chord of side airfoils	d_s	1.5	m
Thickness of foils	t_{kite}	1	cm
KCU mass*	m_{KCU}	8.4	kg
Height of bridle lines*	h_b	4.9	m
Steering bridle line stiffness*	k_b	50	kN/m
Steering bridle line damping*	c_b	300	Ns/m
Power bridle line stiffness*	c_p	∞	N/m
Tether stiffness*	k_t	614.6	kN/m
Tether damping*	c_t	473	Ns/m
Tether diameter*	d_t	4	mm
Tether density*	ρ_t	970	kg/m^3
Tether drag coefficient*	$C_{D,t}$	0.96	–
Initial tether length*	$l_{t,0}$	600	m
Reel-out speed*	$\dot{l}_t$	2.55	m/s
Starting height of kite*	h_0	200	m
Base angle of attack top foil	α_0	4	deg
Base angle of attack side foils	$\alpha_{s,0}$	10	deg

Table 5.1 Design parameters of the kite system. The values that were chosen identical to Fechner's model [7] are marked by an asterisk (*)

Springs	Value (Nm/deg)	Dampers	Value (Nms/deg)
k_δ	200	c_δ	1
$k_{\tau,y}$	1000	$c_{\tau,y}$	1
$k_{\tau,z}$	700	$c_{\tau,z}$	1
		c_{yaw}	0.5

Table 5.2 Optimized springs/dampers of the kite system

tuned to the value in Table 5.1. The gain of the flight controller was set to $K_{ctrl} = -0.2$ m/rad^2.

5.3.2 Simulation Results

To generate the simulation results here presented, a variable step size solver was chosen, with automatically set maximum and minimum step sizes. The used solver was the ODE-45 Dormand-Prince, and the relative tolerance was set to 10^{-3}. The results consist of the deformation angles, angles of attack and traction forces as a function of the de-power settings, the yaw rate and tip displacements, and finally computational speed.

Deformation angles The results for the deformation angles δ, τ_y and τ_z representing bending and torsion are shown in Fig. 5.8. We can note that the deformation angles increase/decrease periodically in very similar, relatively smooth fashion, and reach maximum values at the sides of the trajectory path, in the middle of the turns. As the turn starts, the side of the kite that corresponds to the direction of the turn closes due to the line shortening, slightly less than the amount the side opposite to the turn opens due to the line extension. The magnitude of, and ratios between the values of these angles depend mainly on the magnitude and the ratios between the respective rotational spring constants.

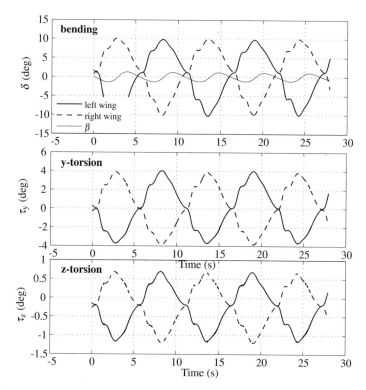

Fig. 5.8 Results for deformation angles: bending, and torsion around y- and z-axes. The kite elevation angle β is added to the top graph as positional reference

Angles of attack The results for the angle of attack of the side wings, along with the steering input of the left line, are shown in Fig. 5.9. The profiles are very similar to the ones found by running the four-point mass model: α_A and α_C vary between $7.5°$ and $15°$, which is approximately tuned to the range found by running Fechner's model [7]. The steering input is centered around the de-power value, and shows maximum amplitudes of around 65 cm. The results for the angle of attack of the main airfoil α_B, together with its velocities (kite velocity and apparent wind), are shown in Fig. 5.10. Again, the model parameters were tuned so that the behavior of α_B approaches that obtained with Fechner's model, although the magnitude of the velocities in Fechner's model were about a factor 1.5 lower. This discrepancy could be due to the different geometry, leading to a different L/D ratio, and could be handled by changing K_D in Eqs. (5.22) to (5.24). Interestingly, the maxima in α_B do not coincide with the maxima in v_B and $v_{a,B}$. This suggests that the asymmetric bending of the kite during the turns causes more increase in α_B than the higher velocity (leading to higher traction force, see Fig. 5.11) when the kite flies straight. This suggestion is supported in Fig. 5.11, where the maxima in α_B and traction force F_{trac} are asynchronous, especially for lower de-power settings.

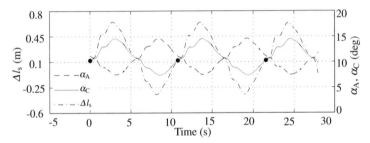

Fig. 5.9 Results of angles of attack of side airfoils and steering input. Dots indicate the start of each orbit

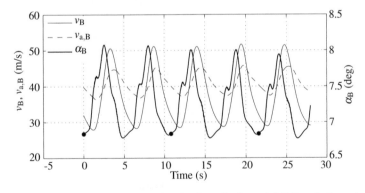

Fig. 5.10 Results for top plane velocity and angle of attack of top airfoil. Dots indicate the start of each orbit

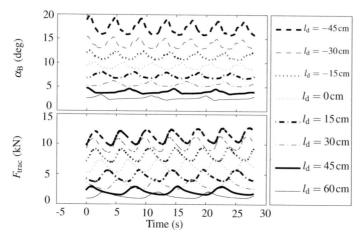

Fig. 5.11 Results for the angle of attack of the top airfoil (top) and traction force (bottom) for different de-power settings

Traction force In Fig. 5.11 the expected behavior with respect to changing the de-power setting is portrayed. Observe that increasing the steering lines (hence more de-power) results in a lower traction force and a lower angle of attack. From the number of orbits it can be noted that the velocity also decreases when increasing the de-power. Moreover, note that the performance of the system shows to be robust with regard to changes in the de-power setting.

Yaw rate and airfoil tip displacements The yaw rate, together with the displacement of the tips of the side airfoils d_s, are shown in Fig. 5.12. The yaw rate v_{yaw} and α_B are both related to the steering input Δl_s, which is sensitive to the controller type, and gain K_{ctrl}. The peaks in d_s and in δ (see Fig. 5.8) coincide, meaning that during the turn, not only asymmetric bending occurs, but the kite is opened more with respect to a straight flight path.

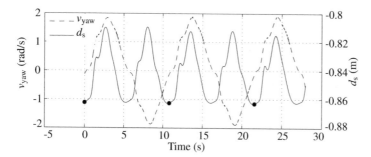

Fig. 5.12 Results for yaw rate and displacement between the tips of the side airfoils

Number of tether segments In this work the tether was modeled as three line segments. It was found through a few tests that the general flight behavior was not changed upon increasing the number of segments. An increase in the number of segments however led to a significant increase in the computational cost. Therefore it was chosen to leave the number of segments at three.

Computational speed The simulation was found to be running close to real-time (75%) on a 5GB RAM, 2.4GHz Intel Core 2 machine, 1066MHz frontside bus. On a more advanced machine, which is common nowadays, the simulation is expected to easily run on real-time.

5.4 Concluding Remarks

In this chapter the development of a dynamic model of a C-shaped power kite for airborne wind energy has been discussed. The motivation was to have a simulation model which could represent more accurately the kite behavior while maintaining the computational complexity low enough to be able to run on real time. The latter criterion is important in order to allow the model to be used for control purposes. The model was developed with the SimMechanics™ toolbox of Matlab®, and consists essentially of 3 rigid plates interconnected by gimbal joints, allowing for three rotational degrees of freedom. These plates are connected by a bridle line system to an airborne kite control unit which, by varying the length of two steering tethers, is responsible for the commands of steering and de-powering the kite. This three-plate model also includes damping and spring elements, and allows for the representation of two basic deformation modes: spanwise bending and torsion. The model parameters were chosen so that the resulting behavior approaches that of an existing 4 point-mass kite model found in the literature [7].

There are several improvements which can be made based on what has been presented in this chapter. For instance, although the system behavior is similar to that of the mentioned reference 4 point-mass model, and the magnitude of the deformations seem realistic based on experience with operation of kites, one important point as future work is the development of a more accurate method for identification and validation of the model parameters. Furthermore, the line control could be improved, especially the extension mechanism. A way to do this would be by breaking up the line into more segments, thus preventing a non-physical compressive force. The same could be done for the number of segments into which the tether is divided.

One could also find an alternative way to include damping in the model, perhaps not via the rotational joints that are present but via translational connections, representing the rotational damping in the physical kite system. This way, the convergence problems associated with the rotational damping may be circumvented. The model could be extended towards more realistic deformations by including more plates and tether segments with relatively low effort. Lastly, the computational speed could be improved by implementing the model in a more efficient engineering language such as Python.

References

1. Archer, C. L., Jacobson, M. Z.: Evaluation of global wind power. Journal of Geophysical Research: Atmospheres **110**(D12), 1–20 (2005). doi: 10.1029/2004JD005462
2. Bosch, A., Schmehl, R., Tiso, P., Rixen, D.: Dynamic nonlinear aeroelastic model of a kite for power generation. AIAA Journal of Guidance, Control and Dynamics **37**(5), 1426–1436 (2014). doi: 10.2514/1.G000545
3. Bosch, H. A.: Finite Element Analysis of a Kite for Power Generation. M.Sc.Thesis, Delft University of Technology, 2012. http://resolver.tudelft.nl/uuid:888fe64a-b101-438c-aa6f-8a0b34603f8e
4. Breukels, J.: An Engineering Methodology for Kite Design. Ph.D. Thesis, Delft University of Technology, 2011. http://resolver.tudelft.nl/uuid:cdece38a-1f13-47cc-b277-ed64fdda7cdf
5. De Lellis, M., Saraiva, R., Trofino, A.: Turning angle control of power kites for wind energy. In: Proceedings of the 52nd Annual Conference on Decision and Control (CDC), pp. 3493–3498, Firenze, Italy, 10–13 Dec 2013. doi: 10.1109/CDC.2013.6760419
6. Fagiano, L.: Control of tethered airfoils for high-altitude wind energy generation. Ph.D. Thesis, Politecnico di Torino, 2009. http://hdl.handle.net/11311/1006424
7. Fechner, U., Vlugt, R. van der, Schreuder, E., Schmehl, R.: Dynamic Model of a Pumping Kite Power System. Renewable Energy (2015). doi: 10.1016/j.renene.2015.04.028. arXiv:1406.6218 [cs.SY]
8. Mathworks: Gimbal: Joint with three revolute joint primitives. http://www.mathworks.com/help/physmod/sm/mech/ref/gimbal.html. Accessed 10 Dec 2014
9. Mathworks: SimMechanics: Model and Simulate Multibody Mechanical Systems. http://www.mathworks.com/products/simmechanics. Accessed 10 Dec 2014
10. Spera, D. A.: Models of Lift and Drag Coefficients of Stalled and Unstalled Airfoils in Wind Turbines and Wind Tunnels. NASA/CR-2008-215434, NASA Glenn Research Center, Cleveland, OH, USA, Oct 2008
11. Til, J. van: Dynamic Model of a Bridled Kite Including Rotational Deformations. In: Schmehl, R. (ed.). Book of Abstracts of the International Airborne Wind Energy Conference 2015, p. 48, Delft, The Netherlands, 15–16 June 2015. doi: 10.4233/uuid:7df59b79-2c6b-4e30-bd58-8454f493bb09. Presentation video recording available from: https://collegerama.tudelft.nl/Mediasite/Play/5b677d6e3d8b4d1691fd01406ef2d53f1d
12. Viré, A.: How to float a wind turbine. Reviews in Environmental Science and Bio/Technology **11**(3), 223–226 (2012). doi: 10.1007/s11157-012-9292-9
13. Vlugt, R. van der, Peschel, J., Schmehl, R.: Design and Experimental Characterization of a Pumping Kite Power System. In: Ahrens, U., Diehl, M., Schmehl, R. (eds.) Airborne Wind Energy, Green Energy and Technology, Chap. 23, pp. 403–425. Springer, Berlin Heidelberg (2013). doi: 10.1007/978-3-642-39965-7_23

Chapter 6
Retraction Phase Analysis of a Pumping Kite Wind Generator

Adrian Gambier

Abstract Airborne wind energy systems have developed very fast in the past five years. One of the most promising systems is the so called pumping kite wind generator, which is based on a cycle of two phases: the traction or generation phase and the retraction or consumption phase. An optimal balance between both phases is crucial in order to obtain an economically viable system. This work is devoted to the investigation of the retraction phase, i.e. the reel-in phase of a pumping kite wind generator, from the theoretical point of view. The most common approaches for the implementation of the retraction phase in the literature are studied from the point of view of the energy as well as time consumption. The first step of this work is the modeling of the dynamic behavior of the system during the tether reel-in process including the aerodynamic coefficients of a ram-air kite and by performing computational simulations. Perfect control is supposed. Hence, assumed that the control system shows its best performance, results of performed simulation experiments confirm that the behavior of the retraction phase is ruled by the system dynamics. The net energy gain of the complete cycle particularly depends on the efficiency of the retraction phase.

6.1 Introduction

The extraction of energy from high-altitude wind, i.e. the wind above the altitudes accessible by conventional wind turbines, is not a new concept [8, 9, 24–26, 32]. Systems to extract energy from high-altitude winds are commonly denoted as airborne wind energy systems (AWES) and can be categorized into flying systems and statically suspended systems [5]. In the case of flying systems, wind energy is

Adrian Gambier (✉)
Fraunhofer Institute for Wind Energy and Energy System Technology IWES, Am Seedeich 45, 27572 Bremerhaven, Germany
e-mail: adrian.gambier@iwes.fraunhofer.de

© Springer Nature Singapore Pte Ltd. 2018
R. Schmehl (ed.), *Airborne Wind Energy*, Green Energy
and Technology, https://doi.org/10.1007/978-981-10-1947-0_6

converted by flying objects as a consequence of their flight. Statically suspended systems have a wind turbine onboard and work relatively stationary in the air. The energy is extracted when the wind blows through the turbine. Therefore, these systems can be seen as airborne wind turbines [30]. The concept proposed by Makani Power [29] is a hybrid design because it also carries wind turbines onboard but instead of staying suspended, it flies.

The flying systems are also named Kite-Based Wind Energy Systems [10] and are characterized by three aspects: a) they generate electric energy, b) the energy is extracted from high-altitude wind and c) the energy conversion is carried out by using a tethered flying object, i.e. the kite.

One of the most used and studied concept of kite-based wind energy systems is the pumping-kite concept [4, 6, 7, 18, 19, 32]. It consists of a crosswind flying kite, which is connected by means of a tether to a drum that, in turn, is coupled with an electric machine. Thus, kite pumping is represented by cycles of alternating traction and retraction phases [4]. The first one is the generation phase, in which power is produced by flying the kite in a crosswind trajectory and using the traction force transmitted through the tether to reel out the drum, which drives the electric machine as a generator. The second one is called retraction or recovery phase. It starts when the tether reaches the maximum length. At that time, the kite is positioned in such way that the traction force is minimal, the operation of the electrical machine is switched from generator to motor and so the tether is reeled in again onto the drum as fast and smooth as possible with the minimum energy consumption. This is illustrated in Fig. 6.1.

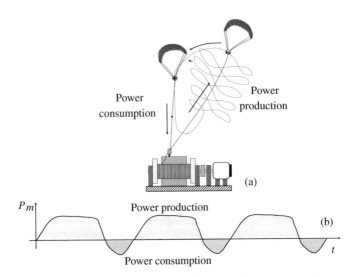

Fig. 6.1 Scheme for the pumping-kite concept: (a) Setup, (b) kite pumping cycle

Because of the fact that pumping kite power is an emergent technology, there are not many experimental setups available and therefore, it is difficult to verify with precision the energy extraction. From the theoretical point of view, several approaches and some formulas have been proposed for an approximated computation of the mean value of the power extraction of a kite in the generation phase [1, 16, 22, 23]. Results obtained by means of these formulas have been compared in [27, 28]. However, these formulas still have not experimentally verified. On the other hand, experimental observations show that the efficiency of the pumping kite systems is strongly dependent on the effectiveness of the retraction phase. A detuned or suboptimal retraction phase results in either low or even negative net power output.

The retraction phase involves at least two aspects. The first aspect consists in which methodology is used to implement it and the second one is related to the control strategy for the implemented methodology.

Regarding the methodology, it was suggested in [4] that the kite is driven to a region where the tethers can be pulled back by spending a small fraction of the energy generated in the traction phase and in [34, 35] two control strategies are presented for this methodology. In [33] it is assumed that the angle of attack of the kite is reduced to a level that maintains the force on the tether at its lower bound when it is reeled in. This is also the approach proposed in [20].

In general, the studies presented in literature include the retraction phase as integrated in the whole cycle and then the total average power is computed. Thus, their interest is centered more in the efficiency of power generation rather than in the retraction phase in particular. This is the case for example of [2, 3, 13, 16, 17, 23].

In the present work, only the retraction phase of the pumping kite wind generator is investigated from the design point of view, i.e. the generating phase and control issues are not considered. The design approaches are analyzed by using simulation studies, which take into account the energy and time consumption during the retraction phase as well as the applicability. For the simulation studies, a dynamic model is used, where the aerodynamic parameters are not obtained experimentally but computed analytically.

The remainder of the chapter is organized as follows: In Sect. 6.2, different approaches, which are proposed in the literature for the retraction phase are schematized and described. Section 6.3 is devoted to derive the dynamic model that is used later for the analysis. Afterwards, in Sect. 6.4, four reference systems are presented in order to carry out simulation experiments, whose results are presented in Sect. 6.5. Finally, conclusions are stated in Sect. 6.6. The preliminary content of the present chapter has been presented at the Airborne Wind Energy Conference 2015 [11].

6.2 Approaches for the Retraction Phase

At present, three approaches have been presented in the literature in order to implement the retraction phase. In the first approach, once the kite is close to reach the maximum tether length, it continues ascending but flies out from the power zone to the position with the minimum resistance to the wind, i.e. the limiting curve of the flight envelope in Fig. 6.2 to start pulling down the kite [4]. A particular point on

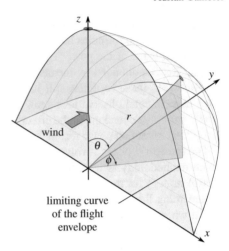

Fig. 6.2 Flight envelope, limiting curve of the flight envelope and the zenith

the limiting curve is the vertical position called the zenith.

It is important here to avoid a flight to the limiting curve at constant tether. Thus, the flight control guides the kite in such a way that the maximum tether length is reached at the same time that the kite reaches the limiting curve. Thus, the generation finishes directly on this curve such that the retraction phase can start immediately.

In the second procedure, the bridle is pulled until the wing's chord is aligned with the tether's direction, and only then, the kite is pulled down [20, 33]. This is the standard procedure for rigid wings and also for LEI kites (Leading Edge Inflatable) because this alignment is only possible for such kites.

Lastly, the third approach used in the simulation of the KiteGen Stem system [21]. This consists in letting one side of the bridling/tether loose and then pulling the kite down. The approach is very illustrative but it can only be implemented with arch kites. If many lines are used, as for example in ram-air kites, this procedure is not applicable. Because of the fact that at present there is no practical system using arch kites, this approach is not included in the study.

The three different approaches will be called in the following as fly to the limiting curve (F2lC), fly to zenith (F2Z), pitch and pull (P&P) and bride and pull (B&P) and they are illustrated in Fig. 6.3, where the angles ϕ and θ are given in Fig. 6.2.

6.3 Modelling the Kite in the Retraction Phase

The modeling process is divided in three subsections. In the first subsection, the dynamic behavior of the kite in the descending movement according to [13] is presented. However, the aerodynamic coefficients of [13] are not applied and, instead

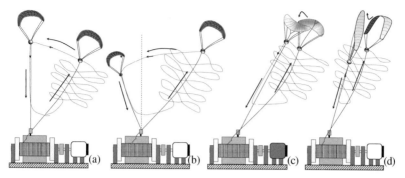

Fig. 6.3 Retraction phase procedures: (a) F2Z: $\theta = 0$ and $\phi = 0$, (b) F2lC: $\theta = 0$ and $\phi \neq 0$, (c) P&P: $\theta \neq 0$ and $\phi \neq 0$ and (d) B&P: $\theta \neq 0$ and $\phi \neq 0$

of these, the used aerodynamic coefficients are presented in the second subsection. Finally, the third subsection is devoted to set the model for the retraction phase.

6.3.1 Dynamic Model for the Retraction Phase

Several models have been proposed to describe the dynamic behavior of kite systems, as described e.g. in [6, 7]. The present analysis uses the simple model of [13]. Notice that this assumes that the kite moves in a vertical plane and therefore it is inadequate for the description of the generating phase (reel-out or traction phase), where the kite ascends in the space following a trajectory in the shape of a horizontal eight. However, the retraction phase is assumed to take place in a vertical plane and in this case, the model can be used. Figure 6.4 illustrates the forces, velocities and angles.

Figure 6.4a describes forces and velocities in the drum, the tether and the kite. The force acting tangent to the drum in the direction of the tether is assumed to be constant and it is calculated as the nominal power of the motor divided by the reel-in speed. Lift, drag and gravitational forces acting on the tether are applied at the middle point. Furthermore, lift, drag and gravitational forces are also present on the kite. The wind velocity vw is assumed to be horizontal, applied at the middle point of the kite and together with the wind velocity relative to the kite v_a and the kite velocity v_k constitute the velocity triangle in the frame of the kite. The corresponding angles are described in Fig. 6.4b, where the most important are the inclination of the tether with the vertical θ and the angle ϕ between v_w and v_a. All angles satisfy the corresponding trigonometric equalities, which lead to more compact equations of motion.

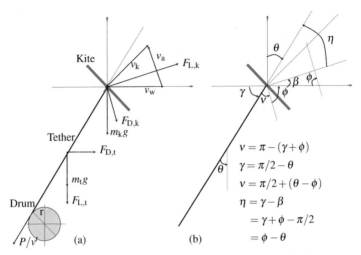

Fig. 6.4 (a) Diagram of forces acting on the kite system. (b) Definition of angles

$$\sin(v) = \sin\left(\frac{\pi}{2} + (\theta - \phi)\right) = \cos(\theta - \phi), \tag{6.1}$$

$$\cos(v) = \cos\left(\frac{\pi}{2} + (\theta - \phi)\right) = -\sin(\theta - \phi), \tag{6.2}$$

$$\sin(\eta) = \sin(\phi - \theta) = -\sin(\theta - \phi), \tag{6.3}$$

$$\cos(\eta) = \cos(\phi - \theta) = \cos(\theta - \phi). \tag{6.4}$$

The derivation of the model as well as the analysis are based under the fulfilment of the following assumptions:

A1. It is possible to control the angle of attack at zero lift all the time.
   This assumption allows the consideration of zero lift forces. Thus, this leads to a theoretical lower bound on the energy consumption, i.e. every other result will elevate the energy consumption.
A2. The electric machine has identical power curves as motor and generator.
   This is required because the nominal power of the generator is used to calculate the nominal force applied by the motor to the tether.
A3. The retraction phase starts without delay.
   This is to avoid times when the electric machine is neither producing nor consuming energy. This leads to a low theoretical boundary for the time consumption, i.e. every other result will elevate the time consumption.
A4. The tether is supposed to be straight.
   This assumption allows modeling of the tether with lumped parameters.
A5. The wind speed is constant at each altitude.
   The assumption leads to a deterministic model, i.e. a model based on ordinary differential equations.

Thus, the dynamic behavior is described by four Newton-Euler equations of motion [13]. The first one describes the motion of the kite in the tether's direction and is given by

$$
\begin{aligned}
(m_k + \mu_t l_t)\dot{v}_r =& F_{L,k}\cos(\theta - \phi) + F_{D,k}\sin(\theta - \phi) + F_{D,t}\cos(\theta) \\
& - F_{L,t}\cos(\theta) - (m_k + \mu_t l_t)g\cos(\theta) \\
& - (m_k + 0.5\mu_t l_t)v_\tau^2/l_t - P_r/v_r.
\end{aligned} \tag{6.5}
$$

The term $(m_k + 0.5\mu_t l_t)v_\tau^2/l_t$ is the force due to the centripetal acceleration of kite. P_r/μ_r is the force on the tether produced by the motor. The other terms are lift, drag and gravitational forces acting in the direction of the tether. The second equation represents the motion perpendicular to the tether

$$
\begin{aligned}
(m_k + \mu_t l_t)\dot{v}_\tau =& F_{L,k}\sin(\theta - \phi) + F_{D,k}\cos(\theta - \phi) + 0.5F_{D,t}\sin(\theta) \\
& - 0.5F_{L,t}\sin(\theta) - (m_k + 0.5\mu_t l_t)g\sin(\theta) \\
& - 2(m_k + \mu_t l_t/3)v_\tau v_r/l_t,
\end{aligned} \tag{6.6}
$$

where the forces act perpendicularly to the tether. The term $2(m_k + \mu_t l_t/3)v_\tau v_r/l_t$ is the Coriolis force. In addition, the equation set is completed by the derivatives

$$
dl_t/dt = v_r, \tag{6.7}
$$
$$
d\theta/dt = v_\tau/l_t. \tag{6.8}
$$

Subscripts r and τ indicate tether direction and the direction perpendicular to the tether, respectively. Moreover, t, k, D and L subscripts are used to represent tether, kite, drag and lift. Symbols m, v, F, P, l, g and μ are mass, velocity, force, power, length, acceleration of gravity and linear density, respectively. For example, m_k is the mass of kite, $F_{L,t}$ the lift force on tether direction and v_τ the velocity component perpendicular to the tether. The forces are computed by using

$$
F_{L,k} = 0.5\rho_a A_k C_{L,k} v_a^2, \tag{6.9}
$$
$$
F_{D,k} = 0.5\rho_a A_k C_{D,k} v_a^2, \tag{6.10}
$$
$$
F_{L,t} = 0.5\rho_a d_t l_t C_{L,t} v_{w,r}^2, \tag{6.11}
$$
$$
F_{D,t} = 0.5\rho_a d_t l_t C_{D,t} v_{w,r}^2, \tag{6.12}
$$

where

$$
v_a^2 = v_{w,r}^2 + v_k^2 - 2v_{w,r}v_k\cos\vartheta, \tag{6.13}
$$
$$
v_{w,r} = v_w\cos\phi, \tag{6.14}
$$
$$
v_k^2 = v_r^2 + v_\tau^2, \tag{6.15}
$$

and ϑ is obtained from

$$
\vartheta = \theta - \tan^{-1}(v_r/v_\tau). \tag{6.16}
$$

Parameters ρ_a, A_k, d_t, C_L and C_D are density of air, area, diameter, lift and drag coefficients, respectively. Variables v_k, v_a and v_w are the kite velocity, the wind velocity relative to the kite and the wind speed, respectively. Moreover, $v_{w,r}$ is the wind speed in the direction of the tether's vertical plane.

6.3.2 Aerodynamic Parameters

Nowadays, ram-air kites have similar designs as paragliders and parachutes and they are known as power kites. Ram air is the air that is forced to pass through an aperture of a moving object such that the created dynamic air pressure inside this object is increased. In the considered case, the moving object is the ram-air kite, i.e. a textile airfoil normally of rip-stop nylon (called parafoil) with a cell structure, which is inflated by the wind such that a wing cross section is produced.

In the following subsections, the glide ratio of a kite will be derived following the way used for rigid wings according to the lifting line theory [15]. A comparative scheme is given in Fig. 6.5.

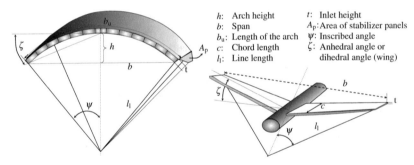

Fig. 6.5 Scheme with geometric parameters of a ram-air glider and a rigid wing (without tethers)

According to the inscribed angle theorem, the anhedral angle is $\zeta = \psi/2$ and due to the fact that the length of the arc is given by $b_u = 2\psi l_1$, it follows

$$\psi = b_u/(2l_1) \quad \text{and} \quad \zeta = b_u/(4l_1).\tag{6.17}$$

The aspect ratio $\mathcal{R}$ is obtained as

$$\mathcal{R} = b_u/c.\tag{6.18}$$

Normally, the aspect ratio and the effective area ($A_k = b_u c$) are used as design parameters. In this case, the chord length c and b_u can be calculated as

$$b_u = \sqrt{A_k \mathcal{R}} \quad \text{and} \quad c = \sqrt{A_k/\mathcal{R}}.\tag{6.19}$$

Parameters l_1, ζ and ψ are now scaled with respect to b_u, i.e.

$$\bar{l}_1 = l_1/b_u, \quad \zeta = 1/\left(4\bar{l}_1\right) \text{ and } \psi = 1/\left(2\bar{l}_1\right). \tag{6.20}$$

The introduced variable is known as relative length.

The total lift for low aspect ratios, i.e. $R \leq 5$ [15], of rectangular wing before stall can be written according to [15] as

$$C_{L,k} = a\left(\alpha - \alpha_{ZL}\right) + k_1 \sin^2\left(\alpha - \alpha_{ZL}\right) \cos\left(\alpha - \alpha_{ZL}\right), \tag{6.21}$$

where α and α_{ZL} are the angle of incidence and the incidence angle for zero lift, respectively. Parameter a is obtained from lifting line theory

$$a = \frac{\pi a_0 k \, R}{\pi R + a_0 k (1 + \tau)}, \tag{6.22}$$

with k given by

$$k = \frac{2\pi R}{a_0} \tanh \frac{a_0}{2\pi R}, \tag{6.23}$$

where a_0 is a constant parameter. Values for the factor k_1 are proposed by [15] as

$$k_1 = \begin{cases} 3.33 - 1.33R & 1 < R < 2.5 \\ 0 & R > 2.5 \end{cases}. \tag{6.24}$$

Parameter τ is a small positive factor which depends on the aspect ratio as shown in Fig. 6.6.

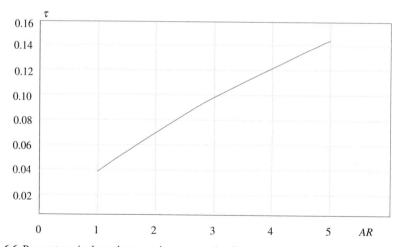

Fig. 6.6 Parameter τ in dependence on the aspect ratio R

This parameter can also be computed by using the polynomial approximation

$$\tau = -0.001649\,\!R^2 + 0.03603\,\!R + 0.003337. \tag{6.25}$$

However, the lift coefficient has to be corrected in the case of anhedral angle ζ as

$$C_{L,k} = C_{L,k,\zeta=0}\cos^2\zeta. \tag{6.26}$$

This equation is also valid for ram-air kites, when the anhedral angle ζ is defined according to Fig. 6.5. Hence, Eq. (6.24) can be modified to introduce the anhedral angle ζ as follows

$$C_{L,k} = a\,(\alpha - \alpha_{ZL})\cos^2\zeta + k_1\sin^2(\alpha - \alpha_{ZL})\cos(\alpha - \alpha_{ZL}). \tag{6.27}$$

The total drag coefficient is given in [15] and consists of the induced drag $C_{D,i}$, the drag of the lines $C_{D,l}$ and the profile drag $C_{D,p}$, i.e.

$$C_{D,i} = a^2\,(\alpha - \alpha_{ZL})(1+\delta)/(\pi\,\!R) + k_1\sin^3(\alpha - \alpha_{ZL}), \tag{6.28}$$

where a, k_1, α and α_{ZL} are already defined, and $C_{D,l}$ is given by

$$C_{D,l} = \frac{n_l d_l l_l \cos^3\alpha}{b_u c}. \tag{6.29}$$

Parameters n_l and d_l are the number of lines and the diameter, and the profile drag $C_{D,p}$ that can be estimated according to [31] as

$$C_{D,p} = 0.015 + 0.004 + 0.5t/c + 0.5A_p/(b_u c), \tag{6.30}$$

where the first term is the basic airfoil drag of a typical section and surface irregularities and the second one is the fabric roughness. Parameter δ is a small factor that depends on the aspect ratio as it is illustrated in Fig. 6.7.

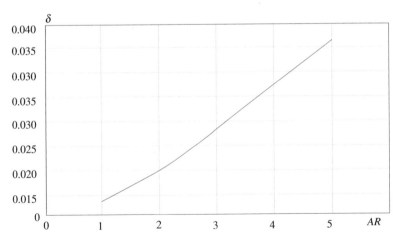

Fig. 6.7 Parameter δ by aspect ratio R

This parameter is computed by using the polynomial approximation

$$\delta = -0.0002519\,\mathscr{R}^3 + 0.002637\,\mathscr{R}^2 + 0.0003589\,\mathscr{R} + 0.0006754. \qquad (6.31)$$

Thus, the total drag of the kite is given by

$$C_{D,k} = 0.019 + 0.5\frac{t}{c} + \frac{a^2(\alpha - \alpha_{ZL})^2(1+\delta)}{\pi\mathscr{R}}$$
$$+ k_1 \sin^3(\alpha - \alpha_{ZL}) + \frac{0.5A_p + n_l d_l l_l \cos^3\alpha}{b_u c}. \qquad (6.32)$$

In order to calculate the coefficient regarding the tether, the equations proposed in [14] for a cable inclined γ with respect to the direction of flow are used, i.e.

$$C_{L,t} = C_{D,0}\sin\gamma\cos^2\gamma, \qquad (6.33)$$
$$C_{D,t} = C_{D,0}\cos^3\gamma + \Delta C_{D,0}. \qquad (6.34)$$

The basic drag coefficient $C_{D,0}$ is obtained by taking into account the number of tethers (n_t) and the number of control lines (n_{cl}) as

$$C_{D,0} = n_t C_{D,0,t} + n_{cl} C_{D,0,cl}, \qquad (6.35)$$

where the coefficients $C_{D,0,t}$ and $C_{D,0,cl}$ are the basic coefficients for a cylinder (at $\gamma = 90\,\text{deg}$) with circular shape, i.e. $C_{D,0} \approx 1.1$. The frictional component $C_{D,0}$ is πC_f, where $C_f \approx 0.004$. Finally, the effective glide ratio [18] is obtained from

$$G_e = \frac{C_{L,k}}{C_{D,k} + C_{D,t}d_t l_t/(4A_k)}. \qquad (6.36)$$

Numerical example The glide radio is computed for several configurations of kite areas, aspects ratios, relative line lengths and angles of attack by using Eq. (6.36) and the results are shown in Fig. 6.8 and Fig. 6.9. For the example, the maximum tether length was 600 m and the number of lines per square meter was 1.5. Moreover, the area of the stabilizer panels A_p was set to zero. Depending on the calculation case, the kite area was chosen between 100 and 1000 m^2, the aspect ratio between 1 and 15, the relative line length between 0.1 and 2. All other parameters were computed internally. For example, the maximum force applied to the tether was calculated for the corresponding kite area and a wind speed of 10 m/s. With these values, the diameters of the tethers and the diameters of the lines were estimated.

Figure 6.8 and Fig. 6.9 show the dependence of G_e by its parameters. Thus, G_e has relatively small changes for large kite areas and increases practically linearly with increasing angle of attack (Fig. 6.8a). However, it is very sensitive with respect to the aspect ratio (Fig. 6.8b). Large aspect ratios produce better G_e (c.f. Fig. 6.9a) but this leads to a long kite with a small chord making the control more difficult (longer control lines, slower reactions, appearance of variable delays).

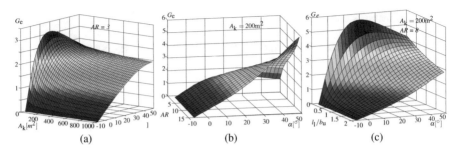

Fig. 6.8 Glide ratio in dependence of: (a) the kite area, (b) the aspect ratio and (c) relative line length and angle of attack

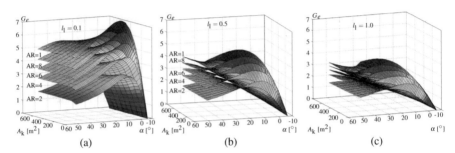

Fig. 6.9 Glide ratio in dependence of the area of the kite and the angle of attack with the aspect ratio as parameter for different relative line lengths: (a) 0.1, (b) 0.5 and (c) 1.0

Augmenting the line length decreases G_e (Fig. 6.8c) and moreover, high flights, which requires longer tether, reduce G_e (Eq. (6.36)). On the other hand, increasing the aspect ratio by making the kite longer with smaller chord also increases G_e, as well. Hence, the relationship between all parameters is sensitive with respect to changing parameters and is nonlinear. Finally, the angle of attack should be maintained small (below 15°) in order to obtain an acceptable linearity. The analysis of G_e for more than one tether should be based on Eq. (6.35).

6.3.3 Model Settings for the Retraction Phase

Depending on which approach is analyzed for the retraction phase, the model equations take different forms. The three cases analyzed in the current subsection (F2Z, F2lC and P&P) are illustrated in Fig. 6.10.

In the case of Fig. 6.10a, angles θ and ϕ are zero. Moreover, the angle of attack α is controlled to be $\alpha = \alpha_{ZL}$ and therefore $C_{L,k} = 0$ (according to assumption 6.3.1). The angles in Fig. 6.10b are $\theta \neq 0$ and $\phi = \pm 90\,\text{deg}$. The angle of attack has also to be maintained here at zero lift. In Fig. 6.10c, the angles are $\theta \neq 0\,\text{deg}$ and $\phi \neq 0\,\text{deg}$. All cases are summarized in Table 6.1.

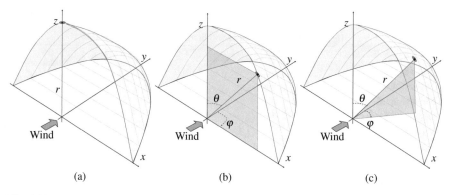

Fig. 6.10 Description of different approaches for the retraction phase: a) F2Z, b) F2IC option 2 and c) P&P

F2Z	F2IC	P&P
$\theta = 0, \varphi = 0 \Rightarrow \dot\theta = 0, \dot\varphi = 0$	$\theta \neq 0°, \varphi \pm 90° \Rightarrow \dot\theta = 0, \dot\varphi = 0$	$\theta \neq 0°, \varphi \neq 0°$
$v_{w,r} = v_w \cos\varphi = v_w$	$v_{w,r} = v_w \cos\varphi = v_w$	$v_{wr} = v_w \cos\varphi$
$v_a = v_w - v_r$	$v_a = v_r$	$v_a = v_w \cos\varphi - v_r$
$\gamma = 0, C_{L,t} = C_{D,0}\sin\gamma\cos^2\gamma = 0$	$\gamma = 90°, C_{L,t} = C_{D,0}\sin\gamma\cos^2\gamma = 0$	$\gamma = \theta, C_{L,t} = C_{D,0}\sin\gamma\cos^2\gamma$
$C_{D,t} = C_{D,0} + \Delta C_{D,0}$	$C_{D,t} = \Delta C_{D,0}$	$C_{D,t} = C_{D,0}\cos^3\gamma + \Delta C_{D,0}$
$F_{L,k} = 0$	$F_{L,k} = 0$	$F_{L,k} = 0$
$F_{D,k} = 0.5\rho_a A_k C_{D,k} v_a^2$	$F_{D,k} = 0.5\rho_a A_k C_{D,k} v_r^2$	$F_{D,k} = 0.5\rho_a A_k C_{D,k} v_a^2$
$F_{L,t} = 0$	$F_{L,t} = 0$	$F_{L,t} = 0.5\rho_a d_t l_t C_{L,t} v_{w,r}^2$
$F_{D,t} = 0.5\rho_a d_t l_t C_{D,t} v_w^2$	$F_{D,t} = 0$	$F_{D,t} = 0.5\rho_a d_t l_t C_{D,t} v_{w,r}^2$

Table 6.1 Particular conditions for all approaches for the retraction phase

6.4 Reference Systems

In order to carry out the study, reference systems are defined. They are based on the ram-air kite-pumping concept with rated powers of 20 kW, 200 kW, 2000 kW and a 34 kW special system. For the design, it was assumed single-tether systems (with two control lines), i.e. three tethers in total operating at an average height of 300 m (i.e. with a cycle between 100 m and 400 m) and average wind speed of 10 m/s. Table 6.2 and Fig. 6.11 summarize the main information of the reference systems. Additional parameter are $A_p = 0$, $l_1 = 0.8 b_u$, $a_0 = 6.89 \, \text{rad}^{-1}$, $n_1 = 1.5$ line per 1.1 m^2, $t = 0.14c$, $d_1 = 2.5$ mm, $\alpha_{ZL} = -7°$, $v_w = 10 \, \text{m/s}$.

Reference systems ($h_m = 250$m; 1 main tether with a maximum length of 600 m)				
Rated power [kW]	20	200	2000	34
Area of kite [m^2]	18	213	2812	23
Aspect ratio	5	5	7	6.7
Number of cells	20	40	80	37
Diameter of tether [mm]	4	12	39	5
Mass [kg] (kite + all tethers)	1.5 + 7.42	22 + 125.25	342 + 848.99	1.37 + 19.74
Linear density of tether [kg/m]	4.122e-03	0.06958	0.47166	0.1097
Max. force [kN]	14.7	147.6	1474.9	25.2

Table 6.2 Description of the reference systems

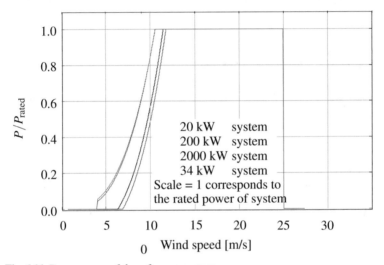

Fig. 6.11 Power curves of the reference systems

6.5 Simulation Experiments and Results

For the experiments, the model presented in Sect. 6.3 was parametrized with the reference systems presented in Sect. 6.4 and simulation experiments for all cases presented in Table 6.1. For the approach P&P, it was assumed that the kites are able to be pitched until the chord aligns the tether direction (like a LEI kite). During the simulation, the tether length was computed from the maximum value of 400 m and the minimum value of 100 m. The initial value for the kite speed was zero. The motor was assumed to be at constant nominal speed. The reel-in time and the energy consumption were also registered.

The dynamic model was implemented in Matlab/Simulink and the solver Ode45 (Dormand-Prince) with adaptive integration step was selected for the simulation.

The main simulation results are summarized in Table 6.3. It is possible to observe that F2Z requires the lower time and P&P the larger. The P&P approach presents the drawback that the angles θ and ϕ are different from zero and this reduces the gravitational forces acting in the direction of the tether. Thus, the advantage of P&P consists in the fact that the retraction phase begins immediately after the maximum length of the tether is reached inside the power zone. F2Z and F2IC are efficient if it is possible to avoid a passive flight, i.e. a flight at constant tether length. On the other hand, F2Z and F2IC have a reduced generation when the kite leaves the power zone in direction to the limiting curve of the flight envelope.

	F2Z		F2IC		P&P	
	time [s]	Energy [Wh]	time [s]	Energy [Wh]	time [s]	Energy [Wh]
20 kW system	27.21	151.17	51.90	288.33	61.24	340.22
34 kW system	23.91	225.82	42.03	396.98	45.15	426.42
200 kW system	52.66	2925.60	85.63	3016.71	95.86	3651.10
2000 kW system	73.40	40777.78	109.54	60855.55	118.68	65933.33

Table 6.3 Time and energy consumption for all reference systems and approaches

Figures 6.12, 6.13 and 6.14 show the evolution of the tether length in the reel-in phase from 400 m to 100 m for all reference systems. With the exception of the 20 and 34 kW systems in the approach F2Z (c.f. Fig. 6.12), all experiments shows similar curve shapes. In the case of F2Z, Eq. (6.5) reduces to

$$\dot{v}_r = \left[-F_{D,k} \sin\varphi - (m_k + \mu_t l_t)g - P_r/v_r \right] / (m_k + \mu_t l_t). \tag{6.37}$$

This equation is nonlinear with respect to v_r and l_t, which decreases with the time. For small kites (small m_k), the second term $\mu_t l_t$ in the mass is important but decreasing and therefore the curve is dominated by $1/l_t$ causing its convex shape. For large m_k, the second term is not relevant. This effect is however not present in the other approaches (i.e. F2IC and P&P). This is because the angle θ for these approaches is different from zero and in this case, the terms change their relative significance. For the present examples, the curve changes from convex to concave for $\theta > 30°$.

An additional observation is the fact that although the system of 34 kW has very similar parameters as the 20 kW system, it performs better. This is because this kite presents a particular optimal design from the aerodynamic point of view and for this reason it has been included in the study.

Notice that the figures show only one instance of the retraction phase, such that after reaching the altitude of 100 m, the kites should begin a new generation phase. Instead of that the altitude is maintained constant at 100 m until the simulation time concludes.

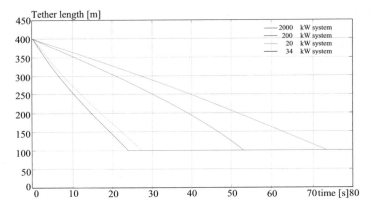

Fig. 6.12 Tether reel-in for all reference systems by using the F2Z approach

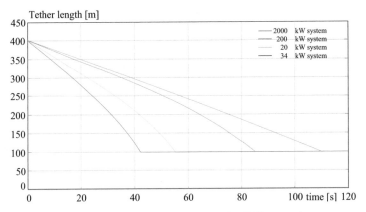

Fig. 6.13 Tether reel-in for all reference systems by using the F2lC approach

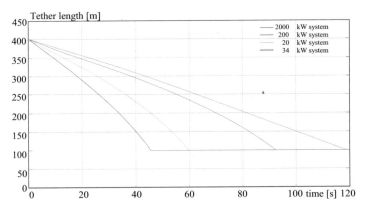

Fig. 6.14 Tether reel-in for all reference systems by using the P&P approach

6.6 Conclusions

The retraction phase is determined by the dynamic behavior of the kite system and also depends on the quality of the control system. The performed study is based on ideal assumptions and therefore, the reached results are very close to the theoretical limit. This means, more realistic assumptions will lead to retraction phases taking more time and higher energy consumption, such that the net energy gain will be reduced.

The future work will include in addition to more realistic assumptions with a combined reel-out and reel-in simulation to establish the theoretical limit for the whole cycle.

Acknowledgements The author acknowledges the funding provided by the Federal Ministry of Economic Affairs and Energy (BMWi) through the project OnKites II, with funding code 0325394A [12].

References

1. Argatov, I., Rautakorpi, P., Silvennoinen, R.: Estimation of the mechanical energy output of the kite wind generator. Renewable Energy **34**(6), 1525–1532 (2009). doi: 10.1016/j.renene.2008.11.001
2. Argatov, I., Silvennoinen, R.: Energy conversion efficiency of the pumping kite wind generator. Renewable Energy **35**(5), 1052–1060 (2010). doi: 10.1016/j.renene.2009.09.006
3. Canale, M., Fagiano, L., Milanese, M.: High Altitude Wind Energy Generation Using Controlled Power Kites. IEEE Transactions on Control Systems Technology **18**(2), 279–293 (2010). doi: 10.1109/TCST.2009.2017933
4. Canale, M., Fagiano, L., Ippolito, M., Milanese, M.: Control of tethered airfoils for a new class of wind energy generator. In: Proceedings of the 45th IEEE Conference on Decision and Control, pp. 4020–4026, San Diego, CA, USA (2006). doi: 10.1109/CDC.2006.376775
5. Cherubini, A., Papini, A., Vertechy, R., Fontana, M.: Airborne Wind Energy Systems: A review of the technologies. Renewable and Sustainable Energy Reviews **51**, 1461–1476 (2015). doi: 10.1016/j.rser.2015.07.053
6. Diehl, M.: Real-time optimization for large scale nonlinear processes. Ph.D. Thesis, University of Heidelberg, 2001. http://archiv.ub.uni-heidelberg.de/volltextserver/1659/
7. Fagiano, L.: Control of tethered airfoils for high-altitude wind energy generation. Ph.D. Thesis, Politecnico di Torino, 2009. http://hdl.handle.net/11311/1006424
8. Fletcher, C. A. J., Honan, A. J., Sappuppo, J. S.: Aerodynamic platform comparison for jet-stream electricity generation. Journal of Energy **7**(1), 17–23 (1983). doi: 10.2514/3.48063
9. Fletcher, C. A. J., Roberts, B. W.: Electricity generation from jet-stream winds. Journal of Energy **3**(4), 241–249 (1979). doi: 10.2514/3.48003
10. Gambier, A.: Projekt OnKites : Untersuchung zu den Potentialen von Flugwindenergieanlagen (FWEA). Final Project Report, Fraunhofer Institute for Wind Energy and Energy System Technology IWES, Bremerhaven, Germany, 2014. 155 pp. doi: 10.2314/GBV:81573428X
11. Gambier, A.: Recovery Phase Analysis of a Pumping Kite Wind Generator. In: Schmehl, R. (ed.). Book of Abstracts of the International Airborne Wind Energy Conference 2015, p. 60, Delft, The Netherlands, 15–16 June 2015. doi: 10.4233/uuid:7df59b79-2c6b-4e30-bd58-8454f493bb09. Presentation video recording available from: https://collegerama.tudelft.nl/Mediasite/Play/88ad32b769b14033bee3fc734bdaf32d1d

12. Gambier, A., Bastigkeit, I., Nippold, E.: Projekt OnKites II : Untersuchung zu den Potentialen von Flugwindenergieanlagen (FWEA) Phase II. Final Project Report, Fraunhofer Institute for Wind Energy and Energy System Technology IWES, Bremerhaven, Germany, June 2017. 105 pp. https://www.tib.eu/de/suchen/id/TIBKAT%3A1002309476/Projekt-OnKites-II-Untersuchung-zu-den-Potentialen/

13. Goela, J. S., Vijaykumar, R., Zimmermann, R. H.: Performance characteristics of a kite-powered pump. Journal of Energy Resource Technology **108**(2), 188–193 (1986). doi: 10. 1115/1.3231261

14. Hoerner, S. F.: Fluid-Dynamic Drag. Bricktown, Brick Town, NJ, USA (1965)

15. Hoerner, S. F., Borst, H. V.: Fluid-dynamic lift. 2nd ed. Mrs. Liselotte A. Hoerner, Brick Town, NJ, USA (1985)

16. Houska, B.: Robustness and Stability Optimization of Open-Loop Controlled Power Generating Kites. M.Sc.Thesis, Ruprecht-Karls-Universität, Heidelberg, 2007. http://sist.shanghaitech. edu.cn/faculty/boris/paper/diploma_thesis.pdf

17. Houska, B., Diehl, M.: Optimal control for power generating kites. In: Proceedings of the 9th European Control Conference, pp. 3560–3567, Kos, Greece, 2–5 July 2007

18. Houska, B., Diehl, M.: Optimal control of towing kites. In: Proceedings of the 45th IEEE Conference on Decision and Control, pp. 2693–2697, San Diego, CA, USA, 13–15 Dec 2006. doi: 10.1109/CDC.2006.377210

19. Ilzhöfer, A., Houska, B., Diehl, M.: Nonlinear MPC of kites under varying wind conditions for a new class of large-scale wind power generators. International Journal of Robust and Nonlinear Control **17**(17), 1590–1599 (2007). doi: 10.1002/rnc.1210

20. Jehle, C., Schmehl, R.: Applied Tracking Control for Kite Power Systems. AIAA Journal of Guidance, Control, and Dynamics **37**(4), 1211–1222 (2014). doi: 10.2514/1.62380

21. KiteGen: KiteGen STEM. http://kitegen.com/en/products/stem (2015). Accessed 20 Jan 2016

22. Loyd, M. L.: Crosswind kite power. Journal of Energy **4**(3), 106–111 (1980). doi: 10.2514/3. 48021

23. Luchsinger, R. H.: Pumping Cycle Kite Power. In: Ahrens, U., Diehl, M., Schmehl, R. (eds.) Airborne Wind Energy, Green Energy and Technology, Chap. 3, pp. 47–64. Springer, Berlin Heidelberg (2013). doi: 10.1007/978-3-642-39965-7_3

24. Oberth, H.: Primer for those who would govern. West Art Pub. (1987)

25. Pocock, G.: The Aeropleustic Art, or, Navigation in the Air by the use of Kites, or Buoyant Sails. Sherwood & Co, London (1827). http://babel.hathitrust.org/cgi/pt?id=mdp. 39015047378875

26. Riegler, G., Riedler, W., Horvath, E.: Transformation of Wind Energy by a High-Altitude Power Plant. Journal of Energy **7**(1), 92–94 (1983). doi: 10.2514/3.62639

27. Sanno, K., Rao, K. V. S.: Estimation of wind power extraction from kites flying at high altitudes. Comparison of five mathematical models. Journal of Chemical and Pharmaceutical Sciences, Special Issue 4, 247–251 (2014). http://www.jchps.com/specialissues/Special% 20issue4/jchps%2087%20kumari%20sanno%20254-256.pdf

28. Sanno, K., Rao, K. V. S.: Estimation of wind power extraction from kites flying at high altitudes. Comparison of two mathematical models. In: 2014 1st International Conference on Non Conventional Energy, 16–17 Jan 2014. doi: 10.1109/ICONCE.2014.6808708

29. Vander Lind, D.: Analysis and Flight Test Validation of High Performance Airborne Wind Turbines. In: Ahrens, U., Diehl, M., Schmehl, R. (eds.) Airborne Wind Energy, Green Energy and Technology, Chap. 28, pp. 473–490. Springer, Berlin Heidelberg (2013). doi: 10.1007/ 978-3-642-39965-7_28

30. Vermillion, C., Glass, B., Rein, A.: Lighter-Than-Air Wind Energy Systems. In: Ahrens, U., Diehl, M., Schmehl, R. (eds.) Airborne Wind Energy, Green Energy and Technology, Chap. 30, pp. 501–514. Springer, Berlin Heidelberg (2013). doi: 10.1007/978-3-642-39965-7_30

31. Ware, G. M., Hassell, J. L.: Wind-tunnel investigation of ram-air-inflated all-flexible wings of aspect ratios 1.0 to 3.0. NASA TM SX-1923, NASA Langley Research Center, Hampton, VA, USA, 1969

32. Williams, P., Lansdorp, B., Ockels, W. J.: Modeling and Control of a Kite on a Variable Length Flexible Inelastic Tether. AIAA Paper 2007-6705. In: Proceedings of the AIAA Modeling and Simulation Technologies Conference and Exhibit, Hilton Head, SC, USA, 20–23 Aug 2007. doi: 10.2514/6.2007-6705
33. Williams, P., Lansdorp, B., Ockels, W.: Optimal Crosswind Towing and Power Generation with Tethered Kites. AIAA Journal of Guidance, Control, and Dynamics **31**(1), 81–93 (2008). doi: 10.2514/1.30089
34. Zgraggen, A. U., Fagiano, L., Morari, M.: Automatic Retraction and Full-Cycle Operation for a Class of Airborne Wind Energy Generators. IEEE Transactions on Control Systems Technology **24**(2), 594–608 (2015). doi: 10.1109/TCST.2015.2452230
35. Zgraggen, A. U.: Automatic Power Cycles for Airborne Wind Energy Generators. Ph.D. Thesis, ETH Zurich, 2014. doi: 10.3929/ethz-a-010350742

Chapter 7
Dynamic Modeling of Floating Offshore Airborne Wind Energy Converters

Antonello Cherubini, Giacomo Moretti and Marco Fontana

Abstract Airborne wind energy converters represent a promising new technology that aims at providing low cost electricity by exploiting airborne systems to harvest energy from high-altitude winds. These plants are interesting for their potential high power density, i.e. ratio between nominal power and weight of required constructions, that makes it possible to forecast extremely low levelized cost for the produced electricity. However, installations of airborne wind energy converters in inland areas might be limited by the required free airspace and by safety problems. For these reasons, marine installations are envisaged, with special interest on the case of floating platforms in deep water locations, that are the most abundantly available. In order to properly address the problem of design and verification of such a kind of system, models that are able to describe the dynamic response of floating platforms to combined kite forces and wave loads have to be developed. This chapter presents a simplified 6 degree-of-freedom model, which couples the linear hydrodynamics of the floating platform with the aerodynamics of the airborne system. A case study is also introduced showing how the dynamic response of the floating platform can affect the performances of the system introducing irregularities in the power output.

7.1 Introduction

Airborne wind energy converters (AWECs) are emerging devices capable of producing electricity from wind energy at altitudes that are currently unreachable by conventional wind turbines. This new technology appears exceptionally promising from the point of view of:

Antonello Cherubini · Giacomo Moretti · Marco Fontana (✉)
PERCRO SEES, Scuola Superiore Sant'Anna, Piazza Martiri 33, 56127 Pisa, Italy
e-mail: marco.fontana-2@unitn.it

© Springer Nature Singapore Pte Ltd. 2018
R. Schmehl (ed.), *Airborne Wind Energy*, Green Energy
and Technology, https://doi.org/10.1007/978-981-10-1947-0_7

- increased power production, because of the high power density and high capacity factors, provided by the level of energy density and persistence of winds that blows at higher layers of the atmosphere [2];
- reduced capital costs, thanks to the tensile slender structure provided by the intrinsic advantageous loading conditions of AWEC foundations.

The combination of the above mentioned positive aspects makes it possible to predict high income and low installation/maintenance costs and consequently reduced levelised cost of electricity [16].

The last decade has seen a remarkable growth of the AWEC sector. A number of prototypes have been built by academic research groups and companies around the world and the associated techno-scientific community is growing fast, trying to close the gap between research and market fit [11]. In the last years, several companies are steadily progressing toward the implementation of full-scale demonstrators.

On the downside, AWECs installations will have to face the problem of available sites. Current AWEC technologies require large airspaces, they can raise safety issues and might face Not In My Back-Yard (NIMBY) effect [48]. This could strongly limit the number of inland areas eligible for the installation of AWECs. As envisaged in [12, 13], one possible option to get around these issues consists in installing AWECs in marine offshore locations - an option that is currently being strongly pursued also by the conventional wind industry [23, 38]. Offshore areas are abundantly available and are not subjected to the cogent regulatory issues of inland sites. Moreover, offshore AWECs could feature typical advantages of conventional offshore wind platforms such as:

- higher wind speeds, which generally increase with distance from the shore;
- high quality of wind with lower turbulence;
- lower wind shear (i.e. thinner surface boundary layer) that makes large power densities available at relative low height;
- availability of large continuous areas suitable for large plants/projects;

On the other hand, offshore AWECs could feature the same problems that characterize conventional offshore wind energy systems, such as:

- expensive marine foundations;
- expensive integration into the electrical network and in some cases a necessary improvement in the capacity of weak coastal grids;
- expensive installation procedures and restricted access during construction owing to weather conditions;
- limited access for operations and maintenance, which results in an additional penalty of reduced availability and hence reduced output.

However, the relatively small size/weight of AWECs foundations, together with a favorable loading scheme characterized by solely traction forces, might be key factors that enable the development of inexpensive/slender offshore platforms. In particular, interest is focused on floating platforms which can be installed in any offshore site, in a wide range of water depth conditions.

The design of an offshore floating platform for AWEC, like other marine structures, is a rather complicated task due to several complex factors. Offshore structures can be subjected and respond to unpredictable loads from ocean waves and currents, combined with external forces (e.g., traction exerted by moorings and AWEC). The understanding of the complex dynamics of these systems is fundamental for their design to resist to extreme loading conditions as well as for the prediction of performance in operating conditions. To this aim, it is extremely useful to have simulation tools that are able to describe the dynamics of floating AWEC systems.

A methodology to model offshore AWECs has been first proposed in [12], where an AWEC mounted on a single Degree-of-Freedom (DoF) heaving platform has been considered. With respect to [12], this chapter introduces a more general modeling approach, considering generic multi-DoF models for the different subsystems (mooring, platform, tethered AWEC). Integrated models are obtained by coupling simplified models for the various subsystems. Attention is focused on AWECs with generators on the lower end of the cable (Ground-Gen [11]) mounted on slack-moored platforms, although the presented models can be easily extended to the case of AWECs with flying generators (Fly-Gen) and to any kind of moored floating platform. The assumed methodology provides a computationally-inexpensive tool that is useful for the preliminary evaluation and conceptual design of floating AWECs. Moreover, the proposed aero-hydrodynamic models allow to assess the influence of the platform motion on the performance of the wind generator, thus providing strategic information for the design of offshore AWECs controllers. It is worth mentioning that the proposed model is only theoretical and that an experimental validation has not been implemented yet.

The chapter is organized as follows. Sect. 7.2 describes the models employed for the different subsystems, i.e. platform, moorings and AWEC. On the basis of such formulations, in Sect. 7.3 a case study is presented in which a dynamic simulator is implemented in a Matlab/Simulink environment. Such a simulator has been conceived as a numerical tool for preliminary assessment of different layouts of floating AWECs and it may serve as a platform to test control strategies and perform feasibility studies. Discussions on practical layouts for the platform and mooring lines and other relevant engineering issues are finally reported in Sect. 7.4. In conclusions Sect. 7.5, a possible roadmap toward full-scale development of offshore AWEC systems is proposed on the basis of well-known methodologies imported from other sectors of offshore renewables. The preliminary content of the present chapter has been presented at the Airborne Wind Energy Conference 2015 [9].

7.2 Model

Accurate mathematical description of the dynamic response of a floating offshore AWEC is an extremely complex problem. The governing physics of the system are characterized by coupled non-linear unsteady hydrodynamic and aerodynamic equations that should be simultaneously solved in order to calculated loads on the platform and predict its time response. Computational Fluid Dynamics (CFD) techniques, such as commonly used Reynolds-Averaged-Navier-Stokes finite-volume solvers, could be employed for this purpose to accurately find solutions but they would result in heavy computational loads and time consuming procedures for setting up simulations.

In this section, a different approach is proposed, that is based on simplified models that are able to grasp the complex multi-DoF dynamic of a floating offshore AWEC with an extremely reduced computational complexity. These models make it possible to run preliminary studies and iterative optimizations with very reduced time-to-solution. On the downside, the proposed models are based on assumptions and approximations that are only valid in operative conditions, i.e. in moderate sea states and wind intensity, far from extreme levels of stress.

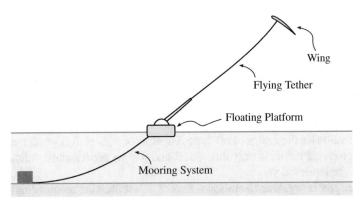

Fig. 7.1 Simplified scheme of floating offshore airborne wind energy converter composed by a wing, a traction tether, a floating platform and a mooring system

A simplified scheme of floating offshore AWEC is shown in Fig. 7.1 where the plant is represented in its main sub-components: a floating platform, a mooring system, a traction tether and a wing.

In this section, a simplified mathematical model that is able to predict the response of each of the sub-components is provided. The approach that is assumed for modeling the airborne components is based on state of the art models from the airborne wind energy (AWE) sector while models for floating elements and mooring are borrowed from the naval/ocean engineering sector [14, 40, 49] and ocean renewable energy [15, 28].

In order to provide a clear description of the models it is important to set appropriate reference frames and coordinates. With reference to Fig. 7.2, the following frames are defined:

- (x, y, z) is a fixed frame with z axis perpendicular to the water plane pointing upwards and having the origin in the position of the center of gravity of the platform in absence of loads from waves and/or tether traction.

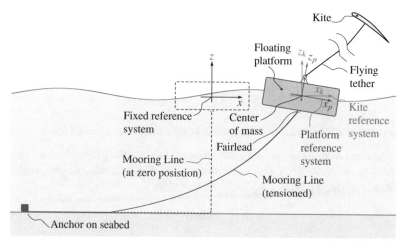

Fig. 7.2 Sketch of offshore AWEC plant sub-components (mooring, platform, tether and kite) and definition of the reference systems

- (x_p, y_p, z_p) is platform-fixed frame having the origin in the center of mass of the platform and whose axes coincide with $x - y - z$ when the platform is not loaded;
- (x_k, y_k, z_k) is the reference system employed to define kite trajectories and has the origin on the center of mass of the platform. The orientation of the axes does not follow the platform-fixed frame of reference, but the x_k axis is along the absolute wind direction and the z_k axis points upwards against gravity.

7.2.1 Floating Platform Dynamic Model

This sub-section presents a 6 DoF hydrodynamic model of the AWEC floating platforms based on the assumptions of potential flow, linear wave theory [36] and small displacements, which make it possible to adopt linear equations [14] for the description of floater dynamics.

The coordinate system assumed to describe the small displacements of the floating platform is provided by a vector $\xi = [\mathbf{x}^\top \; \Theta^\top]^\top = [x \; y \; z \; \theta_x \; \theta_y \; \theta_z]^\top$ with six dimensions, where the first three elements are the linear displacement components of $\mathbf{x}$ and the last three components, Θ, are rotations about the indexed fixed axes (all expressed in the inertial frame of reference). The motion equation can be written in the time-domain according to [1, 30]:

$$\mathbf{M}\ddot{\xi}(t) = \mathbf{f}_H(t) + \mathbf{f}_R(t) + \mathbf{f}_D(t) + \mathbf{f}_E(t) + \mathbf{f}_M(t) + \mathbf{f}_K(t). \qquad (7.1)$$

which represents the balance of the inertia forces, on the right-hand side of the equation and the applied (six-dimension) force-moment, on the left hand-side. The different terms are detailed in the following.

Inertia Matrix The platform inertia matrix $\mathbf{M}$ is represented in the $x - y - z$ frame of reference as

$$\mathbf{M}_{ii} = m \ \text{for } i = 1, 2, 3$$
$$\mathbf{M}_{ij} = \mathbf{I}_{i-3, j-3} \ \text{for } i \geq 4, \ j \geq 4 \qquad (7.2)$$
$$\mathbf{M}_{ij} = 0 \ \text{elsewhere,}$$

being m the platform mass and $\mathbf{I}_{3 \times 3}$ the platform moment of inertia tensor.

Hydrostatic Loads The vector $\mathbf{f}_H$ comprises the forces and moments due to gravity and buoyancy loads. They are a function of the displacement, ξ, and, under the hypothesis of small motion amplitude, they read as

$$\mathbf{f}_H(\xi) \approx \mathbf{f}_{H,0} - \mathbf{G}\xi, \qquad (7.3)$$

where $\mathbf{f}_{H,0}$ is the resultant of gravity and buoyancy loads in the equilibrium configuration, $\xi = \mathbf{0}$, and $\mathbf{G}$ is the so-called buoyancy stiffness matrix, and it is responsible for an elastic-like restoring effect.

Radiation Loads The vector $\mathbf{f}_R$ comprises the force and moments induced by radiated waves that are generated by the platform oscillation, and read as

$$\mathbf{f}_R = -\mathbf{M}_\infty \ddot{\xi} - \int_0^t \mathbf{K}(t - \tau)\dot{\xi}(\tau)\mathrm{d}\tau, \qquad (7.4)$$

where $\mathbf{M}_\infty$ is the added inertia matrix at infinite oscillation frequency, and the convolution integral is a memory term, whose kernel, $\mathbf{K}$, can be expressed in Fourier frequency domain as

$$\widehat{\mathbf{K}}(\omega) = i\omega \left(\widehat{\mathbf{M}}_A(\omega) - \mathbf{M}_\infty \right) + \widehat{\mathbf{B}}_r(\omega). \qquad (7.5)$$

$\mathbf{M}_A$ and $\mathbf{B}_r$ are the frequency-dependent added mass and radiation damping matrices, respectively, ω is the frequency expressed in rad/s, and the superscript $\widehat{}$ indicates the Fourier transform of the labeled quantities. Eq. (7.5) shows that waves radiated by the body influence the total inertia of the system (because of the displaced water volume) and generate a damping effect (i.e., radiated waves propagate at the expense of the platform mechanical energy). The hydrodynamic parameters, i.e., $\mathbf{M}_\infty$, $\widehat{\mathbf{M}}_A(\omega)$, $\widehat{\mathbf{B}}_r(\omega)$, $\widehat{\Gamma}(\omega)$ and $\widehat{\psi}_i(\omega)$ can be computed using potential flow solvers, based on the Boundary Element Method (BEM) [39], with respect to the reference position of the platform, $\xi = \mathbf{0}$. Examples of commercial BEM codes employed for this purpose are WAMIT, ANSYS AQWA, and NEMOH. The convolution term in Eq. (7.4) is computationally inconvenient, and it can be replaced by a state-space approximation, by introducing a state vector, η, of appropriate length:

$$\begin{cases} \dot{\eta} = \mathbf{A}_c \eta + \mathbf{B}_c \dot{\xi} \\ \int_0^t \mathbf{K}(t - \tau)\dot{\xi}(\tau)\mathrm{d}\tau \approx \mathbf{C}_c \eta \end{cases} \qquad (7.6)$$

Matrices $\mathbf{A}_c$, $\mathbf{B}_c$ and $\mathbf{C}_c$ can be chosen using an identification procedure in the frequency domain [50].

Viscous Loads The vector $\mathbf{f}_D$ comprises force and moments produced by drag of viscous friction of the fluid. Although Eq. (7.1) comes from an assumption of inviscid fluid and potential flow, the time-domain formulation allows to include the effects of hydrodynamic drag dissipation. Such contribution can be expressed in a general form as

$$f_{D,i} = -\frac{1}{2}\rho_w C_i A_i l_i^k \, |\dot{\xi}_i| \, \dot{\xi}_i, \quad \text{with} \quad k = 0 \text{ for } i \le 3, \, k = 1 \text{ for } i > 3, \tag{7.7}$$

where ρ_w is the sea water density, A_i and l_i are a characteristic cross sections and lengths, respectively, and C_i are dimensionless drag coefficients.

Wave Excitation Loads The vector $\mathbf{f}_E$ comprises the loads on the floating structure due to sea waves. Thanks to the hypothesis of linear waves, real waves are described as a superimposition of monochromatic waves with different frequencies. With this assumption, wave excitation forces can be approximated by a finite sum of N components as follows [24]:

$$f_{E,i} = \sum_{j=1}^{N} a_{w,j} \widehat{\Gamma}_i(\omega_j) \cos(\omega_j t + \widehat{\psi}_i + \vartheta_j), \tag{7.8}$$

where ω_j are N different angular frequencies, $\widehat{\Gamma}$ is a vector of wave excitation loads (per unit wave amplitude) depending on the wave frequency and direction of propagation, $\widehat{\psi}_i(\omega)$ are angles expressing the phase shift among the different components of the excitation load at each frequency, ϑ_j are random numbers in the interval $[0; 2\pi]$; finally, $a_{w,j}$ are the amplitudes of the different harmonics, given by:

$$a_{w,j} = \sqrt{2 \, \Delta\omega_j \, S_\omega(\omega_j)}. \tag{7.9}$$

In Eq. (7.9), $\Delta\omega_j$ are the differences between consecutive frequencies, and $S_\omega(\omega)$ is the wave spectrum expressed in m^2/(rad/s), which quantifies the distribution of wave energy over pulsation [6]. Besides the action of sea waves, other excitation terms can be kept into account, e.g., the loads induced by sea currents. The latter can be eventually included in the model by means of a term in the same fashion of $\mathbf{f}_D$ in Eq. (7.7).

External Loads The vector $\mathbf{f}_M = \sum_i \mathbf{f}_{M,i}$ is the total load due to the mooring system, and it is the sum of contributions from the single mooring lines (in case of multi-line layout), $\mathbf{f}_{M,i}$; $\mathbf{f}_K$ represents kite loads. Both these contributions will be detailed later on in this section.

7.2.2 Mooring Lines Model

In this section, modeling of platform mooring lines will be discussed, with reference to two traditional classes of computation methods (quasi-static and dynamic).

Offshore renewable energy platforms are moored according to a variety of architectures; a review of the different mooring layouts is presented in [7]. Mooring lines are primarily designed to exert static forces and keep the platform in place and guarantee its stability. However, due to the relevant pulling loads imposed by the wind generator, the dynamic effects of the mooring lines on the platform are non-negligible [30]. The dynamic model of the platform (see Eq. (7.1)) has therefore to be coupled with a model of the mooring system. To this aim, two fundamental approaches exist to model moorings [31]: quasi-static approach and fully-coupled dynamic methods.

Quasi-static methods originate from analytic solutions for the load and shape of continuous homogeneous cables (inextensible or deformable). These methods are only able to account for the gravity/buoyancy forces and, eventually, elastic forces on the cables, thus neglecting any other effect related to inertia and viscous loads. According to these approaches, mooring tensions in any intermediate configuration are a function of the mooring line ends position only. Assuming a quasi-static formulation, the mooring loads on a platform are, in general, a non-linear function of the displacement; nonetheless, it is a common practice to linearize them:

$$\mathbf{f}_M \approx \mathbf{f}_{M0} - \mathbf{G}_M \xi, \tag{7.10}$$

where $\mathbf{f}_{M0} = \sum_i \mathbf{f}_{M0,i}$ is the total mooring load in the static equilibrium position (including the contributions of the single lines), and $\mathbf{G}_M$ is the overall stiffness matrix of the mooring system [33].

Dynamic approaches, on the other hand, account for inertial and dissipative effects. As pointed out in [35], the dynamics of mooring lines is non-negligible when water depth is large or if the line has large-drag elements (e.g., chain moorings). Besides the intrinsic inertia of the mooring line, hydrodynamic parameters (such as added mass, or drag) are crucial in the dynamics of a mooring lines. Such parameters are usually found by means of specialized codes (e.g., ANSYS AQWA or ORCAFLEX). As for the quasi-static case, the load exerted by the mooring on the platform is often expressed in an approximated linear form [8]:

$$\mathbf{f}_M \approx \mathbf{f}_{M0} - \mathbf{G}_M \xi - \mathbf{B}_M \dot{\xi} - \mathbf{M}_M \ddot{\xi}, \tag{7.11}$$

where $\mathbf{B}_M$ and $\mathbf{M}_M$ are overall damping and inertia due to the mooring system.

In this work and in the case study presented in the following, reference is made to a quasi-static non linear model described in [27], which applies for catenary moorings [7] and is based on the assumption of inextensible mooring lines and on the catenary equations.

The geometry of a single mooring line is shown in Fig. 7.3. The line is attached to the platform by means of a fairlead and fixed to the seabed by an anchor. In the

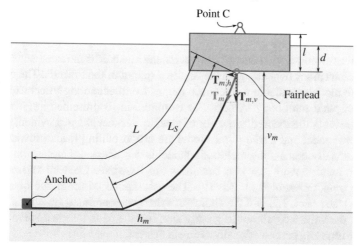

Fig. 7.3 Sketch of a single catenary mooring line, with the main geometric dimensions defined. In the picture, T_m is the force (module) applied by the mooring line to the platform, and $T_{m,h}$ and $T_{m,v}$ its components

picture, L and L_s are respectively the total (constant) length and the length of the suspended portion of the line, while h_m and v_m are respectively the horizontal and vertical distance between the fairlead and the anchor in a generic configuration.

The force exerted by the line on the platform can be decomposed into two vectors, one on the horizontal $x - y$ plane and a vertical one, whose magnitudes ($T_{m,h}$ and $T_{m,v}$) can be expressed as a function of only h_m and v_m (that are directly related to the platform position) [27]:

$$T_{m,h} = \frac{w\, v_m}{2}\left[\left(\frac{L_s}{v_m}\right)^2 - 1\right]$$

$$T_{m,v} = w\, L_s \qquad\qquad (7.12)$$

$$L_s = L - h_m + \frac{T_{m,h}}{w}\operatorname{arccosh}\left(1 + \frac{w\, v_m}{T_{m,h}}\right)$$

where w is the line weight (buoyancy force included) per unit length.

The implicit set of the three Eqs. (7.12), with the three unknowns $T_{m,h}$, $T_{m,v}$ and L_s, applies when $h_m + v_m > L$, otherwise the suspended portion of the line has vertical alignment, horizontal tension is null ($T_{m,h} = 0$) and the vertical force equals the weight (minus the buoyancy) of the suspended part ($T_{m,v} = w\, v_m$). Using these equations on the various mooring lines and computing the associated moments with respect to the platform center of mass, it is possible to express the load vector $\mathbf{f}_M$ as a function of ξ.

7.2.3 Kite Model

In modeling airborne wind energy converters, the kite can be modeled with different degrees of accuracy depending on the required output of the analysis. The complexity of kite models available in literature ranges from simple analytical expressions to sophisticated multibody models. For example, in [34] the basic algebraic formula describing the theoretical power of a cable-free AWEC is provided, in [22] a point mass model and a four point model are proposed, in [4] a sophisticated and computationally-demanding multibody model is described.

In the present work the kite equations are based on the well-known 4 DoF dynamic model presented in [18, 19]. The assumption of the frame of reference (x_k, y_k, z_k), see Sect. 7.2, and the choice of spherical coordinate systems with polar angle on the wind axis allows to greatly simplify the equations and solve a faster-than-real-time and computationally efficient system.

Since tethers are assumed to be simple lines, the kite position and attitude is fully defined with the length of the cables L, the orientation angle ψ and the spherical angles θ and φ. A description of the coordinate system with rotations is as follows: starting with the cables along the x_k axis and the kite pointing upwards, the first rotation has amplitude $-\psi$ around x_k, then $-\theta$ around y_k and finally $+\varphi$ around x_k again. Three examples of wind-polar coordinates are shown in Fig. 7.4.

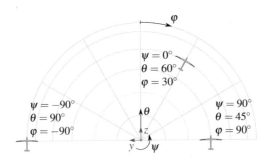

Fig. 7.4 Examples of wind-polar coordinates. Positive wind-polar coordinates (ψ, θ, φ) are indicated by the black arrows. Blue arrows show the kite Cartesian reference frame. By definition, the wind window has the wind going towards the screen, in x direction

The transformation from polar to Cartesian position is therefore:

$$\mathbf{x}_k = \begin{pmatrix} x_k \\ y_k \\ z_k \end{pmatrix} = L \begin{pmatrix} \cos\theta \\ -\sin\theta\sin\varphi \\ \sin\theta\cos\varphi \end{pmatrix} \tag{7.13}$$

The equations of motion of the kite are:

$$\dot{\psi} = g_k v_a \delta + \dot{\varphi}\cos\theta \tag{7.14}$$

$$\dot{\theta} = \frac{v_a}{L}\left(\cos\psi - \frac{\tan\theta}{E}\right) + \frac{v_{yaw}}{L}\tan\theta \qquad (7.15)$$

$$\dot{\varphi} = -\frac{v_a\sin\psi}{L\sin\theta} \qquad (7.16)$$

where the introduced elements assume the following meaning.

Steering response coefficient g_k is a constant that defines the steering response of the kite.

Steering control δ is the non-dimensional control input according to the Turn Rate Law [18].

Apparent wind speed v_a is given by the relation $v_a = v_w E\cos\theta + v_{yaw}E$ (in a realistic system the instantaneous value of this velocity should be measured by a dedicated on-board sensor).

Equivalent aerodynamic efficiency E takes account of all the relevant drag contribution (tether drag, wing drag and where applicable steering unit or fuselage drag).

Drum-platform cable velocity v_{yaw} is the velocity of the kite along the yaw axis aligned towards the floating platform, given by the combination of the drum control input and the motion of the platform.

When the kite is linked to a fixed ground station, as in many AWE prototypes, the component of the kite velocity in the direction of the cables, v_{yaw}, can be assumed to come only from the drum control input on the reeling velocity, $v_{yaw} = -\dot{L}$ [19].

However, if the ground station is a floating platform, the ground ends of the cables of the kite are moved. This may lead to a temporary increase or decrease in the relative wind field at the kite and therefore in the tether tension, depending on the kite position in the wind window and on the direction of motion of the floating platform with respect to the kite position. For example, in case the cables are pulled against the wind, the kite velocity increases and so does the tether force. Likewise, if the platform moves towards the absolute wind direction, the tether tension will decrease. Conversely, a small platform oscillation perpendicular to the cables direction will not generate any effect at the kite. Therefore a kinematic law is needed to couple the airborne system to the floating platform to take into account these effects. In this work, the kinematic link is modeled as:

$$v_{yaw} = -\left(\dot{L} + \dot{\mathbf{c}}\cdot\frac{\mathbf{x}_k}{||\mathbf{x}_k||}\right) \qquad (7.17)$$

$$\mathbf{c} = \mathbf{x} + \mathbf{R}_z\,\mathbf{R}_y\,\mathbf{R}_x\,\mathbf{c}_p \qquad (7.18)$$

where $\mathbf{c}$ and $\mathbf{c}_p$ represent the coordinates of the point where the cables exit from the platform (see Fig. 7.3), expressed in the fixed reference system $(x - y - z)$ and in the platform reference system $(x_p - y_p - z_p)$, respectively. $\mathbf{R}_x$, $\mathbf{R}_y$ and $\mathbf{R}_z$ are the rotation matrices around the x, y and z axis for the transformation from the platform ref. system to the fixed ref. system (see Fig. 7.2). $\mathbf{x}$ and $\mathbf{x}_k$ contain the coordinates of the platform and of the kite defined in Sects. 7.2.1 and 7.2.3, respectively.

In case the platform does not move, the velocity of point C is zero and the coupling equation (7.17) goes back to the case of fixed ground station ($v_{yaw} = -\dot{L}$).

The kite affects the platform motion thanks to the kite aerodynamic loads that generate forces and moments on the platform through the tether.

$$|\mathbf{f}_K| = f_k = \frac{1}{2}\rho_a v_a^2 A_k \sqrt{C_L^2 + C_D^2} \qquad (7.19)$$

where ρ_a is the air density, A_k is the kite aerodynamic area, C_L and C_D are the lift and drag coefficients, respectively.

Even though the lift and drag coefficients change with the angle of attack and other effects, in this analysis C_L and C_D are considered constant, assuming the presence of an inner control loop that keeps the angle of attack in a small range as reported in literature [11, 44, 45]. Stall effects are therefore not captured by this model.

7.3 Case Study

In this section a case study is presented in order to provide an understanding on what kind of analysis can be conducted using the presented models. Specifically, a pumping kite system and its controller are implemented and integrated with a light-weight slack-moored floating platform model. The proposed analysis highlights the effects of the moving/floating platform on the AWEC system and vice versa.

A cylindrical barge with large diameter and short draft is considered, which is assumed to be moored with a single slack mooring line (i.e., the platform can perform large displacements on the water plane). These conditions are considered highly disadvantageous in traditional offshore wind, as both the barge-like platform and slack mooring contribute to large displacements of the generator. Nevertheless, the results of this analysis show that the floating AWEC can preserve stability and good power output in spite of the unfavorable layout, thus providing first positive evidences on the feasibility of offshore AWEC systems based on light-weight floating foundations.

7.3.1 Simulator

The analysis of the case study has been carried out with a dynamic simulator for floating AWECs, which is based on the models presented in Sect. 7.2. The simulator has been developed in a Matlab-Simulink environment and can be adapted to different AWEC architectures with custom layouts of the floating platform, mooring lines and flying kite.

Every time the routine is called, before starting the time-domain simulation, the implicit massless Eqs. (7.12) are solved for the input selection of the system parameters in order to map the force response of the mooring lines as a function of the fairlead position. Then, the simulator calls Simulink routines to solve the dynamic motion of the floating platform (Eq. (7.1)) and of the flying kite (Eqs. (7.14), (7.15) and (7.16)) that are coupled according to Eqs. 7.17 and 7.18 using the control strategy described in Sect. 7.3.3. The modeled physical phenomena are all statically and dynamically stable as there are no negative stiffness and damping coefficients. The simulator showed no numerical stability issues for the range of the parameters set that has been employed. A thorough analysis of the numerical stability would be an interesting topic but it would be out of the scope of this work. The integration of the aforementioned equations is performed with a fixed step explicit solver (ode3 Bogacki-Shampine) with a 0.1 s step size.

A set of three dimensional views showing the animations produced by the simulator for the different subsystems is shown in Fig. 7.5. The source code of the simulator and a sample video is available at [10].

7.3.2 Platform and Mooring

For the sake of clarity, a simplified layout is assumed, which consists in a homogeneous axisymmetric platform moored with a single line (Fig. 7.5). Realistic architectures for offshore AWECs are likely to be more complex (e.g., having multiple mooring lines to keep the platform in place and non-trivial shapes and distribution of ballast masses and buoyant volumes to enhance stability). The platform is assumed to be a cylindrical barge-like buoy as in Fig. 7.3, with the features reported in Table 7.1. The inertia and damping of the mooring lines are neglected in this case study.

In the hypothesis of cylindrical platform with homogeneous density, ρ_p, the inertia matrix $\mathbf{M}$ and the hydrostatic buoyancy matrix $\mathbf{G}$ can be found analytically, with the relations in Table 7.2. Hydrodynamic parameters introduced in Sect. 7.2.1 are obtained with the BEM solver ANSYS AQWA. Platform drag coefficients (C_i in Eq. (7.7)) are all taken equal to one, in order to reasonably estimate the drag loads order of magnitude.

In the present layout, it is assumed that the winch on which the tether is wound and the mooring line fairlead are aligned with the platform centroid along the platform z_p axis direction. The anchor is positioned along the x axis in a way that, in the reference configuration ($\xi = 0$), the suspended length of the line (L_s) is vertical and equal to the vertical distance between the fairlead and the seabed, as shown in Fig. 7.2.

Although the hydrodynamic model proposed in Sect. 7.2.1 comes from a linearization, and is usually valid only for small displacements of the platform, in the case in exam it rigorously holds even for large displacements in x and y direction (not in z direction). Indeed, given the form of hydrodynamic parameters (in particular, hydrostatic stiffness) for this case (Table 7.2), none of the loads in Eq. (7.1) explicitly depends on x and y, except for $\mathbf{f}_M$ and $\mathbf{f}_K$. In the computation of these two

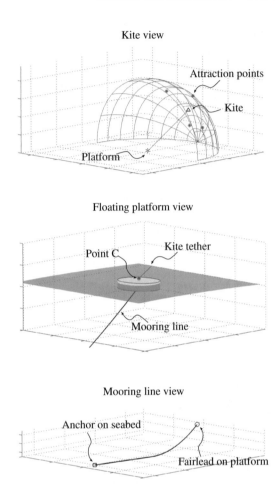

Fig. 7.5 3D view of the simulation environment. The subsystems of a simple offshore AWEC are shown: kite, platform and mooring line

loads general non-linear formulations are used, which apply even for large x and y displacements as well.

For the chosen layout of mooring-barge-tether system and the assumption of small displacements, the platform dynamics is entirely described by 5 DoFs, as no moment can be generated on the z direction, thus the platform yaw rotation, θ_z, is considered identically null.

Table 7.1 Floating platform, mooring line and water data

Platform data	Value
Diameter (m)	10
Draft d (m)	1.2
Height l (m)	1.8
Density ρ_p (kg/m^3)	670
Nominal displacement (ton)	94.7
Drag coefficients, C_i ($i = 1,...,6$) ( )	1
Mooring line data	
Length L (m)	350
Line linear weight w (N/m)	373
Anchor position x (m)	-301.2
Anchor position y (m)	0
Anchor position z (m)	-49.7
Fairlead position x, y (m)	0
Fairlead position z (m)	-0.9
Water data	
Water density ρ_w (kg/m^3)	1025
Water depth h (m)	50

Table 7.2 Inertia matrix and hydrostatic stiffness matrix elements for the case of cylindrical homogeneous platform. Here, d_B=0.6 m is the depth of the barge center of buoyancy and d_G=0.3 m is the depth of the center of gravity below the still water level, in the equilibrium configuration

Inertia Matrix

$$M_{11} = M_{22} = M_{33} = m = \frac{\pi}{4}\rho_p l D^2$$

$$M_{44} = M_{55} = \frac{m}{4}\left(\frac{l^2}{3} + \frac{D^2}{4}\right) \quad , \quad M_{66} = \frac{1}{8}mD^2$$

Hydrostatic Stiffness Matrix

$$G_{11} = G_{22} = G_{66} = 0 \quad , \quad G_{ij} = 0 \quad \text{for } i \neq j$$

$$G_{44} = G_{55} = \frac{\pi}{64}\rho_w g D^4 - \frac{\pi}{4}\rho_w g d D^2 d_B + mg d_G$$

7.3.3 Kite and Controller

Among the many different type of controllers that have been proposed in literature [3, 5, 21, 26, 44, 47], we choose for this case study the simple control strategy illustrated in [18], including improvements proposed in [46].

This controller is based on a simple PID tracking and an algorithm based on four attraction points that makes it possible to reliably implement "side-down" figure-of-eight loops.

With reference to the coordinate system defined in Sect. 7.2.3, it is possible to define the heading, γ, as:

$$\gamma = \arctan\left(\frac{-\dot{\varphi}\sin\theta}{\dot{\theta}}\right) \tag{7.20}$$

The reference heading, γ_0, is computed as

$$\gamma_0 = \arctan\left(\frac{\varphi - \varphi_{0,i}}{\theta_{0,i} - \theta}\right) \tag{7.21}$$

where $\theta_{0,i}$ and $\varphi_{0,i}$ are the constant polar coordinates of the i-th attraction point, P_i, to which the kite is pointing at ($i = 1, 2, 3, 4$).

The non-dimensional steering input of eq. 7.14 is given by a simple proportional control on the reference heading:

$$\delta = k_\gamma(\gamma_0 - \gamma) \tag{7.22}$$

The attraction point P_i is pre-assigned and is changed with the following switch-case algorithm:

Switch i
    Case $i = 1$
        if $\varphi \leq -\varphi_{th}$ then $i = 2$
    Case $i = 2$
        if $\theta \leq \theta_{th}$ then $i = 3$
    Case $i = 3$
        if $\varphi \geq \varphi_{th}$ then $i = 4$
    Case $i = 4$
        if $\theta \leq \theta_{th}$ then $i = 1$

This control algorithm results in a side-down figure-of-eight motion. For example, with reference to Fig. 7.6, viewing the kite from the floating platform with the wind going towards the screen, starting the kite with initial coordinates $\psi_0 = 0$ deg, $\theta_0 = 45$ deg and $\varphi = 0$ deg and initial attraction point P_1, the kite moves upwards and starts steering counterclockwise ($\delta > 0$) while heading to point 1. The φ coordinate becomes negative and keeps decreasing until the threshold $-\varphi_{th}$ is reached and the attraction point switches to point 2. The kite keeps steering counterclockwise to reach point 2 and lowers its θ coordinate. When θ is lower or equal than θ_{th} the attraction point becomes point number 3. The kite steers counterclockwise and starts moving towards greater θ and positive φ until $+\varphi_{th}$ is reached and the attraction point becomes point n. 4. The kite steers clockwise towards lower values of θ and when θ is lower or equal than θ_{th} the attraction point becomes point n. 1 and the cycle starts over.

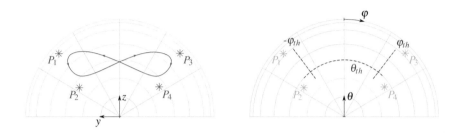

Fig. 7.6 The kite is controlled in to a 'side-down' figure-of-eight motion following the four attraction points marked with a red *. The wind window is shown with the wind going towards the screen

Regarding the control of the reeling drums, different strategies are possible. For example, in [46] the kite is reeled-in after depowering maneuvers, in [25] an optimization of the reeling cycle for a pumping glider is performed and in [20] a detailed control algorithm for soft kites with reel-in at the edge of the wind window is described.

In this work, during reel-out the drum velocity is controlled with a simple proportional controller on the tether force $\dot{L} = k_F(f_k - f_{k0})$. When the maximum tether length is reached, the control mode is switched to reel-in mode and the kite is assumed to fly at constant velocity towards the ground station with zero tension. When the tether length is below the minimum threshold the drum controller is switched to reel-out mode again.

The complete kite data, employed for this case study, together with the control parameters are reported in Table 7.3.

7.3.4 Simulation Results

In order to provide a comparative study, three scenarios have been considered and are presented in this section. Specifically, the same AWEC system have been imagined to be installed: (1) onshore (with fixed ground station), (2) on a floating offshore platform in a calm sea (without waves) and (3) floating offshore platform in presence of intermediate intensity irregular sea waves. In the last case, wave excitation (Eqs. 7.8 and 7.9) is computed assuming a propagation direction along the x axis and a Pierson-Moscowitz distribution [6, 24] to describe the wave spectrum:

$$S_\omega(\omega) = 262.9H_s^2 T_e^{-4}\omega^{-5}\exp(-1054T_e^{-4}\omega^{-4}), \qquad (7.23)$$

Table 7.3 Kite, wind and control data

Kite and wind data	Value
Aerodynamic efficiency E ( )	3.9
Turn rate law constant g_k ( )	0.13
Lift coefficient C_L ( )	0.65
Avg. polar angle θ (deg)	28
Area A_k (m^2)	200
Avg. Tether tension f_k (kN)	81
Avg. reel-out mech. power (kW)	200
Wind speed v_w (m/s)	12
Air density ρ_a (kg/m^3)	1.225
Point C coord. $\mathbf{c}_p$ (m)	$(0,0,1)$
Control parameters	
Attraction point coord. $\varphi_{0,1}$ (deg)	-20
Attraction point coord. $\theta_{0,1}$ (deg)	50
Attraction point coord. $\varphi_{0,2}$ (deg)	-30
Attraction point coord. $\theta_{0,2}$ (deg)	15
Attraction point coord. $\varphi_{0,3}$ (deg)	20
Attraction point coord. $\theta_{0,3}$ (deg)	50
Attraction point coord. $\varphi_{0,4}$ (deg)	30
Attraction point coord. $\theta_{0,4}$ (deg)	15
Heading proportional gain k_γ (rad^{-1})	0.4
Force proportional gain k_F ((m/s)/N)	5e-4
Polar threshold θ_{th} (deg)	50
Longitudinal threshold φ_{th} (deg)	±15
Min. tether length (m)	400
Max. tether length (m)	800
Reel-in speed (m/s)	2.3
Force set point f_{k0} (kN)	76

where H_s and T_e are statistical wave parameters known as 'significant wave height' and 'energy period', and they were assumed to equal H_s=4 m, T_e=10 s.

Results from simulations are reported in Figs. 7.7 and 7.8 that show the time series of the power output and the platform displacements in the three scenarios, respectively.

In Fig. 7.7 the power is zero during reel-in phases (marked in gray on the plot) due to the assumption of null tether tension, although in practice some power is required to reel-in the cables. Comparing the three scenarios, it is clear that when the AWEC is on the floating platform, the power output is quite irregular. i.e. the reeling controller needs to compensate for the platform motion in order to follow the set point of tether tension, thereby generating large oscillations of the instantaneous power around the mean value.

As shown in Fig. 7.8, due to the peculiar layout with single mooring line, the platform undergoes large displacements on the horizontal plane (in x and y direction) in correspondence of both the switching phases between reel-out / reel-in and the periodic lobes of the figure-of-eight path. In particular, the platform oscillates above two different equilibrium positions during reel-out and reel-in phases. In the first case, the mean equilibrium position is set by the force balance among hydrodynamic loads, mooring traction and kite mean tether tension; as the tether tension goes to zero (reel-in), the platform moves in x direction toward the reference equilibrium position due to the restoring mooring force.

Notice also that, for both the scenarios with floating AWEC (with and without waves), the power (Fig. 7.7) becomes negative at the beginning of reel-out cycles, that is, the generator is required to spend power even though the controller is set on reel-out mode. This effect, which is due to the large velocity of the platform in the direction of the tether length, is important and may introduce further complexity in the controller design.

Finally, further considerations can be outlined:

- Heave (z) and pitch (R_y) displacements of the platform in presence of waves are relevantly larger than in calm sea.
- In presence of waves, the hydrodynamic Eq. (7.8) is not rigorously valid immediately after the switching between reel-in and reel-out (i.e. when velocity in x and y direction is large). In these cases, wave frequency should be corrected accounting for the relative velocity between wave and platform.
- Although the considered kite is quite large (with a power in the order of hundreds of kilowatts) and the barge relatively light (the weight is approximately 10 times the mean tether pulling force), the platform does not lose stability, even in presence of relatively tall waves. This is an encouraging result, which demonstrates that offshore AWECs may be deployed with simpler and lighter structures than traditional offshore wind turbines.
- The simulation of the kite retraction phase was done assuming zero tension on the tether. The force during reel-in can be theoretically brought to a near-zero magnitude with a proper flight of a glider [25] and can be practically reduced to roughly one fourth of the reel-out force with a canopy kite [17]. The assumption of zero reel-in force leads to a simple and effective simulation of the transient phase between reel-in and reel-out (see Figs. 7.7 and 7.8), though it could be improved by e.g. choosing more accurate values of the tether force and the reel-in velocity or by taking into account different reel-in strategies.

7.4 Discussion

In this section, a general discussion is provided about technological and engineering issues which cannot be directly observed from the proposed simulator results. Design and operational aspects are discussed on the basis of existing knowledge

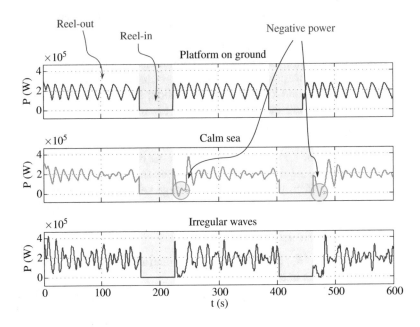

Fig. 7.7 Instant available mechanical power at the generator. Reel-in phases are highlighted in gray

coming from the offshore wind turbines sector. Types of platforms, moorings and very preliminary roadmap towards full-scale devices deployment are outlined. For conventional horizontal-axis wind turbines, it has been suggested [29, 30] that in water deeper than 50 m bottom-fixed support structures are not economically feasible, thus the necessity of floating platform-based solutions arises. The installation of wind turbines on floating platforms presents a number of criticalities related to stability and critical loads due to the large aerodynamic forces applied at large height above the water surface, which are responsible for large pitch overturning moments [42]. Therefore, offshore wind platforms and moorings are generally designed to minimize the wind/wave induced displacements as well as free oscillations [32]. Tethered AWE generators have the intrinsic advantage of generating moderate overturning moments on the platform, thus, it is expectable that design requirements for floating AWE platform may be less stringent than for traditional turbines.

Beside the barge architecture presented in Sect. 7.3 a variety of platform typologies could be considered. Examples of possible configuration are reported in Fig. 7.9 on the basis of envisaged architectures in the wind-offshore sector [29, 32].

The spar buoy (Fig. 7.9(a)) is a platform with deep draft and it is stabilized by lowering the center of mass below the center of buoyancy with a properly sized

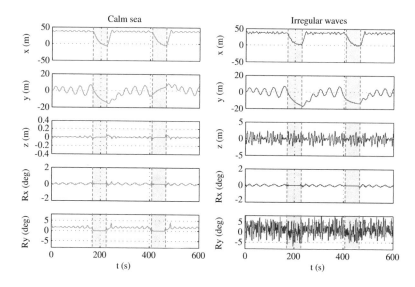

Fig. 7.8 Displacements and rotations of the platform: results of the simulated case study. Reel-in phases are highlighted in gray

ballast. This kind of platform features low vertical wave induced forces, thus undergoing small heave displacements. Due to the small cross section area, its motion is prevalently in roll and pitch [42].

Barge platforms (Fig. 7.9(b)) present large water plane surface (which allows to achieve stability) and relatively shallow draft. According to Roddier et al. [42], this kind of platform is marginally investigated in traditional offshore wind due to its significant angular motions. As shown by the case study in Sect. 7.3, in offshore AWE such angular displacements might be smaller (thanks to the moderate kite-induced moments on the floater), but still relevant. A solution is hypothesized in Fig. 7.9(b) right, where a ring-shaped barge is sketched, which has inertia and buoyant volume located by the platform outer perimeter. The barge central part consists of a lightweight structure (holding the generator), whose structural resistance is possible thanks to the relatively reduced loads generated by the AWE generator. This layout appears to provide a better pitch stability and reduced angular displacements.

Semi-submersible platforms (Fig. 7.9(c)) combine the restoring effects of the two previous types [29]. They are constituted by ballasted pontoons providing buoyancy to the structure, with relatively low wave excitation loads due to the moderate water plane area.

These three platform concepts usually employ catenary lines (like those modeled in Sect. 7.2.2) or taut lines as moorings. Catenary moorings generate a restoring force which is principally due to the line weight, they are slack and approach the

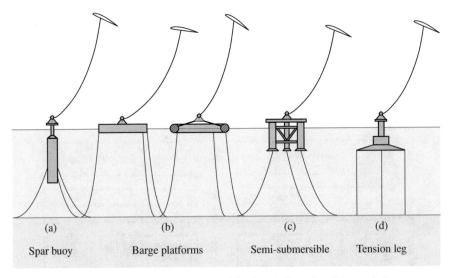

Fig. 7.9 Schematic drawing of four main types of floating platform for offshore wind

sea bed with horizontal tangent. Taut legs arrive at sea bed tense with a certain angle; the main contribution to restoring forces is the line elasticity. Taut moorings have smaller positioning radius, thus requiring a reduced occupied sea bed portion, but they must stand larger tension and their anchor point must resist both horizontal and vertical forces [41].

A different type of floating platform is the so-called Tension Leg Platform (TLP, Fig. 7.9(d)), in which the support tank basement is fully submerged and its excess buoyancy keeps the mooring lines tense and vertically aligned. This architecture minimizes platform displacements, but it is highly costly and presents a number of issues related to mooring installation and lines tension variation as water level changes (due to tidal or incoming waves) [42].

The choice of the platform and mooring type relies on different types of requirement and must be supported by making reference to coupled aero-hydrodynamic models. The key factors for the choice are environmental variables such as water depth (e.g., TLPs and taut moored platforms become advantageous with respect to catenary moored systems as water depth increases [37]), stability, critical loads, displacements. The evaluation of the last two aspects, in particular, is strongly related to the interaction of the floating platform with incoming waves. When the platform resonates (in one of its oscillation modes) with the incoming waves, wave induced loads and/or displacements are maximized. Evaluation of wave spectra and recurring wave energy periods in the installation site is a preliminary step towards the design of platforms with natural frequency outside the range of typical wave frequencies [32]. Moreover, a number of design expedients can be included in order to damp or minimize wave excitation on floating platforms. E.g., pontoons of semi-submersible platforms are usually equipped with horizontal plates at the bottom end, with the aim of increasing the added mass (thus reducing the natural frequency, shifting it away from waves pulsation) and generating viscous damping [42].

Furthermore, it must be remarked that other technical aspects should be kept into account to design and analyze floating AWE plants, i.e.:

- Operating loads and fatigue assessment. This aspect is less crucial than in traditional offshore wind, in which the most critical component is the wind turbine structure itself. In floating AWE, this type of analysis should only be addressed to floating basement and mooring lines.
- Cut-off meteorological conditions. The cut-off wind condition (beyond which a safety landing is necessary) have to be chosen in the light of the platform dynamics. A set of combined wind-wave-current cut-off states for the generator should be defined.
- Extreme loads and survivability. Platform design must guarantee survivability in presence of extreme waves/currents. The modeling methodology proposed in this work is suitable for average operating conditions only (and in presence of small platform displacements). Extreme sea analysis requires different tools, e.g., finite-volume CFD solvers or pool tests of scaled prototypes.
- Non-linear hydrodynamic effects. The relevance of other hydrodynamic effects should be preliminarily assessed with appropriate tools. E.g., Vortex-Induced Vibration (VIV) due to sea currents has to be examined, as it is potentially damaging at frequencies close to the natural frequency of the structure [35]. Moreover, second-order drift hydrodynamic forces can be included in the time domain model of Eq. (7.1), which may cause the platform to oscillate about an offset position with respect to the nominal equilibrium [30].
- Installation and operation. Transportation and anchoring of the platform as well as in-site installation of the tethered generator have to be standardized and automatized. Moreover, take-off and landing phases of the flying generator, which are complex operations still representing a challenge for onshore AWE [11], have to be further improved and adapted to offshore conditions.
- Farms of devices. Similarly to other offshore renewable energy technologies, it is expected that offshore AWE generators become more convenient when installed in farms where operation and maintenance can be more efficient and fixed costs such as electrical connections can be shared among multiple devices [43]. If farms are considered, different layouts can be conceived (e.g., shared platforms and moorings may be employed for different AWECs) and more sophisticated models, keeping into account the aero-hydrodynamic interaction among different devices, should be employed.

Given the high uncertainty related to floating AWECs modeling and design, the above mentioned engineering issues, and the large capital costs involved, it is expectable that a viable roadmap towards implementation of full-scale offshore AWECs should include a series of incremental scale-up prototyping steps, following the methodology of the other offshore renewable energy converters (namely, wave, tidal, offshore wind generators) [42, 43].

After a first phase of concept definition, small-scale tests (e.g., wave tank tests on scaled prototypes of floating platform) are required prior to undertake offshore installation. Such tests are useful to adjust and update previously established numerical models, correct the design choices and assess the device response in operating and extreme load conditions. Final scale-up steps may include sea-tests, monitoring and grid connection of nearly-full-scale (e.g., 1:5) prototypes first, and on full-scale devices finally.

7.5 Conclusions

In this work, the potential of offshore airborne wind energy converters (AWECs) is investigated and numerical models are presented for preliminary modeling and design of floating platforms housing AWECs.

Offshore sites are very promising for wind energy application, as they feature large available airspace and non-turbulent thin wind boundary layers. Installing AWECs offshore, in particular, may open the way to the development of high-altitude wind technology by preventing Not-In-My-Back-Yard (NIMBY) effects that AWECs may induce on the general public. Nevertheless, the implementation of offshore wind generators is a costly and complex operation, which requires a number of progressive steps preceding the installation of a full-scale fully-functional device.

The first step towards deployment is preliminary numerical modeling aimed at understanding the interaction among the different subsystems (platform, wind generator, mooring system) and roughly designing the various parts and the controller. This chapter presents numerical models specifically tailored to this purpose. Existing models for the different subsystems have been collected from literature, adapted, extended and coupled. In particular, a multi-DoF hydrodynamic model for floating platforms is presented, which relies on potential-flow and linear waves theory, a quasi-static model for catenary moorings has been detailed, and a 4-DoF model for kite-type AWECs has been assumed. With respect to other sophisticated numerical tools, which solve fluid continuum equations using a local approach and a domain discretization, this set of models allows faster-than-real-time calculations and is thus preferable at a first stage of analysis.

A Matlab-Simulink simulator that relies on the above mentioned models has been released [10]. An illustrative case study is discussed, which makes reference to a cylindrical barge platform with a single mooring line.

The presented simulator provides a one-of-a-kind numerical platform on which different layouts of floating AWECs can be tested and designed, prior to undergo progressive scale-up prototyping steps towards the installation of full-scale offshore AWECs.

Acknowledgements This work is carried out with the financial support of Kitegen Research Srl and Scuola Superiore Sant'Anna.

References

1. Alves, M.: Numerical simulation of the dynamics of point absorber wave energy converters using frequency and time domain approaches. Ph.D. Thesis, University of Lisbon, 2012
2. Archer, C. L.: An Introduction to Meteorology for Airborne Wind Energy. In: Ahrens, U., Diehl, M., Schmehl, R. (eds.) Airborne Wind Energy, Green Energy and Technology, Chap. 5, pp. 81–94. Springer, Berlin Heidelberg (2013). doi: 10.1007/978-3-642-39965-7_5
3. Baayen, J. H., Ockels, W. J.: Tracking control with adaption of kites. IET Control Theory and Applications 6(2), 182–191 (2012). doi: 10.1049/iet-cta.2011.0037
4. Breukels, J., Ockels, W. J.: Analysis of complex inflatable structures using a multi-body dynamics approach. AIAA Paper 2008-2284. In: Proceedings of the 49th AIAA/ASME/ASCE/AHS/ASC Structures, Structural Dynamics, and Materials Conference, Schaumburg, IL, USA, 7–10 Apr 2008. doi: 10.2514/6.2008-2284
5. Canale, M., Fagiano, L., Ippolito, M., Milanese, M.: Control of tethered airfoils for a new class of wind energy generator. In: Proceedings of the 45th IEEE Conference on Decision and Control, pp. 4020–4026, San Diego, CA, USA (2006). doi: 10.1109/CDC.2006.376775
6. Carter, D. J. T.: Estimation of wave spectra from wave height and period. Report 135, 1982. http://eprints.soton.ac.uk/14556/
7. Castro-Santos, L., González, S. F., Diaz-Casas, V.: Mooring for floating offshore renewable energy platforms classification. In: International Conference on Renewable Energies and Power quality (ICREPQ'13), Bilbao, Spain, 20–22 Mar 2013. http://www.icrepq.com/icrepq'13/277-castro.pdf
8. Cerveira, F., Fonseca, N., Pascoal, R.: Mooring system influence on the efficiency of wave energy converters. International Journal of Marine Energy 3–4, 65–81 (2013). doi: 10.1016/j.ijome.2013.11.006
9. Cherubini, A., Fontana, M.: Modelling and Design of Off-Shore Floating Platform for High Altitude Wind Energy Converters. In: Schmehl, R. (ed.). Book of Abstracts of the International Airborne Wind Energy Conference 2015, p. 42, Delft, The Netherlands, 15–16 June 2015. doi: 10.4233/uuid:7df59b79-2c6b-4e30-bd58-8454f493bb09. Presentation video recording available from: https://collegerama.tudelft.nl/Mediasite/Play/d16d2d1661af47e88f6e13fb8255e2341d
10. Cherubini, A., Moretti, G., Fontana, M.: Offshore AWEC Simulator. http://www.percro.org/AWE (2015). Accessed 1 Feb 2016
11. Cherubini, A., Papini, A., Vertechy, R., Fontana, M.: Airborne Wind Energy Systems: A review of the technologies. Renewable and Sustainable Energy Reviews 51, 1461–1476 (2015). doi: 10.1016/j.rser.2015.07.053
12. Cherubini, A., Vertechy, R., Fontana, M.: Simplified model of offshore Airborne Wind Energy Converters. Renewable Energy 88, 465–473 (2016). doi: 10.1016/j.renene.2015.11.063
13. Coleman, J., Ahmad, H., Pican, E., Toal, D.: Modelling of a synchronous offshore pumping mode airborne wind energy farm. Energy 71, 569–578 (2014). doi: 10.1016/j.energy.2014.04.110
14. Cummins, W. E.: The impulse response function and ship motions. Report 1661, Department of the Navy David Taylor Model Basin, Oct 1962. http://hdl.handle.net/1721.3/49049
15. Day, A. H., Babarit, A., Fontaine, A., He, Y.-P., Kraskowski, M., Murai, M., Penesis, I., Salvatore, F., Shin, H.-K.: Hydrodynamic modelling of marine renewable energy devices: A state of the art review. Ocean Engineering 108, 46–69 (2015). doi: 10.1016/j.oceaneng.2015.05.036
16. Diehl, M.: Airborne Wind Energy: Basic Concepts and Physical Foundations. In: Ahrens, U., Diehl, M., Schmehl, R. (eds.) Airborne Wind Energy, Green Energy and Technology, Chap. 1, pp. 3–22. Springer, Berlin Heidelberg (2013). doi: 10.1007/978-3-642-39965-7_1
17. Erhard, M., Horn, G., Diehl, M.: A quaternion-based model for optimal control of the SkySails airborne wind energy system. ZAMM – Journal of Applied Mathematics and Mechanics (2016). doi: 10.1002/zamm.201500180. arXiv:1508.05494 [math.OC]

18. Erhard, M., Strauch, H.: Theory and Experimental Validation of a Simple Comprehensible Model of Tethered Kite Dynamics Used for Controller Design. In: Ahrens, U., Diehl, M., Schmehl, R. (eds.) Airborne Wind Energy, Green Energy and Technology, Chap. 8, pp. 141–165. Springer, Berlin Heidelberg (2013). doi: 10.1007/978-3-642-39965-7_8

19. Erhard, M., Strauch, H.: Control of Towing Kites for Seagoing Vessels. IEEE Transactions on Control Systems Technology **21**(5), 1629–1640 (2013). doi: 10.1109/TCST.2012.2221093

20. Erhard, M., Strauch, H.: Flight control of tethered kites in autonomous pumping cycles for airborne wind energy. Control Engineering Practice **40**, 13–26 (2015). doi: 10.1016/j.conengprac.2015.03.001

21. Fagiano, L., Milanese, M., Piga, D.: Optimization of airborne wind energy generators. International Journal of Robust and Nonlinear Control **22**(18), 2055–2083 (2011). doi: 10.1002/rnc.1808

22. Fechner, U., Vlugt, R. van der, Schreuder, E., Schmehl, R.: Dynamic Model of a Pumping Kite Power System. Renewable Energy (2015). doi: 10.1016/j.renene.2015.04.028. arXiv:1406.6218 [cs.SY]

23. Henderson, A. R., Morgan, C., Smith, B., Sørensen, H. C., Barthelmie, R. J., Boesmans, B.: Offshore Wind Energy in Europe– A Review of the State-of-the-Art. Wind Energy **6**(1), 35–52 (2003). doi: 10.1002/we.82

24. Henriques, J. C. C., Chong, J. C., Falcão, A. F. O., Gomes, R. P. F.: Latching control of a floating oscillating water column wave energy converter in irregular waves. Paper No. OMAE2014-23260. In: Proceedings of the ASME 2014 33rd International Conference on Ocean, Offshore and Arctic Engineering, San Francisco, CA, USA, 8–13 June 2014. doi: 10.1115/OMAE2014-23260

25. Horn, G., Gros, S., Diehl, M.: Numerical Trajectory Optimization for Airborne Wind Energy Systems Described by High Fidelity Aircraft Models. In: Ahrens, U., Diehl, M., Schmehl, R. (eds.) Airborne Wind Energy, Green Energy and Technology, Chap. 11, pp. 205–218. Springer, Berlin Heidelberg (2013). doi: 10.1007/978-3-642-39965-7_11

26. Jehle, C., Schmehl, R.: Applied Tracking Control for Kite Power Systems. AIAA Journal of Guidance, Control, and Dynamics **37**(4), 1211–1222 (2014). doi: 10.2514/1.62380

27. Johanning, L., Smith, G. H., Wolfram, J.: Mooring design approach for wave energy converters. Proceedings of the Institution of Mechanical Engineers. Part M, Journal of engineering for the maritime environment **220**(4), 159–174 (2006). doi: 10.1243/14750902JEME54

28. Jonkman, J. M.: Dynamics of offshore floating wind turbines—model development and verification. Wind Energy **12**(5), 459–492 (2009). doi: 10.1002/we.347

29. Jonkman, J. M., Matha, D.: Dynamics of offshore floating wind turbines–analysis of three concepts. Wind Energy **14**(4), 557–569 (2011). doi: 10.1002/we.442

30. Jonkman, J. M.: Dynamics modeling and loads analysis of an offshore floating wind turbine. Technical Report NREL/TP-500-41958, National Renewable Energy Laboratory, Golden, CO, USA, Nov 2007

31. Kwan, C. T., Bruen, F. J.: Mooring line dynamics: comparison of time domain, frequency domain, and quasi-static analyses. In: Proceedings of the 23rd Annual Offshore Technology Conference, Houston, TX, USA, 6–9 May 1991. doi: 10.4043/6657-MS

32. Lee, K. H.: Responses of floating wind turbines to wind and wave excitation. Ph.D. Thesis, Massachusetts Institute of Technology, 2005. http://hdl.handle.net/1721.1/33564

33. Loukogeorgaki, E., Angelides, D. C.: Stiffness of mooring lines and performance of floating breakwater in three dimensions. Applied Ocean Research **27**(4), 187–208 (2005). doi: 10.1016/j.apor.2005.12.002

34. Loyd, M. L.: Crosswind kite power. Journal of Energy **4**(3), 106–111 (1980). doi: 10.2514/3.48021

35. Matha, D., Schlipf, M., Cordle, A., Pereira, R., Jonkman, J.: Challenges in simulation of aerodynamics, hydrodynamics, and mooring-line dynamics of floating offshore wind turbines. CP-5000-50544. In: Proceedings of the 21st Offshore and Polar Engineering Conference, Maui, HI, USA, 19–24 June 2011. http://www.nrel.gov/docs/fy12osti/50544.pdf

36. McCormick, M. E.: Ocean engineering wave mechanics. John Wiley & Sons, Chichester (1973)
37. Musial, W., Butterfield, S., Boone, A.: Feasibility of floating platform systems for wind turbines. AIAA Paper 2004-1007. In: Proceedings of the 42nd AIAA Aerospace Sciences Meeting and Exhibit, Reno, NV, USA, 5–8 Jan 2004. doi: 10.2514/6.2004-1007
38. Musial, W., Ram, B.: Large-scale offshore wind power in the United States: Assessment of opportunities and barriers. NREL/TP-500-40745, National Renewable Energy Laboratory, Golden, CO, USA, Sept 2010. http://www.nrel.gov/wind/pdfs/40745.pdf
39. Newman, J. N., Lee, C.-H.: Boundary-element methods in offshore structure analysis. Journal of Offshore Mechanics and Arctic Engineering 124(2), 81–89 (2002). doi: 10.1115/1.1464561
40. Nossen, J., Grue, J., Palm, E.: Wave forces on three-dimensional floating bodies with small forward speed. Journal of Fluid Mechanics 227, 135–160 (1991). doi: 10/d9kr7d
41. Qiao, D., Ou, J., Wu, F.: Design selection analysis for mooring positioning system of deepwater semi-submersible platform. In: Proceedings of the 22nd International Offshore and Polar Engineering Conference, Rhodes, Greece, 17–22 June 2012. https://www.onepetro.org/conference-paper/ISOPE-I-12-012
42. Roddier, D., Cermelli, C., Aubault, A., Weinstein, A.: WindFloat: A floating foundation for offshore wind turbines. Journal of Renewable and Sustainable Energy 2(3), 033104 (2010). doi: 10.1063/1.3435339
43. Ruehl, K., Bull, D.: Wave energy development roadmap: design to commercialization. In: Proceedings of the 2012 Oceans, Hampton Roads, VA, USA, 14–19 Oct 2012. doi: 10.1109/OCEANS.2012.6404795
44. Ruiterkamp, R., Sieberling, S.: Description and Preliminary Test Results of a Six Degrees of Freedom Rigid Wing Pumping System. In: Ahrens, U., Diehl, M., Schmehl, R. (eds.) Airborne Wind Energy, Green Energy and Technology, Chap. 26, pp. 443–458. Springer, Berlin Heidelberg (2013). doi: 10.1007/978-3-642-39965-7_26
45. Vander Lind, D.: Analysis and Flight Test Validation of High Performance Airborne Wind Turbines. In: Ahrens, U., Diehl, M., Schmehl, R. (eds.) Airborne Wind Energy, Green Energy and Technology, Chap. 28, pp. 473–490. Springer, Berlin Heidelberg (2013). doi: 10.1007/978-3-642-39965-7_28
46. Vlugt, R. van der, Peschel, J., Schmehl, R.: Design and Experimental Characterization of a Pumping Kite Power System. In: Ahrens, U., Diehl, M., Schmehl, R. (eds.) Airborne Wind Energy, Green Energy and Technology, Chap. 23, pp. 403–425. Springer, Berlin Heidelberg (2013). doi: 10.1007/978-3-642-39965-7_23
47. Williams, P., Lansdorp, B., Ockels, W. J.: Nonlinear Control and Estimation of a Tethered Kite in Changing Wind Conditions. AIAA Journal of Guidance, Control and Dynamics 31(3) (2008). doi: 10.2514/1.31604
48. Wolsink, M.: Wind power implementation: The nature of public attitudes: Equity and fairness instead of 'backyard motives'. Renewable and Sustainable Energy Reviews 11(6), 1188–1207 (2007). doi: 10.1016/j.rser.2005.10.005
49. Yamamoto, M., Morooka, C. K., Ueno, S.: Dynamic behavior of a semi-submersible platform coupled with drilling riser during re-entry operation in ultra-deep water. In: Proceedings of the ASME 2007 26th International Conference on Offshore Mechanics and Arctic Engineering, pp. 239–248, American Society of Mechanical Engineers, San Diego, CA, USA, 10–15 June 2007. doi: 10.1115/OMAE2007-29221
50. Yu, Z., Falnes, J.: State-space modelling of a vertical cylinder in heave. Applied Ocean Research 17(5), 265–275 (1995). doi: 10.1016/0141-1187(96)00002-8

Chapter 8
Enhanced Kinetic Energy Entrainment in Wind Farm Wakes: Large Eddy Simulation Study of a Wind Turbine Array with Kites

Evangelos Ploumakis and Wim Bierbooms

Abstract Wake effects in wind farms are a major source of power production losses and fatigue loads on the rotors. It has been demonstrated that in large wind farms the only source of kinetic energy to balance the energy extracted by the turbines is the vertical transport of the free-stream flow kinetic energy from above the wind turbine canopy. This chapter explores the possibility to enhance this transport process by introducing kites in steady flight within a small wind turbine array. In a first step, an array of four wind turbines, aligned with the streamwise velocity component, is simulated within a large eddy simulation framework. The turbines are placed in a pre-generated turbulent atmospheric boundary layer and modeled as actuator disks with both axial and tangential inductions, to account for the wake rotation. In a second step an identical turbine configuration with interspersed kites is investigated. The kites are modeled as body forces on the flow, equal in magnitude and opposite in direction to the vector sum of the lift and drag forces acting on the kite surfaces. A qualitative comparison of the mean flow statistics, before and after the introduction of the kites is presented.

8.1 Introduction

Humanity has harnessed the power of the wind for thousands of years, initially through sails and windmills. For more than two thousand years wind-powered machines had been used to pump water and grind grain. By the end of the 19th century pioneers such as Danish scientist, inventor and educator Poul la Cour employed wind energy for generating electricity. Since these days wind energy has progressed

Evangelos Ploumakis (✉) · Wim Bierbooms
Delft University of Technology, Faculty of Aerospace Engineering, Kluyverweg 1, 2629 HS Delft, The Netherlands
e-mail: eploumakis@yahoo.gr, w.a.a.a.bierbooms@tudelft.nl

© Springer Nature Singapore Pte Ltd. 2018
R. Schmehl (ed.), *Airborne Wind Energy*, Green Energy
and Technology, https://doi.org/10.1007/978-981-10-1947-0_8

from being a minor source of electricity to covering 10.2% of the EU's consumption in 2014 [1].

With the rapidly growing global wind energy capacity an increasing number of wind turbines has been installed in large wind farms. These farms are usually organized in patterns of rows and columns and the array configuration is typically chosen according to the dominant wind direction and turbine size. Because of the varying wind direction most turbines in such an array are exposed to the wakes of upstream turbines which leads to significant power losses. For example at Horns Rev, one of the first large-scale offshore wind farms in the world, power losses of more than 40% for certain wind directions have been reported [4]. The wake flows are also associated with increased turbulence levels and higher fatigue loads for exposed turbines. On the other hand, the increased turbulence enhances the flow entrainment from the free stream above the wind turbine canopy, leading to a faster wake recovery and more kinetic energy available for harnessing by downwind turbines. A visualization of simulated turbine wakes in a wind farm is shown in Fig. 8.1.

As wind farms grow larger the asymptotic limit of the fully developed flow inside the farms has been receiving a lot of interest. With the height of the atmospheric boundary layer (ABL) of about 1 km and modern wind farms exceeding 10–20 km in the horizontal direction a fully developed flow regime can be established [21].

Fig. 8.1 Simulated wind farm turbulence: volume rendering of low-velocity wake regions. Visualization generated by D. Bock, National Center for Supercomputing Applications, XSEDE, based on wind farm LES data [26, 27]

This limit state is associated with wind farms on flat terrain whose length exceed the height of the ABL by more than an order of magnitude [9].

From a physical point of view it is important to understand that for large wind turbine arrays entrainment of kinetic energy from the undisturbed flow plays an important role in replenishing the wake. For stand-alone wind turbines the extracted power is related to the difference between the upstream and downstream kinetic energy fluxes. On the contrary, it is shown that for the turbines, operating in a fully developed wind turbine boundary layer, the entrainment of kinetic energy from the free atmosphere into the wind turbine canopy is the only source of kinetic energy to balance that extracted by the wind turbines [8]. The total kinetic energy that is available in the lower parts of the ABL is therefore extracted in two primary ways: from the incoming wind at the leading edge of the wind farm and from above the wind farm [20]. Changes in the streamwise direction can be neglected after the fourth row of turbines and vertical transport of momentum becomes a crucial parameter in determining the overall efficiency of infinitely large wind farms [4].

In view of the realization that for large wind farms it is the vertical entrainment that dominates the availability of power an innovative way to enhance the vertical transport of momentum is proposed. The study of new designs that could potentially assist the wake re-energizing process in wind turbine arrays is a crucial part of the ongoing quest to improve the overall efficiency and lower the cost of energy. In the present study, the possibility to enhance the vertical transport of momentum by introducing kites in steady flight within a wind turbine array is investigated. This concept is visualized in Fig. 8.2. Large eddy simulations (LES) of a small wind

Fig. 8.2 Visualization of the proposed flow entrainment based on interspersed kites [6]

turbine array operating in a turbulent ABL are used to evaluate the effect of the kites on the mean flow statistics.

In Sect. 8.2 the modeling approaches for the simulation of the ABL, the wind turbines and the kites in the computational domain are presented. In Sect. 8.3 an LES of a row of four wind turbines is presented and discussed, while in Sect. 8.4 the same configuration with interspersed kites is analyzed. The main point of interest is the effect of the kites on the recovery characteristics of the wake flow. The preliminary content of the present chapter has been presented at the Airborne Wind Energy Conference 2015 [23] and is published as MSc thesis [22].

8.2 Numerical Setup and Modeling Considerations

Modeling of a wind turbine wake is a key task for the energy yield prediction of operating wind farms as well as the optimization of new wind farm layout configurations. Numerical simulations, instead of experiments, are the focus of scientific research for two main reasons. Full-scale, high-quality experiments are costly and are limited to provide information on the flow field. Wind flow modeling software is nowadays mainly used to extrapolate the flow field data from on-site measurements to locations where poor or no measurements were taken. Most of the modeling software used is based on either micro-scale models derived from measurement campaigns, such as used in WAsP [13], WindPRO [14] and WindFarmer [11], or on computational fluid dynamics (CFD) where the differential governing equations of fluid motion are solved numerically.

8.2.1 Atmospheric Boundary Layer Modeling

When performing CFD simulations a key issue for wind engineers is the accurate representation of the turbulent ABL. Compared to standard Reynolds-averaged-Navier-Stokes (RANS) simulation, LES is generally known to reproduce main turbulence properties with higher accuracy, though it is stressed that further research is needed in sub-grid scale modeling [10, 31]. One of the major difficulties encountered in LES is the definition of realistic upstream conditions at the domain inlet. Several inflow generation techniques have been proposed in the past decade and can be classified into three main categories: synthetic methods, precursor simulations and recycling methods as classified by Keating et al. [16]. To generate the turbulent inflow, also denoted as "numerical wind", the LES model coupled with the Smagorinsky-Lilly sub-grid-scale (SGS) model available in Fluent® was used. More realistic inflow turbulence, with better spatial and temporal correlations, is achieved by running a precursor simulation either before the main simulation or simultaneously with it, usually by extending the domain upstream the area of interest.

A set of general simplifications were considered necessary to make numerical simulations of the ABL possible within the LES framework. Firstly, the ABL is considered neutrally stable meaning that thermal effects are neglected. Since the purpose of the study is to provide a qualitative report of the examined test cases, the assumption of a neutral ABL saves computational time since the additional equation for the transport of potential temperature does not need to be solved. Secondly, the flow is considered to be incompressible, which is valid for low Mach number ($M < 0.3$), at which the flow is quasi-steady and isothermal. For the relatively low wind speeds ($< 20\,\text{m/s}$), typically encountered in our simulations, we can expect minimal compressibility effects on the solution therefore the assumption to keep the density constant in space and time.

In order to generate the fluctuating velocity components in the computational domain the spectral synthesizer method in Fluent® is employed. The synthesized turbulence method is used in our case, to provide an initial perturbation to our LES of turbulent flow and initiate turbulent motions in the flow. For a velocity inlet boundary condition, a random field of fluctuating velocities, based on a random flow generation (RFG) Fourier technique proposed by Smirnov, Shi et al. [25], is superimposed on a specified mean velocity.

The numerical wind field was generated by a precursor simulation in an empty domain without any wind turbines or kites. In this simulation the pressure-driven flow is recycled in the domain allowing the boundary layer and its turbulence to develop naturally. Once the flow approaches the logarithmic profile of the predefined mean velocity

$$\overline{U} = \frac{u_*}{k} \ln\left(\frac{z}{z_0}\right),$$ (8.1)

where z_0 denotes the aerodynamic roughness length, $k = 0.4$ the von Karman constant and u_* the friction velocity, the three velocity components are captured at a specified vertical plane. The friction velocity u_* is related to the level of turbulence in the surface layer, the bottom 5–10% of the mixing layer in which the turbulence is mostly mechanically generated, and typically increases with higher values of the surface roughness z_0.

8.2.2 Rotor Modeling

Simulations of the physical rotor are computationally very expensive because of the fine computational mesh that is required to adequately resolve the different parts of the rotor and the generated flow structures. It is also clear that the high resolution of the near wake is not of major importance when studying the flow in large-scale wind farms. Thus, when modeling wind turbine arrays with CFD, the actuator disk [2, 19] and the actuator line [28, 29] methods are typically used, the actuator surface [12] method has only been recently explored.

Rankine (1865) and Froude (1889) have used one-dimensional momentum theory to predict the performance of ship propellers, later Betz (1919) applied this actuator disk model to assess the energy extraction potential of wind turbines. The theory is depicted in Fig. 8.3 which shows a stream tube that is expanding in flow direction. This expansion effect is caused by the flow resistance of the actuator disk

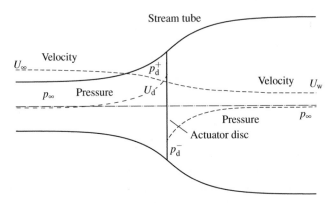

Fig. 8.3 Schematic of the flow passing through the wind turbine rotor modeled as an actuator disk, adapted from [7]

which introduces a pressure drop in the flow. The force that the actuator disk exerts on the flow is added to the momentum equation

$$\frac{D\mathbf{u}}{Dt} = \mathbf{f} - \frac{1}{\rho}\nabla\mathbf{p} + \nu\nabla^2\mathbf{u}. \tag{8.2}$$

The thrust force acting on the wind turbine can also be expressed as the summation of forces acting on both sides of the actuator disk, calculated from the pressure difference across the disk as

$$T = A_d(p_d^+ - p_d^-), \tag{8.3}$$

where $A_d = \pi D^2/4$ denotes the surface area of the actuator disk. Normalization with the dynamic pressure in the flow leads to the non-dimensional thrust coefficient

$$C_T = \frac{T}{\frac{1}{2}\rho U_\infty^2 A_d}. \tag{8.4}$$

However, in LES of wind turbine arrays with significant wind turbine wake interactions the upstream reference velocity U_∞ is not readily known. It is therefore more natural to use the velocity normal to the rotor disk, U_d, to calculate the thrust coefficient at the disks. The thrust force in our simulations is expressed in terms of a modified thrust coefficient [9],

$$C_T' = \frac{C_T}{(1-a)^2}, \tag{8.5}$$

introducing the axial induction factor a as the fractional decrease in wind velocity between the free stream, U_∞, and the rotor plane, U_d, as

$$a = \frac{U_\infty - U_d}{U_\infty}.$$ (8.6)

When representing the turbine rotor by an actuator disk the axial induction factor a is closely related to the power extracted by the wind turbine. In case the wind approaching the turbine is brought to rest ($a = 1$) no power will be extracted by the wind turbine since there will be no flow through the rotor plane. Also, if there is no change in the velocity of the wind passing through the rotor ($a = 0$) no power will be extracted since the kinetic energy of the air before and after the turbine blades remains unchanged. For negative values of the induction factor ($a < 0$) the wind turbine rotor acts as a propeller instead of a generator. The induction factor is related to the power coefficient, $C_P = 4a(1-a)^2$, and thrust coefficient, $C_T = 4a(1-a)$, of a turbine. For an induction factor of $a = 1/3$ one can obtain the maximum power and thrust coefficients of an ideal wind turbine rotor. This condition is well known as the "Betz limit" defining the maximum fraction of kinetic energy that can be extracted from the flow and converted into usable power. It can be shown analytically that this maximum conversion efficiency is $16/27$ (59.3%) independent of the design of the rotor.

Combining Eqs. (8.4) and (8.5) the thrust force is calculated on the basis of the modified thrust coefficient C_T' as

$$T = -\frac{1}{2}\rho C_T' U_d^2 A_d.$$ (8.7)

For flows with significant three-dimensional effects the tangential and radial velocities need to be taken into account to generate swirl in the flow. The induced radial velocity is typically small compared to the axial and tangential velocities and is therefore often neglected in calculations. The tangential velocity is calculated as $U_\theta = U_x \cos\theta + U_y \sin\theta$, where U_x and U_y are the instantaneous velocities in x- and y-direction and θ is the angular coordinate. The change in tangential velocity is expressed in terms of a tangential flow induction factor

$$a' = \frac{1 - 3a}{4a - 1}.$$ (8.8)

The tangential velocity varies along the span of the blades and accordingly is a function of the radial position r. Upstream of the rotor disk the tangential velocity vanishes while immediately downstream the tangential velocity magnitude is $2\omega r a'$, where ω is the angular speed of the rotor. Since it is a reaction of the flow to the motion of the rotor its direction is always opposing the direction of rotation.

Variants of the actuator disk model (ADM) discussed in literature account for thrust and tangential forces (ADM-R) while others account only for thrust forces (ADM-NR). The ADM may simulate wind turbines and the induced wakes but fails to create the tip vortices carried onto the wake [28]. Wu and Porté-Agel [33] compared LES simulations of a wind farm using the ADM-NR and ADM-R approaches with field measurements. It was concluded that the ADM-R yields improved predictions in the wake compared to the ADM-NR which stresses the importance of turbine-induced flow rotation for the accurate prediction of the wake structures. The present study uses the ADM approach with both axial and tangential inductions.

8.2.3 Modeling the Kites

By definition an aerodynamic lifting device uses the relative velocity between wind field and flying device, quantified by the apparent wind velocity vector

$$\mathbf{v}_a = \mathbf{v}_w - \mathbf{v}_k, \tag{8.9}$$

to generate a force component perpendicular to $\mathbf{v}_a$, denoted as lift force

$$L = \frac{1}{2}\rho C_L v_a^2 A_k, \tag{8.10}$$

and a force component aligned with $\mathbf{v}_a$, denoted as drag force

$$D = \frac{1}{2}\rho C_D v_a^2 A_k. \tag{8.11}$$

In these equations, the parameter A_k denotes the projected wing surface area, while C_L and C_D denote the aerodynamic lift and drag coefficients. Both force components constitute the resultant aerodynamic force $\mathbf{F}_a = \mathbf{L} + \mathbf{D}$ with magnitude

$$F_a = \frac{1}{2}\rho C_R v_a^2 A_k. \tag{8.12}$$

Accordingly, the resultant aerodynamic coefficient is given by

$$C_R = \sqrt{C_L^2 + C_D^2} = C_D\sqrt{1 + \left(\frac{L}{D}\right)^2}. \tag{8.13}$$

Because the resultant aerodynamic force $\mathbf{F}_a$ acts on the flying vehicle, the reaction force $-\mathbf{F}_a$ reversely acts on the flow, causing its retardation and deflection.

These fundamental aerodynamic relationships are illustrated schematically in Fig. 8.4 for the cases of untethered gliding flight of a wing at $v_w = 0$ (left), tethered flight at $l_t = $ const. and $v_k > 0$ (center) and tethered flight at $l_t = $ const. and $v_k = 0$ (right). The glide angle ε is defined on the basis of the depicted case of untethered gliding flight as the angle measured from the horizontal plane to the flight velocity vector $\mathbf{v}_k$ or, equivalently, to the apparent wind velocity vector $\mathbf{v}_a = -\mathbf{v}_k$. Because $\mathbf{F}_a$ is perpendicular to the horizontal plane and $\mathbf{L}$ is perpendicular to $\mathbf{v}_k$, the angle ε also characterizes the tilt rotation of the lift vector from the vertical, which

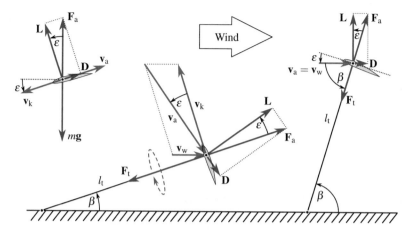

Fig. 8.4 Velocity vectors (red) and force vectors (blue) for an untethered wing in steady gliding flight at $v_k = 0$ (left), for a massless kite exposed to a wind velocity and flying at constant tether length l_t in a continuous loop (center) and at a static position (right), adapted from [15]. The elevation angle β is measured from the ground plane to the tether. The two tethered configurations depict the special case in which the tether and the wind velocity $\mathbf{v}_w$ are in the illustration plane

leads to following relationships between kinematic properties and force components

$$\frac{1}{\tan \varepsilon} = \frac{L}{D}, \tag{8.14}$$

$$\frac{1}{\sin \varepsilon} = \sqrt{1 + \left(\frac{L}{D}\right)^2}. \tag{8.15}$$

These imply that the gliding angle ε depends only on the lift-to-drag ration L/D of the wing and is unaffected by its mass m. By analyzing the vertical equilibrium between aerodynamic force $\mathbf{F}_a$ and gravitational force $m\mathbf{g}$ we can find the gliding velocity v_k of the wing as a function of its mass. We find that the heavier the wing, the faster it descends on the straight gliding trajectory described by the angle ε.

Tethered flight introduces an additional tether force $\mathbf{F}_t$ which, in contrast to the gravitational force, is not constant but adjusts itself to the aerodynamic performance and flight mode of the wing. Especially when the wind velocity v_w or the flight velocity v_k are high, the wing experiences a high apparent wind velocity v_a according to Eq. (8.9) and, as consequence, the steady force equilibrium is dominated by the aerodynamic force $\mathbf{F}_a$, calculated from Eq. (8.12), and the induced tether force $\mathbf{F}_t$. This condition is also typical for fabric membrane wings with a low mass m and for this reason we simplify the analyses of tethered flight by neglecting the gravitational force.

Loyd [18] was among the first to recognize the potential of kites to produce large traction forces that can be used for energy generation with a minimal material effort. In his analysis he distinguishes the two different modes of operation that are schematically illustrated in Fig. 8.4 for the special case of constant tether length. Loyd concludes that operation in crosswind maneuvers, as illustrated in Fig. 8.4 (center), achieves by far larger tether forces than flight at a static position, as shown in Fig. 8.4 (right). It can be shown that for operation in crosswind flight maneuvers at an elevation angle β the apparent wind speed is calculated by [15, 24]

$$v_a = v_w \cos \beta \sqrt{1 + \left(\frac{L}{D}\right)^2} = v_w \frac{\cos \beta}{\sin \varepsilon}. \tag{8.16}$$

In this equation the angle ε is used to quantify the lift-to-drag ratio according to Eq. (8.15) in generalization of the original kinematic definition for steady gliding flight. When the force equilibrium is dominated by $\mathbf{F}_a$ and $\mathbf{F}_t$, the angle ε can be interpreted kinematically as the angle between the tether normal plane and the apparent wind velocity vector $\mathbf{v}_a$. Equation (8.16) shows that the apparent wind velocity and with this also the aerodynamic force according to Eq. (8.12) are maximum for a horizontal tether ($\beta = 0$) and decreasing for increasing elevation angle. For this reason, the elevation angle can be used as an operational parameter to adjust the generated traction force of the kite and the resulting flow deflection towards the ground. The equation further indicates the substantial increase of v_a and F_a with increasing L/D.

For $v_k = 0$, the kite assumes a static position with β reaching its maximum value, v_a its minimum value

$$v_a = v_w, \tag{8.17}$$

and, consequently, also the aerodynamic force F_a reaching its minimum value. The steady force equilibrium illustrated in Fig. 8.4 (right) leads to

$$\tan \beta = \frac{L}{D}, \tag{8.18}$$

$$\cos \beta = \frac{1}{\sqrt{1 + \left(\frac{L}{D}\right)^2}}. \tag{8.19}$$

For $v_k \rightarrow 0$, Eq. (8.16) converges towards Eq. (8.17), which can be shown by inserting the limiting value of $\cos \beta$ given by Eq. (8.19) into Eq. (8.16).

From the above considerations we can conclude that the statically positioned kite can be used as a baseline solution to deflect the flow from the free stream towards the ground. For example, by assuming a glide angle of $\varepsilon = 10°$, which corresponds to a lift-to-drag ratio of $L/D = 5.67$, we can use Eq. (8.18) to calculate the elevation angle of $\beta = 80°$ for static flight. To further increase the traction force the kite can be flown in crosswind flight maneuvers at lower elevation angles. This technique, which requires additional flight control subsystems for the kites, will be used in Sect. 8.4 to intensify the entrainment of flow from the free stream.

To quantify the flow entrainment effect of the kite we define the kite power density as the product of aerodynamic force F_a and apparent wind velocity v_a divided by the projected wing surface area A_k,

$$P' = \frac{F_a v_a}{A_k}. \tag{8.20}$$

It should be noted that this definition differs from the traction power density, which is based on the reeling velocity v_t of the tether [24].

Because the geometrical dimensions of the kites by far exceed the resolution of the computational mesh they are taken into account in the flow simulations as discontinuous pressure jumps over infinitely thin surfaces specified as a function of the instantaneous inflow velocity. The aerodynamic characteristics of a three-dimensional wing section of a ram-air kite were obtained from de Wachter [30] who performed measurements of the inflated wing shape in a wind tunnel using photogrammetry and laser scanning, followed by CFD analysis of the flow past the determined shape. The computed pressure distribution is used to determine the total force on the flow for each kite. The pressure difference between the upper and lower surfaces of the wing is translated into a resultant aerodynamic force that is decomposed into lift and drag force components. The resultant aerodynamic coefficient is evaluated by numerical integration of the pressure difference between the upper and lower wing surfaces over the chord c

$$C_R = \frac{1}{c} \int_0^c \left(C_{p,l} - C_{p,u} \right) dx. \tag{8.21}$$

8.3 Numerical Simulations of an Array of Four Wind Turbines

In this section we present numerical simulations for a wind turbine array in a neutral ABL using the LES framework available in the CFD solver Fluent®. The velocity profiles generated in the precursor simulation are used as inflow conditions for all wind turbine simulations. The turbines are arranged along the main flow direction as illustrated in Fig. 8.5. To model the effect of the wind turbines on the flow field we use the fan boundary condition, which is formulated as a discontinuous pressure jump across an infinitely thin surface and is specified as a function of the normal velocity at the actuator discs. Accordingly, the thrust force of each turbine is calculated from Eq. (8.7) and applied to the flow field as indicated in Fig. 8.5. More details of this procedure are available in [22, Sect. 6-1-1]. The employed ADM-R approach accounts for both axial and tangential inductions and allows for an accurate prediction of the wind turbine wakes at a reasonable computational cost [2, 19]. The wake flow characteristics are the main focus of this study.

The effect of wind turbine loading on wake evolution is studied by applying two different thrust coefficients, corresponding to sub-optimal and to optimal loading of the turbines. The sub-optimal loading, also denoted as partial loading, is defined by $C_T' = 0.85$ and $a = 0.17$ while the optimal loading is defined by $C_T' = 2$ and

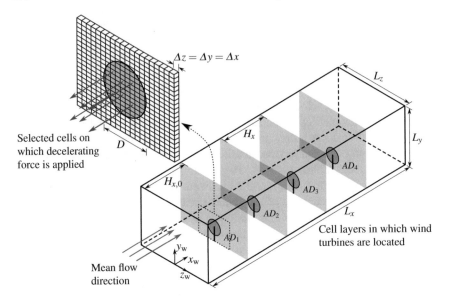

Fig. 8.5 Schematic view of the computational domain (not to scale) with the actuator disks AD_1, AD_2, AD_3 and AD_4 arranged along the main flow in x_w-direction, modified from [17]

$a = 0.33$. For smaller thrust coefficients the performance of the upstream turbines is sub-optimal which allows the downstream turbines to capture more power. Sub-optimal power extraction has already been treated in the literature with the aim to coordinate wind turbine controllers to optimize wind farm performance [3, 5]. This mode of operation was developed to account for the aerodynamic coupling by wake interaction in a group of turbines to maximize the total captured power. The geometry and mesh parameters are summarized in Table 8.1.

Table 8.1 Geometry and mesh parameters for the considered turbine load cases $C_T' = 0.85$ and $C_T' = 2$. The total number of mesh nodes is 2,894,441

Parameter name	Symbol	Value	Unit
Number of turbines	N	4	
Diameter actuator disks	D	80	m
Hub height	H_h	80	m
Inter-turbine spacing	H_x	6D	
Inflow section length	$H_{x,0}$	4D	
Length flow domain	L_x	28D	
Width flow domain	L_y	10D	
Height flow domain	L_z	10D	
Cell size	$\Delta x, \Delta y, \Delta z$	D/10	

The size of the flow domain is identical to the one used to generate the wind profiles in the precursor simulation. The wind reference frame x_w, y_w, z_w depicted in Fig. 8.5 is used to describe absolute positions in the flow domain. A uniform cell size is chosen as compromise between the required resolution of the relevant large vortex structures and the computational cost. In general, three-dimensional unsteady LES is computationally demanding and to finish within a practical timeframe the simulations need to be processed in parallel on high-performance computer clusters. Using 20 Intel® Xeon® CPU E5-2670v2 cores with 4GB of RAM per core, the time required to compute the flow statistics for a sample domain of $8 \times 8 \times 8$ cells was approximately 100 hours.

Results are presented in the normalized local reference frames $x/D, y/D, z/D$ of the individual turbines. The streamwise coordinates x/D are defined such that $x/D > 0$ refers to positions downstream and $x/D < 0$ to positions upstream of the respective turbine. The wake flow is characterized by vertical profiles of the mean streamwise velocity and the kinetic energy flux at relative downstream positions $x/D = 5$. In general, a higher wind turbine loading is associated with a larger thrust coefficient C'_T and increased energy extraction from the wind. This reduces the kinetic energy in the wake flow which can be recognized from the velocity profiles of the load cases $C'_T = 2$ and 0.85 illustrated in Fig. 8.6.

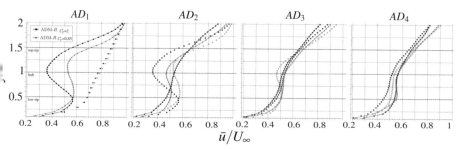

Fig. 8.6 Vertical profiles of the normalized streamwise velocity. The profiles represented by lines and symbols are computed at turbine downstream locations $x/D = 5$. The profiles represented only by symbols are computed at turbine upstream locations $x/D = -1$, with crosses denoting the load case $C'_T = 2$ and dots the load case $C'_T = 0.85$. The maximum tip height of the wind turbines is at $y/D = 1.5$, the hub height at $y/D = 1$ and the minimum tip height at $y/D = 0.5$

The wake velocity recovery reveals the gradual development of the flow towards an equilibrium state as it develops along the wind turbine array. As expected, the wake of actuator disk AD_1 shows a more pronounced velocity deficit for higher turbine loading. From AD_2 onwards the wake velocity recovers much faster, as can be seen when comparing the upstream velocity profiles at $x/D = -1$, represented by only symbols, with the downstream profiles at $x/D = 5$, represented by symbols and lines. For AD_3 and AD_4 the simulations predict a complete recovery of the wake velocity. Interestingly, this recovery is not significantly affected by the turbine loading although the amount of energy extracted from the flow strongly differs for the two load cases. The underlying reason for the accelerated wake velocity recovery

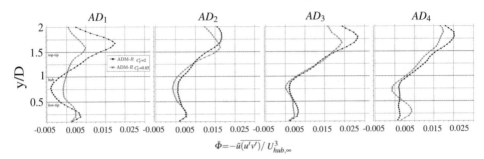

Fig. 8.7 Vertical profiles of the normalized vertical flux of mean flow kinetic energy $\overline{\Phi}$. The profiles are computed at turbine downstream locations $x/D = 5$

at higher turbine loading is the intensified mixing between the undisturbed wind field and the wind turbine canopy. This can be concluded from the profiles of the normalized vertical flux of mean flow kinetic energy $\overline{\Phi}$ depicted in Fig. 8.7 which show that for higher turbine loading the vertical momentum transport in the flow region above the turbines is increased significantly [8].

The power extracted by each turbine is calculated as $\overline{P} = T\overline{U}_d$ with the thrust force T estimated from the surface integral of the mean static pressure immediately upstream and downstream of the actuator disks. Approximately 20% more power was extracted from the flow by the turbines under optimal loading conditions.

8.4 Numerical Simulations of a Wind Turbine Array with Kites

In this section we present simulation results for an array of four wind turbines, interspersed with kites which are modeled as distributed body forces. The domain size, inter-turbine spacing, type of actuator disk model, size, positioning and aerodynamic properties of the kites are kept constant for all evaluated cases. Each of the four kites has a total wing surface area of $A_k = 470\,\mathrm{m}^2$, which corresponds to a relative size $A_k/A_d = 1/11$. A glide angle $\varepsilon = 10°$ is assumed, which, according to Fig. 8.4 (center) and Eq. (8.16), leads to a lift-to-drag ratio of $L/D = 5.67$. To model the interaction with the wind field the double-curved wing is geometrically simplified as a planar elliptical wing with a surface area equivalent to the projected area of the physical wing. Assuming an aspect ratio of $\mathcal{R} = 2$ the dimensions of the elliptical wing are calculated as 16×32 m. Each wing is placed at a vertical distance $D/4$ above and a horizontal distance $3D$ downstream of the corresponding actuator disk. The resulting configuration is illustrated in Fig. 8.8, indicating how several levels of mesh refinement are used to approximate the kite surfaces. The regions of mesh refinement are also shown in the top and isometric views. The elliptical wings can be recognized in the center of these regions.

As mentioned above the kites are not represented as solid flow boundaries, because this would require very fine meshes to resolve viscous boundary layers, but

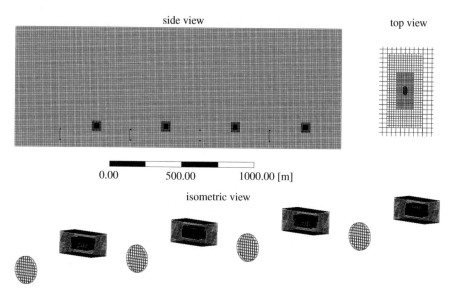

Fig. 8.8 Computational mesh of the flow domain showing several levels of refinement in the regions containing the kite surfaces. The hexahedral cells directly at the kite surface have a uniform side length of $\Delta x = 1$ m

instead are modeled as distributed body forces. These are generated by pressure jumps over infinitely thin surfaces, as derived from de Wachter [30]. The purpose of the local mesh refinement is to better approximate the desired surface area and eventually the total force on the flow (surface integral of static pressure). The four actuator discs (ADM-R) are modeled as circular infinitely thin surfaces without applying any local mesh refinement. The evaluated cases are summarized in Table 8.2.

Case	C_T'	β [°]	ΔP [Pa]	v_a [m/s]	F_a [kN]	P' [W/m^2]
(a)	0.85	80	39	7.8	18.5	307
(b)	2	80	39	7.8	18.5	307
(c)	2	75.7	79	11.1[a]	37	873[a]
(d)	2	69.7	157	15.6[a]	74	2456[a]

[a] Values that kites would experience for crosswind operation

Table 8.2 Simulation cases and parameter combinations for a wind turbine array with interspersed kites operating at $v_w = 7.8$ m/s. The kite power density is defined by Eq. (8.20) as $P' = F_a v_a / A_k$, using the listed values of F_a and v_a as well as $A_k = 470$ m^2

In cases (a) and (b) the kites are positioned statically and the effect of partial and optimal loading of the turbines, $C_T' = 0.85$ and 2, on the wake flow is studied. In cases (c) and (d) the turbines are operated at optimal loading and the effect of increasing kite power density on the wake flow is studied. To double the aerodynamic force of the individual kites and by that also the reaction forces acting on the flow from $F_a = 18.5\,\mathrm{kN}$ to $37\,\mathrm{kN}$ and in a second step further to $74\,\mathrm{kN}$ the kites are considered to perform crosswind flight maneuvers at lower elevation angles. To avoid the computationally demanding direct modeling of flight maneuvers the kites are positioned statically in the flow domain and the flow boundary condition at the wings is prescribed such that the aerodynamic force for true crosswind operation is reproduced. This is achieved by increasing the pressure jump ΔP across the wing surface from $39\,\mathrm{Pa}$ to $79\,\mathrm{Pa}$ and further to $157\,\mathrm{Pa}$.

8.4.1 Effect of Varying Wind Turbine Load

In a first step, the effect of different wind turbine loading on the wake flow is investigated and compared to the reference cases without kites, presented in Figs. 8.6 and 8.7. Similar to these, a higher wind turbine loading is associated with increased energy extraction from the flow. Figure 8.9 shows that the resulting velocity deficit

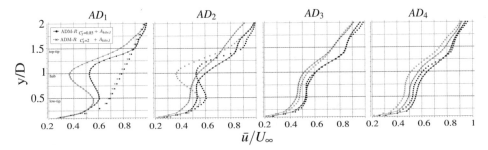

Fig. 8.9 Vertical profiles of the normalized streamwise velocity. The profiles represented by lines and symbols are computed at turbine downstream locations $x/D = 5$. The profiles represented only by symbols are computed at turbine upstream locations $x/D = -1$, with crosses denoting the load case $C_T' = 2$ and dots the load case $C_T' = 0.85$

in the wake of AD_1 is particularly pronounced for the case of higher turbine loading. Further downstream, the turbine wake flows approach an equilibrium state with less pronounced velocity differences. Wake velocity recovery for partial turbine loading stabilizes to about $0.6\overline{U}_{\infty,\mathrm{h}}$ and for optimal loading to about $0.5\overline{U}_{\infty,\mathrm{h}}$, where $\overline{U}_{\infty,\mathrm{h}}$ represents the mean free stream velocity at hub height. Interestingly, for the optimal turbine loading the introduction of kites does not affect the velocity recovery in the wakes of AD_3 and AD_4. However, for the partial loading the effect of the kites is much more pronounced and the wake velocity recovery is enhanced also for AD_4.

The profiles of the normalized vertical flux of mean flow kinetic energy $\overline{\Phi}$ are shown in Fig. 8.10. The diagrams indicate that the wind turbine loading does not significantly affect the vertical transport of momentum in the wakes of the actuator

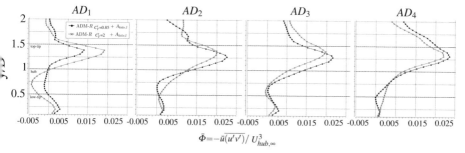

Fig. 8.10 Vertical profiles of the normalized vertical flux of mean flow kinetic energy $\overline{\Phi}$. The profiles are computed at turbine downstream locations $x/D = 5$

disks AD_2, AD_3 and AD_4. When compared to the reference cases shown in Fig. 8.7, the magnitude of the kinetic energy flux is on average four times larger in the wake region, independent of the load cases.

At partial loading of the turbines the use of kites increases the energy extraction from the flow by 14% and at optimal loading by 5.5%. Correspondingly, the turbine array efficiency increases by 8.2% and 2.5%, respectively. Because of the low efficiency gain at optimal loading we investigate the effect of increasing kite power density on the wake flow of the turbine array.

8.4.2 Effect of Varying Kite Power Density

To enhance the flow deflection towards the ground surface and, as consequence, to intensify the flow entrainment into the turbine canopy the kites are considered to perform crosswind flight maneuvers at decreased elevation angles. In this way, the resulting force per kite is doubled in two steps from $F_a = 18.5\,\mathrm{kN}$, which is the value for static flight, to $37\,\mathrm{kN}$ and then to $74\,\mathrm{kN}$, while operating the turbine at optimal loading and maintaining a constant wind velocity of $v_w = 7.8\,\mathrm{m/s}$.

As illustrated in Fig. 8.8 the kites are represented as distributed body forces and are resolved by several levels of local mesh refinement. A direct modeling of flight maneuvers would substantially increase the complexity of the simulations. Because the main interest of this study is the integral force that the kites exert on the turbine wake flows we model the flying kites as static objects and adjust the flow boundary condition such that the prescribed force per kite, as listed in Table 8.2, is generated. Calculation basis of this approach, which is outlined in [22, Appendix C], is the discrete pressure distribution of a ram-air kite adopted from [30]. The integral aerodynamic force of each wing is determined from the discrete pressure distribution along its upper and lower surfaces according to Eq. (8.21). To reach the target values $F_a = 37\,\mathrm{kN}$ and $74\,\mathrm{kN}$ we use an iterative procedure to determine the apparent velocities of $v_a = 11.1\,\mathrm{m/s}$ and $15.6\,\mathrm{m/s}$. The fan boundary condition in Fluent®is used with constant pressure jumps of $\Delta P = 79\,\mathrm{Pa}$ and $157\,\mathrm{Pa}$, respectively. Keep-

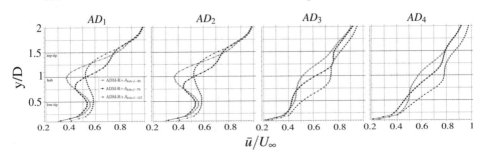

Fig. 8.11 Vertical profiles of the normalized streamwise velocity computed at turbine downstream locations $x/D = 5$ for the actuator disks under optimal load $C_T' = 2$. The three profiles show the effect of different kite power densities on the wake velocity. They are indexed by the different values of the pressure jump ΔP listed in Table 8.2

ing the glide angle $\varepsilon = 10°$ constant and, as consequence also the lift-to-drag ratio $L/D = 5.67$, the required values of the elevation angle β are calculated from Eq. (8.16) as $\beta = 75.7°$ and $69.7°$.

The influence of increasing kite power density on the wake velocity is illustrated in Fig. 8.11. As expected, a stronger deflection of the flow towards the ground surface also leads to a faster recovery of the wake velocity. Doubling the aerodynamic forces of the kites results in a 20% faster velocity recovery at the hub height $y/D = 1$ and downstream $x/D = 5$ of the actuator disks AD_3 and AD_4. Quadrupling the aerodynamic forces results in a 50% faster velocity recovery.

Interesting conclusions can be drawn from the normalized vertical flux of mean flow kinetic energy depicted in Fig. 8.12. The diagrams indicate that increased kite power densities intensify the fluctuations of the vertical momentum transport and as consequence expand the affected flow cross section. Similar trends are expected for the shear stress and the turbulence kinetic energy transport. For higher kite power densities, horizontal oscillations in the magnitude of vertical momentum flux are identified as regions of high turbulent kinetic energy production (oscillations above hub height) and dissipation (oscillations below hub height) which is shown to en-

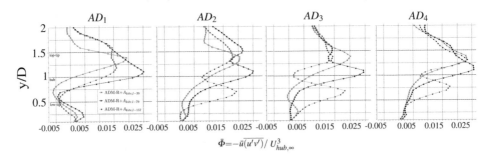

Fig. 8.12 Vertical profiles of the normalized vertical flux of mean flow kinetic energy $\overline{\Phi}$ computed at turbine downstream locations $x/D = 5$ for the actuator disks under optimal load $C_T' = 2$. The three profiles show the effect of different kite power densities

hance the re-energizing of the wake flow [32]. A significantly higher kinetic energy flux in the wake region is translated into more energy available for extraction by the downstream turbines.

At the maximum kite power density $P' = 2456\,\text{W}/\text{m}^2$, a 24% increase in power production and a 15% increase in conversion efficiency is predicted compared to the base case. The use of kites with power density $P' = 873\,\text{W}/\text{m}^2$ results in a 12% increase of extracted power and a conversion efficiency increase of approximately 10% relative to the base case.

8.5 Conclusions

The presented computational study investigates the flow entrainment effect of large kites on the operational characteristics of a conventional wind farm. The considered farm configurations are simplified and many practical aspects concerning the flight operation of kites and wind turbine control schemes are neglected. The large eddy simulations clearly show that the positioning of tethered wings within the wind turbine array significantly affects the spatial distribution of the mean velocity deficit and the vertical kinetic energy flux in the turbine wake flows. The stronger the traction forces of the kites and, accordingly, the reaction forces acting on the flow above the wind turbine canopy, the higher the vertical kinetic energy flux, the faster the wake velocity recovery and the higher the energy extraction efficiency of the wind farm. In all investigated cases the highest values of the turbulent flux are found in the upper regions of the wake flows where the mean shear is maximum. This finding is in line with existing literature [32]. Due to the promising results it is expected that further studies will investigate the effect of altering the wake flow characteristics using downwash-generating devices such as kites.

Acknowledgements The authors would like to thank the anonymous reviewers for their helpful comments and gratefully acknowledge the contribution of Dhruv Mehta and Lorenzo Lignarolo for their expert advice on the wide range of topics treated in the present work.

References

1. Ahmed, M., Hably, A., Bach, S.: High Altitude Wind Power Systems: A Survey on Flexible Power Kites. In: Proceedings of the 2012 XXth International Conference on Electrical Machines, pp. 2085–2091, Marseille, France, 2–5 Sept 2012. doi: 10.1109/ICElMach.2012.6350170
2. Ammara, I., Leclerc, C., Masson, C.: A viscous three-dimensional differential/actuator-disk method for the aerodynamic analysis of wind farms. Journal of Solar Energy Engineering **124**(4), 345–356 (2002). doi: 10.1115/1.1510870
3. Annoni, J., Seiler, P., Johnson, K., Fleming, P., Gebraad, P.: Evaluating wake models for wind farm control. In: Proceedings of the 2014 American Control Conference, pp. 2517–2523, IEEE, Portland, OR, USA, 4–6 June 2014. doi: 10.1109/ACC.2014.6858970

4. Barthelmie, R. J., Hansen, K., Frandsen, S. T., Rathmann, O., Schepers, J. G., Schlez, W., Phillips, J., Rados, K., Zervos, A., Politis, E. S., Chaviaropoulos, P. K.: Modelling and measuring flow and wind turbine wakes in large wind farms offshore. Wind Energy 12(5), 431–444 (2009). doi: 10.1002/we.348
5. Bitar, E., Seiler, P.: Coordinated control of a wind turbine array for power maximization. In: Proceedings of the 2013 American Control Conference, pp. 2898–2904, IEEE, Washington, DC, USA, 17–19 June 2013. doi: 10.1109/ACC.2013.6580274
6. Boonman, D., Broich, C., Deerenberg, R., Groot, K., Hamraz, A., Kalthof, R., Nieuwint, G., Schneiders, J., Tang, Y., Wiegerink, J.: Wind Farm efficiency. S12, Delft University of Technology, Delft, Netherlands, 2011. http://resolver.tudelft.nl/uuid:9ec5078c-c02a-43db-99b9-00bde8871b58
7. Burton, T., Jenkins, N., Sharpe, D., Bossanyi, E.: Wind Energy Handbook. 2nd ed. John Wiley & Sons, Ltd, Chichester (2011). doi: 10.1002/9781119992714
8. Cal, R. B., Lebrón, J., Castillo, L., Kang, H. S., Meneveau, C.: Experimental study of the horizontally averaged flow structure in a model wind-turbine array boundary layer. Journal of Renewable and Sustainable Energy 2(1), 013106 (2010). doi: 10.1063/1.3289735
9. Calaf, M., Meneveau, C., Meyers, J.: Large eddy simulation study of fully developed wind-turbine array boundary layers. Physics of Fluids 22, 015110 (2010). doi: 10.1063/1.3291077
10. Castro, I.: CFD for external aerodynamics in the built environment. The QNET-CFD Network Newsletter 2(2), 4–7 (2003)
11. DNV GL: Wind farm design tool WindFarmer. https://www.dnvgl.com/services/windfarmer-3766. Accessed 1 Nov 2016
12. Dobrev, I., Massouh, F., Rapin, M.: Actuator surface hybrid model. In: Journal of Physics: Conference Series, vol. 75, 1, p. 012019, (2007). doi: 10.1088/1742-6596/75/1/012019
13. DTU Wind Energy: Wind Atlas Analysis and Application Program (WAsP). http://www.wasp.dk/. Accessed 1 Nov 2016
14. EMD International A/S: Software package windPRO. http://www.emd.dk/windpro/. Accessed 1 Nov 2016
15. Fritz, F.: Application of an Automated Kite System for Ship Propulsion and Power Generation. In: Ahrens, U., Diehl, M., Schmehl, R. (eds.) Airborne Wind Energy, Green Energy and Technology, Chap. 20, pp. 359–372. Springer, Berlin Heidelberg (2013). doi: 10.1007/978-3-642-39965-7_20
16. Jarrin, N., Benhamadouche, S., Addad, Y., Laurence, D.: Synthetic turbulent inflow conditions for large eddy simulation. In: Proceedings of the 4th International Turbulence, Heat and Mass Transfer Conference, Antalya, Turkey, 12–17 Oct 2003
17. Jimenez, A., Crespo, A., Migoya, E., Garcia, J.: Advances in large-eddy simulation of a wind turbine wake. In: Journal of Physics: Conference Series, vol. 75, p. 012041, (2007). doi: 10.1088/1742-6596/75/1/012041
18. Loyd, M. L.: Crosswind kite power. Journal of Energy 4(3), 106–111 (1980). doi: 10.2514/3.48021
19. Madsen, H. A.: A CFD analysis of the actuator disc flow compared with momentum theory results. In: Proceedings of the 10th IEA Symposium on the Aerodynamics of Wind Turbines, pp. 109–124, Edinburgh, UK, 16–17 Dec 1996
20. Meyers, J., Meneveau, C.: Large eddy simulations of large wind-turbine arrays in the atmospheric boundary layer. AIAA Paper 2010-827. In: Proceedings of the 48th AIAA Aerospace Sciences Meeting, Orlando, FL, USA, 4–7 Jan 2010. doi: 10.2514/6.2010-827
21. Meyers, J., Meneveau, C.: Optimal turbine spacing in fully developed wind farm boundary layers. Wind Energy 15(2), 305–317 (2012). doi: 10.1002/we.469
22. Ploumakis, E.: Improving the Wind Farm efficiency by simple means – LES study of a wind turbine array with tethered kites. M.Sc.Thesis, Delft University of Technology, 2015. http://resolver.tudelft.nl/uuid:a273b26a-b4bd-4388-8eee-193c6d22626b
23. Ploumakis, E., Mehta, D., Lignarolo, L., Bierbooms, W.: Enhanced Kinetic Energy Entrainment in Wind FarmWakes – LES Study of a Wind Turbine Array with Tethered Kites. In: Schmehl, R. (ed.). Book of Abstracts of the International Airborne Wind Energy Conference

2015, p. 52, Delft, The Netherlands, 15–16 June 2015. doi: 10.4233/uuid:7df59b79-2c6b-4e30-bd58-8454f493bb09. Presentation video recording available from: https://collegerama.tudelft.nl/Mediasite/Play/383929c21f734297ba296717de3d17371d

24. Schmehl, R., Noom, M., Vlugt, R. van der: Traction Power Generation with Tethered Wings. In: Ahrens, U., Diehl, M., Schmehl, R. (eds.) Airborne Wind Energy, Green Energy and Technology, Chap. 2, pp. 23–45. Springer, Berlin Heidelberg (2013). doi: 10.1007/978-3-642-39965-7_2

25. Smirnov, A., Shi, S., Celik, I.: Random flow generation technique for large eddy simulations and particle-dynamics modeling. Journal of Fluids Engineering **123**(2), 359–371 (2001). doi: 10.1115/1.1369598

26. Stevens, R. J. A. M., Gayme, D. F., Meneveau, C.: Effects of turbine spacing on the power output of extended wind-farms. Wind Energy **19**(2), 359–370 (2016). doi: 10.1002/we.1835

27. Stevens, R. J. A. M., Gayme, D. F., Meneveau, C.: Large eddy simulation studies of the effects of alignment and wind farm length. Journal of Renewable and Sustainable Energy **6**(2) (2014). doi: 10.1063/1.4869568

28. Troldborg, N.: Actuator line modeling of wind turbine wakes. Ph.D. Thesis, Technical University of Denmark, 2009. http://orbit.dtu.dk/files/5289075/niels_troldborg.pdf

29. Troldborg, N., Sorensen, J. N., Mikkelsen, R.: Actuator Line Simulation of Wake of Wind Turbine Operating in Turbulent Inflow. Journal of Physics: Conference Series **75**, 012063 (2007). doi: 10.1088/1742-6596/75/1/012063

30. Wachter, A. de: Deformation and Aerodynamic Performance of a Ram-Air Wing. M.Sc.Thesis, Delft University of Technology, 2008. http://resolver.tudelft.nl/uuid:786e3395-4590-4755-829f-51283a8df3d2

31. Wright, N. G., Easom, G. J.: Non-linear k-ε turbulence model results for flow over a building at full-scale. Applied Mathematical Modelling **27**(12), 1013–1033 (2003). doi: 10.1016/S0307-904X(03)00123-9

32. Wu, Y.-T., Porté-Agel, F.: Atmospheric turbulence effects on wind-turbine wakes: An LES study. Energies **5**(12), 5340–5362 (2012). doi: 10.3390/en5125340

33. Wu, Y.-T., Porté-Agel, F.: Simulation of turbulent flow inside and above wind farms: Model validation and layout effects. Boundary-layer meteorology **146**(2), 181–205 (2013). doi: 10.1007/s10546-012-9757-y

Part II
Control, Optimization & Flight State Measurement

Chapter 9
Automatic Control of Pumping Cycles for the SkySails Prototype in Airborne Wind Energy

Michael Erhard and Hans Strauch

Abstract The efficient and economic operation of tethered kites for accessing high-altitude winds as a renewable source of energy requires fully automated setups. During the last decade the SkySails kite systems have been developed for applications in marine propulsion and energy generation. In this chapter we give a descriptive overview of the flight control of the tethered kite and of the control of the tether reeling speed leading to pumping cycles for energy generation. This chapter focuses on the discussion and justification of the overall design choices, functional dependencies and the presentation of the complete system in a self-contained way. For details of the mathematical modeling and control theoretical aspects references for further reading are provided. After an introduction, the dynamical model for the tethered dynamics is briefly summarized. Subsequently, the estimation and sensor system is presented. Then, the control approach is discussed and the parts of the control system are reviewed in detail. Finally the implemented system is briefly compared to other control approaches.

9.1 Introduction

More than thirty years ago [15] energy generation using tethered wings has been proposed for the first time. Since then a great interest in this kind of renewable energy source emerged, especially during the last decade. The introduction of this vision on the market has to take into consideration that the economic operation of

Michael Erhard (✉) · Hans Strauch
SkySails GmbH, Luisenweg 40, 20537 Hamburg, Germany
e-mail: michael.erhard@haw-hamburg.de

Michael Erhard
Systems Control and Optimization Laboratory, Department of Microsystems Engineering
(IMTEK), University of Freiburg, Georges-Köhler-Allee 102, 79110 Freiburg, Germany

© Springer Nature Singapore Pte Ltd. 2018
R. Schmehl (ed.), *Airborne Wind Energy*, Green Energy
and Technology, https://doi.org/10.1007/978-981-10-1947-0_9

airborne wind energy plants demands for reliable and fully automatic operation of the power generation process. Thus, numerous theoretical control proposals [1–3, 10, 12] as well as experimental implementations have been published [6, 7, 11, 13, 14]. However, the robust autonomous operation of complete energy production cycles turns out to be quite challenging, especially as optimization of energy output, i.e. performance and robustness, often appear as opposing design goals. Furthermore, the design process for feasible control systems is strongly influenced by consideration of limitations dictated by real world circumstances. We are convinced, that simplicity, separation of problems and modular structure, grounded in a clear understanding of the physical basis of the controlled plant, are keys to success in mastering the high perturbations and significant uncertainties, which are inevitably coming along with the natural energy resource wind.

This chapter reports on results achieved at the SkySails company, which was founded in 2001 and started with the development of an auxiliary propulsion system for seagoing vessels as shown in Fig. 9.1 in order to save fuel. From the first experiments with kites of 6-10 m^2 up to the latest product generation with a nominal kite size of 320 m^2, several 100 hours of operational experience have been gained during the last decade.

Fig. 9.1 SkySails marine propulsion system. The vessel on this figure with a length of 132 m utilizes kites of sizes up to 320 m^2

In 2011, SkySails started a further business segment using the developed kite systems for airborne wind energy generation. For that purpose, a functional prototype based on the nominal kite size of 30 m^2 was set up as depicted in Fig. 9.2. The kite is controlled by steering lines, which are pulled by an actuator placed in a control pod. The pod is directly located under the kite. This geometry allows for a *single* main towing line, consisting of 6 mm diameter high-performance Dyneema® rope, which connects the flying system to the ground station and transfers the aerodynamic forces. The prototype features 450 m of tether length on the main winch, which is attached to a 50 kW electrical motor/generator-combination.

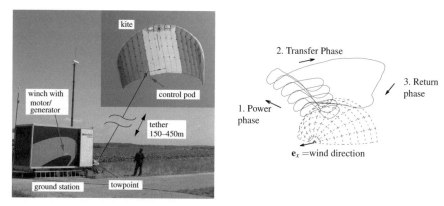

Fig. 9.2 a SkySails functional prototype setup for power generation based on pumping cycles using kites of sizes ranging from 20 to 40 m^2 (30 m^2 shown here). A tether of length 150–400 m transfers forces from the airborne system to a ground-based winch, which is attached to a motor/generator. **b** Curve of an experimentally flown pumping cycle. The control task of fully autonomous operation of such pumping cycles is the main subject of this chapter

    The principle of energy generation is depicted in Fig. 9.2 and shall be briefly summarized: during the power phase, the kite is flown dynamically in lemniscates (figures of eight). This dynamical crosswind flight leads to high tether forces. By reeling out the tether, electrical energy is produced. At a certain line length, the kite is steered towards the neutral zenith position. At this low-force position, the tether is reeled in (return phase) consuming a certain portion of the previously generated energy. Finally, a reasonable amount of generated net power remains. It should be emphasized that the functional prototype is a vehicle optimally suited for carrying out fast development cycles in order to achieve full automation of efficient power generation. However, further future development steps clearly aim at scaling up the system and at utilizing the existing large kite systems for energy generation.

    This chapter aims at providing an overview of the system architecture and the control structures which have been established during the last 10 years while developing the SkySails systems. The focus is put on drawing a conceptual picture and summarizing the whole control design comprising model, sensors, estimation and feedback control structure. Especially relations between these parts and implementation challenges shall be elaborated in order to give an overall idea of design choices to the reader. As a consequence, we will restrict ourselves to a bird eye's view on the established system omitting many technical details and in-depth discussion of experimental data. The interested reader is kindly referred to [5–8] for further information.

    The manuscript is organized as follows: first, the dynamical model for the tethered kite dynamics is depicted and summarized by discussion of the equations of motion. Subsequently, an overview of the sensor system is given and key algorithms are briefly explained. Then, the structure of the complete control system is justified before explaining details of the single stages of the control system. Finally, a classification with respect to alternative control approaches shall be given. The preliminary content of the present chapter has been presented at the Airborne Wind Energy Conference 2015 [9].

9.2 Kite System Dynamics

A dynamic model of the tethered kite dynamics is a prerequisite for an adequate design of the control system. The approach was to keep the model as simple as possible, but still covering the main dynamics. This strategy was chosen for the following reasons: The simplicity allows for an intuitive interpretation of the dynamics and provides motivation for the selected approaches and design decisions of the control system. Further, measurement data from operational flight trials are subject to non-stationary and stochastic disturbances coming along with the wind, mainly gusts. As a consequence, only models with few parameters can be reliably estimated and validated.

In the following, the coordinate system of the model shall be introduced. Then a descriptive explanation of the dynamics will be provided before the equations of motion (EQM) are discussed in detail.

9.2.1 Coordinate System

A definition of the coordinate system for the model is depicted in Fig. 9.3. The

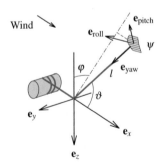

Fig. 9.3 Coordinate system definitions of the fixed inertial reference vectors $\mathbf{e}_x, \mathbf{e}_y, \mathbf{e}_z$ and the body frame vectors $\mathbf{e}_{roll}$, $\mathbf{e}_{pitch}$ and $\mathbf{e}_{yaw}$. The position is determined by the two angles φ, ϑ and the tether length l. The wind vector is given in $\mathbf{e}_x$-direction.

position of the kite is given by

$$\mathbf{r} = l \begin{bmatrix} \cos \vartheta \\ \sin \varphi \sin \vartheta \\ -\cos \varphi \sin \vartheta \end{bmatrix} \tag{9.1}$$

and is defined by the two angles φ, ϑ and the tether length l. As mathematical definitions based on rotation matrices [5] are quite abstract, a comparison to the familiar earth coordinate system shall be given as depicted in Fig. 9.4.

Fig. 9.4 Demonstrative explanation of the defined Euler angles by comparing them to angles used for navigation on earth. For the chosen coordinate definition, the earth's rotation axis has been tilted by 90 degrees with the north pole towards the wind direction. The angles φ, ϑ correspond to the position on earth and ψ to the bearing. The exact mapping of angles is given in Table 9.1

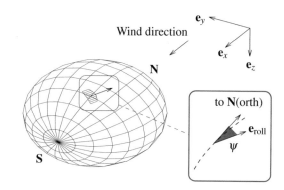

Neglecting changes in tether length for a first consideration, the tethered dynamics could be related to a motion of a vehicle on the earth surface. The two angles φ, ϑ determine the position and correspond to longitude λ_{earth} and latitude ϕ_{earth}, respectively. The ψ angle corresponds to the heading δ_{earth} of the vehicle with respect to the north direction.

In the given geometry, the wind is blowing from north to south, hence 'heading north' means heading against the wind in this comparison. For sake of completeness, a mapping of the angles is summarized in Table 9.1.

Angle in model	'earth' navigation quantity	exact relation
φ	Earth longitude	$-\lambda_{earth}$
ϑ	Earth latitude	$\phi_{earth} + \pi/2$
ψ	Bearing (direction w.r.t. north pole)	δ_{earth}

Table 9.1 Mapping of the Euler angles of the tethered system to earth navigation quantities (compare Fig. 9.4)

Finally, a justification for the specific choice of the coordinate system's symmetry axis shall be given. Choosing the symmetry axis in wind direction has two advantages: first, this leads to a simple set of equations of motion as the main dynamics is induced by the wind. Second, the singularity at the north pole lies on the water or earth surface and is therefore never reached as long as the kite control system is working at all by keeping the system in the air.

9.2.2 Overview on Dynamics

In this subsection, the dynamics shall be illustrated in a pictorial representation before discussing the EQM in the next subsection. Regarding the system as control plant, there are two control inputs: the steering command δ and winch speed $v^{(\text{winch})}$. The dynamics can schematically be represented as shown in Fig. 9.5.

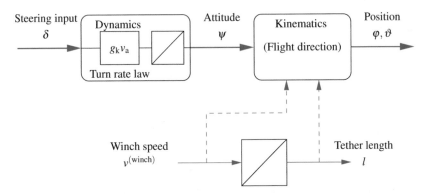

Fig. 9.5 Overview on the kite system dynamics, which can be split up into two subsystems: the first, given by the turn rate law, describes the response in kite yaw-attitude due to an steering deflection δ. The second is the kinematics and determines the change of position as a function of attitude. The winch behavior is simply given by an integration as drawn in the lower row

The kite system dynamics can be split up into two subsystems: first, the steering input δ leads to a turn rate and thus change of yaw-attitude or yaw-orientation. This dynamics is given by the turn rate law (TRL), which basically describes an integrator scaled by the air path speed. The TRL will be further elaborated in the next subsection.

Subsequently, the attitude ψ determines the flight direction as given by the kinematics. Basically, the flight direction is approximately determined by the $\mathbf{e}_{\text{roll}}$-axis of the kite. The deviation between these two quantities due to the ambient wind will be discussed in the next section. Furthermore, the control input $v^{(\text{winch})}$ determines the change of tether length.

Flight speed and forces can be understood by the wind window concept as explained in Fig. 9.6. The reeling speed can then be regarded as a deformation of the wind window. This effect can be explained by the additional apparent wind vector component in tether direction due to the winch speed, see Fig. 9.7.

It is important to note, that 'winching in' allows the kite to enter geometric regions, which are not reachable in the 'no winch' case. A sudden stop of the winch in such a situation would lead to an angle of attack outside of the safe aerodynamic range of operation and most likely to a collapse of the kite. Hence, either the winch speed has to be decreased slowly such that the kite can follow in an quasi-static manner or the kite has to be to be flown actively into the nominal wind window

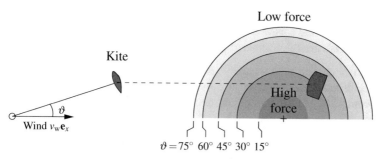

a Side view   **b** View in wind direction

Fig. 9.6 Interpretation of the wind window. **a** The wind window position angle ϑ is defined by the angle between the tether (vector in direction to the kite) and the ambient wind direction vector. **b** From the view in wind direction, different values for the wind angle correspond to semicircles. For the outer circle or higher values of ϑ, flight speed and tether force are low. The inner regions feature high flight speeds and tether forces

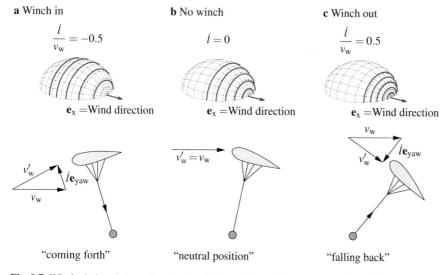

a Winch in   **b** No winch   **c** Winch out

$$\frac{i}{v_w} = -0.5 \qquad i = 0 \qquad \frac{i}{v_w} = 0.5$$

$\mathbf{e}_x$ =Wind direction $\mathbf{e}_x$ =Wind direction $\mathbf{e}_x$ =Wind direction

"coming forth"   "neutral position"   "falling back"

Fig. 9.7 Wind window deformation due to winching. The wind speed vector $\mathbf{v}'_w$, which determines the dynamics in the polar angles φ, ϑ is composed by the ambient wind vector $\mathbf{v}_w$ and the apparent wind vector due to winching $i\mathbf{e}_{yaw}$. The solid rings on the spheres correspond to apparent wind window angles (angles between tether and direction $\mathbf{v}'_w$) of $\vartheta' = 15°, 30°, 45°, 60°$ and $75°$. **a** Winching in increases the geometric size of flyable wind window and leads to a coming forth against the wind of a stationary flown kite. **b** No winch for reference purpose. **c** Winching out decreases the size of the flyable wind window and leads to a falling back of a stationary flown kite

before stopping the winch. In our case the latter is done in order to exit the return phase.

9.2.3 Equations of Motion

The derivation of the equations of motion (EQM) is based on the following assumptions, which are briefly summarized here:

- Aerodynamic forces are usually large compared to gravitational forces. Masses and their acceleration effects are therefore neglected. This assumption reduces the complexity of the model and results in a four dimensional state vector $\mathbf{x} = [\varphi, \vartheta, \psi, l]^\top$ with two controls $\mathbf{u} = [\delta, v^{(\text{winch})}]^\top$.
- The kite is assumed to always fly in its aerodynamic equilibrium, which means that the ratio between the air flow components in $\mathbf{e}_{\text{roll}}$- and $\mathbf{e}_{\text{yaw}}$-directions is given by the lift-to-drag (glide) ratio E.
- No side slip occurs, i.e. absence of an air flow component in $\mathbf{e}_{\text{pitch}}$-direction.
- Steering behavior is added by the turn rate law (TRL) as explained below.

Using these assumptions together with the introduced geometry of Chap. 9.2.1, the following EQM can be derived as conducted in detail in [5, 6]:

$$\dot{\psi} = g_k\, v_a\, \delta + \dot{\varphi} \cos \vartheta \tag{9.2}$$

$$\dot{\varphi} = -\frac{v_a}{l \sin \vartheta} \sin \psi \tag{9.3}$$

$$\dot{\vartheta} = -\frac{v_w}{l} \sin \vartheta + \frac{v_a}{l} \cos \psi \tag{9.4}$$

$$\dot{l} = v^{(\text{winch})} \tag{9.5}$$

Here, the air path speed of the kite v_a, defined as the apparent wind in opposition to the $\mathbf{e}_{\text{roll}}$-direction, is given by

$$v_a = v_w E \cos \vartheta - \dot{l} E \tag{9.6}$$

The set of equations involves three parameters: lift-to-drag (glide) ratio E, steering response proportionality constant g_k and ambient wind speed v_w. The tether force reads

$$F_{\text{tether}} = \frac{\rho A C_R}{2} \frac{1+E^2}{E^2} v_a^2 \tag{9.7}$$

with density of air ρ, projected kite area A and aerodynamic force coefficient C_R.

In the following, the EQM shall be interpreted based on the block structure given in Fig. 9.5. Equation (9.2) basically contains the turn rate law $\dot{\psi}' = g_k v_a \delta$, which states, that the turn rate around the $\mathbf{e}_{\text{yaw}}$-axis is proportional to the steering δ and the air path speed v_a of the kite. It should be noted, that this TRL has been experimentally shown and validated for different kites [6, 11, 14]. A theoretic derivation from first principles can be found in [11]. The term $\dot{\varphi} \cos \vartheta$ incorporates a correction due to the fact of tethered dynamics as explained in [5, 6].

The equations (9.3) and (9.4) describe the dynamics on the sphere. For a proper interpretation, we introduce the kinematic flight direction on the sphere (respective in the tangent plane) as velocity angle γ as shown in Fig. 9.8 with

$$\gamma \doteq \arctan(-\dot{\varphi}\sin\vartheta, \dot{\vartheta}) \tag{9.8}$$

Inserting the equations of motion Eqs. (9.3–9.4) yields

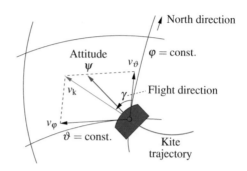

Fig. 9.8 Definitions of attitude ψ and flight direction γ, which is defined as angle in the tangent plane. The relation between attitude and flight direction is given by Eq. (9.9). For crosswind flight $\gamma \approx \psi$ can be assumed

$$\gamma = \arctan\left(\sin\psi, \cos\psi - \frac{1}{E}\frac{v_w\sin\vartheta}{(v_w\cos\vartheta - \dot{l})}\right) \tag{9.9}$$

This equation quantifies the relation between attitude ψ and velocity angle γ. It can be observed, that for crosswind flight, i.e. for small ϑ values, the fraction term in Eq. (9.9) can be neglected, which results in $\gamma \approx \psi$, or in other words: for crosswind flight, the flight direction coincides with the attitude of the kite.

It should be mentioned, that unmodeled dynamics, which are caused by kite inertia, non-stationary aerodynamic effects and delays due to the number of involved hard- and software components, sum up to a combined effect of a delay like phenomena with a typical time-scale of some tenths of a second. These experimentally measured effects are neglected in Eq. (9.2), but they are explicitly considered in the control design as limits on the achievable closed loop bandwidth. Additionally, an appropriate delay is taken into account in the feedforward blocks, compare Fig. 9.15 and Sect. 9.4.1.

We would like to emphasize, that regardless of the model simplicity, most of the experimentally observed effects are reproduced by the given model [5]. This can easily be verified by comparing the 20 kN design load of a typical airborne system with a kite size of 30 m^2 with its mass, which is well below 20 kg including tether. Hence for crosswind flight, the contribution of gravitational forces is a few percent. For the neutral flight phases of the pumping cycles discussed in this paper, the model accuracy is still adequate for control design purposes. At very short tether lengths and in order to extend the operational range towards lower wind speeds, extensions and corrections to the model may be necessary. However, these are neglected in this overview for sake of clarity.

As a final note, it should be remarked, that the model state can be represented based on quaternions [4], which are advantageous for applications in simulations and optimal control problems.

9.3 Sensors and Estimation

The successful automation and control of complex systems strongly depends on a robust and accurate sensing and estimation of the controlled variables. This fact is often underestimated as can be recognized by the scarcity of publications on sensor estimation for airborne wind energy systems. Those are opposed by a huge number of publications presenting sophisticated control proposals, often based on the assumption, that a high-fidelity estimate of a complex state vector is directly available.

A major criterion for the choice of sensors and for the estimation algorithms is robustness in rough conditions, to which the kite systems are steadily exposed in their outdoor and offshore sites of operation. In the following, the sensor and filtering system, successfully applied during the last 10 years, will be presented and some details of the estimation algorithms and design challenges discussed.

9.3.1 Sensor Setup

The complete sensor system is schematically depicted in Fig. 9.9. The sensor can be grouped into airborne and ground sensors. The heart of airborne sensors, which are located in the control pod, is the inertial measurement unit (IMU), consisting of three turn rate sensors (ω_s) and three accelerometers ($\mathbf{a}_s$). The air path speed v_a is measured by an impeller anemometer. Further, the tether force is sensed by a strain gauge and flight altitude by a barometer. The ground sensors comprise angular sensors sensing the tether direction at the tow point and an anemometer for ambient wind speed v_w and direction. The reeled out tether length l is determined by a rotary encoder at the winch drum. A more detailed description of hardware and software architecture can be found in [17].

In order to compute quantities for the controller, the raw sensor values have to be processed in estimation algorithms and combined by sensor fusion. The general and simplified structure for the angles φ, ϑ and ψ is depicted in Fig. 9.10. The ψ-angle is computed by inertial navigation referenced to an inertial frame, spanned by the gravity in $\mathbf{e}_z$ and the estimated wind direction in $\mathbf{e}_x$. Details are explained in the next subsection. In order to obtain φ and ϑ, the angles sensed at the tow point have to be referenced to the wind direction, too. In principle, the wind vector is measured at the ground unit. However, at flight altitude, the wind speed and direction determining the dynamics of the system, might deviate significantly. Thus, a wind estimator is used to provide an appropriate reference.

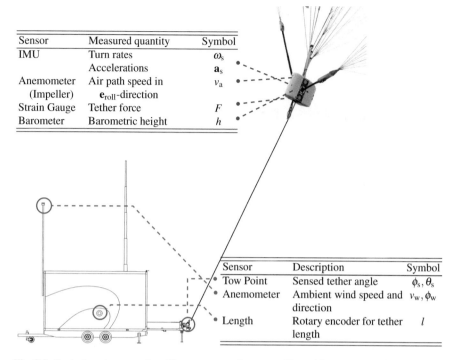

Sensor	Measured quantity	Symbol
IMU	Turn rates	ω_s
	Accelerations	$\mathbf{a}_s$
Anemometer (Impeller)	Air path speed in $\mathbf{e}_{roll}$-direction	v_a
Strain Gauge	Tether force	F
Barometer	Barometric height	h

Sensor	Description	Symbol
Tow Point	Sensed tether angle	ϕ_s, θ_s
Anemometer	Ambient wind speed and direction	v_w, ϕ_w
Length	Rotary encoder for tether length	l

Fig. 9.9 Control system overview. The sensors can be grouped into airborne sensors located in the control pod and sensors installed at the ground unit. It should be kept in mind, that sensor information and actuator commands have to be synchronized in a consistent way between distributed components of the overall control system

9.3.2 Estimation

A sophisticated feature is the estimation of the orientation based on the IMU sensor values. The heart of the algorithm is the integration of turn rates to an orientation in the inertial frame, implemented in quaternion formulation. However, biases in turn rate sensors accumulate to a drift of the estimation. Therefore, referencing of the orientation to an external reference is an important issue. Referencing to the down-direction can be accomplished by making use of the accelerometer measurements. Since the kite system is tethered and hence restricted to a certain area $\|\mathbf{r}\| < r_{max}$, the mean acceleration over longer times must be zero $\langle \ddot{\mathbf{r}} \rangle = \mathbf{0}$. This can be easily justified by consideration of a non-zero mean acceleration value leading to an increasing velocity violating the area restriction. As the accelerometers measure the sum of kinematic acceleration and gravity, transforming the accelerometer measurements to the inertial system (INS) and time averaging yields the gravity vector, i.e. $\langle R\mathbf{a}_s \rangle \approx -g\mathbf{e}_z$. This average can be used in the estimation algorithm for referencing and estimation of the gyro offset rates [8]. A major challenge are acceleration values of multiple of earth gravity g occurring in $\mathbf{e}_{pitch}$-direction during curve flights.

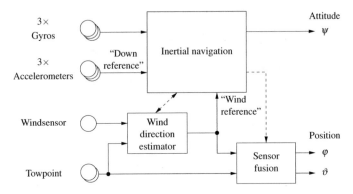

Fig. 9.10 Simplified illustration of sensor fusion and estimation. The central algorithm is the inertial navigation, which estimates the attitude ψ of the kite. The position $[\varphi, \vartheta]$ is determined based on the towpoint sensors. As all angles are referenced to the wind direction, a wind direction estimator provides a reliable reference. In addition, some information is exchanged between the modules as indicated by the dashed lines

These have to be averaged to determine the direction of a vector of length g as depicted in Fig. 9.11. A major design issue is to find the optimum between sensor drifts

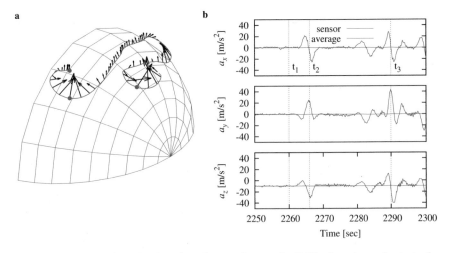

Fig. 9.11 Averaging of accelerations in order to reference the IMU orientation estimate to the down direction. Note, that the given accelerations are the senors measurements transformed to the INS. **a** 3d curve with arrows indicating measured accelerations. **b** Time series for measured accelerations. The averaged values result in $\approx 0\,\text{m/s}^2$ for the components in $\mathbf{e}_x$, $\mathbf{e}_y$-directions and in $\approx -10\,\text{m/s}^2$ for the component in $\mathbf{e}_z$-direction. This demonstrates the correct operation of the estimation algorithm. The times t_1–t_3 indicated by the dashed vertical lines are marked by red dots on the 3d trajectory

and averaging time. Information on referencing of the wind direction and other im-

plementation details are given in [8]. Further, a validation of the IMU algorithms
by comparing outputs to line angle sensor measurements is also presented in this
reference.

In addition to sensor drift and noise, the huge perturbations, coming along with
the wind, pose ambitious requirements to the estimation algorithms. An example of
a typical situation is depicted in Fig. 9.12. A negative gust, i.e. a decreased wind

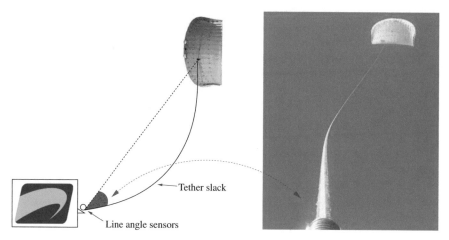

Fig. 9.12 Wind gusts might lead to a temporary untethered situation coming along with slack line,
which has to handled in a proper way by the estimation and control systems

speed for some seconds, may induce a quasi-untethered situation, which leads to
a significant position error in the line angle sensor measurements. This means a
breakdown of the simple model and hence model-based estimation techniques are
in danger to fail in these situations. A solution would be to include more involved
models or to implement some adaptive algorithms detecting such situations and
taking appropriate actions, e.g. temporarily neglecting some measurements.

For our application, a different way to tackle this challenge has been chosen.
The estimation is implemented without an explicit dynamic kite model as given by
Eqs. (9.2–9.5), but uses a more general model for integrating rates and averaging
accelerations, instead. The resulting structure is simple and its behavior for off-
nominal situations, e.g. slack line, can be easily accessed in an interpretative way.
It should be noted, that the approach uses no explicit dynamic model except basic
inertial kinematics. Albeit this restriction leads to higher estimation errors, the stability and robustness of the algorithm over some 100 flight hours is impressive and
convincing. Although improvements of the estimation accuracy by extended models
are subject to current research activities and seem to be promising, the roll-out of
such algorithms into regular operation has to be well-thought-out.

Finally, we would like to comment on two sensor types which are often proposed
in discussions and are suggested for application to tethered kites, but come along
with certain drawbacks for application in operational systems.

First, the global positioning system (GPS), especially when building up a differential GPS with a second stationary receiver, can provide position and velocity vectors of the kite with sufficient accuracy. However, our experience is that common receivers are not capable of handling fast changes in antenna orientation, which occur during dynamic flight of the airborne system. As a consequence, position losses or—even worse—jumps to phantom positions for some seconds occur every few minutes making GPS unusable as major input to the operational control system. This drawback arises from the GPS measurement technique, i.e. correlation tracking of satellites. Hence feasible solutions would involve multiple antennas and multi-channel receivers.

Second, optical sensing systems based on stationary cameras or an airborne camera and stationary landmarks could be used to estimate position and orientation of the kite. However, these systems usually require a huge technical and computational effort. In addition, optical detection becomes difficult or even unfeasible during night and in presence of low-hanging clouds. Furthermore, optical systems are difficult to operate in rough conditions as they occur for airborne wind energy systems: rain, water spray and dust on sensing windows have to be avoided or steadily removed.

Yet, it must be noted that these sensors, even when not suitable for operational use, provide valuable data for research and development purposes, e.g. as independent validation of the performance of estimation algorithms.

9.4 Control Setup

The task of setting up an appropriate control system for the plant dynamics presented in Chap. 9.2 and taking into account available sensors and estimation values described in Chap. 9.3 demands for some design choices, which shall be summarized in advance:

- The general philosophy was to base the design on 'classical' control system approaches. The structure of the plant, as given in Fig. 9.5, should be reflected in the control design and motivates a cascaded control structure. In addition, these single cascaded feedbacks can be engineered separately by loop-shaping based on analytic expressions.
- The occurrence of huge perturbations coming along with the wind has to be considered in the design. This is accomplished basically by two measures. First, the sensor and estimated values used in the various stages of the cascade have to be selected adequately. Second, we discarded the idea of tracking exactly a predefined trajectory. Instead, by allowing the controller a certain degree of freedom in following the kite dynamics instead of forcing a particular trajectory in a tight way, the robustness against larger perturbations can be increased.
- The plant has significant delays and nonlinear limitations, which can be attributed to the control actuator. In order to achieve sufficient bandwidth for dynamical

flight, nonlinear feedforward blocks are employed, which can be easily added to the classical control structure.

An overview on the control structure is shown in Fig. 9.13. The single components

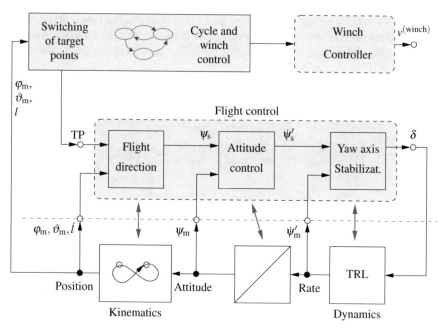

Fig. 9.13 Overview on the feedback structure and its relation with the kite dynamics. Note, that the plant structure is mirrored into the flight controller design resulting in a cascaded structure of the controller as indicated by the red arrows. In addition, the overall process is controlled by a cycle and winch control module. The winch controller is discussed in Sect. 9.4.3 and the target point switching in Sect. 9.4.2

will be discussed in the following subsections.

9.4.1 Attitude Control

The dynamics of the attitude is described by the turn rate law $\dot{\psi}' = g_k v_a \delta$ (see Eq. (9.2) for a more refined version). The turn rate is proportional to the deflection and the airspeed. When regarding v_a as a known (measured) gain the plant dynamics are of integrator type. Following the textbook approach, this suggests that the control of the attitude ψ' can be realized by a proportional feedback controller. The open loop transfer function of controller and plant will then have a slope of 20 dB/decade at the cross-over frequency, which automatically leads to good robustness features [18]. An additional lowpass filter for noise suppression with a sufficient distance to the cross-over frequency is added. By selecting the cross-over frequency as the main tuning knob the parameters of the controller can be readily expressed analytically, thus allowing for an implementation which can be tuned easily during sea trials.

Keeping these design principles in mind and going further into details, the plant can be split into two elements, characterized by the dynamic part expressed in the turn rate law and the integrator part $\dot{\psi} \rightarrow \psi$. As will be explained below, it is advantageous to mirror this structure by the controller architecture, which then is realized in a cascaded structure as depicted in Fig. 9.14. The particular choice is motivated

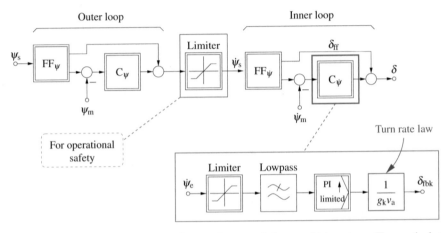

Fig. 9.14 Implementation of the attitude control as cascaded setup with two stages. Two particular features are highlighted here: First, e.g. limiters for operational safety can be easily inserted for exception handling (glitches in measurements) on the physical meaningful state at the connection between two control stages. Second, the turn rate law (TRL) is implemented as nonlinear element in the inner control loop

by the fact, that the turn rate $\dot{\psi}'_m$ is measured by a single sensor directly and hence is a good candidate for an inner loop control quantity. The compensation of errors in attitude, which is estimated by a sophisticated algorithm as depicted in Fig. 9.10, can be considered on a slower time scale in the outer loop. The higher bandwidth of the inner loop deals with controlling the turn rate and as such is responsible for keeping the kite stable by its capability to react fast to disturbances. In addition, the cascade allows for steady-state accuracy in presence of offsets in the rate due to e.g. gravitational effects [5, 6]. Further, the two stage cascade allows for non-linear constraints, e.g. limitation of the commanded rate $\dot{\psi}_s$ to acceptable values from an operational and safety point of view, see Fig. 9.14.

A distinguishing feature of the setup is the handling of the nonlinearity of the turn rate law, which comes from the fact the the rate change is determined by the deflection δ as well as by v_a. As already mentioned, by dividing the control by $g_k v_a$, as highlighted in Fig. 9.14, the plant dynamics is proportional to δ and thus linear. It should be emphasized, that this kind of linearization is acceptable as v_a changes slowly and thus quasi-statically compared to the controller bandwidth.

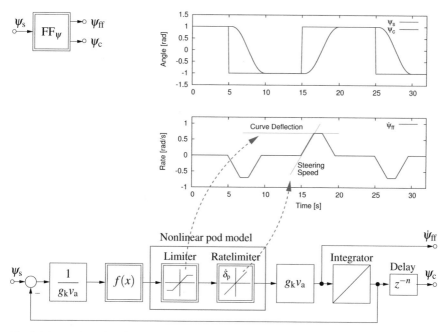

Fig. 9.15 Nonlinear feedforward block FF_ψ (see Fig. 9.14 for context) with embedded control actuator model for curve shaping [6, 7]. The implemented loop shapes a square signal on ψ_s as indicated in the time series for computed rate $\dot\psi_{ff}$ and expected behavior ψ_c. Note, that the implementation exploits explicitly the given constraints of curve deflection and steering speed as indicated in the figure

Additionally, in order to obtain a sufficient bandwidth in the presence of significant actuator non-linearities (rate limits, saturation, delays), feedforward paths are added. Basically, the feedforward elements are the inverse of the respective plant dynamics. The feedforward/feedback structure generates commands immediately in response to set-point changes, while a mere feedback architecture is naturally limited by avoiding gains which are too high. In addition non-linearities, like actuator restrictions in rate and magnitude, can easily be incorporated as inverse elements. A demonstrative example for incorporating non-linearities into the control design is the feedforward block for the attitude (FF_ψ) as depicted in Fig. 9.15.

It should be emphasized, that this feedforward block is even capable of processing steps in the set point ψ_s in such a way, that given constraints of limited steering deflection and speed are not only met, but can even be exploited for shaping the trajectory. This feature is essential for flying narrow curves at shorter tether lengths of the pumping cycle.

9.4.2 Guidance

The guidance control loop determines the flight pattern by providing the set point for the attitude ψ_s based on position and some internal states. In the following, two approaches for the guidance of lemniscates are presented.

Bang-Bang

As is shown in [5, 6], commanding a constant ψ_s ends up at a certain stationary angle ϑ. Hence the amplitude of ψ_s can be considered as an indirect handle for force control. Based on this heuristic, a very simple, but effective control scheme can be proposed [6]. The principle is depicted in Fig. 9.16. The idea is to command

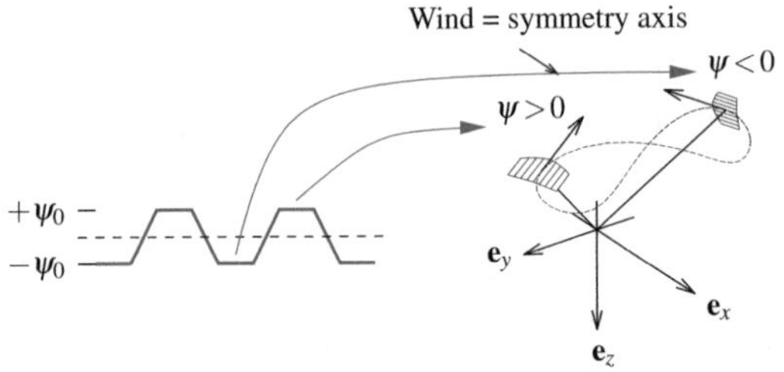

Fig. 9.16 Principle of bang-bang control. The geometrically triggered switching of ψ between $\pm\psi_0$ is the easiest way to fly lemniscates

a constant $\psi_s = \psi_0$ with $\psi_0 > 0$ to fly the kite in one direction. When a certain threshold is reached, i.e. $\varphi_m < -\varphi_{trigger}$, the set point is switched to $\psi_s = -\psi_0$ making the kite flying into the opposite direction until $\varphi_m > \varphi_{trigger}$ is fulfilled. Then, the cycle repeats from the beginning resulting in the lemniscates as sketched in Fig. 9.16.

The repeatability of the lemniscates over hours is really surprising despite the simplicity of the control structure. In addition, the scheme is quite robust in presence of perturbations in the φ measurement. This possibility of handling very rudimentary tether angle information is advantageous for the marine systems, which are subject to wave induced motions. However, wind speeds at the lower limit or significant side gusts, which exist but occur rather rarely, can induce critical flight situations, which may not be recovered by the bang-bang control. The problem can be dealt with by a different type of guidance which allows for stronger position feedback. This guidance is based on target points and is introduced in the subsequent section.

Target Points

A simple guidance with target points was reported for the first time in [11]. A slightly modified scheme can by used for flight control of eight-down patterns for power generation [7] as depicted in Fig. 9.17. The principle is based on a switching

Fig. 9.17 Target point guidance based on alternating
between two target points.
The switching of the active
target point has to be triggered
before the respective point is
reached. This is accomplished
by a trigger region around the
target point as indicated by
the red circle

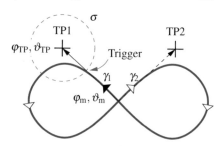

between two target points. The guidance computes the flight direction γ_s towards the
active target point using standard formula as they are used for navigating aircraft or
vessels on earth. The commanded attitude ψ_s is computed by inversion of Eq. (9.9).
As soon as the distance to the target point falls below a certain threshold σ, i.e.

$$(\varphi_{TP} - \varphi_m)^2 \sin^2 \vartheta_{TP} + (\vartheta_{TP} - \vartheta_m)^2 \leq \sigma^2 \qquad (9.10)$$

the other target point becomes active and determines the new flight direction and
thus kite attitude. Note, that this switching logic produces a step in ψ_s, which is
properly shaped by the feedforward block in Fig. 9.15. The direction of the flown
curve can be defined by a proper extension of the standard range of $[-\pi; \pi]$ for the
ψ_s-angle by adding or subtracting 2π as required. In addition, the estimated angle
ψ_m has to be made steady by unwinding it in the same way.

Although the target point guidance needs a position estimation of higher accuracy
compared to the bang-bang scheme, it benefits from a higher robustness especially
in unsteady and low wind situations. This is due to the fact, that even temporary
stalls, which lead to a sagging, are easily recovered as the kite has to fly up into
the trigger zone around the target point before the pattern continues as indicated in
Fig. 9.18. Finally, it should be remarked, that during the retraction phase a single
stationary target point can be used, which is never reached [7].

9.4.3 Winch Control

The positions for flight patterns are more or less given by the desired force and geometric constraints as the kite has to be kept clear off the surface for safety reasons.
In contrast, the winch speed has to chosen in order to maximize the average power
output $\bar{P}$ over complete pumping cycles with cycle time T as given by

$$\bar{P} = \frac{1}{T} \int_0^T l F_{\text{tether}} \, dt \qquad (9.11)$$

For continuously reeling out, maximal power is obtained for $v^{(\text{winch})} = (v_w \cos \vartheta)/a$
for $a = 3.0$ as has been shown by Loyd [15]. For pumping cycles, which also in-

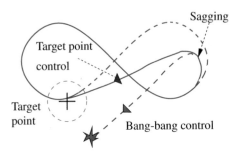

Fig. 9.18 Robustness against sagging. The target point control permanently corrects the flight direction towards the target point. Hence, a sagging has been compensated for when approaching the next target point as shown by the solid blue curve. In contrast, the bang-bang control scheme continues the pattern in a deeper region of the wind window and needs a bunch of lemniscates before being back on the original trajectory. This might lead to critical situations of overload as indicated by the dashed curve. Note that the drawn scheme shows the target point equivalent for the bang-bang case where the pattern is flown in the oposite direction compared to Fig. 9.17. However the same robustness arguments apply for both cases. Further details on flight directions are discussed in [7]

clude retraction phases, the optimal winch speed is a bit lower (i.e. $a > 3$) [16]. The transfer and return phase is a little bit more tricky as it starts in the crosswind region of high forces and ends up in a windward low force position. Hence, it seems reasonable to reel out in the beginning of the transfer phase and to go over continuously to a winching in for the return phase. In order to find out an optimal winch speed profile for the transfer phase, an optimal control problem (OCP) involving complete pumping cycles can be solved [4, 7]. The idea is to reproduce the results of the OCP by classical control structures.

Winch control can be implemented as depicted in Fig. 9.19. For the power phase, a simple proportional feedback of the air path speed v_a on the winch speed is used as follows

$$v_{\text{power}}^{(\text{winch})} = \frac{1}{(a-1)E} v_a \tag{9.12}$$

The closed loop behavior results in $v^{(\text{winch})} = (v_w \cos \vartheta)/a$ as can be easily shown by using Eq. (9.6). Typically $a \approx 3.5$ (instead of the 3.0 of Loyd) is chosen. It should be noted, that this controller automatically adapts to varying wind speed without the need for a direct measurement of the ambient wind speed. The transfer phase is implemented on base of a solution of the OCP. An interesting finding is the linear dependence between winch speed and wind window angle ϑ. As a consequence, a simple feedback of the measured ϑ_m to the winch speed $v^{(\text{winch})}$ can be implemented as shown in Fig. 9.19. These two basic controllers allow for operation of quite efficient pumping cycles. However, it should be mentioned, that in addition to the discussed winch controllers in this chapter, the internal control loops of the motion control hardware are parameterized in order to limit maximal power and maximal tether force of the system. Furthermore sophisticated feedback loops have

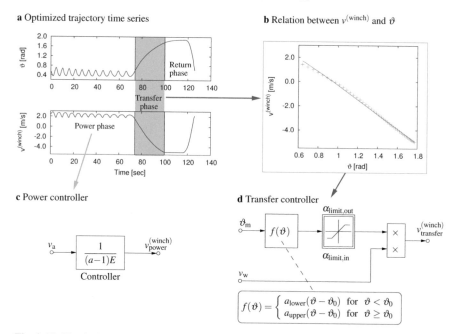

Fig. 9.19 Classical winch controller design based on the solution of an optimal control problem (OCP). **a** Time series for an optimized pumping cycle [4]. **b** Relation between $v^{(\mathrm{winch})}$ and ϑ for the transfer phase. **c** Controller for the power phase. **d** Transfer controller implementing a position feedback based on the solution of the OCP

been added in order to improve the behavior during the return phase and in exceptional flight situations.

9.4.4 Cycle and Process Control

After introduction of the feedback control elements, it has to be mentioned, that a higher-level control structure is needed in order to operate pumping cycles and to adapt parameters to changes in environmental conditions, e.g. wind speed. For the bang-bang control in Sect. 9.4.2, the amplitude ψ_0 has to be chosen and adapted during operation. A well-tested principle is to start with a safe value for ψ_0 and then run a kind of stepping controller, which evaluates the amplitudes of occurring force peaks and increases ψ_0 if the force peaks are below a certain threshold and decreases ψ_0 if the force peaks are above a certain threshold, respectively. As a consequence, the system adapts smoothly to varying wind conditions.

For the pumping cycle system, a state machine as depicted in Fig. 9.20 is used. Note, that this state machine contains part of the guidance given in Sect. 9.4.2 extended by the retraction phase. The main purpose of this overall state machine is to control parameters and determine active controller modes and elements. It should be

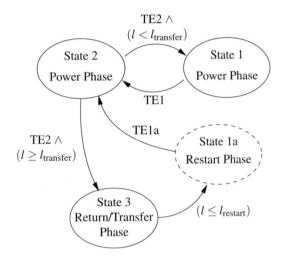

Fig. 9.20 State machine for operating pumping cycles as a simplified example of an overall process control for airborne wind energy systems. Transitions between states are driven by trigger events (TEx), which are activated when reaching a certain region around the target point TPx, compare Fig. 9.17. Note, that for the return phase, an additional target point or phase '1a' is used compared to Fig. 9.17

emphasized, that switching between control elements has to be done carefully in a steady and smooth manner. Furthermore, a general layer is added, which supervises proper operation of the control loops and monitors environmental conditions. As a consequence, parameters as positions of target points are continuously adapted or emergency actions are taken in case of failures.

Finally, it has to be mentioned, that similar state machines exist for automated launch and recovery. These and the underlying control loops are subject to current development and beyond the scope of this manuscript.

9.5 Conclusive Summary

It should be emphasized that the discussed flight control concept has emerged over a period of about 10 years by an iterative approach. Prototypes have been repeatedly put to test on sea trials already at a very early stage. It was paramount to have a control architecture which allowed for quick changes and adaptions during tests. All algorithms are implemented in an analytic fashion and with a close match to basic physical parameters which can easily be estimated. This structure allowed to tune the controller directly while the kite was airborne checking which bandwidth is acceptable, what noise filtering is needed, etc. The cascaded structure also served the purpose of early test needs. The inner loop of yaw rate control was of course the first thing to be mastered in order to keep the kite reliably up in the air. Once sufficient knowledge was gained in the rate dynamics, the next elements of the cascade could be tackled. In fact, the guidance scheme can also be regarded as a further stage in the cascade on top of the attitude control.

Beyond above sketched elements, the most distinguishing feature of the SkySails flight concept is its scalability with ambient wind speed. Regarding the EQM as given in Eqs. (9.2–9.6), time evolution speed of the complete state can be scaled with v_w if the control $v^{(winch)}$ is scaled accordingly. This behavior is anticipated in the controller structure by incorporating the TRL as shown in Fig. 9.14 and by scaling the winch controllers with v_a and v_w as indicated in Fig. 9.19. It is important to note, that the controller parts for the power phase are scaled by v_a and thus take into account the wind conditions at flight altitude in an direct way. This adds crucial robustness against gusts compared to solutions based on models, which depend from an exact knowledge of the ambient wind.

9.6 Classification and Outlook

In this section the above outlined flight control scheme, as operated by SkySails, shall be put into relation to other possible control and design methods. This will highlight the specifics of the approach and further clarify the design choices made.

The selected guidance scheme is also well suited to illustrate the development approach. In the very beginning an architecture was tested which more reflected the classical Guidance Navigation and Control (GNC) concept as commonly known from aerospace control applications. In principle it is possible to generate continuous set-points as function of time for an lemniscate type trajectory and build a controller structure with these set-points as input and the deflection command as output. This is a multiple input/single output (MISO) control problem (inputs are: rate, attitude, azimuth, elevation, wind speed and line length). The feedback can be realized by application of different MIMO design techniques, like linear quadratic control (LQ), pole placement with mode shaping or non-linear dynamic inversion. In all cases one monolithic MIMO feedback transfer function will realize the complete control loop. Such an implementation of the control algorithm has several drawbacks for an operational as well as developmental point of view.

The above sketched type of guidance is widely used in different control applications, but it relies on a sufficiently accurate model of the overall plant. We abandoned this approach after a couple of tests early in the development, mainly due to the limited knowledge available for tethered kite systems and in particular due to the lack of a detailed classical aerodynamic data base, which cannot be easily generated in wind tunnel tests. The described "bang bang" guidance was one answer to this problem. At the core of this scheme is the following idea: If it is difficult to compute a flyable lemniscate in the form of azimuth and elevation angles as function of time, which is consistent with the true kite dynamic, then do not do this at all and let the kite "find" its own pattern by just commanding attitude into the right direction. The target pattern based guidance is a further refinement, but it still stops short of specifying the commanded target trajectory completely.

Even when above issues of modeling can be overcome, the implementation of the flight control in separate blocks, with a clear physical underpinning, provides the capability of easily implementing the non-linear elements (saturation, rate limiting, delays). A MIMO design, rooted in linear control theory, cannot easily handles such types of non-linearities. The exception is model predictive control (MPC). Handling saturation and other limitations on states in a natural way is one of the main features of MPC. Yet, a high reliance on a good model is necessary and therefore this approach was not deemed advisable for pioneering the field of tethered kite control. It should be mentioned, that the present day, improved understanding of system dynamics and progress in model validation reveals the application of MPC in a different light. As a consequence, research and development efforts have been started recently in implementing and evaluating the MPC control approach on the functional model in cooperation with academic institutions.

Acknowledgements Michael Erhard gratefully acknowledges funding from ERC ST HIGH-WIND (259166) and valuable discussions within this collaboration.

References

1. Baayen, J. H., Ockels, W. J.: Tracking control with adaption of kites. IET Control Theory and Applications **6**(2), 182–191 (2012). doi: 10.1049/iet-cta.2011.0037
2. De Lellis, M., Saraiva, R., Trofino, A.: Turning angle control of power kites for wind energy. In: Proceedings of the 52nd Annual Conference on Decision and Control (CDC), pp. 3493–3498, Firenze, Italy, 10–13 Dec 2013. doi: 10.1109/CDC.2013.6760419
3. Diehl, M.: Real-time optimization for large scale nonlinear processes. Ph.D. Thesis, University of Heidelberg, 2001. http://archiv.ub.uni-heidelberg.de/volltextserver/1659/
4. Erhard, M., Horn, G., Diehl, M.: A quaternion-based model for optimal control of the Sky-Sails airborne wind energy system. ZAMM – Journal of Applied Mathematics and Mechanics (2016). doi: 10.1002/zamm.201500180. arXiv:1508.05494 [math.OC]
5. Erhard, M., Strauch, H.: Theory and Experimental Validation of a Simple Comprehensible Model of Tethered Kite Dynamics Used for Controller Design. In: Ahrens, U., Diehl, M., Schmehl, R. (eds.) Airborne Wind Energy, Green Energy and Technology, Chap. 8, pp. 141–165. Springer, Berlin Heidelberg (2013). doi: 10.1007/978-3-642-39965-7_8
6. Erhard, M., Strauch, H.: Control of Towing Kites for Seagoing Vessels. IEEE Transactions on Control Systems Technology **21**(5), 1629–1640 (2013). doi: 10.1109/TCST.2012.2221093
7. Erhard, M., Strauch, H.: Flight control of tethered kites in autonomous pumping cycles for airborne wind energy. Control Engineering Practice **40**, 13–26 (2015). doi: 10.1016/j.conengprac.2015.03.001
8. Erhard, M., Strauch, H.: Sensors and Navigation Algorithms for Flight Control of Tethered Kites. In: Proceedings of the European Control Conference (ECC13), Zurich, Switzerland, 17–19 July 2013. arXiv:1304.2233 [cs.SY]
9. Erhard, M., Strauch, H., Diehl, M.: Automatic Control of Optimal Pumping Cycles in Airborne Wind Energy. In: Schmehl, R. (ed.). Book of Abstracts of the International Airborne Wind Energy Conference 2015, p. 55, Delft, The Netherlands, 15–16 June 2015. doi: 10.4233/uuid:7df59b79-2c6b-4e30-bd58-8454f493bb09. Presentation video recording available from: https://collegerama.tudelft.nl/Mediasite/Play/569b89b60037439bbd9b6e3773588af51d
10. Fagiano, L., Milanese, M., Piga, D.: Optimization of airborne wind energy generators. International Journal of Robust and Nonlinear Control **22**(18), 2055–2083 (2011). doi: 10.1002/rnc.1808
11. Fagiano, L., Zgraggen, A. U., Morari, M., Khammash, M.: Automatic crosswind flight of tethered wings for airborne wind energy:modeling, control design and experimental results. IEEE Transactions on Control System Technology **22**(4), 1433–1447 (2014). doi: 10.1109/TCST.2013.2279592

12. Ilzhöfer, A., Houska, B., Diehl, M.: Nonlinear MPC of kites under varying wind conditions for a new class of large-scale wind power generators. International Journal of Robust and Nonlinear Control **17**(17), 1590–1599 (2007). doi: 10.1002/rnc.1210
13. Jehle, C.: Automatic Flight Control of Tethered Kites for Power Generation. M.Sc.Thesis, Technical University of Munich, Germany, 2012. https://mediatum.ub.tum.de/doc/1185997/1185997.pdf
14. Jehle, C., Schmehl, R.: Applied Tracking Control for Kite Power Systems. AIAA Journal of Guidance, Control, and Dynamics **37**(4), 1211–1222 (2014). doi: 10.2514/1.62380
15. Loyd, M. L.: Crosswind kite power. Journal of Energy **4**(3), 106–111 (1980). doi: 10.2514/3.48021
16. Luchsinger, R. H.: Pumping Cycle Kite Power. In: Ahrens, U., Diehl, M., Schmehl, R. (eds.) Airborne Wind Energy, Green Energy and Technology, Chap. 3, pp. 47–64. Springer, Berlin Heidelberg (2013). doi: 10.1007/978-3-642-39965-7_3
17. Maaß, J., Erhard, M.: Software System Architecture for Control of Tethered Kites. In: Ahrens, U., Diehl, M., Schmehl, R. (eds.) Airborne Wind Energy, Green Energy and Technology, Chap. 35, pp. 599–611. Springer, Berlin Heidelberg (2013). doi: 10.1007/978-3-642-39965-7_35
18. Ruth, M., Lebsock, K., Dennehy, C.: What's new is what's old: use of Bode's integral theorem (circa 1945) to provide insight for 21st century spacecraft attitude control system design tuning. AIAA Paper 2010-8428. In: AIAA Guidance, Navigation, and Control Conference, p. 8428, Toronto, ON, Canada, 2–5 Aug 2010. doi: 10.2514/6.2010-8428

Chapter 10
Attitude Tracking Control of an Airborne Wind Energy System

Haocheng Li, David J. Olinger and Michael A. Demetriou

Abstract We consider attitude tracking control for an airborne wind energy system, which generates electricity through a turbine mounted on a tethered glider flying at higher altitude than conventional wind turbines. The airborne wind energy system, which efficiently harnesses energy due to high-speed crosswind motion, consists of a rigid glider (also referred as a rigid kite) and constant length tether connected to the ground. Full aircraft dynamics are modeled including a rotational equation of motion. The resulting dynamical system is an under-actuated mechanical system with only rotational control inputs. We first propose an attitude tracking theorem that provides desired tracking signals for rotational motion. A feedback linearization controller and a real time differentiator are designed and implemented on the full glider dynamics to try to achieve the desired angle of attack and sideslip angle. A comparison study is conducted between a Lyapunov-based and attitude tracking control for the same baseline conditions for the airborne wind energy system.

10.1 Introduction

To develop a tracking control algorithm for airborne wind energy (AWE) systems, mathematical models to predict the kite motion are needed. The existing kite models can be roughly put into three categories: Multi-Body Model, Point Mass Model and Aircraft Model. In [3–5, 14, 15], the flexible kite is modeled as a multibody system where each body is interconnected by spring elements. Both theoretical and experimental techniques were used in [8, 11, 34], where an empirical yaw rate law is combined with the point mass dynamic for kite motion description. For rigid AWE systems, the aircraft model is used in [18, 29].

Haocheng Li · David J. Olinger (✉) · Michael A. Demetriou
Aerospace Engineering Program, Worcester Polytechnic Institute, 100 Institute Road, 01609 Worcester, MA, USA
e-mail: olinger@wpi.edu

© Springer Nature Singapore Pte Ltd. 2018
R. Schmehl (ed.), *Airborne Wind Energy*, Green Energy
and Technology, https://doi.org/10.1007/978-981-10-1947-0_10

Kite motion control is the key issue in both stabilizing kite motion and optimizing AWE system power output. For flexible kite AWE systems, the yaw rate law proposed in [8, 11, 12] was used to derive a control oriented linear time varying (LTV) system. In [2, 20, 32, 33], methods such as PID, feedback linearization and adaptive control were employed for kite trajectory tracking. Two-layer control structures were developed in this approach where the high-level control algorithm assigned the kite navigation target while different techniques are used in the lower level for tracking. Numerical control methods, such as model predictive control (MPC), were also studied for the LTV system derived from the yaw rate law [6,11,13]. For the rigid glider AWE systems, the equation of motion derived from classical mechanics are usually highly nonlinear so that the analytical control techniques are difficult to use. Various cost functions have been proposed and different methods used to optimize the system behavior [10, 16, 17, 19, 28, 30, 31].

In summary, there are three major kinds of kite models, i.e. Multi-Body Model, Point Mass Model and Aircraft Model and two major kinds of control strategies, i.e. Analytical Control and Numerical Control in the existing literature. For flexible kite systems, both control approaches are developed. However, for rigid kite systems, most of the literature focuses on the numerical approaches. The multi-body flexible kite model is more suitable for simulating kite motion as opposed to control design due to large number of degrees of freedom. The point mass flexible kite model is very suitable for control design. However, it depends on the assumption of stability of crucial aerodynamic parameters, like angle of attack. The numerical control approach applied to rigid kite systems does not provide direct control on kite translation, nor the stability of key aerodynamic parameters.

Previously, the boundedness of kite motion is established in [24, 25] for AWE systems using physics and a comparison principle. Lyapunov-based PD type controllers were developed and baseline simulation were discussed. We proposed the attitude tracking trajectory to try to achieve the constant angle of attack in [23] and a baseline simulation was conducted for an AWE system. The attitude trajectory was extended in order to track kite angle of attack and side slip angle simultaneously in [22], and an AWE system was simulated for baseline conditions. The outline of the chapter is as follow:

- Kinematic relations and the aerodynamic model used in this chapter are first provided.
- A five DOF AWE dynamical system for the tethered kite is derived using Lagrange methods;
- A Lyapunov-based rotational control is designed for an unperturbed AWE system and its asymptotic stability is proven.
- To achieve desired angle of attack and sideslip angles, two types of attitude trajectories are proposed.
- Feedback linearization is applied to design a control signal that achieves the proposed attitude trajectories.
- A baseline simulation is conducted for both Lyapunov-based and attitude tracking control design. Significant improvement in power output is achieved.

10.2 Equations of Motion

In this section the AWE system dynamics are derived, and important kinematic relations are provided. Classical aerodynamics is used to model the fluid structure interaction during kite motion. A Lagrange method is applied to derived the kite translational and rotational dynamics. An analysis of the AWE system power output is presented to justify the control design in the following sections.

10.2.1 Coordinate Systems and Transformations

To describe the kite motion, the following three coordinate systems are first introduced, as shown in Fig. 10.1:

- Inertial Cartesian frame $\mathbf{C}$ $(\mathbf{i}_E, \mathbf{j}_E, \mathbf{k}_E)$,
- *Inertial spherical frame* $\mathbf{S}$ $(\mathbf{i}_q, \mathbf{j}_q)$
- *Body frame* $\mathbf{B}$ $(\mathbf{i}_B, \mathbf{j}_B, \mathbf{k}_B)$.

In this paper, the $\mathbf{i}_E$ is set in the opposite direction of the wind (upwind direction), $\mathbf{k}_E$ is pointing upward and $\mathbf{j}_E$ forms a right-handed coordinate system. The kite CG position in $\mathbf{C}$ frame is denoted as $\mathbf{r}_E = (x_E, y_E, z_E)$. The body frame $\mathbf{B}$ is set to be a conventional North-East-Down (NED) frame [9] as shown in Fig. 10.1. The corresponding spherical coordinate is $\mathbf{q} = (q_1, q_2)$. The kite motion in the q_1 direction is referred as the crosswind and the motion in the q_2 direction as the inclination. Furthermore, assume that the fixed tether length is r. The transformations between these two coordinate systems are

$$(\mathbf{S} \rightarrow \mathbf{C}) \quad (x_E \; y_E \; z_E) = r \left(\cos q_1 \sin q_2 \; \sin q_1 \; \cos q_1 \cos q_2 \right). \qquad (10.1)$$

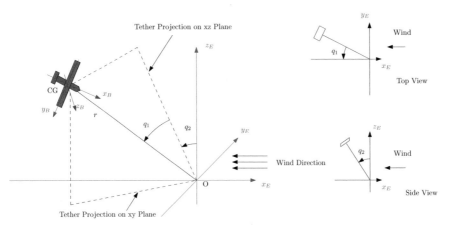

Fig. 10.1 Translational coordinate systems

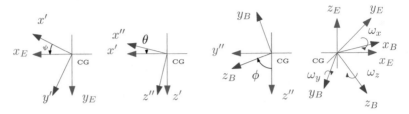

Fig. 10.2 Euler angles

The spherical coordinates can be obtained from the Cartesian coordinate via

$$(\mathbf{C} \rightarrow \mathbf{S}) \quad \begin{pmatrix} q_1 \\ q_2 \end{pmatrix} = \begin{pmatrix} \arctan\left(\frac{y_E}{\sqrt{x_E^2+z_E^2}}\right) \\ \arctan\left(\frac{x_E}{z_E}\right) \end{pmatrix}. \tag{10.2}$$

To derive the equations of motion, the transformation matrices of the infinitesimal displacement between different frames are also needed. The velocity transformation matrix $\mathbf{P}$ between $\mathbf{C}$ and $\mathbf{S}$ can be found by differentiating (10.1), i.e. the kite CG velocity $\mathbf{V}_E$ in the $\mathbf{C}$ frame is

$$(\mathbf{S} \rightarrow \mathbf{C}) \quad \mathbf{V}_E = \dot{\mathbf{r}}_E = \mathbf{P}\dot{\mathbf{q}}. \tag{10.3}$$

Matrix $\mathbf{P}$ can be expressed in terms of q_1 and q_2 and we use s and c for short-hand notation of sine and cosine functions, respectively:

$$\mathbf{P} = \begin{pmatrix} -s_{q_1}s_{q_2} & c_{q_1}c_{q_2} \\ c_{q_1} & 0 \\ -s_{q_1}c_{q_2} & -c_{q_1}s_{q_2} \end{pmatrix}.$$

The attitude of $\mathbf{B}$ with respect to $\mathbf{C}$ can be represented by the Euler angles $\Theta = (\phi, \theta, \psi)$ [9]. The angle ϕ is also referred as the roll angle, θ and ψ as the pitch and yaw angles respectively. By definition, frame $\mathbf{B}$ can be obtained from frame $\mathbf{C}$ by consecutive rotation as shown in Fig. 10.2. The velocity transformation between $\mathbf{C}$ and $\mathbf{B}$ can be found as

$$(\mathbf{C} \rightarrow \mathbf{B}) \quad \mathbf{V}_B = \mathbf{L}_{BE}\mathbf{L}_{EA}\mathbf{V}_E, \tag{10.4}$$

where $\mathbf{L}_{BE}$ is the rotation transformation matrix. $\mathbf{L}_{EA}$ is the flipping matrix for upward $\mathbf{k}_E$ and downward $\mathbf{k}_B$.

$$\mathbf{L}_{BE} = \begin{pmatrix} c_\theta c_\psi & c_\theta s_\psi & -s_\theta \\ s_\phi s_\theta c_\psi - c_\phi s_\psi & s_\phi s_\theta s_\psi + c_\phi c_\psi & s_\phi c_\theta \\ c_\phi s_\theta c_\psi + s_\phi s_\psi & c_\phi s_\theta s_\psi - s_\phi s_\psi & c_\phi c_\theta \end{pmatrix} \quad \mathbf{L}_{EA} = \begin{pmatrix} 1 & 0 & 0 \\ 0 & -1 & 0 \\ 0 & 0 & -1 \end{pmatrix}$$

The inverse transformation from $\mathbf{B}$ to $\mathbf{C}$ is

$$(\mathbf{B} \to \mathbf{C}) \quad \mathbf{V}_E = \mathbf{L}_{EA}\mathbf{L}_{EB}\mathbf{V}_B, \tag{10.5}$$

where $\mathbf{L}_{EB}$ is the inverse of $\mathbf{L}_{BE}$ and the matrix $\mathbf{L}_{EB}$ is orthogonal, i.e. $\mathbf{L}_{EB} = (\mathbf{L}_{BE})^{-1} = (\mathbf{L}_{BE})^T$.

Suppose the rotational velocity in frame $\mathbf{B}$ is $\omega = (\omega_x, \omega_y, \omega_z)$ which can also be specified by the Euler angles and their time rate of change as

$$\omega = \mathbf{R}\dot{\Theta}, \tag{10.6}$$

where the transformation matrix $\mathbf{R}$ is:

$$\mathbf{R} = \begin{pmatrix} 1 & 0 & -s_\theta \\ 0 & c_\phi & c_\theta s_\phi \\ 0 & -s_\phi & c_\theta c_\phi \end{pmatrix}.$$

Therefore, the states of AWE system can be described by the generalized position vector $\mathbf{h} = (q_1, q_2, \phi, \theta, \psi)$ and the generalized velocity $\dot{\mathbf{h}}$.

10.2.2 Kite Aerodynamics Model

Assume the wind velocity $\mathbf{W}$ is constant, then the kite apparent velocity is [1]

$$\mathbf{V}_a = \mathbf{L}_{BE}\mathbf{L}_{EA}(\mathbf{V}_E - \mathbf{W}). \tag{10.7}$$

Denoting the components of the apparent velocity as $\mathbf{V}_a = (u_a, v_a, w_a)$, the angle of attack and side slip angle are

$$\alpha = \arctan_2 \left(\frac{w_a}{u_a} \right) \qquad \beta = \arcsin \left(\frac{v_a}{\|\mathbf{V}_a\|} \right), \tag{10.8}$$

where $\arctan_2$ is the four-quadrant inverse tangent. Generally, kite lift and drag coefficients are function of the angle of attack α

$$C_L = C_L(\alpha), \qquad C_D = C_D(\alpha)$$

By inspection of Fig. 10.3, the aerodynamic coefficients in the $\mathbf{i}_B$ and $\mathbf{z}_B$ directions are

$$\begin{pmatrix} C_x \\ C_z \end{pmatrix} = \begin{pmatrix} 1 & 0 \\ 0 & -1 \end{pmatrix} \begin{pmatrix} \sin\alpha & -\cos\alpha \\ \cos\alpha & \sin\alpha \end{pmatrix} \begin{pmatrix} C_L(\alpha) \\ C_D(\alpha) \end{pmatrix}. \tag{10.9}$$

The aerodynamic coefficient in the $\mathbf{j}_B$ direction is a function of β

$$C_y = C_y(\beta). \tag{10.10}$$

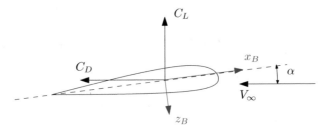

Fig. 10.3 Aerodynamic lift and drag of the airfoil

Turbine drag coefficient C_t is approximated in terms of the induction factor a [26] as

$$C_t = 4a(1-a)S_t/S, \tag{10.11}$$

where S and S_t are the kite and turbine area, respectively. Assuming that the turbine is mounted parallel to the $\mathbf{i}_B$ axis, then the aerodynamic coefficients in frame **B**, $\mathbf{C}_B$, can be computed from (10.9), (10.10) and (10.11):

$$\mathbf{C}_B = \left(C_x \ C_y \ C_z\right)^\top - \left(C_t \ 0 \ 0\right)^\top, \tag{10.12}$$

The aerodynamic force in frame **B** is then

$$\mathbf{A}_B = \frac{1}{2}\rho_{air}\|\mathbf{V}_a\|^2 S\mathbf{C}_B. \tag{10.13}$$

Applying the virtual work principle, $\mathbf{A}^\top \dot{\mathbf{q}} = \mathbf{A}_B^\top \mathbf{V}_B$ and substituting (10.3), (10.4) and (10.13) yield:

$$\mathbf{A}^\top \dot{\mathbf{q}} = \mathbf{A}_B^\top \left(\mathbf{L}_{BE}\mathbf{L}_{EA}\mathbf{P}\dot{\mathbf{q}}\right).$$

Then the generalized aerodynamic force on kite translation is

$$\mathbf{A} = \frac{1}{2}\rho_{air}\|\mathbf{V}_a\|^2 S\mathbf{P}^\top \mathbf{L}_{EA}\mathbf{L}_{EB}\mathbf{C}_B. \tag{10.14}$$

Since the wind velocity **W** is constant, the generalized aerodynamic force is a function of both translational and rotation states, i.e. $\mathbf{A} = \mathbf{A}(\mathbf{q}, \dot{\mathbf{q}}, \Theta)$. In another words, the generalized aerodynamic force represents the coupling between kite translation and rotation.

10.2.3 Airborne Wind Energy System Dynamics

Assume the kite mass is m and moment of inertial about CG is **J**. Denote the kite kinetic energies as K, and gravitational potential energy as U. Substituting (10.1) and (10.3) into expression of kinetic and potential energy yields:

$$K_t = \tfrac{1}{2}(m+\tfrac{1}{3}\rho r)\|\mathbf{V}_E\|^2 = \tfrac{1}{6}(\rho r + 3m)r^2(\dot{q}_1^2 + \dot{q}_2^2 \cos^2 q_1) + \tfrac{1}{2}\omega^\top \mathbf{J}\omega$$

$$U = (m+\tfrac{1}{2}\rho r)gz_E = (m+\tfrac{1}{2}\rho r)gr\cos q_1 \cos q_2,$$

where ρ is the tether line density. Apply the Euler-Lagrangian equation

$$\frac{d}{dt}\left(\frac{\partial L}{\partial \dot{h}_i}\right) - \frac{\partial L}{\partial h_i} = A_i + u_i, \quad i = 1,\ldots,5$$

where $L = K - U$ is the Lagrangian of the AWE system, A_i are the aerodynamic moments and u_i are the control inputs. The equations of translation and rotation can be expressed in matrix form [29] as

$$\mathbf{M}(\mathbf{q})\ddot{\mathbf{q}} + \mathbf{C}(\mathbf{q},\dot{\mathbf{q}})\dot{\mathbf{q}} + \mathbf{G}(\mathbf{q}) = \mathbf{A} \tag{10.15}$$

$$\mathbf{R}^T(\mathbf{J}\dot{\omega} + \omega \times \mathbf{J}\omega) = \mathbf{u}. \tag{10.16}$$

Explicitly, the matrices in (10.15) and (10.16) are

$$\mathbf{M}(\mathbf{q}) = \tfrac{1}{3}(\rho r + 3m)r^2 \begin{pmatrix} 1 & 0 \\ 0 & \cos^2 q_1 \end{pmatrix},$$

$$\mathbf{C}(\mathbf{q},\dot{\mathbf{q}}) = \tfrac{1}{6}(\rho r + 3m)r^2 \sin(2q_1) \begin{pmatrix} 0 & \dot{q}_2 \\ -\dot{q}_2 & -\dot{q}_1 \end{pmatrix},$$

$$\mathbf{G}(\mathbf{q}) = -(\tfrac{1}{2}\rho r + m)rg \begin{pmatrix} \cos q_2 \sin q_1 \\ \cos q_1 \sin q_2 \end{pmatrix}.$$

In the current work, the direct access to the rotational control is assumed. Practically, such control input can be generated by deflection of control surfaces on the glider (or kite). Although the actuation dynamics are a very important aspect of control system design, the assumption allows us to focus on higher level control system design.

There are two important features of the system dynamics (10.15) and (10.16):

- The system is under-actuated: there is no control on kite translation. (10.15)
- The system has one-way coupling: no translational states appear in the kite rotation. (10.16)

The first feature causes great difficulty in control system design for both flexible and rigid AWE system. The second feature indicates that the open loop system can be put into cascade form as shown in Fig. 10.4. In [11], the correlation between translation motion parameter and rotational control input is established using experimental data. In the current work, the controllability of key aerodynamic parameter α and β is provided using rotational motion. Since the angles α and β have great influence on aerodynamic forces on kite, the current work lays down a foundation for controllability of aerodynamic forces on kite. The aircraft model is also applied on

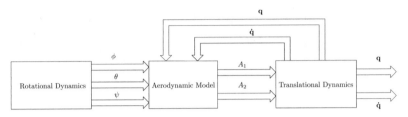

Fig. 10.4 Open loop system diagram

the Tethered Undersea Kite system (TUSK) which is a current power system using undersea kite. Olinger and Wang [27] used a six DOF system to simulate the TUSK trajectory, and Wang and Olinger improved the kite hydrodynamic model in [27].

10.2.4 Power Production

Assuming only one turbine mounted on the rigid glider along the body frame axis i_B, then the power produced by the turbine can be calculated in the following way. The turbine power coefficient C_p is the ratio between extracted power and power available in the airflow [26] which can be estimated using the induction factor a:

$$C_p = 4a(1-a)^2.$$

The power generated by the turbine can then be calculated as:

$$P_G = \frac{1}{2}C_p\rho_{air}(\|\mathbf{V}_a\|\cos\alpha\cos\beta)^3 S_t. \tag{10.17}$$

The theoretical limit of AWE system power production is given in [7]. The resultant coefficient C_R and resultant to drag ratio C_R/C_D are key parameters in determining theoretical maximum power production of AWE systems:

$$C_R = \sqrt{C_L^2 + (C_D + C_t)^2}.$$

The theoretical maximum power output of the AWE system is given by:

$$P_{max} = \frac{2}{27}\rho_{air}V_w^3 SC_R\left(\frac{C_R}{C_D}\right)^2. \tag{10.18}$$

where V_w is the local wind speed at the kite altitude. According to (10.9), (10.10) and (10.11), both C_R and C_R/C_D are function of α and β.

The importance of the controllability of α and β is illustrated by noting that

$$P_G \propto (\|\mathbf{V}_a\|\cos\alpha\cos\beta)^3.$$

Therefore, α and β need to be kept small to extract more power. On the other hand, small α results in small lift forces and C_R/C_D values which reduce the theoretical maximum power output. Additionally, small side slip angles result in small side forces, hence small cross wind motion (q_1 direction) and lower power output. As a result, both α and β need to be controlled to proper values to increase the overall power production.

10.3 Lyapunov-Based Rotational Control

Due to the complexity of the aerodynamic model, the aerodynamic effects on kite motion are neglected and the resulting system is referred to as unperturbed AWE system. In this section, a rotational control signal is designed based on a Lyapunov analysis on the unperturbed AWE system.

10.3.1 Stability of Unperturbed AWE System

The major difficulty in controlling the AWE system is due to the complexity of the coupling of translational and rotational motion in the aerodynamic moment $\mathbf{A}$ in (10.15). Therefore, the control design can be greatly simplified by ignoring the aerodynamic moment in (10.15). The Lyapunov-based controller is designed considering the unperturbed AWE system:

$$\mathbf{M}(\mathbf{q})\ddot{\mathbf{q}} + \mathbf{C}(\mathbf{q}, \dot{\mathbf{q}})\dot{\mathbf{q}} + \mathbf{G}(\mathbf{q}) = \mathbf{0}, \tag{10.19}$$

$$\mathbf{R}^T(\mathbf{J}\dot{\omega} + \omega \times \mathbf{J}\omega) = \mathbf{u}. \tag{10.20}$$

The stability of the unperturbed AWE system can be established using Lyapunov analysis [21].

Theorem 10.1. *For positive definite design matrices* $\mathbf{K}_\Theta$ *and* $\mathbf{K}_\Omega$, *and trim kite attitude* Θ_d, *the unperturbed AWE system, (10.19) and (10.20), is Lyapunov stable under the PD-type control signal*

$$\mathbf{u} = -\mathbf{K}_\Theta(\Theta - \Theta_d) - \mathbf{K}_\Omega\dot{\Theta}. \tag{10.21}$$

Additionally, the rotational dynamics is asymptotically stable under control signal (10.21), *i.e.*

$$\lim_{t \to \infty} \Theta = \Theta_d \tag{10.22}$$

Proof. It can be shown from (10.15) and (10.16) that the matrix $\dot{\mathbf{M}}(\mathbf{q}) - 2\mathbf{C}(\mathbf{q}, \dot{\mathbf{q}})$ is skew-symmetric. Explicitly,

$$\dot{\mathbf{M}}(\mathbf{q}) - 2\mathbf{C}(\mathbf{q}, \dot{\mathbf{q}}) = \frac{1}{3}(\rho r + 3m)r^2 \sin(2q_1)\begin{pmatrix} 0 & \dot{q}_2 \\ -\dot{q}_2 & 0 \end{pmatrix},$$

therefore, the following property holds

$$\left(\dot{\mathbf{M}}(\mathbf{q}) - 2\mathbf{C}(\mathbf{q}, \dot{\mathbf{q}})\right)^\top = -\left(\dot{\mathbf{M}}(\mathbf{q}) - 2\mathbf{C}(\mathbf{q}, \dot{\mathbf{q}})\right). \tag{10.23}$$

Choose the Lyapunov function candidate as

$$V = \frac{1}{2}\dot{\mathbf{q}}^\top \mathbf{M}(\mathbf{q})\dot{\mathbf{q}} + \frac{1}{2}\omega^\top \mathbf{J}\omega + U + \frac{1}{2}(\Theta - \Theta_d)^\top \mathbf{K}_\Theta(\Theta - \Theta_d). \tag{10.24}$$

Take the derivative of (10.23) along the system trajectory (10.19) and (10.20)

$$\dot{V} = \dot{\mathbf{q}}^\top \mathbf{M}(\mathbf{q})\ddot{\mathbf{q}} + \frac{1}{2}\dot{\mathbf{q}}^\top \dot{\mathbf{M}}(\mathbf{q})\dot{\mathbf{q}} + \omega^\top \mathbf{J}\dot{\omega} + \dot{\mathbf{q}}^\top \mathbf{G}(\mathbf{q}) + \dot{\Theta}^\top \mathbf{K}_\Theta(\Theta - \Theta_d)$$

$$= \frac{1}{2}\dot{\mathbf{q}}^\top \left(\dot{\mathbf{M}}(\mathbf{q}) - 2\mathbf{C}(\mathbf{q}, \dot{\mathbf{q}})\right)\dot{\mathbf{q}} + \dot{\Theta}^\top \mathbf{K}_\Theta(\Theta - \Theta_d) + \omega^\top (\mathbf{R}^{-1})^\top \mathbf{u}.$$

The derivative can be further simplified using (10.6) and (10.23)

$$\omega^\top (\mathbf{R}^{-1})^\top \mathbf{u} = \dot{\Theta}^\top \mathbf{u}$$

$$\dot{\mathbf{q}}^\top \left(\dot{\mathbf{M}}(\mathbf{q}) - 2\mathbf{C}(\mathbf{q}, \dot{\mathbf{q}})\right)\dot{\mathbf{q}} = 0$$

Therefore, the derivative of the Lyapunov function becomes:

$$\dot{V} = \dot{\Theta}^\top \left(\mathbf{u} + \mathbf{K}_\Theta(\Theta - \Theta_d)\right). \tag{10.25}$$

Substituting the control signal (10.21), the derivative becomes

$$\dot{V} = -\dot{\Theta}^\top \mathbf{K}_\Omega \dot{\Theta} \leqslant 0. \tag{10.26}$$

Applying the Lyapunov theorem in [21], $V > 0$ and $\dot{V} \leqslant 0$ implies the stability of the unperturbed AWE system. Additionally, by choosing the rotational Lyapunov function as

$$V_r = \frac{1}{2}\omega^\top \mathbf{J}\omega + \frac{1}{2}(\Theta - \Theta_d)^\top \mathbf{K}_\Theta(\Theta - \Theta_d)$$

and control signal (10.21), the asymptotic stability of the rotational motion can be established by LaSell's invariant principle. Taking the time derivative of rotational Lyapunov function,

$$\dot{V}_r = \dot{\Theta}^\top \mathbf{K}_\Theta(\Theta - \Theta_d) + \omega^\top (\mathbf{R}^{-1})^\top \mathbf{u}$$

$$= \dot{\Theta}^\top \left(\mathbf{u} + \mathbf{K}_\Theta(\Theta - \Theta_d)\right)$$

Substituting the control signal (10.21) gives

$$\dot{V} = -\dot{\Theta}^{\top} \mathbf{K}_{\Omega} \dot{\Theta} \leqslant 0.$$

Therefore, combining the kinematic relation (10.6), the invariant set contains the system trajectory satisfies

$$\dot{\Theta} = 0 \qquad \omega = 0 \qquad \dot{\omega} = 0 \qquad (10.27)$$

Substituting the condition (10.27) into rotational dynamics (10.16) gives

$$\mathbf{K}_{\Theta}(\Theta - \Theta_d) = 0 \qquad (10.28)$$

The design matrix $\mathbf{K}_{\Theta}$ is invertible, hence there is only one system trajectory $\Theta = \Theta_d$ in the invariant set.

10.3.2 Control Implementation

In order to form a figure-eight kite CG trajectory, a switching law for trim attitude, depending on the kite cross wind position y_E (or q_1 direction), is proposed:

$$\Theta_d = \begin{cases} \left(\phi_d^+ \ \theta_d^+ \ \psi_d^+\right) & y_E > y_E^+ \\[2mm] \left(\phi_d^- \ \theta_d^- \ \psi_d^-\right) & y_E < y_E^- \end{cases} \qquad (10.29)$$

The Lyapunov-based control system diagram, which uses the switching trim angle and the PD control signal, is shown in Fig. 10.5. As shown in (10.21), the Lyapunov-based controller possess simple PD structure which is very easy to implement. However, the Lyapunov-based controller does not guarantee the stability of α and β which is very important in power generation.

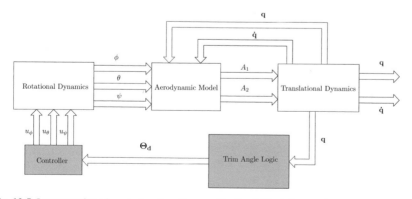

Fig. 10.5 Lyapunov-based control system diagram (feedback block in green)

10.4 Attitude Tracking

To increase the power output of the AWE system, control of the angle of attack and sideslip angle are considered in this section. Based on the trigonometric transformation, two kite attitude trajectories are proposed to achieve the desired angle of attack and sideslip angle. Feedback linearization is applied to achieve the desired attitude tracking and a real time differentiator is used for signal feedback.

10.4.1 Velocity Angles and Tracking Trajectory

Instead of ignoring the aerodynamic moment $\mathbf{A}$ in the translational dynamics (10.15), higher power production can be achieved by tracking the angle of attack α and side-slip angle β, (10.8). Define the kite CG velocity in a flipped inertial Cartesian frame to be:

$$\mathbf{V}^\top = \begin{pmatrix} u & v & w \end{pmatrix} = \mathbf{L}_{EA}(\mathbf{P}\dot{\mathbf{q}} - \mathbf{W})^\top. \tag{10.30}$$

The velocity angles are defined as [23]

$$\gamma_1 = \arctan\left(\frac{v}{u}\right) \tag{10.31}$$

$$\gamma_2 = \arctan\left(\frac{w}{\sqrt{u^2 + v^2}}\right), \tag{10.32}$$

where the inverse sine and tangent function take their principle value in $[-\pi/2, \pi/2]$. Assume that the kite apparent velocity is positive in the $\mathbf{i}_B$ direction, i.e. $u_a > 0$, then the angle of attack and side slip angles are equivalent to

$$\alpha = \arctan\left(\frac{w_a}{u_a}\right) \tag{10.33}$$

$$\beta = \arcsin\left(\frac{v_a}{\|\mathbf{V}_a\|}\right). \tag{10.34}$$

Now, given desired α_d and β_d, a desired rotational trajectory can be found.

Theorem 10.2. *(Tracking) For a desired $\alpha_d, \beta_d \in (-\pi/2, \pi/2)$ and $\alpha_d \neq 0$, there exists two rotational trajectories*

$$
\text{Type I} \quad \Theta_d = \begin{pmatrix} \frac{|u|}{u} \arctan\left(\frac{\tan\beta_d}{\sin\alpha_d}\right) \\ \frac{\alpha_d}{|\alpha_d|} \arccos(\cos\alpha_d \cos\beta_d) - \frac{|u|}{u}\gamma_2 \\ \gamma_1 \end{pmatrix} \tag{10.35}
$$

$$
\text{Type II} \quad \Theta_d = \begin{pmatrix} \frac{|u|}{u} \arctan\left(\frac{\tan\beta_d}{\sin\alpha_d}\right) \\ \arctan\left(\tan\alpha_d \sec\phi\right) - \frac{|u|}{u}\gamma_2 \\ \gamma_1 \end{pmatrix} \tag{10.36}
$$

such that angle of attack α and side slip angle β track the desired value, i.e.

$$
\alpha = \alpha_d; \qquad \beta = \beta_d.
$$

Specifically, if the tracking target is $\alpha_d = 0$ and $\beta_d = 0$, then the desired rotational trajectory becomes

$$
\psi = \gamma_1, \quad \theta = -\frac{|u|}{u}\gamma_2. \tag{10.37}
$$

The detailed proof of the attitude tracking theorem is shown in the Appendix.

10.4.2 Feedback Linearization

In type I tracking, the desired pitch angle θ_d is determined by the desired angle of attack and side slip angle α_d and β_d; in type II tracking, θ_d is computed using the current roll angle ϕ. Therefore, a smoother tracking performance can be achieved by type II signal.

Define the tracking error to be $\mathbf{e} = \Theta - \Theta_d$ and let the desired error dynamics be

$$
\mathbf{R}^\top \mathbf{J}\mathbf{R}(\ddot{\mathbf{e}} + \mathbf{K}_1\dot{\mathbf{e}} + \mathbf{K}_2\mathbf{e}) = \mathbf{0}, \tag{10.38}
$$

where $\mathbf{K}_1$ and $\mathbf{K}_2$ are positive definite design matrices. Assuming that the matrix $\mathbf{R}^\top \mathbf{J}\mathbf{R}$ is positive definite, i.e. $\phi, \theta \in \left(-\frac{\pi}{2}\ \frac{\pi}{2}\right)$, then system in (10.38) can be written as

$$
\ddot{\mathbf{e}} + \mathbf{K}_1\dot{\mathbf{e}} + \mathbf{K}_2\mathbf{e} = \mathbf{0} \quad \text{provided that} \quad \mathbf{R}^\top \mathbf{J}\mathbf{R} > 0
$$

which is locally asymptotically stable. Therefore, the tracking error is locally asymptotically stable, $\mathbf{e}, \dot{\mathbf{e}} \to 0$ as $t \to \infty$. Substituting (10.6) into (10.16) the rotational dynamics becomes

$$
\mathbf{R}^\top \left(\mathbf{J}(\dot{\mathbf{R}}\dot{\Theta} + \mathbf{R}\ddot{\Theta}) + \mathbf{R}\dot{\Theta} \times \mathbf{J}\mathbf{R}\dot{\Theta}\right) - \mathbf{u} = \mathbf{0}. \tag{10.39}
$$

The feedback linearizing control signal can be solved by equating (10.38) to (10.39):

$$\mathbf{R}^\top \left(\mathbf{J} (\dot{\mathbf{R}}\dot{\Theta} + \mathbf{R}\ddot{\Theta}) + \mathbf{R}\dot{\Theta} \times \mathbf{J}\mathbf{R}\dot{\Theta} \right) - \mathbf{u} = \mathbf{R}^\top \mathbf{J}\mathbf{R} (\ddot{\mathbf{e}} + \mathbf{K}_1 \dot{\mathbf{e}} + \mathbf{K}_2 \mathbf{e})$$

$$\mathbf{u} = \mathbf{R}^\top \mathbf{J}\mathbf{R} (-\mathbf{K}_1 \dot{\mathbf{e}} - \mathbf{K}_2 \mathbf{e} + \ddot{\Theta}_d) + \mathbf{R}^\top \left(\mathbf{J}\dot{\mathbf{R}}\dot{\Theta} + (\mathbf{R}\dot{\Theta} \times \mathbf{J}\mathbf{R}\dot{\Theta}) \right). \tag{10.40}$$

Notice that the derivatives of the desired attitude trajectory $\dot{\Theta}_d$ and $\ddot{\Theta}_d$ are required to generate the control signal and therefore the assumption that $\Theta(t) \in C^3$ is required.

10.4.3 Control Implementation

To obtain the derivative of the tracking signal in (10.40), a high-gain observer is applied [21] to generate the derivatives in real time. More generally, if $v_i \in C^3(\mathbb{R})$, then the first and second order derivatives of v_i are estimated by the following observer

$$\dot{\hat{\mathbf{x}}}_1 = \hat{\mathbf{x}}_2 + \frac{\sigma_1}{\varepsilon}(v_i - \hat{\mathbf{x}}_1)$$

$$\dot{\hat{\mathbf{x}}}_2 = \hat{\mathbf{x}}_3 + \frac{\sigma_2}{\varepsilon}(v_i - \hat{\mathbf{x}}_1) \tag{10.41}$$

$$\dot{\hat{\mathbf{x}}}_3 = \frac{\sigma_3}{\varepsilon}(v_i - \hat{\mathbf{x}}_1)$$

where ε is a small number and $\sigma_i, i = 1,2,3$ are the coefficients of the Hurwitz polynomial $s^3 + \sigma_1 s^2 + \sigma_2 s + \sigma_3$. The states of the system (10.41), $\hat{\mathbf{x}} = [\hat{\mathbf{x}}_{1,i}, \hat{\mathbf{x}}_{2,i}, \hat{\mathbf{x}}_{3,i}]^\top$ are the estimation of zeroth, first and second order derivatives of a signal v_i. The estimation error of the observer approaches zero as $\varepsilon \to 0$.

For type I tracking, a constant α_d and alternating β_d according to kite CG position is proposed:

$$\beta_d = \begin{cases} \beta^+ & \text{if } y_E > y_E^+ \\ \\ \beta^- & \text{if } y_E < y_E^- \end{cases} \tag{10.42}$$

For type II tracking, the trim roll angle is switched instead:

$$\phi_d = \begin{cases} \phi^+ & \text{if } y_E > y_E^+ \\ \\ \phi^- & \text{if } y_E < y_E^- \end{cases} \tag{10.43}$$

Moreover the angles β_d and ϕ_d satisfy the following relation:

$$\phi_d = \frac{|u|}{u} \arctan \left(\frac{\tan \beta_d}{\sin \alpha_d} \right).$$

The system diagram for attitude tracking is shown in Fig. 10.6. To meet the continuity requirement of the tracking signal, the switching laws are interpolated with

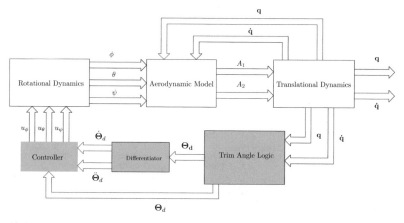

Fig. 10.6 Attitude tracking control system diagram (feedback block in green)

a cosine function. Suppose the switch times are t_i^+ and t_i^- in ith cycle for trim roll angle ϕ^+ and ϕ^-. Defining the transient time to be ΔT, then for switching occurring at t_i^-, the interpolation function is

$$\phi_d(t) = \frac{\phi^+ + \phi^-}{2} + \frac{\phi^+ - \phi^-}{2} \cos \frac{\pi}{\Delta T}(t - t_i^-)$$

where $t \in (t_i^-, t_i^- + \Delta T)$ and $\Delta T < |t_i^+ - t_i^-|$. The smoothing law for ϕ^+ can be obtained similarly.

10.5 Baseline Simulations

In this section, a comparison study between Lyapunov-based control and attitude tracking control is conducted. First, some crucial system parameters are listed. Detailed lift and drag models for the kite are provided, and baseline simulation results are presented.

10.5.1 System Inputs

Table 10.1 lists the key input parameters for the baseline simulation. Tables 10.2 and 10.3 list the gains used for Lyapunov-based controller (10.21) and attitude tracking controller (10.40). The desired kite attitude and crosswind position in (10.29), (10.42) and (10.43) are also provided. Additionally, the gains of the observer (10.41) is given in table 10.3.

Parameters	Value	Parameters	Value
Kite Mass	$40kg$	Tether Density	$0.03kg/m$
Kite Area	$30m^2$	Tether Length	$300m$
Turbine Mass	$20kg$	Tether Diameter	$0.02m$
Turbine Area	$6m^2$	$\bar{W}$	$6m/s$
Aspect Ratio	3.3	C_t	0.072
Induction factor	0.1	C_p	0.324
τ	0.1	C_{L_f}	1.934
k	0.106		

$\bar{W}$: Nominal Wind Speed C_t: Turbine Drag Coefficient
C_p: Turbine Power Coefficient τ: Glauert coefficient
C_{L_f}: vertical tail lift coefficient k: induced drag coefficient

Table 10.1 System input parameters

Parameters	Value	Parameters	Value
$\mathbf{K}_\Theta$	$2592\mathbf{I}$	$\mathbf{K}_\Omega$	$2592\mathbf{I}$
$[y_E^-, y_E^+]$	$[-50, 50]$m	$[\phi_d^-, \phi_d^+]$	$[-25°, 25°]$
$[\theta_d^-, \theta_d^+]$	$[10°, 10°]$	$[\psi_d^-, \psi_d^+]$	$[-65°, 65°]$

Table 10.2 Lyapunov-based control input parameters

Parameters	Value	Trim Angles	Value
$\mathbf{K}_1$	$100\mathbf{I}$	α_d	$12°$
$\mathbf{K}_2$	$10\mathbf{I}$	$[y_E^-, y_E^+]$	$[-30, 30]$m
$[\sigma_1, \sigma_2, \sigma_3]$	$[24, 192, 512]$	$[\phi^-, \phi^+]^a$	$[-37°, 37°]$
ε	0.1	$[\beta^-, \beta^+]^b$	$[-8.9°, 8.9°]$

^{a} Type II Tracking ^{b} Type I Tracking

Table 10.3 Attitude tracking control input parameters

To make the simulation result more realistic, the following aspect is also considered in the baseline case. First, the nominal wind speed $\bar{W}$ is measured at 10m altitude and the wind speed at CG altitude is estimated using a power law with

$$W = \bar{W} \left(\frac{z_E}{10} \right)^{\frac{1}{7}}.$$

Additionally, the aerodynamic force on the tether is estimated using the method proposed in [29]. The 2D lift coefficient of the kite is calculated from interpolating the data of airfoil NACA 0015:

$$C_{L,2D} = \begin{cases} -2.27 \times 10^{-4}\alpha^3 + 0.123\alpha + 0.2 & \text{if } |\alpha| \le 20° \\ 5.15 \times 10^{-10}\alpha^5 - 9.06 \times 10^{-6}\alpha^3 + 0.0405\alpha + 0.2 & \text{else} \end{cases}$$

where α takes the degree values. The 3D effects on the aerodynamic coefficient is also considered:

$$C_L = \frac{1}{1 + \frac{2(1+\tau)}{R}} C_{L,2D}$$

where R is the kite aspect ratio and τ is the Glauert coefficient. The kite drag and side force coefficients are estimated using the following equations:

$$C_D = C_{D0} + kC_L^2 \qquad C_y = -C_{L_f}\beta$$

where C_{D0} is the parasitic drag coefficient, k is the induced drag coefficient and C_{L_f} is the vertical tail lift coefficient.

Denoting the control moments applied on the kite as $\mathbf{u} = (u_\phi \ u_\theta \ u_\psi)$, then the power consumption of the rotational controller is

$$P_C = |u_\phi \dot{\phi}| + |u_\theta \dot{\theta}| + |u_\psi \dot{\psi}| \tag{10.44}$$

Therefore, the net power generated by the AWE system can be calculated from (10.17) and (10.44) as:

$$P_N = P_G - P_C \tag{10.45}$$

10.5.2 Simulation Results

Figures 10.7 and 10.8 show the comparisons of kite translational and rotational motions for the Lyapunov-based and attitude tracking (Type I and II) controllers. All three controllers result in a figure-eight kite CG trajectory. Compared to the Lyapunov-based controller (10.21), the attitude tracking controllers (10.40) result

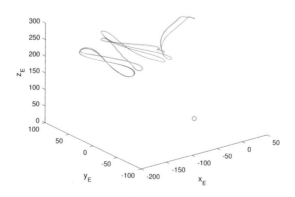

Fig. 10.7 Comparison of kite translation; blue: Lyapunov-based; red: type I tracking; green: type II tracking

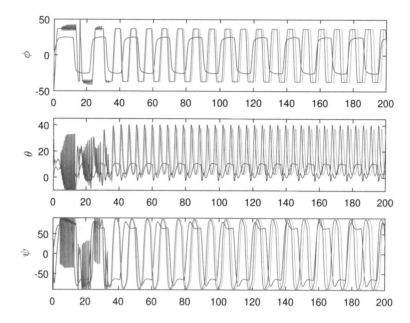

Fig. 10.8 Comparison of kite rotation; blue: Lyapunov-based; red: type I tracking; green: type II tracking

in larger oscillation in pitch motion. The two types of attitude tracking controller behave very similar in terms of kite translation and rotation.

The first two plots of Fig. 10.9 show the time record of the effective angle of attack α and sideslip angle β. The attitude tracking controllers achieve desired values after a short period of oscillation, while the Lyapunov-based controller yields oscillations in these parameters. Compared to type I tracking controller, the type II tracking controller results in smoother behavior of α and β. Hence, the attitude tracking controller achieves more stable time record of these two aerodynamic coefficients as shown in the last two plots of Fig. 10.9.

The first three plots of Fig. 10.10 show that the attitude tracking controller achieve higher kite apparent speed and net power output while consuming less control power compared to the Lyapunov-based controller. The net power output of the Lyapunov-based controller is 6.0kW, while both of the attitude tracking controllers yield 34.4kW, a 473% increase compared to the PD controller for the baseline conditions.

However, due to the high C_R and C_R/C_D, the ratio between net power output and the theoretical maximum output of the attitude tracking controller is fairly low in Fig. 10.10. Further study is needed to optimize power output using the attitude tracking controllers.

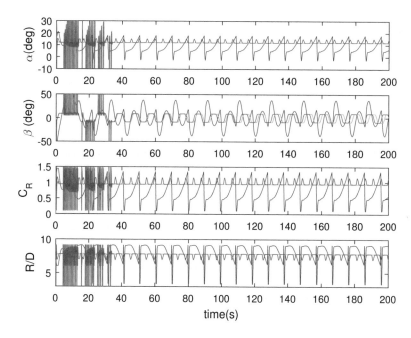

Fig. 10.9 Comparison of kite aerodynamic coefficients; blue: Lyapunov-based; red: type I tracking; green: type II tracking

10.6 Conclusion

In this chapter we have presented Lyapunov-based and attitude tracking control methods for AWE systems. A five-DOF full aircraft model that incorporates translational and rotational dynamics has been used to describe the kite motion. The stability of an unperturbed AWE system is established by Lyapunov methods. An attitude tracking theorem for the simultaneous tracking effective angle of attack and sideslip angle is proposed. Special cases of local wind and effective angle of attack tracking have also been discussed. The effective angle of attack tracking (Type II) can be viewed as an improvement of the general effective angle of attack and sideslip angle tracking (Type I) since its aerodynamic performance is more stable. A feedback linearization controller is designed for tracking desired kite rotation trajectories. A comparison study between the attitude tracking and Lyapunov-based control of AWE system is conducted and discussed in detail. The attitude tracking controller increases the power output significantly compared to Lyapunov control under the same baseline conditions. However, compared to the theoretical maximum power output, the power generated by attitude tracking controller is still fairly low and further optimization is needed. The robustness of the designed control structure under varying environmental conditions also needs further study.

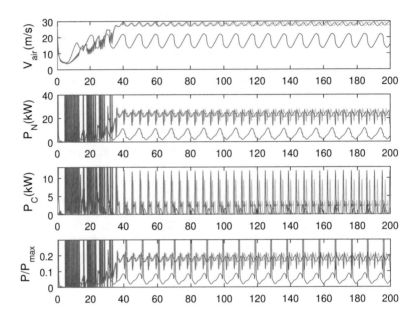

Fig. 10.10 Comparison of power output; blue: Lyapunov-based; red: type I tracking; green: type II tracking

Appendix

According to the kinematic relation (10.30), the kite apparent velocity is

$$
\begin{pmatrix} u_a \\ v_a \\ w_a \end{pmatrix} = \begin{pmatrix} c_\theta c_\psi & c_\theta s_\psi & -s_\theta \\ s_\phi s_\theta c_\psi - c_\phi s_\psi & s_\phi s_\theta s_\psi + c_\phi c_\psi & s_\phi c_\theta \\ c_\phi s_\theta c_\psi + s_\phi s_\psi & c_\phi s_\theta s_\psi - s_\phi s_\psi & c_\phi c_\theta \end{pmatrix} \begin{pmatrix} u \\ v \\ w \end{pmatrix}.
$$

Rearrange the trigonometric functions

$$
\begin{pmatrix} u_a \\ v_a \\ w_a \end{pmatrix} = \begin{pmatrix} (vs_\psi + uc_\psi)c_\theta - ws_\theta \\ [(vs_\psi + uc_\psi)s_\theta + wc_\theta]s_\phi - (us_\psi - vc_\psi)c_\phi \\ [(vs_\psi + uc_\psi)s_\theta + wc_\theta]c_\phi + (us_\psi - vs_\psi)s_\phi \end{pmatrix}
$$

Using trigonometric identities

$$
A\sin\psi + B\cos\psi = \sqrt{A^2 + B^2}\sin\left(\psi - \arctan_2\frac{B}{A}\right)
$$

the following trigonometric functions can be simplified as

$$vs_\psi + uc_\psi = \frac{|u|}{u}\cos(\psi - \gamma_1)\sqrt{u^2 + v^2}$$

$$us_\psi - vc_\psi = -\frac{|u|}{u}\sin(\psi - \gamma_1)\sqrt{u^2 + v^2}$$

By letting $\psi = \gamma_1$, then the kite apparent velocity becomes

$$\begin{pmatrix} u_a \\ v_a \\ w_a \end{pmatrix} = \begin{pmatrix} |u|/u\sqrt{u^2+v^2}c_\theta - ws_\theta \\ [|u|/u\sqrt{u^2+v^2}s_\theta + wc_\theta]s_\phi \\ [|u|/u\sqrt{u^2+v^2}s_\theta + wc_\theta]c_\phi \end{pmatrix}$$

Applying the same trigonometric identities, the expression can be further simplified as:

$$\begin{pmatrix} u_a \\ v_a \\ w_a \end{pmatrix} = \frac{|u|}{u}\|\mathbf{V}\| \begin{pmatrix} \sin\left(\theta + \frac{|u|}{u}\gamma_2\right) \\ \sin\left(\theta + \frac{|u|}{u}\gamma_2\right)\sin\phi \\ \cos\left(\theta + \frac{|u|}{u}\gamma_2\right)\cos\phi \end{pmatrix} \tag{10.46}$$

By substitution of (10.33) and (10.34) and notice $\|\mathbf{V}_a\| = \|\mathbf{V}\|$, the angle of attack α and side slip angle β become

$$\tan\alpha_d = \tan\left(\theta + \frac{|u|}{u}\gamma_2\right)\cos(\phi) \tag{10.47}$$

$$\sin\beta_d = \frac{|u|}{u}\sin\left(\theta + \frac{|u|}{u}\gamma_2\right)\sin(\phi). \tag{10.48}$$

Therefore, by setting $\theta = -\frac{|u|}{u}\gamma_2$, the desired angle of attack $\alpha_d = 0$ and desired sideslip angle $\beta_d = 0$ can be achieved. Additionally, if the control target is to achieved the desired angle of attack $\alpha = \alpha_d$ only, then the desired pitch angle is

$$\theta_d = \arctan\left(\tan\alpha_d\sec\phi\right) - \frac{|u|}{u}\gamma_2 \tag{10.49}$$

More generally, for control target $\alpha_d \neq 0$ and $\beta_d \neq 0$, (10.47) and (10.48) imply

$$\cos\left(\theta + \frac{|u|}{u}\gamma_2\right)\tan\alpha_d = \sin\left(\theta + \frac{|u|}{u}\gamma_2\right)\cos(\phi) \tag{10.50}$$

$$\sin\beta_d = \frac{|u|}{u}\sin\left(\theta + \frac{|u|}{u}\gamma_2\right)\sin(\phi). \tag{10.51}$$

Taking square sum of (10.50) and (10.51) and applying the trigonometric identity $\sin^2\phi + \cos^2\phi = 1$

$$\sin^2\beta_d + \cos^2\left(\theta + \frac{|u|}{u}\gamma_2\right)\tan^2\alpha_d = 1 - \cos^2\left(\theta + \frac{|u|}{u}\gamma_2\right).$$

Rearranging the equation above and using the trig identity $1 + \tan^2 \alpha_d = \sec^2 \alpha_d$:

$$\cos^2\left(\theta + \frac{|u|}{u}\gamma_2\right) = \cos^2 \alpha_d \cos^2 \beta_d. \tag{10.52}$$

For $\alpha_d, \beta_d \in (-\frac{\pi}{2}, \frac{\pi}{2})$, the desired pitch can be solved as

$$\theta = \frac{\alpha_d}{|\alpha_d|}\arccos(\cos \alpha_d \cos \beta_d) - \frac{|u|}{u}\gamma_2,$$

where the inverse cosine takes its value in $[0, \frac{\pi}{2}]$. Suppose that $\alpha \neq 0$, the squared ratio of (10.50) to (10.51) is

$$\frac{\sin^2 \beta_d}{\tan^2 \alpha_d} = \cos^2\left(\theta + \frac{|u|}{u}\gamma_2\right)\tan^2 \phi \tag{10.53}$$

Substitute align (10.52) into (10.53)

$$\tan^2 \phi = \frac{\tan^2 \beta_d}{\sin^2 \alpha_d}.$$

Therefore, the roll trajectory can be obtained as

$$\phi = \frac{|u|}{u}\arctan\left(\frac{\tan \beta_d}{\sin \alpha_d}\right),$$

where the inverse tangent take its value from $[\frac{-\pi}{2}, \frac{\pi}{2}]$.

For type II attitude trajectory, the desired pitch angle needs to satisfy the following equation:

$$\theta = \frac{\alpha_d}{|\alpha_d|}\arccos(\cos \alpha_d \cos \beta_d) - \frac{|u|}{u}\gamma_2 = \arctan\left(\tan \alpha_d \sec \phi\right) - \frac{|u|}{u}\gamma_2.$$

The desired pitch Θ_d is equivalent if and only if

$$\frac{\alpha_d}{|\alpha_d|}\arccos(\cos \alpha_d \cos \beta_d) = \arctan\left(\tan \alpha_d \sec \phi\right).$$

Substituting the desired pitch $\phi = \frac{|u|}{u}\arctan\left(\frac{\tan \beta_d}{\sin \alpha_d}\right)$ yields:

$$\frac{\alpha_d}{|\alpha_d|}\tan\left(\arccos(\cos \alpha_d \cos \beta_d)\right) = \tan \alpha_d \sec\left(\frac{|u|}{u}\arctan\left(\frac{\tan \beta_d}{\sin \alpha_d}\right)\right).$$

Since the inverse tangent is taken value from $(-\frac{\pi}{2}, \frac{\pi}{2})$, the equivalence can be further simplified as

$$\frac{\alpha_d}{|\alpha_d|} \tan\left(\arccos(\cos\alpha_d \cos\beta_d)\right) = \tan\alpha_d \sec\left(\arctan\left(\frac{\tan\beta_d}{\sin\alpha_d}\right)\right)$$

Using the trigonometric identities:

$$\tan(\arccos\xi) = \frac{\sqrt{1-\xi^2}}{\xi}, \qquad \sec\left(\arctan\xi\right) = \sqrt{1+\xi^2}$$

Then the left hand side becomes

$$\frac{\alpha_d}{|\alpha_d|} \tan\left(\arccos(\cos\alpha_d \cos\beta_d)\right) = \frac{\alpha_d}{|\alpha_d|} \frac{\sqrt{1-\cos^2\alpha_d \cos^2\beta_d}}{\cos\alpha_d \cos\beta_d}$$

Additionally, for $\alpha_d, \beta_d \in \left(-\frac{\pi}{2}\ \frac{\pi}{2}\right)$, we have $\cos\alpha_d \cos\beta_d > 0$ and

$$\begin{aligned}
\frac{\alpha_d}{|\alpha_d|} \tan\left(\arccos(\cos\alpha_d \cos\beta_d)\right) &= \frac{|\alpha_d|}{\alpha_d}\sqrt{\frac{1-\cos^2\alpha_d \cos^2\beta_d}{\cos^2\alpha_d \cos^2\beta_d}} \\
&= \frac{|\alpha_d|}{\alpha_d}\sqrt{\sec^2\alpha_d \sec^2\beta_d - 1} \\
&= \frac{|\alpha_d|}{\alpha_d}\sqrt{\sec^2\alpha_d(\sec^2\beta_d - 1) + \tan^2\alpha_d} \\
&= \frac{|\alpha_d|}{\alpha_d}\sqrt{\tan^2\alpha_d + \sec^2\alpha_d \tan^2\beta_d}
\end{aligned}$$

On the right hand side,

$$\begin{aligned}
\tan\alpha_d \sec\left(\arctan\left(\frac{\tan\beta_d}{\sin\alpha_d}\right)\right) &= \tan\alpha_d \sqrt{1 + \left(\frac{\tan\beta_d}{\sin\alpha_d}\right)^2} \\
&= \frac{|\alpha_d|}{\alpha_d}\sqrt{\tan^2\alpha_d + \sec^2\alpha_d \tan^2\beta_d}
\end{aligned}$$

Therefore, the equivalence readily follows.

References

1. Anderson, J. D.: Fundamentals of Aerodynamics. 5th ed. McGraw-Hill (2014)
2. Baayen, J. H., Ockels, W. J.: Tracking control with adaption of kites. IET Control Theory and Applications 6(2), 182–191 (2012). doi: 10.1049/iet-cta.2011.0037
3. Bosch, A., Schmehl, R., Tiso, P., Rixen, D.: Dynamic nonlinear aeroelastic model of a kite for power generation. AIAA Journal of Guidance, Control and Dynamics 37(5), 1426–1436 (2014). doi: 10.2514/1.G000545
4. Bosch, A., Schmehl, R., Tiso, P., Rixen, D.: Nonlinear Aeroelasticity, Flight Dynamics and Control of a Flexible Membrane Traction Kite. In: Ahrens, U., Diehl, M., Schmehl, R. (eds.) Airborne Wind Energy, Green Energy and Technology, Chap. 17, pp. 307–323. Springer, Berlin Heidelberg (2013). doi: 10.1007/978-3-642-39965-7_17

5. Breukels, J., Schmehl, R., Ockels, W.: Aeroelastic Simulation of Flexible Membrane Wings based on Multibody System Dynamics. In: Ahrens, U., Diehl, M., Schmehl, R. (eds.) Airborne Wind Energy, Green Energy and Technology, Chap. 16, pp. 287–305. Springer, Berlin Heidelberg (2013). doi: 10.1007/978-3-642-39965-7_16
6. Canale, M., Fagiano, L., Milanese, M.: High Altitude Wind Energy Generation Using Controlled Power Kites. IEEE Transactions on Control Systems Technology 18(2), 279–293 (2010). doi: 10.1109/TCST.2009.2017933
7. Diehl, M.: Airborne Wind Energy: Basic Concepts and Physical Foundations. In: Ahrens, U., Diehl, M., Schmehl, R. (eds.) Airborne Wind Energy, Green Energy and Technology, Chap. 1, pp. 3–22. Springer, Berlin Heidelberg (2013). doi: 10.1007/978-3-642-39965-7_1
8. Erhard, M., Strauch, H.: Control of Towing Kites for Seagoing Vessels. IEEE Transactions on Control Systems Technology 21(5), 1629–1640 (2013). doi: 10.1109/TCST.2012.2221093
9. Etkin, B., Reid, L. D.: Dynamics of Flight: Stability and Control. John Wiley & Sons, New York (1996)
10. Fagiano, L., Milanese, M., Piga, D.: Optimization of airborne wind energy generators. International Journal of Robust and Nonlinear Control 22(18), 2055–2083 (2011). doi: 10.1002/rnc.1808
11. Fagiano, L., Zgraggen, A. U., Morari, M., Khammash, M.: Automatic crosswind flight of tethered wings for airborne wind energy:modeling, control design and experimental results. IEEE Transactions on Control System Technology 22(4), 1433–1447 (2014). doi: 10.1109/TCST.2013.2279592
12. Fagiano, L., Huynh, K., Bamieh, B., Khammash, M.: On sensor fusion for airborne wind energy systems. IEEE Transactions on Control Systems Technology 22(3), 930–943 (2014). doi: 10.1109/TCST.2013.2269865
13. Fagiano, L., Milanese, M., Razza, V., Bonansone, M.: High Altitude Wind Energy for Sustainable Marine Transportation. IEEE Transactions on Intelligent Trasportation Systems 13(2), 781–791 (2012). doi: 10.1109/TITS.2011.2180715
14. Fechner, U., Vlugt, R. van der, Schreuder, E., Schmehl, R.: Dynamic Model of a Pumping Kite Power System. Renewable Energy (2015). doi: 10.1016/j.renene.2015.04.028. arXiv:1406.6218 [cs.SY]
15. Groot, S. G. C. de, Breukels, J., Schmehl, R., Ockels, W. J.: Modeling Kite Flight Dynamics Using a Multibody Reduction Approach. AIAA Journal of Guidance, Control and Dynamics 34(6), 1671–1682 (2011). doi: 10.2514/1.52686
16. Gros, S., Zanon, M., Diehl, M.: A relaxation strategy for the optimization of Airborne Wind Energy systems. In: Proceedings of the 2013 European Control Conference (ECC), pp. 1011–1016, Zurich, Switzerland, 17–19 July 2013
17. Gros, S., Zanon, M., Diehl, M.: Control of Airborne Wind Energy Systems Based on Nonlinear Model Predictive Control & Moving Horizon Estimation. In: Proceedings of the European Control Conference (ECC13), Zurich, Switzerland, 17–19 July 2013
18. Gros, S., Diehl, M.: Modeling of Airborne Wind Energy Systems in Natural Coordinates. In: Ahrens, U., Diehl, M., Schmehl, R. (eds.) Airborne Wind Energy, Green Energy and Technology, Chap. 10, pp. 181–203. Springer, Berlin Heidelberg (2013). doi: 10.1007/978-3-642-39965-7_10
19. Ilzhöfer, A., Houska, B., Diehl, M.: Nonlinear MPC of kites under varying wind conditions for a new class of large-scale wind power generators. International Journal of Robust and Nonlinear Control 17(17), 1590–1599 (2007). doi: 10.1002/rnc.1210
20. Jehle, C., Schmehl, R.: Applied Tracking Control for Kite Power Systems. AIAA Journal of Guidance, Control, and Dynamics 37(4), 1211–1222 (2014). doi: 10.2514/1.62380
21. Khalil, H.: Nonlinear Systems. 3rd ed. Prentice Hall, Upper Saddle River, NJ (2001)
22. Li, H., Olinger, D. J., Demetriou, M. A.: Attitude Tracking Control of a GroundGen Airborne Wind Energy System. In: Proceedings of the 2016 American Control Conference, Boston, MA, USA, 6–8 July 2016. doi: 10.1109/ACC.2016.7525565

23. Li, H., Olinger, D. J., Demetriou, M. A.: Attitude Tracking Control of an Airborne Wind Energy System. In: Proceedings of the 2015 European Control Conference, Linz, Austria, 15–17 July 2015. doi: 10.1109/ECC.2015.7330752
24. Li, H., Olinger, D. J., Demetriou, M. A.: Control of a Tethered Undersea Kite Energy System Using a Six Degree of Freedom Model. In: Proceedings of the 2015 IEEE International Conference on Decision and Control (CDC), Osaka, Japan, 15–18 Dec 2015. doi: 10.1109/CDC.2015.7402309
25. Li, H., Olinger, D. J., Demetriou, M. A.: Control of an Airborne Wind Energy System using an Aircraft Dynamics Model. In: Proceedings of the 2015 American Control Conference, Chicago, IL, USA, 1–3 July 2015. doi: 10.1109/ACC.2015.7171090
26. Manwell, J. F., McGowan, J. G., Rogers, A. L.: Wind Energy Explained: Theory, Design and Application. 2nd ed. John Wiley & Sons, Ltd., Chichester (2009). doi: 10.1002/9781119994367
27. Olinger, D., Wang, Y.: Hydrokinetic energy harvesting using tethered undersea kites. Journal of Renewable and Sustainable Energy 7(4), 043114 (2015). doi: 10.1063/1.4926769
28. Williams, P., Lansdorp, B., Ockels, W. J.: Nonlinear Control and Estimation of a Tethered Kite in Changing Wind Conditions. AIAA Journal of Guidance, Control and Dynamics 31(3) (2008). doi: 10.2514/1.31604
29. Williams, P., Lansdorp, B., Ockels, W.: Optimal Crosswind Towing and Power Generation with Tethered Kites. AIAA Journal of Guidance, Control, and Dynamics 31(1), 81–93 (2008). doi: 10.2514/1.30089
30. Zanon, M., Gros, S., Andersson, J., Diehl, M.: Airborne Wind Energy Based on Dual Airfoils. IEEE Transactions on Control Systems Technology 21(4), 1215–1222 (2013). doi: 10.1109/TCST.2013.2257781
31. Zanon, M., Horn, G., Gros, S., Diehl, M.: Control of Dual-Airfoil Airborne Wind Energy systems based on nonlinear MPC and MHE. In: European Control Conference, pp. 1801–1806, Strasbourg, France, 17–19 July 2013. doi: 10.1109/ECC.2014.6862238
32. Zgraggen, A. U., Fagiano, L., Morari, M.: Automatic Retraction Phase of Airborne Wind Energy Systems. Proceedings of the 19th IFAC World Congress 47(3), 5826–5831 (2014). doi: 10.3182/20140824-6-ZA-1003.00624
33. Zgraggen, A. U., Fagiano, L., Morari, M.: On Modeling and Control of the Retraction Phase for Airborne Wind Energy Systems. In: Proceedings of the 53rd IEEE Conference on Decision and Control, pp. 5686–5691, Los Angeles, CA, USA, 15–17 Dec 2014. doi: 10.1109/CDC.2014.7040279
34. Zgraggen, A. U., Fagiano, L., Morari, M.: Real-Time Optimization and Adaptation of the Crosswind Flight of Tethered Wings for Airborne Wind Energy. IEEE Transactions on Control Systems Technology 23(2), 434–448 (2015). doi: 10.1109/TCST.2014.2332537

Chapter 11
Nonlinear DC-link PI Control for Airborne Wind Energy Systems During Pumping Mode

Korbinian Schechner, Florian Bauer and Christoph M. Hackl

Abstract During pumping mode, airborne wind energy systems are operated in two phases: A power generating reel-out phase and a power dissipating reel-in phase. The ground winch is connected via a DC-link voltage source converter to the grid. The control of its DC-link voltage is a challenging task due to the bidirectional power flow over the DC-link. Two PI controller designs are discussed: the classical PI controller with constant parameters and a nonlinear PI controller with online parameter adjustment. Based on a worst-case analysis of the physical properties, bounds on the constant parameters of the classical PI controller are derived leading to a conservative design to assure a stable operation also during the reel-in phase where the system dynamics are non-minimum phase. To overcome these limitations in the closed-loop bandwidth, a nonlinear PI controller is proposed which adjusts its parameters online. For controller design, the linearized system model is used and the controller parameters are computed via "online pole placement". Simulation results illustrate robustness, stability and improved control performance of the proposed nonlinear PI controller in comparison to the classical PI controller.

11.1 Introduction

Kites are a promising approach to harvest wind energy at high altitudes (see [5, 10, 14] and references therein): As shown in Fig. 11.1, the kite is tethered to a ground winch which is connected to an electric drive. Electric energy is generated in a

Korbinian Schechner · Christoph M. Hackl (✉)
Technische Universität München, Munich School of Engineering, Research group "Control of renewable energy systems (CRES)", Lichtenbergstr. 4a, 85748 Garching, Germany
e-mail: christoph.hackl@hm.edu

Florian Bauer
Technische Universität München, Institute for Electrical Drive Systems and Power Electronics, Arcisstr. 21, 80333 München, Germany

© Springer Nature Singapore Pte Ltd. 2018
R. Schmehl (ed.), *Airborne Wind Energy*, Green Energy
and Technology, https://doi.org/10.1007/978-981-10-1947-0_11

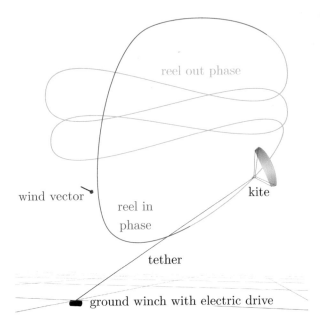

Fig. 11.1 Pumping mode power generation with a kite

pumping process: During the "reel-out phase", the kite is flown in fast crosswind motions like figure eights with a high lift force. The kite pulls the tether which is reeled out slowly. Energy is generated by operating the electric drive in generator mode, i.e. generative braking. In the "reel-in phase", the kite is flown in a low force position like the zenith, or is pitched down, and is reeled back in while only a fraction of the generated energy is dissipated by operating the winch drive in motor mode. Compared to conventional wind turbines, this technology promises to harvest wind energy at higher altitudes using less material. Hence, it promises to have a higher capacity factor, lower capital investments, and, therefore, a lower levelized cost of electricity.

Several challenges of this technology have to be solved for deployment in the power generation industry. In this chapter, we consider a ground winch with electric drive which is connected to the grid via a DC-link voltage source converter (or power converter). This topology allows for independent control of active and reactive power flow to and from the grid (bidirectional power flow). The DC-link dynamics are highly nonlinear and, during the reel-in phase (motor mode), are non-minimum phase which imposes a challenge on controller design. We discuss this challenge of DC-link voltage control under the influence of a highly fluctuating bi-directional power flow to and from the ground winch drive for a given reactive power demand by the grid operator.

From a control point of view, non-minimum phase systems are particularly interesting. In 1940 H.W. Bode was one of the first to discuss the phenomenon of non-minimum phase systems (see [4]). For classical output feedback control, the closed-loop system bandwidth is drastically limited. High gains are not admissible and so a very conservative controller (mostly a proportional-integral (PI) controller) must be designed if constant controller parameters are used.

Although there exists a tremendous number of papers (over 8.000, see [7]) which deal with the subject of DC-link control, only quite few papers (see [6, 7, 12, 16–18, 23, 25, 27, 29–31]) do explicitly address the non-minimum phase behavior of the DC-link dynamics in their metadata (such as abstract, title or indexing words). There is quite a variety of proposed control strategies for the DC-link voltage control problem in power converters such as model predictive control strategies (see [8, 15, 28] and references therein), flatness-based methods, linearization-based or passivity-based approaches (see [11, 22] and references therein) or state-feedback controller designs (see [7, 21, 26]) to name a few. For airborne wind energy systems, DC-link control is conceptually explained in [2] and [1] in the context of grid integration of such renewable energy systems.

In this chapter (for first results see [3]), a nonlinear DC-link controller with online adjustment of its controller parameters for a grid-connected voltage source converter of an airborne wind energy system with bidirectional power flow (pumping mode operation) is proposed. In addition, we investigate the classical PI controller design with constant parameters. The focus on PI controllers is motivated by their widespread use in industry. The contributions of this chapter are:

- Precise problem formulation and detailed modeling of the nonlinear DC-link dynamics of a three-phase grid-connected voltage source converter,
- Linearization of the nonlinear DC-link dynamics around a general equilibrium,
- Illustration and physical explanation of the non-minimum phase property (which depends on the operation point),
- Description of classical PI controller design based on a physical worst-case analysis of the non-minimum phase behavior of the linearized system dynamics,
- Introduction of a nonlinear PI controller design where the controller parameters are continuously adjusted online with respect to the actual "operating point". To ease implementation, analytical expressions to adjust the controller parameters online are derived based on the physical properties of the system dynamics, and
- Simulation results to illustrate and compare the control performance of the classical and the proposed nonlinear PI controller design. To show realistic results, the simulation comprises nonlinear and realistic models of the voltage source converter with pulse width modulation, underlying current control loops and nonlinear power flow and nonlinear DC-link dynamics. Moreover, as realistic input to the simulation model, the measured bi-directional (mechanical) power flow of a real airborne wind energy demonstrator during pumping mode is used (see Fig. 11.2, Courtesy of Roland Schmehl, TU Delft).

We do not present a thorough stability analysis of the proposed nonlinear controller. However, as a proof of concept, the simulation results illustrate that the closed-loop system is stable and robust to (bounded) parameter uncertainties.

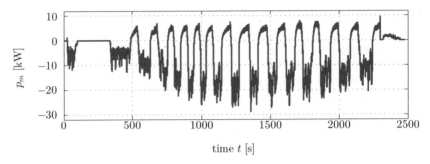

Fig. 11.2 Machine power $p_m(\cdot)$ measured by the TU Delft Kite Power group with their demonstrator on 23rd June 2012 kindly provided for our analysis. Negative is power generation (generator mode), positive is power demand (motor mode)

11.2 Problem Formulation

We consider a grid-connected power converter as shown in Fig. 11.3. It shares its DC-link with at least one electrical drive (electrical machine and voltage source inverter). The electrical drive is the actuator of the electrical drive train of the airborne wind energy (AWE) system which consists of the winch, the electrical machine (e.g. a permanent-magnet, reluctance or electrically-excited synchronous machine under four-quadrant control) and a voltage source inverter. The electrical drive converts mechanical power to electrical (machine[1]) power[2] p_m [W] which is exchanged with the DC-link via the DC-link power p_{dc} [W] and the grid-connected converter via the grid-side converter output power p_g [W]. During pumping mode (see Fig. 11.2), the machine power p_m changes its sign: During the reel-out phase, energy is generated (i.e. $p_m < 0$) and, during the reel-in phase, energy is dissipated (i.e. $p_m > 0$) in the ground winch drive system. Due to the DC-link with capacitance C_{dc} [F], machine and grid side are electrically coupled via the electrical power flow over the DC-link (for more details see Sect. 11.3) but for an almost constant DC-link voltage u_{dc} [V] both sides can be considered separately.

The grid-connected converter generates the voltages $\mathbf{u}_f^{abc} = (u_f^a, u_f^b, u_f^c)^\top [\text{V}]^3$ which are applied to the RL-filter.[3] At the point of common coupling (PCC, i.e. the point of the grid connection), a current $\mathbf{i}_f^{abc} = (i_f^a, i_f^b, i_f^c)^\top [\text{A}]^3$ will flow through the RL-filter with resistance R_f [Ω] and inductance L_f [H] into the balanced (ideal) grid with voltage $\mathbf{u}_g^{abc} = (u_g^a, u_g^b, u_g^c)^\top [\text{V}]^3$. To control the power flow on the grid side, the power converter requires a sufficiently large (positive) and almost constant DC-link voltage $u_{dc} \geq u_{dc,min}$, where $u_{dc,min} > 0$ [V] is the required minimum DC-link volt-

---

[1] Actually, the electrical power $p_e = \eta p_m$ is exchanged with the power converter. But, for simplicity, we assume that the electrical winch drive system has an efficiency of one, i.e. $\eta = 1$. This simplification is justified, since in real world, a non-ideal efficiency $\eta < 1$ would simply scale down the electrical power but will hardly affect the dynamics of the DC-link system.

[2] For details on the nomenclature of this chapter see p. 274.

[3] Another common filter topology is a LCL-filter (see [24, Chap. 11]).

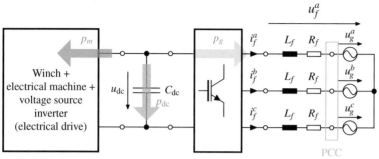

Fig. 11.3 Grid-connected converter with DC-link, filter and electrical drive

age for reasonable operation. The control objective is to achieve a stable set-point tracking of a given constant reference value $u_{dc,ref} \geq u_{dc,min} > 0$ for unknown and possibly bidirectional but bounded mechanical power flows $p_m(\cdot) \in \mathscr{L}^{\infty}(\mathbb{R}_{\geq 0}; \mathbb{R})$.

We consider the reference currents $\mathbf{i}_{f,ref}^{abc} = (i_{f,ref}^a, i_{f,ref}^b, i_{f,ref}^c)^\top \,[\text{A}]^3$ as control inputs, i.e. the underlying current control-loops (with decoupled controllers, voltage source converter and pulse width modulation or space vector modulation; for details see [9]) are already adequately designed. As feedback variables the currents $\mathbf{i}_f^{abc}$ and the DC-link voltage u_{dc} are available (full state-feedback).

11.3 Modeling and System Analysis of the Power Converter

For balanced three-phase systems (see Assumption (A.1) below), the system dynamics reduce to a two-phase system which is represented in a rotating $k = (d, q)$-reference frame or a fixed $s = (\alpha, \beta)$-reference frame instead of the (a, b, c)-reference frame. In general, for $\boldsymbol{\xi} \in \{\mathbf{u}_f, \mathbf{i}_f, \mathbf{i}_{f,ref}, \mathbf{u}_g\}$, we write

$$\boldsymbol{\xi}^k(t) := \left(\xi^d(t), \xi^q(t)\right)^\top := \mathbf{T}_p(\phi_g(t))^{-1} \underbrace{\mathbf{T}_c \boldsymbol{\xi}^{abc}(t)}_{=:\boldsymbol{\xi}^s(t)}$$

where $\phi_g(t)\,[\text{rad}]$ is the angle of the grid voltage,

$$\mathbf{T}_p(\phi_g) := \begin{bmatrix} \cos(\phi_g) & -\sin(\phi_g) \\ \sin(\phi_g) & \cos(\phi_g) \end{bmatrix}, \; \mathbf{J} := \begin{bmatrix} 0 & -1 \\ 1 & 0 \end{bmatrix} \text{ and } \mathbf{T}_c := \frac{2}{3}\begin{bmatrix} 1 & -\frac{1}{2} & -\frac{1}{2} \\ 0 & \frac{\sqrt{3}}{2} & -\frac{\sqrt{3}}{2} \end{bmatrix} \quad (11.1)$$

are Park transformation matrix, rotation matrix (by $\frac{\pi}{2}$ counter-clock wise) and (simplified) Clarke transformation matrix, respectively (for details see [9, 24]). In the remainder of this chapter, we align the $k = (d, q)$-reference frame with the grid voltage ("grid voltage orientation"). For modeling, we impose the following assumptions:

Assumption (A.1) *The grid is balanced with constant angular frequency* $\omega_g > 0\,[\mathrm{rad/s}]$ *and the* $k = (d,q)$ *reference frame is aligned with the grid voltage having magnitude* $\widehat{u}_g > 0\,[\mathrm{V}]$, *i.e.* $u_g^a(t) + u_g^b(t) + u_g^c(t) = 0$ *and*[4]

$$\forall t \geq 0: \qquad \mathbf{u}_g^k(t) = (u_g^d(t),\, u_g^q(t))^\top = (\widehat{u}_g, 0)^\top = \mathbf{T}_p(\phi_g(t))^{-1}\mathbf{T}_c\mathbf{u}_g^{abc}(t).$$

Assumption (A.2) *Power converter and DC-link are lossless (see Fig. 11.3), i.e.*

$$\forall t \geq 0: \qquad p_{\mathrm{dc}}(t) = -p_m(t) - p_g(t). \tag{11.2}$$

Assumption (A.3) *For current control-loop time constant* $T_{\mathrm{app}} > 0\,[\mathrm{s}]$, *the current dynamics are approximated by*

$$\frac{\mathrm{d}}{\mathrm{d}t}\mathbf{i}_f^k(t) = \frac{1}{T_{\mathrm{app}}}\big(-\mathbf{i}_f^k(t) + \mathbf{i}_{f,\mathrm{ref}}^k(t)\big), \quad \mathbf{i}_f^k(0) = \mathbf{i}_{f,0}^k := \mathbf{T}_p(\phi_{g,0})^{-1}\mathbf{T}_c\mathbf{i}_{f,0}^{abc} \in \mathbb{R}^2. \tag{11.3}$$

Assumption (A.4) *Reactive power reference and machine power are unknown but bounded, i.e.* $q_{\mathrm{pcc,ref}}(\cdot) \in \mathscr{L}^\infty(\mathbb{R}_{\geq 0};\mathbb{R})$ *and* $p_m(\cdot) \in \mathscr{L}^\infty(\mathbb{R}_{\geq 0};\mathbb{R})$, *respectively.*

Assumption (A.5) *The magnitude* $\widehat{u}_g$ *of the grid voltage is large compared to the voltage drop over the filter resistance, i.e.* $2R_f i_f^d(t) + \widehat{u}_g > 0$ *for all* $t \geq 0$.

Remark 11.1. For unbalanced (non-symmetric) grids, the situation becomes more difficult and positive, negative and zero sequence components must be considered (see [24]). For the symmetric case, the voltage orientation of the $k = (d,q)$ reference frame is achieved by the use of an adequate phase-locked loop algorithm (see [24, Chap. 8]). Although modern power converters have an efficiency up to 98 %, switching losses depend on the switching frequency. So Assumption (A.2) is a simplification. Assumption (A.3) is a standard assumption for current control-loops and holds for sufficiently high switching and current control frequencies [19, Sect. 13.4]. Assumption (A.4) is reasonable from a physical point of view. For most practical applications, we have $\widehat{u}_g \gg 1$ and $R_f \ll 1$, so Assumption (A.5) should hold. We will show that Assumption (A.5) is crucial for feasibility of any DC-link voltage controller.

11.3.1 Nonlinear DC-link Dynamics

By invoking Kirchhoff's current and voltage laws, the dynamics of the grid-side electrical circuit (as shown in Fig. 11.3) can be derived in the (a,b,c)-reference frame as follows

$$\mathbf{u}_f^{abc}(t) = R_f\mathbf{i}_f^{abc}(t) + L_f\frac{\mathrm{d}}{\mathrm{d}t}\mathbf{i}_f^{abc}(t) + \mathbf{u}_g^{abc}(t), \qquad \mathbf{i}_f^{abc}(0) = \mathbf{i}_{f,0}^{abc}. \tag{11.4}$$

---

[4] Note that $\mathbf{u}_g^k(t) = \mathbf{T}_p(\phi_g(t))^{-1}\mathbf{T}_c\mathbf{u}_g^{abc}(t)$ where $\phi_g(t) = \omega_g t + \phi_{g,0}$ for constant angular grid frequency $\omega_g > 0$ and initial angular position $\phi_{g,0} \in \mathbb{R}$.

Applying the (simplified) Clarke transformation as in Eq. (11.1) to Eq. (11.4) allows to rewrite the dynamics in the stationary $s = (\alpha, \beta)$-reference frame as follows

$$\mathbf{u}_f^s(t) = \mathbf{T}_c \mathbf{u}_f^{abc}(t) \overset{(11.4)}{=} \mathbf{T}_c R_f \mathbf{i}_f^{abc}(t) + \mathbf{T}_c L_f \frac{d}{dt} \mathbf{i}_f^{abc}(t) + \mathbf{T}_c \mathbf{u}_g^{abc}(t)$$

$$= R_f \mathbf{i}_f^s(t) + L_f \frac{d}{dt} \mathbf{i}_f^s(t) + \mathbf{u}_g^s(t), \qquad \mathbf{i}_f^s(0) = \mathbf{T}_c \mathbf{i}_{f,0}^{abc}. \ (11.5)$$

Then, in view of Assumption (A.1), utilizing the Park transformation as in Eq. (11.1) and the product rule[5] yield the system dynamics

$$\mathbf{u}_f^k(t) = \mathbf{T}_p(\phi_g(t))^{-1} \mathbf{u}_f^s(t)$$

$$\overset{(11.5)}{=} \mathbf{T}_p(\phi_g(t))^{-1} R_f \mathbf{i}_f^s(t) + \mathbf{T}_p(\phi_g(t))^{-1} L_f \frac{d}{dt} \left(\mathbf{T}_p(\phi_g(t)) \mathbf{i}_f^k(t) \right) + \mathbf{T}_p(\phi_g(t))^{-1} \mathbf{u}_g^s(t)$$

$$= R_f \mathbf{i}_f^k(t) + L_f \frac{d}{dt} \mathbf{i}_f^k(t) + \omega_g L_f \mathbf{J} \mathbf{i}_f^k(t) + \mathbf{u}_g^k(t) \tag{11.6}$$

in the rotating $k = (d, q)$-reference frame (with grid voltage orientation). In view of Assumption (A.2), the power balance in Eq. (11.2) holds at the DC-link which, for

$$p_{dc}(t) = u_{dc}(t) C_{dc} \frac{d}{dt} u_{dc}(t) \quad \text{and} \quad p_g(t) = \mathbf{u}_f^{abc}(t)^\top \mathbf{i}_f^{abc}(t) = \frac{3}{2} \mathbf{u}_f^k(t)^\top \mathbf{i}_f^k(t),$$

[24, Sect. 9.2] and, in view of Assumption (A.3), leads to the nonlinear DC-link dynamics in the rotating $k = (d, q)$-reference frame (for details see [9])

$$\frac{d}{dt} u_{dc}(t) \overset{(11.2)}{=} \frac{1}{C_{dc} u_{dc}(t)} \left[-p_m(t) - \frac{3}{2} \mathbf{u}_f^k(t)^\top \mathbf{i}_f^k(t) \right], \qquad u_{dc}(0) = u_{dc,0}$$

$$\overset{(11.6)}{=} \frac{3}{2 C_{dc} u_{dc}(t)} \left[-\frac{2}{3} p_m(t) - R_f \left\| \mathbf{i}_f^k(t) \right\|^2 - L_f \mathbf{i}_f^k(t) \frac{d}{dt} \mathbf{i}_f^k(t) \right.$$

$$\left. - \hat{u}_g \mathbf{i}_f^d(t) - \underbrace{\omega_g L_f \mathbf{i}_f^k(t)^\top \mathbf{J}^\top \mathbf{i}_f^k(t)}_{=0} \right] \tag{11.7}$$

$$\overset{(11.3)}{=} \frac{3}{2 C_{dc} u_{dc}(t)} \left[-\frac{2}{3} p_m(t) - \left(R_f - \frac{L_f}{T_{app}} \right) \left\| \mathbf{i}_f^k(t) \right\|^2 \right.$$

$$\left. - \frac{L_f}{T_{app}} \mathbf{i}_f^k(t)^\top \mathbf{i}_{f,ref}^k(t) - \hat{u}_g \mathbf{i}_f^d(t) \right] \tag{11.8}$$

with initial value $u_{dc}(0) = u_{dc,0} \geq u_{dc,min} > 0$, which is positive due to the flyback diodes in the power converter [20, Sect. 8.3].

The reactive power at the point of common coupling (PCC) is given by $q_{pcc}(t) = -\frac{3}{2} \hat{u}_g \mathbf{i}_f^q(t)$ [24, Sect. 9.2] and the reactive power reference $q_{pcc,ref}(\cdot)$ will be provided

---

[5] Note that $\frac{d}{dt} \mathbf{T}_p(\phi_g(t)) = \omega_g \mathbf{T}_p(\phi_g(t)) \mathbf{J}$ holds for all $t \geq 0$ and $\phi_g(t) = \omega_g t + \phi_{g,0}$.

by the grid operator. Hence, in view of Assumption (A.4), the current reference is

$$i_{f,\mathrm{ref}}^q(t) = -\frac{2q_{\mathrm{pcc,ref}}(t)}{3\hat{u}_g}. \tag{11.9}$$

So, the q-component of the filter current is, for all $t \geq 0$, given by

$$|i_f^q(t)| = \left| e^{-\frac{t}{T_{\mathrm{app}}}} i_{f,0}^q - \int_{t_0}^t e^{-\frac{1}{T_{\mathrm{app}}}(t-\tau)} \frac{2q_{\mathrm{pcc,ref}}(\tau)}{3T_{\mathrm{app}}\hat{u}_g} \right| \stackrel{(A.4)}{\leq} |i_{f,0}^q| + \frac{2\|q_{\mathrm{pcc,ref}}\|_\infty}{3\hat{u}_g} \tag{11.10}$$

and can be regarded as time-varying but bounded disturbance to the DC-link dynamics given in Eq. (11.8). For the following, we define state vector, input and disturbance by

$$\mathbf{x} := \begin{pmatrix} x_1 \\ x_2 \end{pmatrix} := \begin{pmatrix} u_{\mathrm{dc}} \\ i_f^d \end{pmatrix}, \quad u := i_{f,\mathrm{ref}}^d, \quad d := \tfrac{2}{3}p_m + \left(R_f - \frac{L_f}{T_{\mathrm{app}}}\right)(i_f^q)^2 - \frac{2L_f}{3T_{\mathrm{app}}\hat{u}_g}i_f^q q_{\mathrm{pcc,ref}} \tag{11.11}$$

(with i_f^q as in Eq. (11.10)), respectively. Note that, by Assumptions (A.4), $d(\cdot) \in \mathscr{L}^\infty(\mathbb{R}_{\geq 0}; \mathbb{R})$ holds. The (reduced) system dynamics can be written in standard form as follows

$$\left. \begin{aligned} \tfrac{\mathrm{d}}{\mathrm{d}t}\mathbf{x}(t) &= \mathbf{f}(\mathbf{x}(t), u(t), d(t)), \quad \mathbf{x}(0) = \begin{pmatrix} u_{\mathrm{dc},0} \\ i_{f,0}^d \end{pmatrix} \\ y(t) &= \underbrace{(1 \ 0)}_{=:\mathbf{c}^\top}\mathbf{x}(t) \end{aligned} \right\} \tag{11.12}$$

where

$$\mathbf{f}: \mathbb{R}^2 \times \mathbb{R} \times \mathbb{R} \to \mathbb{R}^2, \quad (\mathbf{x}, u, d) \mapsto \mathbf{f}(\mathbf{x}, u, d) :=$$

$$\begin{pmatrix} f_1(\mathbf{x}, u, d) \\ f_2(\mathbf{x}, u, d) \end{pmatrix} := \begin{pmatrix} \frac{3}{2C_{\mathrm{dc}}x_1}\left(-\left(R_f - \frac{L_f}{T_{\mathrm{app}}}\right)x_2^2 - \frac{L_f}{T_{\mathrm{app}}}x_2 u - \hat{u}_g x_2 - d\right) \\ \frac{1}{T_{\mathrm{app}}}(-x_2 + u) \end{pmatrix}. \tag{11.13}$$

11.3.2 Equilibrium and Linearization

For the following denote state, control input and disturbance at an equilibrium by

$$\left. \begin{aligned} \mathbf{x}^\star &:= (x_1^\star, x_2^\star)^\top = (u_{\mathrm{dc}}^\star, i_f^{d,\star})^\top \\ u^\star &:= i_{f,\mathrm{ref}}^{d,\star} \\ d^\star &:= \tfrac{2}{3}p_l^\star + \left(R_f - \frac{L_f}{T_{\mathrm{app}}}\right)(i_f^{q,\star})^2 - \frac{2L_f i_f^{q,\star}}{3T_{\mathrm{app}}\hat{u}_g}q_{\mathrm{PCC,ref}}^\star \stackrel{(11.9)}{=} \tfrac{2}{3}p_l^\star + R_f(i_f^{q,\star})^2. \end{aligned} \right\} \tag{11.14}$$

At equilibrium given in Eqs. (11.14), the following must hold $\frac{\mathrm{d}}{\mathrm{d}t}\mathbf{x} = \mathbf{f}(\mathbf{x}^\star, u^\star, d^\star) = \mathbf{0}_2$, which gives

$$u^\star = x_2^\star \qquad \text{and} \qquad R_f(x_2^\star)^2 + \widehat{u}_g x_2^\star = -d^\star, \tag{11.15}$$

where the second condition in Eq. (11.15) has the solution(s)

$$x_2^\star = -\frac{\widehat{u}_g}{2R_f}\left(1 \mp \sqrt{1 - \frac{4d^\star R_f}{\widehat{u}_g^2}}\right).$$

Only, for $d^\star \leq \frac{\widehat{u}_g^2}{4R_f}$ (which holds since $\widehat{u}_g \gg 1$ and $R_f \ll 1$ in real world), the solution is physically meaningful (non-complex roots). By denoting the small signals by

$$\mathbf{x}_l := \mathbf{x} - \mathbf{x}^\star, \quad u_l := u - u^\star, \quad y_l := y - y^\star, \quad \text{and} \quad d_l := d - d^\star, \tag{11.16}$$

a linearization of system in Eqs. (11.12) around the equilibrium $(\mathbf{x}^\star, \mathbf{u}^\star, d^\star)$ yields

$$\left.\begin{array}{l} \frac{d}{dt}\mathbf{x}_l(t) = \mathbf{A}^\star \mathbf{x}_l(t) + \mathbf{b}^\star u_l(t) + \mathbf{b}_d^\star d_l(t) \\ y_l(t) = \mathbf{c}^\top \mathbf{x}_l(t) \end{array}\right\} \tag{11.17}$$

where higher order terms are neglected and[6]

$$\mathbf{A}^\star := \left.\frac{\partial \mathbf{f}(\mathbf{x},u,d)}{\partial \mathbf{x}}\right|_{(\mathbf{x}^\star,u^\star,d^\star)} \overset{(11.15)}{=} \begin{bmatrix} 0 & \overbrace{\frac{3}{2C_{dc}u_{dc}^\star}\left(\left(\frac{L_f}{T_{app}} - 2R_f\right)i_f^{d,\star} - \widehat{u}_g\right)}^{=:a_{12}^\star} \\ 0 & -\frac{1}{T_{app}} \end{bmatrix} \in \mathbb{R}^{2\times2}, \tag{11.18}$$

$$\mathbf{b}^\star := \left.\frac{\partial \mathbf{f}(\mathbf{x},u,d)}{\partial u}\right|_{(\mathbf{x}^\star,u^\star,d^\star)} = \begin{pmatrix} \overbrace{-\frac{3}{2C_{dc}u_{dc}^\star}\frac{L_f}{T_{app}}i_f^{d,\star}}^{=:b_1^\star} \\ \frac{1}{T_{app}} \end{pmatrix} \in \mathbb{R}^2 \tag{11.19}$$

and

$$\mathbf{b}_d^\star := \left.\frac{\partial \mathbf{f}(\mathbf{x},u,d)}{\partial d}\right|_{(\mathbf{x}^\star,u^\star,d^\star)} = \begin{pmatrix} -\frac{3}{2C_{dc}u_{dc}^\star} \\ 0 \end{pmatrix} \in \mathbb{R}^2. \tag{11.20}$$

In the following, for brevity, we use $\star$ as superscript to indicate that the corresponding variable (matrix, vector, coefficient) depends on the operation point (e.g. $\mathbf{A}^\star = \mathbf{A}^\star(\mathbf{x}^\star,u^\star,d^\star)$ or $a_{12}^\star = a_{12}^\star(\mathbf{x}^\star)$), whereas variables without $\star$ do not depend on the operation point as given in Eqs. (11.14).

---

[6] Note that $\left.\dfrac{\partial f_1(\mathbf{x},u,d)}{\partial x_1}\right|_{(\mathbf{x}^\star,u^\star,d^\star)} = -\frac{1}{x_1^\star}\underbrace{f_1(\mathbf{x}^\star,u^\star,d^\star)}_{=0} = 0.$

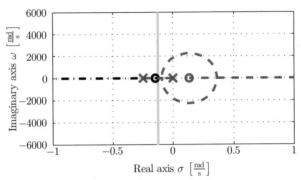

Fig. 11.4 Root locus of Eq. (11.21) for the cases (○: zero, ×: pole): —— $T_0^\star = 0$, - - - $T_0^\star > 0$ and
····· $T_0^\star < 0$ (non-minimum phase case)

11.3.3 Non-Minimum Phase Dynamics

To illustrate the non-minimum phase dynamics of the (linearized) system given
in Eqs. (11.17) during reel-in phase, we compute the transfer function (for details
see [9])

$$F_S(s) := \frac{y_l(s)}{u_l(s)} = \frac{(u_{dc} - u_{dc}^\star)(s)}{(i_{f,ref}^d - i_f^d)(s)} = \mathbf{c}^\top \left(s\mathbf{I}_2 - \mathbf{A}^\star\right)^{-1} \mathbf{b}^\star = -\frac{V_S^\star(1 + sT_V^\star)}{s(1 + sT_{app})}. \qquad (11.21)$$

where system gain $V_S^\star$ [V/A] and numerator time constant $T_V^\star$ [s] depend on the current $i_f^{d,\star}$ and are defined as follows

$$V_S^\star := V_S^\star(i_f^{d,\star}, u_{dc}^\star) := \frac{3(\widehat{u}_g + 2R_f i_f^{d,\star})}{2C_{dc} u_{dc}^\star} \quad \text{and} \quad T_V^\star := T_V^\star(i_f^{d,\star}) := \frac{L_f i_f^{d,\star}}{\widehat{u}_g + 2R_f i_f^{d,\star}}. \qquad (11.22)$$

For different operating points $(\mathbf{x}^\star, u^\star, d^\star)$, the numerator time constant $T_V^\star$ can
either be zero, positive or negative (note that, in view of Assumption (A.5), we have
$\widehat{u}_g + 2R_f i_f^{d,\star} > 0$). During the reel-in phase (motor mode), we have a zero in the right
(unstable) complex half-plane, since power is drawn from the grid and transferred
to the DC-link (see Fig. 11.3), i.e. $i_f^{d,\star} < 0$ and, hence, $T_V^\star < 0$: The system is *non-
minimum phase*. In Fig. 11.4, the root loci of Eq. (11.21) are plotted for the three
cases $T_V^\star = 0$, $T_V^\star > 0$ and $T_V^\star < 0$. If $T_V^\star < 0$, too large gains will render the closed-
system unstable which necessitates a rather conservative controller design.

11.3.4 Physical Explanation of the Non-Minimum Phase Behavior

In this section, we will explain the non-minimum phase behavior from a physical point of view. In particular, we will discuss the often observed initially reversed system response due to step-like reference changes.

First note that the DC-link voltage is (normally) larger than the grid voltage amplitude, i.e. $u_{dc} > \hat{u}_g$. Hence, during the reel-in phase, the power converter operates as boost converter to transfer energy from the grid to the DC-link which requires that energy is stored in the filter inductance before it can be pushed into the DC-link. For our analysis, we will consider a time instant $t \geq 0$ with the following properties

$$\left.\begin{array}{ll} \text{(i)} & i_f^d(t) < 0 \text{ (i.e. current flows from grid to DC-link)}, \\ \text{(ii)} & p_m(t) > 0 \text{ (i.e. reel-in phase, motor mode), and} \\ \text{(iii)} & \frac{d}{dt} i_f^q(t) = i_f^q(t) = 0 \text{ (i.e. no reactive power)}. \end{array}\right\} \quad (11.23)$$

For Eqs. (11.23), the nonlinear DC-link dynamics in Eq. (11.7) simplify to

$$\frac{d}{dt} u_{dc}(t) = \underbrace{\frac{1}{u_{dc}(t)C_{dc}}}_{\substack{> 0, \text{ see} \\ \text{Sect. 11.3.1}}} \left[\underbrace{-p_m(t)}_{\substack{(11.23) \\ < 0}} \underbrace{-\tfrac{3}{2}R_f i_f^d(t)^2 - \tfrac{3}{2}L_f i_f^d(t)\frac{d}{dt}i_f^d(t)}_{<0} \underbrace{-\tfrac{3}{2}\hat{u}_g i_f^d(t)}_{\substack{(11.23) \\ > 0}} \right].$$

$$(11.24)$$

We will only consider the case of a positive DC-link voltage reference change, i.e.

$$u_{dc,ref}(t) > u_{dc}(t) \implies i_{f,ref}^d(t) < i_f^d(t) \overset{(11.23)}{<} 0 \overset{(11.3)}{\implies} \frac{d}{dt} i_f^d(t) < 0. \quad (11.25)$$

The other case follows analogously. To (immediately) increase the DC-link voltage, $\frac{d}{dt} u_{dc}(t) > 0$ must hold and, from Eq. (11.24), it follows that this is feasible if and only if the time derivative of magnetic energy (in the filter inductance) satisfies

$$\tfrac{3}{2}L_f i_f^d(t)\frac{d}{dt}i_f^d(t) < \Big(\underbrace{-p_m(t)}_{\substack{(11.23) \\ < 0}} \underbrace{-\tfrac{3}{2}R_f i_f^d(t)^2 - \tfrac{3}{2}\hat{u}_g i_f^d(t)}_{\substack{(11.23) \\ > 0}} \Big) =: \alpha(t). \quad (11.26)$$

There exist two scenarios when the DC-link voltage will initially decrease, i.e. the typical non-minimum phase behavior with initially reversed system response:

(S$_1$) For a very large machine power $p_m(t) \gg 1$ (energy dissipation), we might have $\alpha(t) < 0$. Then, due to $i_f^d(t) < 0$ and $\frac{d}{dt}i_f^d(t) < 0$ in Eq. (11.25), the change in the magnetic energy $\tfrac{3}{2}L_f i_f^d(t)\frac{d}{dt}i_f^d(t)$ is positive which contradicts Eq. (11.26) and $u_{dc}(\cdot)$ will decrease until $\alpha(\tau) > 0$ will change its sign at some $\tau > t \geq 0$.

(S$_2$) For a small machine power $p_m(t) > 0$ and a large grid voltage $\hat{u}_g \gg 1$, we might have $\alpha(t) > 0$. But very fast current dynamics in Eq. (11.3) might yield $\tfrac{3}{2}L_f i_f^d(t)\frac{d}{dt}i_f^d(t) \geq \alpha(t)$ which also contradicts Eq. (11.26) and leads to an initial decrease of $u_{dc}(\cdot)$ until Eq. (11.26) holds again for some $\tau > t \geq 0$ with $\alpha(\tau) > \alpha(t)$.

Note that such a time instant $\tau > t \geq 0$ does exist, since the active power drawn from the grid, i.e. $-\frac{3}{2}\hat{u}_g i_f^d(t)$ in Eq. (11.26), will become larger and larger for more and more negative currents $i_f^d(\cdot)$ (as result of $\frac{d}{dt}i_f^d(t) < 0$).

Concluding, the non-minimum phase behavior of the DC-link voltage control problem arises from the change of the magnetic energy $\frac{3}{2}L_f i_f^d(t)\frac{d}{dt}i_f^d(t)$ in the filter inductance which might constrain the time derivative of the DC-link voltage for positive changes of the machine power (see Experiment (E_1) in Fig. 11.8 at $t = 0.2$ s) or for positive set-point changes of the DC-link reference voltage (see Experiment (E_3) in Fig. 11.12 at $t = 0.2$ s).

11.4 Classical DC-Link PI Controller Design

In this section, we discuss the classical PI controller design with *constant* controller parameters for the DC-link voltage set-point tracking problem.

For this classical approach the controller parameters are set after a reasonable tuning has been performed. The controller design is based on a (local) analysis of the linearized closed-loop system invoking the Hurwitz criterion. Applying a PI controller with transfer function

$$F_{PI}(s) = \frac{i_{f,ref}^d(s)}{u_{dc,ref}(s) - u_{dc}(s)} = -V_R \frac{1+sT_n}{sT_n}, \tag{11.27}$$

with controller gain $V_R\,[\mathrm{A/V}]$ and controller time constant $T_n\,[\mathrm{s}]$, to the (linearized) system in Eq. (11.21) yields the closed-loop transfer function

$$F_{CL,PI}(s) = \frac{F_{PI}(s)F_S(s)}{1+F_{PI}(s)F_S(s)} = \frac{V_R V_S^\star \frac{(1+sT_n)(1+sT_V^\star)}{s^2 T_n(1+sT_{app})}}{1+V_R V_S^\star \frac{(1+sT_n)(1+sT_V^\star)}{s^2 T_n(1+sT_{app})}} = \frac{\frac{V_R V_S^\star}{T_{app}T_n}(1+sT_n)(1+sT_V^\star)}{s^3 q_3^\star + s^2 q_2^\star + s q_1^\star + q_0^\star} =: \frac{N_{CL,PI}(s)}{D_{CL,PI}(s)}$$

$$\tag{11.28}$$

with the coefficients

$$q_3^\star = 1, \quad q_2^\star = \frac{1}{T_{app}} + \frac{V_R V_S^\star T_V^\star}{T_{app}}, \quad q_1^\star = V_R V_S^\star\left(\frac{1}{T_{app}} + \frac{T_V^\star}{T_n T_{app}}\right), \quad q_0^\star = \frac{V_R V_S^\star}{T_n T_{app}} \tag{11.29}$$

of the denominator polynomial $D_{CL,PI}(s)$. Now, the controller parameters V_R and T_n have to be specified (and tuned) to guarantee a stable closed-loop system behavior for all three operation points $T_V^\star = 0$, $T_V^\star > 0$ and $T_V^\star < 0$.

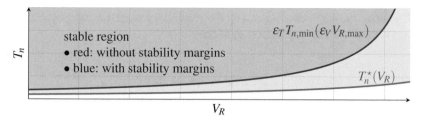

Fig. 11.5 Admissible region for controller gain V_R and controller time constant T_n to guarantee local stability (see sufficient conditions $(\mathbf{C}_1)$ and $(\mathbf{C}_2)$ in Eq. (11.30))

11.4.1 Local Stability Analysis Using the Hurwitz Criterion

In [9], the following two conditions for local stability were derived using the Hurwitz criterion:

$$(\mathbf{C}_1)\ 0 < V_R < \underbrace{\frac{1}{|T_V^\star||V_S^\star|}}_{:=V_R^\star} \quad \text{and} \quad (\mathbf{C}_2)\ T_n > \underbrace{\frac{T_{\text{app}}}{1 - V_R V_S^\star |T_V^\star|} + |T_V^\star|}_{:=T_n^\star(V_R)} > 0. \qquad (11.30)$$

These two conditions are *sufficient* and guarantee *local* stability in the sense that the Hurwitz criterion (i.e. $q_0^\star$, $q_1^\star$, $q_2^\star$, $q_3^\star > 0$ and $q_2^\star q_1^\star - q_3^\star q_0^\star > 0$ [13, Theorem 3.4.71]) is satisfied locally. The region for choosing admissible controller parameters to assure local stability is shown in Fig. 11.5.

Remark 11.2 (Controller sign). Note the minus sign of the PI controller in Eq. (11.27) which is crucial to compensate for the minus sign of the linearized system dynamics in Eq. (11.21).

11.4.2 Worst-Case Analysis

The upper and lower bounds in Eq. (11.30) on the controller gain V_R and the controller time constant T_n depend on the actual operating point in Eq. (11.14) (i.e., in particular, $i_f^{d,\star}$ and $u_{\text{dc}}^\star$). A worst-case analysis is beneficial such that the chosen controller parameters satisfy Eq. (11.30) for a wide range of different operation points. The goal of this section is to determine bounds $V_{R,\text{max}}$ and $T_{n,\text{min}}$ for the conditions in Eq. (11.30) such that the following holds for the complete operation range of the closed-loop system:

$$\forall i_f^{d,\star} \in [i_{f,\text{min}}^d, i_{f,\text{max}}^d]\ \forall u_{\text{dc}}^\star \in [u_{\text{dc,min}}, u_{\text{dc,max}}]:$$
$$0 < V_R < V_{R,\text{max}} \leq V_R^\star(i_f^{d,\star}, u_{\text{dc}}^\star) \quad \text{and} \quad T_n > T_{n,\text{min}} \geq T_n^\star(i_f^{d,\star}) > 0. \qquad (11.31)$$

To derive the worst-case bounds on the controller parameters, the physical limits of the system in *steady state* are computed.

11.4.2.1 DC-Link Voltage Limits (in Steady State)

The steady state DC-link voltage is constrained by the lower (positive) limit

$$u_{dc,min} > \max \left\{ \sqrt{\frac{4\omega_g^2 L_f^2 \hat{u}_g^2}{R_f^2 + \omega_g^2 L_f^2}} ; \frac{3\sqrt{3}}{\pi} \hat{u}_g \right\} > 0, \qquad (11.32)$$

which is due to the flyback diodes (which act as rectifier and, in continuous conduction mode, give the DC-link voltage $\frac{3\sqrt{3}}{\pi} \hat{u}_g$, see [20, pp. 85-90]), and the upper (positive) limit

$$u_{dc,max} > u_{dc,min} > 0 \qquad (11.33)$$

which is set by the user to protect the physical system (e.g. capacitance or switches).

11.4.2.2 Current Limits (in Steady State)

To ease computation of the physical upper and lower limits on the current i_f^d, the derivation is shown for steady state, i.e. $\frac{d}{dt}(\cdot) = 0$, and for $i_f^q = 0$ (which gives a maximal/minimal i_f^d). For this case, the system dynamics of Eq. (11.6) simplify to

$$u_f^d = R_f i_f^d + \hat{u}_g \qquad \text{and} \qquad u_f^q = \omega_g L_f i_f^d. \qquad (11.34)$$

Moreover, for a regularly sampled, symmetrical pulse width modulation scheme, the maximal magnitude of the admissible voltage vector (see [20, pp. 658-720]) is

$$\|\mathbf{u}_f^k\| = \sqrt{(u_f^d)^2 + (u_f^q)^2} \le \frac{u_{dc}}{2},$$

which leads to the following inequality constraint

$$\left\| \mathbf{u}_f^k \right\|^2 - \frac{u_{dc}^2}{4} \stackrel{(11.34)}{=} \left(R_f^2 + \omega_g^2 L_f^2 \right) (i_f^d)^2 + 2R_f \hat{u}_g i_f^d + \hat{u}_g^2 - \frac{u_{dc}^2}{4} \le 0. \qquad (11.35)$$

Solving Eq. (11.35) for i_f^d and inserting $u_{dc} = u_{dc}^\star$ gives the two solutions

$$i_f^{d,\star}(u_{dc}^\star) := \frac{-R_f \hat{u}_g \pm \sqrt{\left(R_f^2 + \omega_g^2 L_f^2 \right) \frac{(u_{dc}^\star)^2}{4} - \omega_g^2 L_f^2 \hat{u}_g^2}}{R_f^2 + \omega_g^2 L_f^2}. \qquad (11.36)$$

Considering the maximally admissible DC-link voltage, i.e. $u_{dc}^\star = u_{dc,max}$, allows to compute the upper (positive) current limit

$$i_{f,\max}^d := \frac{-R_f \widehat{u}_g + \sqrt{\left(R_f^2 + \omega_g^2 L_f^2\right) \frac{(u_{\mathrm{dc,max}})^2}{4} - \omega_g^2 L_f^2 \widehat{u}_g^2}}{R_f^2 + \omega_g^2 L_f^2} > 0. \tag{11.37}$$

and the lower (negative) current limit

$$i_{f,\min}^d := \frac{-R_f \widehat{u}_g - \sqrt{\left(R_f^2 + \omega_g^2 L_f^2\right) \frac{(u_{\mathrm{dc,max}})^2}{4} - \omega_g^2 L_f^2 \widehat{u}_g^2}}{R_f^2 + \omega_g^2 L_f^2} < 0 \ \text{ and } \ |i_{f,\min}^d| > i_{f,\max}^d. \tag{11.38}$$

11.4.2.3 Worst-Case Selection of Controller Gain V_R

To derive the upper limit $V_{R,\max}$ for the controller gain V_R, it is necessary to identify the minimal value of $V_R^\star$ which can be done as follows

$$\forall i_f^{d,\star} \in [i_{f,\min}^d, i_{f,\max}^d] \ \forall u_{\mathrm{dc}}^\star \in [u_{\mathrm{dc,min}}, u_{\mathrm{dc,max}}]:$$

$$V_R^\star (u_{\mathrm{dc}}^\star) \overset{(11.22),(A.5)}{=} \frac{2 C_{\mathrm{dc}} u_{\mathrm{dc}}^\star}{3 L_f |i_f^{d,\star}|} \overset{(11.36)}{=} \frac{2 C_{\mathrm{dc}} \left(R_f^2 + \omega_g^2 L_f^2\right) u_{\mathrm{dc}}^\star}{3 L_f \left(R_f \widehat{u}_g + \sqrt{\left(R_f^2 + \omega_g^2 L_f^2\right) \frac{(u_{\mathrm{dc}}^\star)^2}{4} - \omega_g^2 L_f^2 \widehat{u}_g^2}\right)} > 0. \tag{11.39}$$

To characterize the curve $V_R^\star(\cdot)$, its derivative with respect to $u_{\mathrm{dc}}^\star$ is computed

$$\frac{\mathrm{d}}{\mathrm{d}u_{\mathrm{dc}}^\star} V_R^\star (u_{\mathrm{dc}}^\star) = \frac{2 C_{\mathrm{dc}} \left(R_f^2 + \omega_g^2 L_f^2\right)}{2 L_f \left(R_f \widehat{u}_g + \sqrt{\left(R_f^2 + \omega_g^2 L_f^2\right) \frac{(u_{\mathrm{dc}}^\star)^2}{4} - \omega_g^2 L_f^2 \widehat{u}_g^2}\right)^2} \cdot$$

$$\cdot \left[R_f \widehat{u}_g + \sqrt{\left(R_f^2 + \omega_g^2 L_f^2\right) \frac{(u_{\mathrm{dc}}^\star)^2}{4} - \omega_g^2 L_f^2 \widehat{u}_g^2} - \frac{R_f^2 + \omega_g^2 L_f^2}{4 \sqrt{\left(R_f^2 + \omega_g^2 L_f^2\right) \frac{(u_{\mathrm{dc}}^\star)^2}{4} - \omega_g^2 L_f^2 \widehat{u}_g^2}} u_{\mathrm{dc}}^{\star \, 2} \right], \tag{11.40}$$

which shows that $V_R^\star(\cdot)$ is a monotonically decreasing function until its minimum is reached at (see Fig. 11.6)

$$u_{\mathrm{dc}}^\star = u_{\mathrm{dc}}^{\mathrm{opt}} := \sqrt{\frac{4 \omega_g^2 L_f^2 \widehat{u}_g^2}{R_f^2 + \omega_g^2 L_f^2} \left(\frac{\omega_g^2 L_f^2}{R_f^2} + 1\right)},$$

since the following holds true

$$\forall \, u_{\mathrm{dc}}^\star \in \left(u_{\mathrm{dc,min}}, u_{\mathrm{dc}}^{\mathrm{opt}}\right): \qquad \frac{\mathrm{d}}{\mathrm{d}u_{\mathrm{dc}}^\star} V_R^\star (u_{\mathrm{dc}}^\star) < 0. \tag{11.41}$$

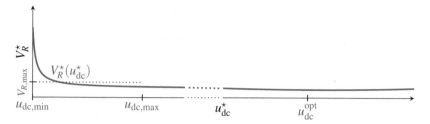

Fig. 11.6 Evolution of the function $u_{dc}^{\star} \mapsto V_R^{\star}(u_{dc}^{\star})$ for $u_{dc}^{\star} \geq u_{dc,min}$

Note that $u_{dc,max} \leq u_{dc}^{\star} < u_{dc}^{opt}$ is *not* feasible, hence the minimum of $V_R^{\star}(\cdot)$ on the admissible interval $[u_{dc,min}, u_{dc,max}]$ is given by

$$V_R^{\star}(u_{dc,max}) = \frac{2C_{dc}\left(R_f^2 + \omega_g^2 L_f^2\right)u_{dc,max}}{3L_f\left(R_f\hat{u}_g + \sqrt{\left(R_f^2 + \omega_g^2 L_f^2\right)\frac{(u_{dc,max})^2}{4} - \omega_g^2 L_f^2 \hat{u}_g^2}\right)} \stackrel{(11.38)}{=} \boxed{\frac{2C_{dc}u_{dc,max}}{3L_f\left|i_{f,min}^d\right|} =: V_{R,max}},$$

$$(11.42)$$

which represents a worst-case upper limit for the choice of the controller gain V_R of the *classical* PI controller in Eq. (11.27) with constant parameters. To satisfy the inequality in Eq. (11.31), we introduce a safety margin ε_V and choose the controller gain as follows

$$\boxed{V_R = \varepsilon_V V_{R,max} \qquad \text{with} \qquad 0 < \varepsilon_V < 1.} \qquad (11.43)$$

Remark 11.3 (Simplified worst case analysis). Note that a simplified worst case analysis yields

$$V_R^{\star} \stackrel{(11.22),(A.5)}{=} \frac{1}{|T_V^{\star}||V_S^{\star}|} = \frac{2C_{dc}u_{dc}^{\star}}{3L_f\left|i_f^{d,\star}\right|} \stackrel{(11.38)}{\geq} \frac{2C_{dc}u_{dc,min}}{3L_f\left|i_{f,min}^d\right|} =: \widetilde{V}_{R,max}, \qquad (11.44)$$

(i.e. using $u_{dc,min}$ in the nominator instead of $u_{dc,max}$) which gives even a more conservative upper bound on the PI controller gain V_R. For the simulated system in Sect. 11.6, this would cause a reduction of $V_{R,max}$ by 37.5 % and, hence, an even more conservative controller design.

11.4.2.4 Worst-Case Selection of Controller Time Constant T_n

To select the controller time constant T_n as requested in Eq. (11.31), we need to derive the lower bound $T_{n,min}$. Straight-forward calculations show that the following holds for all $i_f^{d,\star} \in [i_{f,min}^d, i_{f,max}^d]$ and for all $u_{dc}^{\star} \in [u_{dc,min}, u_{dc,max}]$

$$T_n^\star \overset{(11.30)}{=} \frac{T_{\mathrm{app}}}{1-V_R V_S^\star |T_V^\star|} + |T_V^\star| \overset{(11.30),(11.22)}{=} \frac{T_{\mathrm{app}}}{1-\frac{V_R}{V_R^\star}} + \left| \frac{L_f \dot{i}_f^{d,\star}}{\widehat{u}_g + 2R_f i_f^{d,\star}} \right|$$

$$\overset{(11.39)}{\leq} \frac{T_{\mathrm{app}}}{1-\frac{V_R}{V_{R,\max}}} + \frac{L_f |\dot{i}_f^{d,\star}|}{\widehat{u}_g - 2R_f |i_f^{d,\star}|} \overset{(11.38),(11.43)}{\leq} \boxed{\frac{T_{\mathrm{app}}}{1-\varepsilon_V} + \frac{L_f |\dot{i}_{f,\min}^d|}{\widehat{u}_g - 2R_f |i_{f,\min}^d|} =: T_{n,\min}} .$$

$$(11.45)$$

Now, for any stability margin $\varepsilon_T > 1$, we choose the controller time constant to

$$\boxed{T_n = \varepsilon_T T_{n,\min} > T_{n,\min} \geq T_n^\star(V_R) \qquad \text{with} \qquad \varepsilon_T > 1,} \qquad (11.46)$$

and, therefore, will assure that Eq. (11.30) holds true.

11.5 Nonlinear DC-link PI Controller Design

Due to the possible non-minimum phase behavior of the DC-link dynamics and its *constant* controller parameters, the *classical* PI controller must be tuned in a very conservative fashion (recall Sect. 11.4.2) which leads to a very slow closed-loop system response for most operation points. In this section, we propose a nonlinear PI controller design which instantaneously adjusts its controller parameters to an (approximate) actual operation point ("online parameter adjustment"). The nonlinear controller has the following state space representation

$$\left. \begin{aligned} \frac{\mathrm{d}}{\mathrm{d}t} x_i(t) &= u_{\mathrm{dc,ref}}(t) - u_{\mathrm{dc}}(t) , \qquad\qquad\qquad\qquad\qquad x_i(0) = 0 \\ i_{f,\mathrm{ref}}^d(t) &= V_R(i_f^d, u_{\mathrm{dc}}) \left(u_{\mathrm{dc,ref}}(t) - u_{\mathrm{dc}}(t) \right) + \frac{V_R(i_f^d, u_{\mathrm{dc}})}{T_n(i_f^d, u_{\mathrm{dc}})} x_i(t) \end{aligned} \right\} \quad (11.47)$$

and requires feedback of the actual d-component $i_f^d(t)$ of the filter current and the DC-link voltage $u_{\mathrm{dc}}(t)$ (both are measured and, therefore, available for feedback).

For controller tuning, we specify a desired (local) closed-loop system response via a given Hurwitz polynomial and implement an "online pole placement" strategy to adjust the controller parameters online. Recalling the system order of the closed-loop system in Eq. (11.28), *three* poles have to be specified. More precisely, there is one *real* pole $\lambda_1 \in \mathbb{R}$ and a (possibly) *conjugate-complex* pole pair $\lambda_R \pm \iota \lambda_I \in \mathbb{C}$ which defines the desired closed-loop system polynomial

$$D_{\mathrm{CL,PI}}^{\mathrm{desired}}(s) := (s - \lambda_1)(s - \lambda_R - \iota \lambda_I)(s - \lambda_R + \iota \lambda_I) = s^3 p_3^\star + s^2 p_2^\star + s p_1^\star + p_0^\star$$

$$(11.48)$$

with coefficients

$$p_3^\star = 1, \quad p_2^\star = -2\lambda_R - \lambda_1, \quad p_1^\star = \lambda_R^2 + \lambda_I^2 + 2\lambda_1 \lambda_R, \quad p_0^\star = -\lambda_1(\lambda_R^2 + \lambda_I^2).$$

$$(11.49)$$

Clearly, the desired polynomial in Eq. (11.48) must be a Hurwitz polynomial which is satisfied if and only if $\lambda_1 < 0$ and $\lambda_R < 0$ or $p_0^\star, p_1^\star, p_2^\star, p_3^\star > 0$ and $p_1^\star p_2^\star - p_3^\star p_0^\star > 0$ hold true. Important to note that, due to the use of a PI controller, we only have *two* design parameters (i.e. V_R and T_n) and, so, the problem is under-determined. Therefore, we will only specify (or fix) λ_R and λ_I and leave λ_1 free. It will depend on λ_R, λ_I, V_R and T_n and must be negative which has to be assured by an appropriate choice of λ_R and λ_I.

11.5.1 Pole Placement

Comparing the coefficients of the desired polynomial $D_{\mathrm{CL,PI}}^{\mathrm{desired}}(s)$ in Eq. (11.48) and the denominator polynomial $D_{\mathrm{CL,PI}}(s)$ of the linearized closed-loop system dynamics in Eq. (11.28) allows to solve for the controller parameters V_R and T_n and for the free pole λ_1 as follows (details are omitted)

$$
\begin{aligned}
V_R &= -\frac{2\lambda_R + T_V^\star \lambda_I^2 + T_V^\star \lambda_R^2 - T_{\mathrm{app}}\lambda_I^2 + 3T_{\mathrm{app}}\lambda_R^2 + 2T_V^\star T_{\mathrm{app}}\lambda_R^3 + 2T_V^\star T_{\mathrm{app}}\lambda_I^2 \lambda_R}{V_S^\star\left((T_V^\star)^2\lambda_R^2 + (T_V^\star)^2\lambda_I^2 + 2T_V^\star \lambda_R + 1\right)} \\
&= -\frac{2\lambda_R\left(T_V^\star(\lambda_R^2+\lambda_I^2) + 2\lambda_R + \frac{1}{T_{\mathrm{app}}}\right) + \left(\frac{T_V^\star}{T_{\mathrm{app}}} - 1\right)\left(\lambda_R^2 + \lambda_I^2\right)}{\frac{V_S^\star}{T_{\mathrm{app}}}\left((T_V^\star)^2\left(\lambda_R^2 + \lambda_I^2\right) + 2T_V^\star \lambda_R + 1\right)},
\end{aligned}
\tag{11.50}
$$

$$
\begin{aligned}
T_n &= -\frac{2\lambda_R + T_V^\star \lambda_I^2 + T_V^\star \lambda_R^2 - T_{\mathrm{app}}\lambda_I^2 + 3T_{\mathrm{app}}\lambda_R^2 + 2T_V^\star T_{\mathrm{app}}\lambda_R^3 + 2T_V^\star T_{\mathrm{app}}\lambda_I^2 \lambda_R}{\left(\lambda_R^2 + \lambda_I^2\right)\left(T_V^\star T_{\mathrm{app}}\lambda_I^2 + T_V^\star T_{\mathrm{app}}\lambda_R^2 + 2T_{\mathrm{app}}\lambda_R + 1\right)} \\
&= -\frac{2\lambda_R\left(T_V^\star(\lambda_R^2+\lambda_I^2) + 2\lambda_R + \frac{1}{T_{\mathrm{app}}}\right) + \left(\frac{T_V^\star}{T_{\mathrm{app}}} - 1\right)\left(\lambda_R^2 + \lambda_I^2\right)}{\left(\lambda_R^2 + \lambda_I^2\right)\left(T_V^\star(\lambda_R^2 + \lambda_I^2) + 2\lambda_R + \frac{1}{T_{\mathrm{app}}}\right)}
\end{aligned}
\tag{11.51}
$$

and

$$
\lambda_1 = -\frac{T_V^\star\left(\lambda_R^2 + \lambda_I^2\right) + 2\lambda_R + \frac{1}{T_{\mathrm{app}}}}{(T_V^\star)^2\left(\lambda_R^2 + \lambda_I^2\right) + 2T_V^\star \lambda_R + 1}.
\tag{11.52}
$$

Remark 11.4. Solving for λ_R or λ_I (so one of those is free) instead of λ_1 would yield an infinite closed-loop pole or an infinite controller gain if $T_V^\star = 0$. Therefore, λ_1 is considered as free pole.

11.5.2 Sufficient Condition for (Local) Stability

For a stable behavior of the closed-loop system in Eq. (11.28), the real parts of all poles must be negative, i.e. $\lambda_1 < 0$ and $\lambda_R < 0$. Clearly, λ_R can be chosen negative, but value and sign of λ_1 depend on the time constant $T_V^\star$ of the linearized system in Eq. (11.21), the time constant T_{app} of the converter, and the real pole λ_R and the imaginary part λ_I of the conjugate-complex pole pair. Moreover, it has to be assured that the *varying* controller parameters $V_R = V_R(T_V^\star, V_S^\star)$ and $T_n = T_n(T_V^\star)$ will remain positive over the complete operation range (i.e. $T_V^\star = 0$, $T_V^\star > 0$ and $T_V^\star < 0$).

11.5.2.1 Assuring a Negative Real Pole λ_1

To assure that λ_1 in Eq. (11.52) is negative, we will derive bounds on the choices of $\lambda_R < 0$ and $\lambda_I \in \mathbb{R}$. First note that, we may rewrite the real pole as follows

$$\lambda_1 = -\frac{T_V^\star(\lambda_R^2 + \lambda_I^2) + 2\lambda_R + \frac{1}{T_{\mathrm{app}}}}{(T_V^\star)^2(\lambda_R^2 + \lambda_I^2) + 2T_V^\star\lambda_R + 1} =: -\frac{N_{\lambda_1}(T_V^\star)}{D_{\lambda_1}(T_V^\star)} \tag{11.53}$$

in compact form. Analyzing the denominator yields

$$\begin{aligned}
D_{\lambda_1}(T_V^\star) &:= (T_V^\star)^2\left(\lambda_R^2 + \lambda_I^2\right) + 2T_V^\star\lambda_R + 1 \\
&= \begin{cases} 1 & T_V^\star = 0 \\ (T_V^\star)^2\left(\lambda_R + \frac{1}{T_V^\star}\right)^2 + \lambda_I^2(T_V^\star)^2, & T_V^\star \neq 0 \end{cases} \\
&\Longleftrightarrow \quad \forall T_V^\star \in \mathbb{R}: \quad D_{\lambda_1}(T_V^\star) > 0,
\end{aligned} \tag{11.54}$$

which shows that the denominator D_{λ_1} is positive over the whole operation range. Hence, to achieve $\lambda_1 < 0$, the numerator of Eq. (11.53) must also be positive, i.e. $N_{\lambda_1}(T_V^\star) > 0$ for all operation points $T_V^\star = 0$, $T_V^\star > 0$, and $T_V^\star < 0$. The numerator can be written as

$$\begin{aligned}
N_{\lambda_1}(T_V^\star) &:= T_V^\star(\lambda_R^2 + \lambda_I^2) + 2\lambda_R + \frac{1}{T_{\mathrm{app}}} \\
&= \begin{cases} 2\lambda_R + \frac{1}{T_{\mathrm{app}}} & T_V^\star = 0 \\ T_V^\star\left(\lambda_R + \frac{1}{T_V^\star}\right)^2 + T_V^\star\lambda_I^2 - \frac{1}{T_V^\star} + \frac{1}{T_{\mathrm{app}}}, & T_V^\star \neq 0, \end{cases}
\end{aligned} \tag{11.55}$$

which might change its sign with $T_V^\star$ and the choices of λ_R and λ_I. To check the sign of numerator N_{λ_1}, the three following cases have to be investigated to derive bounds on λ_R and λ_I, respectively:

• Case $T_V^\star = 0$:

$$N_{\lambda_1} \overset{(11.52)}{=} 2\lambda_R + \frac{1}{T_{\mathrm{app}}} > 0 \quad \Longleftarrow \quad \lambda_R > -\frac{1}{2T_{\mathrm{app}}} \tag{11.56}$$

- Case $T_V^\star > 0$:

$$N_{\lambda_1} \overset{(11.52)}{=} \underbrace{T_V^\star \lambda_I^2}_{>0} + \underbrace{T_V^\star \lambda_R^2}_{>0} + 2\lambda_R + \frac{1}{T_{\text{app}}} > 0 \quad \Longleftarrow \quad \lambda_R > -\frac{1}{2T_{\text{app}}} \qquad (11.57)$$

- Case $T_V^\star < 0$:

$$N_{\lambda_1} = T_V^\star \left(\lambda_R + \frac{1}{T_V^\star} \right)^2 + T_V^\star \lambda_I^2 - \frac{1}{T_V^\star} + \frac{1}{T_{\text{app}}} > 0$$

$$\Longleftarrow \quad |\lambda_R| > -\frac{1}{T_V^\star} \pm \sqrt{-\lambda_I^2 + \frac{1}{(T_V^\star)^2} - \frac{1}{T_V^\star T_{\text{app}}}} \quad \text{and} \quad |\lambda_I| < \sqrt{\frac{1}{T_{\text{app}}|T_V^\star|}} \qquad (11.58)$$

Evaluating and combining the results above and imposing the necessary condition $\lambda_R < 0$, we obtain the following sufficient condition

$$\boxed{\begin{aligned} \max\left\{ -\frac{1}{T_V^\star} - \sqrt{-\lambda_I^2 + \frac{1}{(T_V^\star)^2} - \frac{1}{T_V^\star T_{\text{app}}}}, \; -\frac{1}{2T_{\text{app}}} \right\} &< \lambda_R < 0 \quad \text{and} \quad |\lambda_I| < \sqrt{\frac{1}{T_{\text{app}}|T_V^\star|}} \\ \Longrightarrow \qquad \forall T_V^\star \in \mathbb{R}: \quad N_{\lambda_1}(T_V^\star) > 0 \quad &\text{and} \quad D_{\lambda_1}(T_V^\star) > 0 \\ \Longrightarrow \qquad \forall T_V^\star \in \mathbb{R}: \quad \lambda_1 = -\frac{N_{\lambda_1}(T_V^\star)}{D_{\lambda_1}(T_V^\star)} &< 0. \end{aligned}}$$

$$(11.59)$$

which assures local stability of the closed-loop system in Eq. (11.28).

Remark 11.5 (Comments on stability). Clearly, the nonlinear PI controller design is based on the linearized system in Eq. (11.17), hence pole placement will only hold locally. The drawback of a local result, we try to overcome by online adjustment of the controller parameters ("online pole placement"). However, by online adjustment, the controller parameters in Eq. (11.60) and in Eq. (11.61) of the linearized closed-loop system in Eq. (11.28) become "time-varying" or, more precisely, nonlinear. So global stability can not be deduced by checking negativity of the real parts of the poles of the linearized closed-loop system in Eq. (11.28).

11.5.2.2 Assuring Positive Controller Parameters

In addition to conditions in Eqs. (11.59), we check whether the controller parameters will remain positive over the whole operation range (otherwise positive feedback might endanger stability). First note that, by invoking N_{λ_1} as in Eq. (11.55) and D_{λ_1} as in Eq. (11.54), we may rewrite the controller parameters as follows

$$V_R = -\frac{2\lambda_R N_{\lambda_1}(T_V^\star) + \left(\frac{T_V^\star}{T_{\text{app}}} - 1 \right)(\lambda_R^2 + \lambda_I^2)}{\frac{V_S^\star}{T_{\text{app}}} D_{\lambda_1}(T_V^\star)} \quad \text{and} \quad T_n = -\frac{2\lambda_R N_{\lambda_1}(T_V^\star) + \left(\frac{T_V^\star}{T_{\text{app}}} - 1 \right)(\lambda_R^2 + \lambda_I^2)}{(\lambda_R^2 + \lambda_I^2) N_{\lambda_1}(T_V^\star)}.$$

In view of the sufficient conditions in Eqs. (11.59) for local stability and the additional but *physically reasonable* assumption:

Assumption (A.6) *For the whole operation range, the following holds true*

$$\forall i_f^{d,\star} \leq i_{f,\text{max}}^d : \qquad T_V^\star(i_f^{d,\star}) < T_{\text{app}} \quad \Longleftrightarrow \quad \frac{T_V^\star(i_f^{d,\star})}{T_{\text{app}}} < 1,$$

it is easy to see that the numerators are negative and the denominators are positive, i.e. $2\lambda_R N_{\lambda_1}(T_V^\star) + \left(\frac{T_V^\star}{T_{\text{app}}} - 1\right)\left(\lambda_R^2 + \lambda_I^2\right) < 0$, $V_S^\star D_{\lambda_1}(T_V^\star) > 0$ and $\left(\lambda_R^2 + \lambda_I^2\right)N_{\lambda_1}(T_V^\star) > 0$ for all $T_V^\star \in \mathbb{R}$, which implies positivity of the controller parameters, i.e. $V_R(T_V^\star) > 0$ and $T_n(T_V^\star) > 0$ for all $T_V^\star \in \mathbb{R}$.

11.5.3 Online Adjustment of the Controller Parameters

This far the controller design was based on the linearized closed-loop dynamics assuming that an equilibrium exists. For implementation and online parameter adjustment the actual d-component current $i_f^d(t)$ and the actual DC-link voltage $u_{\text{dc}}(t)$ measurements will be used. Using the approximations

$$V_S^\star(i_f^d, u_{\text{dc}}) = \frac{3(\hat{u}_g + 2R_f i_f^d)}{2C_{\text{dc}} u_{\text{dc}}} \approx V_S^\star(i_f^{d,\star}, u_{\text{dc}}^\star) \quad \text{and} \quad T_V^\star(i_f^d) \approx T_V^\star(i_f^{d,\star}) := \frac{L_f i_f^d}{\hat{u}_g + 2R_f i_f^d}$$

the controller parameters become functions of the measured values as follows

$$V_R(i_f^d(t), u_{\text{dc}}(t)) = -\frac{2\lambda_R\left(T_V^\star(i_f^d(t))(\lambda_R^2 + \lambda_I^2) + 2\lambda_R + \frac{1}{T_{\text{app}}}\right) + \left(\frac{T_V^\star(i_f^d(t))}{T_{\text{app}}} - 1\right)(\lambda_R^2 + \lambda_I^2)}{\frac{V_S^\star(i_f^d(t), u_{\text{dc}}(t))}{T_{\text{app}}}\left((T_V^\star)^2(\lambda_R^2 + \lambda_I^2) + 2T_V^\star\lambda_R + 1\right)}$$

$$\tag{11.60}$$

and

$$T_n(i_f^d(t)) = -\frac{2\lambda_R\left(T_V^\star(i_f^d(t))(\lambda_R^2 + \lambda_I^2) + 2\lambda_R + \frac{1}{T_{\text{app}}}\right) + \left(\frac{T_V^\star(i_f^d(t))}{T_{\text{app}}} - 1\right)(\lambda_R^2 + \lambda_I^2)}{(\lambda_R^2 + \lambda_I^2)\left(T_V^\star(i_f^d(t))(\lambda_R^2 + \lambda_I^2) + 2\lambda_R + \frac{1}{T_{\text{app}}}\right)}. \tag{11.61}$$

Note that the integrator time "constant" $T_n(i_f^d(t))$ does not depend on $u_{\text{dc}}(t)$.

11.6 Simulation Results

In this section, the overall grid-connected voltage source power converter (including switching behavior, pulse width modulation underlying current control-loops) with the *classical* and the *nonlinear* DC-link PI controllers is implemented using Malab/Simulink. The goal is to investigate and illustrate (i) closed-loop system sta-

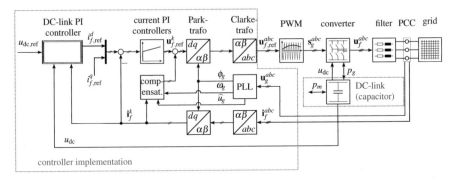

Fig. 11.7 Block diagram of the implementation of the overall DC-link control system with underlying current control-loops, switching behavior of the converter, pulse width modulation (PWM), and phase-locked loop (PLL) for grid synchronization. More details can be found in [9] (with similar notation) or [24, Chap. 9]

bility, (ii) control performance of the controllers and (iii) impact of parameter uncertainties on the control performance of the nonlinear PI controller.

11.6.1 Implementation

Fig. 11.7 shows the block diagram of the implementation of controller and DC-link system in Matlab/Simulink. Filter and grid are implemented as three-phase systems in the (a,b,c)-reference frame (instead of Eq. (11.6), for details see [9]). The DC-link dynamics are as in Eq. (11.2). For given filter voltage reference $\mathbf{u}_{f,\mathrm{ref}}^{abc}\,[\mathrm{V}]^3$ (coming from the current PI controllers), the pulse width modulation (PWM) generates the corresponding switching patterns $\mathbf{s}_g^{abc}\,[1]^3$ for the converter. To estimate angle $\phi_g\,[\mathrm{rad}]$, angular velocity $\omega_g\,[\mathrm{rad/s}]$ and amplitude $\widehat{u}_g\,[\mathrm{V}]$ of the three-phase grid voltage $\mathbf{u}_g^{abc}\,[\mathrm{V}]^3$, a phase-locked loop (PLL) is implemented (see [9] or [24, Chap. 8]). The angle $\phi_g\,[\mathrm{rad}]$ is required for the grid voltage orientation of the $k=(d,q)$-reference frame. Angular velocity $\omega_g\,[\mathrm{rad/s}]$ and voltage amplitude $\widehat{u}_g$ are needed for the compensation of the cross-coupling (see Eq. (11.6)) in the current control-loops to decouple the $i_f^d\,[\mathrm{A}]$- and $i_f^q\,[\mathrm{A}]$-dynamics.[7] With the Park and Clarke transformation in Eq. (11.1), the three-phase signals are transformed to the $k=(d,q)$-reference frame and vice versa (grid voltage orientation). The current PI controllers are tuned according to the *Magnitude Optimum* which, with current decoupling feedforward control, allows to approximate the current control-loop dynamics by Eq. (11.3) (see [9] and Assumption (A.3)). Implementation and system data is collected in Table 11.1.

---

[7] Note that an ideal decoupling is not feasible e.g. due to delays and non-causal compensation terms. For details see [9].

description	symbols & values (with unit)
Implementation data in Matlab/Simulink	
solver (fixed step)	ode4 (Runge-Kutta)
fixed-step size	$h = 2 \times 10^{-6}$ s (fundamental sample time)
System data	
grid	$\widehat{u}_g = 250$ V, $\omega_g = 2\pi 50 \frac{\text{rad}}{\text{s}}$ (balanced)
filter	$R_f = 5 \times 10^{-3} \Omega$, $L_f = 3.6 \times 10^{-3}$ H
converter	$f_{\text{pwm}} = 8 \times 10^3$ Hz, $C_{\text{dc}} = 400 \times 10^{-6}$ F
	$u_{\text{dc,min}} = 500$ V, $u_{\text{dc,max}} = 800$ V
current control	$T_{\text{app}} = 1.25 \times 10^{-4}$ s (implementation as in [9])
	$i_{f,\text{min}}^d = -277$ A, $i_{f,\text{max}}^d = +275$ A
Controller design data	
classical PI (11.27)	$V_R = \varepsilon_V V_{R,\text{max}}$ as in (11.43), $T_n = \varepsilon_T T_{n,\text{min}}$ as in (11.46)
	with $\varepsilon_V = 0.8$ and $\varepsilon_T = 1.25$
nonlinear PI (11.47)	$V_R(i_f^d, u_{\text{dc}})$ as in (11.60), $T_n(i_f^d)$ as in (11.61)
	with $\lambda_I = -200 \frac{\text{rad}}{\text{s}}$ and $\lambda_R = -450 \frac{\text{rad}}{\text{s}}$

Table 11.1 Implementation, system, and controller design data (if not stated otherwise)

The classical DC-link PI controller in Eq. (11.27) is implemented in state space. The controller parameters are listed in Table 11.1. The factors $\varepsilon_V = 0.8$ and $\varepsilon_T = 1.25$ are the stability margins as introduced in Eq. (11.43) and in Eq. (11.46), respectively (see Fig. 11.5).

The nonlinear DC-link PI controller is implemented as in Eq. (11.47). Its varying gains as in Eq. (11.60) and in Eq. (11.61) are adjusted online with respect to the actual measurements of $i_f^d(t)$ and $u_{\text{dc}}(t)$. For "online pole placement", the desired poles were chosen as listed in Table 11.1.

Remark 11.6. In stand-alone operation of the airborne wind energy system (AWES), the AWES usually operates as voltage source (not as current source as described above). Therefore, the grid-side voltage source inverter comes with an LC-filter and the filter output voltage is controlled by an outer control loop. In this case, the DC-link controller must be implemented on the machine side. DC-link controller design on machine side is slightly more complex (due to the nonlinearity of the machine and the aerodynamical torque) but, in principle, very similar to the presented results; in particular, the possible non-minimum phase behavior of the DC-link dynamics remains and imposes the most severe challenge to controller design and stability.

11.6.2 Simulation Experiments

To illustrate and evaluate the control performance of classical and nonlinear DC-link PI controller, *four* simulation experiments are implemented in Matlab/Simulink:

(E$_1$) Comparison of the control performance of the classical PI controller in Eq. (11.27) and the nonlinear PI controller in Eq. (11.47) for decreasing values of the DC-link capacitance $C_{dc} \in \{800 \times 10^{-6}\,\text{F}, 600 \times 10^{-6}\,\text{F}, 400 \times 10^{-6}\,\text{F}\}$ (see Fig. 11.8).

(E$_2$) Disturbance rejection capability of the nonlinear PI controller given in Eq. (11.47) *under parameter uncertainties*:

- $\pm 30\%$ parameter uncertainty in the DC-link capacitance C_{dc} (see Fig. 11.9),
- $\pm 30\%$ parameter uncertainty in filter resistance R_f (see Fig. 11.10), and
- $\pm 30\%$ parameter uncertainty in filter inductance L_f (see Fig. 11.11).

(E$_3$) Set-point tracking performance of the nonlinear PI controller in Eq. (11.47) *under parameter uncertainties*:

- $\pm 30\%$ parameter uncertainty in the DC-link capacitance C_{dc} (see Fig. 11.12),
- $\pm 30\%$ parameter uncertainty in filter resistance R_f (see Fig. 11.13), and
- $\pm 30\%$ parameter uncertainty in filter inductance L_f (see Fig. 11.14).

(E$_4$) Control performance of the nonlinear PI controller in Eq. (11.47) *for a real (measured) machine power flow* (see Fig. 11.15).

11.6.2.1 Discussion of Experiment (E$_1$)

Experiment (E$_1$) compares the disturbance rejection capabilities of the *classical* and *nonlinear* DC-link PI controllers. The simulation results for the experiment are depicted in Fig. 11.8. The following signals are shown: machine power p_m (with changing sign acting as disturbance, see first sub-plot) and the DC-link voltage u_{dc} for three different values of the DC-link capacitor $C_{dc} = 800 \times 10^{-6}\,\text{F}$ (see second sub-plot), $C_{dc} = 600 \times 10^{-6}\,\text{F}$ (see third sub-plot) and $C_{dc} = 400 \times 10^{-6}\,\text{F}$ (see fourth sub-plot). For all three values of C_{dc}, the control performance of the *nonlinear* PI controller is superior to the *classical* PI controller. Its disturbance rejection capability is (much) faster and exhibits (much) smaller under-/overshoots after a step-like change of the machine power. Although the classical PI controller is re-tuned for each value of C_{dc}, for $C_{dc} = 400 \times 10^{-6}\,\text{F}$, it is no longer capable to stabilize the closed-loop system. It becomes unstable after 0.2 s, whereas the *nonlinear* PI controller is able to compensate for the rapid changes in the machine power for all three capacitances. The online adjustment of the controller parameters results in a faster and more accurate disturbance rejection even for the smallest DC-link capacitance $C_{dc} = 400 \times 10^{-6}\,\text{F}$.

Remark 11.7. Due to the *unstable* closed-loop system behavior for the capacitance $C_{dc} = 400 \times 10^{-6}\,\text{F}$, the *classical* PI controller will *no longer* be considered. In the upcoming experiments, *solely* the smallest DC-link capacitance $C_{dc} = 400 \times 10^{-6}\,\text{F}$ will be used.

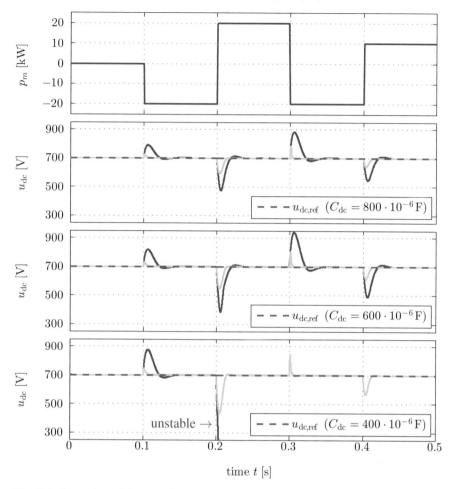

Fig. 11.8 Comparison of the control performance of the ——— classical PI controller in Eq. (11.27) and the ——— nonlinear PI controller in Eq. (11.47) for decreasing values of the DC-link capacitance: $C_{dc} = 800 \times 10^{-6}$ F (second sub-plot), $C_{dc} = 600 \times 10^{-6}$ F (third sub-plot) and $C_{dc} = 400 \times 10^{-6}$ F (fourth sub-plot). The classical PI controller is tuned for each value of C_{dc} separately

11.6.2.2 Discussion of Experiment (E₂)

Experiment (E₂) investigates the disturbance rejection capability of the *nonlinear* PI controller under $\pm 30\%$ parameter uncertainties in DC-link capacitance, filter resistance and filter inductance for step-like changes in machine power p_m and reactive power q_{pcc} (both act as disturbances on the DC-link dynamics). The parameter uncertainties are implemented in such a way that the nonlinear PI controller parameters in Eq. (11.60) and in Eq. (11.61) use the (estimated) values C_{dc}, R_f and L_f whereas the physical system is modeled with the "real" values given by

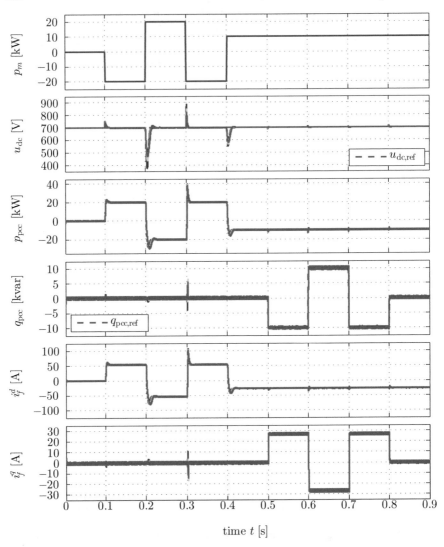

Fig. 11.9 Disturbance rejection of nonlinear DC-link PI controller in Eq. (11.47) for ±30% uncertainties in the DC-link capacitance $C_{\mathrm{dc,real}} = \gamma C_{\mathrm{dc}}$ where —— $\gamma = 0.7$, —— $\gamma = 1.0$, —— $\gamma = 1.3$

$C_{\mathrm{dc,real}} = \gamma C_{\mathrm{dc}}$, $R_{f,\mathrm{real}} = \gamma R_f$ and $L_{f,\mathrm{real}} = \gamma L_f$. The factor $\gamma \in \{0.7, 1, 1.3\}$ is varied. Each value has its own color: —— $\gamma = 0.7$, —— $\gamma = 1.0$, —— $\gamma = 1.3$.

The simulation results for uncertainties in $C_{\mathrm{dc,real}} = \gamma C_{\mathrm{dc}}$, $R_{f,\mathrm{real}} = \gamma R_f$ and $L_{f,\mathrm{real}} = \gamma L_f$ are shown in Fig. 11.9, Fig. 11.10 and Fig. 11.11, respectively. The depicted signals are (from top to bottom) machine power p_m, DC-link voltage u_{dc}, electrical power p_{pcc} at the point of common coupling (PCC), reactive power q_{pcc} at the PCC, and filter currents i_f^d and i_f^q.

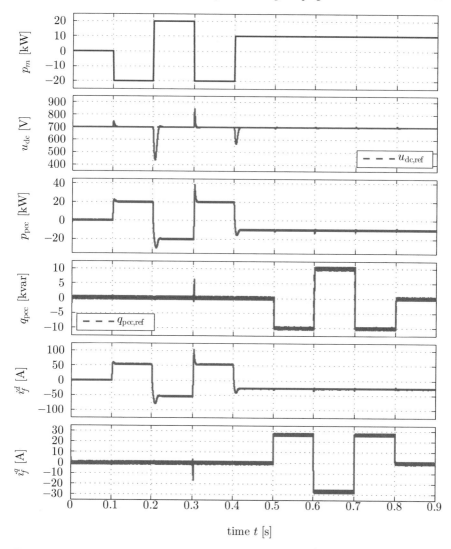

Fig. 11.10 Disturbance rejection of nonlinear DC-link PI controller in Eq. (11.47) for $\pm 30\%$ uncertainties in the filter resistance $R_{f,\text{real}} = \gamma R_f$ where —— $\gamma = 0.7$, —— $\gamma = 1.0$, —— $\gamma = 1.3$

The nonlinear PI controller performs well for all three cases. The closed-loop system remains stable. The disturbances are rejected quickly. The step-like changes in the reactive power have (almost) no effect on the DC-link voltage. Parameter uncertainties in C_{dc} (see Fig. 11.9) affect the set-point tracking control performance. For the case $C_{\text{dc,real}} = 0.7 C_{\text{dc}}$, the DC-link voltage exhibits the largest deviations (over-estimation of the capacitance). The other signals are (almost) not influenced.

Parameter uncertainties in R_f (see Fig. 11.10) are negligible. For the three cases $\gamma \in \{0.7, 1, 1.3\}$, all depicted signals are (almost) identical.

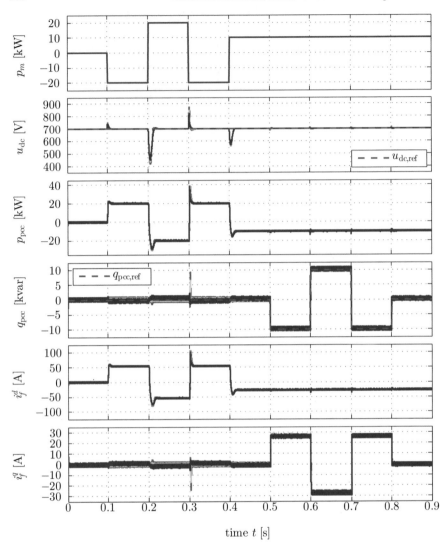

Fig. 11.11 Disturbance rejection of nonlinear DC-link PI controller in Eq. (11.47) for $\pm 30\%$ uncertainties in the filter inductance $L_{f,\mathrm{real}} = \gamma L_f$ where —— $\gamma = 0.7$, —— $\gamma = 1.0$, —— $\gamma = 1.3$

Parameter uncertainties in L_f (see Fig. 11.11) affect the set-point tracking control performances slightly, whereas reactive power and q-component of the current show significant deviations. Here, for $L_{f,\mathrm{real}} = 1.3 L_f$ (under-estimation of the inductance), the largest peaks are visible.

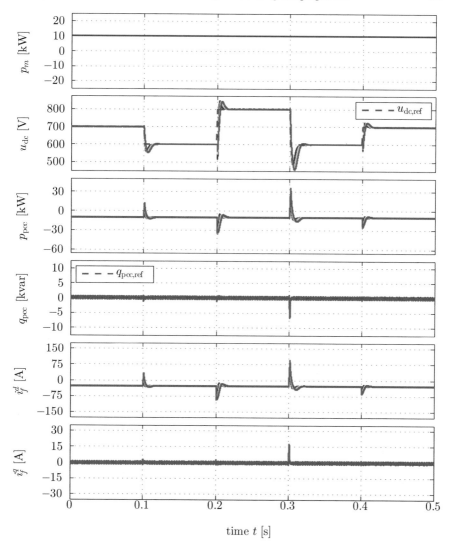

Fig. 11.12 Set-point tracking performance of the nonlinear PI controller in Eq. (11.47) under $\pm 30\%$ uncertainties in the DC-link capacitance $C_{\text{dc,real}} = \gamma C_{\text{dc}}$ where ⸻ $\gamma = 0.7$, ⸻ $\gamma = 1.0$, ⸻ $\gamma = 1.3$

11.6.2.3 Discussion of Experiment (E_3)

Experiment (E_3) illustrates the set-point tracking performance of the *nonlinear* PI controller under $\pm 30\%$ parameter uncertainties in C_{dc}, R_f and L_f for a constant but positive machine power (i.e. the non-minimum phase case with $p_m > 0$; motor mode during reel-in phase). The parameter uncertainties are implemented in the identical manner as for Experiment (E_2), i.e. the physical system is modeled with

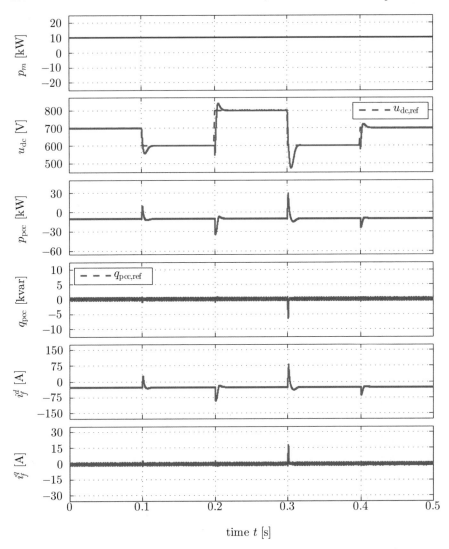

Fig. 11.13 Set-point tracking performance of the nonlinear PI controller in Eq. (11.47) under $\pm 30\%$ uncertainties in the filter resistance $R_{f,\text{real}} = \gamma R_f$ where —— $\gamma = 0.7$, —— $\gamma = 1.0$, —— $\gamma = 1.3$

the "real" values given by $C_{\text{dc,real}} = \gamma C_{\text{dc}}$, $R_{f,\text{real}} = \gamma R_f$ and $L_{f,\text{real}} = \gamma L_f$ where $\gamma \in \{—— 0.7, —— 1.0, —— 1.3\}$ is varied. The values of C_{dc}, R_f and L_f are used for controller implementation and tuning.

The simulation results for uncertainties in $C_{\text{dc,real}} = \gamma C_{\text{dc}}$, $R_{f,\text{real}} = \gamma R_f$ and $L_{f,\text{real}} = \gamma L_f$ are shown in Fig. 11.12, Fig. 11.13 and Fig. 11.14, respectively. The plotted signals represent (from top to bottom) machine power p_m, DC-link voltage

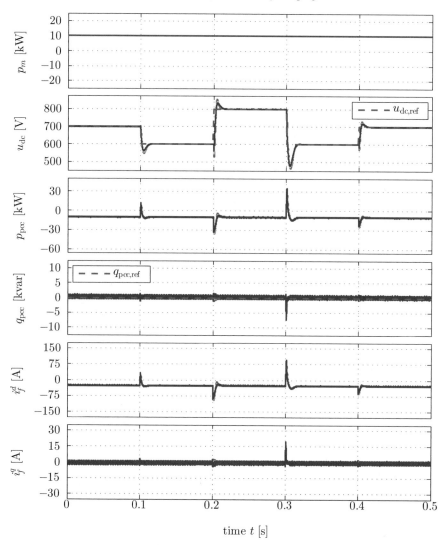

Fig. 11.14 Set-point tracking performance of the nonlinear PI controller in Eq. (11.47) under $\pm 30\%$ uncertainties in the filter inductance $L_{f,\text{real}} = \gamma L_f$ where ⎯⎯ $\gamma = 0.7$, ⎯⎯ $\gamma = 1.0$, ⎯⎯ $\gamma = 1.3$

u_{dc}, electrical power p_{pcc} at the point of common coupling (PCC), reactive power q_{pcc} at the PCC, and filter currents i_f^d and i_f^q.

The set-point tracking performance of the nonlinear PI controller performs is acceptable. Most important, the closed-loop system is stable for all three cases (see Fig. 11.12, Fig. 11.13 and Fig. 11.14). The step-like changes in the reference voltage $u_{\text{dc,ref}}$ are followed quickly with asymptotic accuracy. However, for positive set-point changes at 0.2 s and 0.4 s, the non-minimum phase property of the closed-

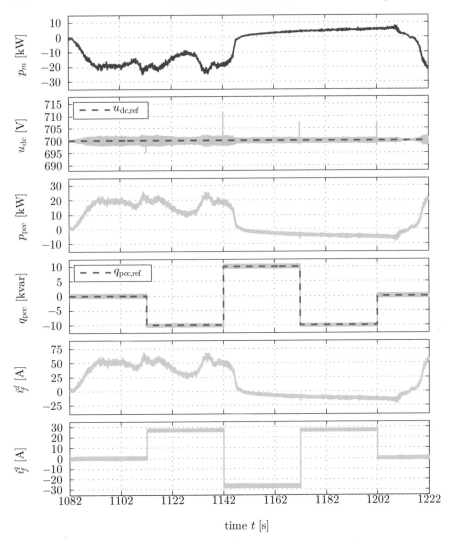

Fig. 11.15 Control performance of the nonlinear PI controller in Eq. (11.47) for a realistic (measured) machine power flow p_m (acting as unfiltered input to the grid-side electrical system)

loop system can be clearly observed: The DC-link voltage u_{dc} decreases before it increases. Moreover, reference changes affect active and reactive power control during transients.

Parameter uncertainties in C_{dc} (see Fig. 11.12), in R_f (see Fig. 11.13) and in L_f (see Fig. 11.14) have only small influence on the set-point tracking performance. Stability is not affected at all.

11.6.2.4 Discussion of Experiment (E$_4$)

Experiment (E$_4$) illustrates the control performance of the *nonlinear* PI controller under the most realistic conditions. The utilized machine power p_m was measured by the TU Delft Kite Power group with their demonstrator on 23rd June 2012 (see Fig. 11.2). The simulation results are shown in Fig. 11.15 (from top to bottom): machine power p_m, DC-link voltage u_{dc}, electrical power p_{pcc} at the point of common coupling (PCC), reactive power q_{pcc} at the PCC, filter currents i_f^d and i_f^q.

Both operation modes are simulated: (a) generator mode during the reel-out phase with $p_m(t) < 0$ for $t \in [1\,082\,\text{s}, 1\,150\,\text{s})$ and (b) motor mode during reel-in phase with $p_m(t) > 0$ for $t \in [1\,150\,\text{s}, 1\,218\,\text{s}]$. The DC-link voltage stays within a 2% band around its reference of $u_{dc,ref} = 700\,\text{V}$. At 1 142 s, due to the high, step-like change in the reactive power q_{pcc}, the DC-link voltage spikes up to $\approx 712\,\text{V}$ which gives the largest deviation of 12 V from $u_{dc,ref} = 700\,\text{V}$ (i.e. a relative error of $\approx 1.7\%$). Concluding, the nonlinear PI controller achieves a very fast and accurate control performance, also, for real data (measured machine power).

Remark 11.8. Note that the noise in Fig. 11.15 is not due to the online adjustment of the controller gains. The noise is induced by (i) the noisy machine power p_m (see top of Fig. 11.15: the provided measurement data was not filtered and directly used as input to the simulation model) and (ii) the switching behavior of the voltage source inverter which leads to ripples in current and power.

11.7 Conclusion

This chapter discusses two different PI controllers for DC-link voltage control: the classical PI controller with constant parameters and a nonlinear PI controller with online parameter adjustment. DC-link voltage control is a non-trivial task due to the nonlinear and possibly non-minimum phase DC-link dynamics (when power flows from the grid to the DC-link). For both PI controllers, the nonlinear system behavior gives different bounds on the choice of the controller parameters. The bounds are derived based on physical system properties (such as admissible currents and DC-link voltages). A comparison of the controllers shows that the classical PI controller becomes unstable for decreasing DC-link capacitances whereas the nonlinear PI controller remains stable. Moreover, the nonlinear DC-link PI controller is (very) robust to parameter uncertainties in filter resistance, filter inductance and DC-link capacitance. Concluding, the implementation of the nonlinear PI controller, compared to the classical PI controller design, seems promising since it is more robust and stable and allows the installation of smaller capacitances which brings economical benefit.

Nomenclature

$\mathbb{R}, \mathbb{C}$	real, complex numbers.
$\mathbf{x} := (x_1, \ldots, x_n)^\top \in \mathbb{R}^n$	column vector, $n \in \mathbb{N}$ where $^\top$ and $:=$ mean 'transposed' (interchanging rows and columns of matrix or vector) and 'is defined as'.
$\mathbf{0}_n \in \mathbb{R}^n$	zero vector.
$\mathbf{a}^\top \mathbf{b} := a_1 b_1 + \cdots + a_n b_n$	scalar product of the vectors $\mathbf{a} := (a_1, \ldots, a_n)^\top$ and $\mathbf{b} := (b_1, \ldots, b_n)^\top$.
$\|\mathbf{x}\| := \sqrt{\mathbf{x}^\top \mathbf{x}} = \sqrt{x_1^2 + \cdots + x_n^2}$	Euclidean norm of $\mathbf{x}$.
$\mathbf{A} \in \mathbb{R}^{n \times n}$	(square) matrix with n rows and columns.
$\mathbf{A}^{-1}$	inverse of $\mathbf{A}$ (if exists).
$\det(\mathbf{A})$	determinant of $\mathbf{A}$.
$\text{spec}(\mathbf{A})$	spectrum of $\mathbf{A}$ (eigenvalues of $\mathbf{A}$).
$\mathbf{I}_n \in \mathbb{R}^{n \times n} := \text{diag}(1, \ldots, 1)$	identity matrix.
$\mathscr{L}^\infty(I; Y)$	space of (essentially) bounded functions with norm $\|\mathbf{f}\|_\infty := \text{ess-sup}_{t \in I} \|\mathbf{f}(t)\|$ (essential supremum). Simple example: For a piecewise continuous function $\mathbf{f}(\cdot) \in \mathscr{L}^\infty(I; Y)$, there exists a positive constant $c_f > 0$, such that $\sup_{t \in I} \|\mathbf{f}(t)\| \leq c_f$ for all $t \in I$. Hence, $\mathbf{f}(\cdot)$ is bounded for all $t \in I$.
$\alpha \overset{(\#)}{=} \beta$	equivalence of α and β follows directly by invoking Eq. (#) (same notation is also used for relations, e.g. $\overset{(\#)}{<}, \overset{(\#)}{\leq}, \overset{(\#)}{\geq}$ and $\overset{(\#)}{>}$).
$\mathbf{x}[X]^n$	physical quantity $\mathbf{x} \in \mathbb{R}^n$, each of the n elements has SI-unit X.
$\boldsymbol{\xi}^{abc} := \left(\xi^a, \xi^b, \xi^c\right)^\top \in \mathbb{R}^3$	signal $\boldsymbol{\xi}^{abc}$ (may represent currents and voltages, i.e. $\boldsymbol{\xi} \in \{\mathbf{i}, \mathbf{u}\}$) in the three-phase (a,b,c)-reference frame.
$\boldsymbol{\xi}^s := \left(\xi^\alpha, \xi^\beta\right)^\top \in \mathbb{R}^2$	signal $\boldsymbol{\xi}^s$ in the stator-fixed (α, β)-reference frame.
$\boldsymbol{\xi}^k = \left(\xi^d, \xi^q\right)^\top \in \mathbb{R}^2$	signal $\boldsymbol{\xi}^k$ in the arbitrarily rotating $k = (d,q)$-reference frame.

References

1. Ahmed, M. S., Hably, A., Bacha, S.: Kite Generator System Modeling and Grid Integration. IEEE Transactions on Sustainable Energy **4**(4), 968–976 (2013). doi: 10.1109/TSTE.2013. 2260364
2. Ahmed, M., Hably, A., Bacha, S., Ovalle, A.: Kite generator system: Grid integration and validation. In: Proceedings of the 40th Annual Conference of the IEEE Industrial Electronics Society (IECON 2014), pp. 2139–2145, Dallas, TX, USA, 29 Oct–1 Nov 2014. doi: 10.1109/ IECON.2014.7048798
3. Bauer, F., Hackl, C. M., Schechner, K.: DC-link control for airborne wind energy systems during pumping mode. In: Schmehl, R. (ed.). Book of Abstracts of the International Airborne Wind Energy Conference 2015, p. 60, Delft, The Netherlands, 15–16 June 2015. doi: 10.4233/ uuid:7df59b79-2c6b-4e30-bd58-8454f493bb09. Presentation video recording available from: https://collegerama.tudelft.nl/Mediasite/Play/599c28511f704d7898d785f584150fb31d
4. Bode, H. W.: Relations Between Attenuation and Phase in Feedback Amplifier Design. Bell System Technical Journal **19**(3), 421–454 (1940). doi: 10.1002/j.1538-7305.1940.tb00839.x
5. Diehl, M.: Airborne Wind Energy: Basic Concepts and Physical Foundations. In: Ahrens, U., Diehl, M., Schmehl, R. (eds.) Airborne Wind Energy, Green Energy and Technology, Chap. 1, pp. 3–22. Springer, Berlin Heidelberg (2013). doi: 10.1007/978-3-642-39965-7_1
6. Ding, X., Qian, Z., Yang, S., Cui, B., Peng, F.: A direct DC-link boost voltage PID-like fuzzy control strategy in Z-source inverter. In: Proceedings of the IEEE Power Electronics Specialists Conference, pp. 405–411, Rhodes, Greece, 15–19 June 2008. doi: 10.1109/PESC.2008. 4591963
7. Dirscherl, C., Hackl, C. M., Schechner, K.: Pole-placement based nonlinear state-feedback control of the DC-link voltage in grid-connected voltage source power converters: A preliminary study. In: Proceedings of the 2015 IEEE Conference on Control Applications (CCA), pp. 207–214, Sydney, Australia, 21–23 Sept 2015. doi: 10.1109/CCA.2015.7320634
8. Dirscherl, C., Hackl, C., Schechner, K.: Explicit model predictive control with disturbance observer for grid-connected voltage source power converters. In: Proceedings of the 2015 IEEE International Conference on Industrial Technology, pp. 999–1006, Seville, Spain, 17–19 Mar 2015. doi: 10.1109/ICIT.2015.7125228
9. Dirscherl, C., Hackl, C., Schechner, K.: Modellierung und Regelung von modernen Windkraftanlagen: Eine Einführung. In: Schröder, D. (ed.) Elektrische Antriebe – Regelung von Antriebssystemen, Chap. 24, pp. 1540–1614. Springer, Berlin Heidelberg (2015). doi: 10. 1007/978-3-642-30096-7_24
10. Fagiano, L., Milanese, M.: Airborne Wind Energy: an overview. In: Proceedings of the 2012 American Control Conference, pp. 3132–3143, Montréal, QC, Canada, 27–29 June 2012. doi: 10.1109/ACC.2012.6314801
11. Gensior, A., Sira-Ramirez, H., Rudolph, J., Guldner, H.: On Some Nonlinear Current Controllers for Three-Phase Boost Rectifiers. IEEE Transactions on Industrial Electronics **56**(2), 360–370 (2009). doi: 10.1109/TIE.2008.2003370
12. Heidary Yazdi, S., Fathi, S., Gharehpetian, G., Ma'ali Amiri, E.: Regulation of DC link voltage in VSC-HVDC to prevent DC voltage instability based on accurate dynamic model. In: Proceedings of the 4th Power Electronics, Drive Systems and Technologies Conference, pp. 394–400, Tehran, Iran, 13–14 Feb 2013. doi: 10.1109/PEDSTC.2013.6506740
13. Hinrichsen, D., Pritchard, A. J.: Mathematical Systems Theory I – Modelling, State Space Analysis, Stability and Robustness. Texts in Applied Mathematics, vol. 48. Springer-Verlag, Berlin Heidelberg (2005). doi: 10.1007/b137541
14. Loyd, M. L.: Crosswind kite power. Journal of Energy **4**(3), 106–111 (1980). doi: 10.2514/3. 48021
15. Muslem Uddin, S. M., Akter, P., Mekhilef, S., Mubin, M., Rivera, M., Rodriguez, J.: Model predictive control of an active front end rectifier with unity displacement factor. In: Proceed-

ings of the 2013 IEEE International Conference on Circuits and Systems, pp. 81–85, Kuala Lumpur, Malaysia, 18–19 Sept 2013. doi: 10.1109/CircuitsAndSystems.2013.6671612

16. Olalla, C., Leyva, R., El Aroudi, A., Queinnec, I.: LMI control applied to non-minimum phase switched power converters. In: Proceedings of the IEEE International Symposium on Industrial Electronics, pp. 154–159, Cambridge, United Kingdom, 30 June–2 July 2008. doi: 10.1109/ISIE.2008.4676993

17. Pérez, M. A., Fuentes, R., Rodríguez, J.: Predictive control of DC-link voltage in an active-front-end rectifier. In: Proceedings of the 2011 IEEE International Symposium on Industrial Electronics, pp. 1811–1816, Gdansk, Poland, 27–30 June 2011. doi: 10.1109/ISIE.2011. 5984432

18. Rodriguez, H., Ortega, R., Escobar, G.: A robustly stable output feedback saturated controller for the Boost DC-to-DC converter. In: Proceedings of the 38th IEEE Conference on Decision and Control, vol. 3, pp. 2100–2105, Phoenix, AZ, USA, 7–10 Dec 1999. doi: 10.1109/CDC. 1999.831229

19. Schröder, D.: Elektrische Antriebe - Regelung von Antriebssystemen. 3rd ed. Springer, Berlin Heidelberg (2009). doi: 10.1007/978-3-540-89613-5

20. Schröder, D.: Leistungselektronische Schaltungen: Funktion, Auslegung und Anwendung. Springer-Verlag, Berlin Heidelberg (2012). doi: 10.1007/978-3-642-30104-9

21. Shukla, A., Ghosh, A., Joshi, A.: State Feedback Control of Multilevel Inverters for DSTAT-COM Applications. IEEE Transactions on Power Delivery 22(4), 2409–2418 (2007). doi: 10.1109/TPWRD.2007.905271

22. Song, E., Lynch, A., Dinavahi, V.: Experimental Validation of Nonlinear Control for a Voltage Source Converter. IEEE Transactions on Control Systems Technology 17(5), 1135–1144 (2009). doi: 10.1109/TCST.2008.2001741

23. Sosa, J. M., Martínez-Rodríguez, P. R., Vázquez, G., Nava-Cruz, J. C.: Control design of a cascade boost converter based on the averaged model. In: Proceedings of the 2013 IEEE International Autumn Meeting on Power, Electronics and Computing, pp. 1–6, Morelia, Michoacán, Mexico, 13–15 Nov 2013. doi: 10.1109/ROPEC.2013.6702718

24. Teodorescu, R., Liserre, M., Rodríguez, P.: Grid Converters for Photovoltaic and Wind Power Systems. John Wiley & Sons, Ltd., Chichester, United Kingdom (2011)

25. Thakur, R. K.: Analysis and control of a variable speed wind turbine drive system dynamics. In: Proceedings of the International Conference on Power Systems, pp. 1–5, Kharagpur, India, 27–29 Dec 2009. doi: 10.1109/ICPWS.2009.5458447

26. Vasiladiotis, M., Rufer, A.: Dynamic Analysis and State Feedback Voltage Control of Single-Phase Active Rectifiers With DC-Link Resonant Filters. IEEE Transactions on Power Electronics 29(10), 5620–5633 (2014). doi: 10.1109/TPEL.2013.2294909

27. Wen-Lei, L.: Adaptive dynamic surface tracking control for DC-DC boost converter. In: Proceedings of the 31st Chinese Control Conference, pp. 750–755, Hefei, China, 25–27 July 2012

28. Yaramasu, V., Wu, B.: Predictive Control of a Three-Level Boost Converter and an NPC Inverter for High-Power PMSG-Based Medium Voltage Wind Energy Conversion Systems. IEEE Transactions on Power Electronics 29(10), 5308–5322 (2014). doi: 10.1109/TPEL. 2013.2292068

29. Zhang, L., Nee, H.-P., Harnefors, L.: Analysis of Stability Limitations of a VSC-HVDC Link Using Power-Synchronization Control. IEEE Transactions on Power Systems 26(3), 1326–1337 (2011). doi: 10.1109/TPWRS.2010.2085052

30. Zhang, Y., Liu, J., Ma, X., Feng, J.: Model and control of diode-assisted buck-boost voltage source inverter. In: Proceedings of the 1st International Future Energy Electronics Conference, pp. 734–739, Tainan, Taiwan, 3–6 Nov 2013. doi: 10.1109/IFEEC.2013.6687599

31. Zhang, Y., Liu, J., Ma, X., Feng, J.: Multi-loop controller design for diode-assisted buck-boost voltage source inverter. In: Proceedings of the 2014 International Power Electronics Conference, pp. 835–842, Hiroshima, Japan, 18–21 May 2014. doi: 10.1109/IPEC.2014. 6869685

Chapter 12
Control of a Magnus Effect-Based Airborne Wind Energy System

Ahmad Hably, Jonathan Dumon, Garrett Smith and Pascal Bellemain

Abstract This chapter studies the control of an airborne wind energy system that is operated in pumping cycles and uses a rotating cylinder to provide aerodynamic lift with the Magnus effect. The proposed control strategy aims at stabilizing the output power production which can be used for off-grid applications, for example. In a first case study, the wind tunnel setup of a small-scale system is investigated experimentally and by means of numerical simulation. The proposed controller works well to effectively manage the tether length. However, a comparison of the results demonstrates the penalizing effects of wind turbulence with a factor of three difference in power production. In a second case study, the control strategy is used for the numerical simulation of a medium scale prototype with a potential power rating of 50 kW. The results show that the control strategy is very effective to track the desired power production even in the presence of wind velocity fluctuations. In a third case study, the scalability of the system is evaluated by applying the control scheme to the numerical simulation of a MW scale platform. The results show that the system with a span equal to the diameter of a conventional wind turbine can generate an equivalent amount of power.

12.1 Introduction

The concept of airborne wind energy (AWE) has attracted a lot of interest in the last few years [3]. In an interview in June 2015, Bill Gates said that he is planning to invest more than two billion dollars in green technologies and highlighted promising

Ahmad Hably (✉) · Jonathan Dumon · Pascal Bellemain
Université Grenoble Alpes, CNRS, Grenoble INP*, GIPSA-lab, 38000 Grenoble, France
*Institute of Engineering Université Grenoble Alpes
e-mail: ahmad.hably@gipsa-lab.grenoble-inp.fr

Garrett Smith
Wind Fisher S.A.S., 2 allee du Vivarais, 31770 Colomiers, France

© Springer Nature Singapore Pte Ltd. 2018
R. Schmehl (ed.), *Airborne Wind Energy*, Green Energy
and Technology, https://doi.org/10.1007/978-981-10-1947-0_12

areas of research that "involve among others kites, kite-balloon hybrids known as kytoons or flying turbines" [8].

AWE systems replace the blades of conventional wind turbines by a controlled flying wing that captures the energy of the wind. EnerKite claims that their AWE system doubles the output whilst saving 95 % of resources [4]. This concept of producing more energy with less material goes in the direction of recent movement of frugal innovation that promotes "How to do more with less".

AWE systems can be divided into two main classes depending on the location where energy is produced:

- Systems using the lift mode as introduced in [10] where the mechanical power is transferred to the ground. Energy is produced during a production phase, in which the aerodynamic lifting device produces a traction force which is used to pull the tether from a drum which drives an electrical generator. This phase is followed by a recovery phase that begins when the tether reaches its predefined maximum length, and hence needs to be reeled-in, an operation that consumes energy. These on-ground generation systems are studied, for example by Kitegen [2], and Ampyx Power [1].
- On-board production using the drag mode as introduced in [10]. The generator is embedded in the airborne structure and electric energy is produced in-flight and transferred to the ground using conducting tethers. This type of systems is investigated, for example by Makani Power [12].

Most of the aforementioned systems use either flexible or rigid wings as aerodynamic lifting devices. To increase the traction force during reel-out or the energy harvesting in drag mode, the wing is operated in crosswind maneuvers.

On the other hand, Omnidea Lda has used a Magnus effect-based system in its High Altitude Wind Energy project (HAWE) [14]. The operation principle of their platform is based on the traction force of a rotating cylindrical balloon employing both aerostatic as well as aerodynamic lift mechanisms [16, 17]. Magnus effect-based airborne wind energy systems generate an aerodynamic lift that depends on the apparent wind speed at the Magnus cylinder and its angular speed. Electrical energy is produced as for on-ground systems, with the difference that the balloon is not operated in crosswind flight maneuvers. The fundamental reason is that a cylindrical body will always have a significant aerodynamic drag because of its flow cross section. As a consequence, the lift-to-drag ratio is limited to comparatively low values which leads to relatively low performance when operated in crosswind mode despite of its high lift coefficient.

However, the Magnus effect-based system offers a huge advantage of being lighter than air, which greatly simplifies the takeoff and landing phases when the wind is insufficient. In addition, as opposed to the AWE systems using soft or rigid wings whose lift and drag vector magnitudes depend on the angle of attack measured between airfoil chord and apparent wind velocity, Magnus effect-based systems only change lift and drag vector direction, and not magnitude, when there is a change in apparent wind direction.

In [18], the feasibility of the Magnus effect-based concept has been studied. In [13], the control variables were optimized for an airborne wind energy production system showing optimal vertical trajectories. In the present chapter, using similar vertical trajectories, the proposed control strategy aims at controlling the power produced by a Magnus effect-based AWE system. The on-ground generator controls the tether length and the cycle period. The performance of the control strategy is satisfying even in the presence of highly perturbed wind speeds.

The chapter is organized as follows. In Sect. 12.2 the system model is introduced. The control strategy presented in Sect. 12.3 applies to the experimental platform described in Sect. 12.4 where both simulation and experimental results are shown. A numerical application of the proposed control strategy on a medium scale system based on Omnidea's experimental platform is presented in Sect. 12.5. Some numerical results on a MW scale system are presented in Sect. 12.6. The chapter ends with some conclusions and perspectives in Sect. 12.7. The preliminary content of the present chapter has been presented at the Airborne Wind Energy Conference 2015 [6].

12.2 System Modeling

The airborne wind energy system studied is composed of a light-weight rotating cylindrical lifting device, called hereafter the Magnus rotor, that supplies a traction force to an on-ground generator through a tether. This device generates lift and drag forces, as shown in Fig. 12.1. The lift mechanisms are aerodynamic lift, the well-known Magnus effect, and aerostatic lift, by using Helium as a filling gas for the balloon. The resultant traction force is transferred via the tether to the on-ground generator where a drum is used to convert the linear motion of the tether into shaft power, which is used to drive a generator. For recovery, this operation is reversed, i.e. the generator is operated as a motor and the aerodynamic force of the Magnus rotor is reduced by switching off the rotation. Acting on the Magnus rotor are the aerodynamic force $\mathbf{F}_a$, which can be split up into a aerodynamic lift $\mathbf{L}$ and a drag $\mathbf{D}$ component, the gravitational force on the Magnus rotor $\mathbf{P}$, the buoyancy force $\mathbf{B}$ and the tensile force $\mathbf{T}$ in the tether.

Aerodynamic lift and drag forces can be expressed by:

$$L = \frac{1}{2}\rho S v_a^2 C_L, \quad D = \frac{1}{2}\rho S v_a^2 C_D \tag{12.1}$$

where ρ is the air density, v_a is the apparent wind velocity, S is the Magnus rotor projected surface area in the direction of the apparent wind velocity, C_L and C_D are aerodynamic lift and drag coefficients respectively. For Magnus effect-based systems, aerodynamic lift coefficient C_L and drag coefficient C_D are functions of the spin ratio X [19]. The Magnus rotor spin ratio is given by the following equation:

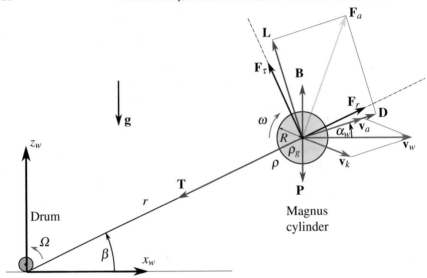

Fig. 12.1 The different forces acting on the Magnus effect-based airborne wind energy system, the translation velocity of the Magnus rotor v_k and its angular velocity ω, the angular velocity of the drum $\Omega = \dot{r}/R_d$ and the wind velocity with respect to the ground v_w

$$X = \frac{\omega R}{v_a} \tag{12.2}$$

where ω is the Magnus rotor angular velocity and R is the Magnus rotor radius. The buoyancy force can be calculated from Archimedes' principle:

$$B = \rho V_0 g \tag{12.3}$$

where V_0 is the volume of the Magnus rotor and g is the gravitational acceleration. The combined mass of all airborne system components (Magnus rotor, contained gas, rotor drive train plus tether and bridle lines) is denoted by M_M:

$$M_M = M + V_o \rho_g + M_l r \tag{12.4}$$

where M is the mass of the airborne structure, M_l denotes the mass per tether length and ρ_g is the gas density. The wind velocity $\mathbf{v}_w$ is assumed to be parallel to ground. The apparent wind velocity v_a is defined by:

$$\mathbf{v}_a = \mathbf{v}_w - \mathbf{v}_k \tag{12.5}$$

where $\mathbf{v}_k$ is the translation velocity of the Magnus rotor. In this study, the movement of the Magnus rotor is assumed to be in the vertical plane. It is also assumed that the tether of length r is always in tension and forms a straight line. It has an elevation angle β with respect to the ground plane. Furthermore, the cylindrical lifting device does not allow the definition of an angle of attack in the cross sectional plane. In order to find the dynamic model, fundamental dynamic equations are used. Considering the two translational degrees of freedom of the system, r and β, $\mathbf{v}_k$ can be decomposed into a radial velocity component $v_{k,r} = \dot{r}$ and a tangential velocity component $v_{k,\tau} = r\dot{\beta}$. Differentiation of $\mathbf{v}_k$ with respect to time yields the radial acceleration component $dv_{k,r}/dt = \ddot{r} - r\dot{\beta}^2$ and the tangential acceleration component $dv_{k,\tau}/dt = r\ddot{\beta} + 2\dot{r}\dot{\beta}$. The resultant forces F_r and F_τ on the Magnus rotor are respectively the radial and tangential force components according to the polar coordinate system (r, β) as shown in Fig. 12.1.

$$F_r = -T + L\sin(\beta - \alpha_w) + D\cos(\beta - \alpha_w) - P\sin\beta + B\sin\beta \qquad (12.6)$$
$$F_\tau = L\cos(\beta - \alpha_w) - D\sin(\beta - \alpha_w) - P\cos\beta + B\cos\beta \qquad (12.7)$$

where α_w is the angle that the apparent wind velocity forms with the horizontal. The dynamic model can then be derived in 2D polar coordinates:

$$\ddot{\beta} = \frac{1}{r}\left[-2\dot{\beta}\dot{r} + \frac{F_\tau}{M_M}\right] \qquad (12.8)$$

$$\ddot{r} = \frac{1}{M_M + M_D}\left[r\dot{\beta}^2 M_M + F_r\right] \qquad (12.9)$$

where $M_D = I/R_d^2$ with moment of inertia of the on-ground generator I and its radius R_d. In addition to these equations, we add the dynamics of the on-ground generator:

$$\dot{T} = \beta_T\left(u_T - T\right) \qquad (12.10)$$

where u_T is the desired traction force and β_T, homogeneous to a frequency, represents its dynamic response modeled here as a first order dynamic system.

12.3 Control Strategy

The control strategy to be applied to the Magnus-based system aims at stabilizing the mean output power produced during a given cycle (recovery phase then production phase). The tether traction force T and its speed $\dot{r}$ are forced to track some reference signals related to a desired reference power P_{ref} to be produced. For simplicity, P_{ref} is assumed to be constant, however the control strategy can be adapted to varying P_{ref} as shown later.

During the cycle, the Magnus rotor moves from a minimum radial position r_{min} to a maximum radial position r_{max} at a reel-out speed $\dot{r}_{prod}$ during production phase and from r_{max} to r_{min} at a negative reel-in speed $\dot{r}_{rec}$ during the recovery phase. Since $\dot{r}_{prod}$ and $\dot{r}_{rec}$ are assumed to be constant, the proposed algorithm tracks P_{ref} by controlling the traction force T. A given cycle is defined by the time period from the beginning of the recovery phase to the end of the production phase. The recovery

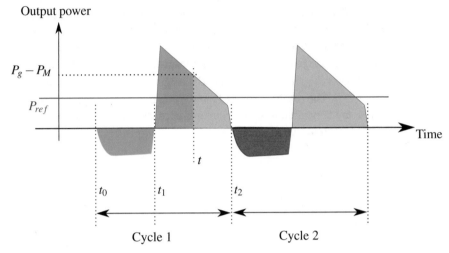

Fig. 12.2 A sketch of the instantaneous output power as a function of time covering several pumping cycles. For a cycle, the red area E_{rec} represents the energy consumed during the recovery phase between time t_0 and time t_1 and the green area E_{prod} represents the energy produced during the production phase between time t_1 and time t_2. The blue area represents the output energy at time t for the beginning of the cycle. Desired reference power P_{ref} is assumed to be constant

phase starts at time t_0 and ends at time t_1. Then the production phase starts at time t_1 and ends at time t_2 (see Fig. 12.2). The time t_1 can be calculated by

$$t_1 = t_0 + \frac{r_{max} - r_{min}}{-\dot{r}_{rec}} \tag{12.11}$$

and the time t_2 can be calculated by

$$t_2 = t_1 + \frac{r_{max} - r_{min}}{\dot{r}_{prod}} \tag{12.12}$$

To produce a net output power equals to P_{ref}, the output energy to be produced during a cycle E_{ref} is given by $P_{ref}(t_2 - t_0)$. During the cycle, the output energy produced from time t_0 to time t is calculated by

$$E = \int_{t_0}^{t} (P_g - P_M)dt \tag{12.13}$$

where P_g is the produced output power of the on-ground generator and P_M is the power consumed by the Magnus motor. In order to satisfy $E = E_{ref}$ at the end of the cycle, the remaining energy to be produced E_{prod} from time t to time t_2 has to satisfy:

$$E_{prod} = E_{ref} - E \tag{12.14}$$

Subsequently, the reference traction force has to satisfy for $t \in [t_0, t_2]$

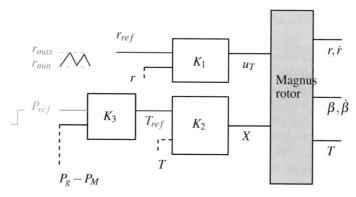

Fig. 12.3 An overview of the proposed control system. Three controllers K_1, K_2, and K_3 are used. The Magnus rotor moves from minimum radial position r_{min} to a maximum radial position r_{max}. T_{ref} is obtained by Eq. (12.15). X is the spin ratio and u_T is the desired traction force

$$T_{ref} = \frac{E_{prod}}{\dot{r}(t_2 - t)} \tag{12.15}$$

As T cannot be negative, T_{ref} is set to zero for $t \in [t_0, t_1]$.

To implement the proposed control strategy, two other controllers K_1 and K_2 are used as shown in Fig. 12.3. The tether length is controlled by K_1 through the desired traction force u_T of the on-ground generator. In order to track T_{ref} obtained from controller K_3, K_2 controls spin ratio X of the Magnus rotor. Controller K_3 is given in Eq. (12.15). Controllers K_1 and K_2 are classical PID controllers in parallel form whose parameters are tuned empirically with the following constraints:

- K_2 is set to have a fast response time to get $T_{ref} = T$.
- K_1 is set to have a faster response time than K_2 in order to have a decoupled control between the tether length r and the traction force T.

In the rest of this chapter, three case studies will be presented. In the first case, the control will be applied numerically and experimentally on a small-scale indoor system. In the second case, we will numerically study a medium scale system. In the last case, the control strategy will be numerically applied to a MW scale system.

12.4 Control of a Small Scale Laboratory Test Setup

In this section, the control of a small-scale wind tunnel setup will be presented. It is based on Gipsa-lab's experimental setup. This experimental setup has been used in [7] and [11]. This setup gives us some flexibility and allows us to test our prototypes and the proposed control strategies independently of the weather conditions. It is composed of a wind tunnel, the Magnus rotor, and the ground station.

12.4.1 Wind Tunnel

The fan section of the wind tunnel is composed of nine brushless electrical motors equipped with two-blade fans of 0.355 m diameter, see Fig. 12.4 (top left). These motors, 800 W each, are distributed on a tunnel cross section area of 1.85 m^2. The air flow first passes through a honeycomb then in a tunnel of 1.8 m length in order to stabilize it. A hot wire wind speed sensor[1] is used to measure the airspeed. The output air flow speed can reach 9 m/s with a typical standard deviation of 0.18 measured at 1 Hz. As there are fast variations, the air flow can be better characterized with a smaller sample time.

We choose to use nine hobbyist propulsion sets of electrical motors because this option turns out to be cheaper than the use of a single 7.2 kW motor with a driver and a propeller. This design of an open wind tunnel was chosen based on economic reasons at the expense of flow quality that can be provided by a closed wind tunnel architecture.

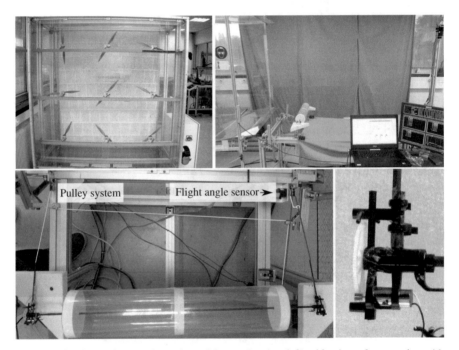

Fig. 12.4 Wind tunnel setup: front view of fan section (top left), side view of test section with computer hardware in foreground (top right), Magnus rotor (bottom left) and DC motor used to rotate the Magnus rotor (bottom right)

---

[1] Measurement frequency of 1 Hz achievable with serial interface.

12.4.2 Magnus Rotor

The Magnus rotor used in the experimental setup is a light-weight cylinder built with carbon rods, polystyrene and transparent plastic paper, see Fig. 12.3 (bottom left). The rotation of the Magnus rotor is provided by a mini DC motor mounted on one end of it. Its current control and speed sensor are implemented with a homemade driver. The parameters of the Magnus rotor are given in Table 12.1. The size of this Magnus rotor does not allow us to have a lighter-than-air system so it is not filled with Helium. The rigid frame design proposed allows us to have a better balanced symmetrical structure than if we have used a textile structure.

Symbol	Name	Value
M_M	Mass of airborne subsystem	0.11 kg
M_l	Mass per tether length	neglected
R	Magnus rotor radius	0.047 m
L_m	Magnus rotor length	0.45 m
M_D	Ground station rotor mass	0.0481 kg
R_d	Drum radius	0.05 m
ρ	Air density	1.225 kg/m^3
β_T	Inverse of time constant of motor current loop	14.28 s^{-1}
Re	Reynolds number	4×10^4

Table 12.1 Parameters of the test setup

12.4.3 Ground Station

The ground station is composed of a dynamo-motor system Maxon 2260L DC 100W driven by a four-quadrants amplifier Maxon ADS 50/10 that controls current through the motor, see Fig. 12.5. Two incremental encoders provide measurement of the elevation angle β and tether length r. Control references of DC motors are sent to drivers with a DAC PCI DAS1200 from Measurement Computing and a torque sensor provides an accurate measurement of tether tension. Controllers are implemented on the experimental setup using the xPC target real-time toolbox of Matlab, see Fig. 12.3 (top right).

12.4.4 Identification

First, the response time of the DC motor used to rotate the Magnus rotor is identified (see Fig12.6). The second step is to identify, by regression of the measured data, the

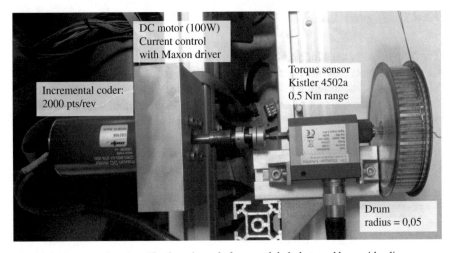

Fig. 12.5 The ground station. The drum is made for a tooth belt, but used here with a line

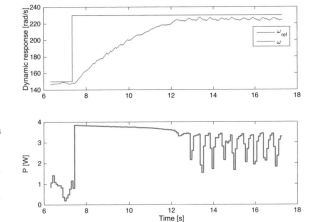

Fig. 12.6 The dynamic response of the DC motor to a step signal for a wind speed $v_w = 0$ m/s. In blue is the measured angular speed and in red the reference angular speed (top). The power consumed for this step signal is also shown (bottom)

aerodynamic lift and drag coefficients as functions of the spin ratio. The obtained results are comparable to the theoretical results [21] for Reynolds number $Re = 3.8 \times 10^4$ and used in [13] (see Fig. 12.7). The identified model for the aerodynamic lift and drag coefficients function of spin ratio X is:

$$C_D = 0.73X^2 - 1.2X + 1.2131 \tag{12.16}$$

$$C_L = 0.0126X^4 - 0.2004X^3 + 0.7482X^2 + 1.3447X - 0.2 \tag{12.17}$$

The last step in the identification phase is to find the operational limits of our platform. We have noticed that friction in the pulleys is significant. The increase of mechanical friction forces is a well known physical phenomenon when scaling down. We have measured the tension in the tether as a function of the tether length r

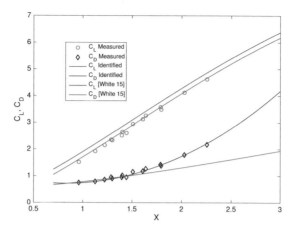

Fig. 12.7 Aerodynamic Lift and drag coefficients identification as functions of the spin ratio X. Wind speed v_w varies from 4.76 m/s to 7.26 m/s

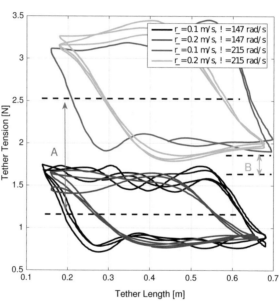

Fig. 12.8 The measured tension in the tether as a function of the tether length r for different angular speeds ω of the Magnus rotor and tether speed $\dot{r}$, for a wind speed $v_w = 6.2$ m/s. The zone A is the possible force difference that can be used to produce energy. This zone is reduced to zone B due to the pulleys friction. This gives an idea of the feasibility of a positive power production cycle and what one could potentially get if this friction is reduced

for different angular speeds ω of the Magnus rotor and the tether speed $\dot{r}$. We have found our platform can provide a limited difference of traction force that one can use to produce energy. This is shown in the difference between the upper and lower zones of Fig. 12.8.

12.4.5 Simulation Results

In this section, the proposed control strategy is tested numerically in a simulation environment using Eqs. (12.8), (12.9), and (12.10). Our objective is to test the control strategy and to have a cycle with a positive net energy output. The following conditions are used.

- The Reynolds number is 4×10^4 and the wind speed $v_w = 6.28$ m/s.
- The minimum tether length is $r_{min} = 0.1$ m and its maximum is $r_{max} = 0.7$ m. These limits are imposed by the wind window of the wind tunnel.
- The tether speed in the production phase is $\dot{r}_{prod} = 0.1$ m/s and in the recovery phase $\dot{r}_{rec} = -0.1$ m/s. These values, even far from the optimal value $(v_w/2)$ according to the main theory of simple kite, [10, Fig. 2], were imposed again by the limits of the wind window.
- Friction identified in Fig. 12.8 leads to a limited range of power production. For this reason, we choose to deactivate the controllers K_2 and K_3 using constant values of spin ratio X.
- We have chosen to control only ω which gives an average value of X. A constant angular speed of the Magnus rotor is used in the production phase $\omega_{prod} = 200$ rad/s and in the recovery phase $\omega_{rec} = 140$ rad/s. A value of $\omega_{rec} = 0$ cannot be set due to limits of the experimental setup. Considering that $\dot{r}$ is negligible with respect to v_w and simple kite conditions ($\mathbf{v_k}$ and $\mathbf{T}$ are colinear), this leads to $v_a \approx v_w = 6.28$ m/s. The spin ratio is then $X_{prod} = 1.5161$ in the production phase and $X_{rec} = 1.0613$ in the recovery phase.

The tether length follows the desired radial position as shown in Fig. 12.9. As expected, the traction force increases as the angular Magnus rotor speed increases. The movement in the vertical plane is shown in Fig. 12.10. The application of this control strategy enables us to produce a positive net energy output as shown in Fig. 12.11.

12.4.6 Experimental Results

Using the same conditions as in the previous simulation section, the control strategy is applied to the experimental setup (Figs. 12.12–12.14). A movie that shows the experiment can be found on our website [5].

12.4.7 Results Discussion

From the simulation results, one can see that after stabilizing the angular velocity ω, the elevation angle β also stabilizes confirming simple kite conditions. In the experimental results, there are significant oscillations of β due to wind turbulence of the used wind tunnel. As a consequence, simple kite conditions are not verified.

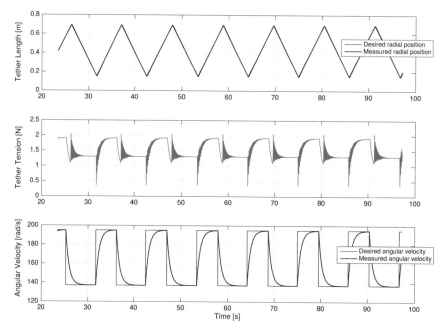

Fig. 12.9 Tether length, tether tension and the angular speed of the Magnus rotor as function of time in the simulation of the small-scale system for a wind speed $v_w = 6.28$ m/s. The oscillation in the tether tension is due to the choice of controller parameters

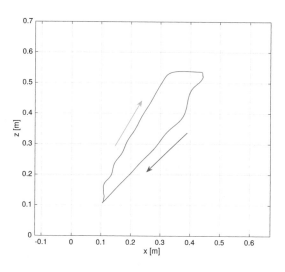

Fig. 12.10 The production cycles in the simulation of the small-scale system for a wind speed $v_w = 6.28$ m/s. The direction of the arrows indicates the movement of the Magnus rotor: Green for the production phase and red for the recovery phase

Moreover, these oscillations impact the tether tension as shown in Fig. 12.12 and the produced power shown in Fig. 12.14, which is reduced by a factor of three compared to the simulation results. Nevertheless, the similar shapes of the cycles shown in Fig. 12.13 gives us an idea on the validity of the proposed model. For this

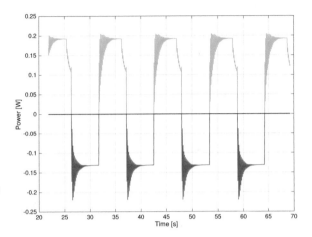

Fig. 12.11 The power produced in the simulation of a small-scale system. The net output power is 0.0327 W. Wind speed $v_w = 6.28$ m/s

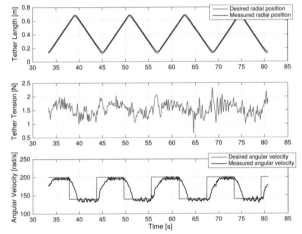

Fig. 12.12 Tether length, tether tension and the angular speed of the Magnus rotor as function of time in the experimentation on the small-scale system. The measured wind speed $v_w = 6.28$ m/s with a standard deviation of 0.184

small-scale system, the power consumed by the Magnus motor P_M is much larger than the power produced by the system due, among others, to the significant effect of frictions.

For larger scale systems, frictions become less important compared to aerodynamic forces. Wind turbulence will also produce less impact on elevation angle β due to length of the tether and to larger inertia of the airborne subsystem. One has to adjust the dynamics of the on-ground generator and the Magnus motor in order to match the time scale of production cycles. In addition, a model of aerodynamic lift and drag coefficients must be adapted to the considered Reynolds number. Finally, as the volume increases with the cube of the cylinder dimensions and the mass increases with the square since it is related to the surface, scaling up the system allows to get a lighter-than-air structure by filling it with lighter-than-air gas like Helium.

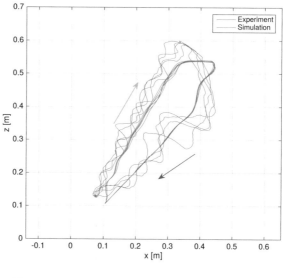

Fig. 12.13 The measured production cycles of the small-scale system versus the simulation production cycles using the same wind data. The measured wind speed $v_w = 6.28$ m/s with a standard deviation of 0.184. The direction of the arrows indicates the movement of the Magnus rotor: Green for the production phase and red for the recovery phase

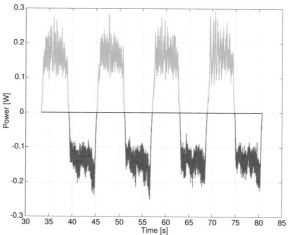

Fig. 12.14 The power produced in the experimentation on the small-scale system. The net output power is 0.0099055 W. The measured wind speed $v_w = 6.28$ m/s with a standard deviation of 0.184

12.5 Numerical Application to a Medium Scale System

The complete control strategy has been numerically applied to Omnidea's system [9, 15]. We have used the dimensions of the Magnus rotor presented in [16]. This Magnus rotor is filled with Helium. Its parameters are listed in Table 12.2. The aerodynamic lift and drag coefficients used are those presented in [21] for Reynolds number $Re = 3.8 \times 10^4$:

$$C_D = -0.0211X^3 + 0.1837X^2 + 0.1183X + 0.5 \tag{12.18}$$
$$C_L = 0.0126X^4 - 0.2004X^3 + 0.7482X^2 + 1.3447X \tag{12.19}$$

Note that for a wind speed $v_w = 10$ m/s, Reynolds number is Re $= 1.7 \times 10^6$. For higher values of Re, Eq. (12.18) can be slightly different.

Symbol	Name	Value
M_M	Mass of airborne subsystem	91.22 kg
R	Magnus rotor radius	1.25 m
L_m	Magnus rotor length	16 m
ρ_{He}	Helium density	0.1427 kg/m^3
ρ_{air}	Air density	1.225 kg/m^3
M_l	Mass per tether length	0.2 kg/m
M_D	Ground station rotor mass	2000 kg
$u_{T_{max}}$	Saturation on traction actuator	65 kN
v_w	Wind speed	10 m/s
Re	Reynolds number	1.7×10^6

Table 12.2 Parameters of the medium scale Magnus rotor

In order to implement the proposed control strategy, we choose to reproduce a vertical trajectory similar to those suggested in [13] that we will reduce its efficiency in order to stabilize a desired power produced P_{ref}. We have determined the feasibility regions for $r_{min} = 200$ m and $r_{max} = 300$ m. For a wind speed $v_w = 10$ m/s, the tether speed during the production phase $\dot{r}_{prod}$ and during the recovery one $\dot{r}_{rec}$ are found numerically offline. One gets $\dot{r}_{rec} = -0.52 v_w$ and $\dot{r}_{prod} = 0.33 v_w$ which is not the optimal value given by the main theory of simple kite [10]. This is because the colinearity condition of v_k and T is not satisfied in vertical trajectories. By simulating this system at a wind speed $v_w = 10$ m/s, we get the net output power produced during a full cycle as a function of X during the production phase (Fig. 12.15). X is then set to 0 during the recovery phase. The maximum net output power equals

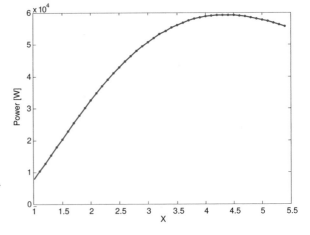

Fig. 12.15 The variation of the net output power as a function of the spin ratio X during the production phase for the medium scale system. During the recovery phase, X is set to zero. Wind speed $v_w = 10$ m/s

59.23 kW for $X = 4.3$. The proposed control strategy will therefore use this nominal production cycle and vary the spin ratio X between 0 and 4.3 in order to stabilize the desired power produced.

For this nominal production cycle, the energetic performance is 1.48 kW/m^2 which is consistent with 1.25 kW/m^2 found in [13] where a similar sized system is used. Note that we do not consider here the motor consumption that actuates the Magnus rotor. An estimation of this consumption can be computed as follows: Based on the C_{M_z} Magnus parameter of [18] for Re $= 10^6$, the torque exerted on the Magnus rotor is:

$$M_z = 0.5\rho\pi R^2 L_m v_a^2 C_{M_z} \qquad (12.20)$$

and the motor power consumption can be calculated by:

$$P_M = \omega M_z = v_w \frac{X}{R} M_z = 0.5 X \rho \frac{\pi}{2} S v_w^3 C_{M_z} \qquad (12.21)$$

If one considers a spin ratio of $X = 4.3$ and $v_w = 10$ m/s, one can estimate $C_{M_z} = 0.0055$, and $P_M = 910$ W for the production phase (61.2% of the time). The consumption of the motor is 556.7 W for the whole cycle which is 0.9% of the 59.23 kW produced.

12.5.1 Nominal Production Cycle

In this section, the results of the nominal production cycle are presented. In order to have a smooth movement of the Magnus rotor, the reference tether length r_{ref} is filtered by $1/(\tau_R s + 1)^2$ with $\tau_R = 2$s.

The PID controller K_1 parameters are $K_p = 8250$ N/m, $K_i = 1.32$ N/(ms), $K_d = 45 \times 10^3$ Ns/m. We find that the apparent wind speed increases thanks to the temporal evolution of elevation angle β (Fig. 12.16) which produces the cycle of Fig. 12.17 with a maximum of $v_a = 14.26$ m/s in the production phase and

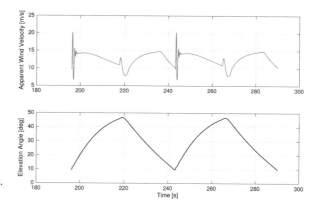

Fig. 12.16 Apparent wind speed in [m/s] (top) and elevation angle β in [deg] (bottom) as function of time for the medium scale system. Wind speed $v_w = 10$ m/s

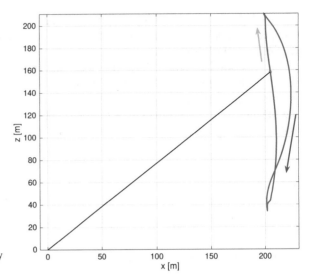

Fig. 12.17 The production cycles of the medium scale system. Wind speed $v_w = 10$ m/s. The direction of the arrows indicates the movement of the Magnus rotor: Green for the production phase and red for the recovery phase

$v_a = 14.79$ m/s in the recovery phase. Following the simple kite theory, one can get an elevation angle $\beta = 0$ for the recovery phase and $\beta = 52.6$ deg for the production phase. This type of cycle is composed of the succession of transition phases between these two values of β. In Fig. 12.18, we show the temporal evolution of the tether length, tether tension and angular speed of the Magnus rotor. One can find the maximum tension on the tether is $T_{max} = 42.4$ kN, the maximum angular speed $\omega_{max} = 49.02$ rad/s. The production phase reel-out speed is 3.3 m/s with an over-shoot measured at 8 m/s, the recovery phase speed is set to -5.2 m/s, without any observed overshoot. Omnidea's current system cannot completely meet these values since the announced maximum force has been 5 kN with a maximum angular speed of 9.42 rad/s [16].

12.5.2 Energy Control

In this section, the complete control strategy has been applied. To find the control parameters of the controller K_2 (PD controller), we have chosen the increasing line slope of Fig. 12.15 between $X = 1$ and $X = 4.3$. The control parameters are then $K_p = 6.4 \times 10^{-3} N^{-1}$ and $K_d = 6.4 \times 10^{-3}$ s/N. One can clearly see the performance of the proposed control strategy (Fig. 12.19). The produced power will follow the desired one even in the presence of noise on the wind speed. It is worth noting that if the output of PD is saturated, one can simply apply a very large reference to achieve the nominal production cycle, with $X = 4.3$ throughout the production phase.

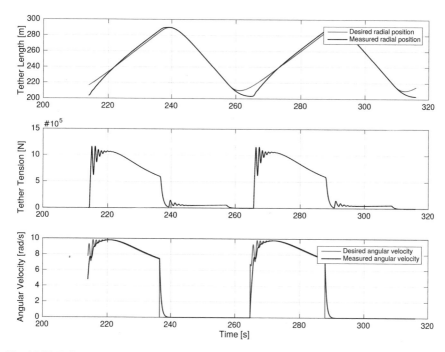

Fig. 12.18 Tether length, tether tension and the angular speed of the Magnus rotor as function of time for the medium scale system. Wind speed $v_w = 10$ m/s

12.5.3 Energy Control with Real Wind Data

The energy control algorithm is also applied using real wind data taken on October 2015 at the Bard station of the Loire region in France [20].[2] Only the wind magnitude is considered given that we are studying the movement in the vertical plane. The wind speed varies from 7 m/s to 20 m/s. Three power reference levels are considered (Fig. 12.21):

- $P_{ref} = 20$ kW: In this case, the system succeeds to track the desired power reference by limiting the energy produced even in the presence of wind turbulence. These variations in the wind speed generate a traction force that exceeds the on-ground generator saturation which causes an error on the control of r but does not affect the power produced.
- $P_{ref} = 50$ kW: The system succeeds to track the desired power reference when the available wind speed is enough. A short-term storage system can be used to ensure that the system catches up with the remaining energy of the previous cycle and thus obtains the desired average power in the presence of such fast changes in the wind.

---

[2] The measurement sampling period is five seconds.

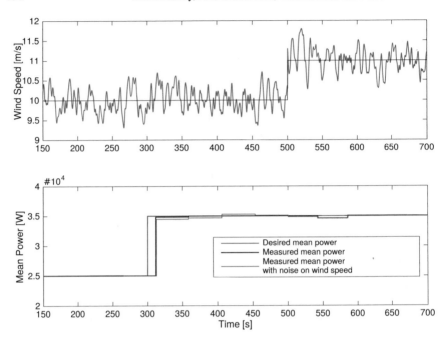

Fig. 12.19 A noise is added to the wind speed to test the performance of the control strategy (top). The net output power produced as a function of the desired level of power reference with a change in wind speed (from 10 m/s to 11 m/s) for the medium scale system (bottom)

- $P_{ref} = 90$ kW: In this case, the wind speed is not high enough and the desired power reference is never attained.

12.6 Numerical Application to a Future MW Scale System

In order to evaluate the feasibility and scaling behavior of this kind of system, numerical simulations for a MW scale system have been performed. Its parameters are listed in Table 12.3 and correspond to a factor 25 from the medium scale system of the previous section. For $v_w = 10$ m/s, Reynolds number reaches Re $= 8.6 \times 10^6$.

By scaling up, the volume of the Magnus rotor increases with the cube of the rotor dimension while the mass increases with the square, because it is related to the Magnus rotor surface. The gas used to fill the Magnus rotor can be more dense, keeping the whole system lighter-than-air without using pure Helium. As in the previous section, the cycle parameters are set in order to get a nominal production cycle with vertical trajectories. We have determined the feasibility regions for $r_{min} = 200$ m and $r_{max} = 300$ m. For a wind speed $v_w = 10$ m/s, the tether speed in the production phase $\dot{r}_{prod}$ and in the recovery phase $\dot{r}_{rec}$ are found numerically offline.

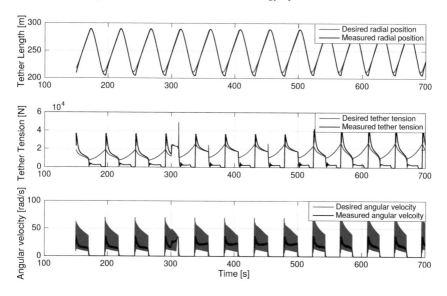

Fig. 12.20 Tether length, tether tension and the angular speed of the Magnus rotor as function of time in absence of noise for the medium scale system. The oscillations in the rotation speed are due to the choice of the control parameters. Wind speed $v_w = 10$ m/s

Symbol	Name	Value
M_M	Mass of airborne subsystem	1.133×10^4 kg
R	Magnus rotor radius	6.25 m
L_m	Magnus rotor length	80 m
ρ_g	Buoyant gas density	0.95 kg/m^3
ρ_{air}	Air density	1.225 kg/m^3
M_l	Mass per tether length	5 kg/m
M_D	Magnus rotor mass	50000 kg
$u_{T_{max}}$	Saturation on traction actuator	2×10^6 N
v_w	Wind speed	10 m/s
Re	Reynolds number	8.6×10^6

Table 12.3 Parameters of the MW scale Magnus rotor

One gets $\dot{r}_{prod} = 0.31 v_w$ and $\dot{r}_{rec} = -0.46 v_w$ which are slightly different form those found for the medium scale system.

By simulating this system at a wind speed $v_w = 10$ m/s, with the same method of the previous section, the net output power is found to be 1.37 MW for $X = 4.3$, which corresponds to an energetic performance of 1.37 kW/m^2. This is consistent with the results of the medium scale system 1.48 kW/m^2 and 1.25 kW/m^2 found in [13].

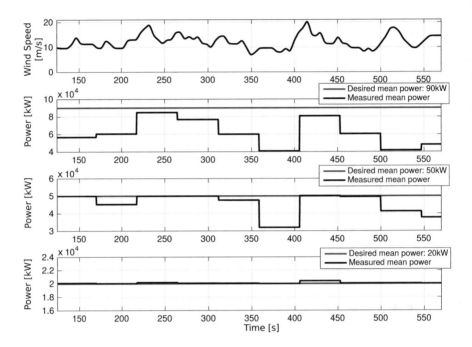

Fig. 12.21 Energy control of the medium scale system. Real wind speed data is used (top). Three levels of power reference are considered: 20 kW, 50 kW, and 90 kW. The wind speed varies from 7 m/s to 20 m/s

PID controller K_1 parameters are $K_p = 5.16 \times 10^5$ N/m, $K_i = 82.5$ N/ms, $K_d = 2.81 \times 10^6$ Ns/m. These control parameters are chosen empirically.

In Fig. 12.22, tether length, tether tension and the angular speed of the Magnus rotor as function of time are shown. One can find the maximum tension in the tether $T_{max} = 1.16 \times 10^6$ N and the maximum angular speed $\omega_{max} = 9.8$ rad/s. The production phase reel-out speed is 3.1 m/s with an overshoot measured at 7.4 m/s, the recovery phase reel-in speed is set to -4.6 m/s, without any observed overshoot.

In Fig. 12.23, one can see the vertical trajectory of the MW scale system. We also present a comparison with an equivalent conventional wind turbine. Even though the Magnus effect-based system is less efficient to capture mechanical energy from wind, it produces the same amount of power as an 80 m diameter wind turbine (around 1.4 MW for 10 m/s wind speed) since it works on a larger area. In other words, an 80 m diameter wind turbine works on 5000 m² with a power coefficient $c_p = 0.45$ where the Magnus effect-based system works on 13940 m² with a power coefficient $c_p = 0.157$. With the same method used in Sect. 12.5, the Magnus motor consumption can be estimated by $P_M = 22.7$ kW for $C_M = 0.0055$, $X = 4.3$ and $v_w = 10$ m/s. Knowing that production phase is 59% of the time, the net output power of the Magnus motor over the whole cycle is 13.56 kW which is about 1% of the power produced.

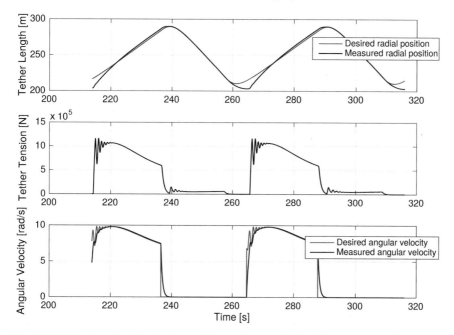

Fig. 12.22 Tether length, tether tension and the angular speed of the Magnus rotor as function of time for the MW scale system. Wind speed $v_w = 10$ m/s

12.7 Conclusions and Perspectives

The control of an airborne wind energy system based on a Magnus rotor have been presented. The small-scale system wind tunnel experiments have enabled us to test different aspects of the system and to validate part of the proposed control strategy. The Magnus effect-based model was validated for a spin ratio ranging from 1 to approximately 2.3. Our goal for a future work is the experimentation of such models for a spin ratio greater than 5.5 in order to increase efficiency. The small size of our wind tunnel does not allow to reach tether reel-out speeds that would achieve the simulated performance of 1.48 kW/m^2, but faster dynamics of the actuators would allow to achieve vertical trajectories.

Nominal production cycles have been studied for a medium scale systems with vertical trajectories achieving the simulated performance of 1.48 kW/m^2. For these vertical trajectories, the simple kite case described in [10] cannot be considered since the translation velocity of the Magnus rotor $\mathbf{v_k}$ is not colinear with the traction force $\mathbf{T}$. This type of 2D cycles has to be formally studied in order to optimize its performance. The complete control strategy has been applied. The system succeeds to track the desired power reference even in the presence of wind turbulence. This strategy can be applied in a future work on other types of AWE systems by adapting the control variables.

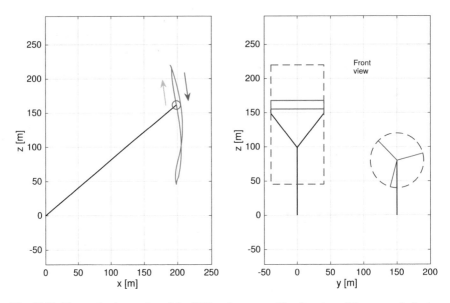

Fig. 12.23 The production cycles of the MW scale system. The direction of the arrows indicates the movement of the Magnus rotor: Green for the production phase and red for the recovery phase. On the right side figure, the MW scale system and an equivalent conventional wind turbine are compared

Finally, in order to study the feasibility of a MW scale system, numerical simulation has been performed. The production cycle gives 1.38 kW/m^2 based on vertical trajectories. The problem of scalability due to the structure's resistance to important forces is not treated here and must be addressed later to ensure that the Magnus rotor can withstand such mechanical stress.

Acknowledgements The authors of this chapter would like to thank the technical staff of Gipsa-lab and the trainees Azzam Alwann, Alexandre Kajiyama and Pierre Estadieu. They also thank the editor and the anonymous reviewers for their constructive comments, which helped to improve the quality of the chapter.

References

1. Ampyx Power B.V. http://www.ampyxpower.com/. Accessed 24 May 2013
2. Canale, M., Fagiano, L., Milanese, M.: KiteGen: A revolution in wind energy generation. Energy **34**(3), 355–361 (2009). doi: 10.1016/j.energy.2008.10.003
3. Cherubini, A., Papini, A., Vertechy, R., Fontana, M.: Airborne Wind Energy Systems: A review of the technologies. Renewable and Sustainable Energy Reviews **51**, 1461–1476 (2015). doi: 10.1016/j.rser.2015.07.053
4. Enerkite GmbH: Products. http://www.enerkite.de/en/products. Accessed 30 Oct 2015

5. GIPSA-lab. http://www.gipsa-lab.grenoble-inp.fr/recherche/plates-formes.php?id_plateforme=70. Accessed 30 Oct 2015
6. Hably, A., Dumon, J.: Éoliennes Volantes: Airborne Wind Energy Activities at the Gipsa-Lab. In: Schmehl, R. (ed.). Book of Abstracts of the International Airborne Wind Energy Conference 2015, p. 41, Delft, The Netherlands, 15–16 June 2015. doi: 10.4233/uuid: 7df59b79-2c6b-4e30-bd58-8454f493bb09. Presentation video recording available from: https://collegerama.tudelft.nl/Mediasite/Play/5068e380738143bbb8cc8aa59fe677481d
7. Hably, A., Lozano, R., Alamir, M., Dumon, J.: Observer-based control of a tethered wing wind power system: indoor real-time experiment. In: Proceedings of the 2013 American Control Conference, Washington, DC, USA, 17–19 June 2013. doi: 10.1109/ACC.2013.6580368
8. Johnston, I.: Bill Gates calls for Manhattan Project-style renewable energy drive. The Independent, 26 June 2015. http://www.independent.co.uk/news/people/bill-gates-calls-for-manhattan-project-style-renewable-energy-drive-10346752.html Accessed 30 Oct 2015
9. Loctier, D.: The search for a high flying clean energy generator. http://www.euronews.com/2016/02/22/the-search-for-a-high-flying-clean-energy-generator (2016). Accessed 6 Dec 2017
10. Loyd, M. L.: Crosswind kite power. Journal of Energy 4(3), 106–111 (1980). doi: 10.2514/3.48021
11. Lozano, R., Dumon, J., Hably, A., Alamir, M.: Energy production control of an experimental kite system in presence of wind gusts. In: Proceedings of the 2013 IEEE/RSJ International Conference on Intelligent Robots and Systems, pp. 2452–2459, IEEE, Tokyo, Japan, 3–7 Nov 2013. doi: 10.1109/IROS.2013.6696701
12. Makani Power/Google. http://www.google.com/makani. Accessed 14 Jan 2016
13. Milutinović, M., Čorić, M., Deur, J.: Operating cycle optimization for a Magnus effect-based airborne wind energy system. Energy Conversion and Management 90, 154–165 (2015). doi: 10.1016/j.enconman.2014.10.066
14. Omnidea, Lda. http://www.omnidea.net/hawe/. Accessed 28 June 2013
15. Omnidea, Lda: Omnidea High Altitude Wind Energy with Magnus effect. https://www.youtube.com/watch?v=Ne_aEa__svo (2015). Accessed 6 Dec 2017
16. Pardal, T., Silva, P.: Analysis of Experimental Data of a Hybrid System Exploiting the Magnus Effect for Energy from High Altitude Wind. In: Schmehl, R. (ed.). Book of Abstracts of the International Airborne Wind Energy Conference 2015, pp. 30–31, Delft, The Netherlands, 15–16 June 2015. doi: 10.4233/uuid:7df59b79-2c6b-4e30-bd58-8454f493bb09. Presentation video recording available from: https://collegerama.tudelft.nl/Mediasite/Play/e51a679525fe491990de3a55a912f79d1d
17. Penedo, R. J. M., Pardal, T. C. D., Silva, P. M. M. S., Fernandes, N. M., Fernandes, T. R. C.: High Altitude Wind Energy from a Hybrid Lighter-than-Air Platform Using the Magnus Effect. In: Ahrens, U., Diehl, M., Schmehl, R. (eds.) Airborne Wind Energy, Green Energy and Technology, Chap. 29, pp. 491–500. Springer, Berlin Heidelberg (2013). doi: 10.1007/978-3-642-39965-7_29
18. Perković, L., Silva, P., Ban, M., Kranjčević, N., Duić, N.: Harvesting high altitude wind energy for power production: The concept based on Magnus' effect. Applied Energy 101, 151–160 (2013). doi: 10.1016/j.apenergy.2012.06.061
19. Seifert, J.: A review of the Magnus effect in aeronautics. Progress in Aerospace Sciences 55, 17–45 (2012). doi: 10.1016/j.paerosci.2012.07.001
20. Station Météo Bard. http://www.meteobard.fr/. Accessed 30 Oct 2015
21. White, F.: Fluid mechanics. 8th ed. McGraw-Hill Education (2015)

Chapter 13
Optimization-Inspired Control Strategy for a Magnus Effect-Based Airborne Wind Energy System

Milan Milutinović, Mirko Čorić and Joško Deur

Abstract An optimization study has been conducted and the corresponding control strategy developed for the lighter-than-air airborne wind energy system. The linchpin of the system is an airborne module in the form of a buoyant, rotating cylinder, whose rotation in a wind stream induces the Magnus effect-based aerodynamic lift, thereby facilitating traction power generation. The optimization is aimed at maximizing the average power produced at the ground-based generator during a continuously repeatable operating cycle. This chapter provides a recap of the optimization methodology, results, and their physical interpretation, and builds on this foundation to develop control strategies aimed at approaching the optimization results. Comparative analysis of the two proposed control strategies and the optimization results shows that the simpler and more robust strategy can approach the performance of the more sensitive strategy that closely matches the optimization results.

13.1 Introduction

The research field of airborne wind energy has given rise to a multitude of innovative systems, featuring different techniques for harnessing the wind at altitudes beyond the reach of conventional, ground-based turbine systems [1, 5, 7]. Whereas most of the systems that aim to produce traction power (ground-based generator systems) employ wings or kites to induce aerodynamic forces, this chapter is concerned with a quite distinct system, based on a lighter-than-air cylindrical balloon exploiting the Magnus effect [18, 19].

The ground-based-generator (GBG) system in question, proposed by Omnidea and developed by the HAWE consortium [16], consists of an airborne module (ABM) connected by a single tether to the winch-generator system (Fig. 13.1). A

Milan Milutinović (✉) · Mirko Čorić · Joško Deur
University of Zagreb, Faculty of Mechanical Engineering and Naval Architecture, Zagreb, Croatia
e-mail: milan.milutinovic@fsb.hr

© Springer Nature Singapore Pte Ltd. 2018
R. Schmehl (ed.), *Airborne Wind Energy*, Green Energy
and Technology, https://doi.org/10.1007/978-981-10-1947-0_13

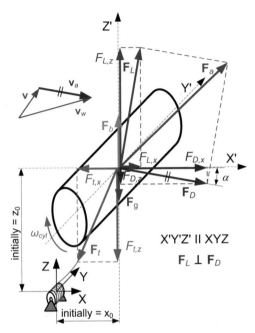

Fig. 13.1 Reference coordinate system, forces and velocities of the basic two-dimensional ABM dynamics model [14]

wind stream produces the Magnus effect-based aerodynamic lift on a rotating cylindrical balloon, thereby driving the generator in an ascending-descending pumping mode. The rotation of the cylinder is accomplished using an electric drive attached to the ABM, with the tether serving as both the mechanical and the electrical link with the ABM. The intended motion of the ABM is contained in a vertical plane. Previously conducted theoretical studies [19] and, in particular, preliminary proof-of-concept experimental results [18] have shown that the energy production based on the proposed concept is viable.

The key functionality of the system is to achieve continuous cyclical operation accompanied by power production. Using the control-oriented 2D dynamics model initially outlined in [19] and slightly expanded in Sect. 13.2, it is reasonably straightforward to define the underlying control system structure, including the elementary winch-ABM coordination logic and the basic supervisory control strategy used to facilitate this goal. This is done in Sect. 13.3, which also presents realistic, low-level control loops of the generator and cylinder motor speeds.

To develop a more fruitful control strategy, aiming to maximize the average mechanical power transferred by the tether to the winch, an open-loop (off-line) control optimization study has been conducted [14]. Similar studies have thus far been conducted mainly for GBG AWE systems that are kite or wing-based [6, 9–11], which are aerodynamically quite different from the Magnus effect-based concept presented herein, and for which it is well known that a crosswind motion is preferable [13].

Section 13.4 recapitulates this optimization study and its results, which provide an insightful basis and a benchmark for further control strategy development, and compares the optimized power production with the basic control strategy. This content is the first major topic of the chapter.

The optimization is conducted using the dynamically ideal low-level control, and has been compared with basic control strategy simulation using equivalent low-level control. However, when realistic dynamic properties of the system are introduced, pronounced irregularities in responses of tether force and power occur (such as spikes or sudden drops). This is mitigated by refining the winch-ABM coordination during reversing. Also, an adaptation is needed to limit the tether force. These refinements, taken up in Sect. 13.5 in relation to the basic control strategy and in Sect. 13.6 in relation to the optimization-inspired control strategy, form the second major topic of this chapter.

The final major topic of the chapter, presented in Sect. 13.7, is the analysis of the ABM initial position influence on produced power, since the operating position has important consequences for the choice of the control strategy.

This chapter does not deal with power transfer beyond the generator, i.e. with grid supply and the related requirement for an electric storage subsystem. For an assessment of various kinds of storage appropriate for AWE systems, motivated by the system described in this chapter, the reader is referred to [17].

13.2 Process Model

A model used for the purpose of ABM control system design should be simple, but also capable of reflecting the basic dynamics of the system. For this purpose, a simple 2D ABM state-space dynamics model derived from [19] is used throughout this chapter. The model uses Cartesian coordinates, as illustrated in Fig. 13.1.

The forces acting on the ABM include: the aerodynamic drag force $\mathbf{F}_D$, the aerodynamic lift force $\mathbf{F}_L$, the buoyancy force $\mathbf{F}_b$, the ABM weight force $\mathbf{F}_g$, and the ABM-side tether force $\mathbf{F}_t$. The drag and lift forces form the resultant aerodynamic force $\mathbf{F}_a$. The force components are considered positive when oriented as in Fig. 13.1, and negative for opposite orientation. The velocities include: the ABM velocity $\mathbf{v}$, the wind velocity $\mathbf{v}_w$, and the relative velocity between the wind and the ABM (apparent wind velocity), $\mathbf{v}_a = \mathbf{v}_w - \mathbf{v}$. In case of velocities, positive directions of their components always correspond to positive directions of the coordinate axes. Reference direction of the cylinder angular velocity ω_{cyl} is chosen so that for the given reference direction of wind velocity $\mathbf{v}_w$, the Magnus effect causes a lift force $\mathbf{F}_L$ that points upward (i.e. that has positive z-component).

The corresponding model equations, starting with velocities, are:

$$v_{a,x} = v_{w,x} - v_x, \tag{13.1}$$

$$v_{a,z} = v_{w,z} - v_z, \tag{13.2}$$

$$v_a = \sqrt{v_{a,x}^2 + v_{a,z}^2} > 0. \tag{13.3}$$

The drag force is calculated in its x- and z-direction components as:

$$F_{D,x} = C_D \rho \, r_{cyl} l_{cyl} v_{a,x} v_a, \tag{13.4}$$

$$F_{D,z} = -C_D \rho \, r_{cyl} l_{cyl} v_{a,z} v_a. \tag{13.5}$$

Similarly, the lift force components read:

$$F_{L,x} = -\text{sign}(\omega_{cyl}) C_L \rho \, r_{cyl} l_{cyl} v_{a,z} v_a, \tag{13.6}$$

$$F_{L,z} = \text{sign}(\omega_{cyl}) C_L \rho \, r_{cyl} l_{cyl} v_{a,x} v_a. \tag{13.7}$$

It is assumed that the wind velocity $\mathbf{v}_w$ is oriented in x-direction, with constant speed of 10 m/s, and that the air density ρ is constant (1.18 kg/m^3), i.e. the wind speed and air density height dependence is disregarded, mainly to reduce the computational complexity of the subsequent optimization problem.

The aerodynamic drag and lift coefficients, C_D and C_L, are given as functions of the cylinder circumferential speed to apparent wind speed ratio:

$$X = |\omega_{cyl}| r_{cyl} / v_a. \tag{13.8}$$

The C_D and C_L maps are derived from [20, 22] and additional research by HAWE consortium, and conveniently approximated by polynomials:

$$C_D(X) = -0.0211X^3 + 0.1837X^2 + 0.1183X + 0.5, \tag{13.9}$$

$$C_L(X) = 0.0126X^4 - 0.2004X^3 + 0.7482X^2 + 1.3447X. \tag{13.10}$$

Note that the aerodynamic coefficients vary considerably with the speed ratio, which is an important distinguishing feature compared to wings, where they are far less variable. According to Eqs. (13.9) and (13.10), maximizing the lift-to-drag ratio C_L/C_D by maintaining the associated X is not particularly rewarding for this specific system, because it occurs at low values of the aerodynamic coefficients.

The buoyancy force acting on the ABM cylinder and the weight of the entire ABM including the gas inside the cylinder are given by the following expressions, with radius $r_{cyl} = 1.75$ m, length $l_{cyl} = 21$ m and ABM mass $m_{ABM} = 150$ kg:

$$F_b = \rho \, r_{cyl}^2 \pi l_{cyl} g, \tag{13.11}$$

$$F_g = m_{ABM} g. \tag{13.12}$$

The tether force acting on the ABM, $\mathbf{F}_t$, is obtained from the straight-line elastic tether model with neglected aerodynamic drag, as shown in Fig. 13.2(a) (for a more

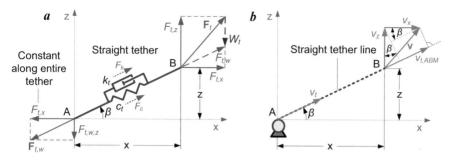

Fig. 13.2 Straight line elastic tether model (a) and calculation of tether speed (b)

accurate tether model, which takes into account the aerodynamic drag and inertial effects of tether mass, see [15]).

The tether model amounts to calculating the winch-side tether force, $\mathbf{F}_{t,w} = \mathbf{F}_c + \mathbf{F}_k$ (a sum of tether elastic and damping forces $\mathbf{F}_c$ and $\mathbf{F}_k$, Fig. 13.2(a)) as described below, and adding tether weight W_t to get the ABM-side force x- and z-components:

$$F_{t,x} = F_{t,w} \cos \beta = F_{t,w} \frac{x}{\sqrt{x^2 + z^2}}, \tag{13.13}$$

$$F_{t,z} = F_{t,w} \sin \beta + W_t = F_{t,w} \frac{z}{\sqrt{x^2 + z^2}} + m_{t,rel} g l_t. \tag{13.14}$$

The tether weight $W_t = m_{t,rel} g l_t$ is calculated from the tether mass-per-meter, $m_{t,rel} = 0.3$ kg/m, and the variable wound-out tether length l_t, obtained from the winch dynamics model (described below).

$\mathbf{F}_{t,w}$ is calculated using tether stretch according to the spring-damper model in Fig. 13.2(a). The tether longitudinal speed as observed at the ABM, $v_{t,ABM}$, which is required for stretch speed calculation, is modeled as the projection of ABM velocity v on the straight tether line, i.e. as the radial component of velocity v, as in Fig. 13.2(b):

$$v_{t,ABM} = v_x \cos \beta + v_z \sin \beta = v_x \frac{x}{\sqrt{x^2 + z^2}} + v_z \frac{z}{\sqrt{x^2 + z^2}}. \tag{13.15}$$

The tether stretch speed is then defined as the difference between the tether speed at the ABM, $v_{t,ABM}$, given by Eq. (13.15), and the tether unwinding speed $v_t = \omega_w r_w$, obtained from the winch dynamics model, i.e. as $v_{t,ABM} - v_t$ (see Figs. 13.2(b) and 13.3). Tether stretch is the time integral of thus calculated stretch speed. The tether stiffness $c_t = A_t E_t / l_t \approx 52$ kN/m is calculated from tether representation as an axially loaded, 600 m long rod (tether material is Dyneema® [4]), while its damping $k_t = 500$ Ns/m is estimated from damping of steel and Kevlar ropes [8].

The winch dynamics model is presented in Fig. 13.3. The winch, of radius r_w and moment of inertia J_w, is driven by the winch-side tether force $F_{t,w}$, and loaded by the winch generator torque τ_g. The resulting winch rotation speed is ω_w. The

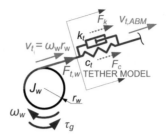

Fig. 13.3 Winch dynamics model derivation

corresponding tether winding or unwinding speed, v_t, and the length of the wound-out tether, l_t, represent inputs of the tether dynamics model.

The winch dynamics is therefore described by the following equations:

$$\dot{l}_t = v_t,$$ (13.16)

$$\dot{v}_t = r_w \dot{\omega}_w = \frac{r_w}{J_w} \left(r_w F_{t,w} - \tau_g \right).$$ (13.17)

Note that J_w is variable since the length of stored tether is variable, although this change in inertia is small for low tether mass per meter $m_{t,rel}$.

The cylinder dynamics is described by a very basic model (moment of inertia accelerated by the cylinder motor torque multiplied through a gearbox, with aero-dynamic drag torque neglected), and is therefore omitted for brevity.

The ABM dynamics is finally described by the following state equations:

$$\dot{x} = v_x,$$ (13.18)

$$\dot{z} = v_z,$$ (13.19)

$$\dot{v}_x = \frac{F_x}{m_{ABM}} = \frac{F_{D,x} + F_{L,x} - F_{t,x}}{m_{ABM}},$$ (13.20)

$$\dot{v}_z = \frac{F_z}{m_{ABM}} = \frac{F_{L,z} - F_{D,z} - F_{t,z} + F_b - F_g}{m_{ABM}}.$$ (13.21)

13.3 Control System Structure and Basic Functions

The overall control system structure (Fig. 13.4(a)) consists of the low-level ABM and winch feedback control subsystems coordinated by the high-level supervisory control strategy. The winch speed control essentially controls the unwinding speed of the tether v_t, which is a more meaningful quantity for ABM dynamics than the winch rotation speed $\omega_w = v_t/r_w$ (see the process model in the previous section). Thus, the two controlled variables in the system are taken to be the tether unwinding speed v_t and the cylinder speed ω_{cyl}, as indicated in Fig. 13.4(a).

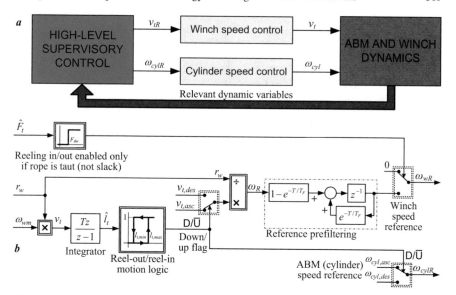

Fig. 13.4 Functional block diagram of ABM and winch control subsystems coordination (a) and elementary coordination logic of high-level supervisory control strategy (b)

13.3.1 Elementary Coordination Logic

To coordinate the ABM cylinder and winch drives, the elementary coordination logic shown in Fig. 13.4(b) is applied. This logic is the foundation for the development of the supervisory control strategy. It first determines the current cycle phase (ascending or descending) by comparing the unwound tether length l_t with the tether length thresholds $l_{t,max}$ and $l_{t,min}$, and accordingly setting the down-up flag $D/\overline{U}$ to TRUE or FALSE (the $D/\overline{U}$ flag carries the command "pull the ABM down" or "do not pull it down"). The logic sets $D/\overline{U}$ to TRUE if l_t exceeds the maximum allowed $l_{t,max}$, and to FALSE if l_t falls below the minimum allowed $l_{t,min}$. The tether length l_t can be either reconstructed by integrating the tether unwinding speed v_t, or measured using the winch position sensor. The $D/\overline{U}$ flag then provides a rudimentary coordination between the winch and the ABM, notwithstanding the actual values of the speed references and tether length thresholds, whose calculation is the task for the particular control strategy that is employed.

13.3.2 Low-Level Winch Speed and ABM Cylinder Speed Control

The structure of the winch/generator speed control loop is shown in Fig. 13.5. It includes a PI speed controller tuned according to the symmetrical optimum tuning procedure [12], which gives a fast and well-damped response of the speed control loop.

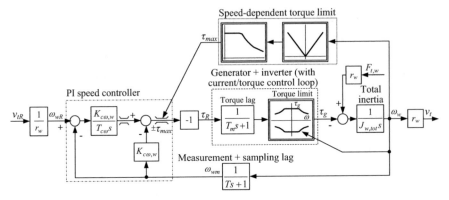

Fig. 13.5 Block diagram of the winch generator speed control system

The associated parameters are $T_{c\omega} = 4(T + T_m)$ and $K_{c\omega} = 0.5J_{w,tot}(T + T_m)$, where T_m is the generator torque/current control lag time constant, $J_{w,tot} = J_w + m_{ABM}r_w^2$ is the total equivalent winch inertia referred to the generator output shaft (assuming negligible tether mass and elasticity), and T is the sampling time. As explained above, the controlled variable is essentially the tether unwinding speed, so the reference is originally defined as reference unwinding speed v_{tR}.

It should be noted that the generator+inverter dynamic model in Fig. 13.5 also includes the speed-dependent torque limit, based on the maximum torque static curve of the chosen electrical machine. In order to account for this torque limit, the maximum torque static map $\tau_{max}(\omega_w)$ is used within the PI controller saturation algorithm, which also includes the so-called reset-integrator logic of the controller integral term saturation [2].

The cylinder speed control system has the same structure as the winch speed control system, with a formal difference related to the sign of torque variables since the cylinder machine is considered a motor, while the winch machine operates mainly as a generator. Specifically, compared to Fig. 13.5, cylinder speed block diagram would not include reversal of the motor torque (no "-1" block), and the torque would enter the summation as positive, while the load, aerodynamic drag torque M_D scaled down by the gear ratio $i_{ABM} = 45$ (and thus neglected in the model) would enter it as negative. Therefore, the PI controllers for the two systems are tuned in the same way, only the obtained parameters are different.

13.3.3 Basic Control Strategy

The basic supervisory control strategy, first outlined in [19], is presented here as an important first step in achieving sustainable operation, while the presentation of its results is postponed until the comparison with the optimization benchmark (Sect. 13.4.2). In this strategy, the tether unwinding speed v_t is controlled so that its maximum magnitudes are maintained during ascending phase (positive speed $v_{t,max}$) and descending phase (negative speed $v_{t,min}$), in order to try to maximize the ascending power and minimize the descending duration. During ascending, the cylinder rotation speed, ω_{cyl}, is controlled to obtain a desirable speed ratio $X_R = \omega_{cyl}r_{cyl}/v_a$, for instance the one that gives the maximum magnitude of the total aerodynamic force $\mathbf{F}_a$, see Fig. 13.1. During descending, ω_{cyl} is controlled so that the ABM does not get too close to the ground, but without causing too large lift and thereby a heavy motor load. This is accomplished by observing the elevation angle β or the ratio between the x- and z-coordinates of the ABM (see Fig. 13.2). The described control strategy is therefore (cf. Fig. 13.4(b)):

$$v_{tR} = \begin{cases} v_{t,max}, \text{ ascending,} \\ v_{t,min}, \text{ descending,} \end{cases} \tag{13.22}$$

$$\omega_{cylR} = \begin{cases} \dfrac{v_a X_R}{r_{cyl}}, \text{ ascending,} \\ K_c \left| \cot \beta \right| = K_c \left| \dfrac{x}{z} \right|, \text{ descending.} \end{cases} \tag{13.23}$$

The appropriate values of gain K_c depend on the chosen trade-off between low consumed power and safe ABM height during descending.

This is a simple and robust strategy, but it is not obvious whether its power yield is near optimum. This motivates an optimization study aimed at finding a strategy that maximizes the power production.

13.4 Optimization and Optimization-Based Control Strategy

The goal of numerical optimization is to find control variables' time-responses that maximize energy production during continuous system operation.

To ease the computational burden while still capturing the most important features of the system, the optimization uses a simplified version of the model. The order of the model is reduced by (i) excluding the winch submodel (Fig. 13.3) altogether while treating the tether as inelastic, and (ii) disregarding the cylinder dynamics (inertia). Accordingly, one of the two control variables is changed from tether unwinding speed v_t to the magnitude of the winch-side tether force, $F_{t,w}$. Namely, for an inelastic tether (that does not stretch), one cannot calculate the tether force from stretch and stretch speed as explained in Sect. 13.2, so $F_{t,w}$ is now commanded directly, while Eqs. (13.13) and (13.14) continue to hold. The resulting state variables' vector is $\mathbf{x} = [x\, z\, v_x\, v_z]^\top$ while the control variables' vector is $\mathbf{u} = [F_{t,w}\, \omega_{cyl}]^\top$.

13.4.1 Optimization problem formulation

The optimization problem needs to be formulated by defining: (i) the cost function, (ii) the constraints and (iii) boundary (initial/final) values of state variables. Once formulated, the problem is solved numerically using the TOMLAB/PROPT software tool [21], which uses a pseudo-spectral collocation method of optimization. Since the method does not guarantee global optimality, a convenient approach (as opposed to analytic methods [3], which are mathematically intractable for this problem) is to use TOMLAB solvers and eventually perturb the main problem formulation parameters (e.g. initial solution guess and number of grid points) for multi-run optimizations and check their sensitivity to local optima to obtain a more accurate result (for details, see [14]).

The chosen approach to optimization problem formulation is to gradually refine it. This procedure is described in detail in [14], where four optimization problem formulations were employed, whereas this section focuses on the final formulation, summarized in Table 13.1. The simpler formulations from which it evolved gave

Cost function, J	$J = -\frac{1}{t_c}\int_0^{t_c} F_{t,w} v_t \mathrm{d}t + \frac{c_{pen}}{t_c}\int_0^{t_c}\left(\dot{F}_{t,w}^2 + \dot{\omega}_{cyl}^2\right)\mathrm{d}t$	
Constraints	$z \geq 32$ m $v_t \leq 4$ m/s $v_t \geq -6$ m/s $F_{t,w} \geq 0$ $F_t < 40000$ N	$X \leq 7$ $l_t < 800$ m $x_0 = x_f = x_b$ (x_b defined in next row) $\mathbf{u}_0 = \mathbf{u}_f$
Initial/final state $x_b=[x_b\ z_b\ v_{xb}\ v_{zb}\ ]^\top$	[free free free free]$^\top$	
Final time, t_c	free	
Asc./desc. phases	A-D	
Phase constraints	$v_{z,A} > 0$ $v_{z,D} < 0$ $\mathbf{x}_{Af} = \mathbf{x}_{D0}$ $\mathbf{u}_{Af} = \mathbf{u}_{D0}$	

Table 13.1 Optimization problem formulation

practically unfeasible, but theoretically valuable results [14].

The cost function features the mechanical power at the generator, defined by the product of tether force at winch $F_{t,w}$ and the tether unwinding speed v_t. The first term in the cost function is the average power during the cycle of duration t_c. The negative sign appears because the cost function is to be minimized. The second term in the cost function serves to limit the derivatives of the control variables, i.e. to limit sudden changes and oscillations of $F_{t,w}$ and ω_{cyl}, which can be unrealistic and can also cause numerical problems. The level of derivatives suppression can be modulated by the penalty coefficient c_{pen} multiplying the integral.

The constraints consist of several conditions that have to be obeyed during operation, related to altitude z, range of tether speed v_t (matches basic control strategy, Eq. (13.22)), winch-side tether force $F_{t,w}$ that must be positive (cannot exert a push), maximum tether force F_t, speed ratio X (must not exceed range of definition of maps $C_{L,D}(X)$) and unwound tether length l_t. The basic cycle periodicity constraint is the equality of the initial and final state, i.e. $\mathbf{x}_0 = \mathbf{x}_f$. Additional periodicity constraint is the equality of the initial and final control variables, i.e. $\mathbf{u}_0 = \mathbf{u}_f$. Note that the initial/final state $\mathbf{x}_b$ is a free parameter that is optimized together with the control variables (Table 13.1). The same is true for the cycle duration t_c.

The operation is strictly divided into one ascending (A) and one descending (D) phase. Otherwise, the optimization algorithm may find a solution involving several cycles, instead of a single repeatable cycle. To define a single ascending phase followed by a single descending phase, the sign of z-component of velocity, v_z, is constrained to be positive during ascending and negative during descending. For a continuous overall cycle, the state at the end of the ascending phase is constrained to be equal to the state at the start of the descending phase, and the same applies to control variables. This reversal-point state and control values are optimized, as well as the corresponding reversing instant (the time at which the reversal occurs).

13.4.2 Optimization Results

The results obtained using the final formulation with cylinder speed limited to 200 rpm are shown green in Fig. 13.6, while the results of basic control strategy, described in Sect. 13.3.3, are blue. The particular cylinder speed limit was chosen because it roughly equals the observed maximum of the basic strategy.[1] Three important conclusions can be drawn from the results:

1. The trajectory loop is almost vertical, as opposed to slant loops produced by the basic control strategy. Given the horizontal direction of the wind, this suggests that the ABM should fly in the crosswind direction, which was not immediately clear for this system. This was corroborated by conducting the optimization with slanted (non-horizontal) wind direction [14], which also produced a crosswind motion direction.

2. The trajectory position in the xz-plane is quite far from the winch, more specifically as far as possible for the given tether length limit. This suggests the optimal operation is more easily achieved far from the winch and that this location influences power production (in fact, an earlier formulation clearly pointed out that vertical trajectory is practically unsustainable near the winch [14]).

---

[1] The process model and the accompanying control system were simplified to match the model used in the optimization, hence the steady state error visible in the basic control strategy rope speed response. Realistic control system, introduced in Sect. 13.3, will be revisited in later sections.

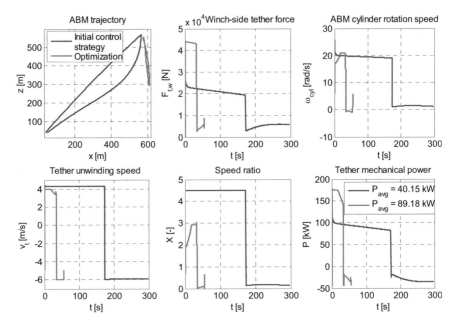

Fig. 13.6 Comparison between basic control strategy-based results and optimization results for cylinder speed limit $\omega_{cyl,max} = 200$ rpm [14]

3. The results of the optimization, providing average power of 89.18 kW, are significantly better than the basic control strategy results, which produced average power of 40.15 kW. This means that an improvement of 122% has been achieved through optimal control. Note that the basic control strategy causes radial lift, in contrast with the vertical lift associated with the crosswind motion of the optimally controlled system.

The optimization study has been extended by analyzing the effects of limited cylinder rotation speed ω_{cyl} (unconstrained in the original formulation). The analysis investigates eight levels of cylinder speed limit, by adding the following constraint to the problem formulation: $\omega_{cyl} < \omega_{cyl,max}$, where $\omega_{cyl,max} \in \{25, 50, 75,$ $100, 125, 150, 175, 200\}$ rpm (as noted above, Fig. 13.6 corresponds to the last of these values). The optimized trajectories and the most relevant time responses are shown in Fig. 13.7, as well as the corresponding dependence of the average produced power on the cylinder speed limit (solid line in the lower-right plot).

The results in Fig. 13.7 show that it remains beneficial to lift the ABM vertically even if the cylinder rotation speed is limited. The shape of other system time responses is not largely influenced by the cylinder rotation speed $\omega_{cyl,max}$. However, the magnitude of these responses during the power production phase increases as $\omega_{cyl,max}$ grows, and it saturates at the speed of around 150 rpm at which the winch-side tether force $F_{t,w}$ becomes saturated for the given system (particularly wind)

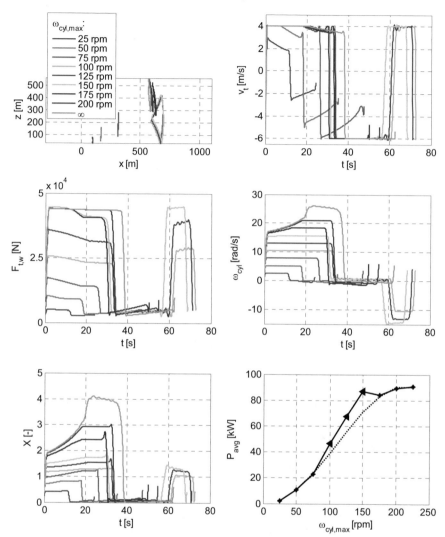

Fig. 13.7 Optimization results for different levels of cylinder speed limit $\omega_{cyl,max}$. Last point in $P_{avg} - \omega_{cyl,max}$ graph, shown in bottom right corner, corresponds to unlimited $\omega_{cyl,max}$ (adapted from [14])

parameters. Correspondingly, the average produced power P_{avg} also saturates[2] at the cylinder speed of approximately 150 rpm.

---

[2] The fact that the average power decreases as $\omega_{cyl,max}$ increases from 150 rpm to 175 rpm (Fig. 13.7) appears to be due to a failure of the optimization algorithm to find a precise solution for $\omega_{cyl,max}$ = 175 rpm (the optimizer apparently got stuck in a local optimum, see Sect. 13.4.1). This may be related to the fact that the optimal triangle-shaped trajectory occurs for $\omega_{cyl,max} \in$ {100, 125, 150} rpm (triangle markers in Fig. 13.7), but not at the apparently sub-optimal result for $\omega_{cyl,max}$ = 175 rpm (and also higher speeds).

The main discrepancy in the shapes of responses in Fig. 13.7 is in the appearance of a triangular form of the ABM trajectory for medium cylinder speeds, which relates to the occurrence of a power production phase at the end of the response. This kind of trajectory incorporates a somewhat counterintuitive form of lowering the ABM, where the tether is actually unwound and the power produced while the ABM is descending, i.e. $v_t > 0$ while $v_z < 0$. This corresponds to the lower cathetus of the triangle-shaped trajectory and the positive power in the last stage of response.

13.4.3 Algebraic Analysis of the Optimization Results

A brief interpretation of the optimization results by means of an algebraic analysis (see [14] for details) is concentrated on the ascending part of the operating cycle. The main aim of the analysis is to explain why the optimal ABM trajectory is nearly perpendicular to the wind direction and executed far away from the winch.

13.4.3.1 Optimal Shape of the ABM Trajectory

The power $P = F_{t,w} v_t$ produced during the ascending phase is predominantly dependent on the winch-side tether force $F_{t,w}$, because the tether speed is preferably kept at or near its maximum value. The tether force $F_{t,w}$ is induced mainly by the aerodynamic force F_a, which is the dominant force in the system. This force is determined from the ABM model as (see Fig. 13.1 and Eqs. (13.4) through (13.7)):

$$
\begin{aligned}
F_a &= \sqrt{F_{a,x}^2 + F_{a,z}^2} = \sqrt{(F_{L,x} + F_{D,x})^2 + (F_{L,z} - F_{D,z})^2}, \\
&= \rho\, r_{cyl} l_{cyl} v_a^2 \sqrt{C_L^2 + C_D^2} = \rho\, r_{cyl} l_{cyl} v_a^2 C_A.
\end{aligned}
\tag{13.24}
$$

Firstly, Eq. (13.24) explains why the optimization results tended to give approximately time-constant aerodynamic coefficients $C_{L,D}$ (as visible in [14]). Namely, it aimed at maintaining the maximum value of the total aerodynamic coefficient C_A, thus maximizing the aerodynamic force F_a (while taking into consideration the tether force limit). Note that, according to Eq. (13.8), the required value of X can be attained by means of controlling the cylinder rotation speed ω_{cyl}.

Secondly, for the approximately constant C_A, and since ρ is assumed constant, the only variable term in Eq. (13.24) is the apparent wind speed, v_a. This speed is described by Eqs. (13.1) through (13.3) and it should be maximized for maximal aerodynamic force F_a. To facilitate further analysis, three characteristic modes of ABM ascending are introduced:

1. Radial lift—elevation angle β is constant, thus $v = v_t$.
2. Vertical lift—ABM x-coordinate is constant, thus $v = v_z$, while $v_x = a_x = 0$
3. Backward lift—ABM ascends against the wind, i.e. "backwards", with $v_x =$ const. < 0

Equations (13.1) through (13.3) and (13.15), with approximation $v_{t,ABM} \approx v_t$ (Fig. 2b), give the following general formula for the apparent speed:

$$v_a = \sqrt{(-v_z)^2 + (v_w - v_x)^2} = \sqrt{\left(\frac{v_x}{\tan\beta} - \frac{v_t}{\sin\beta}\right)^2 + (v_w - v_x)^2}. \qquad (13.25)$$

For the characteristic mode definitions above, this reduces to the following expressions (note that the wind velocity is assumed to have only x-component):

$$v_{a,rad} = \sqrt{(v_t \sin\beta)^2 + (v_w - v_t \cos\beta)^2}, \qquad (13.26)$$

$$v_{a,vert} = \sqrt{\left(\frac{v_x}{\sin\beta}\right)^2 + v_w^2}, \qquad (13.27)$$

$$v_{a,back} = \sqrt{\left(\frac{|v_x|}{\tan\beta} + \frac{v_t}{\sin\beta}\right)^2 + (v_w + |v_x|)^2}. \qquad (13.28)$$

For the same position in the xz-plane, i.e. the same β, clearly $v_{a,back} > v_{a,vert}$. Also, since $\sin\beta$ and $\cos\beta$ are between 0 and 1 and $v_t < v_w$, relation $v_{a,vert} \gg v_{a,rad}$ clearly holds. This yields:

$$v_{a,back} > v_{a,vert} \gg v_{a,rad} \Rightarrow F_{a,back} > F_{a,vert} \gg F_{a,rad}$$
$$\Rightarrow P_{back} > P_{vert} \gg P_{rad}, \qquad (13.29)$$

which confirms that it is favorable to lift the ABM backwards or vertical to the wind, as opposed to sub-optimal radial lift. Note that this analysis is independent of the wind shear, meaning that a control strategy based on it (Sect. 13.6) is independent of wind shear as well.

13.4.3.2 Region of Feasible Vertical Lift

In addition to having constant (maximum) tether unwinding speed $v_t = 4$ m/s (see Table 13.1 and e.g. Fig. 13.6), all lift cases are characterized by a constant value of the ABM velocity x-component, v_x, implying that $a_x = 0$ (see the above three definitions of ascending modes and the kinematical relationships in Fig. 13.2(b)). Taking the time derivative of kinematic equation (13.15), using the usual approximation $v_{t,ABM} \approx v_t$, noting that $dv_t/dt = 0$ and $a_x = d^2x/dt^2 = 0$, and rearranging yields:

$$\frac{\dot{x}^2 + z\ddot{z} + \dot{z}^2}{\sqrt{x^2 + z^2}} - \frac{(x\dot{x} + z\dot{z})^2}{\sqrt{(x^2 + z^2)^3}} = 0. \qquad (13.30)$$

Equation (13.30) can be rewritten as a state-space kinematic model (see [14]) suitable for describing all three characteristic cases of ABM lift. When fed by the inputs $v_t = $ const. and $v_x = $ const., it gives a set of simulation results for $x(t)$, $z(t)$,

$v_{x,z}(t)$ and $a_{x,z}(t)$. In order to verify the feasibility of this "prescribed" motion with respect to the available set of forces acting on the ABM, the simulation results are used as inputs of the dynamic model described in Sect. 13.2. In this way, time response of the required resultant ABM force components F_x and F_z is obtained by inverted dynamic equations (13.18) through (13.21). Similarly, the individual ABM forces, other than tether force at winch, $\mathbf{F}_{t,w}$, can be calculated from "prescribed" kinematic model simulation results (this includes $\mathbf{F}_b$, $\mathbf{F}_g$, $\mathbf{F}_L$, $\mathbf{F}_D$ and W_t, Eqs. (13.1) to (13.12) and (13.14)). Subtracting them from the resultant and taking into account Eq. (13.14) gives the winch-side tether force components $F_{t,w,x} = F_{t,x}$ and $F_{t,w,z}$.

The kinematic model response will be feasible at any instant at which the winch-side tether force $\mathbf{F}_{t,w}$, obtained in the above way, satisfies the tether collinearity condition, which states that the tether force needs to be collinear with the tether direction, as shown in Fig. 13.8.

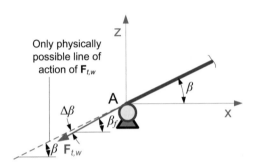

Fig. 13.8 Illustration of difference between angle of winch-side tether force $\mathbf{F}_{t,w}$, β_f, required for desired lift, and actual tether angle, β, which is the only angle $\mathbf{F}_{t,w}$ can assume

That is, the following equation needs to be satisfied, in which β_f is the angle of tether force at winch:

$$\beta_f = \arctan \frac{F_{t,w,z}}{F_{t,x}} = \beta = \arctan \frac{x}{z}. \tag{13.31}$$

A small deviation from the ideal trajectory may be tolerated in the analysis: $\Delta\beta = |\beta_f - \beta| \ll \beta$. The lift feasibility simulation tests can now be performed individually for a succession of ABM initial x-values, i.e. horizontal positions, while the initial height z is always near zero. Figure 13.9(a) shows the feasibility analysis results for the given system parameters, vertical lift ($v_x = 0$), and the collinearity error margin $\Delta\beta = 1°$. Evidently, as the horizontal distance from the winch ($x = x_0$) is increased, the vertical lift feasibility region expands externally, and also the prohibitive internal gap diminishes. This explains why the optimization results showed that the favorable vertical lift trajectory needed to be far from the winch (see Fig. 13.6). If the analysis is extended to the case of backward lift, where a small negative velocity v_x is prescribed, the feasibility region becomes significantly narrower externally (Fig. 13.9(b)). This may explain why the optimization results mostly included vertical ABM trajectories, as opposed to backward ones, despite the fact that the latter can give somewhat higher power.

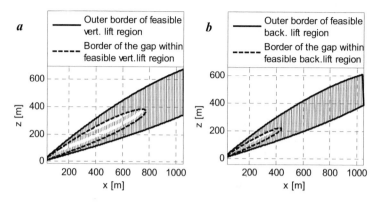

Fig. 13.9 Region where vertical lift is possible (a) and where backward lift with $v_x = -0.25$ m/s is possible (b), both for $\Delta\beta = 1°$ (adapted from [14])

13.5 Refinements of the Basic Control Strategy

Although the basic strategy is clearly subdued in terms of power production, it can be conveniently used to detect difficulties with the realistic control system and adjust it accordingly, and, as described in later sections, may be very useful in itself.

The basic control strategy was initially combined with small minimum length $l_{t,min}$, i.e. the starting position of the cycle (x_0, z_0) was close to the winch [19]. Such approach with $x_0 = z_0 = 40$ m, $l_{t,max} = 800$ m and $K_c = 1$ gives the response shown blue in Fig. 13.10, where two ascending-descending cycles are shown. As previ-

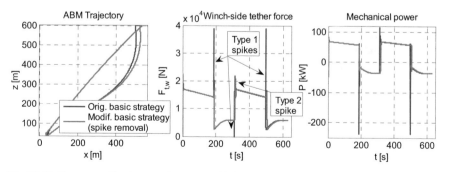

Fig. 13.10 Response of basic control strategy with initial position near winch

ously noted, the trajectory during ascending is approximately radial (a ray from the origin). Average produced mechanical power is P_{avg} = 24.54 kW for the system parameters given in Sect. 13.2.

An obvious issue detected in the results is the appearance of large tether force spikes during reversing (Fig. 13.10, type 1-spikes), as a consequence of insufficient coordination between the fast-response winch drive and the slow-response ABM drive. During up-down reversing, the winch starts pulling the ABM down before the cylinder had time to slow down, i.e. while it still pulls the ABM up, causing a sudden increase in tether force. This is detrimental because of both the tether durability and the energy production quality. During down-up reversing, the winch starts winding the tether out before the cylinder had time to accelerate, i.e. while the ABM is still descending, causing a sudden drop in tether force. This is detrimental primarily to energy production, but the subsequent sudden transition from slack to taut tether may also lead to increased stress.

Type 2 of tether force spikes, which is characterized by large tether forces, is also observed during reversing, but has a different cause: too large cylinder angular speed, i.e. too fast increase in cylinder speed during down-up reversing. These spikes are smaller, but last longer than Type 1 spikes. They are not as detrimental to energy production and are primarily a concern regarding the tether stress they may cause.

In the following sections, two methods are proposed as a systematic and effective solution for avoiding the Type 1 and Type 2 tether force spikes.

13.5.1 Type 1 Spikes Removal—Signal Shaping

The signal shaping approach to coordinating the fast winch response with the slow cylinder response (Fig. 13.11) is to make the winch speed ω_w change in the same proportion relative to winch-speed-reference-jump magnitude $\Delta \omega_{w,tot}$ as the proportion the cylinder speed ω_{cyl} changes relative to the cylinder-speed-reference-jump magnitude $\Delta \omega_{cyl,tot}$.

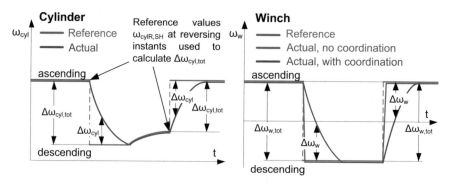

Fig. 13.11 Signal-shaping intervention for Type 1 tether force spikes removal

The equations for the cylinder speed control error $\Delta\omega_{cyl}$ and the difference between the current cylinder speed reference and the reference speed at the start of the reversing, $\Delta\omega_{cyl,tot}$, are defined as follows (Fig. 13.11):

$$\Delta\omega_{cyl} = \omega_{cyl} - \omega_{cylR}, \tag{13.32}$$

$$\Delta\omega_{cyl,tot} = \left|\omega_{cylR} - \omega_{cylR,SH}\right|. \tag{13.33}$$

Equations for the coordinated winch speed follow from those above, as illustrated in Fig. 13.11. Namely, the coordination strategy corrects the winch speed reference according to the following law:

$$\omega_{wR,coor} = \omega_{wR} + \Delta\omega_w = \omega_{wR} + \Delta\omega_{w,tot}\frac{\Delta\omega_{cyl}}{\Delta\omega_{cyl,tot}}. \tag{13.34}$$

13.5.2 Type 2 Spikes Removal—PI Control of Tether Force During Ascending

An additional PI controller (Fig. 13.12) is used to correct the cylinder motor speed reference ω_{mcylR} in order to suppress too large cylinder speeds causing Type 2 spikes of tether force. The PI controller input is the difference between the actual tether force F_t and the maximum allowed tether force $F_{t,max}$.

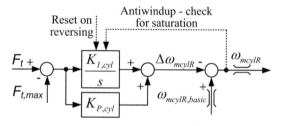

Fig. 13.12 PI controller of tether force (based on cylinder speed reference correction) for tether force Type 2 spikes removal

The PI controller lowers or increases the basic reference $\omega_{mcylR,basic}$, obtained from Eq. (13.23) using gear ratio i_{ABM}, depending on whether the tether force overshoots or undershoots the maximum allowed force, based on the assumption that the aerodynamic force increases with cylinder speed (according to Eqs. (13.9) and (13.10), this is almost always true; note that this provides the additional functionality of maintaining high tether force without explicit knowledge of wind speed). To avoid the integrator windup, the I-action is active only when (a) the final reference signal ω_{mcylR} is not saturated, or (b) the final reference signal is saturated, but the I-action tends to pull it out of saturation. The controller is active only during ascending, so as not to increase the tether force during descending. The I-action is reset to zero at every reversing instant.

The effect of spikes-removing interventions is shown green in Fig. 13.10. After the interventions, P_{avg} = 25.09 kW, which is a 2.2% increase over the original approach, which reflects the slight detrimental effect of spikes to energy production.

To better illustrate the effect of Type 2 spikes-removing approach, the wind speed was increased to 20 m/s, which gives tether forces far larger than the chosen limit of $F_{t,max}$ = 20 kN. Figure 13.13(a) shows the resulting tether forces without the spikes-removing interventions, and Fig. 13.13(b) with them included.

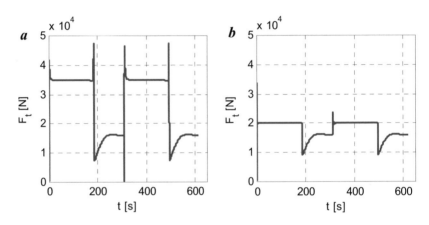

Fig. 13.13 Tether force response of basic control strategy without (a) and with (b) tether force spikes-removing interventions, for wind speed 20 m/s and tether force limit $F_{t,max}$ = 20 kN

13.6 Optimization-Inspired Vertical Lift Control Strategy

To achieve optimal-like vertical ABM operation, the control strategy is modified by changing the tether speed reference v_{tR} that is fed to the winch speed low-level control loop. Generation of the reference v_{tR} is based on Eq. (13.15) for the tether speed at the ABM, v_{tABM}, which needs to be adjusted to obtain the desired x-component of the ABM velocity, v_{xR}, and observation that $v_t \approx v_{tABM}$ (because we cannot directly control v_{tABM}), leading to the following simple control law for the speed unsaturated case:

$$v_{tR,unsat} = v_{xR}\cos\beta + v_z\sin\beta. \tag{13.35}$$

The reference tether speed $v_{tR,unsat}$ is determined from the desired x-component of the ABM velocity, v_{xR}, while the z-component v_z is predominantly influenced by the ABM cylinder rotation speed (v_z is assumed to be measurable either by inertial of GPS-based measurement system). This means that whether ascending or descending ($v_z > 0$ or $v_z < 0$) occurs is determined by control of the cylinder speed

ω_{cyl} alone, through modulating the aerodynamic forces by changing the speed ratio X, see Eqs. (13.4) through (13.10). The extended winch control ensures only that resulting winding or unwinding occurs with the desired x-component of ABM velocity, which in case of vertical motion is $v_{xR} = 0$, so that vertical shape of trajectory is obtained. In this relevant particular case, Eq. (13.35) reduces to $v_{tR,unsat} = v_z \sin \beta$.

Note that the described indirect dependence of $v_{tR,unsat}$ on ω_{cyl} through v_z means that the winch and the cylinder are indirectly coordinated, i.e. there will be no Type 1 tether force spikes (see previous section) even without any further intervention. Type 2 intervention (PI control of $\omega_{m,cylR}$) remains fully applicable to vertical operation.

Since the tether speed is limited by maximum and minimum allowed speeds corresponding to winding-out and winding-in ($v_{t,max}$ and $v_{t,min}$, respectively), Eq. (13.35) is modified to limit v_{tR} to these values:

$$v_{tR} = \begin{cases} v_{xR}\cos\beta + v_z \sin\beta, & v_{t,min} \le v_{xR}\cos\beta + v_z \sin\beta \le v_{t,max}, \\ v_{t,min}, & v_{xR}\cos\beta + v_z \sin\beta < v_{t,min}, \\ v_{t,max}, & v_{xR}\cos\beta + v_z \sin\beta > v_{t,max}. \end{cases} \quad (13.36)$$

An important consideration when applying this modification is the choice of the initial position of the ABM. The initial position was also provided by the optimization study (see Fig. 13.6), and is set here to similar value $x_0 = 550$ m, $z_0 = 350$ m. The result is shown blue in Fig. 13.14, for three simulated cycles (the green responses correspond to the improved vertical operation strategy, introduced below in Sect. 13.6.1). Average produced mechanical power is 31.93 kW, which is a 30% increase over the basic control strategy without refinements (cf. Fig. 13.10).

Although it is clearly better to use the vertical operation strategy from the standpoint of energy production, a potential difficulty with the described original approach can be recognized in Fig. 13.14 (blue). Namely, during the descending phase the absolute values of the tether speed reference $|v_{tR}|$, and therefore the actual tether speed $|v_t|$, which are now variable according to Eq. (13.36), can significantly decrease when compared with the basic control strategy, Eq. (13.22). The extent of departure from the largest allowed absolute value of tether speed during winding-in $|v_{t,min}|$ depends on the chosen initial position (x_0, z_0). If a particularly bad choice of position is made, the tether speed can even decrease to zero, thereby stopping the ABM. Therefore, it should be known in advance which position is suitable. However, this information is not readily available, especially under the realistic conditions of time-variable wind speed. Another difficulty arises from the possibility that after limiting the original reference $v_{tR,unsat}$ according to Eq. (13.36), the final reference v_{tR} cannot provide the desired vertical motion.

13.6.1 Solution to the Low Tether Speed Problem

The cylinder speed reference is modified in order to bring the unsaturated tether speed reference during descending, $v_{tR,unsat}$, close to the minimum allowed tether

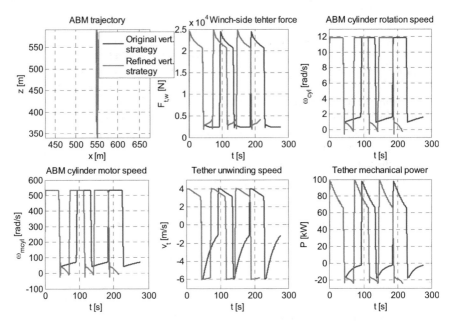

Fig. 13.14 Simulation results for original (blue) and improved (green) version of vertical operation control strategy

speed $v_{t,min} = -6$ m/s, so that descending is as short as possible, and v_{tR} is not saturated. Namely, since $v_{tR,unsat}$ is calculated from the ABM vertical velocity v_z and the elevation angle β by Eq. (13.35), for a constant x-coordinate of the vertical trajectory it will be affected mostly by v_z. This component of the ABM velocity can be influenced by the cylinder rotation speed ω_{cyl} through aerodynamic force $\mathbf{F}_a$. The z-component of aerodynamic force, $F_{a,z}$, generally decreases as ω_{cyl} decreases. This fact can be used to decelerate the ABM in the z-direction, i.e. to modify v_z and in that way $v_{tR,unsat}$. If doing so, attention has to be paid to the magnitude of $\mathbf{F}_a$, so as not to load the winch motor excessively, which is done by appropriately limiting the cylinder speed reference ω_{cylR}.

Figure 13.15 shows a block-diagram for the resulting PI controller of the unsaturated tether unwinding speed reference $v_{tR,unsat}$, based on the above-outlined new way of generating the cylinder speed reference during descending, $\omega_{cylR,des}$.

Note that the unsaturated reference tether speed $v_{tR,unsat}$ does not exceed speed limits substantially any more, meaning that the difference between Eq. (13.35) and Eq. (13.36) is smaller and vertical motion can be approached more closely. Therefore, the following rudimentary modification of the definition of the ABM horizontal velocity reference v_{xR}, used in Eq. (13.36) and nominally equal to zero, suffices to prevent position drift from accumulating over many cycles:

$$v_{xR} = K_{pos}\left(x_0 - x\right). \tag{13.37}$$

Generation of cylinder speed reference during descending

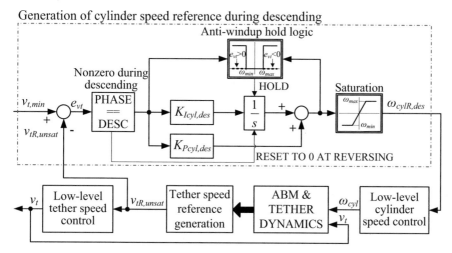

Fig. 13.15 PI controller of unsaturated tether unwinding speed reference $v_{tR,unsat}$, intended to speed up the winding-in of tether during vertical operation

This simple P controller commands a nonzero v_{xR} when there is a position error, i.e. it counters the position drift when x departs from x_0. Finally, note that strategy for generating ω_{cylR} during ascending is unchanged.

The green response in Fig. 13.14 illustrates the effect of this modified control strategy for the same operating conditions as in the original case (whose response is plotted blue in Fig. 13.14). The average mechanical power increases by 27% compared to initial vertical operation strategy, to $P_{avg} = 40.43$ kW, i.e. the accumulated improvement over the refined basic (radial-lift) strategy equals 61%.

Even though these modifications improve the robustness of the vertical operation strategy, it should be noted that for some choices of the initial ABM position the overall system behavior remains unsatisfactory. This is because the modifications are related to descending only. For particularly unfavorable initial positions, ABM may significantly slow down or stop even when it should be ascending (in accordance with the feasibility discussion of Sect. 13.4.3). Solution to this might entail additional modifications, however a simpler solution exists, as shown later in Sect. 13.7. Finally, note that the strategy relies on the assumption that the tether is nearly straight, i.e. on tether speed approximation $v_{t,ABM} \approx v_t$ (see Fig. 13.2 and Eq. (13.15)). The accuracy of this approximation is decreased for increased tether weight and drag, in which case a more detailed tether model is preferable during strategy development, e.g. the one given in [15].

13.6.2 Comparison of Optimization and Simulation Using Idealized Winch and Cylinder Control Systems

The optimization results that were presented in Sect. 13.4 were obtained using the dynamically ideal low-level control systems of the winch and the cylinder, i.e. control systems whose output was equal to their reference input (giving $v_t = v_{tR}$ for the winch, and $\omega_{cyl} = \omega_{cylR}$ for the cylinder). The same intervention is used in this section to compare the realistic closed-loop control strategy for vertical operation with the "idealized" open-loop optimization results used as a benchmark.

The summarized optimization results for the average produced mechanical power P_{avg} as a function of the cylinder speed limit $\omega_{cyl,max}$ were given in Fig. 13.7 (the lower right plot), where the result assigned to 225 rpm corresponds to no speed limit, i.e. to $\omega_{cyl,max} = \infty$. An important observation was an occurrence of a triangle shaped trajectory in some of the optimization results, which is a mode of operation that would require an exceedingly intricate control strategy, and its replication in realistic control strategies was therefore not attempted. Consequently, those three optimization results are replaced by the estimated results that would be obtained for vertical operation, by using a third order polynomial approximation of the P_{avg}-$\omega_{cyl,max}$ curve obtained when the triangle points are removed (the dotted line in the lower right plot in Fig. 13.7).

The comparison between the optimization and closed-loop control results is shown in Table 13.2.

Speed limit, $\omega_{cyl,max}$ [rpm]	Optimization: average mech. power, $P_{avg,opt}$ [kW]	Simulation: average mech. power, P_{avg} [kW]	Relative difference, $P_{avg}/P_{avg,opt} - 1$ [%]
25	2.18	1.95	-10.55
50	10.57	9.90	-6.34
75	23.47	23.04	-1.83
100	39.12	38.33	-2.02
125	55.61	56.71	1.98
150	70.94	73.45	3.54
175	83.96	87.70	4.45
200	89.18	90.62	1.61
∞	90.79	92.60	1.99

Table 13.2 Comparison between the optimization and the vertical operation strategy simulation

The comparison indicates that the optimization is notably better only at low cylinder speed limits $\omega_{cyl,max}$, but as the speed limit is increased to and over 75 rpm, the presented vertical operation strategy approaches, and at high speed limits even exceeds optimization in terms of average produced power P_{avg} (the latter points to the conclusion that optimization algorithm found local optimums there, see Sect. 13.4.2).

13.7 Analysis of ABM Initial Position Influence on Produced Power and Implications for Control Strategy Choice

From the thus far described features of the basic and the vertical operation strategies, it can be concluded that the basic strategy is simple and inherently robust, but yields comparatively little energy, whereas the situation with vertical operation is exactly the opposite. Furthermore, the quality of vertical operation strongly depends on the chosen position in the xz-plane where the operation is to take place.

To gain further insights, a more detailed analysis of the initial ABM position influence has been conducted, including both the vertical and the basic operation strategy. To this end, a set of initial positions was defined in the xz-plane, as given by the nodes of the grid defined by vertical lines at $x \in \{50, 175, 300, 425, 550, 675\}$ m and horizontal lines at $z \in \{50, 150, 250, 350\}$ m. For each point, simulations of the final versions of the basic control strategy and the vertical control strategy were conducted, but with the "idealized" low-level control systems, to facilitate comparison with optimization. To preserve legibility, only nine points were selected for subsequent trajectory plots.

There are two ways of defining the available tether length during a cycle, which determines when the production phase ends and descending begins. The one that has been used thus far is to explicitly set the maximum length of the tether. The other way is to dismiss the initial length and consider only the effective length of the active tether (part of the tether that is cyclically wound out and wound in during continuous operation).

13.7.1 Operation with Maximum Tether Length Prescribed

Reversing from ascending to descending occurs upon reaching the maximum length of wound-out tether $l_{t,max} = 800$ m. Ascending begins when the tether length decreases to initial value. Accordingly, the active tether length is large if the initial position is near the winch, and small if the initial position is far from the winch.

Figures 13.16(a) and 13.16(b) show the average power vs. the initial ABM position plots in the cases of vertical operation strategy and basic control strategy (with $K_c = 0.35$ during descending), respectively. Robustness of the basic control strategy and at the same time sensitivity of the vertical operation control strategy with respect to initial ABM position are apparent. While the vertical operation strategy indeed provides more average power when far from the winch, it fails to provide any power if the initial position is near the winch, because the operation is unsustainable there (the ABM comes to a halt; see Sect. 13.6.1). Moreover, the power gain in the region where the vertical operation has an advantage is not overwhelming. For $x_0 = 650$ m and $z_0 = 250$ m, where the largest average power for vertical operation is achieved, the gain is 5.5%, and it does not exceed 7.3% for any initial position.

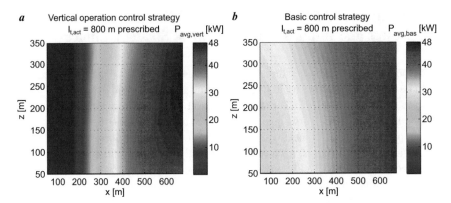

Fig. 13.16 Operation with prescribed maximum tether length: Average produced power vs. initial ABM position for vertical operation control strategy (a) and basic control strategy (b)

Trajectories for a subset of starting positions are shown in Fig. 13.17. In the case of vertical operation, Fig. 13.17(a), the approximate feasible region of vertical lift with tether winding out speed of $v_t = 4$ m/s, obtained in Sect. 13.4.3, is drawn in. The full line is the outside border of the feasible vertical lift region, and the dashed line is the border of the gap inside the feasible vertical lift region. This means that the approximately vertical lift with $v_t = 4$ m/s can take place roughly between the dashed line and the full line. Outside that space, either the lift will not be vertical or the tether speed will significantly depart from 4 m/s. In accordance with this, it can be observed that for the three simulations with $x_0 = 50$ m, the ABM came to a halt (Fig. 13.17(a)), i.e. the cycle could not be completed. This is also visible in Fig. 13.16(a), as the average power is zero for those cases.

On the other hand, the trajectories for the basic control strategy all tend to converge to a single, approximately radial line during ascending. This line is determined by the resultant force on the ABM, which changes negligibly during ascending (by both direction and magnitude). When the tether speed v_t and the speed ratio X are constant (a feature of the basic control strategy) in addition to the constant wind speed v_w and air density ρ, the forces acting on the ABM can find a steady balance where their resultant is approximately zero. Namely, the aerodynamic force $\mathbf{F}_a$ is then constant, see Eqs. (13.1) through (13.10), as well as the ABM buoyancy $\mathbf{F}_b$ and the ABM weight $\mathbf{F}_g$, and the change of the tether weight W_t (Eq. (13.14)) is negligible as it is small compared to the dominant $\mathbf{F}_a$. Then ascending occurs for one distinct elevation angle β, which is the angle of action of the winch-side tether force $\mathbf{F}_{t,w}$ that must be constant in magnitude and direction to balance the constant resultant force to the ABM. The trajectories do not fully converge only for large initial tether lengths, i.e. limited active part of tether.

An important observation is that for the shorter cycles, corresponding to initial ABM positions farther from the winch and, consequently, smaller active tether lengths, the trajectories become increasingly vertical. This, in fact, is the reason that at those positions the basic strategy is only marginally worse than the vertical strategy. Hence, the nearly vertical operation can be approximated rather well by the simple and robust basic control strategy if the cycle is short and occurs far from the winch.

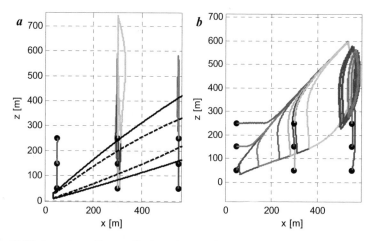

Fig. 13.17 Operation with prescribed maximum tether length: Sample trajectories for vertical operation control strategy, including feasible vertical lift area (a) and basic control strategy (b)

13.7.2 Operation with Active Tether Length Prescribed

In this approach, the reversing from ascending to descending occurs when the active tether length $l_{t,act} = 100$ m is wound out. This means that the maximum length is now variable.

The average power vs. initial position plots for the two control strategies are shown in Fig. 13.18. The corresponding individual trajectories are shown in Fig. 13.19. The area of large average power values for the vertical operation is now much larger (cf. Fig 13.16(a)), which is expected because smaller active tether length prevents the ABM to go far outside the area of feasible vertical lift (cf Fig. 13.19(a) and Fig. 13.17(a)). For this reason, the operation with $x_0 = z_0 = 50$ m manages to complete a cycle, albeit an unsatisfactory one. Hence, prescribing a small active length of the tether increases the robustness of the vertical operation control. Nevertheless, the basic control strategy is still more robust, and the produced power values approach those of the vertical operation strategy in its best operating region; for instance, for $x_0 = 550$ m and $z_0 = 150$ m, where the largest vertical operation average power is achieved (48.63 kW), the gain over basic strategy average power (44.98 kW) is 8.1%, and it is not larger than 9.7% at any position.

Figure 13.19(b) clearly indicates that the basic control strategy trajectories have the following features: (i) they cluster around a single line (the convergence line

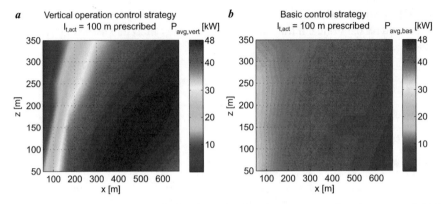

Fig. 13.18 Operation with prescribed active tether length: Average produced power vs. initial ABM position for vertical operation control strategy (a) and basic control strategy (b)

from Fig. 13.17(b)), and (ii) they become increasingly vertical as the active tether length is decreased, i.e. the cycle distance from the winch is increased. Therefore, in the case of prescribed active tether length, the basic strategy performance is rather independent of the initial ABM position (particularly after the initial settling to the final vertical trajectory), which is confirmed by a rather uniform power production plot in Fig. 13.18(b).

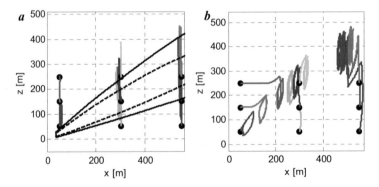

Fig. 13.19 Operation with prescribed active tether length: Sample trajectories for vertical operation control strategy, including feasible vertical lift area (a) and basic control strategy (b)

13.8 Conclusion

The conducted optimization study has shown that the optimal power production cycle of the presented Magnus effect-based airborne wind energy system should take place at a position far from the winch, so that a vertical or slightly upstream-inclined trajectory can be produced for a reasonably long cycle period. If the cylinder rotation speed is not significantly limited and for the particular system parameters that were used, the increase in power production compared with the original, basic control strategy, whose main defining feature is the constancy of tether reel-in/reel-out speeds, can reach over 120%.

The analysis of the optimization results has shown that the physical reason for the optimal, near-vertical shape of the trajectory is the possibility of achieving large air-ABM relative velocities through crosswind motion, which induces large aerodynamic (and consequently tether) forces. The kinematic model-based feasibility analysis has confirmed that the optimal vertical trajectory can be achieved for a practically sustainable period of time only if it is horizontally quite distant from the winch position, especially if reasonably high tether speed is desired.

The subsequent work has proposed a new, optimization-inspired vertical control strategy for the power production cycle. The developed practical implementation of both the vertical and the basic control strategy accounts for potential issues such as the occurrence of tether force spikes due to coordination problems between the winch drive and the ABM cylinder drive, and for the tether force overload in general. Comparison with optimization showed that the optimization-inspired vertical operation provides nearly optimal power output.

The vertical operation strategy requires a variable tether speed and a good knowledge of suitable position for operation in the xz-plane, which can be challenging to predict (especially on-line), making this strategy sensitive to system parameters. Consequently, its power output can drop drastically if the appropriate position of operation is poorly selected. On the other hand, although the basic control strategy is suboptimal in terms of energy production, it is simple to implement and robust. Provided that it is combined with a small active tether length and a relatively large distance from the winch, it approaches the vertical trajectory and only lags the power production quality of the truly vertical, nearly optimal operation by up to 10%. The exact choice of initial position is not critical, because the ABM position will tend to converge regardless of the initial coordinates, and the variability of power production with respect to this position is acceptable, as there are no drastic drops. This makes the basic control strategy a good trade off between the high power yield on one hand and the robustness and simplicity on the other.

Acknowledgements It is gratefully acknowledged that this work was supported by the European Commission through the "High Altitude Wind Energy" FP7 project, grant No. 256714. The authors would also like to express their gratitude to the project coordinator Omnidea Lda for the support extended on the project activities, as well as to project partner DTU Wind Energy for useful discussions on HAWE system modeling.

References

1. Ahrens, U., Diehl, M., Schmehl, R. (eds.): Airborne Wind Energy. Green Energy and Technology. Springer, Berlin Heidelberg (2013). doi: 10.1007/978-3-642-39965-7
2. Åström, K. J., Wittenmark, B.: Computer-Controlled Systems. 3rd ed. Prentice Hall, Upper Saddle River, NJ (1997)
3. Bertsekas, D. P.: Dynamic programming and optimal control. 3rd ed., vol. 1. Athena Scientific, Nashua (2005)
4. Bosman, R., Reid, V., Vlasblom, M., Smeets, P.: Airborne Wind Energy Tethers with High-Modulus Polyethylene Fibers. In: Ahrens, U., Diehl, M., Schmehl, R. (eds.) Airborne Wind Energy, Green Energy and Technology, Chap. 33, pp. 563–585. Springer, Berlin Heidelberg (2013). doi: 10.1007/978-3-642-39965-7_33
5. Cherubini, A., Papini, A., Vertechy, R., Fontana, M.: Airborne Wind Energy Systems: A review of the technologies. Renewable and Sustainable Energy Reviews **51**, 1461–1476 (2015). doi: 10.1016/j.rser.2015.07.053
6. Fagiano, L., Milanese, M., Piga, D.: Optimization of airborne wind energy generators. International Journal of Robust and Nonlinear Control **22**(18), 2055–2083 (2011). doi: 10.1002/rnc.1808
7. Fagiano, L., Milanese, M.: Airborne Wind Energy: an overview. In: Proceedings of the 2012 American Control Conference, pp. 3132–3143, Montréal, QC, Canada, 27–29 June 2012. doi: 10.1109/ACC.2012.6314801
8. Hamilton, J. M.: Vibration-based techniques for measuring the elastic properties of ropes and the added mass of submerged objects. Journal of Atmospheric and Oceanic Technology **17**(5), 688–697 (2000). doi: 10.1175/1520-0426(2000)017<0688:VBTFMT>2.0.CO;2
9. Houska, B., Diehl, M.: Optimal control for power generating kites. In: Proceedings of the 9th European Control Conference, pp. 3560–3567, Kos, Greece, 2–5 July 2007
10. Houska, B., Diehl, M.: Optimal control of towing kites. In: Proceedings of the 45th IEEE Conference on Decision and Control, pp. 2693–2697, San Diego, CA, USA, 13–15 Dec 2006. doi: 10.1109/CDC.2006.377210
11. Houska, B., Diehl, M.: Robustness and Stability Optimization of Power Generating Kite Systems in a Periodic Pumping Mode. In: Proceedings of the IEEE Multi-Conference on Systems and Control, pp. 2172–2177, Yokohama, Japan, 8–10 Sept 2010. doi: 10.1109/CCA.2010.5611288
12. Leonhard, W.: Control of electrical drives. 3rd ed. Springer, Berlin Heidelberg (2001)
13. Loyd, M. L.: Crosswind kite power. Journal of Energy **4**(3), 106–111 (1980). doi: 10.2514/3.48021
14. Milutinović, M., Čorić, M., Deur, J.: Operating cycle optimization for a Magnus effect-based airborne wind energy system. Energy Conversion and Management **90**, 154–165 (2015). doi: 10.1016/j.enconman.2014.10.066
15. Milutinović, M., Kranjčević, N., Deur, J.: Multi-mass dynamic model of a variable-length tether used in a high altitude wind energy system. Energy Conversion and Management **87**, 1141–1150 (2014). doi: 10.1016/j.enconman.2014.04.013
16. Omnidea, Lda: HAWE High Altitude Wind Energy. http://www.omnidea.net/hawe/. Accessed 31 Jan 2016
17. Pavković, D., Hoić, M., Deur, J., Petrić, J.: Energy storage systems sizing study for a high-altitude wind energy application. Energy **76**, 91–103 (2014). doi: 10.1016/j.energy.2014.04.001
18. Penedo, R. J. M., Pardal, T. C. D., Silva, P. M. M. S., Fernandes, N. M., Fernandes, T. R. C.: High Altitude Wind Energy from a Hybrid Lighter-than-Air Platform Using the Magnus Effect. In: Ahrens, U., Diehl, M., Schmehl, R. (eds.) Airborne Wind Energy, Green Energy and Technology, Chap. 29, pp. 491–500. Springer, Berlin Heidelberg (2013). doi: 10.1007/978-3-642-39965-7_29

19. Perković, L., Silva, P., Ban, M., Kranjčević, N., Duić, N.: Harvesting high altitude wind energy for power production: The concept based on Magnus' effect. Applied Energy **101**, 151–160 (2013). doi: 10.1016/j.apenergy.2012.06.061
20. Reid, E. G.: Tests of rotating cylinders. Technical Note NACA-TN-209, Langley Memorial Aeronautical Laboratory, Langley Field, VA, USA, 1924. http://ntrs.nasa.gov/archive/nasa/casi.ntrs.nasa.gov/19930080991.pdf
21. Rutquist, P. E., Edvall, M. M.: PROPT – Matlab optimal control software, TOMLAB Optimization Inc., 26 Apr 2010. http://tomopt.com/docs/TOMLAB_PROPT.pdf
22. White, F. M.: Fluid mechanics. 4th ed. McGraw-Hill, Boston (1998)

Chapter 14
Optimization of Pumping Cycles for Power Kites

Marcelo De Lellis, Ramiro Saraiva and Alexandre Trofino

Abstract The main contribution of this chapter is the formulation of an optimization problem to find the set of parameters of two decentralized control schemes—one for the wing flight and another for the ground winch—that maximizes the cycle power of a pumping kite. The pumping cycle consists of two phases, traction (reel-out) and retraction (reel-in), with predefined trajectories. The optimization takes into account constraints of reel speed saturation and minimum angle of attack, and can be applied to any wing with de-powering capability and given aerodynamic curves. The solution is computed through an iterative algorithm that uses a model of massless kite in dynamic equilibrium for the traction phase, and a dynamic 2D point mass model for the retraction phase. Other contributions are a discussion on the influence of the tether drag on the optimal angle of attack, and how the base angle of attack affects the average angle of attack. All results are validated by simulations with a dynamic 3D point mass kite model.

14.1 Introduction

In 1980 Loyd published his pioneering work in the field of airborne wind energy (AWE), describing how the power in the wind flow could be harvested by a flying kite tethered to the ground in two possible modes: the "lift power" and "drag power" [20]. Since then many variants of AWE systems using tethered wings have been proposed, such as the "laddermill" [19, 23], the "fast-motion transfer system" [15], the "dancing kites" [16], the "rotokites" [26], the "Makani energy kite" [22], the "carousel" [5], and the "pumping kite" (or "yo-yo") [10]. Among these concepts, the pumping kite stands out as one of the simplest and cheapest

Marcelo De Lellis (✉) · Ramiro Saraiva · Alexandre Trofino
Federal University of Santa Catarina, Department of Automation and Systems. Roberto Sampaio Gonzaga, 88040-900, Florianopolis, Brazil
e-mail: marcelo.lellis@posgrad.ufsc.br

© Springer Nature Singapore Pte Ltd. 2018
R. Schmehl (ed.), *Airborne Wind Energy*, Green Energy
and Technology, https://doi.org/10.1007/978-981-10-1947-0_14

ones to experiment with. Employing a single wing, energy is harvested in the traction phase by letting the pulling force in one or more tethers, wound around drum(s), to drive a generator at the ground while the tether(s) are reeled out. The flight control actuators of a pumping kite can be either at the ground [12] or airborne, in a kite control unit (control pod) [3]. To avoid twisting the tether(s) the kite usually flies in a "lying-eight" trajectory. After a predefined tether length has been reached, the retraction phase is executed, when the tether is reeled back in, ideally at only a small expense of time and energy.

Pumping kite implementations can employ rigid, semi-flexible or flexible wings. The latter wing type, such as Leading-Edge-Inflated (LEI) tube or ram-air (foil) kites, are lighter and usually cheaper. Also, because of the weight advantage, flexible kites offer less damaging potential and are more robust in the case of land crashes, which is something to keep in mind, especially during the current stage of research and development. On the other hand, flexible kites usually have a lower aerodynamic efficiency (glide ratio), as well as a lower de-powering capability when going into the retraction phase. As a consequence, for a same kite area and wind speed, flexible kites tend to harvest less wind energy than other airfoil types.

Probably due to its mentioned relative simplicity, the pumping kite has been the AWE configuration most investigated in the literature. Flight control has been tackled basically with two approaches. One of them is Nonlinear Model Predictive Control (NMPC) [10, 16, 17] in a centralized topology. It consists of computing the references for the flight control (steering) and the ground winch control (reel speed) by solving an optimization problem to maximize the average power in a pumping cycle (the cycle power). In this case the references are jointly determined resulting in a coupled solution, in the sense that the flight trajectory is not predefined, but results from the optimization problem that must be solved within a sampling time typically smaller than 100 ms. Despite its advantages, the NMPC may be quite a demanding real-time task to solve, especially for more complex, accurate models.

Alternatively, a decentralized control topology has received increasing attention in recent years [1, 7, 9, 11, 18]. Keeping in mind that the complete pumping kite system can be viewed as a connection of two distinct subsystems, the wing flight subsystem and the ground winch subsystem, the idea is to design a decentralized control for the pumping kite system, in the sense that the control loops use only local variables of the subsystems [2]. Decentralized control is normally used to improve robustness of the control strategy against failures in the communication between the subsystems, as e.g. failures in the communication link between the ground station and the airborne control pod. The flight subsystem can be controlled in a cascade scheme comprising two (or more) loops. An advantage of this approach is that it allows for a parametrization of the flight trajectory in the outermost loop, either from a continuous reference or from some points of reference. In the inner loop the steering input of the wing subsystem is manipulated, usually to control the course angle,[1] whose reference is generated in the outer loop based on the wing position relative to the reference trajectory.

---

[1] It represents the instantaneous direction of flight, also referred to as "turning" or "velocity" angle.

Motivated by the advantages of the decentralized control topology, and with the purpose of maximizing the cycle power, the main contribution of this chapter is the formulation of an optimization problem to choose suitable values of the parameters of the flight and ground winch controls, including both traction and retraction phases. The optimization takes into account constraints of reel speed saturation and minimum angle of attack, and applies to any wing with de-powering capability and given aerodynamic curves. The idea is that the optimizer, based on updates of the wind speed, could be running in a supervisory system in a much larger time interval than the sampling period of the control loops. After computation, the new setpoints would then be transmitted to the flight and ground winch controllers, thus contributing to a more efficient operation of the AWE system.

The optimization proposed in this chapter is carried out through an iterative algorithm that uses, for the traction phase, a massless kite model in dynamic equilibrium[2], and a dynamic point mass kite model for the retraction phase. The choice for the relatively simple models is based on a compromise between computational speed and modeling accuracy, especially in the traction phase, where the apparent forces and weight can be neglected due to the high aerodynamic forces. The control setpoints to be optimized are the base angle of attack and average values of polar angle, tether length and reel-out speed for the traction phase, whereas for the retraction phase the parameters are the base angle of attack and traction force. The optimization is validated by simulations with a 3D point mass kite model.

Another contribution of this chapter is to provide answers to some open questions regarding traction power optimization. For instance, is the angle of attack that maximizes the traction power solely dependent on the kite lift and drag curves, as already discussed in the literature [24, 25], or is it significantly affected by the tether drag? Moreover, what should be the base angle of attack in order to operate the kite at such optimal point?

The rest of the chapter is structured as follows: in Sect. 14.2 we recall some related works in the literature. Next, in Sect. 14.3 we present the models used for optimization and validation. In Sect. 14.4 we discuss the influence of the tether drag on the optimal angle of attack, and how the base angle of attack affects the (average) angle of attack. This analysis is followed by Sect. 14.5, where we parametrize the retraction phase in terms of traction force and base angle of attack, and propose a grid search to find the solution that maximizes the cycle power. In Sect. 14.6 we present an iterative algorithm that also adjusts the reel-out speed in the traction phase to further maximize the cycle power. Then we show how the optimal cycle power and corresponding solution changes when the kite aerodynamic efficiency and effective area are varied. Sect. 14.7 concludes the chapter with some final remarks. The preliminary content of the chapter has been presented at the Airborne Wind Energy Conference 2015 [6].

Notation: a is a scalar, $\mathbf{x}$ is a vector. $x = \|\mathbf{x}\|$ is the Euclidean norm of $\mathbf{x}$. $\mathbf{x} \cdot \mathbf{y}$ and $\mathbf{x} \times \mathbf{y}$ are the dot (scalar) and cross (vector) products of $\mathbf{x}$ and $\mathbf{y}$, respectively. $\mathbf{v}_{k,(n)}$ indicates vector $\mathbf{v}_k$ is represented in the coordinate system of subscript $_{(n)}$.

---

[2] "static model" but at constant, non-zero speed.

14.2 Related Works

One of the first attempts to optimize a complete pumping cycle was carried out by [10], who assumed constant values of the polar angle θ, reel-out speed $\dot{r}$, and aerodynamic lift coefficient C_L and drag coefficient C_D for each pumping phase. It was also considered that the tether length variation Δr is negligible in comparison to the average tether length $\bar{r}$, so that r would be approximately constant during the whole pumping cycle. Under these assumptions the tether traction force T could be considered constant for each phase and, consequently, a simplified expression for the cycle power P_{cyc} was found. Two subsets of optimal solutions were proposed, $\mu_o = (\theta_o, \dot{r}_o, r_o)$ for the traction and $\mu_i = (\theta_i, \dot{r}_i, r_i)$ for the retraction phase, and the optimal solution was computed as $\mu^* = \arg \max \{ P_{cyc}(\mu_o, \mu_i) \}$. The solution for the azimuth angle was trivial, $\phi_o^* = \phi_i^* = 0$, for the "wing-glide maneuver".[3]

Some issues with this approach can be mentioned. First, in order for Δr to be negligible in comparison to $\bar{r}$ the traction phase must be short. By choosing to reel out 50 m around the optimal tether length of 611 m, i.e. $\Delta r \approx 8\% r^*$, the traction phase in [10, p. 84] lasted about 23 s. This relatively short duration may give rise to a practical issue: if one takes into account the need for a transition maneuver between the pumping phases, and that such maneuver can take e.g. between 4 s and 10 s, this relatively long transition time may have a negative impact on the cycle power. This is because, during the transition, energy is not being optimally generated nor consumed. Secondly, keeping θ constant during the retraction phase may be a hard task to achieve in practice because pulling the wing "as a flag", as required for the wing-glide maneuver, leaves the elevation angle uncontrolled. It can also happen that the kite, especially if flexible, simply may not be de-powered as much as required to generate the needed low lift and stabilize the elevation. If this is the case, the decrease of θ during the retraction phase significantly affects the power consumption and thus should be taken into account by the optimizer. A third concern, and perhaps more importantly, is that it was not clear what angle of attack was used for the traction phase in order to run the optimization, nor was it justified the choice of the base angle of attack $\alpha_0 = 3.5°$. In other words, the angle of attack α was not (at least explicitly) considered as an argument of the optimization, although the aerodynamic curves $C_L(\alpha)$ and $C_D(\alpha)$ had been declared.

A similar approach and assumptions was used by [21], who approximated the cycle power as a function of the constant values of reel-out and reel-in speeds, aerodynamic coefficients (different values for each pumping phase), and the elevation angle, which was assumed to be the same for the whole pumping cycle. However, the dependency of the system drag on the tether length was left out, as well as the wind shear model. Therefore, given θ, C_L and C_D, the only arguments to be optimized were $\dot{r}_o$ and $\dot{r}_i$.

Besides the works mentioned so far and those dealing strictly with the traction phase, only a few papers have been published focused on the retraction phase. A

---

[3] In [10] it was also considered the "low-power maneuver" for the retraction phase, where the kite is brought to the border of the wind window at $0 < |\phi| < \pi/2$ before the tether is reeled-in.

well-designed reel-in maneuver is essential for a higher cycle power since, regardless of the power generated in the traction phase, if that same amount is spent in the retraction phase, no net power is obtained. We can mention the work of [13], who proposed a reel-in maneuver where the kite flies towards zenith while the azimuth angle is kept at zero. A numerical, iterative procedure based on a simplified model of the kite dynamics was used to compute the reel-in speed and elevation, at each time step, in order to maximize the cycle power. This strategy was chosen because the elevation angle typically increases during the maneuver execution, hence rendering the use of a quasi-steady model inappropriate. The optimization framework could be further explored though, as only the kite trajectory was shown as a result. Furthermore, the assumptions of constant aerodynamic coefficients and constant ratio between reel-in speed and traction force could be relaxed for a more thorough analysis. These are some of the points to be addressed in this chapter. On a different approach, in [28] a retraction maneuver was designed to maintain the wing flying parallel to the ground, i.e. with a constant elevation angle, while the tether was reeled in. However, no criterion for the choice of the reel-in speed nor its impact on the cycle power was presented.

14.3 Models

Many models for tethered wings can be found in the literature, with different levels of complexity and, therefore, for distinct purposes. In this section we will discuss the three models involved in this work: one used to validate the optimization output, and two others used by the optimization algorithm.

14.3.1 3D Point Mass Kite

When it comes to a high level of detail and accuracy for simulating tethered wings, the state-of-the-art can be found in models such as [3, 4]. However, the price for the high accuracy achieved is a high computational load, besides the non-analytical framework. To be able to run simulations of a full pumping cycle in the desired time scale of a minute or less, and thus validate the optimizations carried out in this work, we chose the point mass kite model approach, first presented by [8] and later extended by [10]. In the following we will summarize the model equations, for the sake of completeness, emphasizing the changes made in comparison to [10].

The kite, with projected area A—i.e the area considering a flat, straight wingspan w between the two curved wingtips—, has its mass m concentrated in one point, which is also the kite aerodynamic center. Let us consider two steering/generation scenarios. In the first one each of the two wingtips is actuated (pulled) by a tether reeled around a respective drum, of radius r_d, connected to an electric machine on the ground, with moment of inertia J_e, through a gearbox with transmission factor

κ. By applying a differential steering length $\Delta l = w \sin \psi$ the kite is rolled through an angle ψ, which is used for flight control. These two tethers are also responsible for power generation, which occurs when they are reeled-out while subject to a traction force. Hence, in this scenario we must consider the drag and weight of two steering/traction tethers of length r. In the other scenario the traction force is transmitted from the kite by two secondary tethers which merge into a main tether, close to the kite. This main traction tether is then reeled around a drum connected to an electric machine at the ground. The steering actuator is located inside an airborne control pod, kept in place right above the junction of the traction tethers. Therefore we can take into account the weight and drag of only the main traction tether by assuming that the length of the steering tethers is negligible. To make our model applicable for both cases, let us define the amount of tethers reaching the ground as n_t. The tether has a mass density ρ_t, diameter d_t, drag coefficient $C_{D,t}$, and is considered to be perfectly straight (no sagging) and inelastic. We can now define the tether mass m_t—assumed to be concentrated at half the tether length r—, the tangential mass m_{tan}, and the radial mass m_{rad} as

$$m_t = n_t r (\pi d_t^2 / 4) \rho_t, \tag{14.1a}$$

$$m_{tan} = m + (m_t / 4), \tag{14.1b}$$

$$m_{rad} = (J_d + \kappa^2 J_e)/r_d + m_{tan}. \tag{14.1c}$$

The kite position in the inertial coordinate system, whose origin is the (average) point where the tether leaves the drum at the ground, is represented by the spherical coordinates (θ, ϕ, r), where θ is the polar angle, and ϕ is the azimuth angle. The kite Cartesian coordinates are $\mathbf{r}_{(n)} = [x_k, y_k, z_k]' = r[\sin \theta \cos \phi, \sin \theta \sin \phi, \cos \theta]'$, where the x_k direction is determined by the nominal wind[4] $\mathbf{v}_{w,(n)} = [v_w, 0, 0]'$, the z_k direction is upwards, and the y_k is sideways, completing the orthonormal system, according to Fig. 14.1. Vectors $\mathbf{e}_\theta$, $\mathbf{e}_\phi$ and $\mathbf{e}_r$ form the basis of the local coordinate system and the rotation matrix $\mathbf{R}_l^n$ from the local (l) to the inertial (n) frame:

$$\mathbf{R}_l^n = \begin{bmatrix} \mathbf{e}_\theta & \mathbf{e}_\phi & \mathbf{e}_r \end{bmatrix}_{(n)} = \begin{bmatrix} \cos \theta \cos \phi & -\sin \phi & \sin \theta \cos \phi \\ \cos \theta \sin \phi & \cos \phi & \sin \theta \sin \phi \\ -\sin \theta & 0 & \cos \theta \end{bmatrix}. \tag{14.2}$$

Considering Earth's gravitational acceleration g, the vectors of weight $\mathbf{G}$ and apparent forces $\mathbf{P}$ are

$$\mathbf{G}_{(l)} = g \begin{bmatrix} (m + 0.5 m_t) \sin \theta \\ 0 \\ -(m + m_t) \cos \theta \end{bmatrix} \text{ and } \mathbf{P}_{(l)} = m_{tan} \begin{bmatrix} \dot{\phi}^2 r \sin \theta \cos \theta - 2 \dot{r} \dot{\theta} \\ -2 \dot{\phi} \left(\dot{r} \sin \theta + r \dot{\theta} \cos \theta \right) \\ r \left(\dot{\theta}^2 + \dot{\phi}^2 \sin^2 \theta \right) \end{bmatrix}. \tag{14.3}$$

---

[4] By "nominal wind" we refer to the wind vector averaged in some time interval, e.g. every 2 minutes.

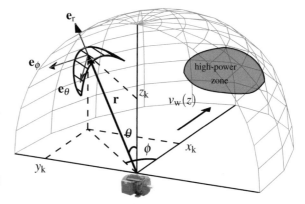

Fig. 14.1 Concept of a power kite including wind window, spherical coordinates (θ, ϕ, r), Cartesian coordinates (x_k, y_k, z_k), and unitary vectors $(\mathbf{e}_\theta, \mathbf{e}_\phi, \mathbf{e}_r)$ of the local coordinate system

Observe how the results obtained for $\mathbf{G}$ and $\mathbf{P}$ are slightly different from those in [10]. First of all, in that work it was considered that the magnitude of the tether weight in the θ and r directions is the same, namely $(m + m_t)g$. However, since the equivalent tether mass is concentrated at half the tether length, it is intuitive to think that only half the tether weight will affect the θ dynamics. Secondly, although considering the tether mass for the system's potential energy (hence weight), the tether mass was disregarded in [10] for the total kinetic energy. Thus, to be rigorous and coherent in physical terms, the factor $m_{\text{tan}} = m + (m_t/4)$ appears in the apparent force vector in Eq. (14.3), instead of only m.

The vectors of kite velocity $\mathbf{v}_k$, apparent wind speed $\mathbf{v}_a$, considering that $\mathbf{R}_n^l = (\mathbf{R}_l^n)'$, and apparent wind speed projected onto the tangent plane $\mathbf{v}_{a,\tau}$, are

$$\mathbf{v}_{k,(l)} = \begin{bmatrix} r\dot{\theta} \\ r\dot{\phi}\sin\theta \\ \dot{r} \end{bmatrix}, \quad \mathbf{v}_{a,(l)} = \mathbf{R}_n^l \mathbf{v}_{w,(n)} - \mathbf{v}_{k,(l)} \quad \text{and} \quad \mathbf{v}_{a,\tau} = \mathbf{v}_a - (\mathbf{v}_a \cdot \mathbf{e}_r)\mathbf{e}_r. \quad (14.4)$$

Given a steering input Δl we obtain the roll angle ψ and the angle η:

$$\psi = \arcsin\left(\frac{\Delta l}{w}\right) \quad \text{and} \quad \eta = \arcsin\left(\frac{\mathbf{v}_a \cdot \mathbf{e}_r}{v_{a,\tau}}\tan\psi\right). \quad (14.5)$$

Let us define the auxiliary vectors

$$\mathbf{e}_w = \mathbf{v}_{a,\tau}/v_{a,\tau} \quad \text{and} \quad \mathbf{e}_o = \mathbf{e}_r \times \mathbf{e}_w. \quad (14.6)$$

Now we can propose an orthonormal basis $(\mathbf{x}_a, \mathbf{y}_a, \mathbf{z}_a)$ for the so-called aerodynamic coordinate system, which is necessary to define the aerodynamic forces:

$$\mathbf{x}_a = -\mathbf{v}_a/v_a, \quad (14.7a)$$
$$\mathbf{y}_a = (-\cos\psi\sin\eta)\mathbf{e}_w + (\cos\psi\cos\eta)\mathbf{e}_o + (\sin\psi)\mathbf{e}_r, \quad (14.7b)$$
$$\mathbf{z}_a = \mathbf{x}_a \times \mathbf{y}_a. \quad (14.7c)$$

Considering the base angle of attack α_0 as a control input, the partial and total angle of attack are, respectively,

$$\Delta\alpha = \arcsin\left(\frac{\mathbf{v}_a \cdot \mathbf{e}_r}{v_a}\right) \text{ and } \alpha = \alpha_0 + \Delta\alpha. \tag{14.8}$$

Taking into account the kite and tether drag coefficients, $C_{D,k}(\alpha)$ and $C_{D,t}$, respectively, we can define a total drag coefficient

$$C_D(\alpha) = C_{D,k}(\alpha) + \frac{n_t d_t r \cos\Delta\alpha}{4A} C_{D,t}. \tag{14.9}$$

Knowing that ρ is the air density, the aerodynamic lift $\mathbf{L}$ and drag $\mathbf{D}$ forces are

$$\mathbf{L} = -(1/2)\rho A C_L(\alpha)v_a^2 \mathbf{z}_a, \tag{14.10a}$$
$$\mathbf{D} = -(1/2)\rho A C_D(\alpha)v_a^2 \mathbf{x}_a. \tag{14.10b}$$

Finally, we have all the definitions needed to state the equations of motion. In the tangent plane $(\mathbf{e}_\theta, \mathbf{e}_\phi)$ to the kite position, we have

$$\begin{bmatrix}\ddot{\theta}\\\ddot{\phi}\end{bmatrix} = \begin{bmatrix}(1/(m_{\tan}r))(\mathbf{G}+\mathbf{P}+\mathbf{L}+\mathbf{D})\cdot\mathbf{e}_\theta\\(1/(m_{\tan}r\sin\theta))(\mathbf{G}+\mathbf{P}+\mathbf{L}+\mathbf{D})\cdot\mathbf{e}_\phi\end{bmatrix}. \tag{14.11}$$

The radial dynamics we can treat in two ways. We can either assume we control the traction force T and have the tether acceleration $\ddot{r}$ as a result, or vice-versa:

$$\ddot{r} = (1/m_{\text{rad}})((\mathbf{G}+\mathbf{P}+\mathbf{L}+\mathbf{D})\cdot\mathbf{e}_r - T), \text{ or} \tag{14.12a}$$
$$T = (\mathbf{G}+\mathbf{P}+\mathbf{L}+\mathbf{D})\cdot\mathbf{e}_r - m_{\text{rad}}\ddot{r}. \tag{14.12b}$$

To represent how the nominal wind speed varies with altitude due to friction with the ground—a behavior assumed to hold up to about 600 m of altitude—we use the logarithmic wind shear model

$$v_w(z) = v_{w,\text{ref}}\frac{\ln(z/z_0)}{\ln(z_{\text{ref}}/z_0)}, \tag{14.13}$$

where z_{ref} is the reference height, $v_{w,\text{ref}}$ is the wind speed at z_{ref}, and z_0 is the surface roughness. All nominal parameter values are listed in Table 14.1. We also consider the curves of kite lift C_L, drag $C_{D,k}$, and aerodynamic efficiency $E_k = C_L/C_{D,k}$ shown in Fig. 14.2 (left), obtained from [14]. We highlight that these curves were not validated through experimental wind tunnel nor CFD data. Instead, they were obtained from empirical modifications on a wind turbine airfoil in order to reproduce the expected behavior of a LEI tube kite. We chose these curves because they are defined over a broader span of angle of attack, allowing simulations in more distinct scenarios.

Description	Symbol	Value	Unit
Air density	ρ	1.2	kg/m^3
Gravitational acceleration	g	9.82	m/s^2
Wing mass	m	7	kg
Projected wing area	A	12	m^2
Projected wingspan	w	7.75	m
Electric machine inertia	J_e	0.25	kg m^2
Drum inertia	J_d	0.1	kg m^2
Drum radius	r_d	0.2	m
Ground winch transmission ratio	κ	9	-
Number of main tether(s)	n_t	1	-
Tether density	ρ_t	970	kg/m^3
Tether drag coefficient	$C_{D,t}$	1.2	-
Main tether diameter	d_t	5	mm
Wind model reference height	z_{ref}	15	m
Wind model reference speed	$v_{w,ref}$	7	m/s
Wind model roughness coefficient	z_0	0.05	m

Table 14.1 Nominal model parameters (constants)

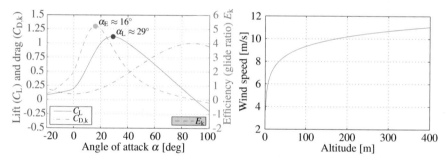

Fig. 14.2 Aerodynamic characteristics (not validated) of a LEI tube kite, reproduced from [14] (left) and logarithmic wind shear model (right)

Regarding flight control, we used the cascaded approach with two loops, similar to [7]. In short, we used Bernoulli's lemniscate in the outer loop as a reference for the "lying-eight" flight pattern of the traction phase. There are two control parameters in this loop: the lemniscate focus a_1, and the length δ of the normalized tangent vector $\mathbf{r}_{t,n}^\star$ to the lemniscate. This vector is computed at the lemniscate angular coordinate $\omega^\star$ of minimal distance (norm of) $\mathbf{r}_d^\star$ between the kite position $\mathbf{r}_k$ and the lemniscate position $\mathbf{r}_l$, as shown in Fig. 14.3. The course angle is γ, and its reference is given by the direction of $\mathbf{r}_r$. We used in our simulations $a_1 = 10°$ and $\delta = 1.25°$.

In the inner loop, the kite steering input Δl is manipulated as an attempt to make the course vector $\mathbf{v}_k$ converge to $\mathbf{r}_r$, i.e. to make the control error—defined as

$$e_\gamma = \arctan 2 \left(\frac{\|\mathbf{r}_r \times \mathbf{v}_{k,\tau}\|}{\mathbf{r}_r \cdot \mathbf{v}_{k,\tau}} \mathrm{sign}(b_3) \right), \text{ where } \mathbf{r}_r \times \mathbf{v}_{k,\tau} = (b_1, b_2, b_3) \quad (14.14)$$

and $\arctan 2()$ is the four-quadrant version of the $\arctan()$ function—, converge to zero. As a trade-off between simplicity and performance we chose a proportional

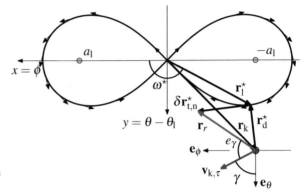

Fig. 14.3 Bernoulli's lemniscate as a reference for the lying-eight trajectory

controller for the course angle, $\Delta_l = \beta e_\gamma$. Based on the open-loop dynamics of the course angle proposed by [11], we set the feedback gain $\beta = 2$ to ensure closed-loop stability. We must here omit the stability demonstrations due to space restrictions.

During the traction phase we control $v_t = \dot{r}$, and the direction of flight is going down on the lemniscate edges (x-axis extremities). This offers two advantages. First, when flying down along the edges, the acceleration due to gravity reduces the valley caused in the traction force (and power) due to deviating more from the high-power zone. Second, it makes it easier for the kite to enter the retraction phase already closer to the reference trajectory at $\phi = 0$. The retraction phase starts as soon as the following three conditions are met: (a) the maximum tether length r_{max} is achieved; (b) the azimuth falls below a threshold, $|\phi| < \phi_{thr}$; and (c) the polar angle falls below the lemniscate center coordinate, $\theta < \theta_1$. Then, while in the retraction phase, the zenith is the reference for the kite course angle, and the traction force T is used as the tether control input instead of v_t. The reason for this will be explained in Sect. 14.3.3. As soon as the minimal tether length $r_{min} = r_{max} - \Delta r$ is achieved a new traction phase begins by switching back to the lemniscate trajectory and ramping up the reel-speed during 5 s.

14.3.2 Massless Kite in Dynamic Equilibrium

During the traction phase the kite flies in the high-power zone, in approximate crosswind conditions. Hence it is reasonable to assume that the high magnitude of the aerodynamic forces causes the weight and apparent forces to be negligible, and the kite is in dynamic equilibrium. We will further assume that the elevation angle remains approximately constant, and $\Delta l \approx 0$, so that we have no "loss" of lift force due to steering in order to make turns. This scenario is depicted in Fig. 14.4

Considering this model, we will borrow some results from [25]. Knowing that $f = v_t/v_w$ is the reel-out factor,

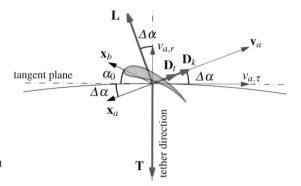

Fig. 14.4 Massless kite in dynamic equilibrium, with null steering input and subject to no apparent forces

$$CR = \sqrt{C_L{}^2 + C_D{}^2} \text{ , and } C = C_R \left(\frac{C_R}{C_D}\right)^2, \qquad (14.15)$$

the tether traction force can be approximated as

$$T = \frac{1}{2}\rho A C \left(\sin\theta\cos\phi - f\right)^2 v_w^2, \qquad (14.16)$$

while the traction power is

$$P = T v_t = T f v_w. \qquad (14.17)$$

We can calculate the kite tangential speed $v_{k,\tau}$ with the factor $\lambda = v_{k,\tau}/v_w$. Knowing that γ is the course angle (see Fig. 14.3) the tangential factor is

$$\lambda = a + \sqrt{a^2 + b^2 + 1 + E^2(b-f)^2}, \text{ where}$$
$$a = \cos\theta\cos\phi\cos\gamma - \sin\phi\sin\gamma \quad \text{and} \quad b = \sin\theta\cos\phi. \qquad (14.18)$$

To Eqs. (14.15) to (14.18) we add our models of $C_L(\alpha), C_D(\alpha,r),$ and $v_w(z(r,\theta))$.

14.3.3 2D Point Mass Kite at Zero Azimuth

To achieve a smooth transition into the retraction phase we must ramp down the base angle of attack from the value used in the traction phase, $\alpha_{0,o}$, to the value $\alpha_{0,i}$. Because α_0 usually cannot be low (negative) enough, the polar angle typically decreases during the reel-in maneuver and, as discussed by [13], we must work with a dynamic model instead of a "static" (in equilibrium) one. To obtain this dynamic model let us establish some assumptions. Firstly, we replace the lemniscate in the outer control loop by a trajectory with $\phi = 0$. By doing this the airborne weight contributes to decreasing the traction force, thus allowing us to save some reel-in power. Secondly, we would like to gradually decrease $v_t = \dot{r}$ (increase the reel-in speed) during the retraction phase. By doing so we avoid spending too much power while the kite is still in the high-power zone. One way to pursue this, as in [27], is to work with a traction force setpoint: as the kite leaves the high-power zone, the reel-in speed must increase in order for the desired traction force to be maintained. To achieve a smooth transition we also ramp down the traction force setpoint from the traction phase value T_o to the retraction phase value T_i. Therefore our control inputs will vary in the retraction phase according to

$$\alpha_0(t) = \max\{\alpha_{0,o} + c_\alpha t, \alpha_{0,i}\} \quad \text{and} \quad T(t) = \max\{T_o + c_T t, T_i\}, \quad (14.19)$$

where the ramp inclinations are constant values $c_\alpha < 0$ and $c_T < 0$, and the time t is counted from the moment the retraction phase starts. Applying the assumptions made here to Eqs. (14.11) and (14.12) we arrive at

$$\ddot\theta = (1/m_{\text{tan}} r)\left[(\mathbf{G}+\mathbf{P})\cdot\mathbf{e}_\theta + (1/2)\rho A\,(C_D v_{a,\theta} - C_L v_{a,r})v_a\right], \quad (14.20a)$$

$$\ddot r = (1/m_{\text{rad}})\left[(\mathbf{G}+\mathbf{P})\cdot\mathbf{e}_r + (1/2)\rho A\,(C_D v_{a,r} + C_L v_{a,\theta})v_a - T\right], \quad (14.20b)$$

where the apparent wind vector is simplified to

$$\mathbf{v}_a = \begin{bmatrix} v_{a,\theta} \\ v_{a,r} \end{bmatrix} = \begin{bmatrix} v_w \cos\theta - r\dot\theta \\ v_w \sin\theta - \dot r \end{bmatrix}. \quad (14.21)$$

14.4 Traction Power Optimization

We start our effort to maximize the cycle power by maximizing the traction power[5], approximated in Eq. (14.17) by the function $P(f,\theta,\phi,r,\alpha)$. To this end, the optimal (instantaneous) reel-out factor is well known to be $f^* = (1/3)\sin\theta\cos\phi$ (see e.g. [25]). Keep in mind that the arguments θ, ϕ and α typically suffer cyclic variations inside a lying-eight orbit. Therefore we will attempt to optimize their average values within a single orbit, as well as the average tether length, since r constantly increases during the traction phase. By inspection in Eq. (14.17) we conclude that the optimal azimuth is $\phi^* = 0$.

Regarding the remaining three arguments observe that, by decreasing θ the kite is able to reach a higher altitude $z = r\cos\theta$, where the wind is stronger. However, this positive effect on P is counterbalanced by displacing the system from the ideal crosswind condition, represented by the $(\sin\theta)$ factor in Eq. (14.17). There is also a trade-off with the tether length optimization: by increasing r the kite gains altitude where, again, the wind is stronger, nevertheless the total airborne drag (Eq. (14.9)) increases due to the longer tether(s). In fact, we will here predetermine the tether length variation $\Delta r = r_{\text{max}} - r_{\text{min}}$ and have it centered around the optimized value of r, i.e. the initial tether length of the traction phase will be $r_{\text{min}} = r^* - (\Delta r)/2$. Regarding the angle of attack note that, if we neglect the tether drag, the optimal

---

[5] As we will discuss in Sect. 14.6, maximizing the power in the traction phase does not necessarily mean maximizing the cycle power.

value would be the one that maximizes the C term in Eq. (14.15), as discussed by [24]. However, because we are taking the tether drag into account, there may be another value of α that, combined with a certain value of r, further maximizes P, even though C is not maximum. With this in mind we come to realize that the traction power depends on nonlinear combinations (multiplications) between θ, r and α and, therefore, these three arguments must be numerically and simultaneously optimized.

Before we formulate the optimization problem, observe that we intend to use Bernoulli's lemniscate as the lying-eight trajectory reference. Hence we could also treat the lemniscate focus a_l as an optimization argument. Basically, if a_l is too small the harvested power would be small due to the large steering needed to execute the tight curves and, consequently, due to the loss of the lift force decomposition onto the tether direction. On the other hand, if a_l is too big the kite would deviate too much from the high-power zone (at $\phi^* = 0$), hence P is also small. This leads us to conclude there should be an intermediate value of a_l which is optimal. However, this is still work in progress and therefore it will not be addressed here.

Aiming at a safe operation, we should also limit the traction force to a maximum value T_{max}, defined by the tether minimum breaking load. The polar angle should also be constrained to a maximum in order to ensure the kite does not fly too close to the ground. We establish a minimum altitude $z_{min} = 3.5w$, i.e. the kite wingtip should keep a least distance of 3 wingspans from the ground.[6]. Based on what has been presented so far, the optimization solution is

$$v_o^* = (\theta^*, r^*, \alpha^*) = \arg\max \left\{ C(\alpha, r) \left[v_w(\theta, r) \sin\theta \right]^3 \right\}$$

subject to

$$T \leq T_{max} \tag{14.22}$$

$$\theta \leq \arccos \left(\frac{z_{min}}{r - (\Delta r)/2} - \frac{a_l}{2} \right).$$

Once we obtain the optimal solution, we need to find the corresponding base angle of attack α_0^* that yields the desired angle of attack α^*. To this end let us consider once more the flight in dynamic equilibrium of Fig. 14.4. The partial angle of attack $\Delta\alpha$ can be computed as the solution of the force equilibrium equation in the tangent plane, while α_0 follows by using the definition of α:

$$\Delta\alpha = \arg\left\{ C_L(\alpha)\sin\Delta\alpha - C_D(\alpha, r)\cos\Delta\alpha = 0 \right\},$$
$$\alpha_0 = \alpha - \Delta\alpha. \tag{14.23}$$

Considering the system parameters of Table 14.1 and Fig. 14.2, the constraints of minimum altitude $z_{min} = 27\,\text{m}$, and maximum tether traction force $T_{max} = 15\,\text{kN}$, and setting the tether length variation as $\Delta r = 200\,\text{m}$, the optimal solution obtained is $v_o^* = (70.6°, 456\,\text{m}, 20.9°)$, yielding a mechanical power $P = 14.5\,\text{kW}$ at a traction

---

[6] considering no tracking error on the lemniscate trajectory.

force $T = 7.2\,\text{kN}$, achieved with a base angle of attack $\alpha_0 = 6.4°$ and at a reel-out speed $v_t = 3.1\,\text{m/s}$.

Note how $\alpha^* = 20.9°$ differs from the angle of attack of maximum kite aerodynamic efficiency $\alpha_E = 16°$, and of maximum lift $\alpha_L = 29°$ (see Fig. 14.2), as expected. The optimal angle of attack is, in fact, very close to the value that maximizes the C term without considering the tether drag, $\alpha = 19.4°$, the theoretical optimum discussed by [24]. To investigate how α^* varies with the tether drag, we ran the optimization of Eq. (14.22) for three different values of the tether drag coefficient $C_{D,t}$. For each value we changed α around the optimal value. We also applied the setpoints to the validation model of Sect. 14.3.1. The results are presented in Fig. 14.5.

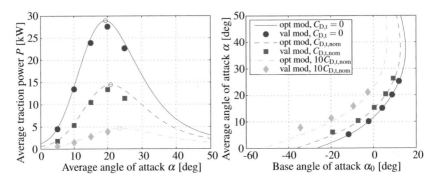

Fig. 14.5 Traction phase optimization results with the optimization (opt mod) and validation (val mod) models for different values of tether drag coefficient $C_{D,t}$

Observe in Fig. 14.5 (left) that, by varying $C_{D,t}$ from 0 to 10 times the nominal value, the optimal angle of attack (indicated by the small circle on the peak of each power curve) only slightly varies inside the interval $\alpha^* \in [19.4°, 23.4°]$. This leads us to conclude that, although the tether drag has a strong impact on the harvested power (peak value of the P curves), it does not significantly influence the optimal angle of attack. Therefore it is indeed a good approximation to compute $\alpha^* = \arg\max\{C(\alpha)\}$, as proposed by [24], instead of having to use Eq. (14.22).

We can also see that the results obtained with the validation model (filled circles) have a high correlation with the optimization results (continuous curves). This is also the case in Fig. 14.5 (right), where we show the angle of attack obtained as a function of the base angle of attack. For the optimization model we used Eq. (14.23), whereas for the validation model we took the average value of α. The validation experiments were carried out starting at $\alpha = 5°$, in increments of $5°$. Note that at $\alpha = 30°$ onwards no more validation results are shown. The reason is that, after the lift peak at $\alpha_L = 29°$ the kite eventually comes to a stall and a complete traction phase is no more possible. We can here draw two conclusions. First, that the dynamic equilibrium model of Sect. 14.3.2 produces results which are very close to those obtained with a more complex model (of Sect. 14.3.1), hence the simpler model is a good choice between computational speed and modeling accuracy (the optimization takes about 3 s). The second conclusion is that we should avoid operating close to the angle of attack of maximum lift, to keep away from a stall condition.

Back to Fig. 14.5 (right), observe that the angle of attack becomes more sensitive on the base angle of attack as the latter is increased. Moreover, for high enough values of α_0 there are two equilibria of α, one of them beyond $35°$. This is something to keep in mind, especially under turbulent wind conditions: a wind gust may perturb the angle of attack in such a way that it is attracted to the high-value equilibrium, causing the kite to stall. A more detailed study on the α-equilibria is yet to be carried out before we can draw further conclusions in this regard.

We conclude this section by emphasizing the importance of choosing the proper base angle of attack for maximizing the traction power, in face of the results in Fig 14.5. For instance, if we apply this optimization approach to the $500\,\mathrm{m}^2$ kite considered in [7] we obtain the solution $v_0^* = (74°, 408\,\mathrm{m}, 4.2°)$, with corresponding $\alpha_0 = 1.1°$. Operation at this point results in 4.3 MW of electric power, i.e. a roughly 20% increase with respect to the solution $v_0 = (79.8°, 652\,\mathrm{m}, 7.8°)$ presented in that work, where it was used $\alpha_0 = 3.5°$, the same as in [10].

14.5 Retraction Phase Optimization

Our goal here is to take a more realistic look into the retraction phase, obtaining results which could be actually tested with current prototype technology. Therefore we assumed a smooth transition from traction to retraction phase by ramping down—instead of abruptly changing—the base angle of attack and traction force to constant setpoints, which will be optimally determined in this section. Intuition on how to design a more efficient reel-in maneuver led us to establish the flight trajectory at $\phi = 0$, and use the traction force as control input for the electric machine at the ground, as explained in Sect. 14.3.3. For a realistic maneuver we should also take into account system constraints. Firstly, the kite, especially if made of flexible material, as LEI tube kites—or even more critically for ram-air kites—, should not fly at too low angles of attack. Otherwise the wing loading can be very low, the kite may lose its proper inflated shape, and thus lose steering capability. Hence let us consider a lower limit α_{min} on the angle of attack. Secondly, there is in practice a speed limitation $v_{t,sat}$ of the electric machine, imposed by factors such as the maximum centrifugal force withstood by the windings, the number of poles, and the transmission ratio κ between drum and electric machine.

Differently from Sect. 14.4, our goal now is to maximize the cycle power. Considering that $P_o(t)$ and $P_i(t)$ are the instantaneous mechanical power in the traction and retraction phases, respectively, the cycle power is

$$P_{cyc} = \frac{\int P_o(t)dt + \int P_i(t)dt}{\Delta t_o + \Delta t_i}. \tag{14.24}$$

Having optimized the traction phase a priori, we know $\int P_o(t)dt$, Δt_o, and can promptly set the following initial conditions of the retraction phase: $\dot{r}_i(0) = v_{t,o}^* = f^* v_w$, $r_i(0) = r_o^* + (\Delta r)/2$ and $\theta_i(0) = \theta_o^*$. If we consider the kite flying towards zenith, i.e. $\gamma = \pm \pi$, the tangential velocity is solely in the θ direction, $v_{k,\tau} = r\dot{\theta}$. By replacing this in Eq. (14.18) the initial angular condition $\dot{\theta}_i(0)$ can be computed. The initial value of the traction force can be obtained by solving Eq. (14.20b) for T with $\ddot{r} = 0$ and all other variables as already discussed. Then, with the model of Sect. 14.3.3, including given ramp inclinations c_α and c_T of the control input curves of Eq. (14.19), we can simulate the retraction phase and compute the instantaneous power $P_i(t) = T_i(t)v_{t,i}(t)$ and cycle power in Eq. (14.24). We are now in conditions of optimizing the constant setpoints of base angle of attack and traction force to be used during the retraction phase. To this end we execute a grid search and obtain

$$v_i^* = (\alpha_{0,i}^*, T_i^*) = \arg\max \left\{ P_{cyc}\left(\alpha_{0,i}, T_i, \int P_o(t)dt, \Delta t_o\right) \right\}$$

$$\text{subject to} \hspace{4cm} (14.25)$$

$$\alpha \geq \alpha_{min}$$

$$v_t \geq -v_{t,sat}.$$

It is intuitive to think we would like to de-power the kite as fast as possible, but there is also a speed limitation to execute that. Let us assume we are using the (maximum) de-powering speed $c_\alpha = -10°/s$. For ramping down the traction force, we will consider a ramp inclination $c_T = 982 \, N/s$. Moreover, let us assume the optimization constraints are $\alpha_{min} = -5°$ and $v_{t,sat} = 10 \, m/s$. To have a more comprehensive look into the optimization, we define a broad grid interval, composed by $\alpha_{0,i} \in [-100, 0]°$ and $T_i \in [0, 1000] \, N$, with resolution $\delta\alpha_{0,i} = 5°$ and $\delta T_i = 50 \, N$. Using a time-integration step $\delta t = 50 \, ms$ to simulate the retraction phase, the resulting cycle power surface is shown in Fig. 14.6.

Observe that, for intermediate values of T_i and $\alpha_{0,i}$, the retraction phase is completed without violating the constraints on the angle of attack and reel-in speed. In this case we say the solution $v_i = (\alpha_{0,i}, T_i)$ is inside the feasible region. Let us take a grid point for our analysis somewhere in the middle of this region. Starting there, if T_i is increased there may be initially an increase in P_{cyc} because the tether must be reeled-in faster. However, at a certain point reel-in saturation may happen. If not, we may reach a certain value of T_i beyond which the benefit of operating at a faster reel-in speed—thus with a shorter retraction phase—is overshadowed by the increase in the power expense to reel-in the tether, hence P_{cyc} starts to decrease. Back to the starting point of our analysis and moving towards lower values of T_i the tendency is for the cycle power to decrease because the retraction phase must be carried out at a reel-in speed closer to zero, hence Δt_i increases in Eq. (14.24). If T_i becomes too low, the retraction maneuver is not possible anymore because the tether must be actually reeled-out to allow for the low traction force.

Once again in the middle of the feasible region and then decreasing $\alpha_{0,i}$, the cycle power increases monotonically, regardless of T_i. The problem is that, below a certain value of base angle of attack, the constraint on the minimum angle of attack

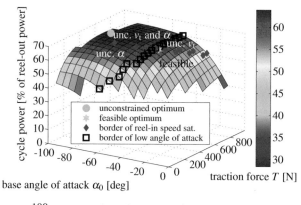

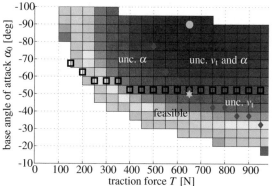

Fig. 14.6 Cycle power as a function of constant references of traction force T_i and base angle of attack $\alpha_{0,i}$ used during the retraction phase

is violated at some point during the maneuver. Looking into the other direction, when increasing $\alpha_{0,i}$ it comes a point when the retraction phase cannot be completed anymore for the same reason as for a traction force too low: the tether must be reeled out. Grid points for which the retraction phase cannot be completed were omitted (blank/white squares) in Fig. 14.6.

Observe that the maximum cycle power $P_{\text{cyc,unc}}^* = 0.639 P_0$ is obtained with $v_{i,\text{unc}}^* = (-90°, 650\,\text{N})$, in a region where both constraints $v_{t,\text{sat}}$ and $\alpha_{\min}$ are violated. We refer to $v_{i,\text{unc}}^*$ as the unconstrained optimum. To comply with the constraints we must look for the optimal solution inside the feasible region. Thus our choice becomes $v_i^* = (-50°, 650\,\text{N})$, the feasible optimum, which yields $P_{\text{cyc}}^* = 0.556 P_0$. As can be seen in Fig. 14.6, one could think of the optimal solution more as a "plateau region", since the cycle power remains approximately constant in the vicinity of $v_{i,\text{unc}}^*$. In fact note that, relative to the unconstrained optimum, the decrease in cycle power to reach the feasible optimum is below 10%.

If this optimization were applied to a rigid wing, the constraint on the minimum angle of attack could probably be disregarded. However, when considering a foil (ram-air) kite, the $\alpha_{\min}$ constraint could be more restrictive. In this case the border of $\alpha_{\min}$ violation—indicated by the black squares in Fig. 14.6—would move

towards higher values of α_0, "squeezing" the feasible region. This would also be
the case if we had a ground station with less reel speed capability: the feasible region would be squeezed to the left by the $v_{t,sat}$ line (blue diamonds in Fig. 14.6).
Also note that, not only power kites have a limited de-powering speed c_α, but also
the maximum de-powering itself may be limited, i.e. α_0 may not be arbitrarily low,
especially for flexible kites. This constraint can be easily taken into account by reducing the $\alpha_{0,i}$ search domain. Similarly, a minimum value of traction force T_{min},
which is important to ensure the kite maintains steering capability during the reel-in
maneuver, can be taken into account by reducing the search domain to $T_i \geq T_{min}$.

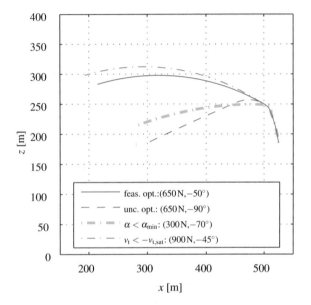

Fig. 14.7 Examples of retraction phase trajectories in the $\phi = 0$ plane: (i) feasible optimum, (ii) unconstrained optimum, (iii) violation of α_{min}, and (iv) violation of $v_{t,sat}$

Legend in figure:
- feas. opt.:$(650\,N, -50°)$
- unc. opt.: $(650\,N, -90°)$
- $\alpha < \alpha_{min}$: $(300\,N, -70°)$
- $v_t < -v_{t,sat}$: $(900\,N, -45°)$

In Fig. 14.7 we can compare flight trajectories obtained with different values of
v_i. Note that, in the first 4 s, when α_0 and T are being ramped down in all cases,
the trajectories are the same because of the same c_α and c_T. Furthermore, we can
see that the unconstrained optimum, which requires $\alpha_{0,i} = -90°$, causes the kite
to follow more of a straight trajectory towards the ground station—a behavior we
already expected since it makes the kite operate at a very low angle of attack. To
verify this, and understand how these trajectories are generated, let us take a look at
some system variables, presented in Fig. 14.8. Note how the unconstrained optimum
causes both the $v_{t,sat}$ and α_{min} constraints to be violated. In fact, the angle of attack
becomes extremely low, closing in on $\alpha = -20°$, below which the aerodynamic
curves of our validation model are not defined (see Fig. 14.2). We can also see
that the maneuver with $(-70°, 300\,N)$ is the longer one, which is mostly due to
the relatively low traction force, requiring very little power to be spent, but also
violating α_{min} because of the high de-powering.

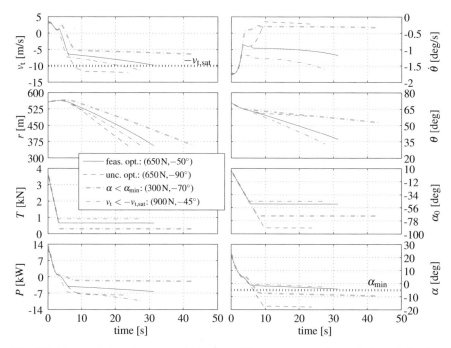

Fig. 14.8 Time evolution of system variables for 4 different retraction phase solutions: (i) feasible optimum, (ii) unconstrained optimum, (iii) violation of α_{min}, and (iv) violation of $v_{t,sat}$

14.6 Iterative Pumping Cycle Optimization

As stated in Sect. 14.1, the end goal of this chapter is to propose a pumping cycle parametrization that maximizes the cycle power. This was already considered in the retraction phase optimization of Sect. 14.5. However, in Sect. 14.4 the maximized function was the traction power P, whose maximum does not necessarily coincide with that of the cycle power P_{cyc}. The reason is simple: imagine we decrease the traction phase reel-out speed $v_{t,o}$. It can be shown that the traction power, approximated by Eq. (14.17), will decrease, yet the duty cycle $\Delta t_o / (\Delta t_o + \Delta t_i)$ will increase—i.e. the kite will spend a greater ratio of the pumping cycle duration harvesting energy. While the decrease of P has a negative effect on P_{cyc} (see Eq. (14.24)), the increase in the duty cycle has a positive effect on P_{cyc}. We will see in this section that if $v_{t,o}$ is decreased to a certain amount, we can further maximize the cycle power.

The idea is, once the traction power and retraction phase are optimized in a first iteration, to run a second iteration where the traction phase is optimized to maximize the cycle power, this time considering the reel-out speed as an argument. Because we have discussed it is a good approximation to optimize the angle of attack by maximizing $C(\alpha)$ (Eq. (14.15)) regardless the tether drag, we will keep the value of α^* found in the first iteration. Also, since the retraction phase initial conditions depend on the optimal solution of the traction phase, we must complete the iteration with a new optimization of the retraction phase as well. We will execute these iterations until the increase in P_{cyc} falls below a tolerance, meaning we reached a new optimum. This iterative algorithm is represented in pseudo code as:

---

while *increase in $P_{cyc} \geq$ tolerance* **do**
   if *1st iteration* **then**
$$(\theta_o^*, r_o^*, \alpha_o^*) = \arg\max\left\{ C(\alpha, r)\left[v_w(\theta, r)\sin\theta \right]^3 \right\};$$
$$v_{t,o}(t) = (1/3)\sin\theta(t)\cos\phi(t);$$
   else
$$(\theta_o^*, r_o^*, v_{t,o}^*) = \arg\max\left\{ P_{cyc}\left(\theta, r, v_t, \int P_i(t)dt, \Delta t_i\right) \right\};$$
$$v_{t,o}(t) = v_{t,o}^*\cos\phi(t);$$
   end
$$\alpha_{0,o}^* = \alpha_o^* - \Delta\alpha(\alpha_o^*, r_o^*);$$
$$(\alpha_{0,i}^*, T_i^*) = \arg\max\left\{ P_{cyc}(\alpha_{0,i}, T_i, \int P_o(t)dt, \Delta t_o) \right\};$$
end

Algorithm 1: Iterative algorithm for pumping cycle optimization (constraints have been omitted for a compact representation).

Let us consider the de-powering limit of the kite is $\alpha_0 = -60°$, set the convergence tolerance to 100 W, and shorten the retraction phase optimization grid to intervals in which the optimal values are more likely to be found. For the 1st iteration, we search inside $\alpha_{0,i} \in [-60, -20]°$ ($\delta\alpha_{0,i} = 2.5°$) and $T_i \in [200, 900]$ N ($\delta T_i = 100$ N). For the following iterations we still consider these intervals, but center the grid around the solution previously found 5 times the resolution in each direction. The resolution is then set to $\delta\alpha_{0,i} = 2°$ and $\delta T_i = 25$ N. The values of all other involved parameters remain as presented so far. With these settings, the solution found is $v^* = (\theta_o^*, r_o^*, v_{t,o}^*, \alpha_{0,o}^*, \alpha_{0,i}^*, T_i^*) = (70.4°, 400\,\text{m}, 2.3\,\text{m/s}, 6.8°, -50°, 600\,\text{N})$, yielding an optimal cycle power $P_{cyc}^* = 9.2\,\text{kW} = 0.672 P_0$. The algorithm took 1 minute and 55 seconds to converge in 3 iterations, with more than 90 % of the duration required for simulating the retraction phases. Note that, because the optimal reel-out speed found was 25.8% lower than the value that maximizes the traction power (1st iteration), we obtained a 9.3% increase in the optimal value of cycle power, accompanied by a 27.8% increase in the traction force. The optimized kite trajectory is shown in Fig. 14.9.

The time course of some system variables are shown in Fig. 14.10. Note how the design and validation results of the retraction phase, with the 2D and 3D models of Sects. 14.3.3 and 14.3.1, respectively, closely match. In the traction phase we can see the cyclic variations in the traction force, power, polar angle, angle of attack, and in the steering input Δl. The discontinuity in Δl is due to the transition from the reel-in trajectory with $\phi = 0$ back to the lemniscate, when a new pumping cycle begins.

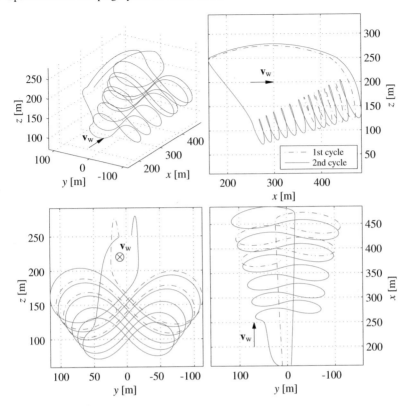

Fig. 14.9 Optimized flight trajectory during two pumping cycles. The 1st cycle begins in the middle of the traction phase

To show how the optimal pumping cycle solution behaves when system parameters are varied we ran a sensitivity analysis with respect to the nominal solution obtained in this section. Because the traction force is influenced by the kite glide ratio E_k through the C coefficient, whereas T depends linearly on the kite area A (see Eq. (14.16)), we adjusted the ramp inclination from the nominal value as $c_T = c_{T,nom}(C/C_{nom})$ when varying E_k, and $c_T = c_{T,nom}(A/A_{nom})$ when altering A. The results are shown in Fig. 14.11. Variations of the kite mass were not considered because the model used for the traction power optimization is massless. Also keep in mind that the tether diameter remained unaltered, although in reality it should vary according to the traction force variations.

Observe the strong influence of the kite glide ratio on the cycle power: a variation of 30% on E_k almost doubled P_{cyc}, only possible because the base angle of attack increased by 50%, to $\alpha_{0,i} = -27.5°$. Less de-powering results in a higher traction force during the reeling-in, which is indeed observed. The higher glide ratio causes the kite tangential speed to increase, and therefore the reel-in maneuver must end around zenith. This situation may require a more elaborate strategy to maintain the traction force above a minimum safety level when making the transition back to the

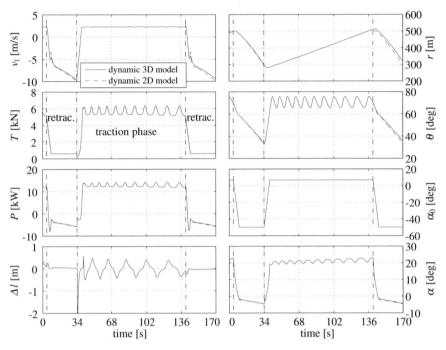

Fig. 14.10 Time evolution of system variables in a pumping cycle

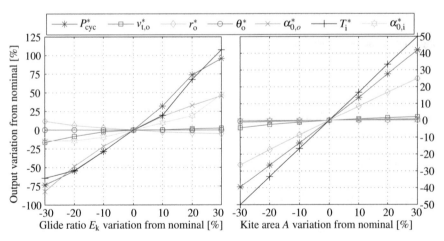

Fig. 14.11 Sensitivity analysis by varying the kite glide ratio $E_k = C_L/C_{D,k}$, and mass m, around the nominal values

traction phase. In any case, the high sensitivity of P_{cyc} on E_k justifies the efforts in developing more aerodynamically efficient materials for power kites, especially for flexible (ram-air and LEI tube) ones. In the right-hand side plot we can see that the base angle of attack is not affected by variations of the kite area, as one could have expected. On the other hand, the inclination of the P_{cyc} curve is higher than one, which does not correspond to the expected linear behavior modeled by Eqs. (14.16) and (14.17). This is partially because the tether diameter was kept constant, but especially due to variations of the average tether length r_o^*, and the corresponding changes on the operating altitude and nominal wind, according to Eq. (14.13).

14.7 Final Remarks

In this chapter we formulated an optimization problem to choose suitable setpoints of the flight and ground winch controls of a pumping kite system. The optimization takes into account constraints of reel speed saturation and minimum angle of attack, and applies to any wing with de-powering capability and given aerodynamic curves. The optimization goal is to maximize the cycle power by adjusting four setpoints of the traction phase—the base angle of attack and average values of polar angle, tether length and reel-out speed—, and two setpoints of the retraction phase—the base angle of attack and traction force. This optimizer could be constantly running in a supervisory system, e.g. based on updates of the nominal (average) wind, whose update frequency is much smaller than the control sampling frequency.

Among our main findings is that, when optimizing the traction phase by attempting to maximize the cycle power P_{cyc} instead of the traction power P_o, the increase in P_{cyc} was about 12% of P_o, jumping from $0.556P_o$ to $0.672P_o$. This gain in cycle power was almost solely due to a the decrease in the reel-out speed of about 26% with respect to the value that maximizes the traction power. This result corroborates previous findings from [21] with a simpler model. Furthermore we found that, although the traction power P_o is very sensitive on the tether drag $C_{D,t}$, the latter does not significantly affect the angle of attack α that maximizes P_o. We also showed the nonlinear dependency of α on the base angle of attack α_0, especially in the traction phase, where there may be more than one equilibrium of α.

Through a sensitivity analysis we observed that, when increasing by 30% the wing glide ratio, the cycle power almost doubled. This highlights how important it is to pursue more aerodynamically efficient materials for power kites. Also, when varying the projected wing area by 30%, the retraction force setpoint increased linearly, but the cycle power gain was more than one, about 40%. This behavior was mostly due to an increase in the optimal (average) tether length, resulting in a higher operating altitude and stronger wind.

We can mention a few points for investigation in future works. Firstly, the lemniscate focus a_1 is a parameter that is still to be optimized. Also, the ramp inclinations of the base angle of attack c_α and traction force c_T were set empirically. Therefore an interesting question is what would be the best values of c_α and c_T to be considered or, more generally, how to optimize these parameters. Alternatively, it would be interesting to relax the requirement of constant setpoints and linear ramping down of the base angle of attack and traction force, and use optimization tools to find trajectories of these parameters that result in a higher cycle power. In this chapter we have used a simple grid search method to optimize these setpoints.

Another improvement would be to model and include in the optimization the transition from the retraction back to the traction phase in order to guarantee a safe maneuver. Also, for the cases when e.g. the wing aerodynamic curves and the wind model are not well known, an alternative modeling and optimization procedure based on more easily estimated (available) parameters could be developed. Finally, regarding the aerodynamics and structural dynamics of flexible kites, it would be valuable to know where the constraint on the minimum angle of attack is, since we showed how a relaxation of this constraint yields a higher cycle power.

Acknowledgements The authors would like to thank the reviewers for the helpful comments. This work was supported by the Brazilian government through CNPq under grants 480931/2013-5 and 406996/2013-0, and CAPES.

References

1. Baayen, J. H., Ockels, W. J.: Tracking control with adaption of kites. IET Control Theory and Applications 6(2), 182–191 (2012). doi: 10.1049/iet-cta.2011.0037
2. Bakule, L.: Decentralized control: An overview. Annual Reviews in Control 32(1), 87–98 (2008). doi: doi:10.1016/j.arcontrol.2008.03.004
3. Bosch, A., Schmehl, R., Tiso, P., Rixen, D.: Dynamic nonlinear aeroelastic model of a kite for power generation. AIAA Journal of Guidance, Control and Dynamics 37(5), 1426–1436 (2014). doi: 10.2514/1.G000545
4. Breukels, J.: An Engineering Methodology for Kite Design. Ph.D. Thesis, Delft University of Technology, 2011. http://resolver.tudelft.nl/uuid:cdece38a-1f13-47cc-b277-ed64fdda7cdf
5. Canale, M., Fagiano, L., Milanese, M.: KiteGen: A revolution in wind energy generation. Energy 34(3), 355–361 (2009). doi: 10.1016/j.energy.2008.10.003
6. De Lellis, M., Saraiva, R., Trofino, A.: On the Optimisation of Pumping Kites for Wind Power. In: Schmehl, R. (ed.). Book of Abstracts of the International Airborne Wind Energy Conference 2015, p. 63, Delft, The Netherlands, 15–16 June 2015. doi: 10.4233/uuid: 7df59b79-2c6b-4e30-bd58-8454f493bb09. Presentation video recording available from: https://collegerama.tudelft.nl/Mediasite/Play/bde2b8dbc269407cbfa987b1c680eec71d
7. De Lellis, M., Saraiva, R., Trofino, A.: Turning angle control of power kites for wind energy. In: Proceedings of the 52nd Annual Conference on Decision and Control (CDC), pp. 3493–3498, Firenze, Italy, 10–13 Dec 2013. doi: 10.1109/CDC.2013.6760419
8. Diehl, M.: Real-time optimization for large scale nonlinear processes. Ph.D. Thesis, University of Heidelberg, 2001. http://archiv.ub.uni-heidelberg.de/volltextserver/1659/
9. Erhard, M., Strauch, H.: Flight control of tethered kites in autonomous pumping cycles for airborne wind energy. Control Engineering Practice 40, 13–26 (2015). doi: 10.1016/j.conengprac.2015.03.001
10. Fagiano, L.: Control of tethered airfoils for high-altitude wind energy generation. Ph.D. Thesis, Politecnico di Torino, 2009. http://hdl.handle.net/11311/1006424
11. Fagiano, L., Zgraggen, A. U., Morari, M., Khammash, M.: Automatic crosswind flight of tethered wings for airborne wind energy:modeling, control design and experimental results. IEEE Transactions on Control System Technology 22(4), 1433–1447 (2014). doi: 10.1109/TCST.2013.2279592

12. Fagiano, L., Marks, T.: Design of a Small-Scale Prototype for Research in Airborne Wind Energy. IEEE/ASME Transactions on Mechatronics **20**(1), 166–177 (2015). doi: 10.1109/ TMECH.2014.2322761
13. Fechner, U., Schmehl, R.: Model-Based Efficiency Analysis of Wind Power Conversion by a Pumping Kite Power System. In: Ahrens, U., Diehl, M., Schmehl, R. (eds.) Airborne Wind Energy, Green Energy and Technology, Chap. 14, pp. 249–269. Springer, Berlin Heidelberg (2013). doi: 10.1007/978-3-642-39965-7_14
14. Fechner, U., Vlugt, R. van der, Schreuder, E., Schmehl, R.: Dynamic Model of a Pumping Kite Power System. Renewable Energy (2015). doi: 10.1016/j.renene.2015.04.028. arXiv:1406. 6218 [cs.SY]
15. Goldstein, L.: Theoretical analysis of an airborne wind energy conversion system with a ground generator and fast motion transfer. Energy, 987–995 (2013). doi: 10.1016/j.energy. 2013.03.087
16. Houska, B., Diehl, M.: Optimal control for power generating kites. In: Proceedings of the 9th European Control Conference, pp. 3560–3567, Kos, Greece, 2–5 July 2007
17. Ilzhöfer, A., Houska, B., Diehl, M.: Nonlinear MPC of kites under varying wind conditions for a new class of large-scale wind power generators. International Journal of Robust and Nonlinear Control **17**(17), 1590–1599 (2007). doi: 10.1002/rnc.1210
18. Jehle, C., Schmehl, R.: Applied Tracking Control for Kite Power Systems. AIAA Journal of Guidance, Control, and Dynamics **37**(4), 1211–1222 (2014). doi: 10.2514/1.62380
19. Lansdorp, B., Williams, P.: The Laddermill - Innovative Wind Energy from High Altitudes in Holland and Australia. In: Proceedings of Global Windpower 06, Adelaide, Australia, 18–21 Sept 2006. http://resolver.tudelft.nl/uuid:9ebe67f0-5b2a-4b99-8a3d-dbe758e53022
20. Loyd, M. L.: Crosswind kite power. Journal of Energy **4**(3), 106–111 (1980). doi: 10.2514/3. 48021
21. Luchsinger, R. H.: Pumping Cycle Kite Power. In: Ahrens, U., Diehl, M., Schmehl, R. (eds.) Airborne Wind Energy, Green Energy and Technology, Chap. 3, pp. 47–64. Springer, Berlin Heidelberg (2013). doi: 10.1007/978-3-642-39965-7_3
22. Makani Power Inc. http://www.makanipower.com. Accessed 4 July 2013
23. Ockels, W. J.: Laddermill, a novel concept to exploit the energy in the airspace. Journal of Aircraft Design **4**(2-3), 81–97 (2001). doi: 10.1016/s1369-8869(01)00002-7
24. Paulig, X., Bungart, M., Specht, B.: Conceptual Design of Textile Kites Considering Overall System Performance. In: Ahrens, U., Diehl, M., Schmehl, R. (eds.) Airborne Wind Energy, Green Energy and Technology, Chap. 32, pp. 547–562. Springer, Berlin Heidelberg (2013). doi: 10.1007/978-3-642-39965-7_32
25. Schmehl, R., Noom, M., Vlugt, R. van der: Traction Power Generation with Tethered Wings. In: Ahrens, U., Diehl, M., Schmehl, R. (eds.) Airborne Wind Energy, Green Energy and Technology, Chap. 2, pp. 23–45. Springer, Berlin Heidelberg (2013). doi: 10.1007/978-3-642-39965-7_2
26. Vergnano, G., Don Bosco, C.: Ultralight Airfoils for Wind Energy Conversion. US Patent 8,113,777, Feb 2012
27. Vlugt, R. van der, Peschel, J., Schmehl, R.: Design and Experimental Characterization of a Pumping Kite Power System. In: Ahrens, U., Diehl, M., Schmehl, R. (eds.) Airborne Wind Energy, Green Energy and Technology, Chap. 23, pp. 403–425. Springer, Berlin Heidelberg (2013). doi: 10.1007/978-3-642-39965-7_23
28. Zgraggen, A. U., Fagiano, L., Morari, M.: Automatic Retraction and Full-Cycle Operation for a Class of Airborne Wind Energy Generators. IEEE Transactions on Control Systems Technology **24**(2), 594–608 (2015). doi: 10.1109/TCST.2015.2452230

Chapter 15
Flight Path Planning in a Turbulent Wind Environment

Uwe Fechner and Roland Schmehl

Abstract To achieve a high conversion efficiency and at the same time robust control of a pumping kite power system it is crucial to optimize the three-dimensional flight path of the tethered wing. This chapter extends a dynamic system model to account for a realistic, turbulent wind environment and adds a flight path planner using a sequence of attractor points and turn actions. Path coordinates are calculated with explicit geometric formulas. To optimize the power output the path is adapted to the average wind speed and the vertical wind profile, using a small set of parameters. The planner employs a finite state machine with switch conditions that are highly robust towards sensor errors. The results indicate, that the decline of the average power output of pumping kite power systems at high wind speeds can be mitigated. In addition it is shown, that reeling out towards the zenith after flying figure eight flight maneuvers significantly reduces the traction forces during reel-in and thus increases the total efficiency.

15.1 Introduction

Converting the traction power of kites into electricity is a potential low cost wind energy solution. A minimal implementation is the pumping kite power system which harvests wind energy in a cyclic pattern, alternating between traction and retraction phases. During the traction phase the kite is flown in crosswind flight maneuvers which generates a high traction force that is used to drive a generator. During the retraction phase the kite is depowered and the generator is used as a motor to pull the kite back towards the ground station. The net energy output per cycle essentially

Uwe Fechner (✉) · Roland Schmehl
Delft University of Technology, Faculty of Aerospace Engineering, Kluyverweg 1, 2629 HS Delft, The Netherlands
e-mail: fechner@aenarete.eu

© Springer Nature Singapore Pte Ltd. 2018
R. Schmehl (ed.), *Airborne Wind Energy*, Green Energy
and Technology, https://doi.org/10.1007/978-981-10-1947-0_15

depends on the values of the traction force, reeling speed and time duration of both phases [5].

A key advantage of this energy harvesting technique is the possibility to adjust and modify these operational parameters and the three-dimensional flight path of the wing within a broad range. In practice the operational envelope is constrained by hardware limits of the wing, the tether and the control system, limits of the governing flight physics and safety limits. The purpose of the flight path planner is to design the operation for optimal performance while complying to the imposed constraints. It is an indispensable functional unit of a kite power system, translating the technical capabilities of the system and the characteristics of the deployment scenario into an optimal energy harvesting process.

Sophisticated methods for maximizing the energy output from a closed flight path, combining reel-in and reel-out within a single figure eight flight maneuver, have been described in literature [14]. With some restrictions these methods can also be applied for a flight path concatenating multiple circular maneuvers within a reel-out phase [15]. However, the described methods have two practical limitations. Firstly, only very simple system models can be used during the optimization process. This makes it impossible to realistically account for operational constraints in a turbulent wind field, such as, for example, the maximum tether force. This limiting force strongly depends on the tether sag which is generally neglected by fast, analytic system models. Secondly, the publications cited above do not include a crest factor of the power. This means that the power was optimized without accounting for important operational limits such as maximum power or force. As consequence the trajectories are of limited practical relevance.

In [29, 30] the optimal average position of the kite in the wind window is determined, i.e. the average azimuth and elevation angles which maximize the energy output. The proposed algorithm to find the average wind direction is effective and can in principle be combined with the flight path planner described in the present chapter. Not investigated in the cited publications is how to limit the power at high wind speeds and how a realistic wind profile affects the optimal average elevation angle.

In [27, 28] the control of the kite during retraction is investigated, proposing a flight path at the side of the wind window at an azimuth angle of $45°$. The authors do not provide any evidence that this is the optimal path to reel in the kite. The focus of the cited publications is the control of the retraction phase and not the planning of the flight path. Furthermore, the required transitioning from retraction to traction and the impact on the power production is not investigated.

The focus of [21] is the control and flight performance of a tethered rigid wing system. The flight path planning is based on a lemniscate curve which is discretized by a large number of attractor points. As consequence, a large number of control commands have to be executed per flight maneuver and control loop delays can accordingly be more problematic. During retraction the aircraft pitches down and flies a waypoint track directed towards the ground station, while during traction a constant elevation angle is used. These design decisions are partially related to the rigid wing designed used in the study.

The chapter is organized as follows. In Sect. 15.2 we analyze the kinematics of tethered flight and the mechanism of steering, introduce a suitable kinematic model for path planning and improve the modeling quality of the wind resource by accounting for a realistic vertical profile and turbulence characteristics of the wind velocity. In Sect. 15.3 we present a planning approach for constructing the flight path from straight line and circle segments defined in terms of spherical surface coordinates. The approach can be used for both the retraction and traction phases providing a smooth transition and it can be optimized using a small set of parameters. The crest factors for traction force and power were carefully optimized to harvest the maximum average power for a given hardware without compromising the robustness. In Sect. 15.4 the performance of the flight path planner is assessed by simulating pumping cycle operation using a dynamic system model in conjunction with a wind turbulence model. Some of the design objectives are verified by a subsequent quasi-steady analysis. The preliminary content of the present chapter has been presented at the Airborne Wind Energy Conference 2015 [13] and is also published in [11].

15.2 Tethered Flight in a Realistic Wind Environment

The control strategy for the kite power system is closely connected to the available flight and winch control mechanisms, the resulting mathematical description of the control problem and the effect on the kinematics and dynamics of the system. The practical control approach presented in the following is based on the 20 kW demonstrator system designed, built and tested by Delft University of Technology [9, 18, 22, 26]. The flying components of this system are illustrated in Fig. 15.1.

Fig. 15.1 Kite system consisting of a flexible membrane wing with 25 m² total surface area, a bridle line system to transfer the aerodynamic load and a suspended, remote-controlled cable robot (kite control unit) [18]. For this specific flight a piece of barricade tape ❶ is attached to the rear end of the center strut of the kite and another piece ❷ is attached to the end of the depower tape to indicate the direction of the apparent wind (24 November 2010)

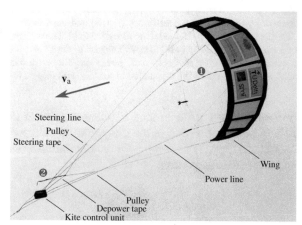

15.2.1 Reference Frames and Kinematics of Tethered Flight

To describe the flight of a tethered wing in a wind field we define the wind reference frame x_w, y_w, z_w as illustrated in Fig. 15.2. The origin $\mathbf{O}$ of this Cartesian reference

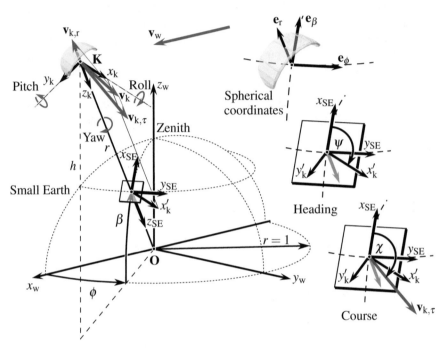

Fig. 15.2 Wind reference frame x_w, y_w, z_w, small earth reference frame x_{SE}, y_{SE}, z_{SE}, kite reference frame x_k, y_k, z_k, spherical coordinates (r, ϕ, β) and corresponding local vector base $\mathbf{e}_r, \mathbf{e}_\phi, \mathbf{e}_\beta$ to describe the flight of a tethered wing in a wind field. The course angle χ describes the orientation of the tangential velocity of the kite with respect to the local x_{SE}-axis. Depicted is the ideal case of a straight tether for which the z_k- and z_{SE}-axes are aligned and the heading angle ψ, also denoted as yaw angle, describes the orientation of the kite with respect to the local x_{SE}-axis

frame is located at the ground attachment point of the tether and the x_w-axis is pointing in the direction of the average wind velocity v_w. We introduce spherical coordinates (r, ϕ, β) to decompose the translational motion of the wing into radial and tangential components

$$\mathbf{v}_k = \dot{r}\mathbf{e}_r + r\dot{\phi}\cos\beta\,\mathbf{e}_\phi + r\dot{\beta}\mathbf{e}_\beta, \tag{15.1}$$

$$= \mathbf{v}_{k,r} + \mathbf{v}_{k,\tau}, \tag{15.2}$$

from which the following kinematic relations can be derived

$$v_{k,r} = \dot{r}, \tag{15.3}$$

$$v_{k,\tau} = r\sqrt{\dot{\phi}^2 \cos^2 \beta + \dot{\beta}^2}. \tag{15.4}$$

Defining the angular velocity of point $\mathbf{K}$ with respect to origin $\mathbf{O}$ as

$$\omega = \sqrt{\dot{\beta}^2 + \dot{\phi}^2 \cos^2 \beta}, \tag{15.5}$$

Eq. (15.4) can be reformulated as

$$v_{k,\tau} = r\omega. \tag{15.6}$$

The radial and tangential motion components are governed by two different control systems. Assuming that the tether is always tensioned, the radial velocity $v_{k,r}$ is determined by the winch controller of the ground station. The direction of the tangential velocity vector $\mathbf{v}_{k,\tau}$ is described by the course angle χ and controlled by the steering system of the wing. The velocity magnitude $v_{k,\tau}$, however, is depending on the angle between tether and wind velocity vector, the reeling velocity of the tether, the aerodynamic properties and the mass of the wing [23].

The geometric similarity of tethered flight at constant radial distance and level flight of an aircraft above the curved surface of the earth motivates the use of a small earth analogy [17, 18]. The corresponding small earth coordinate system is defined on the surface of the unit sphere around the ground attachment point of the tether, using longitude ϕ and latitude β to describe the angular position of the kite. Following common practice in aerospace engineering, the local reference frame x_{SE}, y_{SE}, z_{SE} is defined as a North, East, Down (NED) frame[1] with the local x_{SE}-axis pointing towards the zenith, the local y_{SE}-axis pointing towards East and the local z_{SE}-axis always pointing towards the origin $\mathbf{O}$.

The body-fixed reference frame of the kite has its origin at point $\mathbf{K}$ and is denoted as x_k, y_k, z_k with unit vectors $\mathbf{e}_{k,x}, \mathbf{e}_{k,y}, \mathbf{e}_{k,z}$. As kite we define the entire flying system consisting of wing, bridle line system and suspended control unit [22, 26]. Following aerospace engineering practice, the y_k-axis defines the direction from the left to the right wing tip and the z_k-axis defines the direction from the wing to the suspended control unit, chosen such that the z_k-axis is aligned with the tether at the kite attachment point. The x_k-axis is by definition orthogonal to y_k and z_k. If the tether is not straight the z_k- and z_{SE}-axes are not aligned.

The rotations around the body-fixed x_k-, y_k- and z_k-axes are denoted as roll, pitch and yaw. For the ideal case of a fully tensioned, straight tether the pitch and roll rates, are kinematically coupled to the angular velocity ω of point $\mathbf{K}$, as defined by Eq. (15.5). This specific situation, in which the yaw angle describes the heading of the kite on the unit sphere, is illustrated in Fig. 15.1. For a flexible and thus sagging tether the pitch and roll rotations are not kinematically coupled to the tangential flight motion of the kite.

---

[1] North, East, Down and zenith refer in this context to the small earth

15.2.2 Kinematic Kite Model and Steering Mechanism

For designing the flight path planner we will use an idealized kinematic model based on a straight tether as discussed in the previous section. The first three degrees of freedom of this model are the spherical coordinates (r, ϕ, β) of the kite point **K**. Because of the bridling of the wing, the pitch and roll rotations of the kite are kinematically coupled to the tangential motion of the kite, while the heading angle ψ is a fourth degree of freedom describing the rotation of the kite around the tether.

It is important to note that this kinematic model describes the translation and rotation of the entire kite system consisting of wing, bridle line system and suspended kite control unit. Within this system, the kite control unit actuates the bridle line system to deform and rotate the wing relative to the kite reference frame. The changing aerodynamic forces and moments induce accelerations that adjust the flight motion of the kite. From this perspective, the entire wing functions as an aerodynamic control surface similar to the control surfaces of an aircraft.

To avoid the physical modeling of the complex aeroelastic phenomena which govern the mechanism of steering [3, 4] we use an empirical correlation between the turn rate $\dot{\psi}$ and the steering actuation of the bridle line system. This turn rate law has been established on the basis of experimental data [6–8, 18] and validated by numerical simulations [3]. As a result, the rotation of the kite around the tether is directly coupled to the prescribed control input.

The course angle χ describes the direction of the tangential velocity $\mathbf{v}_{k,\tau}$. For the ideal hypothetical case of a massless kite ($m = 0$) in a stagnant wind field ($v_w = 0$), the heading vector $\mathbf{e}_{k,x}$ and the tangential velocity vector $\mathbf{v}_{k,\tau}$ are aligned at all times which means that heading and the course angle are identical. This because kites and aircraft in general are designed to align with the relative flow velocity $\mathbf{v}_a$, when the flight condition does not require to compensate a lateral inertial or gravitational force component. For a stagnant wind field $\mathbf{v}_a$ is parallel to the flight velocity $\mathbf{v}_k$. Any deviation from this ideal case leads to a misalignment of heading and course, which is quantified by the kinematic side slip angle [18].

For example, a non-vanishing wind velocity $\mathbf{v}_w$, which is a prerequisite for wind energy conversion, will incline the relative flow velocity with respect to the flight velocity according to the definition

$$\mathbf{v}_a = \mathbf{v}_w - \mathbf{v}_k. \tag{15.7}$$

However, kites with a lift-to-drag ratio $L/D \gg 1$ operate at crosswind flight speeds $v_{k,\tau} \gg v_w$ and it is thus reasonable to neglect the misalignment of $\mathbf{v}_a$ and $\mathbf{v}_k$ in the traction phase.

For a real kite ($m > 0$) inertial and gravitational force effects transverse to the tether have to be balanced by an aerodynamic side force $\mathbf{F}_{a,s}$. Given the constrained pitch and roll rotations this side force can only be generated by inclining the heading of the kite with respect to its course, thus enforcing a kinematic side slip angle. This mechanism is illustrated in Fig. 15.3 for a kite flying a left turn (perspective from ground) and including the effects of the gravitational force mg and centrifugal force

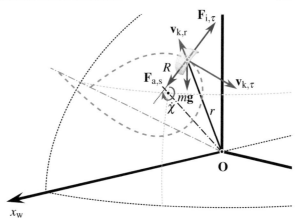

Fig. 15.3 Kite performing a left turn ($\dot{\chi} > 0$) while flying a downloop figure eight maneuver and generating traction power ($v_{k,r} > 0$). Only one half of the maneuver is displayed. The dash-dotted line passing through the center of the turn is the turning axis, while the corresponding turn radius is denoted as R and the non-dimensional turning radius is defined as $\varrho = R/r$

due the turning maneuver,

$$F_{i,\tau} = m \frac{v_{k,\tau}^2}{R}, \tag{15.8}$$

where the radius of curvature R characterizes the tangential motion component.

In the described kinematic framework, the turning radius R links the rate of change of the course angle $\dot{\chi}$ to the tangential kite velocity $v_{k,\tau}$ and, using Eq. (15.6), to the angular velocity ω of the kite point

$$\dot{\chi} = \frac{v_{k,\tau}}{R} = \frac{r}{R}\omega = \frac{\omega}{\varrho}. \tag{15.9}$$

The relation between the heading of the kite and the relative flow is illustrated in Fig. 15.4 for three typical flight modes during pumping cycle operation. The flow direction is indicated by two pieces of red/white striped barricade tape, as described in the caption of Fig. 15.1. The end point of the depower tape, to which the two

Fig. 15.4 Wing and bridle line system during a downloop figure eight maneuver performing a right turn (left) and following straight flight diagonally upwards (center) and during retraction of the kite (right). Video stills taken from the kite control unit (24 November 2010)

steering tapes and the barricade tape in the foreground are tied, is highlighted by a dot. The barricade tape attached to the kite is indicated by an arrow.

When in crosswind flight (left & center) the wing is powered by reeling in the depower tape. This tensions the steering lines and consequently the highlighted knot stays in the line of sight from control unit to wing. When retracting the kite towards the ground station (right) the wing is depowered by reeling out the depower tape. This relaxes the steering lines which are deflected substantially by the relative flow.

To fly the turn, the kite needs to balance a centrifugal acceleration of 2 to 4g by generating an aerodynamic side force towards the turning axis. The strength of this force is indicated by the pronounced sideslip angle (the angle between center strut and attached barricade tape) in the left photo. The steering mechanism is illustrated in more detail in Fig. 15.5 including the generated aerodynamic steering forces acting on the wing tips. The sketch shows how the asymmetric steering input of reeling

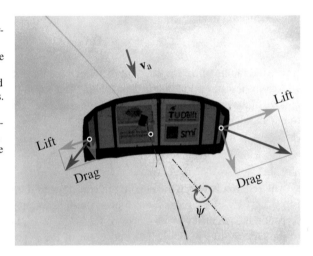

Fig. 15.5 Aerodynamic steering forces acting on a kite flying a right turn (perspective from ground). The forces at the wing tips are decomposed into lift and drag components. The dash-dotted line is in extension of the tether and indicates the yaw rotation axis. The center point indicated the approximate location around which the aerodynamic moment of the wing tips act. Photo of the kite, which has a total wing surface area of 25 m^2, is taken from the ground (24 November 2010)

in the right and reeling out the left steering tape warps the entire wing, which leads to a resultant side force to the right which is required for the right turn. The importance of actively controlled wing warping for the excellent turning characteristics of C-shaped flexible membrane wings has been confirmed by computational analysis [4].

Because the kite control unit represents almost 50% of the total airborne mass it experiences a strong centrifugal force during the turns of the figure eight maneuvers. As can be seen in Fig. 15.4 (left) this leads to a substantial outwards deflection of the unit which distorts not only the entire bridle line system but also affects the video recording from this point of observation.

When flying diagonally upwards the kite needs to balance only the gravitational acceleration g. This is indicated by the comparatively small sideslip angle in the center photo. During retraction the kite asymptotically approaches a steady flight state which is characterized by a descent at constant elevation angle. As consequence, the tangential velocity $v_{k,\tau}$ and angular velocity ω can become very small towards the end of the retraction phase. In this situation of undefined course angle, the heading of the kite is planned to keep the nose of the kite pointing towards zenith.

To summarize the above considerations we conclude that an asymmetric steering input leads to an aerodynamic side force. Instead of using a dynamic model we couple the turn rate $\dot{\psi}$ of the kite directly to the steering input by using an empirical turn rate law. On the other hand, the rate of change of the course angle $\dot{\chi}$ is governed by the tangential flight velocity $v_{k,\tau}$ of the kite, which is a dependent variable, and by the turn radius R required for a specific maneuver.

The objective of the steering system is to control the course angle to fly the kite towards any feasible attractor point or on a turn with well-defined radius. This is achieved by controlling the steering signal to the kite control unit. By introducing the small earth analogy and using the course angle as the controlled variable the kite control problem is reduced to a Single Input Single Output (SISO) problem [2].

15.2.3 Wind Resource

Within the scope of this study we limit the pumping cycle operation to a maximum altitude of 600 m and a minimum tether length of 300 m. We further assume that the maximum altitude and maximum tether length are identical, although a safety margin will have to be applied for any practical use of the results. To cover the operational altitude range we use wind data measured with the 213 m high KNMI-mast in Cabauw, The Netherlands, in 2011. The publicly available CESAR database provides data for different altitudes, sampled with 10 min resolution for at least a full year [24].

The average ground wind speed, measured at a height of 10 m, is 4.26 m/s. The corresponding estimated cumulative probability distribution function (CDF) is shown in Fig. 15.6. The performance simulations are based on a kite power system which is designed to reach its nominal power at a wind speed that is exceeded about 20% of the time. According to the CDF this threshold is at a ground wind speed of 6 m/s. For this choice a capacity factor of about 40% is expected.

The measured vertical profile of the average wind speed is shown in Fig. 15.7. The diagram includes the fitted power law [1]

$$v_w = v_{w,g} \left(\frac{\bar{z}}{z_{ref}} \right)^{\alpha}, \tag{15.10}$$

with the ground wind speed $v_{w,g}$ at $z_{ref} = 10$ m height, and the average height $\bar{z}$ of the kite during the traction phase. Average wind speeds of 7.28 and 8.56 m/s at 98.7 and 131.6 m altitude were estimated. These heights correspond to the average operational altitude of the kite during the traction phase, as will be shown in Sect. 15.4.

Wind turbulence is characterized by the relative intensity I of the turbulent velocity fluctuations. To account for the sampling interval of the experimental data,

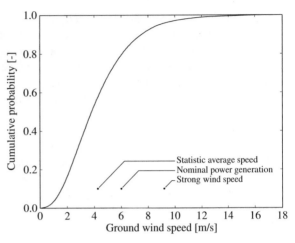

Fig. 15.6 Cumulative probability distribution function (CDF) of the ground wind speed at Cabauw, The Netherlands, in 2011. Data from [24]

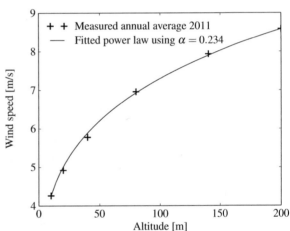

Fig. 15.7 Vertical profile of the wind speed measured at Cabauw in 2011 [24] and fitted power law with an exponent $\alpha = 0.234$

we define this intensity as the ratio of the standard deviation σ_1 of the wind speed within 10min intervals and the corresponding 10min average of the wind speed. Three-dimensional turbulence is simulated using the approach described in [19, 20] for the three different ground wind speeds displayed in Fig. 15.6 and listed in Table 15.1.

Table 15.1 Simulation scenarios based on Cabauw data [24]. I_{99} and I_{197} are intensities at 98.7m and 197.4m

Ground wind speed	$v_{w,g}$	I_{99}	I_{197}
Annual average	4.26 m/s	8.5%	6.3%
Nominal power generation	6.00 m/s	9.7%	7.2%
Reel-out depower required	9.20 m/s	9.8%	7.9%

Parameter	Value	Description
γ	3.9	anisotropy parameter, isotropic turbulence $\gamma = 0$
σ_1	see Table 15.1	standard deviation of the wind component at the average height during reel-out in the mean wind direction
σ_{iso}	0.55 σ_1	standard deviation of wind speed for isotropic turbulence
l	33.6 m	turbulence length scale for an average height > 60 m

Table 15.2 Parameters of the Mann model used to generate the three-dimensional wind field, from IEC 61400 [16]

The homogeneous velocity field is obtained by three-dimensional Fast Fourier Transformation (FFT) of the spectral tensor. A white noise vector is used to give the wave numbers a random phase and amplitude. The model parameters are listed and described in Table 15.2. The ground wind speed of 9.20 m/s marks the upper limit of pumping cycle operation without the need to depower the wing during tether reel-out in the traction phase.

For each value of the ground wind speed, a turbulent wind field is computed on a grid of $4050 \times 100 \times 500$ nodes, using a uniform spatial resolution of 2 m. The size of the computational domain in x-direction is large enough to cover 10 min of kite operation with uniquely simulated wind data, as illustrated in Fig. 15.8.

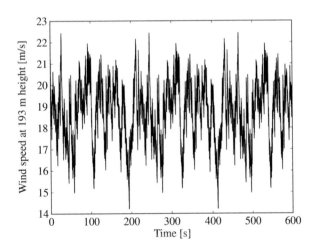

Fig. 15.8 Simulated wind speed as function of time for the scenario $v_{w,g} = 9.2$ m/s. The wind speed varies between 14.2 and 22.5 m/s and changes can be as fast as 4 m/s²

It should be noted that he kite is substantially larger than the spatial discretization of the wind field. We thus use a third-order spline interpolation to determine the three-dimensional wind velocity vector at any given position in the computational domain. The time dependency of the wind is taken into account by an advection correction, adding the product of simulation time and average wind speed at the height of the kite to its x-position before determining the wind velocity vector by interpolation of pre-computed grid data.

15.3 Flight Path Planner

The objective of the planning algorithm is to provide an optimal flight path as a function of the variable wind conditions and the various operational constraints of the system. If multiple systems are arranged in a wind park configuration, as analyzed in Chap. 16, additional constraints due to joint operation have to be considered.

15.3.1 General Design Considerations

Path planning can be based on an ordered list of positions, so-called waypoints, that describe the path to be tracked by the kite. Alternatively, the desired path can also be described as a continuous curve [2, 17, 18]. It is the task of the flight control algorithm to generate steering commands that maneuver the kite from any deviating position back to this reference path. A variant is to steer the kite sequentially towards attractor points without the goal to actually reach and pass these points. For example, the original flight path planner of the 20 kW demonstrator of Delft University of Technology uses four attractor points to describe a complete figure eight flight maneuver [26], while the path planner developed by Ampyx Power uses a substantially larger number of attractor points to describe this maneuver [21].

In the present study we adopt this alternative approach, however, instead of using attractor points only, we introduce additional turning maneuvers, with well-defined center point and radius, to advance from one attractor point to the next. This planning scheme is illustrated schematically in Fig. 15.9, combining two attractor points P_3 and P_4 with two turns around points C_2 and C_3 to describe the figure eight maneuvers. An important design constraint is the minimum turning radius R_{min} of the kite which is not only determined by the inherent maneuverability of the wing, but also by the operational limits of integrated sensors. One example are the GNSS sensors which tend to fail when the turning radius is too small.

It is important to note that the depicted representation by linear and circular segments in the $\phi\beta$-plane is only a geometric approximation of the planned flight path. This true path is constructed in three-dimensional Cartesian space according to basic flight-physical considerations. For example, we assume that for straight flight at constant tether length the kite moves along a great circle. Also denoted as a geodesic, this circular arc segment represents the shortest distance between two positions on the spherical surface. Similarly we assume that for a turn with constant radius and at constant tether length the kite moves along a small circle. Both type of arc segments can be used to construct a figure eight flight maneuver.

However, because the angular coordinates ϕ and β describe a spherical coordinate surface in Cartesian space, the mapping of these segments to the $\phi\beta$-plane introduces a geometric distortion. This distortion increases with the elevation angle and reaches a maximum at the zenith. For the purpose of simplicity and because the planned path is not explicitly tracked we neglect this effect in Fig. 15.9 using straight lines and circle segments to illustrate the flight path.

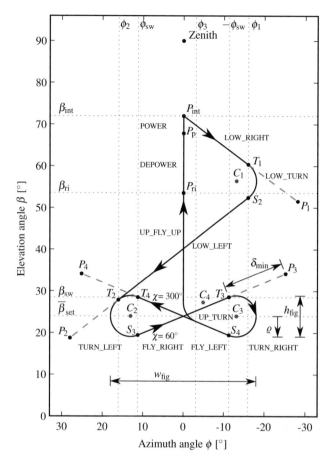

Fig. 15.9 Planned pumping cycle with downloop figure eight maneuvers at an average elevation angle of $\overline{\beta}_{set} = 24°$, represented in the $\phi\beta$-plane. The entire cycle is described by five attractor points (P_1 to P_4 and zenith) and four turning points (C_1 to C_4). The minimum angular distance between kite and an active attractor point is $\delta_{min} = 15°$. This threshold value for δ is used to trigger the turning maneuvers. Adapted from [11, 12]

Following the crosswind operation, the kite performs a turn around point C_4 into the x_wz_w-plane to fly towards zenith. When reaching the elevation angle β_{ri} the kite is depowered by reeling out the depower tape. As consequence, the tether force drops and the retraction phase can begin. When the minimum tether length is reached at P_p the depower tape is reeled in to power the kite again. When the target power setting is reached the transition phase begins in which the kite is steered towards P_1. The position of P_p is not planned by the algorithm, but measured and used as input for the planned flight path. In the first cycle P_p is identical to the parking position.

Reeling in while steering towards zenith has the advantages that for one, the wind speed in the direction of the tether and thus the retraction force is low. For the other, the kite is rising at the end of the traction phase, increasing its potential energy, which later helps to power the retraction phase. The use of potential energy is more efficient than using electrical storage only. If the traction energy is converted into electricity, stored in a battery and then a portion of it is used to drive a motor to retract the kite, this has an efficiency of about $\eta = 90\% \cdot 80\% \cdot 95\% = 68.4\%$. The three contributions are the estimated generator, motor and battery efficiencies. Potential energy can be used without any losses to reduce the tether forces during retraction. Within certain limits a heavier kite thus increases the overall efficiency. On the other hand a higher mass of the kite makes launching and landing more difficult and increases the minimum wind speed required for operation.

Flying towards an intermediate point after the retraction phase helps to mitigate the tether force peak which can otherwise occur when the kite is diving too rapidly towards the ground. We concluded that a combination of flying towards attractor points and turn actions allows planning of an any technically feasible flight path. Our tests have show that when limiting the duration of the turn actions in time the flight control is very robust against sensor or communication failures.

To conclude the above considerations we summarize the following goals for the design of the flight path planner:

1. a high projected wind speed $\mathbf{v}_w \cdot \mathbf{e}_r$ at the height of the kite in the traction phase;
2. a low crest factor (ratio of peak value to effective value) of the reel-out force and reel-out power;
3. a large turn radius R to limit the steering effort and the additional aerodynamic drag caused by steering;
4. a low steering effort also when flying straight;
5. a retraction phase with low projected wind speed;
6. short transition phases between retraction and traction;
7. low tether force overshoot at the transition from retraction to traction;
8. high robustness with respect to sensor errors and delays;
9. good controllability of the maximum and minimum height;
10. good controllability of the projected wind speed and thus reel-out power, especially at high wind speeds, but also at medium wind speeds with high turbulence.

The design goals (2) and (3) are competing, as do the goals (6) and (7), and a good design accordingly requires a compromise solution.

15.3.2 Supervisory Control for Automated Power Production

The proposed supervisory control for operation in pumping cycles is shown in Fig. 15.10. The path is planned in the two-dimensional plane spanned by the azimuth and elevation angles. The initial condition for automated power production is

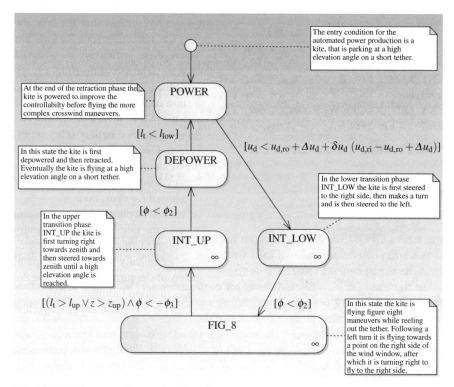

Fig. 15.10 Finite state diagram detailing the supervisory controller for automatic operation in pumping cycles. To start the traction phase the minimum tether force for the winch controller is reduced to about 10% of the maximum tether force. This results in an attenuation of the reel-out speed and a reversal of the drum rotation from reel in to reel out, because the winch controller now tracks a set speed, that is positive and proportional to the square-root of the force. Adapted from [11]

a parking position of the kite at a high elevation angle on a short tether ($l_t = l_{low}$). When activated, the supervisory controller enters the state POWER. In this state, the angle of attack of the wing is increased to the set value for tether reel out. When this target is reached to at least δu_d percent the controller is switching into the lower transition phase INT_LOW. In this state the kite is first flying to the right, then makes a turn and is finally flying to point P_2 on the left of the wind window. When an azimuth angle $\phi > \phi_2$ on the figure eight trajectory is reached the kite enters the state FIG_8.

In this state the kite is forced to fly figure eight maneuvers, the tether is reeled out and power is harvested. When either the height of the kite or the tether length reach an upper limit and the kite is near the center of the wind window, then the controller switches into the upper transition state INT_UP.

The upper transition phase begins with a turn towards zenith. Then the kite is flying straight upwards, slowing down while the kite is still harvesting energy. When the wind speed dependent elevation angle β_{ri} is reached the controller switches into

State	Next state	$u_{d,set}$	Condition
PARKING	POWER	$u_{d,ro} + \Delta u_d$	Event START_POWER_PRODUCTION
POWER	INT_LOW	$u_{d,ro} + \Delta u_d$	$u_d < u_{d,ro} + \Delta u_d + \delta u_d \, (u_{d,ri} - u_{d,ro} - \Delta u_d)$
INT_LOW	FIG_8	$u_{d,ro} + \Delta u_d$	Event EXIT(INT_LOW)
FIG_8	INT_UP	$u_{d,ro} + \Delta u_d$	Event EXIT(FIG_8)
INT_UP	DEPOWER	$u_{d,ro} + \Delta u_d$	Event EXIT(INT_UP)
DEPOWER	POWER	$u_{d,ri}$	$l_t < l_{low}$

Table 15.3 Finite states and state transitions of the supervisory flight path controller during automated power production. This controller activates one of the sub-controllers for the planning of the flight sections. In addition it changes the angle of attack of the kite by changing the depower setting u_d

the state DEPOWER. In this state the angle of attack of the wing is reduced for depowering and the set value of the tether force is reduced. After a short transition time the winch begins to reel in.

In the state DEPOWER the elevation angle increases. When the lower tether length is reached ($l_t = l_{low}$) the controller switches to the lower transition state INT_LOW and the next pumping cycle begins.

The state transition table for the supervisory control is shown in Table 15.3. The settings of the winch depend on the state of the supervisory flight path controller. The value Δu_d is calculated by the winch controller. It is an additional depowering of the kite for limiting the power output at high wind speeds. δu_d is a constant in the order of 70% and determines to which degree the powering has to be finished before flying to the side.

15.3.3 Lower Transition Phase

The design of the lower transition phase aims at the following objectives:

- a low tether force overshoot. This overshoot is caused by the gravitational acceleration of the kite flying downwards,
- a minimum impact on total efficiency. If the kite is flying too far to the side of the wind window to limit the force overshoot, too much time and energy is lost,
- a low undershoot of the minimum elevation angle. This prevents that the kite flies too close to the ground, which is a safety risk, and also reduces the power output because of the lower wind velocities towards the ground.

The proposed layout of the flight path is illustrated schematically in Fig. 15.9 by the segments LOW_RIGHT, LOW_TURN and LOW_LEFT. The implementation keeps the rate of change of the elevation angle identical to the value during the straight flight segments of the figure eight maneuvers. The state transition table is shown in Table 15.4. Because the path construction depends on the desired change rate of β, we explain the calculation of ϕ_1 at the end of the next section.

State	Next state	$\mathbf{p}^{SE}_{k,set}$	$\dot{\chi}_{set}$	Condition
Initial	LOW_RIGHT	P_1	from PID	always
LOW_RIGHT	LOW_TURN	–	$\dot{\chi}_{turn}$	$\phi < \phi_1$
LOW_TURN	LOW_LEFT	P_2	from PID	$\chi < 180 + \delta\chi_{int}$
LOW_LEFT	Final	–	$-\dot{\chi}_{turn}$	$\phi > \phi_2$

Table 15.4 Finite sub-states of the lower transition phase INT_LOW. The parameter $\delta\chi_{int}$ is introduced to compensate the delay of the steering actuator and the inertia of the kite at the end of the turn. It is chosen such that the kite does not turn more than required before flying straight towards the attractor point P_2

15.3.4 Traction Phase and Crosswind Flight Maneuvers

The four-step flight path planner for flying figure eight maneuvers during the traction phase is shown in Fig. 15.9. Flying past point T_2 the kite first turns left, then steers towards P_3, then turns right and finally steers towards P_4, after which the sequence is repeated. Table 15.5 describes the outputs and the switch conditions of the six different sub-states for flying figure eight maneuvers.

State	Next state	$\mathbf{p}^{SE}_{k,set}$	$\dot{\chi}_{set}$	Condition
Initial	TURN_LEFT	$--$	$\dot{\chi}_{turn}$	always
FLY_LEFT	TURN_LEFT	$--$	$\dot{\chi}_{turn}$	$\phi > \phi_{sw}$
TURN_LEFT	FLY_RIGHT	P_3	from PID	$\chi > 270° - \delta\chi$
FLY_RIGHT	TURN_RIGHT	–	$-\dot{\chi}_{turn}$	$\phi < -\phi_{sw}$
TURN_RIGHT	FLY_LEFT	P_4	from PID	$\chi < 90° + \delta\chi$
FLY_LEFT	LAST_LEFT	$--$	from PID	$(l_t > l_{up} \vee z > z_{up}) \wedge \phi \leq \phi_3$
LAST_LEFT	Final	$--$	$-\dot{\chi}_{turn}$	$\phi > \phi_3$
FLY_RIGHT	LAST_RIGHT	$--$	from PID	$(l_t > l_{up} \vee z > z_{up}) \wedge \phi \geq -\phi_3$
LAST_RIGHT	Final	$--$	$\dot{\chi}_{turn}$	$\phi < -\phi_3$

Table 15.5 Finite sub-states of the figure eight flight path planner. Flying these maneuvers is not finished before the upper height z_{up} or the upper tether length l_{up} is reached. The final up-turn is always started such that it ends at $\phi = 0$. The parameter $\delta\chi$ is introduced to compensate the delay of the steering actuator and the inertia of the kite at the end of the turn. It is chosen such that the kite does not turn more than required before flying straight towards one of the attractor points

The path planner has the following inputs and outputs:

- IN: set value of the average elevation angle $\overline{\beta}_{set}$,
- IN: course angle χ and heading angle ψ,
- IN: azimuth angle ϕ,
- OUT: boolean value PID_{active},
- OUT: set value of the position $\mathbf{p}^{SE}_{k,set}$ when the PID is active.
- OUT: set value for the turn rate $\dot{\psi}_{set}$ when the PID is not active.

The path is parametrized by the angular width w_{fig} and height h_{fig} of the figure eight and the minimum angular distance δ_{min} between kite and an active attractor point. When the angular distance δ from the currently active attractor point drops below this threshold value the next turning maneuver is triggered. When these values are given, $P_3, P_4, \psi_{\text{turn}}$ and ϕ_{sw} can be calculated.

As a first step we calculate the non-dimensional turning radius $\varrho = h_{\text{fig}}/2$ as a linear function of the average elevation angle $\overline{\beta}_{\text{set}}$

$$\varrho = \varrho_{\text{max}} - (\varrho_{\text{max}} - \varrho_{\text{min}}) \frac{\overline{\beta}_{\text{set}} - \beta_{\text{min}}}{\beta_{\text{max}} - \beta_{\text{min}}}, \tag{15.11}$$

using the constant values

$$\beta_{\text{min}} = 20°, \qquad \beta_{\text{max}} = 60°, \qquad \varrho_{\text{min}} = 3°, \qquad \varrho_{\text{max}} = 5°, \tag{15.12}$$

that have been determined empirically for the 20 kW demonstrator system illustrated in Fig. 15.1. Equation (15.11) decreases the turning radius with increasing average elevation angle of the flight maneuvers. This suppresses the kinematically induced variation of the effective wind speed $\mathbf{v}_w \cdot \mathbf{e}_r$ which, for constant turning radius, would increase towards the zenith. Using the measured angular velocity ω of the kite the required turn rate $\dot{\chi}_{\text{turn}}$ can then be calculated from Eq. (15.9).

In a next step, the azimuth angle of the turning point C_2 is calculated as

$$\phi_{C2} = \frac{w_{\text{fig}}}{2} - \varrho \tag{15.13}$$

and the angular coordinates of the switch points $T_3 = (-\phi_{\text{sw}}, \beta_{\text{sw}})$ and $T_4 = (\phi_{\text{sw}}, \beta_{\text{sw}})$ are determined from the equations of the right turning circle and the tangent as

$$\phi_{\text{sw}} = \phi_{C2} - \frac{\varrho^2}{\phi_{C2}}, \qquad \beta_{\text{sw}} = \overline{\beta}_{\text{set}} + \sqrt{\varrho^2 - (\phi_{\text{sw}} - \phi_{C2})^2}. \tag{15.14}$$

The slope k of the straight line from S_4 via T_4 to P_4 can now be calculated as

$$k = \sqrt{\frac{\phi_{C2} - \phi_{\text{sw}}}{\phi_{\text{sw}}}} \tag{15.15}$$

and the angular coordinates of the attractor points $P_3 = (-\phi_P, \beta_P)$ and $P_4 = (\phi_P, \beta_P)$ are determined as

$$\phi_P = \phi_{\text{sw}} + \delta_{\text{min}} \sqrt{\frac{1}{1+k^2}}, \qquad \beta_P = \beta_{\text{sw}} + \delta_{\text{min}} k \sqrt{\frac{1}{1+k^2}}. \tag{15.16}$$

We chose the azimuth angle of C_1 to be equal to the azimuth angle of C_3

$$\phi_{C1} = \phi_{C3} = -\phi_{C2}, \tag{15.17}$$

and accordingly calculate its elevation angle as

$$\beta_{C1} = \beta_{int} - k_1 + k_2 \overline{\beta}_{set}. \tag{15.18}$$

The coefficients k_1 and k_2 are determined empirically, such that

- for $\overline{\beta}_{set} = \beta_{max}$ the turning points C_1 and C_3 coincide to avoid an overshoot of the tether force during the first pumping cycle and
- for $\overline{\beta}_{set} = \beta_{min}$ the worst case tether force overshoot is negligible while the time for the lower transition phase is still as short as possible.

Within the scope of this chapter this is achieved by the following values

$$k_1 = 18.6°, \qquad k_2 = 0.11. \tag{15.19}$$

In a next step, ϕ_1 is calculated as

$$
\begin{aligned}
\phi_1 = \frac{1}{\phi_{C1} \left(\beta_{C1}^2 - 2\beta_{C1}\beta_{int} + \beta_{int}^2 + \phi_{C1}^2 \right)} & \left(\beta_{C1}^2 \phi_{C1}^2 - 2\beta_{C1}\beta_{int}\phi_{C1}^2 \right. \\
- \beta_{C1} \sqrt{\phi_{C1}^2 r^2 \left(\beta_{C1}^2 - 2\beta_{C1}\beta_{int} + \beta_{int}^2 + \phi_{C1}^2 - \varrho^2 \right)} + \beta_{int}^2 \phi_{C1}^2 & \\
+ \beta_{int} \sqrt{\phi_{C1}^2 \varrho^2 \left(\beta_{C1}^2 - 2\beta_{C1}\beta_{int} + \beta_{int}^2 + \phi_{C1}^2 - r^2 \right)} + \phi_{C1}^4 & \left. - \phi_{C1}^2 \varrho^2 \right).
\end{aligned}
\tag{15.20}
$$

The elevation angle β_{int} at the beginning of the lower transition phase is measured. For the schematic shown in Fig. 15.9 we assume a value of $\beta_{int} = 72°$. To determine ϕ_2 we use the geometric fact that the flight path during the state LOW_LEFT must cross the line $\phi = 0$ at the average elevation angle of the turning points C_1 and C_2

$$\beta_M = \frac{1}{2} \left(\beta_{C1} + \beta_{C2} \right). \tag{15.21}$$

With these parameters the azimuth angle ϕ_2 is calculated as

$$
\begin{aligned}
\phi_2 = \frac{1}{\beta_M^2 - 2\beta_M\beta_{C1} + \beta_{C1}^2 + \phi_{C1}^2} & \left(-\beta_M^2 \beta_{C1} + 2\beta_M \beta_{C1}^2 - \beta_M \varrho^2 - \beta_{C1}^3 \right. \\
& \left. -\beta_{C1}\phi_{C1}^2 + \beta_{C1}\varrho^2 + \sqrt{\phi_{C1}^2 \varrho^2 \left(\beta_M^2 - 2\beta_M\beta_{C1} + \beta_{C1}^2 + \phi_{C1}^2 - \varrho^2 \right)} \right).
\end{aligned}
\tag{15.22}
$$

Using the basic rules of geometry the azimuth angle, needed in the exit condition of Table 15.5, is evaluated as

$$\phi_3 = \varrho - \sqrt{\frac{k^2 \varrho^2}{k^2 + 1}}. \tag{15.23}$$

15.3.5 Upper Transition Phase

The design of the upper transition phase aims at the following objectives:

- bringing the kite out of the power zone, while still harvesting energy,
- implementing a fast and smooth transition.

The state transition table of the upper transition phase is shown in Table 15.6. The optimal elevation angle β_{ri} at which the traction phase is terminated depends on the wind conditions. A first estimate can be calculated as

$$\beta_{ri} = k_5 + k_6 \overline{\beta}_{set} \tag{15.24}$$

with

$$k_5 = 37.5^\circ, \qquad k_6 = 0.5. \tag{15.25}$$

To further improve the path planner both constants can be optimized using a dynamic system model [10]. Instead of using a pre-calculated value for β_{ri} it is also possible to use a switch condition depending on the traction power. For example, the traction phase can be terminated when the traction power drops to 40% of the average mechanical power in the state FIG_8. Such a dynamic switch condition is less dependent on the average wind speed or on the aerodynamic characteristics of the kite. Most of these options for improvement are further assessed in [11, 12].

At the end of the upper transition phase the kite is depowered and the set force of the winch changed, which is not shown in Table 15.6.

State	Next state	$\mathbf{p}^{SE}_{k,set}$	$\dot{\chi}_{set}$	Condition
Initial	UP_TURN	–	χ_{turn}	always
UP_TURN	UP_FLY_UP	Zenith from PID		$\psi > 360^\circ - \delta\psi \lor \psi < \delta\psi$
UP_FLY_UP	Final	Zenith from PID		$\beta > \beta_{ri}$

Table 15.6 Finite sub-states of the upper transition phase INT_UP. An offset of $\delta\psi \approx 60^\circ$ is needed to compensate for the time delay δt_{up} between the command to stop turning and the kite actually stopping to turn

15.3.6 Influence of the Elevation Angle

The planned flight path is depicted in Fig. 15.11 for different set values of the average elevation angle. For the limiting case $\overline{\beta}_{set} = \beta_{max} = 60^\circ$ the turning points C_1 and C_3 as well as the attractor points P_2 and P_4 coincide, as defined in Sect. 15.3.4. As consequence the states LOW_TURN and LOW_LEFT disappear and the kite directly transitions from the retraction into the traction phase.

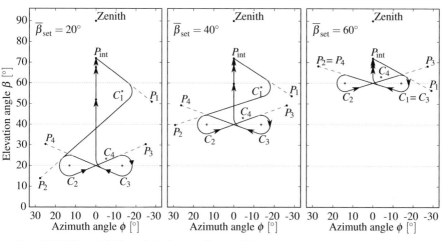

Fig. 15.11 Planned flight path for low, medium and high set values of the average elevation angle

15.4 Planning Performance

The performance of the planning approach is first assessed by a dynamic simulation of the 20 kW demonstrator system of Delft University of Technology in a realistic wind environment and subsequently investigated by a quasi-steady analysis.

15.4.1 Dynamic System Simulation

The dynamic system model [10] is used with a combination of flight path and winch controllers [11, 12] for simulating pumping cycle operation. The key parameters of the model are listed in Table 15.7. The wind data for the onshore location Cabauw is used, as described in Sect. 15.2.3.

Table 15.7 Parameters of the simulation model including all relevant system components and accounting for site-specific wind shear profile and turbulent fluctuations

Parameter	Value
Total wing surface area A_k [m^2]	25.0
Projected wing surface area A [m^2]	20.36
Relative side area A_{side}/A [%]	30.6
Wing mass including sensors m_k [kg]	10.58
Mass of kite control unit m_{KCU} [kg]	11.0
Maximum tether force $F_{t,max}$ [N]	8000.0

The computed flight path of the kite for a nominal ground wind speed of 6 m/s is shown in Fig. 15.12. The influence of the turbulence intensity $I_{197} = 7.2\%$ is hardly

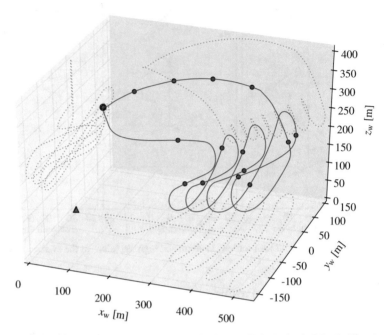

Fig. 15.12 Simulated flight path of the kite at the nominal ground wind speed of 6 m/s. The simulation starts at the enlarged red dot. Smaller dots are placed on the flight path in 10 s intervals. Positions are relative to the ground station which is represented as a red triangle

visible. The corresponding tether force at the ground station and the tether reel-out speed are shown in Fig. 15.13. Figure 15.14 shows the flight path at an increased ground wind speed of 9.2 m/s. The average traction elevation angle $\overline{\beta}_{set}$ has been increased to limit the maximum power. Retraction and traction are at nearly the same

Fig. 15.13 Tether force at the ground and tether reel-out speed for two power cycles at the nominal ground wind speed of 6 m/s. The tether force is close to the maximum value of 8000 N. The reel-out speed is quite constant. Only when the force exceeds the value of 7600 N the upper force controller becomes active and the reel-out speed increases to limit the tether force

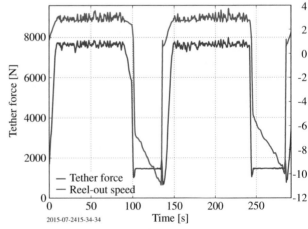

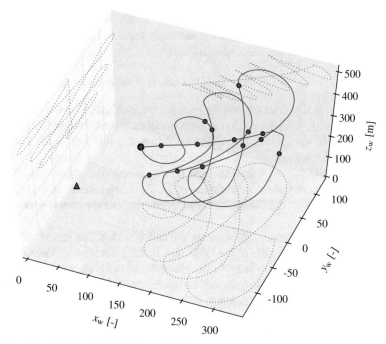

Fig. 15.14 Simulated flight path of the kite at a higher ground wind speed of 9.2 m/s. The kite is reeled out at an average elevation angle of 58° to limit the maximum power to 30 kW. It is reeled in at about the same elevation angle, skipping the lower transition phase

elevation angle. As a result of the slower retraction phase the cycle time increased to 170 s. The corresponding tether force at the ground station and the reel-out speed are shown in Fig. 15.15.

Fig. 15.15 Tether force at the ground and reel-out speed for two pumping cycles at a higher ground wind speed of 9.2 m/s. The variations of the force and the reel-out speed are much higher than at lower wind speeds. The reason for this is, that the kite is flying at a high elevation angle and the effective wind speed is strongly varies during the figure eight maneuvers. Nevertheless the traction force and traction power stay within the allowed limits

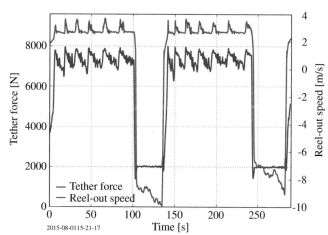

To achieve these results it was necessary to adapt the depower settings during the traction phase, depending on the elevation angle and the mechanical power. The depower settings are calculated as follows

$$u_d = u_{d,ro} + u_{d,add}K_1 - u_{d,sub}K_2, \qquad (15.26)$$

where $u_{d,ro}$ is a parameter which depends on the estimated wind speed and turbulence intensity. A look-up table is used to obtain this parameter. The values in this look-up table were optimized offline for different combinations of wind speed and turbulence intensities.

Empirically derived values are used for the additional depower contribution $u_{d,add}$ and the additional power contribution $u_{d,sub}$. The integer values K_1 and K_2 depend on the mechanical power, the elevation angle, the turbulence intensity and the ground wind speed.

The simulation results are summarized in Table 15.8. The crest factor of the traction force, defined as ratio of the maximum force to the average force in the traction phase, should be close to unity to maximize the power generation of a given system. Further listed are the duty cycle, defined as the ratio of the retraction time to the total cycle time, the pumping efficiency, defined as the ratio of the net mechanical energy to the energy generated in the traction phase, and the cycle efficiency, defined as the ratio of the average mechanical power of the cycle to the average mechanical power during the traction phase. For wind speeds above the nominal wind speed of the simulated system, which is 6 m/s, the traction power increases while the traction power stays constant. Therefore the average power slightly drops.

The computed power curve is shown in Fig. 15.16 demonstrating the advantage of adjusting the average elevation angle during traction compared to just depowering the kite at constant elevation angle.

Table 15.8 Numerical results for operating the kite described in Table 15.7 at a tether length between 300 and 600 m at ground wind speeds of 6 and 9.2 m/s, respectively. For a detailed definition of the listed parameters see [9]

Ground wind speed	6 m/s	9.2 m/s
Average mechanical power [W]	11953.2	10523.0
Lift to drag ratio, traction [-]	4.97	5.01
Lift to drag ratio, retraction [-]	2.01	1.96
Duty cycle [%]	72.2	73.0
Pumping efficiency [%]	80.1	74.6
Cycle efficiency [%]	57.8	54.5
Crest factor traction power [-]	1.30	1.43
Maximum traction power [W]	27738.2	28781.8
Crest factor traction force [-]	1.08	1.11
Maximum traction force [N]	7899.7	7914.3

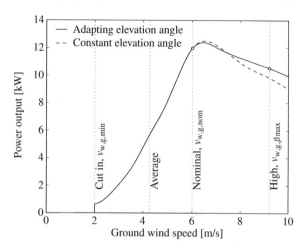

Fig. 15.16 Average mechanical power of a kite power system, using a 20 kW generator with a maximum tether force of 8000 N. Increasing the average elevation angle for larger than nominal wind speeds increases the power output because of the shorter transition phase, but also increases the wear of the depower actuator. See Figs. 15.6 and 15.7 as references for the available wind resource

15.4.2 Quasi-Steady Analysis

Some of the design goals mentioned in Sect. 15.3.1 can be verified on the basis of a quasi-steady modeling framework as described in [9, 23, 25]. Accordingly the analysis presented in the following has the objective to describe the sensitivity of the power output with respect to key problem parameters, such as the average elevation angle $\overline{\beta}_{\text{set}}$ during the traction phase and the exponent α characterizing the wind speed profile.

To extrapolate the ground wind speed $v_{\text{w,g}}$ to the operating altitude of the kite we use the power law given by Eq. (15.10) with an exponent $\alpha = 0.234$ to describe the wind resource at Cabauw, the Netherlands. Projecting the wind velocity vector $\mathbf{v}_{\text{w}}$ at the kite onto the direction vector $\mathbf{e}_{\text{r}}$ pointing from the ground station to the kite leads to the effective wind speed

$$v_{\text{w,e}} = \mathbf{v}_{\text{w}} \cdot \mathbf{e}_{\text{r}} = v_{\text{w}} \cos \beta \cos \phi. \tag{15.27}$$

Combining Eqs. (15.27) and (15.10) and representing the average height as $\overline{z} = l_{\text{t}} \sin \beta$ we can formulate the dimensionless wind speed gain as

$$\mu = \frac{v_{\text{w,e}}}{v_{\text{w,g}}} = \cos \beta \cos \phi \left(\frac{l_{\text{t}} \sin \beta}{z_{\text{ref}}} \right)^{\alpha}. \tag{15.28}$$

Based on the quasi-steady theory of tethered flight the normalized tether force for vanishing mass of the airborne system can be evaluated as [23]

$$\frac{F_{\text{t}}}{q_{\text{g}}S} = C_{\text{R}} \left[1 + \left(\frac{L}{D} \right)^2 \right] (\mu^2 - f_{\text{g}}^2), \tag{15.29}$$

where $q_g = 1/2\rho v_{w,g}^2$ is the dynamic wind pressure at the ground and $f_g = v_{k,r}/v_{w,g}$ is a nondimensional reeling factor. Accordingly, the crest factor of the traction force can be evaluated for flying figure eight maneuvers at a given tether length and average elevation angle as [9]

$$CF_{f,q} = \left(\frac{\mu_{max}}{\mu_{av}} \right)^2. \tag{15.30}$$

The wind speed gain and the crest factor are illustrated in Fig. 15.17 as functions of the average elevation angle $\overline{\beta}_{set}$. At lower elevation angles the wind speed gain can

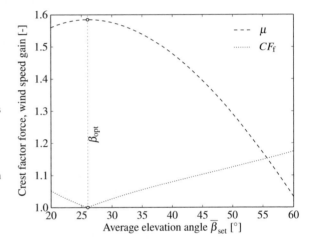

Fig. 15.17 The dimensionless wind speed gain μ and the crest factor CF_f of the tether force as functions of the average elevation angle $\overline{\beta}_{set}$. The tether length is $l_t = 300\,\mathrm{m}$ and the wind shear exponent is $\alpha = 0.234$ to approximate the wind speed profile at Cabauw

reach a value of up to 1.53, while the crest factor does not exceed a value of 1.2. Keeping the crest factor below this limit is a design choice: up to this value electrical machines still work efficiently. We find from Fig. 15.17 that the crest factor reaches its minimum of $CF_f \simeq 1$ at an elevation angle $\overline{\beta}_{set} \approx 26°$.

This minimum is a result of two competing mechanisms. In the lower half of the figure eight maneuver the effective wind speed increases because of a dominating factor $\cos\beta$, while in the upper half it increases because of a dominating effect of the wind speed profile. The elevation angle $\overline{\beta}_{set} = 26°$ is thus the optimal choice for operating at the nominal wind speed $v_{w,g,nom}$ at which the nominal power output is just reached (neglecting the tether drag within the scope of this analysis). Figure 15.18 quantifies how the optimal elevation angle β_{opt} varies as a function of the wind shear exponent α.

As discussed in the context of Fig. 15.16 the tether force in the traction phase reaches its maximum value $F_{t,max}$ for the nominal wind speed $v_{w,g,nom}$. For higher wind speeds the tether force can efficiently be limited to $F_{t,max}$ by increasing the average elevation angle $\overline{\beta}_{set}$ without the need to additionally depower the wing. This planning strategy can be used until the maximum elevation angle β_{max} is reached. For even higher wind speeds the kite must additionally be depowered to keep the tether force during the traction phase below $F_{t,max}$.

To determine this threshold wind speed, which is an operational characteristic of the system, we define the velocity ratio

$$v = \frac{\mu_{av,max}}{\mu_{av,min}}, \qquad (15.31)$$

where $\mu_{av,max}$ is the average wind speed gain at $\overline{\beta}_{set} = \beta_{opt}$ and $\mu_{av,min}$ is the average wind speed gain at $\overline{\beta}_{set} = \beta_{max}$. The ratio v and the optimum elevation angle β_{opt} are depicted in Fig. 15.18 as functions of the wind shear exponent. Given its definition

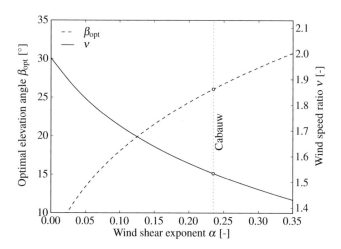

Fig. 15.18 Optimal elevation angle β_{opt} and wind speed ratio v both as functions of the wind shear exponent α. For a uniform wind field ($\alpha = 0$) a value of $v = 1.98$ can be achieved by adjusting the average elevation angle to β_{max}. For the wind speed profile at Cabauw ($\alpha = 0.234$) a value of $v = 1.53$ is achievable

by Eq. (15.31) the wind speed ratio v can be used in two different ways. For a specific fixed ground wind speed $v_{w,g}$ it quantifies the ratio of the effective wind speeds at nominal operation with β_{opt} and at operation with maximum elevation angle β_{max}. Alternatively, v quantifies the ratio of the maximum ground wind speed $v_{w,g,\beta max}$ at which the system can be operated at β_{max}, without additionally depowering the wing during traction, and the nominal ground wind speed $v_{w,g,nom}$ for operation with β_{opt}. Both interpretations follow from the definitions by Eqs. (15.28) and (15.31).

For the wind speed profile in Cabauw we find from Fig. 15.18 a ratio of $v = 1.53$. From this we derive that for a nominal ground wind speed $v_{w,g,nom} = 6$ m/s the system can be operated up to a ground wind speed of $v_{w,g,\beta max} = 9.2$ m/s without increasing the depower settings during traction, assuming a negligible turbulence. According to the measured wind speed distribution displayed in Fig. 15.3 this means that when neglecting turbulence, in total about 96% of the time it is not necessary to change the angle of attack of the wing to limit the maximum power.

15.5 Conclusions and Outlook

This chapter comprises three distinct contributions. Firstly, we investigate the physics of tethered flight in a realistic wind environment with a particular focus the steering and depowering of a flexible membrane wing by a suspended cable robot as well as on the modeling of wind shear and turbulence. Introducing spherical coordinates we separate the radial motion, managed by the winch controller, from the tangential motion, managed by the flight path controller. We derive the kinematic relations describing straight flight along great circle segments as well as turning flight with constant radius, along small circle segments. These path segments are used to compose an entire pumping cycle, consisting of figure eight flight maneuvers during the traction phase and retraction and transition maneuvers to close the cycle. The effect of gravity on the flight dynamics during straight and turning flight is shown qualitatively by photographic data.

As a second contribution a flight path planning scheme for automatic power generation in pumping cycles is presented. The path is described in the plane spanned by the azimuth and elevation angles by a concise set of parameters: the width w_{fig}, height h_{fig} and average elevation angle $\overline{\beta}_{\text{set}}$ of the figure eight flight maneuvers, the elevation angle β_{ri} for starting the depower phase and the minimum attractor point distance $\delta_{\min}$. To reduce the power fluctuations at higher elevation angles the height of the figure eight is decreased linearly with increasing $\overline{\beta}_{\text{set}}$. To compensate the steering delay we propose to use three empiric parameters, $\delta\chi$, $\delta\chi_{\text{int}}$ and $\delta\psi$, which are tuned manually to minimize the error between the planned and the actual flight paths.

The third contribution of the chapter is a performance assessment and sensitivity analysis of the planning scheme. To assess the performance we numerically simulate the pumping operation in a realistic turbulent wind environment. Using a dynamic model of the kite power system in conjunction with the Mann turbulence model we consider nominal operation at a ground wind speed of 6 m/s and operation at a higher wind speed of 9.2 m/s. Within this speed range the generated power can be kept nearly constant by adjusting the average elevation angle $\overline{\beta}_{\text{set}}$ during traction. This force control strategy has proven to be more effective than just depowering the kite in the traction phase. Based on a quasi-steady analysis we demonstrated that increasing $\overline{\beta}_{\text{set}}$ up to a maximum value of $\beta_{\max} = 60°$ significantly reduces the loss of efficiency above the nominal wind speed. It does however require a higher level of activity of the depower actuator.

Although a first important step has been achieved the flight path can be optimized further. The current system model predicts a maximum average power output for a minimum width w_{fig} of the figure eight maneuver. It is clear though that this does not correspond to reality because the steering-induced aerodynamic drag has been neglected in the model. Based on our experiences from test flights a good compromise between maximum power output and high robustness towards sensor errors is achieved by a value of $w_{\text{fig}} = 36°$. Extending the flight path planner to a control scheme that allows retraction of the kite at the side of the wind window is also a future goal. This might improve the power output particularly for deployment scenarios with low-altitude limit.

Acknowledgements The financial support of the European Commission through the projects AWESCO (H2020-ITN-642682) and REACH (H2020-FTIPilot-691173) is gratefully acknowledged.

References

1. Archer, C. L., Jacobson, M. Z.: Evaluation of global wind power. Journal of Geophysical Research: Atmospheres **110**(D12), 1–20 (2005). doi: 10.1029/2004JD005462
2. Baayen, J. H., Ockels, W. J.: Tracking control with adaption of kites. IET Control Theory and Applications **6**(2), 182–191 (2012). doi: 10.1049/iet-cta.2011.0037
3. Bosch, A., Schmehl, R., Tiso, P., Rixen, D.: Dynamic nonlinear aeroelastic model of a kite for power generation. AIAA Journal of Guidance, Control and Dynamics **37**(5), 1426–1436 (2014). doi: 10.2514/1.G000545
4. Breukels, J.: An Engineering Methodology for Kite Design. Ph.D. Thesis, Delft University of Technology, 2011. http://resolver.tudelft.nl/uuid:cdece38a-1f13-47cc-b277-ed64fdda7cdf
5. Cherubini, A., Papini, A., Vertechy, R., Fontana, M.: Airborne Wind Energy Systems: A review of the technologies. Renewable and Sustainable Energy Reviews **51**, 1461–1476 (2015). doi: 10.1016/j.rser.2015.07.053
6. Erhard, M., Strauch, H.: Theory and Experimental Validation of a Simple Comprehensible Model of Tethered Kite Dynamics Used for Controller Design. In: Ahrens, U., Diehl, M., Schmehl, R. (eds.) Airborne Wind Energy, Green Energy and Technology, Chap. 8, pp. 141–165. Springer, Berlin Heidelberg (2013). doi: 10.1007/978-3-642-39965-7_8
7. Erhard, M., Strauch, H.: Control of Towing Kites for Seagoing Vessels. IEEE Transactions on Control Systems Technology **21**(5), 1629–1640 (2013). doi: 10.1109/TCST.2012.2221093
8. Fagiano, L., Zgraggen, A. U., Morari, M., Khammash, M.: Automatic crosswind flight of tethered wings for airborne wind energy:modeling, control design and experimental results. IEEE Transactions on Control System Technology **22**(4), 1433–1447 (2014). doi: 10.1109/TCST.2013.2279592
9. Fechner, U., Schmehl, R.: Model-Based Efficiency Analysis of Wind Power Conversion by a Pumping Kite Power System. In: Ahrens, U., Diehl, M., Schmehl, R. (eds.) Airborne Wind Energy, Green Energy and Technology, Chap. 14, pp. 249–269. Springer, Berlin Heidelberg (2013). doi: 10.1007/978-3-642-39965-7_14
10. Fechner, U., Vlugt, R. van der, Schreuder, E., Schmehl, R.: Dynamic Model of a Pumping Kite Power System. Renewable Energy (2015). doi: 10.1016/j.renene.2015.04.028. arXiv:1406.6218 [cs.SY]
11. Fechner, U.: A Methodology for the Design of Kite-Power Control Systems. Ph.D. Thesis, Delft University of Technology, 2016. doi: 10.4233/uuid:85efaf4c-9dce-4111-bc91-7171b9da4b77
12. Fechner, U., Schmehl, R.: Flight Path Control of Kite Power Systems in a Turbulent Wind Environment. In: Proceedings of the 2016 American Control Conference (ACC), pp. 4083–4088, Boston, MA, USA, 6–8 July 2016. doi: 10.1109/ACC.2016.7525563
13. Fechner, U., Schmehl, R.: Flight Path Planning in a Turbulent Wind Environment. In: Schmehl, R. (ed.). Book of Abstracts of the International Airborne Wind Energy Conference 2015, pp. 56–57, Delft, The Netherlands, 15–16 June 2015. doi: 10.4233/uuid:7df59b79-2c6b-4e30-bd58-8454f493bb09. Presentation video recording available from: https://collegerama.tudelft.nl/Mediasite/Play/0856a922984242c58bd46c84db3d320f1d

14. Gros, S., Zanon, M., Diehl, M.: A relaxation strategy for the optimization of Airborne Wind Energy systems. In: Proceedings of the 2013 European Control Conference (ECC), pp. 1011–1016, Zurich, Switzerland, 17–19 July 2013

15. Horn, G., Gros, S., Diehl, M.: Numerical Trajectory Optimization for Airborne Wind Energy Systems Described by High Fidelity Aircraft Models. In: Ahrens, U., Diehl, M., Schmehl, R. (eds.) Airborne Wind Energy, Green Energy and Technology, Chap. 11, pp. 205–218. Springer, Berlin Heidelberg (2013). doi: 10.1007/978-3-642-39965-7_11

16. International Electrotechnical Commission: Power performance measurements of electricity producing wind turbines, IEC Standard 61400-12-1:2005(E)

17. Jehle, C.: Automatic Flight Control of Tethered Kites for Power Generation. M.Sc.Thesis, Technical University of Munich, Germany, 2012. https://mediatum.ub.tum.de/doc/1185997/1185997.pdf

18. Jehle, C., Schmehl, R.: Applied Tracking Control for Kite Power Systems. AIAA Journal of Guidance, Control, and Dynamics 37(4), 1211–1222 (2014). doi: 10.2514/1.62380

19. Mann, J.: The spatial structure of neutral atmospheric surface-layer turbulence. Journal of Fluid Mechanics 273, 141 (1994). doi: 10.1017/S0022112094001886

20. Mann, J.: Wind field simulation. Probabilistic Engineering Mechanics 13(4), 269–282 (1998). doi: 10.1016/S0266-8920(97)00036-2

21. Ruiterkamp, R., Sieberling, S.: Description and Preliminary Test Results of a Six Degrees of Freedom Rigid Wing Pumping System. In: Ahrens, U., Diehl, M., Schmehl, R. (eds.) Airborne Wind Energy, Green Energy and Technology, Chap. 26, pp. 443–458. Springer, Berlin Heidelberg (2013). doi: 10.1007/978-3-642-39965-7_26

22. Schmehl, R., Vlugt, R. van der, Fechner, U., Wachter, A. de, Ockels, W.: Airborne Wind Energy System. Dutch Patent Application 2,009,528, Mar 2014

23. Schmehl, R., Noom, M., Vlugt, R. van der: Traction Power Generation with Tethered Wings. In: Ahrens, U., Diehl, M., Schmehl, R. (eds.) Airborne Wind Energy, Green Energy and Technology, Chap. 2, pp. 23–45. Springer, Berlin Heidelberg (2013). doi: 10.1007/978-3-642-39965-7_2

24. The Royal Netherlands Meteorological Institute (KNMI): Cabauw Tower Meteorological Profiles. CESAR Database. http://www.cesar-database.nl (2011). Accessed 30 May 2016

25. Vlugt, R. van der, Bley, A., Schmehl, R., Noom, M.: Quasi-Steady Model of a Pumping Kite Power System. Submitted to Renewable Energy (2017). arXiv:1705.04133 [cs.SY]

26. Vlugt, R. van der, Peschel, J., Schmehl, R.: Design and Experimental Characterization of a Pumping Kite Power System. In: Ahrens, U., Diehl, M., Schmehl, R. (eds.) Airborne Wind Energy, Green Energy and Technology, Chap. 23, pp. 403–425. Springer, Berlin Heidelberg (2013). doi: 10.1007/978-3-642-39965-7_23

27. Zgraggen, A. U., Fagiano, L., Morari, M.: Automatic Retraction and Full-Cycle Operation for a Class of Airborne Wind Energy Generators. IEEE Transactions on Control Systems Technology 24(2), 594–608 (2015). doi: 10.1109/TCST.2015.2452230

28. Zgraggen, A. U., Fagiano, L., Morari, M.: Automatic Retraction Phase of Airborne Wind Energy Systems. Proceedings of the 19th IFAC World Congress 47(3), 5826–5831 (2014). doi: 10.3182/20140824-6-ZA-1003.00624

29. Zgraggen, A. U., Fagiano, L., Morari, M.: On real-time optimization of airborne wind energy generators. In: Proceedings of the 52nd IEEE Conference on Decision and Control, pp. 385–390, IEEE, Firenze, Italy, 10–13 Dec 2013. doi: 10.1109/CDC.2013.6759912

30. Zgraggen, A. U., Fagiano, L., Morari, M.: Real-Time Optimization and Adaptation of the Crosswind Flight of Tethered Wings for Airborne Wind Energy. IEEE Transactions on Control Systems Technology 23(2), 434–448 (2015). doi: 10.1109/TCST.2014.2332537

Chapter 16
Design and Economics of a Pumping Kite Wind Park

Pietro Faggiani and Roland Schmehl

Abstract The development of airborne wind energy is steadily progressing towards the market introduction of the technology. Even though the physical foundations of the various conversion concepts are well understood, the actual economic potential of distributed small-scale and centralized large-scale power generation under real-world conditions is still under investigation. In the present chapter we consider the clustering of units into a large kite wind park, specifically the spatial arrangement and collective operation. The analysis starts from a quasi-steady flight model of the kite to estimate the power production in pumping cycle operation. From the surface area and aerodynamic properties of the kite all other system parameters are determined. A genetic algorithm is used to optimize the operation of a single unit and to derive its power curve. Based on this information multiple interconnected units are simulated and an economic model is added. The results show that a coordinated collective operation not only achieves a continuous net electricity output, but also decreases the LCOE from 106 to 96 €/Mwh as consequence of economic scale effects. The prediction supports the substantial economic potential of pumping kite wind parks for large-scale power generation.

16.1 Introduction

A common feature of airborne wind energy (AWE) is the use of tethered flying devices for harvesting the kinetic energy of wind. Replacing the foundation and rigid tower of conventional wind turbines by lightweight tethers and control technology, AWE systems can potentially achieve lower energy costs and access wind at higher altitudes. However, apart from this common feature, the technical details and de-

Pietro Faggiani (✉) · Roland Schmehl
Delft University of Technology, Faculty of Aerospace Engineering, Kluyverweg 1, 2629 HS Delft, The Netherlands
e-mail: p.faggiani@kitepower.nl

© Springer Nature Singapore Pte Ltd. 2018
R. Schmehl (ed.), *Airborne Wind Energy*, Green Energy
and Technology, https://doi.org/10.1007/978-981-10-1947-0_16

signs of the currently pursued conversion concepts can be quite different [5]. In view of the current development activities, the pumping kite power system (PKPS) with either flexible or rigid wings seems to be a clear industry favorite because of its conceptual simplicity and scalability.

The presented study is based on the PKPS concept. The considered implementation is using a leading edge inflatable (LEI) tube kite operated on a single tether and steered by a remote-controlled suspended kite control unit (KCU), as illustrated in Fig. 16.1 and described in more detail in [25]. To maximize the tether tension

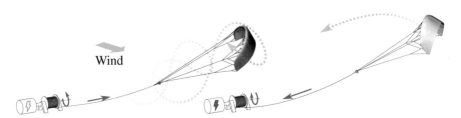

Fig. 16.1 Schematic representation of the pumping cycle: traction phase with crosswind flight maneuvers (left) and retraction phase with de-powered wing (right). Adapted from [25]

during the traction phase the kite is steered in crosswind flight maneuvers while the tether is being reeled from a drum. This rotational motion drives the connected generator. Reaching a predefined altitude, the kite is de-powered to minimize the tether tension. The tether is then reeled back onto the drum consuming a fraction of the energy produced in the previous phase. When reaching the minimum tether length the next traction phase is started. Energy is thus generated in pumping cycles. Because the flight motion of a lightweight tethered wing is dominated by the equilibrium of aerodynamic, tether and gravitational forces a quasi-steady theoretical model can be used to efficiently predict the mechanical power production or consumption of the kite at different wind speeds and in different operational phases [7, 22, 24]. Extending this framework to all components of the kite power system, the global energy conversion efficiency is broken down to the efficiencies of the individual system components [12].

The objective of the present study is to analyze the economic potential of the pumping kite power technology under real-world conditions, considering a wind park configuration. Specific elements of such analysis have already been treated, for example, the economics of single systems [2, 15, 17] or the spacial arrangement in a wind park [14]. However, only few recent studies have quantitatively compared the characteristics of conventional wind turbines and AWE systems and extended this to large-scale park configurations [6, 8]. Starting point of the present analysis is the approach described in [17], which governs the economics of a single PKPS and which we develop into a framework to assess a kite wind park in terms of the achievable levelized cost of energy (*LCOE*). This measure quantifies the cost per unit of produced energy in €/MWh throughout the project lifetime, allowing a consistent comparison with other energy technologies. The *LCOE* is evaluated as the ratio of the discounted costs C_t of the installation, accumulated over the years $t = 1, \ldots, n$ of its lifetime, and the discounted energy E_t produced, equally accumulated over the years

$$LCOE = \frac{Cost}{Energy} = \frac{\sum\limits_{t=1}^{n} \dfrac{C_t}{(1+i)^t}}{\sum\limits_{t=1}^{n} \dfrac{E_t}{(1+i)^t}}, \qquad (16.1)$$

where the parameter i denotes the discount rate. Costs can be divided into operational and maintenance costs OMC, expressed in €/y, and initial capital costs ICC, expressed in €. If the annual energy production AEP, expressed in MWh/y, is constant we can write

$$LCOE = \frac{ICC \times CRF + OMC}{AEP}, \qquad (16.2)$$

using the capital recovery factor CRF, which takes into account the time value of money. This parameter can be computed for the lifetime of the system as

$$CRF = \frac{i(1+i)^n}{(1+i)^n - 1}. \qquad (16.3)$$

Evaluating Eq. (16.1) requires detailed knowledge of the system performance at the specific deployment location and of the associated cost components. Because all commercial development programs are still in a prototype stage, the scale effects of mass production are taken into account by reasonably estimated cost reductions.

The chapter is organized as follows. In Sect. 16.2 a quasi-steady flight model of the kite is developed to derive the power curve of a single PKPS. The approach is based on [22, 24] but several aspects of the analytical framework have been simplified to reduce the computational effort without considerably affecting the result quality. In Sect. 16.3 a genetic algorithm is used to optimize the main operational parameters of the system for maximizing the energy production at every wind speed. In Sect. 16.4 multiple PKPS are used in a wind park configuration, investigating the effects of the spacial arrangement, the modes of operation depending on the wind direction, the control strategy and the electrical interconnection of the units. In Sect. 16.5 a basic cost model is used to determine the LCOE of the wind park configuration. The influence of the initial parameter choices and assumption is investigated by a sensitivity analysis. The preliminary content of the present chapter has been presented at the Airborne Wind Energy Conference 2015 [11] and is described in detail in [10].

16.2 Quasi-Steady Flight Model

The flight motion of a kite operated in pumping cycles can be described, for most of the time, as a quasi-steady transitioning through equilibrium states. This observation can be used to formulate an efficient model to predict the traction power and energy production over a pumping cycle as function of the system design and operational parameters. To account for the different kinematics and force balances in the retraction, transition and traction phases the cycle is generally discretized along the flight trajectory.

16.2.1 Theoretical Framework

The present study is based on the quasi-steady flight model developed in [22] and further detailed, extended to pumping cycle operation and validated experimentally in [21, 24]. Starting point is the Cartesian wind reference frame x_w, y_w, z_w, which is centered at the tether ground attachment point $\mathbf{O}$, has its x_w-axis oriented along the wind velocity vector $\mathbf{v}_w$ and is assumed to be an inertial frame. The kite position $\mathbf{K}$ is described in spherical coordinates (r, θ, ϕ) as illustrated in Fig. 16.2. Assuming that

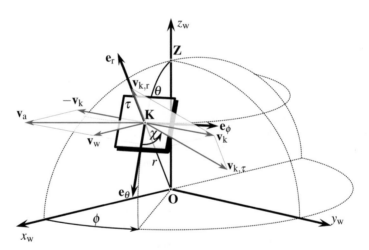

Fig. 16.2 Definition of the apparent wind velocity $\mathbf{v}_a = \mathbf{v}_w - \mathbf{v}_k$. Decomposition of the kite velocity $\mathbf{v}_k$ into radial and tangential components $\mathbf{v}_{k,r}$ and $\mathbf{v}_{k,\tau}$, respectively. The course angle χ is measured in the tangential plane τ, the spherical coordinates (r, θ, ϕ) are defined with respect to the wind reference frame x_w, y_w, z_w. The tether elevation angle is defined as $\beta = 90° - \theta$. Figure and caption from [22]

the tether is straight, the flight motion can then be decomposed into a radial velocity component $\mathbf{v}_{k,r}$, which is controlled by the ground station, and a tangential velocity component $\mathbf{v}_{k,\tau}$. The course angle χ describes the direction of $\mathbf{v}_{k,\tau}$ with respect to the local unit vector $\mathbf{e}_\theta$ and it is controlled by the steering system of the kite. However, the magnitude $v_{k,\tau}$ of the tangential velocity component is a dependent problem variable and not a kinematic degree of freedom [22]. The corresponding non-dimensional velocity components are denoted as reeling factor f and tangential velocity factor λ defined as

$$f = \frac{v_{k,r}}{v_w} \quad \text{and} \quad \lambda = \frac{v_{k,\tau}}{v_w}. \tag{16.4}$$

The mass of the kite, its control unit and part of the tether are taken into account as a lumped mass located at point **K**. Similarly the resultant aerodynamic force generated by the kite and part of the aerodynamic drag acting on the tether are lumped to point **K**. The quasi-steady flight behavior is governed by the equilibrium of the resultant aerodynamic force, gravitational force and tether force. Each pumping cycle is divided into a sequence of traction, retraction and transition phases and the aerodynamic properties of the kite are assumed to be constant for each phase.

To account for the varying kinematics and forces the flight path $\mathbf{r}_k$ is advanced in discrete time steps Δt according to the finite difference scheme

$$\mathbf{r}_k(t + \Delta t) = \mathbf{r}_k(t) + \mathbf{v}_k(t)\Delta t. \tag{16.5}$$

The control strategy for the simulation is based on set values for the tether force F_t which are achieved by adjusting the reeling factor according to [22]

$$f = \sin\theta \cos\phi - \sqrt{\frac{F_t}{qS_k C_R (1 + \kappa^2)}}, \tag{16.6}$$

where q denotes the dynamic wind pressure

$$q = \frac{1}{2}\rho v_w^2, \tag{16.7}$$

the resultant aerodynamic coefficient is evaluated as

$$C_R = \sqrt{C_L^2 + C_D^2}, \tag{16.8}$$

the kinematic ratio is given by

$$\kappa = \frac{v_{a,\tau}}{v_{a,r}}, \tag{16.9}$$

and S_k denotes the projected area of the kite. For vanishing mass of the airborne components, κ is identical to the lift-to-drag ratio C_L/C_D. For real systems this idealization does not hold anymore and Eq. (16.6) has to be solved iteratively [22, 24].

16.2.2 Retraction Phase

The simulation of the pumping cycle starts with the retraction phase because it is only at the start of this phase that the kite position is fully defined by the model settings. The kite is fully de-powered to its minimum lift-to-drag ratio C_L/C_D to consume as little energy as possible for the retraction flight maneuver. Adjusting the course angle to $\chi = 180°$ the kite flies against the wind, with azimuth angle $\phi = 0°$, while the tether elevation angle β continuously increases. The tether length is at its maximum at the start of this phase and at its minimum when the end is reached.

For low reel-in velocity $v_{k,r}$ the kite can reach a steady flight state on a radial trajectory descending towards the ground station. For higher reel-in velocity, as generally used in practice, this steady-state flight condition with $\lambda = 0$ and constant β_{max} is approached asymptotically but not reached before switching to the transition phase. This is clearly visible in Fig. 16.3 which shows a representative computed trajectory. Because of the high reel-in velocity in this particular case the kite overflies the ground station in upwind direction to positions $x < 0$.

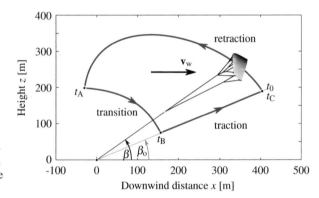

Fig. 16.3 Two-dimensional flight trajectory computed with the quasi-steady model. The radial line segment $\beta_o = $ const. representing the traction phase does not resolve the crosswind flight maneuvers but is computed on the basis of an averaged flight state. The time integration starts at t_0, the transition phase at t_A, the traction phase at t_B and the cycle ends at t_C

16.2.3 Transition Phase

At the end of the retraction phase the elevation angle is much larger than the design value for the traction phase. The purpose of the transition phase is to perform a flight maneuver that brings the kite back to the elevation angle that governs the traction phase. For this maneuver the kite is again fully powered such that it has the aerodynamic properties of the traction phase. The kite flies in downwind direction with course angle $\chi = 0$ until it reaches the target elevation angle for the traction phase.

The control strategy during this phase is not based on the tether force but on the reeling velocity. The aim is here to fly the maneuver at constant tether length, which means that Eq. (16.6) needs to be solved for F_t, setting $f = 0$. However, any implemented AWE system will need to maintain a certain minimum tension in the tether to ensure operational stability. For the present simulation, the minimum tension limit is applied to the entire pumping cycle by means of adjusting the reeling velocity. For example, when the tension drops below the limiting the tether is reeled in such that the tension increases again.

16.2.4 Traction Phase

In the traction phase the kite is flown in crosswind maneuvers to maximize the apparent wind velocity at the wing and correspondingly also the traction force. Because circular flight maneuvers can lead to torsion of the tether and entanglement of the bridle line system it is common to use flight maneuvers that track a horizontal figure of eight. To maximize the traction power the maneuvers are generally centered at $\phi = 0$, as illustrated in Fig. 16.4.

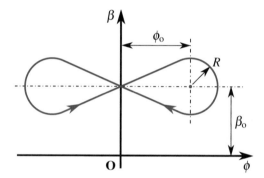

Fig. 16.4 Representative figure of eight flight maneuver in the $\phi\beta$-plane [13]. In Cartesian space this plane represents a spherical surface around the origin

Instead of resolving the actual physical flight path of a crosswind maneuver, the present approach uses a constant average flight state to compute the generated traction force. Accordingly, the varying tangential motion in the $\phi\beta$-space is represented by a constant average angular position (ϕ_o, β_o) and flight velocity (λ_o, χ_o). This approach substantially reduces the computational effort because the crosswind maneuver can be optimized separately and the flight path can be integrated in two dimensions, as illustrated in Fig. 16.3. The values ϕ_o, β_o and χ_o are determined as time averages of the real flight state over a crosswind maneuver. Because the traction power is a function of the product term $\cos\phi\cos\beta$ we define the positional averages by

$$\cos\phi_o = \overline{\cos\phi} \quad \text{and} \quad \cos\beta_o = \overline{\cos\beta}. \tag{16.10}$$

These definitions result in an average angular position (ϕ_o, β_o) coinciding with the center of the lobe of the figure of eight, as illustrated in Fig. 16.4. Because of gravity the kite is flying slower in upward than in downward direction the average course angle χ_o is larger than $90°$, which is the value characterizing horizontal flight. The traction phase is completed when the tether reaches its maximum length at t_C. The position of the kite at this time coincides with its initial position at t_0.

16.2.5 Complete Cycle

With the averaging of the crosswind flight trajectory in the traction phase the pumping cycle can be analyzed in a two-dimensional framework. The side view of a representative computed trajectory is illustrated in Fig. 16.3. The mean mechanical net power is computed as time average of the power provided or required by the system over one cycle

$$P_{\mathrm{m}} = \frac{\overline{P}_{\mathrm{o}}\Delta t_{\mathrm{o}} + \overline{P}_{\mathrm{x}}\Delta t_{\mathrm{x}} + \overline{P}_{\mathrm{i}}\Delta t_{\mathrm{i}}}{\Delta t_{\mathrm{o}} + \Delta t_{\mathrm{x}} + \Delta t_{\mathrm{i}}}, \qquad (16.11)$$

where the subscripts refer to traction (o), transition (x) and retraction phases (i). Equation (16.11) is maximized by an optimization procedure that is discussed in the following section.

16.3 Optimization

The present analysis considers the size of the kite to be a prescribed design parameter which is not varied during the optimization process. All other design parameters are scaled accordingly following system-level engineering practices to minimize losses, while meeting the specific technical requirements and complying with physical and regulatory limitations. Once the design parameters are set, the operational parameters of the system are determined by systematic optimization. In its outermost loop the computational framework steps through the range of expected wind speeds in discrete increments to determine the power curve of the system.

16.3.1 Methodology

To maximize the power production of any wind energy system it is crucial to adjust the operational parameters to the available wind resource. Analyzing the potential of kites for power generation, Loyd [18] found that the tether of a kite flying in crosswind direction should be reeled out with 1/3 of the wind speed to maximize the produced power. Although this idealized theory neglects the effect of gravity on kite and tether as well as the effect of aerodynamic drag on the tether it provides a fundamental understanding of the mechanism of traction power generation and thus represents a first basic guideline for optimization.

More accurate models have been developed subsequently to describe the influence of a broader set of problem parameters and also of gravitational and inertial force contributions that can significantly affect the operation of the kite [1, 19, 22]. However, with increasing mathematical complexity an explicit analytical solution is not possible anymore and as consequence numerical solution techniques are required. The work of Grete [15] is used as reference to choose the most important operational parameters to be optimized. Those are the tether forces during traction and retraction phases, $F_{t,o}$ and $F_{t,i}$, the minimum and maximum tether lengths, $l_{t,min}$ and $l_{t,max}$ as well as the average elevation angle in the traction phase, β_o.

The tether reeling speed is continuously adjusted by the winch control system to meet the constant set values of the tether force for each cycle phase. This radial velocity has a dominating influence on the instantaneous traction power and the system reacts very sensitively to deviations from its optimal value. The optimization of the minimum and maximum tether length is motivated by the observation that the wind power density generally increases with flight altitude while the aerodynamic drag and gravitational forces acting on the tether increase with the deployed length. The competing effect on the traction power leads to an altitude range which maximizes the power production of the kite power system. The sensitivity of the power output to the average elevation angle in the traction phase is rather low. While the power output does not change notably within a range of $\pm 5°$ it does decrease rapidly for values far away from the optimum value.

Because the power output of the pumping cycle is the result of a numerical integration which depends on several operational parameters that are optimization variables, a Monte Carlo genetic algorithm is used. The approach starts from parameter sets that are chosen randomly within specified ranges. In genetic algorithm terminology these sets represent families which together form a generation of the population. Among the families only those performing best in terms of power production are retained for the next generation of the population. An effort is made to restrict the parameter ranges to practically suitable limits in order to reduce the computational effort. To achieve this the ranges are derived from the optimization results obtained for the previous wind speed.

It is important to note that the traction power is subject to several physical constraints. The maximum wing loading and the maximum tether loading both impose a limit on the tether force that can be reached. Additional limiting factors are the reeling speed and the nominal power of the electrical machines on the ground. When neither the tether force nor the reeling speed can be increased anymore to compensate for a large wind speed the kite has to be depowered.

16.3.2 Case study

In this section we present a case study to demonstrate the performance of the modelling and optimization framework. Considering a utility-scale energy system we chose a wing surface area of 100 m^2 for the kite. The derived design parameters of the system as well as the aerodynamic properties of the kite are summarized in Table 16.1. For each discrete wind speed in the considered range, the operational parameters are optimized for maximum power. The result is the power curve of the pumping kite power system. The computation of the curve shown in Fig. 16.5 has taken about 30 minutes on a standard Laptop.

Parameter name	Symbol	Value	Unit
Total wing surface area	A_k	100	m²
Projected wing surface area	S_k	72	m²
Kite mass	M_k	48	kg
Kite control unit mass	M_{kcu}	16	kg
Maximum wing loading		450	N/m²
Tether diameter	d_t	12	mm
Aerodynamic lift coefficient	C_L		
• retraction phase		0.3	
• transition & traction phases		0.8	
Aerodynamic drag coefficient	C_D		
• retraction phase		0.1	
• transition & traction phases		0.2	

Table 16.1 Design parameters and aerodynamic parameters of a representative pumping kite power system for utility-scale energy generation

With increasing wind speed the power output reaches a maximum value and then continuously decreases. This behavior at larger wind speeds is not known from conventional wind turbines. It can be explained by the fact that above a certain wind speed the energy required for the retraction keeps increasing, while the energy produced in the traction phase remains constant or decreases due to physical limitations, such as the maximum tether force, for example.

Together with a the probability distribution of the wind speed at the specific location the annual energy production (AEP) of the system can be determined.

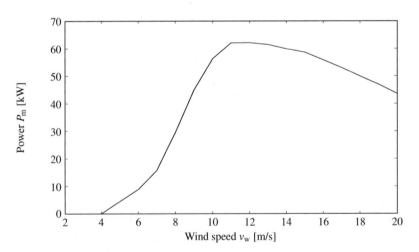

Fig. 16.5 Power curve of a pumping kite power system equipped with a 100 m² kite having aerodynamic properties as listed in Table 16.1

16.4 Wind Park Arrangement

The optimized unit is used as building block for a complete kite wind park. In the following sections we investigate how pumping cycle characteristics affect the collective operation in a park configuration. Specific aspects are the spacial arrangement of the units, the quantity and quality of the generated electricity as well as the economic performance.

16.4.1 Spacing of Units

For conventional wind parks the aerodynamic interaction between turbines strongly depends on the inter-turbine spacing because the energy is harvested from the atmospheric layer close to the ground surface. In contrast to that, a crosswind kite operated in pumping cycles covers a substantially larger airspace and as consequence wake interaction effects are assumed to be negligible. This can be justified by the relatively small wing surface area compared to the swept area of the kite. Moreover, the kites can be flown at different heights and maneuvered in such a way as to avoid perfect alignment with the wind.

In the present study the spacing between units is determined by the requirement of safe collective operation. This requirement has already been applied in previous work on the subject [14, 17]. The flight envelope of each unit is designed in such a way that mechanical interference between the airborne components is avoided. The most restrictive distance constraint is required for units that are aligned with the wind direction, such as illustrated in Fig. 16.6. In this sketch the maximum tether length resulting from the optimization process is denoted as L, the maximum radius of the operational envelope as R, the opening angle of the operational envelope as v and the distance between two units as d_u.

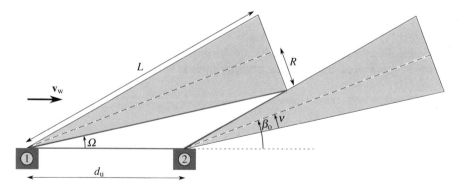

Fig. 16.6 Side view of two units aligned with the wind direction, with the shaded areas defining the operational envelopes

We follow the approach described in [16] to determine the minimum safe distance to avoid collision of airborne components. Starting point is the red triangle highlighted in Fig. 16.6 which can be used to formulate the expression [11]

$$d_u = \frac{L}{\sin(\beta_o - v_1)\left[\frac{1}{\tan(\beta_o - v_1)} + \frac{1}{\tan(v_1 + v_2)}\right]}, \tag{16.12}$$

where indices 1 and 2 refer to the upwind and downwind units, respectively. To minimize Eq. (16.12) the two units have to be operated synchronously, which, for cyclic pumping operation means that they have to be operated in phase. Although smaller deviations from synchronous operation can be covered by application of a safety factor, a robust supervisory control strategy has to be implemented to prohibit larger phase differences. This is of particular importance when the units are aligned with the wind direction.

However, while synchronous operation allows a close spacing of the units it is not favorable from the power production perspective. We will show in the following section that collective operation with different phase shifts has an equalizing effect on the output power which improves the quality of the electricity delivered to the network.

To estimate the maximum radius R of the operational envelope we assume that the kite is operated in figure eight maneuvers, as illustrated in Fig. 16.4. Starting point is the turn rate law [9]

$$\dot{\chi} = g_k v_a \delta, \tag{16.13}$$

which is a mechanistic model describing how the non-dimensional steering input δ and the apparent wind velocity v_a influence the time derivative of the course angle χ. In this equation the maneuverability g_k is regarded as an empirical constant that can be determined experimentally or by high-resolution computational simulation of the flexible ad deforming kite [4]. The turn rate is coupled to the radius R and the tangential flight velocity $v_{k,\tau}$ by the kinematic relation (see Fig. 15.9 in this book)

$$v_{k,\tau} = R\dot{\chi}, \tag{16.14}$$

noting that for crosswind maneuvering $v_{k,\tau}$ can be calculated as shown in [22]. Knowing the maximum radius of the circular trajectory segments and the maximum tether length we can determine the opening angle from

$$v = \arcsin\frac{R}{L} \tag{16.15}$$

and the minimal distance from Eq. (16.12).

For the considered kite size of 100 m^2 we calculate a turning radius of approximately 50 m, which results in a minimal distance of 100 to 150 m between units. Figure 16.7 shows the result of a parametric analysis, investigating the influences of the elevation angle and the maximum radius of the operational envelope.

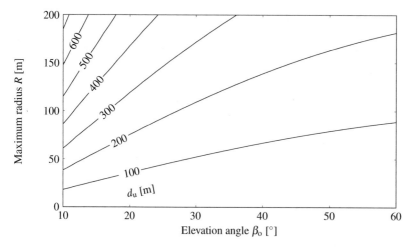

Fig. 16.7 Isolines of the minimum distance d_u between two units as function of the elevation angle β_o during traction and the maximum radius R of the operational envelope. In this specific case a maximum tether length of $L = 1500$ m has been assumed

16.4.2 Quality of Electricity Output

For simplicity we consider a square array layout of the farm. Neither the temporal variability of the wind direction nor the flow interaction between kites is taken into account at this stage. The wind direction is used as a reference to define columns and rows of kite power systems. The units roughly aligned with the wind direction are grouped into columns, while the units roughly aligned in perpendicular direction are grouped into rows. The two extreme inflow scenarios are depicted in Fig. 16.8

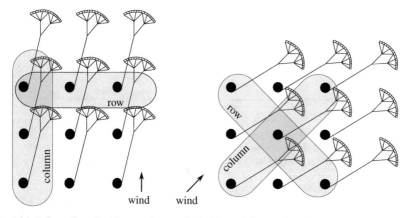

Fig. 16.8 Inflow aligned with array layout of wind farm (left) and diagonal to array layout (right)

indicating that diagonal inflow leads to a maximum asymmetry of the distribution of units into columns and rows.

In the previous section we have shown that units in columns need to be operated synchronously, without significant phase shift, to allow a close spacing. Across the columns, on the other hand, units can be operated safely with phase shifts to internally balance the electricity output of the farm. From these considerations it is clear that the inflow direction plays an important role, affecting the collective operation and production characteristics of the wind farm. In the following we detail the operational strategy on farm level.

For inflow aligned with the array, the phase shift between the columns is calculated as the cycle period divided by the number of columns. For diagonal inflow, the phase shift is calculated as twice this value, which ensures that the outer columns with fewer units are synchronized and in opposite phase to the inner columns with more units. To account for imperfect control a small phase shift is applied between units in the same column. The minimal distance is determined as a function of the maneuverability of the kite and the maximum phase shift of the units in the same column.

The key parameters influencing the power output of a wind farm are the number of units as well as the direction and the magnitude of the wind speed. The instantaneous power output of a single pumping kite power system and two farm configurations of different sizes is illustrated in Fig. 16.9. The simulations show that with the

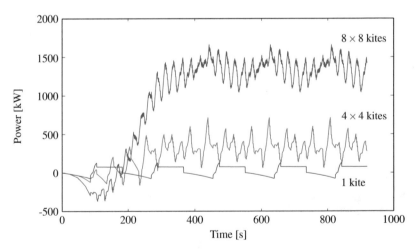

Fig. 16.9 Instantaneous electricity generation of kite wind farms with a wing surface area of 100 m^2 per kite. Inflow diagonal to the array with a wind velocity of 8 m/s

number of contributing units the fluctuation frequency as well as the average output power increase. We further conclude that the farm configuration with the largest number of units can be operated over the broadest wind speed range [10]. Because of the internal load balancing of the units the need for temporary energy storage to retract the kites can be substantially reduced for larger farms.

The generated traction power and the consumed retraction power of the individual units increase strongly with the wind speed and as a result also the fluctuation amplitude of the instantaneous power output of the farm increases. To quantify the quality of the electricity output we use the normalized standard deviation. The simulation results show that by introducing phase shifts, as discussed above, the deviation can be reduced to the minimum for all combinations of wind speed and direction.

Figure 16.10 shows the result of a parametric analysis for array-aligned and diagonal inflow. The standard deviation decreases with increasing wind speed as long as the average cycle power increases. The latter is evident from the power curve of the single unit, illustrated in Fig. 16.5, which peaks at a wind speed of 12 m/s. Above this value the average cycle power decreases because the retraction power further increases while the traction power is limited by the maximum loading constraint. As consequence the standard deviation increases because it is inversely proportional to the average power.

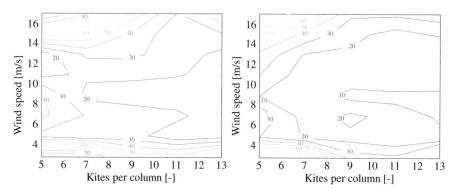

Fig. 16.10 Isolines of the normalized standard deviation for inflow aligned with the array layout of the farm (left) and diagonal to the array layout (right)

With increasing number of kites the standard deviation decreases. The effect is stronger for diagonal inflow because the non-uniform distribution of units to columns allows a better internal balancing of the power contributions.

16.5 Economic Performance

To estimate the levelized cost of energy a simple cost model is added to the simulation framework. The different cost items, their parametric dependencies and the resulting values used for this analysis are listed in Table 16.2. The costs functions are adapted from previous works on wind turbine farms [26] or from specific literature on kite power systems [15, 16]. The specific cost values refer to a wind farm based using kites of $A_k = 100$ m^2 surface area, for which the power curve has been presented in Fig. 16.5. The rated continuous power of the individual units, after balancing internally with an energy storage system or on park level among the units, is $P_{rat} = 60$ kW. The nominal power of the electrical machines of the individual units is $P_{nom} = 100$ kW. We assume a square array layout of the farm with 7×7 individual units and a discount rate of 5% [20].

Hardware			[€/unit]
Electrical machines	C_{em}	$= c_{em}\omega_{nom}^{-0.6}P_{nom}$	15000
Drum	C_{dr}	$= c_{dr,1}M_{dr} + c_{dr,2}d_{dr}$	3200
Power electronics	C_{pe}	$= c_{pe}P_{nom}$	2300
Transformer	C_{tr}	$= (c_{tr,1}P_{rat} + c_{tr,2})e^{c_{tr,3}r_{tr}}$	4200
Tether handling and bearings	C_{thb}	$= c_{thb}F_{t,max}^{0.5}$	9000
Cover frame	C_{cf}	$= c_{cf,1}P_{nom}^{0.85} + c_{cf,2}$	300
Launching and landing	C_{ll}	$= c_{ll}M_kA_k^{0.5}$	4800
Kite	C_k	$= c_kA_k^{0.75}$	22000
Kite Control Unit	C_{kcu}	$= c_{kcu,1} + c_{kcu,2}A_k^{0.5}$	3000
Tether	C_t	$= c_tL\pi d_t^2/4$	9000
Electrical connections	C_{ec}	$= c_{ec}d_u$	23000
Controls	C_{co}	$= c_{co}P_{nom}^{0.2}$	3000
Total	C_{unit}		98800

Operation and Maintenance		[€/unit/y]
Consumables	C_{cons}	17000
O&M	$C_{om} = c_{om,1}AEP + c_{om,2}$	4000
Insurance	$C_{ins} = c_{ins}C_{unit}$	1300
Land lease	$C_{land} = c_{land}AEP$	300

Installation and Decommissioning		[€]
Transport	$C_{mov} = c_{mov}P_{nom}n_u$	196000
Civil works	$C_{cw} = c_{cw}d_un_u$	241000
Cables installation	$C_{ci} = c_{ci}d_u^{0.5}n_u$	6555000
Farm design	$C_{fd} = c_{fd}P_{rat}n_u$	55000
Units removal	$C_{ur} = c_{ur}Mn_u$	241000
Cables removal	$C_{cr} = c_{cr}d_un_u$	6555000

Table 16.2 Cost items taken into account by the model as functions of the total wing surface area A_k, system mass M and rated continuous power output P_{rat} per unit, component masses M_k and M_{dr}, drum diameter d_{dr}, nominal power P_{nom} and rotational speed ω_{nom} of the electrical machines, winding ratio r_{tr} of the transformer, number of units in the farm n_u, distance between the units d_u and their individual annual energy production AEP [10]. The symbols c denote constants. The values assigned are for an array of 7×7 units powered by kites of 100 m^2 surface area

Of particular interest are the scale effects on the costs. The results indicate that the step from single unit to wind farm reduces the cost of energy by 5%. Increasing the number of units, the combined effects of increasing energy production and scale effect on the installation and cable costs of the farm, reduce the investment asymptotically.

The computed LCOE of the kite wind farm is illustrated in Fig. 16.11 as function of the number of units. This prediction is in line with the cost of comparable renewable energy technologies, specifically it is in between the cost of conventional onshore and offshore wind energy. The diagram shows that for a wind farm of 49 units the cost of energy is just below 120 €/MWh.

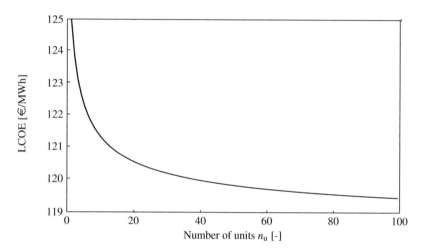

Fig. 16.11 Levelized cost of energy as function of the number of units in the farm based on 100 m^2 wing surface area per unit

The annual electricity production (AEP) of a single unit is computed as 162 MWh/y. When considering operation of the unit at the rated power of 60 kW in a wind environment that can be described by a Weibull probability distribution with parameters $k = 2$ and $A = 12$ a capacity factor of 54% can be achieved. For a wind energy system in the 100 kW range this factor is remarkably high. It is caused by the the low cut-in wind speed which enables the system to produce energy already at very low wind speeds and to access more steadier and stronger winds at higher altitudes.

The areal power density of 6 W/m^2 is comparable to the values of conventional wind turbines farms. This is a remarkable finding considering the much smaller nominal power of the kite power systems. The high power density is the result of a close spacial arrangement, assuming the availability of a suitable control strategy. However, the present analysis has not accounted for possible flow interaction effects between kites, which is left for investigation by follow-up studies.

The sensitivity analysis shows a strong influence of the wing loading and wing surface area on the LCOE. Increasing the wing loading also increases the annual energy production and therefore lowers the LCOE. The maximum wing loading is a design parameter that depends on material properties and the specific design of the kite, including its bridle system, however, these aspects are not in the scope of the present study. As illustrated in Fig. 16.12, increasing the wing size has the same effect until the higher price of larger kites outbalances the gain in terms of energy production. Given the presented cost model the optimum kite size is at around 250 m^2.

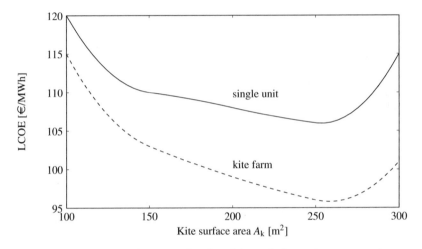

Fig. 16.12 Levelized cost of energy as function of the total wing surface are per unit

16.6 Conclusion

The presented computational approach uses the size of the kite as a starting point to dimension all other functional components of the pumping kite power system. To maximize the harvesting performance the key operational parameters are optimized for the entire range of expected wind speeds. Arranging multiple systems in a wind farm and synchronizing their operation in dependence of the wind direction it is possible to internally balance the collective power generation to create a more uniform electricity output. To assess the economic performance of the wind farm the simulation framework is complemented by a cost model that accounts for the different parametric relations of cost items.

The analysis reveals several scale effects with increasing number of kites, most notably the decreasing cost of energy and the increasing quality of the electrical power. Considering a square array layout of the farm, a minimum cost of 96 €/MWh is achieved for units equipped with kites of 250 m^2 surface area. The corresponding cost for a single kite power power system is 105 €/MWh.

Within the scope of the study it was not possible to cover all options for further optimization. For example, we did not investigate the effect of different kite designs, such as semi-rigid or rigid wings. These are generally heavier and more expensive than flexible membrane wings, but in turn have a better aerodynamic performance, can sustain a higher wing loading and are more durable. Not surprisingly, the analysis showed that the wing loading is the most limiting property of the currently analyzed kite power system. Another component with a considerable optimization potential is the tether. The aerodynamic line drag substantially affects the power production and together with the gravitational effect limits the optimal operating altitude.

The offshore deployment of kite wind parks is a particularly interesting solution for large-scale energy generation. The pumping kite power systems are suitable for mounting on floating platforms because of the low mass and negligible bending moment occurring at the ground station. The application is explored further in Chap. 7 of this book and pursued in current industry projects [3, 23].

Acknowledgements The financial support of the European Commission through the projects AWESCO (H2020-ITN-642682) and REACH (H2020-FTIPilot-691173) is gratefully acknowledged.

References

1. Argatov, I., Rautakorpi, P., Silvennoinen, R.: Estimation of the mechanical energy output of the kite wind generator. Renewable Energy **34**(6), 1525–1532 (2009). doi: 10.1016/j.renene. 2008.11.001
2. Argatov, I., Shafranov, V.: Economic assessment of small-scale kite wind generators. Renewable Energy **89**, 125–134 (2016). doi: 10.1016/j.renene.2015.12.020
3. Bloomberg New Energy Finance: Shell will test energy-generating kites this summer. BNEF Blog, 26 May 2017. https://about.bnef.com/blog/shell-will-test-energy-generating-kites-this-summer/ Accessed 30 May 2017
4. Bosch, A., Schmehl, R., Tiso, P., Rixen, D.: Dynamic nonlinear aeroelastic model of a kite for power generation. AIAA Journal of Guidance, Control and Dynamics **37**(5), 1426–1436 (2014). doi: 10.2514/1.G000545
5. Cherubini, A., Papini, A., Vertechy, R., Fontana, M.: Airborne Wind Energy Systems: A review of the technologies. Renewable and Sustainable Energy Reviews **51**, 1461–1476 (2015). doi: 10.1016/j.rser.2015.07.053
6. Coleman, J., Ahmad, H., Pican, E., Toal, D.: Modelling of a synchronous offshore pumping mode airborne wind energy farm. Energy **71**, 569–578 (2014). doi: 10.1016/j.energy.2014.04. 110
7. Costello, S., Costello, C., François, G., Bonvin, D.: Analysis of the maximum efficiency of kite-power systems. Journal of Renewable and Sustainable Energy **7**(5), 053108 (2015). doi: 10.1063/1.4931111

8. De Lellis, M., Mendonça, A. K., Saraiva, R., Trofino, A., Lezana, Á.: Electric power generation in wind farms with pumping kites: An economical analysis. Renewable Energy **86**, 163–
   172 (2016). doi: 10.1016/j.renene.2015.08.002
9. Erhard, M., Strauch, H.: Theory and Experimental Validation of a Simple Comprehensible
   Model of Tethered Kite Dynamics Used for Controller Design. In: Ahrens, U., Diehl, M.,
   Schmehl, R. (eds.) Airborne Wind Energy, Green Energy and Technology, Chap. 8, pp. 141–
   165. Springer, Berlin Heidelberg (2013). doi: 10.1007/978-3-642-39965-7_8
10. Faggiani, P.: Pumping Kites Wind Farm. M.Sc.Thesis, Delft University of Technology, 2015.
    http://resolver.tudelft.nl/uuid:66cddbd2-5f50-4fc7-be0b-468853128f37
11. Faggiani, P., Schmehl, R. S., Vlugt, R. van der: Pumping Kites Wind Farm. In: Schmehl, R.
    (ed.). Book of Abstracts of the International Airborne Wind Energy Conference 2015, pp. 102–
    103, Delft, The Netherlands, 15–16 June 2015. doi: 10.4233/uuid:7df59b79-2c6b-4e30-bd58-
    8454f493bb09. Poster available from: http://www.awec2015.com/images/posters/AWEC15_
    Faggiani-poster.pdf
12. Fechner, U., Schmehl, R.: Model-Based Efficiency Analysis of Wind Power Conversion by a
    Pumping Kite Power System. In: Ahrens, U., Diehl, M., Schmehl, R. (eds.) Airborne Wind
    Energy, Green Energy and Technology, Chap. 14, pp. 249–269. Springer, Berlin Heidelberg
    (2013). doi: 10.1007/978-3-642-39965-7_14
13. Fechner, U., Schmehl, R.: Flight Path Control of Kite Power Systems in a Turbulent Wind
    Environment. In: Proceedings of the 2016 American Control Conference (ACC), pp. 4083–
    4088, Boston, MA, USA, 6–8 July 2016. doi: 10.1109/ACC.2016.7525563
14. Goldstein, L.: Density of Individual Airborne Wind Energy Systems in AWES Farms. http:
    //www.awelabs.com/wp-content/uploads/AWES_Farm_Density.pdf (2014). Accessed
    19 May 2016
15. Grete, C.: The Economic Potential of Kite Power. Journal of the Society of Aerospace Engi-
    neering Students VSV Leonardo da Vinci October, 10–11 (2014). http://resolver.tudelft.nl/
    uuid:f852545f-2946-4556-9ef8-0b5cdbdaf289
16. Heilmann, J.: Technical and Economic Potential of Airborne Wind Energy. M.Sc.Thesis,
    Utrecht University, 2012. http://dspace.library.uu.nl/handle/1874/258716
17. Heilmann, J., Houle, C.: Economics of Pumping Kite Generators. In: Ahrens, U., Diehl, M.,
    Schmehl, R. (eds.) Airborne Wind Energy, Green Energy and Technology, Chap. 15, pp. 271–
    284. Springer, Berlin Heidelberg (2013). doi: 10.1007/978-3-642-39965-7_15
18. Loyd, M. L.: Crosswind kite power. Journal of Energy **4**(3), 106–111 (1980). doi: 10.2514/3.
    48021
19. Luchsinger, R. H.: Pumping Cycle Kite Power. In: Ahrens, U., Diehl, M., Schmehl, R. (eds.)
    Airborne Wind Energy, Green Energy and Technology, Chap. 3, pp. 47–64. Springer, Berlin
    Heidelberg (2013). doi: 10.1007/978-3-642-39965-7_3
20. OECD/NEA/IEA: Projected Costs of Generating Electricity 2010, OECD Publishing, Paris,
    2010. 218 pp. doi: 10.1787/9789264084315-en
21. Schmehl, R.: Traction Power Generation with Tethered Wings - A Quasi-Steady Model for
    the Prediction of the Power Output. In: Schmehl, R. (ed.). Book of Abstracts of the Inter-
    national Airborne Wind Energy Conference 2015, pp. 38–39, Delft, The Netherlands, 15–
    16 June 2015. doi: 10.4233/uuid:7df59b79-2c6b-4e30-bd58-8454f493bb09. Presen-
    tation video recording available from: https://collegerama.tudelft.nl/Mediasite/Play/
    02a6612b8d004580b08681efd10611351d
22. Schmehl, R., Noom, M., Vlugt, R. van der: Traction Power Generation with Tethered Wings.
    In: Ahrens, U., Diehl, M., Schmehl, R. (eds.) Airborne Wind Energy, Green Energy and Tech-
    nology, Chap. 2, pp. 23–45. Springer, Berlin Heidelberg (2013). doi: 10.1007/978-3-642-
    39965-7_2
23. TKI Wind op Zee: Exploratory Research and LCOE of Airborne Offshore Wind Farm, Project
    Number TEWZ116048. https://topsectorenergie.nl/tki-wind-op-zee/exploratory-research-
    and-lcoe-airborne-offshore-wind-farm (2017). Accessed 1 June 2017
24. Vlugt, R. van der, Bley, A., Schmehl, R., Noom, M.: Quasi-Steady Model of a Pumping Kite
    Power System. Submitted to Renewable Energy (2017). arXiv:1705.04133 [cs.SY]

25. Vlugt, R. van der, Peschel, J., Schmehl, R.: Design and Experimental Characterization of a
    Pumping Kite Power System. In: Ahrens, U., Diehl, M., Schmehl, R. (eds.) Airborne Wind
    Energy, Green Energy and Technology, Chap. 23, pp. 403–425. Springer, Berlin Heidelberg
    (2013). doi: 10.1007/978-3-642-39965-7_23
26. Zaaijer, M.: Great Expectations for Offshore Wind Turbines. Ph.D. Thesis, Delft Univeristy
    of Technology, 2013. doi: 10.4233/uuid:fd689ba2-3c5f-4e7c-9ccd-55ddbf1679bd

Chapter 17
Visual Motion Tracking and Sensor Fusion for Kite Power Systems

Henrik Hesse, Max Polzin, Tony A. Wood and Roy S. Smith

Abstract An estimation approach is presented for kite power systems with ground-based actuation and generation. Line-based estimation of the kite state, including position and heading, limits the achievable cycle efficiency of such airborne wind energy systems due to significant estimation delay and line sag. We propose a filtering scheme to fuse onboard inertial measurements with ground-based line data for ground-based systems in pumping operation. Estimates are computed using an extended Kalman filtering scheme with a sensor-driven kinematic process model which propagates and corrects for inertial sensor biases. We further propose a visual motion tracking approach to extract estimates of the kite position from ground-based video streams. The approach combines accurate object detection with fast motion tracking to ensure long-term object tracking in real time. We present experimental results of the visual motion tracking and inertial sensor fusion on a ground-based kite power system in pumping operation and compare both methods to an existing estimation scheme based on line measurements.

17.1 Introduction

In this work we consider ground-based kite power systems as the ones developed in Switzerland within the Autonomous Airborne Wind Energy (A^{2}WE) project [4]. Ground-based airborne wind energy (AWE) systems feature ground-based steering of tethered wings through differential line lengths [5]. Since electrical power is also generated at the ground following a pumping cycle approach, most weight of the AWE system is contained to the ground. This approach allows the ground station (GS) to be constructed using mostly off-the-shelf components, reducing the cost and risk of development, and increasing reliability.

Henrik Hesse (✉) · Max Polzin · Tony A. Wood · Roy S. Smith
ETH Zurich, Automatic Control Laboratory, Physikstrasse 3, 8092 Zurich, Switzerland
e-mail: henrik.hesse@glasgow.ac.uk

© Springer Nature Singapore Pte Ltd. 2018
R. Schmehl (ed.), *Airborne Wind Energy*, Green Energy
and Technology, https://doi.org/10.1007/978-981-10-1947-0_17

Automatic control approaches for AWE rely on the availability of estimated parameters describing the state of the kite, typically given by kite position and heading, for feedback control. Existing estimation schemes for ground-based generators, for example [5, 14], compute estimates of the kite state using measurements of the line length and line angles obtained at the GS. Estimators based on ground-based position measurements are effective if the kite and winch system can ensure sufficient line tension. However, when operating a two-phase generation cycle, we desire low line forces during the retraction phase to improve cycle efficiency. Additionally, when operating at long line lengths to reach higher altitudes, the aerodynamic forces generated by the kite may not sufficiently balance the weight and drag induced by the lines even during the traction phase. Both scenarios lead to situations where degradation of the estimate quality reduces controller performance and imposes a critical limitation on the usable cycle efficiency of ground-based kite power systems.

Visible effects of the decline in estimator performance are line sag and increased delay in the estimation of the kite heading. The latter has been shown to significantly affect the performance of tracking controllers [1]. To actively incorporate system delay in the control design, in [1, 32] the kite steering behavior is modeled as a delayed dynamical system. Identification of the involved model parameters is used in [32] in a predictive manner to account for the delay in the path generation and tracking steps.

In this work we focus instead on sensor fusion to reduce the estimation and overall system delay. Different sensor setups including ground-based line and onboard inertial and position measurements have been explored in [14] using a kinematic model for sensor fusion. Experimental demonstration on a ground-based kite system with relatively short and fixed line lengths showed the benefits of fusion of inertial sensors with line angle measurements. They further found that onboard position measurements from GPS are not usable in kite power applications due to the large accelerations and fast changes in direction inherent to such systems. A similar finding was suggested in [10] and an inertial navigation algorithm based on acceleration and gyroscope measurements is proposed for a fixed line-length system.

From the lessons learned in [10, 13] we develop a filtering scheme to fuse onboard inertial measurements with line data for a ground-based kite power system with pumping operation. In particular, we formulate an extended Kalman filter (EKF) based on a sensor-driven kinematic model to fuse onboard yaw-rate measurements with *delayed* position measurements to correct for the bias in low-cost inertial sensors. Limiting ourselves to one onboard measurement further ensures that we can establish a reliable downlink to the GS. The developed estimation scheme has been implemented and demonstrated in closed-loop pumping operation on the AWE system developed at Fachhochschule Nordwestschweiz (FHNW). As an alternative to noisy line measurement, we can also obtain position measurements from range sensors using ultra-wideband radios. In [22] we have developed a range-inertial estimator specifically for AWE applications based on range measurements between a transceiver fixed to the kite and a number of static range beacons scattered on the ground. Only approximate knowledge of the range beacon locations is required.

Experimental validation of any estimation algorithm however requires true knowledge of the kite dynamics. Video streams of moving kites can provide useful position information which is unaffected by tether dynamics, can be acquired in real time, and requires no transmission of data from kite to GS. Hence, in this work we further develop a visual motion tracking (VMT) approach which produces improved position estimates from ground-based video streams and can be used in postprocessing to evaluate other estimation approaches. The developed VMT approach is demonstrated in this work for tethered wings and conventional soft kites.

The rest of this chapter is structured as follows. In Sect. 17.2 we describe the kite system and derive the model equation required for the sensor fusion scheme in Sect. 17.4. The underlying sensor configuration is defined in Sect. 17.3. In Sect. 17.5 we described the developed VMT approach. In Sect. 17.6 we present experimental results from the VMT and estimators implemented on a ground-based kite system. The preliminary content of the present chapter has been presented at the Airborne Wind Energy Conference 2015 [24].

17.2 System Description and Dynamics

In this work we consider a two-line AWE system with ground-based actuation and generation. The system has been developed at the FHNW as part of the A^2WE project [4] to focus on autonomous operation of kite power generators. The kite is actuated at the GS which contains a drum and motor for each line. The steering of the kite is achieved through differential line length.

This system can operate with variable tether lengths such that full power cycles can be flown using conventional soft kites or tethered wings. In such two-phase operation power is generated in the so-called *traction phase* where the kite is flown in the power zone on a figure-eight trajectory under high aerodynamic forces in crosswind flight. Winching of the lines enables power generation during this phase and controls the forces exerted on the kite system. Once a maximum line length has been reached, the kite is guided to the side of the wind window where it can be stabilized and rewound under low aerodynamic forces. This second phase where the lines are reeled in at low tether force is referred to as *retraction phase*. When a minimum line length is reached the cycle is repeated leading to a net power generation.

17.2.1 Model Equations

We describe the motion of the kite in terms of three right-handed reference frames, as defined in Fig. 17.1, where we follow the definitions as in [14] but denote unit vectors by $\mathbf{e}$ with subscripts indicating the axis. The inertial frame $\{G\} := (\mathbf{e}_{Gx}, \mathbf{e}_{Gy}, \mathbf{e}_{Gz})$ with its origin at the GS is defined such that $\mathbf{e}_{Gx}$ is parallel to the ground and the wind direction, $\mathbf{e}_{Gz}$ is pointing upwards, and $\mathbf{e}_{Gy}$ completes the right hand system. For a given line length r the kite position $\mathbf{p}^G$ is expressed in the inertial frame $\{G\}$ in terms of the spherical coordinates $\phi \in [-\pi, \pi]$ and $\theta \in [0, \pi]$, as shown in Fig. 17.1a, such that

$$\mathbf{p}^G = r \begin{bmatrix} \cos(\phi)\cos(\theta) \\ \sin(\phi)\cos(\theta) \\ \sin(\theta) \end{bmatrix}, \qquad (17.1)$$

where we denote vectors as bold variables with superscripts indicating the reference system for projection of the vectors. We also define the kite position vector in spherical coordinates as

$$\mathbf{p}_{\theta\phi r} = \begin{bmatrix} \theta & \phi & r \end{bmatrix}^\top. \qquad (17.2)$$

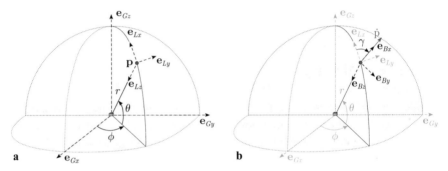

Fig. 17.1 Definition of reference frames: inertial frame $\{G\} = (\mathbf{e}_{Gx}, \mathbf{e}_{Gy}, \mathbf{e}_{Gz})$, local frame $\{L\} = (\mathbf{e}_{Lx}, \mathbf{e}_{Ly}, \mathbf{e}_{Lz})$, and body frame $\{B\} = (\mathbf{e}_{Bx}, \mathbf{e}_{By}, \mathbf{e}_{Bz})$. The wind direction is aligned with $\mathbf{e}_{Gx}$. **a** Kite position $\mathbf{p}$ in spherical components with azimuth angle ϕ, elevation angle θ, and line length r. **b** Definition of velocity vector orientation γ with velocity vector $\dot{\mathbf{p}}$ aligned with $\mathbf{e}_{Bx}$

We can further define a *local* north-east-down (NED) coordinate system $\{L\} := (\mathbf{e}_{Lx}, \mathbf{e}_{Ly}, \mathbf{e}_{Lz})$ with its origin at the kite position $\mathbf{p}$. As shown in Fig. 17.1a, $\mathbf{e}_{Lx}$ and $\mathbf{e}_{Ly}$ define the local tangent plane on a sphere with radius r with $\mathbf{e}_{Lz}$ pointing down from the kite towards the GS. A vector given in the inertial frame $\{G\}$ can be expressed in terms of the local frame $\{L\}$ using the transformation matrix C^{LG} as [14],

$$C^{LG} = \begin{bmatrix} -\cos(\phi)\sin(\theta) & -\sin(\phi)\sin(\theta) & \cos(\theta) \\ -\sin(\phi) & \cos(\phi) & 0 \\ -\cos(\phi)\cos(\theta) & -\sin(\phi)\cos(\theta) & -\sin(\theta) \end{bmatrix}, \qquad (17.3)$$

such that for example $\mathbf{p}^L = C^{LG}\mathbf{p}^G$. The inverse transformation from frame $\{L\}$ to frame $\{G\}$ is given as $C^{GL} = \left(C^{LG}\right)^\top$. Note that the kite position, $\mathbf{p}^G$, defined in Eq. (17.1) can be extracted from C^{GL} as $\mathbf{p}^G = -rC^{GL}\mathbf{e}_{Lz}$.

At last, we define the *body* frame $\{B\} := (\mathbf{e}_{Bx}, \mathbf{e}_{By}, \mathbf{e}_{Bz})$, which is non-inertial, centered at the kite position, and fixed to the kite body, as shown in Fig. 17.1b. We will use frame $\{B\}$ in this work to describe the orientation of the kite and hence assume that the axis $\mathbf{e}_{Bx}$ is always aligned with the kite velocity vector $\dot{\mathbf{p}}$. For the derivation of the filtering equations, we further assume a rigid tether such that axis $\mathbf{e}_{Bz}$ coincides with $\mathbf{e}_{Lz}$, and $\mathbf{e}_{By}$ completes a right handed coordinate system. Hence, the transformation from the body frame $\{B\}$ to the local frame $\{L\}$ is given as

$$C^{LB} = \begin{bmatrix} \cos(\gamma) & -\sin(\gamma) & 0 \\ \sin(\gamma) & \cos(\gamma) & 0 \\ 0 & 0 & 1 \end{bmatrix}, \tag{17.4}$$

where $\gamma \in [-\pi, \pi]$ is the *velocity vector orientation* introduced in [13, 14] as velocity angle. The notion of the velocity vector orientation has been demonstrated in recent publications as a crucial feedback variable to achieve successful autonomous flight during the traction phase [1, 10, 13] but also retraction phase [35]. Assuming small reeling speed, i.e. $\dot{r} \ll \|\dot{\mathbf{p}}\|$, the velocity vector orientation is defined as

$$\gamma := \arctan_2\left(v_\phi, v_\theta\right) = \arctan_2\left(\cos(\theta)\,\dot{\phi}, \dot{\theta}\right). \tag{17.5}$$

where $\arctan_2\left(v_\phi, v_\theta\right) \in [-\pi, \pi]$ is the 4-quadrant arc tangent function and the kite velocity vector expressed in the $\{L\}$ frame is defined as

$$\mathbf{v}^L = \begin{bmatrix} v_\theta \\ v_\phi \\ v_r \end{bmatrix} = \begin{bmatrix} r\dot{\theta} \\ r\cos(\theta)\dot{\phi} \\ -\dot{r} \end{bmatrix}. \tag{17.6}$$

The velocity vector orientation can therefore be interpreted as the angle between the local north, $\mathbf{e}_{Lx}$, and the projection of the kite velocity vector onto the tangent plane of the wind window at the kite position. The velocity vector orientation is particularly suitable as feedback variable as it can be used to deduce the heading of the kite in a single scalar, e.g. for $\gamma = 0$ the kite moves upwards and for $\gamma = \pi/2$ parallel to the ground towards the left (as seen from the GS). More details on the derivation of the velocity vector orientation can be found in [13]. During retraction phases, v_θ can converge to zero and we use a regularized version of the velocity vector orientation, defined as [35]

$$\gamma_{\text{reg}} := \arctan_2\left(\cos(\theta)\dot{\phi} + c\sin(\phi - \phi_W), \dot{\theta} + c\sin\theta\cos(\phi - \phi_W)\right), \tag{17.7}$$

during retraction phases. A tuning parameter $c = 0.02$ was used in this work and ϕ_W can account for misalignment of the wind direction in the $(\mathbf{e}_{Gx}, \mathbf{e}_{Gy})$ plane.

Based on the definition of the kite velocity vector orientation, Eq. (17.5), we can derive the kinematic model equations which will be used to propagate the estimator states in Sect. 17.4.2. Similar to [10, 32] we describe the behavior of the kite as a unicycle on the (θ, ϕ)-plane with heading γ and tangent velocity $v_{\theta\phi} := \sqrt{v_\theta^2 + v_\phi^2}$ as

$$\dot{\theta} = \frac{v_{\theta\phi}}{r}\cos(\gamma), \tag{17.8a}$$

$$\dot{\phi} = \frac{v_{\theta\phi}}{r\cos(\theta)}\sin(\gamma), \tag{17.8b}$$

which we can extend with the tether reeling kinematics,

$$\dot{r} = v_r, \tag{17.8c}$$

to complete the model equations in this work.

17.3 Sensor Configuration and Modeling

The two-line, ground-based AWE system used in this work provides a range of sensors installed on the kite (inertial measurement unit) and the GS (line angles, line lengths, line forces, wind speed, wind direction, and video footage) as shown in Fig. 17.2. A wind sensor, mounted roughly 5 m above the ground, provides measurements of the wind direction, $\tilde{\phi}_W$, in the $(\mathbf{e}_{Gx}, \mathbf{e}_{Gy})$ plane and wind speed, $\tilde{v}_W$, where $(\tilde{\bullet})$ denotes the noise-corrupted, unfiltered measurement of a variable for the remainder of this work. The different sensors, which are relevant to the sensor fusion and VMT approaches in Sects. 17.4 and 17.5, will be presented next together with the corresponding sensor modeling approach. The important sensor parameters are summarized in Table 17.1.

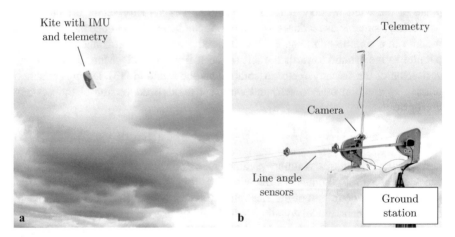

Fig. 17.2 Sensor configuration of the kite power system developed at FHNW. **a** Kite system. **b** Ground station

Sensor	Variable	Description	Properties
Line angle sensor at GS	$\tilde{\theta}$	Elevation angle	$\sigma_\theta^2 = 8 \times 10^{-2}$
	$\tilde{\phi}$	Azimuth angle	$\sigma_\phi^2 = 8 \times 10^{-2}$
Ground station motor	$\tilde{r}$	Line length	$\sigma_r^2 = 10^{-3}$
	$\tilde{v}_r$	Reeling speed	$\sigma_{v_r}^2 = 10^{-3}$
Inertial measurement unit	$\tilde{\omega}$	Yaw rate of kite	$\sigma_\omega^2 = 10^{-4}$

Table 17.1 Overview of sensors with the corresponding variance $\sigma_{(\bullet)}^2$ of the sensor noise.

17.3.1 Line Sensors

The GS is equipped with line angle sensors on both lines, as shown in Fig. 17.2b, with two dedicated encoders on each lead-out sheave to measure azimuth and elevation angles, $\tilde{\theta}$ and $\tilde{\phi}$, respectively. In this work we only consider the measurements from the left line although weighting of left and right line angle measurements with the corresponding line force measurements can lead to improved estimates by favoring readings from the line with higher tension. Combined with line length measurements, $\tilde{r}$, obtained from the motors inside the GS, we model the line angle measurements to provide unbiased measurements of the kite position given in terms of spherical coordinates as

$$\tilde{\mathbf{p}}_{\theta\phi r} = \begin{bmatrix} \tilde{\theta} \\ \tilde{\phi} \\ \tilde{r} \end{bmatrix} = \begin{bmatrix} \theta \\ \phi \\ r \end{bmatrix} + \begin{bmatrix} \zeta_\theta \\ \zeta_\phi \\ \zeta_r \end{bmatrix} = \mathbf{p}_{\theta\phi r} + \zeta_{\theta\phi r}, \tag{17.9}$$

where $\zeta_{\{\theta,\phi,r\}}$ are the measurement noises which we assume to be independent, zero-mean Gaussian processes, i.e. $\zeta_{\{\theta,\phi,r\}} \sim \mathcal{N}(0, \sigma_{\{\theta,\phi,r\}}^2)$. We have determined the variances $\sigma_{\{\theta,\phi,r\}}^2$ from experiments as stated in Table 17.1.

The assumption of unbiased line angle measurements in Eq. (17.9) is clearly questionable for ground-based AWE systems. Especially at long lines we can observe that tether dynamics significantly deteriorate the quality of ground-based measurements introducing bias and measurement delay [1].

17.3.2 Inertial Measurements

To improve the estimation of the kite heading we exploit additional onboard measurements from an inertial measurement unit (IMU) installed on the kite as depicted in Fig. 17.2a. We use the Pixhawk Autopilot which is equipped with a Cortex M4 processor and several redundant sensors including a 3-axis accelerometer, 3-axis gyroscope, magnetometer, and barometer [23]. In this work we will focus on fusing the gyroscope measurements only as they are particularly relevant in the estimation of the kite velocity vector orientation [11].

The 3-axis gyroscope installed in the Pixhawk Autopilot has been set to operate at a bandwidth of 100 Hz providing measurements in the range of $\pm 2.75\pi$ rad/s. In the temperature range of operation we can expect a bias of $\pm 3 \times 10^{-3}$ rad/s and a noise density of $\pm 5 \times 10^{-4}$ rad/(s$\sqrt{\text{Hz}}$). The yaw-rate measurement is streamed via telemetry to the GS, where it is synchronized with the line measurements. Since we only require gyroscope measurements for the developed estimator, a small Baud rate of 57,600 Hz could be selected to establish a reliable connection over 250 m at 100 Hz bandwidth.

Although we can calibrate the gyroscope to remove a static bias, it is common that low-cost inertial sensors tend to drift due to external factors. We will therefore model the gyroscope following a common approach in the literature, e.g. [20], which relates the true angular velocities ω^B to the gyroscope measurements, $\tilde{\omega}$, through

$$\tilde{\omega} = \omega^B + \mathbf{b}_\omega + \eta_\omega, \tag{17.10}$$

where ω^B indicates the angular velocity of the $\{B\}$ frame relative to the $\{G\}$ frame projected in the $\{B\}$ frame. The inertial measurements are corrupted by zero-mean Gaussian noise captured by the vector $\eta_\omega \sim \mathcal{N}(0, \sigma_\omega^2 \, \mathrm{I}_{3\times 3})$. The non-static gyroscope bias is modeled as a continuous-time Gaussian random process,

$$\dot{\mathbf{b}}_\omega = \eta_b, \tag{17.11}$$

with $\eta_b \sim \mathcal{N}(0, \sigma_b^2 \, \mathrm{I}_{3\times 3})$ with $\sigma_b^2 = 10^{-3}$ throughout this work.

Next, we relate the gyroscope measurements in Eq. (17.10) to our model equations in Sect. 17.2 to arrive at an expression for the velocity vector orientation, γ, defined in Eq. (17.5). For this we first link the rate of change of the orientation of the kite,

$$\dot{C}^{GB} = \frac{d}{dt}\left(C^{GL}C^{LB}\right) = \dot{C}^{GL}C^{LB} + C^{GL}\dot{C}^{LB}, \tag{17.12}$$

to the kite angular velocity, ω^B, as

$$\omega^B \times = C^{BG}\dot{C}^{GB} = C^{BG}\dot{C}^{GL}C^{LB} + C^{BL}\dot{C}^{LB}, \tag{17.13}$$

where we denote the skew-symmetric matrix of ω^B as $[\omega^B \times]$ with $[\omega^B \times]\mathbf{b} := \omega^B \times \mathbf{b}$ for $\mathbf{b} \in \mathbb{R}^3$. Hence, a gyroscope fixed to the kite will measure the angular velocities due to the motion on the sphere, $\dot{C}^{GL}$, and a change in heading, $\dot{C}^{LB}$. With

$$C^{BL}\dot{C}^{LB} = \begin{bmatrix} 0 & -\dot{\gamma} & 0 \\ \dot{\gamma} & 0 & 0 \\ 0 & 0 & 0 \end{bmatrix}, \tag{17.14}$$

we can directly relate the time-derivative of the velocity vector orientation, $\dot{\gamma}$, to the measured gyroscope outputs in Eq. (17.10) as

$$\dot{\gamma} = \left(\omega^B\right)^\top \mathbf{e}_{Bz} + \dot{\phi}\sin(\theta) = \omega_{Bz} + \frac{v_{\theta\phi}}{r}\tan(\theta)\sin(\gamma), \qquad (17.15)$$

where $\omega_{Bz} = \left(\omega^B\right)^\top \mathbf{e}_{Bz}$ is the measurable yaw rate of the kite. The effect of the additional term, $\dot{\phi}\sin(\theta)$, due to the motion on the sphere is illustrated (for a different definition of reference frames) in [11].

17.3.3 Visual Measurements

The GS is equipped with a GoPro HD Hero 2 video camera. The camera has been set to capture video streams with a resolution of 1280×960 pixel at 48 frames per second. Although the camera is equipped with a fisheye lens, extending the field of view to nearly π rad in a static setup, the field of view is not sufficient to capture the kite at all times. Especially during retractions the wind window is increased due to the reeling of the tethers. We therefore attached the camera to the lead-out sheave of the right line at the GS, as shown in Fig. 17.2b. The lead-out sheave (and hence the camera) rotates with the azimuth angle, $\tilde{\phi}$, but is fixed in elevation.

The built-in fisheye lens introduces nonlinear distortions to the camera image. To successfully link the true kite position $\mathbf{p}^G$ in the inertial frame $\{G\}$ with a visual position measurement $\tilde{\mathbf{p}}^V$ from the video stream in the camera frame, $\{V\}$, we need to compensate for distortions (intrinsic calibration) and estimate the camera pose (extrinsic calibration). To undistort the video stream from the fisheye effect and calculate the intrinsic camera parameters, we used the calibration procedure from the Computer Vision System Toolbox^{TM} in Matlab®based on a pinhole camera with distortion coefficients for radial and tangential correction [18].

From the undistorted video stream, we extract the camera pose relative to the inertial frame given by the transformation matrix C^{GV} (extrinsic camera parameters). The orientation of the camera, C^{GV}, is computed using linear regression over the measured line angles and visually tracked positions at high line forces where we trust the line angle data. In each experiment, the calculated camera parameters do not change over time, except for an azimuth rotation of the lead-out sheaves which is measured at the GS and added to the tracked position estimate. Depth information is lost in position measurements extracted from a single video stream. However, since the kite is assumed to move on the sphere, we are able to recover position measurements $\tilde{\mathbf{p}}^V$ from the camera stream by taking the line length measurement, $\tilde{r}$, as depth information.

17.4 Filtering Schemes

In this section we present the filtering algorithms to obtain estimates of the feedback variables $\hat{\mathbf{p}}^G$ and $\hat{\gamma}$ where $(\hat{\bullet})$ will denote estimated variables in this work. The first approach presented in Sect. 17.4.1 uses only line measurements at the GS and provides a good starting point for simple kite power systems. It will also serve as a baseline estimator in this work. As line dynamics and filtering of noisy position measurements can deteriorate the control performance, we introduce a sensor fusion algorithm based on onboard inertial measurements in Sect. 17.4.2.

17.4.1 Line-Angle-Based Estimation

Ground-based AWE systems typically provide basic measurements of azimuth/elevation angles and line length, $(\tilde{\theta}, \tilde{\phi}, \tilde{r})$, as described in Sect. 17.3.1. We therefore first present a line-based estimator following the approach in [14] based on an orientation-free kinematic process model. In this work, however, we describe the filter states directly in terms of the measurable spherical coordinates leading to the 6-dimensional state vector $\mathbf{x}_{la} = [\mathbf{p}_{\theta\phi r}^\top \ \dot{\mathbf{p}}_{\theta\phi r}^\top]^\top$ with the vector of spherical position coordinates, $\mathbf{p}_{\theta\phi r}$, defined in Eq. (17.2). The discrete line angle and length measurements, $\mathbf{z}_{la} = \tilde{\mathbf{p}}_{\theta\phi r}^\top$ are modeled as defined in Eq. (17.9). Under the assumption of decoupling the kite motion in azimuth and elevation directions, we can obtained the following simplified process model given as

$$\mathbf{x}_{la}^k = \begin{bmatrix} I_{3\times3} & T_s I_{3\times3} \\ 0_{3\times3} & I_{3\times3} \end{bmatrix} \mathbf{x}_{la}^{k-1} + \eta_{la}^{k-1} \tag{17.16a}$$

$$\mathbf{z}_{la}^k = \begin{bmatrix} I_{3\times3} & 0_{3\times3} \end{bmatrix} \mathbf{x}_{la}^k + \zeta_{la}^k, \tag{17.16b}$$

where we have discretized the process dynamics with the forward Euler method using a constant sampling rate, T_s. In Eq. (17.16) we denote identity and zeros matrices as I and 0, respectively, with subscripted dimensions. The process and measurement noises, η_{la} and ζ_{la}, respectively, are modeled as independent zero-mean Gaussian noise vectors. Writing the filter equations directly in the measurable spherical coordinates ensures that ζ_{la}, as defined in Eq. (17.9), captures the line angle and length sensor noises with the variances defined in Table 17.1.

We can then derive a steady-state *Kalman filter* [2] based on Eq. (17.16) to extract state estimates $\hat{\mathbf{x}}_{la}$. Since the underlying linear process model Eq. (17.16) is autonomous, the Kalman filter equations reduce to,

$$\hat{\mathbf{x}}_{la}^k = \begin{bmatrix} I_{3\times3} & T_s I_{3\times3} \\ 0_{3\times3} & I_{3\times3} \end{bmatrix} \hat{\mathbf{x}}_{la}^{k-1} + K_{la} \left(\tilde{\mathbf{p}}_{\theta\phi r}^k - \hat{\mathbf{p}}_{\theta\phi r}^{k-1} - T_s \hat{\dot{\mathbf{p}}}_{\theta\phi r}^{k-1} \right), \tag{17.17}$$

where we have combined the prediction and measurement update steps in the steady-state Kalman filter [2]. The steady-state Kalman gain K_{la} can be precomputed using the variances of the measurement noise, ζ_{la} defined in Table 17.1. The estimated velocity vector orientation $\hat{\gamma}_{la}^k$, as defined in Eq. (17.5), is finally extracted at each time step from the estimated velocity vector, $\hat{\dot{\mathbf{p}}}_{\theta\phi r}^k$.

17.4.2 Sensor Fusion

Since position-based estimation of the velocity vector orientation, as presented in 17.4.1, involves differentiation of noisy position measurements, we require excessive filtering which introduces estimation delay in addition to the existing system delay. Hence, in this section we derive an estimator that uses onboard yaw-rate measurements, $\tilde{\omega}_{Bz}$, as modeled in Sect. 17.3.2.

Coupling Eq. (17.15) with the unicycle model in Eq. (17.8), we can derive a sensor-driven kinematic process model given in discrete time as

$$
\begin{bmatrix} \gamma \\ v_{\theta\phi} \\ \theta \\ \phi \\ r \\ b_{\omega z} \end{bmatrix}^k = \begin{bmatrix} \gamma \\ v_{\theta\phi} \\ \theta \\ \phi \\ r \\ b_{\omega z} \end{bmatrix}^{k-1} + T_s \begin{bmatrix} v_{\theta\phi}r^{-1}\tan(\theta)\sin(\gamma) + \tilde{\omega}_{Bz} - b_{\omega z} \\ 0 \\ v_{\theta\phi}r^{-1}\cos(\gamma) \\ v_{\theta\phi}(r\cos(\theta))^{-1}\sin(\gamma) \\ \tilde{v}_r \\ 0 \end{bmatrix}^{k-1} + \eta^{k-1},
$$

(17.18)

with the state vector defined as,

$$
\mathbf{x}_\omega = [\gamma \ v_{\theta\phi} \ \theta \ \phi \ r \ b_{\omega z}]^\top.
$$

(17.19)

The time-varying gyroscope bias in yaw, $b_{\omega z} = (b_\omega)^\top \mathbf{e}_{Bz}$, is estimated as part of the filtering which is a common approach in estimation with low-cost IMUs [26]. The measured yaw rate, $\tilde{\omega}_{Bz}$, and reeling speed measured at the GS, $\tilde{v}_r$, are the inputs to the system which we can stack as $\mathbf{u}_\omega = [\tilde{\omega}_{Bz} \ \tilde{v}_r]^\top$. We further stack the individual process noise terms in $\eta = [\eta_\gamma \ \eta_v \ \eta_\theta \ \eta_\phi \ \eta_r \ \eta_{bz}]^\top$ and again use T_s as the sampling rate in the temporal forward Euler discretization. We can then write Eq. (17.18) as

$$
\mathbf{x}_\omega^k = \mathbf{f}(\mathbf{x}_\omega^{k-1}, \mathbf{u}_\omega^{k-1}, \eta^{k-1}).
$$

(17.20)

Estimation of the bias however requires an unbiased measurement. In ground-based AWE systems line angle measurements typically provide unbiased position measurements of the kite position during the traction phase [14]. We therefore define the measurement function, $\mathbf{z}_\omega^k$, to be the outputs of the process model Eq. (17.18) with

$$
\mathbf{z}_\omega^k = \mathbf{h}(\mathbf{x}_\omega^k, \zeta^k) = \mathbf{p}_{\theta\phi r}^k + \zeta_{\theta\phi r}^k,
$$

(17.21)

where $\zeta_{\theta\phi r}$ accounts for the noise in the line measurements as defined in Table 17.1.

17.4.2.1 Extended Kalman Filtering

Due to the nonlinear nature of the process dynamics in Eq. (17.18) we implement an EKF [25], as outlined in Fig. 17.3, to fuse the inertial measurements $\tilde{\omega}_{Bz}$ characterized in Eq. (17.10) with the line measurements $\tilde{\mathbf{p}}_{\theta\phi r}$ defined in Eq. (17.9). In Kalman filtering an estimate is computed at each time step performing a *prior up-*

date step to propagate the estimate and a *measurement update* step to correct the propagated estimate using line angle measurements.

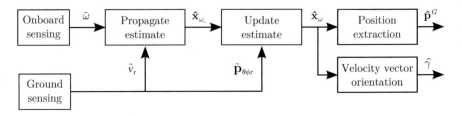

Fig. 17.3 Overview of sensor fusion scheme

To arrive at the estimate propagation equations used in the prior update step, we define the estimate state vector, $\hat{\mathbf{x}}_\omega$, analogous to Eq. (17.19), and take the expectation of Eq. (17.18) to propagate the estimate $\hat{\mathbf{x}}_\omega^{k-1}$ based on the model inputs $\mathbf{u}_\omega^{k-1} = [\bar{\omega}_{Bz}^{k-1} \ \ \tilde{v}_r^{k-1}]^\top$. The propagated estimate, $\hat{\mathbf{x}}_\omega^k$, is updated at each time step with the measurement $\mathbf{z}_\omega^k$ such that the resulting sensor fusion estimate is given as [25]

$$\hat{\mathbf{x}}_\omega^k = \hat{\mathbf{x}}_{\omega_-}^k + K_\omega^k \left(\mathbf{z}_\omega^k - \mathbf{h}(\hat{\mathbf{x}}_{\omega_-}^k, 0) \right), \qquad (17.22)$$

where $\hat{\mathbf{x}}_{\omega_-}^k$ denotes the propagated state at k.

In the EKF scheme the Kalman gain K_ω is time varying due to the nonlinear process dynamics and is a function of the covariance matrices of the process and measurement noises given as $\eta \sim \mathcal{N}(0,Q)$ and $\zeta \sim \mathcal{N}(0,R)$, respectively. We can estimate $Q \in \mathbb{R}^{6\times6}$ and $R \in \mathbb{R}^{3\times3}$ from experiments, as done for R in Sect. 17.3.1, or tune them to effect the performance of the sensor fusion estimator. Unlike the line-based estimator in Sect. 17.4.1, we can directly extract estimates of the position vector and velocity vector orientation, $\hat{\mathbf{p}}^G$ and $\hat{\gamma}$, respectively, from the estimate state vector $\hat{\mathbf{x}}_\omega$.

In the absence of onboard measurements, the proposed estimator reduces to a line-based estimator, similar to the one in Eqs. (17.16)–(17.17), but with a sound definition of the kinematic process model based on spherical coordinates and the definition of the velocity vector orientation in Eq. (17.5).

17.4.2.2 Extended Kalman Filtering with Delayed Measurements

The estimator in Sect. 17.4.2.1 was derived on the assumption that yaw rate, reeling speed and line measurements are obtained simultaneously. From experiments however we observe that line angle readings can be significantly delayed due to line dynamics, especially at low tether tension. Hence, in this section we want to account for a static delay τ in the line angle readings $\tilde{\theta}_\tau^k = \tilde{\theta}^{k-\tau}$ and $\tilde{\phi}_\tau^k = \tilde{\phi}^{k-\tau}$ during the measurement update step in Eq. (17.22), where the subscript denotes the delay in measured variables.

We can augment the estimate state vector with the additional delay states,

$$\hat{\mathbf{x}}_{\omega,\tau} = [\hat{\gamma} \ v_{\theta\phi} \ \hat{\theta} \ \hat{\theta}_1 \ \dots \ \hat{\theta}_\tau \ \hat{\phi} \ \hat{\phi}_1 \ \dots \ \hat{\phi}_\tau \ \hat{r} \ \hat{b}_{\omega z}]^\top, \qquad (17.23)$$

with $\hat{\mathbf{x}}_{\omega,\tau} \in \mathbb{R}^{6+2\tau}$ to arrive at the augmented estimate propagation equations given as

$$\hat{\mathbf{x}}_{\omega,\tau}^k = \hat{\mathbf{x}}_{\omega,\tau_-}^{k-1} + T_s \begin{bmatrix} \hat{v}_{\theta\phi} \hat{r}^{-1} \tan(\hat{\theta}) \sin(\hat{\gamma}) - \hat{b}_{\omega z} \\ 0 \\ \hat{v}_{\theta\phi} \hat{r}^{-1} \cos(\hat{\gamma}) \\ 0_{\tau \times 1} \\ \hat{v}_{\theta\phi} (\hat{r} \cos(\hat{\theta}))^{-1} \sin(\hat{\gamma}) \\ 0_{(\tau+2) \times 1} \end{bmatrix}^{k-1} + T_s \mathbf{u}_\omega^{k-1}, \qquad (17.24)$$

where $\mathbf{u}_\omega^{k-1} = [\tilde{\omega}_{Bz}^{k-1} \ \tilde{v}_r^{k-1}]^\top$ accounts again for the simultaneous measurements of the kite yaw rate and line reeling speed. The propagated augmented estimate is updated at time step k with the delayed measurements

$$\mathbf{z}_{\omega,\tau}^k = [\tilde{\theta}_\tau^k \ \tilde{\phi}_\tau^k \ \tilde{r}^k]^\top. \qquad (17.25)$$

The introduction of additional delay states in Eq. (17.23) increases the system size and hence numerical cost of the filtering scheme. To facilitate real-time operation we therefore assume the line length measurement $\tilde{r}$ in Eq. (17.25) to be simultaneous. Based on the EKF scheme [25] and using the augmented estimate propagation model Eq. (17.24) and measurement model Eq. (17.25) we can obtain estimates where we actively account for the delay in the line angle measurements.

17.5 Visual Motion Tracking of Kites

Next we aim to extract the kite position from a video stream obtained from a camera at the GS as described in Sect. 17.3.3. Such visual position measurements are not affected by line dynamics and can help to assess characteristics of line dynamics in a typical pumping operation. Kite power systems, however, commonly operate at long line lengths leading to very small target sizes in a video stream. Additionally, we require from VMT applied to kite applications to be able to cope with:

- dynamic backgrounds due to camera motion,
- highly cluttered backgrounds and target occlusions due to clouds and sunlight,
- varying illumination and changing appearances over one pumping cycle, and
- high frame rates to track fast moving rigid wings.

In summary, tracking of kites in real time from ground-based video streams involves most challenges for modern motion trackers as described in [33]. Hence, despite the large development in motion tracking, as surveyed in [29], there still exists no solution tailored to AWE systems. In this work we therefore adapt existing methodologies for motion tracking and object detection to develop a novel VMT approach.

The algorithm developed here reaches the low computational complexity to achieve high frame rates while producing accurate tracking results. Long-term robust object tracking is accomplished in real time.

Localization of a target object in a sequence of consecutive frames is generally defined as *motion tracking*. From the performance evaluation of state-of-the-art motion trackers in [29], there currently exists no approach that achieves the high frame rates ($\geq$ 48 fps) and long-term tracking capabilities ($\geq$ 2 h) required for kite applications. Therefore, a simple classical motion tracker is not applicable in this scenario. Alternatively, *object detection* provides very accurate detection performance for long-term tracking. However, the numerical burden to achieve this increase in accuracy leads to low frame rates which are too low to achieve real-time tracking capabilities with object detection alone.

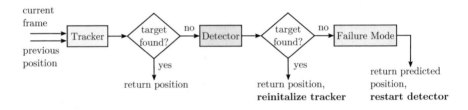

Fig. 17.4 Flowchart of the presented VMT approach

By combining state-of-the-art algorithms for object detection and motion tracking, we overcome existing drawbacks in both and achieve low failure rates and good recovery capabilities. The approach followed in this work to couple the motion tracker [19] and object detector [8] is shown in Fig. 17.4. To switch between tracker and detector we require a measure that we can use as a threshold. We have therefore extended the approach in [19], to return a quality measure for the localization performance of the tracker. Hence, based on the last known position, the tracker processes the current frame and locates the target. If the tracker succeeds with sufficient certainty, it returns the new target position. Otherwise, the detector assists by processing a sub-image of the current frame extracted at the last known kite position. The kite motion is assumed to be continuous and sufficiently slow such that the target is likely to be present in the sub-image. After successful target detection in the processed sub-image, the motion tracker is reinitialized. To address the situation when both the tracker and the detector fail to locate the target we have added three selectable *failure modes* to predict the most likely positions for reappearance: (a) remain at last tracked position, (b) predict new position based on a motion model and appropriate filtering, or (c) use an external source such as line angle measurements. Once the detector has successfully relocated the kite, the VMT algorithm continues with motion tracking.

To start the VMT, the motion tracker requires a single frame for initialization. The object detector can be initialized with an arbitrary number of labeled samples. During the initialization phase, we collect such training samples from the first frames of the video. Once enough training samples have been acquired, the detector is trained and the automatic tracking process starts. Next, we provide a brief overview of the implementation of the tracker and detector.

17.5.1 Motion Tracker

The implemented tracker is a modified version of a kernelized correlation filter [19] which has been shown to be fast while maintaining high tracking performance in recent benchmark tests [34]. Core of the tracker is a discriminating classifier obtained from a kernel ridge regression problem. We briefly summarize the method of [19] with our modifications in this section.

The tracker is limited to training data extracted from a single frame. We extract edges from a section of the video frame with the kite in the center to generate the feature vector, $\mathbf{x} \in [0,1]^n$, by using Sobel edge detection [21]. The extracted feature will be referred to as base training feature. It is the only training sample that is labeled to be positive. Permutations of the base sample corresponding to horizontal and vertical cyclic shifts of the frame section serve as negative samples and complete the training set,

$$\mathscr{X} := \left\{ P^{l-1}\mathbf{x} \mid l = 1,\dots,n \right\},$$

where P is a shift generating permutation matrix. The classifier, $H_t(\cdot) = \mathbf{w}^\top \psi(\cdot)$, is obtained from a kernel ridge regression problem,

$$\mathbf{w} = \arg\min_{\mathbf{q}} \sum_i \left(\mathbf{q}^\top \psi(\mathbf{x}_i) - \mathbf{y}_i \right)^2 + \lambda \|\mathbf{q}\|^2, \qquad (17.26)$$

where the regularized squared error between the training samples, $\mathbf{x}_i \in \mathscr{X}$, mapped to an implicit feature space by the kernel mapping, $\psi(\cdot) : [0,1]^n \to \mathbb{R}^m$, and their corresponding labels, $\mathbf{y}_i \in [0,1]$, is minimized. The ridge parameter, λ, penalizes over-fitting.

The optimization problem defined by Eq. (17.26) is linear in the dual space and its solution can be written as a linear combination of the samples, $\mathbf{x}_i$, mapped to the feature space [27],

$$\mathbf{w} = \sum_i \alpha_i \psi(\mathbf{x}_i),$$

where the dual variables, α_i, are elements of the vector, α, obtained from the stacked label vector, $\mathbf{y}$, by

$$\alpha = (K + \lambda I_{n \times n})^{-1} \mathbf{y}. \qquad (17.27)$$

The elements of the kernel matrix, K, correspond to the inner products of all training samples in the feature space,

$$K_{ij} = \langle \psi(\mathbf{x_i}), \psi(\mathbf{x_j}) \rangle = \kappa(\mathbf{x}_i, \mathbf{x}_j),$$

and can be calculated by the kernel function, $\kappa(\mathbf{x}_i, \mathbf{x}_j)$, without instantiating any sample in the feature space. The interested reader is referred to [28] for more information on kernel methods.

For the particular structure of the training set, the kernel matrix becomes circulant. Exploiting the diagonalization property of the Fourier transform of any circulant matrix [17], the computation of Eq. (17.27) can be simplified to element-wise operations in the Fourier domain,

$$F\alpha = \frac{F\mathbf{y}}{F\mathbf{k_{xx}} + \lambda}, \tag{17.28}$$

where F denotes the discrete Fourier transform (DFT) matrix of the unitary DFT, $\mathbf{k_{xx}}$ is the first column of the kernel matrix K, and the fraction denotes element-wise division. For arbitrary samples, $\mathbf{x}_i, \mathbf{x}_j \in [0,1]^n$, and a linear kernel, $\kappa(\mathbf{x}_i, \mathbf{x}_j) = \mathbf{x}_i^\top \mathbf{x}_j$, the computation of the *kernel phase correlation*, $\mathbf{k_{x_i x_j}}$, can be computed efficiently in the Fourier domain as,

$$F\mathbf{k_{x_i,x_j}} = \frac{F^{-1}\mathbf{x}_i \odot F\mathbf{x}_j}{\left| F^{-1}\mathbf{x}_i \odot F\mathbf{x}_j \right|}, \tag{17.29}$$

where $\odot$ denotes the element-wise product. The magnitude normalization is an added modification from the derivation in [19]. We consider here a linear kernel for its simplicity but the method can by extended to more complex kernels potentially resulting in more discriminating classifiers.

In each frame a base candidate feature, $\mathbf{z} \in [0,1]^n$, is extracted given the previous position estimate. Similar to the training set, we consider a candidate set consisting of the base candidate feature, $\mathbf{z}$, and its relative shifts,

$$\mathscr{Z} := \left\{ P^{l-1}\mathbf{z} \mid l = 1, \ldots, n \right\}. \tag{17.30}$$

The new position estimate is given by the shift corresponding to the element of the candidate set that maximizes the classifier response,

$$\mathbf{z}^* = \arg\max_{\mathbf{s} \in \mathscr{Z}} H_t(\mathbf{s}).$$

The evaluation of the classifier, $H_t(\cdot)$, on all elements of the candidate set, $\mathscr{Z}$, is efficiently computed in the Fourier domain by,

$$F\mathbf{h}_t = F\mathbf{k_{zx}} \odot F\alpha, \tag{17.31}$$

where $\mathbf{h}_t \in [0, 1]^n$ is a vector of the stacked classifier outputs applied to the elements in $\mathscr{Z}$, $\mathbf{k}_{\mathbf{zx}}$ denotes the kernel phase correlation between the base training feature, $\mathbf{x}$, and base candidate feature, $\mathbf{z}$. The values of the elements of $\mathbf{h}_t$ serve as similarity measures between the candidate features and the base training feature, $\mathbf{x}$, and allow to assess the quality of the location estimate. Note that given the normalization in Eq. (17.29) the output of the classifier $H_t(\cdot)$, is bound to be in $[0, 1]$.

17.5.2 Object Detector

The object detector is based on the work described in [3, 8, 31] which has been shown to achieve high performances on benchmark tests in [12]. In this section, we illustrate the concept of the object detector by summarizing its individual components.

We consider labeled training samples manually collected from a video stream. Multiple feature channels are extracted from the data to train the detection classifier, $H_d(\cdot)$. The features consist of color information, the histogram of oriented gradients, and the gradient magnitude as suggested in [7]. The used classifier is a cascade of boosted decision trees introduced in [31]. Boosting methods are used in machine learning to construct strong classifiers by combining multiple weak classifiers. An illustrative example of boosting is shown in Fig. 17.5. In particular, we consider the discrete AdaBoost method discussed in [16] with binary decision trees as weak classifiers. The boosted decision trees are coupled in a cascade with a constant rejection threshold as in [6, 9]. In such a cascade candidate samples are discarded if the sum of weak classifiers drops below a rejection threshold.

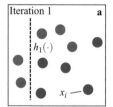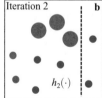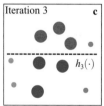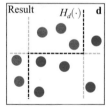

Fig. 17.5 Illustration of boosting [15]. In each iteration j a new weak classifier, $h_j(\cdot)$, (vertical or horizontal dashed line) is trained which minimizes the current classification error of the weighted positive (red), and negative (blue) samples. The resulting boosted classifier, $H_d(\cdot)$, is a weighted sum over all weak classifiers. **a** Equal weight is assigned to each feature image, x_i. **b** Weights of incorrectly classified samples are increased. **c** Subset of the heaviest samples is sufficient for optimal learning [3]. **d** Combination of all weak classifiers to form the boosted classifier, $H_d(\cdot)$

The object detector is applied when the motion tracker fails to locate the target and is initiated at the last position that was tracked, as illustrated in Fig. 17.6. The position is then obtained from the highest scoring detection. Successful detection initializes re-training of the tracker. If the detector fails to locate the target, a location

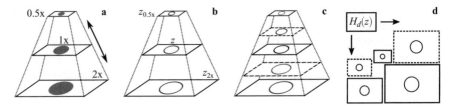

Fig. 17.6 Illustration of sliding window detection with aggregated channel features over an approximated feature pyramid. Approximating features is computationally less expensive then the extraction. Detection over a feature pyramid enables scale-invariant detections. The best detection of $H_d(z)$ is returned. **a** Resample extracted candidate patch at higher ($2x$) and lower scales ($0.5x$). **b** Extract feature images z, z_{2x}, $z_{0.5x}$ at all scales. **c** Approximate feature images at scales in between, $z_{0.75x}$, $z_{1.5x}$ [8]. **d** Run sliding window over all extracted and approximated feature images [3]

estimate is obtained from the failure mode. Note that limiting the detection area enables detection in real time.

17.6 Experimental Results

In this section we demonstrate VMT and the presented filtering approaches on experimental data which was obtained using the AWE platform developed at FHNW. In Sect. 17.6.1, we first compare VMT against (unfiltered) line angle measurements from experimental data of a tethered wing flight. This case aims to demonstrate the potential of VMT for rigid tethered wings with depower capabilities where line-based approaches would fail especially during retraction phases. Although the VMT approach has been implemented for high sampling rates of 100 Hz, due to hardware limitations it currently only provides vision-based estimates in post-processing. We therefore focused on a (real-time) sensor fusion approach which is evaluated in Sect. 17.6.2 using vision-based results as a reference solution.

17.6.1 Visual Motion Tracking of Tethered Wings

In this work we apply VMT as a tool to verify other estimation approaches for ground-based AWE systems. We have demonstrated the VMT approach for a range of kites in a series of visually challenging videos containing occlusions, camera motion, appearance changes, and long video duration. Here, we compare ground-based line angle measurements (Sect. 17.3.1) from experimental data of a rigid tethered wing flight to the positions tracked in video recordings from a ground-based fisheye camera (see Sect. 17.3.3). To obtain the tracked wing position from the video image we apply the VMT approach in Sect. 17.5. The tethered wing used in this section consists of a 3 m^2 main wing and an elevator for depower during retractions.

Figure 17.7 shows a frame from the tracking results during the retraction phase of a small-scale tethered wing flight at 200 m line length. In Fig.s 17.8 and 17.9 the tracked image positions are compared against line-based position measurements. Note that the presented VMT approach provides 2D information and results are therefore presented in the (θ,ϕ)-plane only. We further see that VMT tends to lose the object regularly but the tracking failure is detected correctly and limited to short instances. As a result target loss is indicated correctly and the VMT returns no erroneous tracked positions.

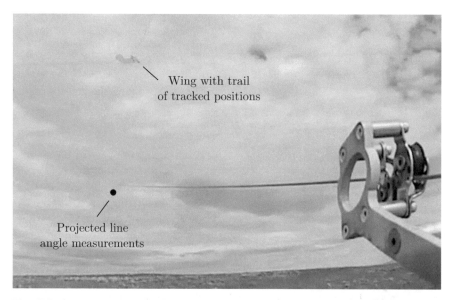

Wing with trail
of tracked positions

Projected line
angle measurements

Fig. 17.7 Overlay of last 20 tracked positions of a tethered wing using VMT in green with the instantaneous projected line angle measurement in black

The overlay in Fig. 17.7 hints at the weakness of line-based position measurements during retractions where we desire low line forces to improve cycle efficiency. Drops in tether tension can significantly deteriorate the quality of line-based measurements leading to increased estimation delay and large errors due to line sag. From Fig. 17.8 we can see that such events are not limited to the retraction phase but can also occur in turns flying a figure-eight trajectory. While the effect of line dynamics is less obvious in the azimuth measurements, we can clearly see large errors in elevation measurements of up to 0.1 rad even during the traction phase. This is further illustrated in Fig. 17.9 which shows the tracked image positions over one figure-eight. The markers in Fig. 17.9 also demonstrate the effect of delay in the line angle measurements. One would generally expect the delay to increase with increasing line lengths [32] but Fig. 17.8 clearly indicates variations of delay in the line-based elevation measurements.

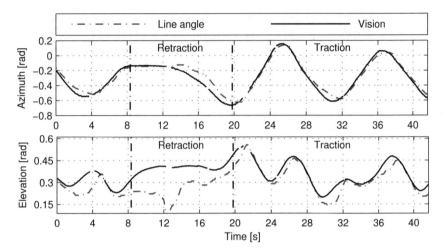

Fig. 17.8 Unfiltered position measurements of a tethered wing over a complete pumping cycle between 130m and 200m line lengths tracked at 24fps

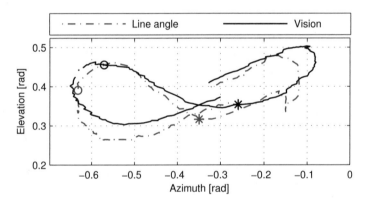

Fig. 17.9 Tracked image positions mapped in (θ, ϕ)-plane after calibration and correcting for the relative motion of the camera. Markers illustrate the instantaneous tracked positions at two different time instances

In summary, line dynamics introduce varying lag, especially during low-line-tension scenarios such as uploop curves and retractions. This demonstrates the importance of effective winch control in ground-based AWE approaches to not only optimize the cycle power but also ensure sufficient estimation capabilities of the wing heading and position. The inertial sensor fusion approach, as demonstrated next, can help to reduce the estimation delay allowing for lower line forces and hence improve system efficiency.

17.6.2 Inertial Sensor Fusion for Soft Kites

In this section we demonstrate the inertial sensor fusion for a HQ Apex III 10 m^2 kite at wind speeds of 3-6 m/s measured 5 m above ground. All estimators have been implemented on a Speedgoat Real-Time Target Machine [30] and demonstrated at 100 Hz in closed-loop operation on the ground-based FHNW system. The IMU data is streamed to the GS as described in Sect. 17.3.2. Note that the current FHNW system with two lines is designed for rigid wings with onboard depower capabilities. The soft kite in this experiment therefore remained fully powered during retractions and no significant line sag was observed. This test case nonetheless serves well to demonstrate the potential of limited onboard measurements in the proposed sensor fusion approach.

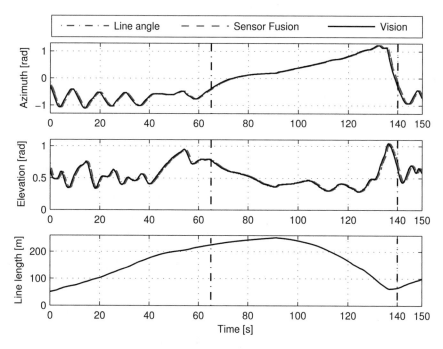

Fig. 17.10 Reference vision-based kite position estimate (solid) with sensor fusion (dashed) and line-angle (dotted) estimates over one pumping cycle. Retraction occurs between vertical lines

We apply the VMT approach in Sect. 17.5 to obtain a reference solution that allows comparison of the different estimators presented in Sect. 17.4. An example of the resulting position measurements over one full pumping cycle is presented in spherical coordinates in Fig. 17.10, where the retraction phase (65-140 s) is indicated by vertical lines. From the video recording we can also compute estimates of the kite velocity vector orientation, $\hat{\gamma}$, as presented in Fig. 17.11, where the regularized definition of the velocity vector orientation, Eq. (17.7), was used during the retraction phase.

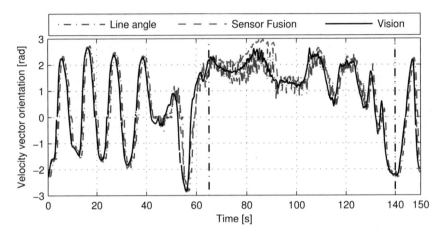

Fig. 17.11 Reference vision-based velocity vector orientation (solid) with sensor fusion (dashed) and line-angle (dotted) estimates over one pumping cycle. The regularized velocity vector orientation $\hat{\gamma}_{reg}$ is shown during retraction between vertical lines

Figures 17.10 and 17.11 also show estimates of the position and velocity vector orientation obtained using sensor fusion (Sect. 17.4.2.2) and the line-based filtering scheme (Sect. 17.4.1) with the variances in Table 17.1. We tuned the covariance matrix of the process noise, Q, in the EKF equations of the sensor fusion approach to compromise between filter performance during traction and retraction phases. A delay of $\tau = 0.3$ s in the line angle measurements was assumed in Eq. (17.25).

The initial observation that no significant line sag was apparent during the experiment is confirmed in Fig. 17.10. Since the kite could not be depowered during the retraction phase, line dynamics have no significant effect on the estimation of the kite position in this demonstration case. This is in sharp contrast to the results obtained with a tethered wing in Sect. 17.6.1 which requires no line tension to remain airborne. A delay of 0.6 s and 0.4 s is introduced in the estimation of the line angles using the line-based estimator and inertial sensor fusion, respectively.

During the traction phase a similar behavior can be observed in the estimation of the velocity vector orientation in Fig. 17.11 where the line-based estimator and the sensor fusion results follow the vision-based reference with a slight delay. The exact estimation delay can be deduced from the root mean square (RMS) errors of line-based and sensor fusion estimates presented in Fig. 17.12 for the traction phase ($t \leq$ 65 s). The vision-based solution serves as reference to compute the respective errors. By shifting the reference solution in the RMS error computation by the indicated delay, we can see that incorporating inertial measurements can reduce the estimation delay from 0.9 s to 0.2 s. This is a significant reduction in estimation delay compared to a common overall system delay of 1.5 s [1].

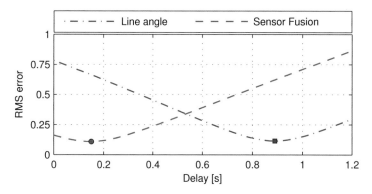

Fig. 17.12 RMS errors between the line-based and sensor fusion estimate of the velocity vector orientation $\hat{\gamma}$ during the traction phase. The vision-based solution shifted by the indicated delay is used as reference

Note that tuning of the respective filter is a compromise between noisy feedback variables and estimation delay. We can generally see however that the yaw-rate measurements require less filtering, compared to line-based estimators, to arrive at a satisfactory estimate of the velocity vector orientation. A similar trend can be seen in Fig. 17.11 during the retraction phase ($t > 65$ s) where the noise on line-based measurements is usually magnified due to the reduced kite velocity. Such oscillations combined with an increased system delay during this critical phase can significantly affect the performance of pumping cycle controllers, e.g. [35]. Sensor fusion in comparison provides smoother estimates closer to the vision-based reference solution. This is also reflected in Table 17.2 summarizing the RMS errors of both estimators during traction and retraction phases.

Estimator	Traction		Retraction	
	$\hat{\gamma}$ (rad)	$\|\hat{\mathbf{p}}_{\theta\phi}\|$ (rad)	$\hat{\gamma}$ (rad)	$\|\hat{\mathbf{p}}_{\theta\phi}\|$ (rad)
Line angle	0.78	0.09	0.39	0.09
Sensor fusion	0.16	0.04	0.38	0.06

Table 17.2 RMS errors in the estimated velocity vector orientation, $\hat{\gamma}$, and position in (θ, ϕ)-plane given as $\hat{\mathbf{p}}_{\theta\phi} = \begin{bmatrix} \hat{\phi} & \hat{\theta} \end{bmatrix}^{\top}$ during traction and retraction phases. The vision-based solution is taken as reference.

17.7 Conclusion

Automatic control of ground-based kite power systems commonly relies on ground-based measurements of line angles and lengths to estimate the kite position and heading for feedback control. In a two-phase operation line dynamics can significantly deteriorate the performance of flight controllers due to increased estimation delay and line sag. We have therefore proposed a sensor fusion scheme to reduce the estimation delay in ground-based kite power systems in pumping operation by fusing yaw-rate and line measurements. Limiting the extent of incorporated onboard information to one variable enables robust telemetry over long tether lengths demonstrated up to 280 m. The developed extended Kalman filtering scheme for sensor fusion has been demonstrated on experimental data to significantly reduce the estimation delay compared to an estimator relying solely on ground-based line angle and length measurements. Although the demonstration case was limited to soft kites, which showed little influence of line dynamics, we expect the strength of inertial-based sensor fusion to be exposed for longer line lengths.

We further developed a visual motion tracking approach to obtain reference solutions, or a ground truth, of the kite position and heading from ground-based video streams. Combining fast motion tracking with accurate object detection allows long-term tracking of kites in real time. In this work we demonstrated the visual motion tracking capabilities on experimental data of a tethered wing in pumping operation. The vision-based position estimates showed large discrepancies to ground-based line measurements during uploops and retractions when the tether forces are low. Line dynamics effectively limits the achievable cycle efficiency of rigid wings with ground-based estimation. The proposed sensor fusion can help to reduce the detrimental effects of line dynamics even in the presence of delayed line measurements.

The vision-based results revealed the complex nature of line dynamics leading to time-varying delay over a whole pumping cycle and even a single figure-eight. The current sensor fusion approach is however limited to static measurement delay of up to 50 time steps in real-time operation due to the augmentation of the filter state vector to incorporate the delay states. Future development of estimators for ground-based kite systems will therefore focus on fusion of spatially and temporally misaligned sensors in a computationally efficient manner.

We can further characterize the line dynamics as a time-varying bias which can be modeled or identified using the visual motion tracking results as a reference solution. Modeling of the line dynamics ensures that ground-based measurements can be fully exploited, but the proposed sensor fusion scheme could further benefit from additional onboard measurements. In particular, measurement of the kite velocity, which is a crucial quantity in the unicycle model employed in the developed extended Kalman filter, would improve the state estimation in the prediction step.

Acknowledgements This research was support by the Swiss National Science Foundation (Synergia) No. 141836 and the Swiss Commission for Technology and Innovation (CTI). We further acknowledge Corey Houle, Damian Aregger, and Jannis Heilmann from the Fachhochschule Nordwestschweiz (FHNW) for their test support and providing all the hardware used in the experiments.

Development of the hardware architecture enabling onboard measurements was done by Martin Rudin and Alexander Millane (ETH Zurich). We are grateful for their support. The authors acknowledge the SpeedGoat Greengoat program.

References

1. A., W. T., Hesse, H., Zgraggen, A. U., Smith, R. S.: Model-based Identification and Control of the Velocity Vector Orientation for Autonomous Kites. In: Proceedings of the 2015 American Control Conference, Chicago, IL, USA, 1–3 July 2015. doi: 10.1109/ACC.2015.7171088
2. Anderson, B. D. O., Moore, J. B.: Optimal Filtering. English. Dover Publications, Mineola, N.Y., USA (2005)
3. Appel, R., Fuchs, T., Dollár, P.: Quickly Boosting Decision Trees – Pruning Underachieving Features Early. In: International Conference on Machine Learning (ICML), vol. 28, pp. 594–602, Atlanta, USA, June 2013
4. Autonomous Airborne Wind Energy Project (A^2WE). http://a2we.skpwiki.ch. Accessed 31 Oct 2015
5. Bormann, A., Ranneberg, M., Kövesdi, P., Gebhardt, C., Skutnik, S.: Development of a Three-Line Ground-Actuated Airborne Wind Energy Converter. In: Ahrens, U., Diehl, M., Schmehl, R. (eds.) Airborne Wind Energy, Green Energy and Technology, Chap. 24, pp. 427–437. Springer, Berlin Heidelberg (2013). doi: 10.1007/978-3-642-39965-7_24
6. Bourdev, L., Brandt, J.: Robust object detection via soft cascade. In: Conference on Computer Vision and Pattern Recognition, vol. 2, pp. 236–243, San Diego, CA, USA, June 2005. doi: 10.1109/CVPR.2005.310
7. Dalal, N., Triggs, B.: Histograms of oriented gradients for human detection. In: Conference on Computer Vision and Pattern Recognition, vol. 1, pp. 886–893, San Diego, CA, USA (2005). doi: 10.1109/CVPR.2005.177
8. Dollár, P., Appel, R., Belongie, S., Perona, P.: Fast Feature Pyramids for Object Detection. 36(8), 1532–1545 (2014). doi: 10.1109/TPAMI.2014.2300479
9. Dollár, P., Appel, R., Kienzle, W.: Crosstalk Cascades for Frame-Rate Pedestrian Detection. In: European Conference on Computer Vision, pp. 645–659, Florence, Italy (2012). doi: 10.1007/978-3-642-33709-3_46
10. Erhard, M., Strauch, H.: Theory and Experimental Validation of a Simple Comprehensible Model of Tethered Kite Dynamics Used for Controller Design. In: Ahrens, U., Diehl, M., Schmehl, R. (eds.) Airborne Wind Energy, Green Energy and Technology, Chap. 8, pp. 141–165. Springer, Berlin Heidelberg (2013). doi: 10.1007/978-3-642-39965-7_8
11. Erhard, M., Strauch, H.: Sensors and Navigation Algorithms for Flight Control of Tethered Kites. In: Proceedings of the European Control Conference (ECC13), Zurich, Switzerland, 17–19 July 2013. arXiv:1304.2233 [cs.SY]
12. Ess, A., Leibe, B., Gool, L. V.: Depth and appearance for mobile scene analysis. In: International Conference on Computer Vision (ICCV), pp. 1–8, Rio de Janeiro, Brazil (2007). doi: 10.1109/ICCV.2007.4409092
13. Fagiano, L., Zgraggen, A. U., Morari, M., Khammash, M.: Automatic crosswind flight of tethered wings for airborne wind energy:modeling, control design and experimental results. IEEE Transactions on Control System Technology 22(4), 1433–1447 (2014). doi: 10.1109/TCST.2013.2279592
14. Fagiano, L., Huynh, K., Bamieh, B., Khammash, M.: On sensor fusion for airborne wind energy systems. IEEE Transactions on Control Systems Technology 22(3), 930–943 (2014). doi: 10.1109/TCST.2013.2269865
15. Freund, Y., Schapire, R. E.: Experiments with a New Boosting Algorithm. In: International Conference on Machine Learning (ICML), pp. 148–156, Bari, Italy, July 1996

16. Friedman, J., Hastie, T., Tibshirani, R.: Additive logistic regression: a statistical view of boosting. The Annals of Statistics **28**(2), 337–407 (2000)
17. Gray, R. M.: Toeplitz and Circulant Matrices: A Review. Foundations and Trends in Communications and Information Theory **2**(3), 155–239 (2006). doi: 10.1561/0100000006
18. Heikkila, J., Silvén, O.: A four-step camera calibration procedure with implicit image correction. In: Conference on Computer Vision and Pattern Recognition, pp. 1106–1112, IEEE, San Juan, Puerto Rico, June 1997
19. Henriques, J. F., Caseiro, R., Martins, P., Batista, J.: High-Speed Tracking with Kernelized Correlation Filters. **37**(3), 583–596 (2015). doi: 10.1109/TPAMI.2014.2345390
20. Lefferts, E. J., Markley, F. L., Shuster, M. D.: Kalman filtering for spacecraft attitude estimation. Journal of Guidance, Control, and Dynamics **5**(5), 417–429 (1982)
21. MATLAB® Computer Vision System Toolbox™ Reference, MathWorks, Inc., Natick, MA, USA, 2015. http://mathworks.com/products/computer-vision
22. Millane, A., Wood, T. A., Hesse, H., Zgraggen, A. U., Smith, R. S.: Range-Inertial Estimation for Airborne Wind Energy. In: Conference on Decision and Control (CDC), pp. 455–460, Osaka, Japan, Dec 2015. doi: 10.1109/CDC.2015.7402242
23. Pixhawk Autopilot. Accessed 31. October 2015. https://pixhawk.org/modules/pixhawk.
24. Polzin, M., Hesse, H., Wood, T. A., Smith, R. S.: Visual Motion Tracking for Estimation of Kite Dynamics. In: Schmehl, R. (ed.). Book of Abstracts of the International Airborne Wind Energy Conference 2015, p. 110, Delft, The Netherlands, 15–16 June 2015. doi: 10.4233/uuid: 7df59b79-2c6b-4e30-bd58-8454f493bb09. Poster available from: http://www.awec2015. com/images/posters/AWEC42_Hesse-poster.pdf
25. Sabatini, A. M.: Kalman-filter-based orientation determination using inertial/magnetic sensors: Observability analysis and performance evaluation. Sensors **11**(10), 9182–9206 (2011)
26. Savage, P. G.: Strapdown Inertial Navigation Integration Algorithm Design Part 2: Velocity and Position Algorithms. Journal of Guidance, Control, and Dynamics **21**(2), 208–221 (1998)
27. Schölkopf, B., Herbrich, R., Smola, A. J.: A Generalized Representer Theorem. In: Computational Learning Theory, pp. 416–426. Springer-Verlag, Berlin, Germany (2001)
28. Shawe-Taylor, J., Cristianini, N.: Kernel methods for pattern analysis. Cambridge University Press (2004)
29. Smeulders, A. W. M., Chu, D. M., Cucchiara, R., Calderara, S., Dehghan, A., Shah, M.: Visual tracking: An experimental survey. **36**(7), 1442–1468 (2014). doi: 10.1109/TPAMI.2013.230
30. Speedgoat User Story: Efficiently harnessing wind power high above the ground using autonomous kites, Speedgoat GmbH, Liebefeld, Switzerland, 2015. https://www.speedgoat.ch/ Portals/0/Content/UserStories/ethz_user_story.pdf
31. Viola, P., Jones, M.: Rapid object detection using a boosted cascade of simple features. In: Conference on Computer Vision and Pattern Recognition, vol. 1, pp. 511–518, (2001). doi: 10.1109/CVPR.2001.990517
32. Wood, T. A., Hesse, H., Zgraggen, A. U., Smith, R. S.: Model-Based Flight Path Planning and Tracking for Tethered Wings. In: Conference on Decision and Control (CDC), pp. 6712–6717, Osaka, Japan, Dec 2015. doi: 10.1109/CDC.2015.7403276
33. Wu, Y., Lim, J., Yang, M. H.: Online object tracking: A benchmark. In: Conference on Computer Vision and Pattern Recognition, pp. 2411–2418, (2013). doi: 10.1109/CVPR.2013.312
34. Wu, Y., Lim, J., Yang, M.-H.: Object Tracking Benchmark. In: IEEE Transactions on Pattern Analysis and Machine Intelligence, vol. 37, 9, pp. 1834–1848, (2015). doi: 10.1109/TPAMI. 2014.2388226
35. Zgraggen, A. U., Fagiano, L., Morari, M.: Automatic Retraction and Full-Cycle Operation for a Class of Airborne Wind Energy Generators. IEEE Transactions on Control Systems Technology **24**(2), 594–608 (2015). doi: 10.1109/TCST.2015.2452230

Part III
Concept Design & Analysis

Chapter 18
Crosswind Kite Power with Tower

Florian Bauer, Christoph M. Hackl, Keyue Smedley and Ralph M. Kennel

Abstract Crosswind kite power replaces the tower and the support structure of a conventional wind turbine by a lightweight tether leading to a potentially lower levelized cost of electricity. However, in this chapter it is shown that tethering the kite to the top of a tower instead of to the ground can have advantages: Most notably, the "cosine loss" is reduced, i.e. the misalignment of the wind velocity vector and the direction of the traction power transfer. Hence, a tower can increase the power and energy yield up to about the double. Even for small tower heights compared to the kite's operation altitude, a significant efficiency increase can be obtained. Further advantages of a tower are highlighted e.g. for the autonomous start and landing and for the wind velocity measurement. Possible tower concepts are illustrated.

18.1 Motivation

Kites, or tethered wings, are promising alternatives to harvest wind energy (see e.g. [1, 5, 11, 20]): As shown in Fig. 18.1, a (rigid) kite is flown in crosswind trajectories resembling figure eights or circles. The kite has onboard turbines and gen-

Florian Bauer (✉) · Ralph M. Kennel
Institute for Electrical Drive Systems and Power Electronics, Technical University of Munich, Arcisstrasse 21, 81477 Munich, Germany
e-mail: florian.bauer@tum.de

Christoph M. Hackl
Munich School of Engineering, Research group "Control of renewable energy systems (CRES)", Technische Universität München, Lichtenbergstr. 4a, 85748 Garching, Germany

Keyue Smedley
The Henry Samueli School of Engineering, Power Electronics Laboratory, University of California, Irvine, CA 92697, USA

© Springer Nature Singapore Pte Ltd. 2018
R. Schmehl (ed.), *Airborne Wind Energy*, Green Energy and Technology, https://doi.org/10.1007/978-981-10-1947-0_18

441

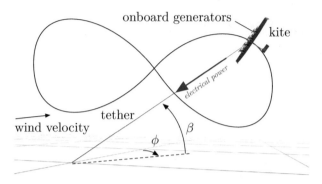

Fig. 18.1 "Drag power": continuous onboard generation of electricity

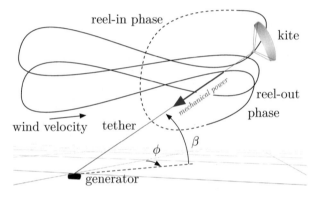

Fig. 18.2 "Lift power": traction power conversion in periodic pumping cycles by ground-based generator

erators to generate electrical power which is transmitted to the ground via electrical cables that are integrated in the tether. Due to the high speed of the kite, the apparent wind speed at the kite is about a magnitude higher than the actual wind speed, so that the onboard turbines can be small. This concept is called "drag power" [20].[1]

A second possibility for crosswind kite power is shown in Fig. 18.2: A kite (from soft materials like a paraglider or alternatively from rigid materials like a glider) is tethered to a winch on the ground which is connected to an electrical drive. The kite is flown in crosswind motions with a high lift force and pulls the tether from the winch. Energy is generated by operating the winch drive as generator (generative braking). When the maximum tether length is reached, the kite is flown to a low force position like the zenith, and/or pitched down, and reeled back in. A rigid kite can also dive towards the ground winch for minimal reel-in time. During the reel-in

---

[1] Also called "onboard-", "continuous power generation" or "fly-gen".

phase, only a fraction of the generated energy is dissipated by operating the winch drive as motor. This concept is called "lift power" [20].[2]

Both concepts can generate the same amount of power [20]. Compared to conventional wind turbines, crosswind kite power promises to harvest wind energy at higher altitudes with stronger and steadier winds, but by needing only a fraction of the construction material. Hence, it promises to have a higher capacity factor, lower capital investments, and in the end a lower levelized cost of electricity (LCOE). Mechanical output powers of two megawatts have already been achieved by a commercial soft kite by the company SkySails [9]. A drag power rigid kite with a rated electrical power of 600 kW is currently under development by the company Makani Power/Google [21], shown in Fig. 18.3 right.

The power P a kite can generate is proportional to (see also e.g. [22, Eq. (2.38)])

$$P \sim \underbrace{\cos^3 \beta}_{=:\, \eta_{\cos}(\beta)} \tag{18.1}$$

where β is the angle between the wind velocity vector and the direction of the traction power transfer, which is the elevation angle if the tether is assumed straight (see Figs. 18.1–18.2), and $\eta_{\cos}(\beta)$ is the cosine efficiency or $1 - \eta_{\cos}(\beta)$ is the cosine loss. If the kite is tethered to the ground, then $\beta \gg 0$. With a typical elevation of $\beta \approx 30°$, the cosine efficiency is already reduced to $\eta_{\cos}(30°) \approx 0.65$. With $\beta \approx 40°$, as used for Makani Power's/Google's Wing 7 demonstrator [24, p. 486, Table 28.7], the cosine efficiency is only $\eta_{\cos}(40°) \approx 0.45$. Consequently, if the kite is attached to the top of a tower—as proposed in this chapter—, whose height is ideally similar to the operation altitude of the kite, up to $1/\eta_{\cos}(40°) \approx 2.22$ times more power and energy, i.e. more than the *double*, can be generated. Even for smaller towers, the cosine efficiency and thus power and energy yield can be increased significantly. A tower can have further advantages e.g. for the autonomous

Fig. 18.3 Two groups experimenting with towers/masts. Left: TU Delft's demonstrator, provided by Roland Schmehl. Right: Makani Power's/Google's 600 kW prototype [25], reprinted with permission of Fort Felker

---

[2] Also called "traction power", "ground-", "pumping mode power generation" or "ground-gen".

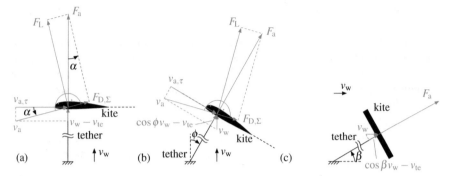

Fig. 18.4 Sketch of a crosswind flying kite from top (a), with azimuth angle $\phi \neq 0$ (b), and seen from the side with elevation angle $\beta \neq 0$ while $\phi = 0$ (c)

start and landing, which is why some groups already experimented with towers, see Fig. 18.3.

The use of a (possibly high) tower is a counter-intuitive approach, as the kite power technology minimizes the construction mass, particularly by avoiding a tower. No publication that details the potentials of a tower was found, so this contribution aims at closing that gap.

This chapter is organized as follows: Section 18.2 derives Eq. (18.1) and reveals the potential of cosine efficiency maximization as function of tether length, tower height and kite operation altitude. Section 18.3 proposes a tower design concept and reveals further advantages of a tower. In Sect. 18.4 the start and landing of a kite from a tower is discussed. Finally, conclusions and an outlook are given in Sect. 18.5.

18.2 Cosine Efficiency and its Maximization

18.2.1 Crosswind Kite Power

The potential of crosswind kite power was first derived by Loyd in [20]. In the following, Loyd's derivation is extended and relies only on the following two assumptions, which are valid for wind speeds above some minimum wind speed and for crosswind flight:

Assumption 18.1: *Gravitational and inertial forces are small compared to aerodynamic forces, i.e.* $m_k + m_{te} \approx 0$ *with kite mass* m_k *and tether mass* m_{te}.

Assumption 18.2: *The tether is straight, so that, in combination with Assumption 18.1, aerodynamic force* F_a *and tether force on ground* F_{te} *are in balance, i.e.* $F_{te} = F_a$, *see Fig. 18.4.*

The following derivation of crosswind kite power can partly be found in a similar way e.g. in [1] and references therein.

Figure 18.4 (a) shows a kite flying perpendicular to the wind (i.e. crosswind) when the kite is exactly in the downwind position and the aerodynamic force and the tether force are in balance. One can find the relation

$$\frac{v_w - v_{te}}{v_a} = \sin \alpha = \frac{F_{D,\Sigma}}{F_a}, \qquad (18.2)$$

where v_w is the wind speed, v_{te} is the tether speed, v_a is the apparent wind speed, α is the angle of attack and $F_{D,\Sigma}$ is the sum of the drag forces. The aerodynamic forces are determined by

$$F_L = \frac{1}{2}\rho A v_a^2 C_L \qquad (18.3)$$

$$F_{D,\Sigma} = \frac{1}{2}\rho A v_a^2 C_{D,\Sigma} \qquad (18.4)$$

$$F_a = \sqrt{F_L^2 + F_{D,\Sigma}^2} \qquad (18.5)$$

with air density ρ, the kite's characteristic (projected wing-) area A, lift coefficient C_L and drag coefficient sum $C_{D,\Sigma}$. The latter is given by

$$C_{D,\Sigma} = \underbrace{C_{D,k} + C_{D,te}}_{=:C_{D,eq}} + C_{D,tu} \qquad (18.6)$$

with the kite's drag coefficient $C_{D,k}$, the tether drag coefficient $C_{D,te}$, which both can be summarized by an equivalent drag coefficient $C_{D,eq}$, and the "drag" coefficient of onboard turbines $C_{D,tu}$. All aerodynamic coefficients are in general functions of time. More specifically, C_L and $C_{D,k}$ depend e.g. on the angle of attack, $C_{D,te}$ lumps the drag forces of the tether to the kite and depends e.g. on the tether length (see also [15] or [10, Chap. 3.4.1, pp. 44]), and $C_{D,tu}$ depends e.g. on the angular speed of the turbines. Inserting Eqs. (18.3)–(18.5) into Eq. (18.2) and solving for v_a leads to

$$v_a = (v_w - v_{te})\frac{\sqrt{C_L^2 + C_{D,\Sigma}^2}}{C_{D,\Sigma}}. \qquad (18.7)$$

Figure 18.4 (b) shows the same situation as Fig. 18.4 (a) if tether and wind velocity have azimuth $\phi \neq 0$. The vector diagram is similar, but compared to Fig. 18.4 (a) the effect of the wind speed is reduced to $\cos \phi\, v_w$ leading to a reduced apparent wind speed and kite speed as well as forces. In Fig. 18.4 (c) the situation is shown from the side with an elevation of $\beta \neq 0$ so that, compared to Fig. 18.4 (a), the effect of the wind speed is reduced to $\cos \beta\, v_w$. Combining both effects, i.e. for arbitrary ϕ and β, leads to the projected wind speed

$$\tilde{v}_w = \cos\phi \cos\beta \, v_w. \tag{18.8}$$

Inserting Eq. (18.8) into Eq. (18.7) gives a "corrected" apparent wind speed (see also e.g. [22, Eq. (2.15)])

$$\tilde{v}_a = (\cos\phi \cos\beta \, v_w - v_{te})\frac{\sqrt{C_L^2 + C_{D,\Sigma}^2}}{C_{D,\Sigma}}. \tag{18.9}$$

In case of drag power $v_{te} = 0$ and $C_{D,tu} \neq 0$. With the turbine's thrust force

$$F_{tu} = \frac{1}{2}\rho A \tilde{v}_a^2 C_{D,tu}, \tag{18.10}$$

the power is given by

$$
\begin{aligned}
P_{tu} &= \tilde{v}_a F_{tu} \\
&= \frac{1}{2}\rho A \cos^3\phi \cos^3\beta \, v_w^3 \frac{\left[C_L^2 + (C_{D,k} + C_{D,te} + C_{D,tu})^2\right]^{\frac{3}{2}}}{(C_{D,k} + C_{d,t} + C_{D,tu})^3} C_{D,tu}
\end{aligned}
\tag{18.11}
$$

which contains the proportionality stated in Eq. (18.1).

In case of lift power $v_{te} \neq 0$ and $C_{D,tu} = 0$. With $F_{te} = F_a$, the power for the reel-out phase is given by (see also e.g. [22, Eq. (2.35)])

$$
\begin{aligned}
P_{te} &= v_{te} F_{te} \\
&= v_{te}\frac{1}{2}\rho A(\cos\phi \cos\beta \, v_w - v_{te})^2 \frac{\left(C_L^2 + C_{D,eq}^2\right)^{\frac{3}{2}}}{C_{D,eq}^2}.
\end{aligned}
\tag{18.12}
$$

By expressing v_{te} in terms of $\tilde{v}_w$ with reeling factor f_{te},

$$v_{te} = f_{te}\cos\phi \cos\beta \, v_w, \tag{18.13}$$

Eq. (18.12) can be rewritten as

$$P_{te} = \frac{1}{2}\rho A \cos^3\phi \cos^3\beta \, v_w^3 f_{te}(1 - f_{te})^2 \frac{\left(C_L^2 + C_{D,eq}^2\right)^{\frac{3}{2}}}{C_{D,eq}^2} \tag{18.14}$$

which also contains the proportionality stated in Eq. (18.1).

Note that

$$\eta_{cos}(\beta) = \cos^3\beta \tag{18.15}$$

is a factor in Eqs. (18.11) and (18.14) and holds for any time instant (if Assumptions 18.1–18.2 hold true).

In [6] it is shown that $P \leq \cos\beta F_{te} v_w$ for an arbitrary tethered object and the term $1 - \cos\beta$ is called cosine loss. It is stated: "Fortunately, for moderate angles, the cosine is still close to one, for example the cosine loss is less than 30% even if the tether goes upwards with an angle of 45 degrees." [6, p. 14] However, as derived above, in crosswind kite power, also the force is proportional to $\cos^2\beta$ leading to Eq. (18.1). Therefore at $45°$ the actual cosine loss is 65%. In other chapters e.g. [22, 24], the same proportionality as Eq. (18.1) is derived.

For sake of completeness, maximizing Eq. (18.11) over $C_{D,tu}$ or maximizing Eq. (18.14) over f_{te} both yield, with the assumption $C_L \gg C_{D,eq}$, to a maximum power of

$$P_{\max} = \frac{2}{27} \rho A \cos^3\phi \cos^3\beta \, v_w^3 \frac{C_L^3}{C_{D,eq}^2}, \tag{18.16}$$

where respectively $C_{D,tu}^* = \frac{1}{2} C_{D,eq}$ or $f_{te}^* = \frac{1}{3}$ are the optimal arguments [20].

18.2.2 Mean Cosine Efficiency

Figure 18.5 shows a plot of Eq. (18.15) for $\beta \in [-30°, 90°]$. Hereby the region $\beta \in [20°, 40°]$ is marked by a bold black line, representative for a mean elevation $\bar\beta \approx 30°$, i.e. for a ground-tethered kite, and the region $\beta \in [-10°, 10°]$ is marked by a bold green line, representative for a mean elevation $\bar\beta \approx 0$, i.e. for a tower tethered kite where the tower height coincides with the kite's mean operation altitude.

The mean cosine efficiency (i.e. over a complete flight path at a given wind speed) is determined by

$$\bar\eta_{\cos} := \frac{1}{T} \int_{t_0}^{t_0+T} \eta_{\cos}(\beta) \, dt \tag{18.17}$$

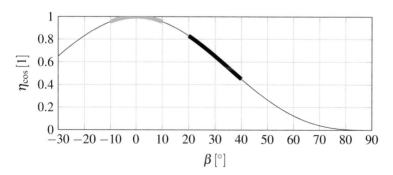

Fig. 18.5 Plot of Eq. (18.15) for $\beta \in [-30°, 90°]$

with initial time t_0 and period time T. Consider that the kite is flown in circles or lying figure eights with $\overline{\beta} - \Delta\beta_{max} \leq \beta \leq \overline{\beta} + \Delta\beta_{max}$ with maximum elevation variation due to the flight path of $\Delta\beta_{max} \approx 10°$ or below. Consider further that the kite's speed is approximately constant. Then the mean elevation (i.e. approximately the elevation of the circle's center or the figure eight's intersection point) is in view of Fig. 18.5 good to estimate $\overline{\eta}_{cos}$, i.e. the following assumption can be made:

Assumption 18.3:   *The mean cosine efficiency $\overline{\eta}_{cos}$ can be approximated with the cosine efficiency at mean elevation $\overline{\beta}$,*

$$\overline{\eta}_{cos} \approx \eta_{cos}\left(\overline{\beta}\right) = \cos^3\overline{\beta}. \tag{18.18}$$

In the following, usually mean values are considered.

18.2.3 Cosine Efficiency With Tower

Derived from an initial solution without tower (case A), Fig. 18.6 sketches modified kite power systems (cases B...D). Hereby, $\overline{x}_k$ is the mean horizontal distance, $\overline{h}_k$ is the mean operation altitude of the kite and h_{to} is the tower height. Cases B–D have the following modifications compared to the initial ground-tethered case A:

B: Tower-tethered; $\overline{h}_k$ and $\overline{x}_k$ unchanged, leading to shorter l_{te} and decreased $\overline{\beta}$.
C: Tower-tethered; $\overline{h}_k$ and l_{te} unchanged, leading to larger $\overline{x}_k$ and decreased $\overline{\beta}$.
D: Ground-tethered with longer l_{te} while $\overline{h}_k$ unchanged, leading to a larger $\overline{x}_k$ and decreased $\overline{\beta}$.

In case B, and stronger in case C, $\overline{\beta}$ is decreased and thus $\overline{\eta}_{cos}$ is increased. A disadvantage of case C compared to case B is the increased $\overline{x}_k$, particularly if a

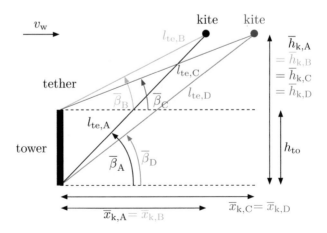

Fig. 18.6 Possible kite power system modifications with a tower or a longer tether

kite power farm or distance restrictions to urban areas are regarded. Moreover, as l_{te} is smaller in case B compared to case C, the mass and drag losses of the tether are reduced (assuming that the tether drag coefficient is proportional to the tether length as in [15]). Case D increases $\overline{\eta}_{cos}$ without a tower while maintaining $\overline{h}_k$, by increasing l_{te}. However, $\overline{x}_k$ and the tether drag and mass losses are increased, and only with $l_{te} \to \infty$ the cosine loss is zero. So this alternative is with limited value, unless a multiple kite system as in [7] is considered, which however would lead to increased complexity and requires further research.

Remark 18.1: *Note that all cases are considered with the same operational altitude $\overline{h}_k$, so that wind shear has no effect on efficiency comparisons. Moreover, as shown in [24, pp. 476], a higher elevation angle is hardly a good choice of tapping stronger winds in higher altitudes for crosswind kite power: "Unless the shear exponent is remarkably high, the best AWT is that which flies at near the minimum practical tether inclination." [24, p. 477]*

Via trigonometric relations, elevation and cosine efficiency are given by

$$\overline{\beta} = \arcsin \frac{\overline{h}_k - h_{to}}{l_{te}} = \arctan \frac{\overline{h}_k - h_{to}}{\overline{x}_k} \tag{18.19}$$

$$\Rightarrow \overline{\eta}_{cos} = \cos^3 \arcsin \frac{\overline{h}_k - h_{to}}{l_{te}} = \cos^3 \arctan \frac{\overline{h}_k - h_{to}}{\overline{x}_k}. \tag{18.20}$$

Obviously, $\overline{\beta} = 0$ and thus $\overline{\eta}_{cos} = 1$ if $\overline{h}_k = h_{to}$.

18.2.4 Numerical Results

Figure 18.7 shows numerical results for increasing tower heights $h_{to} \in [0, \overline{h}_k]$ for two different initial elevations, revealing that $\overline{\eta}_{cos}$ increases almost linearly for small h_{to}. Table 18.1 shows the results for two possible tower heights which are smaller then the operation altitude of the kite: Even if the tower height is only half of the kite's altitude, an efficiency gain of up to 1.89 in case C is possible. If the tower height is only a third of the kite's altitude, still an increase of 1.25 in case B can be achieved. Regarding a mean operation altitude of $\overline{h}_k = 225\,m$ (as projected for the Makani M600 [14]), the tower heights in the two examples of Table 18.1 are $h_{to} \approx 113\,m$ or $h_{to} \approx 74\,m$, respectively. As today's conventional wind turbines have hub heights of up to $\approx 150\,m$, these figures seem feasible. However, as visualized in Fig. 18.7, a considerable efficiency gain is only achievable if the tower height to operation altitude ratio is not too small. Consequently, the efficiency gain and effectiveness of a tower would be rather low for kite operation altitudes above $\overline{h}_k \approx 1000\,m$.

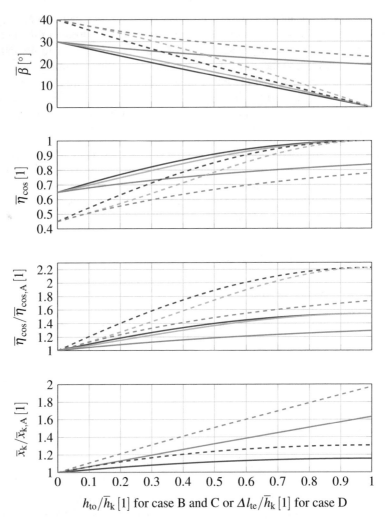

Fig. 18.7 Numerical results for increasing tower heights h_{to} for cases B (——) and C (——), or for increasing tether lengths Δl_{te} for case D (——). From top to bottom: Elevation, cosine efficiency, cosine efficiency gain, horizontal distance normalized to initial horizontal distance. Results for initial elevation $\overline{\beta}_A = 30°$ are drawn-through (——) and for $\overline{\beta}_A = 40°$ are dashed (- - -)

18.2.5 Further Efficiency Increase Effects through a Tower

A tower has further beneficial effects, particularly for drag power where the kite is heavy carrying the generators and the tether is heavy and thick due to integrated cables: As mentioned in Sect. 18.2.3, the tether can be shorter to reach the desired altitude reducing airborne mass and tether drag (assuming that the tether drag coefficient is proportional to the tether length as in [15]). The tether drag reduction

$\frac{h_{to}}{h_k}$ [1] for cases B and C or $\frac{\Delta l_{te}}{h_k}$ [1] for case D	case	$\frac{\overline{\eta}_{cos}}{\eta_{cos,A}}$ [1] for $\overline{\beta}_A = 30°$	$\frac{\overline{\eta}_{cos}}{\eta_{cos,A}}$ [1] for $\overline{\beta}_A = 40°$
	B	1.37	1.74
0.50	C	1.40	1.89
	D	1.19	1.48
	B	1.25	1.47
0.33	C	1.28	1.64
	D	1.13	1.36

Table 18.1 Numerical results for two smaller tower heights for cases B and C, or for increased tether lengths for case D

leads to an increased apparent wind speed which further leads to an increased aerodynamic force, see Eqs. (18.3)–(18.7). As a tower reduces β, the aerodynamic force $F_a \sim \cos^2 \beta$ is additionally increased. All three aforementioned effects can increase the strength to weight ratio $F_a/(m_k g)$ with gravitational acceleration $g = 9.81\,\mathrm{m/s^2}$, leading to a reduced impact of the airborne mass on the efficiency and cut-in wind speed.

18.3 Proposed Tower Concepts

Figure 18.8 illustrates possible tower concepts for different kite power concepts. The proposed tower is a steel framework supported by suspension lines reducing the bending moment absorbed by tower and foundation. In special cases, the suspension lines can absorb the majority of the kite's force. Such a tower can be cost-effective and transported in small parts and mounted on site. Similar to a conventional wind turbine, only a small area is occupied and the area around can be used e.g. for agriculture. Moreover, after its lifetime, such a steel framework tower has a high recyclability. Note, that such a tower concept is not an option for conventional wind turbines, as the rotor disk would intersect with suspension lines, see Fig. 18.9. In crosswind kite power the suspension lines do not intersect with the tether or the kite, even if $\beta \in [-30°, 30°]$ (due to flight path), as the elevation of the suspension lines β_s may be designed e.g. $\beta_s > 60°$. Though, a certain safety distance also for transient situations must be considered in a detailed design.

A steel framework tower with suspension lines is also an option for offshore deployment. Figure 18.10 illustrates a possible concept, which could also be simpler than an offshore tower for conventional wind turbines due to the possibility to absorb a major portion of the tower's moments with suspensions lines.

In all cases in Fig. 18.8, the tower top is yawable for wind alignment. For drag power, a vertical winch on the tower top can be used, as pursued by Makani Power/Google. For lift power, the following solutions are imaginable to avoid tether

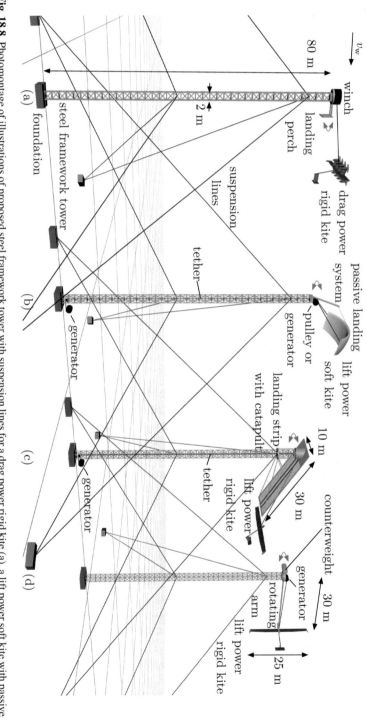

Fig. 18.8 Photomontage of illustrations of proposed steel framework tower with suspension lines for a drag power rigid kite (a), a lift power soft kite with passive start and landing (b), a lift power rigid kite with catapult start and landing strip (c), and a lift power rigid kite with rotating arm start and landing (d)

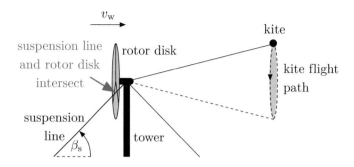

Fig. 18.9 Tower with suspension lines: conventional wind turbine vs. crosswind kite power with tower

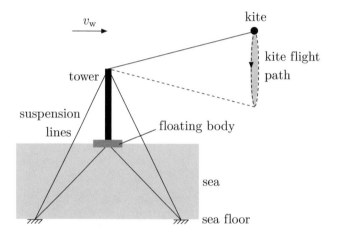

Fig. 18.10 Tower with suspension lines for offshore deployment

twisting during wind alignment: (i) Vertical axis winch with generator on the tower top. (ii) Horizontal axis winch and generator on yawing system on the tower top. (iii) Vertical axis winch and generator inside the tower on ground with a respective tether guidance system. (iv) Horizontal axis winch and generator on ground, pulley system for wind alignment on tower top and tether that allows a twist (inside the tower) of at least ±180°. (v) Same as (iv) without tether twist, but whole tower with winch and generator on ground is yawable (though probably most expensive and only feasible for small towers). More detailed studies on tether wear and tether guidance are required to evaluate the best solution for lift power.

Figure 18.11 depicts the force diagram for a tower for a drag power system with the assumption that the tower only absorbs compression forces. This simplification can be made for an offshore tower on a floating platform, but an onshore tower with foundation would also absorb a portion of the tether force. In this simplified 2D consideration, tower force and suspension line force are given via trigonometric

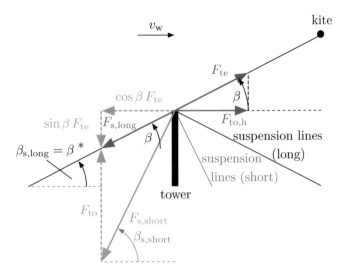

Fig. 18.11 Force distribution with long (red) and short (orange) suspension lines for a drag power system with the assumption that the tower only absorbs compression forces

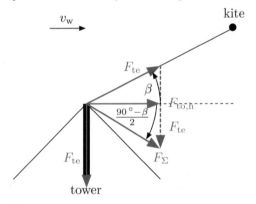

Fig. 18.12 Force distribution for a lift power system with ground-based generator

relations by $F_s = \cos\beta / \cos\beta_s F_{te}$ and $F_{to} = (\tan\beta_s \cos\beta - \sin\beta)F_{te}$. Consequently, the compression force of the tower and the force in the suspension lines increase with the elevation of the suspension lines. For a lift power system, the situation is more complex and unfortunately less beneficial: If the generator is placed on the tower top, the forces are the same as in Fig. 18.11 (with the assumption that the tower only absorbs compression forces), but the generator introduces a torque that tower, suspension lines and foundation have to withstand—this is the torque with which the actual power is generated. It translates either in a bending moment of the tower for a horizontal axis generator or into a twisting moment of the tower for a vertical axis generator. If the generator is placed on ground, a pulley has to direct the tether force downwards to the generator and leads to a different resultant force which needs to be balanced by tower and suspension lines, as shown in Figure 18.12. In the extreme case of $\beta = 0$, the angle of the resultant force is $45°$ downwards. As the suspension lines also introduce a downward force component, the compression force of the tower can be much higher than compared to Fig. 18.11.

If the elevation angle of suspension lines and kite coincide (superscript * in Fig. 18.11), it might be questioned, why the tower is not replaced by a longer tether. The advantage of the tower of this (unlikely) case would be, that the suspension lines do not move and are not airborne, i.e. they do not contribute to drag nor to the airborne mass and can thus be made cost-efficient from arbitrary materials.

Besides the stated advantages, a tower would have more:

- Autonomous start and landing of the kite is a major challenge, particularly for lift power. This challenge is apparently simplified if the kite is started and landed from a tower, as discussed below in Sect. 18.4.
- Started from the top of the tower, the kite already has a portion of the potential energy of the operation altitude, which reduces energy and time required for the start.
- A wind sensor can be mounted on the tower top. Additionally, several wind sensors can be attached along the height of the tower to allow for a low-cost measurement/estimation of the wind shear and the wind velocity at the kite's altitude. If the wind sensors are mounted on booms similar to a meteorological tower, the measured data would be almost undisturbed, contrary to a conventional wind turbine where a wind sensor is placed on the nacelle behind the rotor. Moreover, as the wind sensors are in upwind direction as seen from the kite, provisions to mitigate gusts (such as the "50-year gust" [4, pp. 214]) can be made and the power output for the near future (magnitude of seconds) can be predicted.
- Another advantage is that a tower allows kite power deployments also in/over forests. If the kite would be anchored to the ground, this would only be possible in a (rather large) cleared area.
- The tower can have additional not kite power-related functions, e.g. a weather station or antennas can be mounted to it.

However a disadvantage is the higher construction mass and costs compared to ground-tethered kites. Moreover, the tower has to withstand about twice the load as a tower of a conventional wind turbine of same power rating: Assume $C_L \gg C_{D,eq}$ and consider the maximum power point $\phi, \beta = 0$. Assume also, that the tower has to withstand only the tether force, so the horizontal tower force is $F_{to,h} = F_{te}$ (see Fig. 18.11 or 18.12). Using the aerodynamic force as given by Eq. (18.3) with apparent wind speed

$$v_a^* = \frac{2C_L}{3C_{D,eq}} v_w \tag{18.21}$$

that occurs at the maximum power as given by Eq. (18.16) [6], the ratio of the horizontal tower force $F_{to,h}$ to the power P is

$$\left(\frac{F_{\text{to,h}}}{P}\right)_{\text{kite}} = \frac{F_{\text{L}}\left(v_{\text{a}} = v_{\text{a}}^*\right)}{P_{\text{max}}} = \frac{\frac{1}{2}\rho A\left(\frac{2}{3}\frac{C_{\text{L}}}{C_{\text{D,eq}}}v_{\text{w}}\right)^2 C_{\text{L}}}{\frac{2}{27}\rho A v_{\text{w}}^3 \frac{C_{\text{L}}^3}{C_{\text{D,eq}}^2}} = \frac{3}{v_{\text{w}}}. \tag{18.22}$$

Thrust and power of a conventional wind turbine (CWT) can be formulated by

$$T_{\text{CWT}} = 2\rho A_{\text{s}} v_{\text{w}}^2 a\left(1 - a\right), \tag{18.23}$$

$$P_{\text{CWT}} = 2\rho A_{\text{s}} v_{\text{w}}^3 a\left(1 - a\right)^2, \tag{18.24}$$

where A_{s} is the swept area and a is the inflow or induction factor [18]. At optimal inflow factor $a^* = \frac{1}{3}$ [18], the horizontal tower force to power ratio is

$$\left(\frac{F_{\text{to,h}}}{P}\right)_{\text{CWT}} = \frac{T_{\text{CWT}}\left(a^* = \frac{1}{3}\right)}{P_{\text{CWT}}\left(a^* = \frac{1}{3}\right)} = \frac{2\rho A_{\text{s}} v_{\text{w}}^2 \frac{1}{3}\left(1 - \frac{1}{3}\right)}{2\rho A_{\text{s}} v_{\text{w}}^3 \frac{1}{3}\left(1 - \frac{1}{3}\right)^2} = \frac{1.5}{v_{\text{w}}} \tag{18.25}$$

which is only half of what the tower for a kite has to withstand (compare with Eq. (18.22)).

Nevertheless, considering a kite power system with $P = 5\,\text{MW}$ at (rated) wind speed of $v_{\text{w}} = 10\,\text{m/s}$, the horizontal tower force is

$$F_{\text{to,h}} = \left(\frac{F_{\text{to,h}}}{P}\right)_{\text{kite}} P = \frac{3}{v_{\text{w}}}P = \frac{3}{10\,\text{m/s}} \times 5\,\text{MW} = 1.5\,\text{MN} \approx 153\,\text{t}. \tag{18.26}$$

Considering that today's cranes can reach more than 100 m height and can lift many hundred tons (see e.g. [19]), it seems possible to design and construct a tower which can withstand the loads of a kite.

18.4 Start and Landing from the Top of a Tower

As mentioned, a tower can have advantages for the autonomous start and landing of the kite. In the following, four major starting and landing concepts are discussed (see Fig. 18.8), with focus on the advantages of the use of a tower.

18.4.1 Drag Power Kites and Vertical Take Off and Landing (VTOL) Lift Power Kites

A drag power (rigid) kite can hover to and from a perch mounted on top of the tower, see Fig. 18.8 (a). Particularly, the kite may have a long tail (if the tower is at least as high as the tail long). Consequently, there is no need for tiltable propellers or a tiltable tail as in [2], which were required if the kite should land on the (flat) ground. This concept is indeed pursued by the company Makani Power/Google [21], however with a relatively small tower as visible in Fig. 18.3 right. Regarding an operation altitude of $\bar{h}_k \approx 225\,\text{m}$ and tether length $l_{te} = 440\,\text{m}$ (taken from [14] from "M600 Specs"), i.e. $\bar{\beta}_A \approx 30.75°$, and assuming a tower height of $h_{to} = 30\,\text{m}$, the efficiency gain is already $\overline{\eta}_{\cos}/\overline{\eta}_{\cos,A} \approx 1.13$ (assuming case C in Fig. 18.6) compared to a ground-tethered solution.

If the kite has no tail, as it was pursued by the company Joby Energy [16], the kite could also land on a platform similar to a heliport on top of the tower, or alternatively on ground next to the tower for simpler maintenance from ground.

Similar concepts are also applicable for lift power kites with onboard propellers, which are used for vertical take-off and landing (VTOL) only.

18.4.2 Passive Method for Light (Soft) Lift Power Kites

If a tower is used and the kite is started from the top of the tower, the kite is already exposed to a wind speed that is, depending on the tower height, close to the wind speed of the operation altitude. This allows for a "passive" start and landing just with the help of the wind speed, which is hardly possible if the kite shall be started from ground, due to wind shear. The company SkySails [9, 23] pursues such a concept for offshore applications, which is—with a tower—also possible for onshore sites, see Fig. 18.8 (b).

However, a "passive" concept is only feasible, if the kite is light enough, which is usually the case for soft kites only. The maximum kite mass can be estimated by

$$m_k g \leq F_{L,\max} = \frac{1}{2}\rho A v_{w,\text{cut-in},h_{to}}^2 C_{L,\max}$$

$$\Leftrightarrow \frac{m_k}{A} \leq \frac{\rho v_{w,\text{cut-in},h_{to}}^2 C_{L,\max}}{2g}. \tag{18.27}$$

Regarding a cut-in wind speed at the tower height of $v_{w,\text{cut-in},h_{to}} = 3\,\text{m/s}$, air density $\rho = 1.2\,\text{kg/m}^3$ and a soft kite with maximum lift coefficient of $C_{L,\max} = 1$, the result is $m_k/A \leq 0.55\,\text{kg/m}^2$. For $v_{w,\text{cut-in},h_{to}} = 4\,\text{m/s}$ and otherwise identical values, the result is $m_k/A \leq 0.98\,\text{kg/m}^2$.

18.4.3 Catapult-Method for (Rigid) Lift Power Kites

Similar to a fighter jet on an aircraft carrier, a heavier rigid lift power kite can be started in a catapult launch, as pursued e.g. by Ampyx Power [17] or ABB [12]. The kite is launched with a high acceleration powered by the winch or by a catapult technology, such as linear motors (which are also used e.g. for roller coasters). Small onboard propellers may help to climb to the operation altitude. To allow for a short landing strip, the kite is caught and stopped by the ground winch or an additional braking system such as a hook on the kite and lines on the landing strip.

As shown in Fig. 18.8 (c), a landing strip on a tower can be inclined (quite intensely). Consequently, during launch, the kite is catapulted already with an upward component, instead of purely horizontal. The kite needs to climb to a minimum operation altitude $h_{k,min}$ and minimum operation tether length, i.e. from which it can start to fly crosswind motions and make the remaining way with the wind power. Regarding that the kite has to climb to $h_{k,min}$ alone with the kinetic energy from the catapult, the required start speed $v_{k,start}$ can be approximated by conservation of energy (see also [3, Chap. 4.3])

$$E_{kin,start} = E_{kin,min} + E_{pot,min-op}$$
$$\frac{1}{2}m_k v_{k,start}^2 = \frac{1}{2}m_k v_{k,min}^2 + m_k g(h_{k,min} - h_{to})$$
$$\Leftrightarrow v_{k,start} = \sqrt{v_{k,min}^2 + 2g(h_{k,min} - h_{to})} \tag{18.28}$$

where $E_{kin,start}$ is the kinetic energy at the end of the catapult, $E_{kin,min}$ is the minimum kinetic energy with speed $v_{k,min}$, and $E_{pot,min-op}$ is the minimum operation potential energy. With acceleration a, the catapult length l_c is given by

$$l_c = \frac{v_{k,start}^2}{2a} = \frac{v_{k,min}^2 + 2g(h_{k,min} - h_{to})}{2a} \tag{18.29}$$

where Eq. (18.28) is inserted. Taking the requirements of the Ampyx Power-Plane from [3, p. 29, Table 3.1] as example with $v_{k,min} = 22\,\text{m/s}$, $a = 50\,\text{m/s}^2$, $h_{k,min} = 125\,\text{m}$, and the example tower height of $h_{to} = 80\,\text{m}$, the catapult (or landing strip-) length has to be only $l_c \approx 13.7\,\text{m}$, i.e. less than half as long as sketched in Fig. 18.8 (c). For a 30 m long catapult and otherwise identical values, an altitude of up to $\approx 208\,\text{m}$ can be reached. This implies that the kite might not need further measures like propellers for the start.

Concerning the landing, the kite can approach on a low altitude even below the landing strip and is put into a steep climb shortly before touch-down. As a consequence, a portion or the complete kinetic energy of the kite can be converted into potential energy of the kite itself. This effect can be approximated by conservation of energy

$$E_{pot} = m_k g \Delta h = \frac{1}{2}m_k v_{k,min}^2 = E_{kin,min} \tag{18.30}$$

$$\Leftrightarrow \Delta h = \frac{v_{k,min}^2}{2g} \tag{18.31}$$

where Δh is the maximum height the kite can climb with its kinetic energy $E_{kin,min}$ at $v_{k,min}$. Regarding that the kite approaches the landing strip with speed $v_{k,min} = 22\,\text{m/s}$, the maximum climb height is $\Delta h \approx 25\,\text{m}$. Consider Fig. 18.8 (c) with tower height $h_{to} = 80\,\text{m}$, landing strip inclination $30°$ and length 30 m, i.e. the landing strip starts at altitude $\approx 80\,\text{m} - \sin(30°) \times 30\,\text{m} = 65\,\text{m}$. The kite would be stopped by potential energy alone, if it approaches in approximately 55 m altitude or 10 m below the start of the landing strip. This implies that no further braking system might be required.

The tether can help to guide the kite during landing. However, if during approaching e.g. a crosswind gust hits the kite in the situation shown in Fig. 18.8 (c) with the danger of a crash landing, the landing can hardly be aborted with a subsequent retry, especially if the kite has no propeller. One exit of such a situation could be the separation from the tether and an emergency landing on ground. A small rocket engine on the kite (with fuel just for a few seconds) may help to catapult the kite away and avoid a collision with the tower or suspension lines. By any means, the landing strip should be broad enough for a low probability of the need to perform such an emergency landing.

18.4.4 Rotating Arm-Method for Lift Power Kites

Another start and landing method, is the "rotating arm" method [3, 13]: The kite is rotated at the tip of an arm and slowly released. Operation altitude and -tether length are approached with a helix flight path. The landing is (roughly) the reverse motion.

If such a concept is implemented on ground, the kite would need to start with a (very) small roll angle, i.e. the kite's wings are approximately parallel to the ground (otherwise the outer wing would intersect with the ground), which complicates the design and control and may require a long arm. Moreover, the circular area enclosed with the radius of the arm's length plus half of the wing span cannot be accessed and used e.g. for agriculture. To mitigate these problems, the company EnerKite [8] intends to use a telescoping arm to which the kite is attached for the start and the landing, and retracted during power generation.

If the rotating arm is attached to the top of a tower, as illustrated in Fig. 18.8 (d) (which is also similar to [13, Fig. 1.2 (c)]), these disadvantages are not existent: The kite can be attached to the arm with a high roll angle, even up to 90°, i.e. the wings are parallel to the tower (regarding that the arm is long enough and the elevation of the suspension lines is large enough). The length of the arm has no effect on the occupied ground area. However, the (heavy) generator might need to be placed on the top of the tower and needs to be rotated with the arm to avoid tether twisting during start and landing.

18.5 Conclusions and Outlook

In this chapter, the possible benefits of a tower in crosswind kite power technology are discussed: Although it is a counterintuitive and contrary approach to tether the kite to the top of a tower, as this wind energy technology does not rely on a tower, a significant efficiency increase can be obtained yielding up to more than the double of the power and energy output compared to a ground-tethered kite. A tower design based on a steel framework and suspension lines is proposed. The advantages of a tower can be summarized as follows:

- Significant increase of cosine efficiency, or decrease of cosine loss almost down to its elimination.
- Decrease of tether drag and mass losses as the tether is shorter to reach the same altitude without increase of horizontal distance.
- Increase of the strength to weight ratio.
- (Apparent) Simplification of start and landing for both, drag power and lift power kites.
- Lower potential energy demand for the start, if the kite is started from the top of the tower.
- Simplified wind velocity measurement/estimation for the kite's altitude.
- Possibility to deploy kite power in/over forests (without the need for a clearing).
- Multi-functionality of the tower, e.g. by adding a weather station or antennas.
- Compared to conventional wind turbines, a simpler and more cost-effective tower seems possible.

However, disadvantages compared to ground-tethered kites include the higher material demand, higher construction costs and higher maintenance costs. As with conventional wind turbines, the tower needs to transmit the induced bending moment to the ground, while a ground-tethered kite requires only a lightweight tensile structure.

In a future work, dynamic simulations should be carried out to identify the cosine efficiency and loads on tower and suspension lines more accurately. The results can then be used for a specific tower design. An economical investigation which considers, capital, material, transportation, construction, demolition and recycle costs could then quantify the financial impact of the tower on the LCOE and optimize the tower height for a site. As the tower is not a requirement, it is even possible to build a (higher) tower some time after the start of operation to reduce capital costs. The tower design concept (also e.g. concrete and steel tube towers), and reliable start and landing from the top of the tower, are subject to further studies.

Acknowledgements The authors thank the anonymous reviewers and the editors for their helpful comments. This study was supported by Bund der Freunde der TU München e.V.

References

1. Ahrens, U., Diehl, M., Schmehl, R. (eds.): Airborne Wind Energy. Green Energy and Technology. Springer, Berlin Heidelberg (2013). doi: 10.1007/978-3-642-39965-7
2. Bevirt, J. B.: Apparatus for generating power using jet stream wind power. US Patent 20,100,032,947, Feb 2010
3. Bontekoe, E.: How to Launch and Retrieve a Tethered Aircraft. M.Sc.Thesis, Delft University of Technology, 2010. http://resolver.tudelft.nl/uuid:0f79480b-e447-4828-b239-9ec6931bc01f

4. Burton, T., Sharpe, D., Jenkins, N., Bossanyi, E.: Wind Energy Handbook. John Wiley & Sons, Ltd, Chichester (2001). doi: 10.1002/0470846062
5. Cherubini, A., Papini, A., Vertechy, R., Fontana, M.: Airborne Wind Energy Systems: A review of the technologies. Renewable and Sustainable Energy Reviews **51**, 1461–1476 (2015). doi: 10.1016/j.rser.2015.07.053
6. Diehl, M.: Airborne Wind Energy: Basic Concepts and Physical Foundations. In: Ahrens, U., Diehl, M., Schmehl, R. (eds.) Airborne Wind Energy, Green Energy and Technology, Chap. 1, pp. 3–22. Springer, Berlin Heidelberg (2013). doi: 10.1007/978-3-642-39965-7_1
7. Diehl, M., Horn, G., Zanon, M.: Multiple Wing Systems – an Alternative to Upscaling? In: Schmehl, R. (ed.). Book of Abstracts of the International Airborne Wind Energy Conference 2015, p. 96, Delft, The Netherlands, 15–16 June 2015. doi: 10.4233/uuid:7df59b79-2c6b-4e30-bd58-8454f493bb09. Presentation video recording available from: https://collegerama.tudelft.nl/Mediasite/Play/1065c6e340d84dc491c15da533ee1a671d
8. Enerkite GmbH. http://www.enerkite.com/. Accessed 14 Jan 2016
9. Erhard, M., Strauch, H.: Control of Towing Kites for Seagoing Vessels. IEEE Transactions on Control Systems Technology **21**(5), 1629–1640 (2013). doi: 10.1109/TCST.2012.2221093
10. Fagiano, L.: Control of tethered airfoils for high-altitude wind energy generation. Ph.D. Thesis, Politecnico di Torino, 2009. http://hdl.handle.net/11311/1006424
11. Fagiano, L., Milanese, M.: Airborne Wind Energy: an overview. In: Proceedings of the 2012 American Control Conference, pp. 3132–3143, Montréal, QC, Canada, 27–29 June 2012. doi: 10.1109/ACC.2012.6314801
12. Fagiano, L., Schnez, S.: The Take-Off of an Airborne Wind Energy System Based on Rigid Wings. In: Schmehl, R. (ed.). Book of Abstracts of the International Airborne Wind Energy Conference 2015, pp. 94–95, Delft, The Netherlands, 15–16 June 2015. doi: 10.4233/uuid:7df59b79-2c6b-4e30-bd58-8454f493bb09. Presentation video recording available from: https://collegerama.tudelft.nl/Mediasite/Play/2ebb3eb4871a49b7ad70560644cb3e2c1d
13. Geebelen, K., Gillis, J.: Modelling and control of rotational start-up phase of tethered aeroplanes for wind energy harvesting. M.Sc.Thesis, KU Leuven, June 2010
14. Hardham, C.: Response to the Federal Aviation Authority. Docket No.: FAA-2011-1279; Notice No. 11-07; Notification for Airborne Wind Energy Systems (AWES), Makani Power, 7 Feb 2012. https://www.regulations.gov/#!documentDetail;D=FAA-2011-1279-0014
15. Houska, B., Diehl, M.: Optimal control for power generating kites. In: Proceedings of the 9th European Control Conference, pp. 3560–3567, Kos, Greece, 2–5 July 2007
16. Joby Energy. http://www.jobyenergy.com/. Accessed 14 Jan 2016
17. Kruijff, M., Ruiterkamp, R.: Status and Development Plan of the PowerPlane of Ampyx Power. In: Schmehl, R. (ed.). Book of Abstracts of the International Airborne Wind Energy Conference 2015, pp. 18–21, Delft, The Netherlands, 15–16 June 2015. doi: 10.4233/uuid:7df59b79-2c6b-4e30-bd58-8454f493bb09. Presentation video recording available from: https://collegerama.tudelft.nl/Mediasite/Play/2e1f967767d541b1b1f2c912e8eff7df1d
18. Kulunk, E.: Aerodynamics of Wind Turbines. INTECH Open Access Publisher (2011). doi: 10.5772/17854
19. Liebherr GmbH: Lattice boom mobile crane LG 1750. http://www.liebherr.com/en/deu/products/mobile-and-crawler-cranes/mobile-cranes/lg-lattice-mast-cranes/details/lg1750.html. Accessed 14 Jan 2016
20. Loyd, M. L.: Crosswind kite power. Journal of Energy **4**(3), 106–111 (1980). doi: 10.2514/3.48021
21. Makani Power/Google. http://www.google.com/makani. Accessed 14 Jan 2016
22. Schmehl, R., Noom, M., Vlugt, R. van der: Traction Power Generation with Tethered Wings. In: Ahrens, U., Diehl, M., Schmehl, R. (eds.) Airborne Wind Energy, Green Energy and Technology, Chap. 2, pp. 23–45. Springer, Berlin Heidelberg (2013). doi: 10.1007/978-3-642-39965-7_2
23. Skysails GmbH. http://www.skysails.info. Accessed 14 Jan 2016
24. Vander Lind, D.: Analysis and Flight Test Validation of High Performance Airborne Wind Turbines. In: Ahrens, U., Diehl, M., Schmehl, R. (eds.) Airborne Wind Energy, Green Energy

and Technology, Chap. 28, pp. 473–490. Springer, Berlin Heidelberg (2013). doi: 10.1007/
978-3-642-39965-7_28

25. Vander Lind, D.: Developing a 600 kW Airborne Wind Turbine. In: Schmehl, R. (ed.). Book of
abstracts of the International Airborne Wind Energy Conference 2015, pp. 14–17, Delft, The
Netherlands, 15–16 June 2015. doi: 10.4233/uuid:7df59b79-2c6b-4e30-bd58-8454f493bb09.
Presentation video recording available from: https://collegerama.tudelft.nl/Mediasite/Play/
639f1661d28e483cb75a9a8bdedce6f11d

Chapter 19
Multicopter-Based Launching and Landing of Lift Power Kites

Florian Bauer, Christoph M. Hackl, Keyue Smedley and Ralph M. Kennel

Abstract Crosswind kite power is a promising alternative wind power technology. However, unlike the rotor blades of a conventional wind turbine, a kite needs to be launched prior to power generation and needs to be landed during low-wind conditions or for maintenance. This study proposes multicopter-based concepts for an autonomous solution. Basic system components and different system configurations are discussed. Static and dynamic feasibility analyses are carried out. Results show that such systems are feasible and have advantages compared to other launching and landing concepts. However, also the weaknesses of such systems become apparent e.g. the increased airborne mass.

19.1 Introduction

Crosswind kite power is becoming more and more attractive in both academia and industry (see e.g. [2, 8, 31] and references therein) and is considered as promising alternative wind power technology: Compared to conventional wind turbines, kites can harvest wind power at higher altitudes with stronger and steadier winds, but

Florian Bauer (✉) · Ralph M. Kennel
Institute for Electrical Drive Systems and Power Electronics, Technical University of Munich, Arcisstrasse 21, 81477 Munich, Germany
e-mail: florian.bauer@tum.de

Christoph M. Hackl
Munich School of Engineering, Research group "Control of renewable energy systems (CRES)", Technische Universität München, Lichtenbergstr. 4a, 85748 Garching, Germany

Keyue Smedley
The Henry Samueli School of Engineering, Power Electronics Laboratory, University of California, Irvine, CA 92697, USA

© Springer Nature Singapore Pte Ltd. 2018
R. Schmehl (ed.), *Airborne Wind Energy*, Green Energy
and Technology, https://doi.org/10.1007/978-981-10-1947-0_19

Florian Bauer, Christoph M. Hackl, Keyue Smedley and Ralph M. Kennel

only by needing a fraction of the construction material. Hence, it promises to have a higher capacity factor, lower capital investments, and in the end a lower levelized cost of electricity [2, 31]. Mechanical output powers of two megawatts were already achieved by a commercial product of the company SkySails [11].

Two beneficial concepts to generate power with a kite are also referred as "crosswind kite power" [2, 8, 31]: (i) In "lift power" [31],[1] the kite is tethered to a winch on ground which is connected to an electrical drive that can be operated as motor or generator. The kite is flown in crosswind motions like figure eights with a high speed and a high lift force and pulls the tether from the winch. The winch drive counteracts by generative braking, i.e. it is operated as generator and electric energy is generated. Before the tether is pulled out completely, the kite is flown towards a low force position like the zenith and then reeled in. A rigid kite can also dive towards the ground station. During this reel-in phase the ground winch drive is operated as motor, but only a fraction of the generated energy is dissipated. (ii) In "drag power" [31],[2] onboard wind turbines are attached to a rigid kite or to an airborne unit beneath a soft kite. The kite is also flown in crosswind motions with a high speed, but with constant tether length. The turbines generate electric power which is transmitted to the ground via electrical cables that are integrated in the tether.

Unlike the rotor blades of a conventional wind turbine, the kite needs to be launched prior to power generation and needs to be landed during low-wind conditions or for maintenance. This challenge can be seen as solved for the drag power principle with rigid kites which is pursued e.g. by the company Makani Power/Google [2, 33]: They are developing rigid kites with a tether with integrated electrical cables. The kite is launched and retrieved by using the turbines as propellers in motor mode like a multicopter. This concept seems very successful, robust and autonomous. It is independent of the wind speed near the ground station. Most components required for launch and retrieval are already present. However, in contrast to the lift power principle, the drag power principle incorporates that the masses of all generators are carried by the kite and the need for a tether with integrated electrical cables. The latter results in a more expensive tether which is also heavier and thicker and thus reduces efficiency. To reduce the tether's mass, a high voltage is chosen for the cables but complicates the electrical system. Another challenge is the reduction of the noise of the propellers.

The implementation of completely autonomous and robust launch and retrieval is a major challenge for the lift power principle. Most tested launch and retrieval concepts for lift power kites suffer at least one of the following: (i) A material-intensive and complex ground station, (ii) requirement of strong and constant wind near the ground, or (iii) challenging control. This is contrary to the drag power principle and Makani Power's/Google's successful multicopter launch and retrieval. Another advantage can be seen in a simpler implementation of "dancing kites" [23, 42] as the two kites can hover with a low speed side-by-side for the launching and landing. These are motivations for this study: It considers multicopter-based con-

---

[1] Also called "traction power", "ground-", "pumping mode power generation" or "ground-gen".

[2] Also called "onboard-", "continuous power generation" or "fly-gen".

cepts for lift power-operated kites. In fact, the companies e-kite and TwingTec recently announced in [7, 32] to pursue such a concept. Also the company Kitemill pursues such a concept now [28]. However, no detailed studies can be found. The contributions of this study can thus be summarized as follows: (i) Exploration of multicopter-based launching and landing concepts for lift power kites, (ii) proposal of two new concepts for soft kites and proposals for low-weight solutions through detachable electrical cables, (iii) formulation of simple models for static feasibility analyses with example results and (iv) presentation of results of dynamic feasibility analyses (multi-body simulation) for a soft kite solution.

This chapter is organized as follows: The next section presents previously investigated launch and retrieval concepts. Section 19.3 discusses different concepts of a lift power system with multicopter launching and landing. Sections 19.4–19.5 present static and dynamic feasibility analyses. Section 19.6 discusses the results as well as advantages and disadvantages of such multicopter concepts for lift power kites. Finally, Sect. 19.7 gives conclusions and an outlook. The preliminary content of the present chapter has been presented at the Airborne Wind Energy Conference 2015 [4].

19.2 Other Launching and Landing Methods–Related Works

For first experiments for both, soft kites and rigid kites, many companies and research groups use a conventional winch launch with the ground station winch after the kite is placed at some distance by the testing team. For the landing, a rigid kite may be disconnected from the tether and land like a conventional sailplane while a soft kite may be steered to one side of the wind window for a "soft crash landing". In the following, existing launch and retrieval concepts for a commercial deployment found in literature are summarized.

The company SkySails [38, 41] uses a telescope mast to pull a ram-air kite out of its storage. An extra tether connects the leading edge of the kite with the mast's tip. When the kite is inflated by the wind, it is released upwards. The retrieval is performed in the reverse order. This concept seems successful, robust and autonomous (neglecting the storing of the folded kite). However, a relatively strong and constant wind near the ground and/or a high mast is required. The dutch company e-kite also experimented with a similar mast-launch for ram-air kites [7]. Unlike SkySails, no extra tether is connected to the kite's leading edge. Instead the kite is lifted by a metal rack. However, the concept did only work under "ideal situations" [7].

Geebelen and Gillis [19] propose a centrifugal launch and retrieval. This is also persued by the company EnerKite [10] and is being further investigated e.g. in [17, 18]. A (rigid) kite is attached to a rotating arm. Through its inertia and lift forces the kite is released and finally flies downwind with a helical-like flight path. An advanced control method e.g. nonlinear model predictive control is used to control the kite's trajectory. Bontekoe [5] extended this concept by adding a propeller to the rigid kite to support the launch and retrieval with longitudinal thrust. A centrifugal launch and retrieval is independent of the wind speed near the ground. It works well in simulations and small prototypes. A disadvantage is the relatively complex and material-intensive ground station.

In his master thesis, Haug [22] started from scratch and evaluated several part solutions from a mind map. He also considered a concept in which multicopters lift the kite, but discarded such concepts because an "unexpected reaction of one UAV [unmanned aerial vehicle] can lead to hazardous outcomes" [22, p. 35]. Instead he proposed a mast or crane-like construction on which the soft kite hangs upside down and is launched with the help of the wind. Haug was able to launch and retrieve the kite several times, but the kite also crashed several times into the mast or into suspension lines. A video of a successful launch is online available [36]. A similar concept was pursued by the company KiteGen [27] with the additional support of fans on ground. It is not clear if such a mast concept can be implemented with the required robustness. Another disadvantage of the mast launching and landing is the relatively complex ground station and the need for a relatively strong and constant wind near the ground station. These could be reasons why KiteGen now also seems to experiment with multicopters [25].

The company NTS [2, 35] pursues to use several kites, each tethered to a railroad trolley moving on a circular track. The kite pulls the trolley which counteracts by generative braking similar to an electric train for deceleration. NTS could use the trolley in motor mode and launch and retrieve the kite similar to SkySails' mast concept. Since the trolley can generate enough true air speed, this would particularly also work with no wind near ground and with a small mast. However, a disadvantage of such track- or carousel concepts is the high material demand of the track.

To bring the kite into a starting altitude, Breukels [6] successfully used a helium-filled airship. He did not consider the retrieval of the kite. The company Festo Cyberkite [39] used a helium filled kite such that the kite pulls itself into the air even without wind. In their master theses Bontekoe [5] and Haug [22] also reviewed lighter-than-air concepts to launch the kite. However, only few pursue such concepts today. Reasons may include the helium leakage through any membrane or controllability issues of aerostats.

Alula Energy [3, 40] proposes to catapult a rigid kite into the air. The kite is retrieved by landing slowly on the catapult platform. The launch was tested already on small prototypes, but no references were found if the landing can be performed that way. A disadvantage of Alula's ground station concept is its comparably high material demand. A similar concept is being pursued by the companies Ampyx Power [30] and ABB [14]: Similar to a fighter jet on an aircraft carrier, a rigid kite shall be started in a catapult launch with a high acceleration powered by the winch or by a catapult technology, such as linear motors which are also used e.g. for roller coasters. Small onboard propellers may help to climb to the operation altitude. To allow for a short landing strip, the kite is caught and stopped by the ground winch or an additional braking system such as a hook on the kite and lines on the landing strip. However, the feasibility and robustness of such a concept (without the help of a human pilot) is yet to be demonstrated.

As mentioned in the introduction, most of those concepts suffer at least one of the following: (i) A material-intensive and complex ground station, (ii) requirement of strong and constant wind near the ground, or (iii) challenging control.

19.3 Lift Power Kite Concepts With Multicopter Launching and Landing

In this section the basic system components and different concepts for soft and rigid lift power kites with multicopter launching and landing are explored.

19.3.1 Basic System Components

The basic system components or "building blocks" can be summarized by Fig. 19.1: The kite is attached via one or more force-transmitting tethers to one or more winches on ground. The airborne system (blue in Fig. 19.1) comprises three main functions (from right to left in Fig. 19.1):

- The rigid wing or soft kite generates the aerodynamic forces for power generation.
- If the kite is not solely steered by all ground winches on its flight path during power generation, an airborne control unit is required: In the case of a soft kite, this could be a control pod with winches and steering tethers, whereas a rigid kite

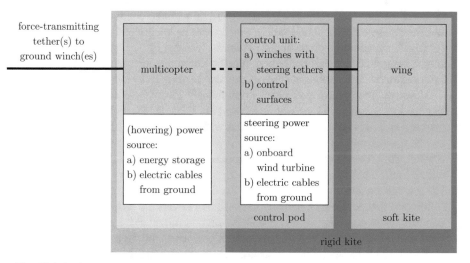

Fig. 19.1 Basic system components or "building blocks" of lift power kites with multicopter launching and landing

can comprise control surfaces. The steering unit needs a steering power source which can either be an onboard wind turbine or electrical cables from ground.

- Finally, a multicopter for launching and landing is required. Its power source can either be an energy storage like lithium-type batteries (which are recharged before/after launching/landing) or electrical cables from ground (which could be identical to the steering power source). The multicopter can be part of the control pod of a soft kite. In the case of a rigid kite, all three functions (wing, control unit, multicopter) can be integrated in one assembly. Note however, that the rated power required by the multicopter should be much lower than the rated system power (i.e. of the ground winch), otherwise it might be more meaningful to use the "drag power" principle.

One or more separate multicopters might be used to launch and land the kite. However, such a concept is not considered here, since it seems too challenging to dock such a drone to the kite for retrieval, particularly in strong and gusty wind conditions. However, a multicopter that can be attached to and detached from the control pod or the rigid kite and moved along the tether, and is thus guided by the tether (light part in Fig. 19.1), is part of the considerations of this study.

19.3.2 Soft Kite Concepts

In the case of a soft kite, a control pod solution, as pursued e.g. by SkySails [38] or TU Delft [15, 29], seems meaningful as one can find the following advantages: (i) Only one winch on ground is required, because the steering is performed by small steering winches onboard the control pod. (ii) The tether drag is minimal, as only one tether connects the airborne system to the ground station. (iii) The steering is direct, as the steering lines are short. (iv) For accurate control a GPS sensor, an inertial measurement unit and further sensors can be integrated into the control pod. (v) Lighting and collision avoidance systems, which might be mandatory in future, can be integrated into the control pod. (vi) A communication system to exchange the information with the onboard systems can also be integrated into the control pod.

However, the control pod needs continuously electric power for steering (see Fig. 19.1). To avoid a tether with integrated electrical cables and its disadvantages, one or more onboard wind turbines (each with electrical drive in generator mode) can be used. One proposal of this study is to use those wind turbines as propellers (by operating their electrical drives in motor mode) for the launching and landing with certain flight maneuvers, and thus turning the control pod into a multicopter. For low mechanical complexity, several fixed pitch propellers are considered. In the following, two possible launch and retrieval maneuvers are proposed, one with up-wind mounted propellers and one with downwind mounted propellers. Hereby up-wind and downwind refers to the propeller's position at the control pod with respect to the relative air velocity vector during normal flight, see Fig. 19.2.

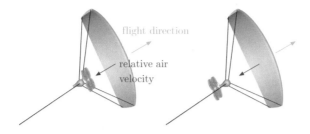

Fig. 19.2 Control pod with upwind mounted propellers (left) and control pod with downwind mounted propellers (right)

Launch and Retrieval Maneuver with Upwind Propellers. Figure 19.3 illustrates the proposed launch maneuver with upwind propellers: The propellers are operated in motor mode with which the pod hovers downwind and into higher altitudes while the kite hangs below and the tether is slack. Then, the tether is pulled from the ground station winch, which should erect the kite. Alternatively or additionally, the pod is steered by its propellers accordingly to erect the kite. The kite then flies towards the zenith, the propellers are turned off and energy generation can be started.

This launch maneuver is similar to a launching paraglider pilot: The pilot unfolds his/her paraglider on ground the same way the kite hangs below the control pod in Fig. 19.3. Then the pilot starts running which erects the kite above the pilot.

Figure 19.4 illustrates the retrieval maneuver with upwind propellers: The kite is flown towards the zenith. At a low altitude, the propellers are powered with full thrust while the tether is slack. In this way the pod accelerates upwards while the kite approximately remains on its position which can be supported by depowering the kite. The pod is partly forced on a circular path by the tethers between pod and kite. As soon as the pod is above the kite, the pod is again operated like a multicopter and hovers back to a landing site.

A disadvantage of the concept with upwind propellers is that during hovering the trailing edge of the kite instead of its leading edge faces towards the wind vector. Consequently, the kite might be out of control. To avoid this risk, a second concept with downwind propellers is proposed.

Launch and Retrieval Maneuver with Downwind Propellers. Figure 19.5 illustrates the launch: Initially the propellers are operated again in motor mode with which the pod hovers downwind and into greater altitudes while the kite hangs below and the tether is slack. Hereby, unlike Figs. 19.3 and 19.4, the leading edge of the kite faces towards the wind vector and thus should be controllable. Then, the tether is pulled from the ground station winch, which should erect the kite. Alternatively or additionally, the pod is steered by its propellers accordingly to erect the

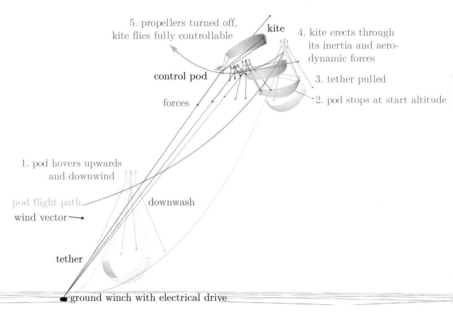

Fig. 19.3 Illustration of the launch maneuver with upwind propellers

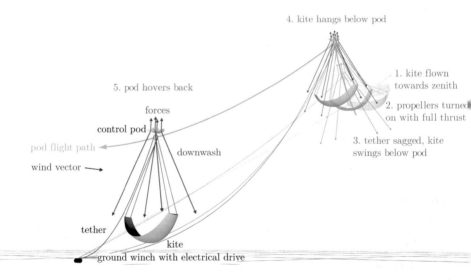

Fig. 19.4 Illustration of the retrieval maneuver with upwind propellers

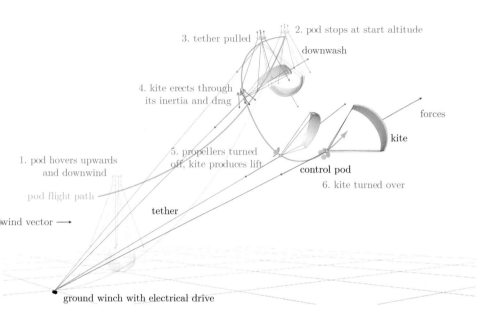

Fig. 19.5 Illustration of the launch maneuver with downwind propellers [4]. Reprinted with permission of R. Schmehl

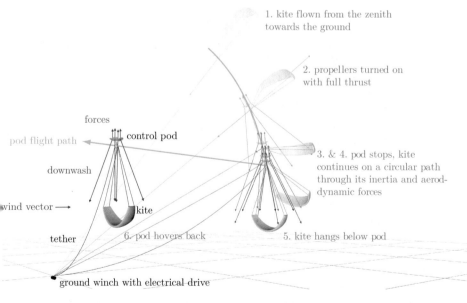

Fig. 19.6 Illustration of the retrieval maneuver with downwind propellers [4]. Reprinted with permission of R. Schmehl

kite. After the propellers are turned off, the kite is turned over and energy generation can be started.

Figure 19.6 illustrates the retrieval maneuver with downwind propellers: The kite is flown from the zenith towards the ground. The kite's speed may be reduced by depowering the kite. Then, the propellers are operated in motor mode with full thrust, while the tether length is kept constant or the tether is slack. The pod stops and the kite swings below. Finally, the pod is again operated as multicopter and hovers back to a landing site.

A disadvantage of the concept with downwind propellers is, that the kite can generate a high aerodynamic force counteracting the propeller forces. Moreover, in both concepts the propeller downwash can hit the kite which is not desirable. Such issues are addressed in the following.

Design Considerations. To solve the challenges imposed by the proposed concepts, certain design criteria are required for both, control pod and kite. In the following the challenges are stated and possible solutions are sketched.

As derived later in Sect. 19.4, the higher the airborne mass and the lower the overall propeller disk (or swept) area (i.e. the higher the disk loading), the higher the power and energy demand of the propellers to hover. Since the propeller disk area and the energy storage size are limited, a combination of the following measures should be taken: (i) The control pod should be made from lightweight materials, e.g. carbon-fiber. Additionally, the kite should be as light as possible. (ii) The energy storage should consist of rechargeable lithium batteries, ultra capacitors and/or other high energy density and high power density storage technologies. (iii) The energy storage should be recharged on ground before launch, such that it needs to be dimensioned only for one launch or one retrieval, respectively. (vi) The propeller disk area should be maximized to obtain a high propeller efficiency (or figure of merit) through low disk loading. However, such big propellers might be disadvantageous during normal flight because the onboard loads and the recharging of the energy storage require less power than large propellers can provide. So the load factor of the propellers in wind turbine mode could be rather low and generate undesired drag during crosswind flight. To avoid a high propeller drag, all or some propellers may be carried out as folding propellers, single blade propellers or variable pitch propellers. Additionally, a small optimized wind turbine could be used at the pod which is the only propeller operated in generator mode. (v) Contra rotating propellers could be a possibility to increase propeller efficiency. (vi) Ducted propellers could be another possibility to increase propeller efficiency and decrease propeller noise. Although the downwind or upwind placement of the propellers with the respective launching and landing maneuvers decreases the possibility of collisions between propeller blades and kite or tethers (see Figs. 19.2–19.6), the ducts would give a further protection. However, the last two measures (v) and (vi) also increase mass, so that a good compromise has to be found.

To avoid a negative effect of the propeller downwash to the kite during hovering, a combination of the following measures is proposed: (i) Change the pitch angle of the kite such that the kite area which is affected by downwash is small. (ii) Change the kite's geometry: The kite might be folded or flagged. The latter means that the right or the left tethers between pod and kite are slack. (iii) Design the thrust system so that downwash passes the kite to the left and to the right side and a downwash free zone occurs below the pod in which the kite can swing in and out. E.g. use propellers which are inclined or can be inclined to the left and the right side, see Figs. 19.3–19.6. Another possibility might be to mount the propellers on (long, telescoping) arms to the right and left side. Yet another possibility is to use baffles and/or design the casing of the pod such that downwash is slightly deflected to create a downwash free zone.

Particularly for the downwind propeller concept where the kite's leading edge faces towards the wind vector while hanging below the pod, see Figs. 19.5 and 19.6, the kite can create a significant aerodynamic force counteracting the propellers' thrust and could lead to a considerably higher power demand to keep the system aloft. To reduce the kite's force a combination of measures (ii) and (iii) of the last paragraph may be applied. However, the controllability would be affected negatively with these solutions. A better solution might be to use a ram air kite and reduce its aerodynamic efficiency while hanging upside down by a combination of the following measures: (i) Partly close the leading edge inlets. (ii) Partly open trailing edge outlets. (iii) Change the airfoil of the kite to a symmetric one or an inverted one to reduce the generated lift. This could be achieved by shortening tethers between leading edge and trailing edge.

The propellers, their drives and particularly the energy storage add significant mass to the control pod. This not only can reduce efficiency during energy generation but also can reduce stability: In dynamic simulations (details in Sect. 19.5) it was observed that with a high mass the control pod could oscillate perpendicularly to the tether. To avoid this oscillation, a combination of the following measures may be applied: (i) The oscillation may be damped actively by steering actuations of kite or ground winch(es). (ii) Small stabilizer wings may be attached to the pod to damp the oscillation passively. The wings may also be rotatable or have flaps to damp the oscillation actively. Additionally, these actuators may be used to generate some lift to counteract the mass of the control pod.

The system needs a set of additional sensors: E.g. to control the hanging kite during hover mode, the control pod could have a camera. Data processing extracts the kite's position and attitude. To enable this concept during night, a spotlight is also required.

19.3.3 Rigid Kite Concepts

Rigid kites (gliders/airplanes) equipped with propellers have been proposed e.g. in [21, 32, 33]. One can imagine a rigid kite with propellers at various locations and with various orientations. The wings can be placed to support the propellers during hovering with aerodynamic lift. The evaluation of a specific design is out of scope of this study. Instead, the aerodynamic forces of the wings of a rigid kite are considered negligible. Thus a worst case is considered where only the thrust of the propellers is available during hovering.

19.3.4 Electrical Cables-Based Hovering Power Sources

So far, only concepts were considered where the multicopter control pod carries an energy storage. Besides its high mass and cost, the energy density of an energy storage diminishes over time. To avoid these disadvantages, concepts which partly or completely replace the energy storage by an electrical cable are discussed in the following: Figure 19.7 illustrates the considered solutions for a soft kite, which are applicable similarly to a rigid kite.

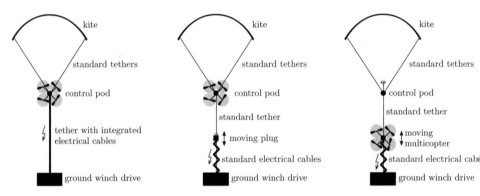

Fig. 19.7 Partly or complete replacement of the onboard energy storage: by a tether with integrated electrical cables (left), by external electrical cables which can move along the tether (the control pod has a small energy storage to buffer fluctuations, middle) and by a separate multicopter that can move along the tether (the control pod has a small wind turbine and a small energy storage, right)

Tether with Integrated Electrical Cables. One solution is using a tether with integrated electrical cables (with all its disadvantages) and thus replace completely the onboard energy storage, see Fig. 19.7 (left). However, this solution has a similar disadvantage as an energy storage solution: Both are designed for the propeller power during launch and retrieval, because they need much more power than the onboard loads during normal flight. Neither a big energy storage nor a high power electrical cable is needed in the latter phase. Moreover, it is questionable if such a tether for a lift power system can be made durable enough to withstand the tear imposed by winding.

External Electrical Cables. A high power electrical cable is only needed during launch and retrieval. Figure 19.7 (middle) visualizes the idea to plug an electrical cable in the control pod for launch and retrieval only: A small carriage moves along the tether from the ground station to the control pod prior to the retrieval. After launch the carriage moves back to the ground station. There it may decouple itself from the tether, because the tether can move fast during the pumping cycles. As a consequence, the mass which is to be lifted is smaller, the providable power should be high and the providable energy is practically unlimited.

Once launched, this solution has the advantage to be a relatively light airborne system without heavy energy storage and a thin, light and inexpensive tether. The additional electrical cable may be as heavy as the kite or the multicopter control pod can carry, i.e. a low voltage level and an inexpensive off-the-shelf electrical cable may be used. While the carriage moves towards the control pod, the kite may stay in the zenith. If the wind is too calm while the carriage is moving, the kite may also be powered mechanically by the ground winch through reverse pumping [5, 20] or winching [5].

Separate Multicopter. Another idea is to split the control pod into two parts, see Fig 19.7 (right): The first part is a conventional control pod only with a small wind turbine to power the onboard devices and with a small energy storage to buffer fluctuations. The second part is a multicopter with greater propellers and is docked at the control pod for launch and retrieval only. To reduce mass, that multicopter is preferably powered by electrical cables from ground rather than an energy storage onboard of the multicopter. To simplify the docking, the multicopter moves along the tether from the ground station towards the control pod by using the propellers or a small onboard winch. The undocking is the reverse.

Contrary to the additional electrical cables solution in Fig 19.7 (middle), no additional mass and drag through propellers are present at the control pod during normal flight. However, the docking and undocking as well as the construction are more challenging, because the mass of a multicopter is higher than the mass of a small carriage which moves just a plug to the control pod.

19.4 Static Feasibility Analyses

In the following, feasibility analyses are carried out to show that the propellers can generate enough thrust to lift a kite, both powered by an energy storage or electrical cables. The flight maneuvers are not considered in these analyses. A simplified model is formulated and solved with example parameters for a soft kite and for a rigid kite.

19.4.1 Model for Energy Storage Powered Solutions

The following assumptions and equations are employed to allow for simplified analyses:

Assumption 19.1: *The power of a single propeller $P_{p,s}$ to generate thrust $T_{p,s}$ with propeller disk area $A_{p,s}$ is given by*

$$P_{p,s} = \sqrt{\frac{T_{p,s}^3}{2\rho A_{p,s}}}, \tag{19.1}$$

where ρ is the air density (actuator disk or momentum theory). [16, pp. 152]

For n propellers and to lift the mass m in hovering flight, each propeller must generate the thrust $T_{p,s} = mg/n$ with gravitational acceleration $g \approx 9.81\,\text{m/s}^2$. Then the total propeller power is

$$P_p = nP_{p,s} = n\sqrt{\frac{\left(\frac{mg}{n}\right)^3}{2\rho A_{p,s}}} = \sqrt{\frac{(mg)^3}{2\rho \frac{n^3 A_{p,s}}{n^2}}} = \sqrt{\frac{(mg)^3}{2\rho \underbrace{nA_{p,s}}_{=A_p}}} = \sqrt{\frac{(mg)^3}{2\rho A_p}}, \tag{19.2}$$

where $A_p = nA_{p,s}$ is the total propeller disk area.

Assumption 19.2: *The energy storage must provide the constant electric power*

$$P_e = s\frac{1}{\eta}P_p, \tag{19.3}$$

where η is the efficiency (i.e. the ratio of the power drawn from the energy storage and the power required to generate thrust mg in Eq. (19.2)) and s is a safety factor which regards e.g. the needed power for steering actuations or other electrical loads.

The airborne mass is

$$m = m_k + m_p + m_e + m_o \tag{19.4}$$

with the kite mass m_k, the propellers' mass m_p, the energy storage mass m_e and the mass of other parts m_o, such as control pod winches, control pod casing, tethers etc. in the case of a soft kite with control pod. The following assumptions are made for the mass portions:

Assumption 19.3: *The kite mass is*

$$m_k = \mu_k A_k, \tag{19.5}$$

where μ_k is the specific kite mass and A_k is the kite's projected area.

Assumption 19.4:   *The mass of the propulsion unit (i.e. propellers and electrical drives) is*

$$m_p = \mu_p P_e, \tag{19.6}$$

where μ_p is the specific mass of the propellers.

Assumption 19.5:   *The mass of the energy storage is*

$$m_e = \max\left\{ \frac{E_e}{\gamma_{e,E}}, \frac{P_e}{\gamma_{e,P}} \right\}, \tag{19.7}$$

where $E_e = P_e t_h$ is the electric energy needed to hover for the time t_h, and $\gamma_{e,E}$ and $\gamma_{e,P}$ are the energy and power densities of the energy storage technology.

Note that the max-function of Eq. (19.7) ensures that the energy storage is large enough to provide the required energy *and* the required power.

Assumption 19.6:   *The mass of other parts is*

$$m_o = \mu_o A_k, \tag{19.8}$$

where μ_o is the specific mass of other parts.

Remark 19.1:   *Assumption 19.1 is a simplification as it assumes a propeller disk (Betz' analysis). Real propellers require more power (i.e. their "actuator disk efficiency" is less then 100%), which can be covered by Assumption 19.2 through an adequate value of η. Also a possibly high power demand due to the kite's aerodynamic force that counteracts propeller thrust during hovering can be covered by Assumption 19.2 through an adequate value for s. Assumption 19.3 with constant μ_k is only valid within tight bounds of A_k, as the kite mass does not increase linearly with its area. Similar limitations apply for Assumptions 19.4 through 19.6.*

Combining Eqs. (19.1) to (19.8) yields

$$P_e = s\frac{1}{\eta} \sqrt{ \frac{\left(A_k[\mu_k + \mu_o] + P_e\left[\mu_p + \max\left\{ \frac{t_h}{\gamma_{e,E}}, \frac{1}{\gamma_{e,P}} \right\}\right] \right)^3 g^3}{2\rho A_p} } \tag{19.9}$$

which was solved numerically for P_e (note that the unknown P_e is on the left hand side and on the right hand side of Eq. (19.10)),[3] i.e.

$$P_e = f(s, \eta, A_k, \mu_k, \mu_o, \mu_p, t_h, \gamma_{e,E}, \gamma_{e,P}, A_p). \tag{19.10}$$

The masses of the propellers and of the energy storage can then be calculated by inserting the numerical result of Eq. (19.10) into Eqs. (19.6) and (19.7).

---

[3] It is also possible to convert Eq. (19.9) to the form $0 = p_0 + p_1 x + p_2 x^2 + p_3 x^3$. So there should be an analytical solution $x = P_e$. However, numerical solving was preferred for sake of simplicity.

Symbol & Value	Comment
Parameters (for both, soft kite or rigid kite).	
$s/\eta = 3$	assumed
$\mu_{\mathrm{o}} = 0.25\,\mathrm{kg/m^2}$	assumed
$\mu_{\mathrm{p}} = 0.2\,\mathrm{kg/kW}$	assumed; taken from [26]
$t_{\mathrm{h}} = 5\,\mathrm{min}$	assumed to be enough for one launch or one retrieval incl. safety factor
$\gamma_{\mathrm{e,E}} = 130\,\mathrm{Wh/kg}$	assumed for lithium batteries; taken from a recent model making battery [9]
$\gamma_{\mathrm{e,P}} = 5\,\mathrm{kW/kg}$	assumed for lithium batteries; taken from a recent model making battery [9]
$A_{\mathrm{p}} = 1\,\mathrm{m^2}$	design decision
$\rho = 1.2\,\mathrm{kg/m^3}$	assumed (for low elevation flight)
Soft kite specific parameters.	
$A_{\mathrm{k}} = 20\,\mathrm{m^2}$	size of a commercial surf kite or paraglider [1]; the system could have a rated power of about 22 kW (from Loyd's analysis [31] with lift coefficient 1.0, drag coefficient 0.2, overall system efficiency 50%, wind speed 10 m/s)
$\mu_{\mathrm{k}} = 0.25\,\mathrm{kg/m^2}$	assumed; typical for commercial paragliders of the regarded size [1]
Rigid kite specific parameters.	
$A_{\mathrm{k}} = 15\,\mathrm{m^2}$	size of a commercial hang glider [34]; the system could have a rated power of about 56 kW (from Loyd's analysis [31] with lift coefficient 1.5, drag coefficient 0.2, overall system efficiency 50%, wind speed 10 m/s)
$\mu_{\mathrm{k}} = 2.5\,\mathrm{kg/m^2}$	assumed; typical for commercial hang gliders of the regarded size [34]
Soft kite solutions.	
$P_{\mathrm{e}} \approx 2.5\,\mathrm{kW}$	solution of Eq. (19.10)
$m_{\mathrm{p}} \approx 0.5\,\mathrm{kg}$	solution of Eq. (19.10) inserted into Eq. (19.6)
$m_{\mathrm{e}} \approx 1.6\,\mathrm{kg}$	solution of Eq. (19.10) inserted into Eq. (19.7)
Rigid kite solutions.	
$P_{\mathrm{e}} \approx 36\,\mathrm{kW}$	solution of Eq. (19.10)
$m_{\mathrm{p}} \approx 7.2\,\mathrm{kg}$	solution of Eq. (19.10) inserted into Eq. (19.6)
$m_{\mathrm{e}} \approx 23.1\,\mathrm{kg}$	solution of Eq. (19.10) inserted into Eq. (19.7)

Table 19.1 Considered parameters and numerical results for a soft kite with battery-powered control pod and a rigid kite with battery-powered propellers

19.4.2 Example Results for Battery-Powered Solutions

Table 19.1 lists possible parameters of a soft kite prototype system as well as the numerical results. Figure 19.8 (left) shows also a plot of the results for varied s/η and A_{p}, since s/η could be higher due to the kite's aerodynamic force that counteracts propeller thrust during hovering. Additionally, μ_{k} is likely to be higher since a commercial paraglider, from which the value of μ_{k} originates, is not designed to generate electricity. Figure 19.8 (right) shows a plot of the results for higher μ_{k} and varied A_{p}. Note that Eq. (19.10) has no solution for high s/η with low A_{p} or for high μ_{k} with low A_{p}, i.e. the propellers cannot generate enough thrust to lift the masses or it can only hover for a time smaller than t_{h}.

Table 19.1 and Fig. 19.9 show the considered parameters and results for a lightweight rigid kite which is designed similarly to a hang glider. The differences to the soft kite concept are that the specific kite mass μ_{k} is higher and the kite area A_{k} is smaller.

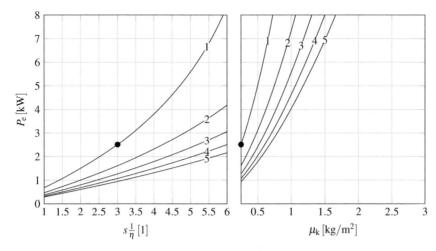

Fig. 19.8 Numerical results for a soft kite with the data of Table 19.1 (black dot) and for varied s/η (left) and for higher μ_k (right) each for $A_p \in [1\,\mathrm{m}^2, 2\,\mathrm{m}^2, \ldots, 5\,\mathrm{m}^2]$ and otherwise unchanged parameters

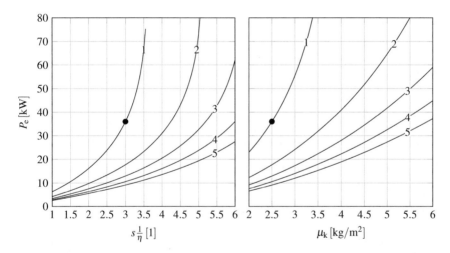

Fig. 19.9 Numerical results for a rigid kite with the data of Table 19.1 (black dot) and for varied s/η (left) and for varied μ_k (right) each for $A_p \in [1\,\mathrm{m}^2, 2\,\mathrm{m}^2, \ldots, 5\,\mathrm{m}^2]$ and otherwise unchanged parameters

19.4.3 Model for Electrical Cables-Powered Solutions

The same assumptions and equations from Sect. 19.4.1 are taken for feasibility analyses of electrical cables-powered solutions. Hereby, Assumption 19.5 is replaced by the following assumption:

Assumption 19.7:   *The mass of the electrical cables is*

$$m_c = \mu_c l_c P_e \qquad (19.11)$$

where μ_c is the specific cable mass and l_c is the carried cables' length.

With that, the required electric power is given by

$$P_e = s\frac{1}{\eta}\sqrt{\frac{\left(A_k[\mu_k + \mu_o] + P_e\left[\mu_p + \mu_c l_c\right]\right)^3 g^3}{2\rho A_p}} \qquad (19.12)$$

which was again solved numerically for P_e, i.e.

$$P_e = f(s, \eta, A_k, \mu_k, \mu_o, \mu_p, \mu_c, l_c, A_p). \qquad (19.13)$$

The masses of propellers and electrical cables are again given by inserting the numerical result of Eq. (19.13) into Eqs. (19.6) and (19.11).

19.4.4 Example Results for Electrical Cables-Powered Solutions

Table 19.2 lists possible parameters for a prototype system similar to the last section as well as numerical results. Figure 19.10 (left) shows a plot of the results for a soft kite for varied s/η and A_p and Fig. 19.11 (right) shows a plot of the results for varied μ_c and A_p. Figure 19.11 shows plots of the results for a rigid kite.

Symbol & Value	Comment
Parameters (for both, soft kite or rigid kite).	
$\mu_c \approx 3 \cdot 10^{-6}\,\text{kg/m/W}$	assumed; taken from two cables used for photovoltaics applications [24]
$l_c = 100\,\text{m}$	assumed to be enough to safely launch and retrieve the kite with the proposed maneuvers
Soft kite solutions.	
$P_e \approx 2.2\,\text{kW}$	solution of Eq. (19.13)
$m_p \approx 0.4\,\text{kg}$	solution of Eq. (19.13) inserted into Eq. (19.6)
$m_c \approx 0.7\,\text{kg}$	solution of Eq. (19.13) inserted into Eq. (19.11)
Rigid kite solutions.	
$P_e \approx 22.7\,\text{kW}$	solution of Eq. (19.13)
$m_p \approx 4.5\,\text{kg}$	solution of Eq. (19.13) inserted into Eq. (19.6)
$m_c \approx 6.8\,\text{kg}$	solution of Eq. (19.13) inserted into Eq. (19.11)

Table 19.2 Considered parameters that differ from Tab 19.1 and numerical results for a soft kite with electrical cables-powered control pod and a rigid kite with electrical cables-powered propellers

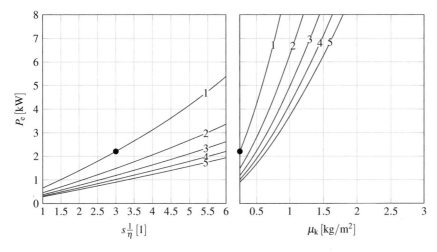

Fig. 19.10 Numerical results for a soft kite with the data of Table 19.2 (black dot) and for varied s/η (left) and for higher μ_k (right) each for $A_p \in [1\,m^2, 2\,m^2, \ldots, 5\,m^2]$ and otherwise unchanged parameters

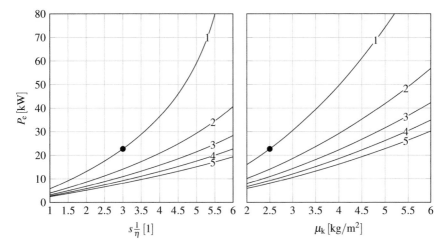

Fig. 19.11 Numerical results for a rigid kite with the data of Table 19.1 (black dot) and for varied s/η (left) and for higher μ_k (right) each for $A_p \in [1\,m^2, 2\,m^2, \ldots, 5\,m^2]$ and otherwise unchanged parameters

19.5 Dynamic Feasibility Analyses

The flight maneuvers of rigid kite solutions are feasible: videos of such maneuvers have been presented by e-kite [7] and TwingTec [32]. Moreover, for a few years Makani Power/Google has successfully been demonstrating the multicopter launching and landing as well as transitions to and from crosswind flight with their drag power rigid kite prototypes. Consequently, only the soft kite solutions need to be investigated, whereby the following analyses focus on the proposed maneuvers of the downwind propeller concept (see Figs. 19.5–19.6). A multicopter control pod with lithium-type battery energy storage without the possibility to separate the multicopter is considered in these first dynamic analyses.

A multi-body modeling approach is employed where kite and control pod are modeled as rigid bodies connected by spring-dampers as tether models. No collisions between tethers, pod nor kite are modeled. The kite follows the turning rate law formulated in [11] and its angle of attack can be changed actively. The shaft of the propellers and of the ground winch are modeled with Newtonian dynamics and the propellers are modeled to generate thrust and torque proportional to the square of the propeller speed. No downwash is modeled which implies the assumption that downwash never effects the kite (due to inclined propellers as in Figs. 19.3–19.6). For the kite a controller similar to [13] and for the control pod a simple multicopter controller is employed. For the overall system control, a state machine is used. Feedback of all for the control necessary quantities are regarded as available without offset, delay, noise or other disturbances. All controller parameters were tuned by hand and are thus not optimized. Model and controller are implemented in C++. Table 19.3 lists important model parameters. Note that the parameters correspond to a soft kite system sketched in Table 19.1. A more detailed description is omitted due to space limitation.

Parameter	Symbol & Value
Logarithmic Wind Model: $v_{\mathrm{w}}(z) = v_{\mathrm{w,ref}}\ln(z/z_0)/\ln(z_{\mathrm{ref}}/z_0)$.	
reference wind speed	$v_{\mathrm{w,ref}} = 6\,\mathrm{m/s}$
reference altitude	$z_{\mathrm{ref}} = 50\,\mathrm{m}$
surface roughness	$z_0 = 0.2\,\mathrm{m}$
Kite and Control Pod.	
lift coefficient during power generation[a]	$C_{\mathrm{L}} \approx 0.7$
drag coefficient during power generation[a]	$C_{\mathrm{D}} \approx 0.15$
lift coefficient during hovering[a]	$C_{\mathrm{L}} \approx 0.1$
drag coefficient during hovering[a]	$C_{\mathrm{D}} \approx 0.1$
kite area	$A_{\mathrm{k}} = 20\,\mathrm{m}^2$
kite mass	$m_{\mathrm{k}} = 5\,\mathrm{kg}$
control pod mass	$m_{\mathrm{pod}} = 7.1\,\mathrm{kg}$
number of propellers	$n = 8$ (4 pairs of contra-rotating)
total propeller disc area	$A_{\mathrm{p}} = 1\,\mathrm{m}^2$

[a] Lift and drag coefficients are simulated as functions of angle of attack.

Table 19.3 Model Parameters

Figure 19.12 (a) shows the 3D flight trajectory of the kite and of the control pod for a launch maneuver. Figure 19.12 (b) shows the sum of the electric (propeller) power. Similar plots for the landing are shown in Figs. 19.12 (c) and (d). The plots show successful maneuvers, as sketched in Figs. 19.5–19.6.

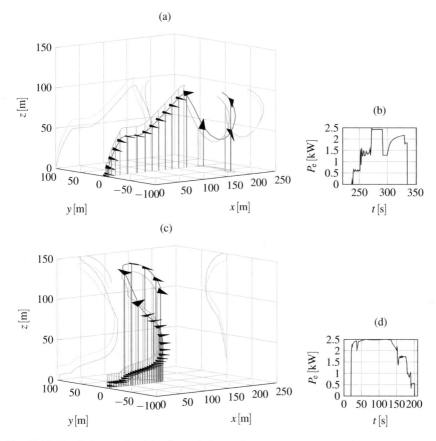

Fig. 19.12 Simulation results of the downwind propeller concept (compare with Figs. 19.5–19.6): 3D flight trajectories during launching from ground into crosswind figure eights (a) with sum of the electric propeller power in (b). The kite trajectory is printed in blue (with kite attitude visualization every five seconds) and control pod trajectory in green. Similar plots are given for the landing in (c) and (d)

Simulations for turbulent and rotating winds were also carried out. To handle the latter case, the control algorithm always turns the kite's leading edge towards the wind. Even under such non-ideal conditions, the simulations show that the airborne system stays aloft remarkably stably, i.e. almost without oscillations and without crashing. Moreover, simulations with a much lighter control pod (i.e. without energy storage and propellers) where carried out to investigate the effect of the added control pod mass on the efficiency, with otherwise unchanged model and controller parameters: With an unrealistic light control pod of $m_{pod} = 0.5\,$kg, the power increase in the reel out phase is $\approx 5.6\,\%$ and the average power increase for the pumping cycle is with $\approx -0.5\,\%$ even negative. The latter can be explained by a more efficient reel-in phase of the heavier system as gravity helps better to bring the airborne system back and a heavier control pod further depowers the kite. Yet another simulation with a stronger wind with $v_{w,ref} = 10\,\mathrm{m/s}$ was carried out. Unfortunately, under such conditions the propellers are not able to generate enough thrust to keep the system aloft, i.e. the kite's downward pulling aerodynamic force is too strong.

19.6 Discussion

General Results of Analyses. The results of P_e, m_p and m_e of the static analyses seem to be feasible figures, for all, battery and cable power solutions, rigid and soft kites. Based on the simulation results with the underlaying simplifications/assumptions, the maneuvers of Figs. 19.5–19.6 also seem to be flyable. The kite is quite stable while hanging under the control pod even under turbulent and rotating wind and even with the hand tuned controller parameters. However, it cannot be excluded that this is just the result of the relatively simple simulation model, so that the control can be much more challenging with a more elaborate kite model or with a real demonstrator. A disadvantage of that soft kite concept is that the kite generates a significant lift force that opposes the propeller thrust. As visible in Figs. 19.12 (b) and (d), the full available power with $s/\eta = 3$ is required for a large portion of the time already at a wind speed of about $6\,\mathrm{m/s}$. For higher wind speeds, even for the arguably optimistically assumed parameters, the available propeller thrust is not sufficient to keep the system aloft. Consequently, more propeller power (and a bigger energy storage) and further measures to reduce the kite's lift, instead of just pitching the kite as in the simulation, are necessary. Contrary to the soft kite solutions, a rigid kite can be designed such that its aerodynamic force does not counteract propeller thrust, with the further advantage of a lower factor s/η. However, the mass of a rigid kite is higher, for which the required hovering power is very sensitive. An interesting observation of the dynamic simulations is that the average power of one pumping cycle with a heavy control pod is higher than the average power of a very light control pod (however, note that all system and controller parameters other than the control pod mass were not changed and that the controller parameters were not optimized). This apparent contradiction was also observed by [37]: An airborne mass greater than $0\,\mathrm{kg}$ results in a more efficient pumping cycle due to a more efficient reel-in phase. Consequently, a higher mass of control pod or rigid kite due to propellers and energy storage might not necessarily lead to critical system efficiency losses.

Hovering Power Source. Particularly the battery-powered solutions are very sensitive to worse s/η and to higher masses, visible in Figs. 19.8 and 19.9. This problem is more severe for larger systems as μ_k actually increases with A_k. Moreover, for a solution with a soft kite and control pod, A_p is limited also for high A_k. As rigid kites are usually heavier than soft kites, a battery-powered solution for such kites is only feasible for very light structures (e.g. through using a combination of soft and rigid materials similar to hang gliders). Note that, in the considered cases $m_e = \max\{E_e/\gamma_{e,E}, P_e/\gamma_{e,P}\} = E_e/\gamma_{e,E}$ is driven by the energy density and not by the power density. Consequently, with today's energy storage technologies, lithium-type batteries are most suitable (compared to e.g. ultra-capacitors which have a higher power density but a lower energy density and are thus not suitable).

Electrical cables-powered solutions are less sensitive to worse s/η and μ_k, compare Figs. 19.8 and 19.9 with Figs. 19.10 and 19.11. For instance, a battery solution for a rigid kite would add $m_p + m_e \approx 30.3\,\text{kg}$ while a solution with detachable electrical cable would only add $m_p = 4.5\,\text{kg}$. However, the propeller power of the rigid kite of this example is already about half of the expected rated power of the system, see Tabs. 19.1 and 19.2.

Comparison of the Detachable Electrical Cables Concept with Drag Power. A rigid lift power kite with detachable electrical cables is similar to the drag power concept developed by Makani Power/Google. The following advantages, but also disadvantages, can be found for such a lift power concept:

+ During power generation, up to about one fourth less airborne mass (due to absence of the conductive portion of the tether, compare with [21]).
+ During power generation, thinner and cheaper tether without integrated electrical cables.
+ More freedom to dimension voltage level, propellers and propeller drives, since (possibly heavy and thick low voltage) electrical cables are only present for launch and retrieval and (most of the) propellers are needed only to hover. This could also result in a lower price of system components.
+ (Almost) no noise during power generation (since a propeller or a small wind turbine is only used to power onboard electronics).
+ Improved control authority due to high power and high force ground winch, i.e. also the ground winch can help for the transition from hover to crosswind flight and back.
− Higher complexity e.g. through a carriage moving the electrical cable along the tether up and down.
− (Most of the) propellers might be unused during power generation, but generate drag. An alternative with foldable propellers (or similar) would (further) increase complexity.
− In a safety critical situation, the kite cannot be brought in a hovering state within seconds unless the electrical cable is already connected.

One must acknowledge that some challenges of the drag power principle might be solved within the coming years and thus diminish some advantages of a multicopter launching and landing lift power solution. As an example, the reduction of the turbine noise might be addressed by using "Bionic Loop Propellers" [12], ducted propellers, a large blade count or a combination of these. Additionally, the development of high force transmitting tethers with integrated high voltage electrical cables has just started. With new materials and higher production capabilities, the mass and the price of such tethers can be expected to be decreasing. Moreover, the mass of the system is only critical for low wind conditions. In high/rated wind conditions (in the magnitude of $\approx 12\,\text{m/s}$) or already medium wind conditions (in the magnitude of $\approx 8\,\text{m/s}$), the lift force generated by the kite is many times higher than its weight, also for the heavier rigid kites. Consequently, the justification of the increased complexity of a multicopter lift power system compared to a drag power system is subject to further studies.

19.7 Conclusions and Outlook

Multicopters are an interesting alternative to autonomously launching and landing kites. It has been successfully implemented by Makani Power/Google for a rigid drag power kite. This was motivation to explore possibilities how the multicopter principle can also be applied to lift power kites. Basic system components were introduced and new soft kite concepts were derived. As a relatively high power is demanded by the propellers to lift the system in hovering flight, electrical cable-based hovering power sources as alternative to an energy storage, like lithium-type batteries, were explored. Since an onboard energy storage or electrical cables are only needed during launch and retrieval but add significant mass to the airborne parts, the idea emerged to use standard electrical cables which are connected only for launch and retrieval, and moved by a carriage along the tether. To gain a high efficiency during kite power generation, also the whole multicopter could be separated from the control pod or from the rigid kite in a similar way.

Static and dynamic feasibility analyses were carried out. The results for considered example systems similar to a paraglider for a soft kite and to a hang glider for a rigid kite are feasible. However, energy-storage-powered solutions are prone to high airborne masses, whereas cables-powered solutions are less sensitive. Compared to other launching and landing concepts for lift power kites, the main advantages can be summarized as follows: (i) The ground station is simple as it primarily consists only of one winch and tether guides. Hence, the overall system complexity could be lower compared to other launching and landing concepts (e.g. rotating arm launching and landing), although the airborne part is more complex. (ii) It is possible to launch and land the system also without wind near the ground, (iii) with relatively simple and proven flight maneuvers in the case of rigid kites.

The investigations also revealed the weaknesses of such concepts which are (i) a high airborne mass due to propellers and particularly due to an energy storage (if such a solution is considered), (ii) higher drag (if the propellers/multicopter cannot be detached), (iii) and/or, in the case of detachable cables or detachable multicopter solutions, a then again increased complexity. Moreover, the soft kite concept has the disadvantage that the kite generates a lift force opposite to the propeller thrust. Subject to further research in this area would be the reduction of this force during the hovering phases. Further studies would focus on more detailed dynamic simulations, the design of a system or carriage with which an external electrical cable can be attached and detached and also economical efficiency analyses. A more elaborate comparison between the multicopter-based lift power rigid kite concept (vertical takeoff and landing, VTOL) with short takeoff and landing concepts (STOL), and with the drag power rigid kite concept are of particular interest.

Acknowledgements The authors thank the anonymous reviewers for their helpful comments. This study was supported by Bund der Freunde der TU München e.V.

References

1. Advance Thun AG: Alpha 6. http://www.advance.ch/de/alpha. Accessed 20 Jan 2016
2. Ahrens, U., Diehl, M., Schmehl, R. (eds.): Airborne Wind Energy. Green Energy and Technology. Springer, Berlin Heidelberg (2013). doi: 10.1007/978-3-642-39965-7
3. Alula Energy Oy: Takeoff and landing system - Airborne Wind Energy and Tethered UAV. http://vimeo.com/78090844. Accessed 20 Jan 2016
4. Bauer, F., Hackl, C. M., Smedley, K., Kennel, R.: On Multicopter-Based Launch and Retrieval Concepts for Lift Mode Operated Power Generating Kites. In: Schmehl, R. (ed.). Book of Abstracts of the International Airborne Wind Energy Conference 2015, pp. 92–93, Delft, The Netherlands, 15–16 June 2015. doi: 10.4233/uuid:7df59b79-2c6b-4e30-bd58-8454f493bb09. Presentation video recording available from: https://collegerama.tudelft.nl/Mediasite/Play/a303417db9114a9f876819208bd889c71d
5. Bontekoe, E.: How to Launch and Retrieve a Tethered Aircraft. M.Sc.Thesis, Delft University of Technology, 2010. http://resolver.tudelft.nl/uuid:0f79480b-e447-4828-b239-9ec6931bc01f
6. Breukels, J.: Kite launch using an aerostat. Technical Report, Delft University of Technology, 21 Aug 2007. http://repository.tudelft.nl/view/ir/uuid%3A1a0c6dfd-6115-461f-ac04-bd8751efd6fb
7. Brink, A. van den: Design of the e-50 Ground Station. In: Schmehl, R. (ed.). Book of Abstracts of the International Airborne Wind Energy Conference 2015, pp. 34–35, Delft, The Netherlands, 15–16 June 2015. doi: 10.4233/uuid:7df59b79-2c6b-4e30-bd58-8454f493bb09. Presentation video recording available from: https://collegerama.tudelft.nl/Mediasite/Play/eec678673e7b4056961269ab59fd4d6b1d
8. Cherubini, A., Papini, A., Vertechy, R., Fontana, M.: Airborne Wind Energy Systems: A review of the technologies. Renewable and Sustainable Energy Reviews **51**, 1461–1476 (2015). doi: 10.1016/j.rser.2015.07.053
9. Conrad Electronic SE: Modellbau-Akkupack (LiPo) 22.2 V 5000 mAh 40 C. https://www.conrad.de/de/modellbau-akkupack-lipo-222-v-5000-mah-40-c-conrad-energy-offene-kabelenden-239016.html. Accessed 20 Jan 2016
10. Enerkite GmbH. http://www.enerkite.com/. Accessed 14 Jan 2016
11. Erhard, M., Strauch, H.: Control of Towing Kites for Seagoing Vessels. IEEE Transactions on Control Systems Technology **21**(5), 1629–1640 (2013). doi: 10.1109/TCST.2012.2221093
12. EvoLogics GmbH: Bionic Loop Propeller. http://www.evologics.de/en/products/propeller/index.html. Accessed 20 Jan 2016
13. Fagiano, L., Zgraggen, A. U., Morari, M., Khammash, M.: Automatic crosswind flight of tethered wings for airborne wind energy:modeling, control design and experimental results. IEEE Transactions on Control System Technology **22**(4), 1433–1447 (2014). doi: 10.1109/TCST.2013.2279592
14. Fagiano, L., Schnez, S.: The Take-Off of an Airborne Wind Energy System Based on Rigid Wings. In: Schmehl, R. (ed.). Book of Abstracts of the International Airborne Wind Energy Conference 2015, pp. 94–95, Delft, The Netherlands, 15–16 June 2015. doi: 10.4233/uuid:7df59b79-2c6b-4e30-bd58-8454f493bb09. Presentation video recording available from: https://collegerama.tudelft.nl/Mediasite/Play/2ebb3eb4871a49b7ad70560644cb3e2c1d
15. Fechner, U., Schmehl, R.: Design of a Distributed Kite Power Control System. In: Proceedings of the 2012 IEEE International Conference on Control Applications, pp. 800–805, Dubrovnik, Croatia, 3–5 Oct 2012. doi: 10.1109/CCA.2012.6402695
16. Filippone, A.: Advanced Aircraft Flight Performance. 1st ed. Cambridge University Press (2012). doi: 10.1017/CBO9781139161893

17. Geebelen, K., Ahmad, H., Vukov, M., Gros, S., Swevers, J., Diehl, M.: An experimental test set-up for launch/recovery of an Airborne Wind Energy (AWE) system. In: Proceedings of the 2012 American Control Conference, pp. 5813–5818, Montréal, QC, Canada, 27–29 June 2012. doi: 10.1109/ACC.2012.6315033

18. Geebelen, K.: Design and Operation of Airborne Wind Energy Systems.Experimental Validation of Moving Horizon Estimation for PoseEstimation. Ph.D. Thesis, KU Leuven, 2015. https://lirias.kuleuven.be/handle/123456789/485714

19. Geebelen, K., Gillis, J.: Modelling and control of rotational start-up phase of tethered aeroplanes for wind energy harvesting. M.Sc.Thesis, KU Leuven, June 2010

20. Gillis, J., Goos, J., Geebelen, K., Swevers, J., Diehl, M.: Optimal periodic control of power harvesting tethered airplanes. In: Proceedings of the 2012 American Control Conference, pp. 2527–2532, Montréal, QC, Canada, 27–29 June 2012. http://ieeexplore.ieee.org/xpls/abs_all.jsp?arnumber=6314924

21. Hardham, C.: Response to the Federal Aviation Authority. Docket No.: FAA-2011-1279; Notice No. 11-07; Notification for Airborne Wind Energy Systems (AWES), Makani Power, 7 Feb 2012. https://www.regulations.gov/#!documentDetail;D=FAA-2011-1279-0014

22. Haug, S.: Design of a Kite Launch and Retrieval System For a Pumping High Altitude Wind Power Generator. M.Sc.Thesis, University of Stuttgart, 2012. doi: 10.18419/opus-3936

23. Houska, B., Diehl, M.: Optimal control for power generating kites. In: Proceedings of the 9th European Control Conference, pp. 3560–3567, Kos, Greece, 2–5 July 2007

24. IBC Solar AG: IBC FlexiSun 2,5/4/6/10/16 mm^2 PV1-F. http://www.photovoltaik-shop.com/downloads/dl/file/id/312/solarkabel_ibc_flexisun_1x2_5_16mm_datenblatt_pdf.pdf. Accessed 20 Jan 2016

25. Ippolito, M.: System and process for starting the flight of power wing airfoils, in particular for wind generator. Patent WO2014199406 A1, Dec 2014

26. Joby Motors, Inc.: JM1. http://www.jobymotors.com/public/views/pages/jm1.php. Accessed 20 Jan 2016

27. KiteGen. http://kitegen.com. Accessed 20 Jan 2016

28. Kitemill. http://kitemill.com. Accessed 20 Jan 2016

29. KitePower. http://www.kitepower.eu. Accessed 29 Apr 2015

30. Kruijff, M., Ruiterkamp, R.: Status and Development Plan of the PowerPlane of Ampyx Power. In: Schmehl, R. (ed.). Book of Abstracts of the International Airborne Wind Energy Conference 2015, pp. 18–21, Delft, The Netherlands, 15–16 June 2015. doi: 10.4233/uuid:7df59b79-2c6b-4e30-bd58-8454f493bb09. Presentation video recording available from: https://collegerama.tudelft.nl/Mediasite/Play/2e1f967767d541b1b1f2c912e8eff7df1d

31. Loyd, M. L.: Crosswind kite power. Journal of Energy **4**(3), 106–111 (1980). doi: 10.2514/3.48021

32. Luchsinger, R. H. et al.: Closing the Gap: Pumping Cycle Kite Power with Twings. In: Schmehl, R. (ed.). Book of Abstracts of the International Airborne Wind Energy Conference 2015, pp. 26–28, Delft, The Netherlands, 15–16 June 2015. doi: 10.4233/uuid:7df59b79-2c6b-4e30-bd58-8454f493bb09. Presentation video recording available from: https://collegerama.tudelft.nl/Mediasite/Play/646b794e7ac54320ba48ba9f41b41f811d

33. Makani Power/Google. http://www.google.com/makani. Accessed 14 Jan 2016

34. Moyes USA: Litespeed RS. http://www.moyesusa.com/products/litespeedRSspecs.html. Accessed 20 Jan 2016

35. NTS Nature Technology Systems. http://www.x-wind.de/en/. Accessed 20 Jan 2016

36. Schmehl, R.: Experimental setup for automatic launching and landing of a 25m2 traction kite. https://www.youtube.com/watch?v=w4oWs_zNpr8. Accessed 20 Jan 2016

37. Schmehl, R.: Traction Power Generation with Tethered Wings - A Quasi-Steady Model for the Prediction of the Power Output. In: Schmehl, R. (ed.). Book of Abstracts of the International Airborne Wind Energy Conference 2015, pp. 38–39, Delft, The Netherlands, 15–16 June 2015. doi: 10.4233/uuid:7df59b79-2c6b-4e30-bd58-8454f493bb09. Presentation video recording available from: https://collegerama.tudelft.nl/Mediasite/Play/02a6612b8d004580b08681efd10611351d

38. Skysails GmbH. http://www.skysails.info. Accessed 14 Jan 2016
39. Stoll, W., Fischer, M., Bormann, A., Skutnik, S.: CyberKite. http://www.festo.com/net/ SupportPortal/Files/42084/CyberKite_en.pdf. Accessed 20 Jan 2016
40. Suominen, I., Berg, T.: Method and System for Towing a Flying Object. Patent WO2013156680 A1, Oct 2013
41. Wortmann, S.: Mast arrangement and method for starting and landing an aerodynamic wing. Patent WO2013164446 A1, Nov 2013
42. Zanon, M., Gros, S., Andersson, J., Diehl, M.: Airborne Wind Energy Based on Dual Airfoils. IEEE Transactions on Control Systems Technology **21**(4), 1215–1222 (2013). doi: 10.1109/ TCST.2013.2257781

Chapter 20
Linear Take-Off and Landing of a Rigid Aircraft for Airborne Wind Energy Extraction

Lorenzo Fagiano, Eric Nguyen Van and Stephan Schnez

Abstract An overview of recent results on the take-off and landing phases of airborne wind energy systems with a rigid aircraft is given. The considered take-off approach employs a linear motion system installed on the ground to accelerate the aircraft to take-off speed and on-board propellers to sustain the climb up to operational altitude. Theoretical analyses are employed to estimate the power, additional on-board mass and land occupation required to realize such a take-off strategy. A realistic dynamical model of the tethered aircraft is then employed, together with a decentralized control approach, to simulate the take-off maneuver, followed by a low-tension flight and a landing maneuver back on the linear motion system. The consequences of different wing loadings for this approach are discussed as well. The simulation results indicate that the take-off and landing can also be accomplished in turbulent wind conditions with good accuracy when the wing loading is relatively small. On the other hand, with larger wing loading values the performance is worse. Possible ways to improve the approach and further research directions are finally pointed out.

20.1 Introduction

The take-off and landing phases of airborne wind energy systems employing rigid aircraft and pumping cycles are among the aspects that received relatively little attention in the last decade. In fact, this functionality has been demonstrated, at least on a small-scale, in systems with on-board generation [16, 17] and in systems based on pumping cycles and flexible wings [9], using a rather compact ground area. For AWE systems with rigid wings and ground-level electric generators, there is also evidence of autonomous take-off [1], however by using a winch launch that requires

Lorenzo Fagiano · Eric Nguyen Van · Stephan Schnez (✉)
ABB Switzerland Ltd, Corporate Research, Segelhofstrasse 1K, 5405 Baden-Dättwil, Switzerland
e-mail: stephan.schnez@ch.abb.com, eric.nguyenvan@protonmail.ch, lorenzo.fagiano@polimi.it

© Springer Nature Singapore Pte Ltd. 2018
R. Schmehl (ed.), *Airborne Wind Energy*, Green Energy and Technology, https://doi.org/10.1007/978-981-10-1947-0_20

a significant space in all directions in order to adapt to the prevalent wind direction during take-off. Regarding the landing phase, to the best of the authors' knowledge this has been demonstrated and documented only by detaching the tether from the aircraft, which can be dangerous and impractical due to the absence of control on the tether behavior after detachment, and the need to re-attach it to the aircraft for the subsequent take-off.

So far, this issue has been addressed to a limited extent within the scientific community, where only the take-off phase has been partially studied. In Ref. [10], a rotational take-off is considered and simulated with a focus on the control and optimization aspects. Reference [2] gives an analysis of several take-off strategies, considering different performance criteria. There, three alternatives are deemed the most promising: buoyant systems, linear ground acceleration plus on-board propeller, and rotational take-off. Then, the rotational take-off is examined in more detail by means of numerical simulations.

At ABB Corporate Research, we recently started to investigate this problem via theoretical, numerical and experimental research. In a first contribution [7], we provide a theoretical study of three possible take-off approaches, aimed to assess and compare their technical and economic viability. Out of this study, we conclude that the most promising approach for the considered type of AWE system is to combine a linear acceleration phase on the ground, using a ground-level linear motion system, with on-board propellers, to sustain the climb of the aircraft to operational altitude. In the AWE community, the company Ampyx Power is currently exploring such an approach for their system [11]. In a second contribution [14], we then study the modeling and control design aspects of this approach, including the landing phase without detaching the tether and aiming to touch ground back on the linear motion system used for the take-off. In the same reference we provide realistic numerical simulations that indicate that a satisfactory landing accuracy can be obtained, even in the presence of wind turbulence, with an aircraft with low wing loading.

In this chapter, we provide an overview of the mentioned findings, and we exploit the developed tools to further comment on the effects of the wing loading on both the take-off and landing phases. In particular, we show how the additional power (both on the ground and on board), the additional on-board mass, and the ground area required for take-off change with the wing loading, and how the flight pattern and wind conditions that can be managed successfully during landing are affected by this parameter.

The chapter is structured as follows. Section 20.2 provides a brief description of the considered take-off and landing strategy. Section 20.3 summarizes the main findings pertaining to the theoretical analysis of the approach, as well as new considerations on the effects of the wing loading. Section 20.4 presents the results concerning modeling and control design of the system. Section 20.5 describes the obtained simulation results. Finally, Sect. 20.6 provides conclusions and further steps in this research. The preliminary content of the present chapter has been presented at the Airborne Wind Energy Conference 2015 [8].

20.2 Linear Take-Off and Landing: System Description and Operation

In Ref. [7], we analyzed four different launching procedures which are conceivable for an AWE concept based on a rigid wing and ground-based electric generation: a vertical take-off with vertical-axis propellers, a rotational take-off on a carousel-like structure, a linear take-off with small on-board propellers, and a standard winch launch. Based on qualitative and quantitative performance criteria, we concluded that the linear take-off is the most appealing approach. Here, we will summarize the most important results from Ref. [7] for the linear take-off and investigate the influence of the wing loading more deeply.

20.2.1 System Description

The system we consider is composed of a ground station equipped with a winch, storing a tether connected to a rigid aircraft, see Fig. 20.1 for a sketch and Fig. 20.2 for a picture of our small-scale experimental setup. Details of the prototype are reported in Ref. [4]. Two electric machines are installed on the ground station: The first one controls the winch in order to achieve, during power production, a repetitive cycle of reeling-out under high load, hence converting the mechanical power into electrical, and of reeling-in under low load, spending a small fraction of the energy to start a new production phase. The second machine controls the movement of a linear motion system composed of a slide on rails. In particular, in the approach we consider here, the slide can be pulled both forward and backwards by two tethers,

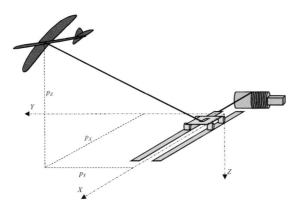

Fig. 20.1 Sketch of the considered system together with the (X, Y, Z) inertial frame and the position $[p_X, p_Y, p_Z]^\top$ of the aircraft. The electrical motor controlling the slide position is not depicted

Fig. 20.2 Photo of the small-scale prototype built at ABB Corporate Research. The numbers in the picture indicate: ① the winch, ② the motor that moves the slide via a pair of additional tethers, ③ the rails, ④ the slide, ⑤ the rigid aircraft, ⑥ the aluminum frame. The rails are 4.5 m long and about 0.5 m wide. The employed aircraft is a commercial model made of Styrofoam, with 1.85-m-wingspan, a mass of about 1 kg, and a propeller connected to a 300 W DC motor. It has been modified to host a tether attach/detach mechanism, an inertial measurement unit and an autopilot. 200 meters of Dyneema® tether with 0.002 m diameter are coiled on the winch. The motor connected to the winch and the one connected to the slide are ABB permanent-magnet, three-phase motors with 2 kW of rated power and 13 Nm of rated torque, controlled by two ABB MicroFlex® drives

which are coiled around a drum ("slide drum") attached to the second electric motor, see Fig. 20.2. A series of pulleys redirects the tether from the winch to the slide and then to the aircraft. The position and speed of both electric machines are measured via encoders and hall-effect sensors. The position of the ground station is determined via global positioning system (GPS). The tether tension is estimated by measuring the compression of a tether tensioning system installed in the ground station. All these measurements can be used for feedback control of the ground station. The available manipulated variables on the ground are the torques of the two electric machines. The tether is made of ultra-high-molecular-weight polyethylene.

We consider a rigid aircraft with on-board electric propellers. The aircraft's attitude, absolute position, angular rates and linear velocity vector are measured with an inertial measurement unit (IMU) and a GPS. The incoming airspeed along the longitudinal body axis is also measured using an air speed sensor. The available control surfaces are the ailerons (for roll control), elevator (for pitch control), rudder (for yaw control), and flaps (to increase lift and drag during take-off and landing). Together with the propellers' thrust, these form the five manipulated variables available to influence the aircraft's motion. In our experimental setup (see Fig. 20.2) the aircraft is a commercially available model made of foam, which is inexpensive, easy to modify and resilient to impacts.

20.2.2 System Operation

We can divide the desired system operation in three phases:

Take-off The aircraft is initially attached to the slide. The slide is accelerated to take-off speed and decelerated down to rest within the length of the rails. The aircraft starts its on-board propulsion during the acceleration and detaches from the slide when the peak speed has been reached.

Flight Since power generation is not our objective, transition from normal flight to pumping cycles is not considered here. Instead, the aim is to control the flight at relatively low speed, notwithstanding the perturbation induced by the tether and by wind turbulence, and to prepare for the landing procedure. To this end, a roughly rectangular path is executed before approaching the ground station for landing.

Landing The landing strategy we consider consists in reeling in the tether to guide the aircraft to the rails. When it is close enough to the ground station, the winch is stopped and the slide starts to accelerate, hence reeling-in the remaining part of the tether and engaging again with the aircraft. Finally, the slide is slowed down and the tether is used to keep the aircraft on it during the braking.

In the considered concept, the take-off is the most energy-intensive phase and determines the system requirements in terms of power (both on the ground and on board), additional on-board mass, and ground area. On the other hand, the landing phase requires accurate and fast control, but it does not consume additional energy in principle. Rather, the kinetic and potential energy of the aircraft have to be dissipated in a controlled way.

Thus, in the following we first focus our attention on the take-off phase and we analyze its requirements and its impact on the design of the whole AWE generator (Sect. 20.3), and then we consider the control problem for the whole cycle of take-off, flight and landing (Sect. 20.4).

20.3 Power, Mass and Ground Area Required for the Take-Off Phase

The considered AWE system generates energy by means of a pumping operating principle composed of the power-generation (or traction) phase, the retraction phase, and the transition phases linking them [6]. During the traction phase, the on-board control system steers the aircraft into figure-of-eight patterns under crosswind conditions. The generated aerodynamic forces exert a large traction load on the tether, which is reeled out from the winch. The winch drives an electric machine and thus generates electricity. Under perfect crosswind conditions (i.e. the tether is parallel to the incoming wind) and a reel-out speed equal to one third of the wind speed, the maximum mechanical power is generated [12]:

$$P_m^* = \frac{2}{27}\rho A \frac{C_L^3}{C_{D,eq}^2} v_w^3,$$ (20.1)

where ρ is the air density, A the effective area of the aircraft, C_L and $C_{D,eq}$ the aerodynamic lift and drag (including tether drag) coefficients, and v_w the absolute wind speed.

For the sake of estimating the generated power, the mass of the airborne components is irrelevant in a first approximation, since the weight and apparent forces of the aircraft and of the tether are significantly smaller than the force acting on the tether during the traction phase. On the other hand, this parameter clearly plays a crucial role when discussing take-off approaches. In order to evaluate a given take-off technique on a quantitative basis, the total mass of the aircraft m has to be linked to the system's capability in terms of force and power. Such a link is given by the so-called wing loading w_l, i.e. the ratio between m and A:

$$m = w_l A.$$ (20.2)

The total mass of the aircraft is then the sum of m and of the additional mass Δm required for the take-off capability, as further discussed below.

20.3.1 Acceleration Phase on the Ground

The acceleration phase on the ground lasts until the take-off speed v^* is reached:

$$v^* = \sqrt{\frac{2(m+\Delta m)g}{\rho A C_L}},$$ (20.3)

computed by setting $F_L = (m+\Delta m)g$ and using $F_L = \frac{1}{2}\rho A C_L v^{*2}$. Assuming that this speed shall be reached after a horizontal acceleration distance L, the required acceleration is $a = v^{*2}/(2L)$. The corresponding required force is then $F_g = (m+\Delta m)a$. The other forces acting at take-off are significantly smaller, but not negligible, namely the drag force $F_D = \frac{1}{2}\rho C_{D,eq} A v^{*2}$ and the viscous resistance $F_v = c_v v^*$, where c_v is the viscous friction coefficient of the system employed for the linear acceleration. Hence, the required maximal power on the ground, $\bar{P}_g$, is

$$\bar{P}_g = v^* \left(F_g + F_D + F_v\right)$$

$$= \frac{\sqrt{2}(m+\Delta m)^{5/2}g^{3/2}}{L(\rho A C_L)^{3/2}} + \frac{C_{D,eq}(2g(m+\Delta m))^{3/2}}{\sqrt{\rho A C_L^{3/2}}} + \frac{2g(m+\Delta m)c_v}{\rho A C_L}.$$ (20.4)

As we comment later on, the additional mass Δm results to be proportional to the mass m of the aircraft without considering the take-off equipment. Hence, Eqs. (20.3) and (20.4) reveal that, for given wing loading and aerodynamic coefficients, the take-off speed is independent of the wing's size and that, for a fixed travel distance, the peak required ground power is proportional to the effective area in a first approximation. Moreover, Eq. (20.4) clearly shows that, if the take-off distance L is small, e.g. of the same order of magnitude of the aircraft wingspan, the inertia of the plane (and of the linear motion system) is the dominating term in the power requirement on the ground, since $C_{D,eq}$ and c_v are small.

As regards the land occupation A_g, we choose to fix the acceleration distance L on the rails, such that it is independent from the wing size, and we assume that the system shall be able to adapt to the widest possible range of prevalent wind conditions, i.e. the linear acceleration phase can be carried out in all directions. At the same time, the area spanned by the wings throughout the ground launching phase is considered to be occupied by the system. Thus, we obtain

$$A_g \simeq \frac{\pi L^2}{4} + \frac{\pi d^2}{4}, \tag{20.5}$$

where d is the wingspan of the aircraft.

20.3.2 Powering the Plane During the Ascend

We analyze the climbing phase assuming the worst conditions possible, i.e. with zero prevalent wind speed, which yields the required peak on-board power. It turns out that the required propeller thrust is approximately equal to

$$F_T \approx (m + \Delta m)g \left(\frac{C_{D,eq}}{C_L} + c_r \right), \tag{20.6}$$

where c_r denotes the ratio of the vertical speed v_c and the horizontal (forward) speed v_{fwd} (see Fig. 20.3). We refer to Ref. [7] for a detailed derivation. The required on-board power is then (see [7])

$$\overline{P}_{ob} = \frac{F_T}{\eta} \left(\sqrt{\frac{F_T}{2\rho A_{prop}} + \frac{v_{fwd}^2}{4}} + \frac{1}{2} v_{fwd} \right), \tag{20.7}$$

Here, we consider two propellers with a total area A_{prop}, each having a diameter of half the chord (where the chord length is the distance from the leading to the trailing edge of the wing), with horizontal axes, and with an efficiency of η.

Finally, as regards the additional on-board mass Δm, this is mainly determined by the on-board batteries and the electric motors that drive the propellers. The required battery mass is calculated from the energy density of lithium-polymer batteries E_{batt} and the required power P_{ob}, target altitude h and climb speed v_c (i.e. the climb du-

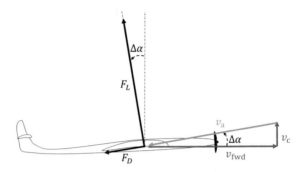

Fig. 20.3 Schematic representation of an airplane with horizontal speed of v_{fwd} (assuming no wind) and a vertical speed of v_c. The lift force has a component opposite to the thrust and the drag force has a component which adds to the gravitational pull

ration is h/v_c). The power density of an electric motor is indicated by E_{mot}. The resulting equation for the additional on-board mass is:

$$\Delta m = P_{ob}\left(\frac{h}{v_c\, E_{batt}} + \frac{1}{E_{mot}}\right). \tag{20.8}$$

We solve Eqs. (20.6) to (20.8) to compute the required on-board power, accounting also for the additional mass.

20.3.3 Results and Discussion

We evaluate expressions given by Eqs. (20.1) to (20.8) for the input parameters given in Table 20.1 for three different wing sizes, namely 5, 10 and 20 m wingspan. The results are reported in bold numbers in the Table. It turns out that the required power values—be it on the ground or on board—are small compared to the generated (mechanical) power: the required on-ground power is about 11% of the peak mechanical power (at a wind speed of 15 m/s and with take-off travel $L = 12\,\text{m}$); the on-board power only 3%. The rather small power which is required on board results in a weight increase of 5%.

Some aspects should be pointed out:

- The required power on the ground could, in principle, be provided by the electric machine connected to the winch: one could envision a solution where this machine is also employed in the initial phase of the take-off, e.g. by means of a clutch to (dis-) engage a linear motion system to accelerate the aircraft. In our prototype, we use an additional electric motor for simplicity.
- The required ground occupation is dominated by the wing size when scaling up. Hence, it turns out to be quite favorable.

Parameter	Aircraft 1	Aircraft 2	Aircraft 3
Wing span d (m)	5	10	20
Aspect ratio $\mathcal{R}$		10	
Chord $d/\mathcal{R}$ (m)	0.5	1	2
Wing area A (m^2)	2.5	10	40
Wing loading $w_l = m/A$ (kg/m^2)		15	
Mass m	37.5	150	600
Lift coefficient C_L		1	
Drag coefficient $C_{D,eq}$		0.1	
Desired vertical velocity v_c (m/s)		1	
Propeller efficiency η		0.7	
Peak mech. power P_m^* at $W = 15$ m/s (kW)	75	300	1200
Ground travel distance L (m)		12	
Target height h (m)		100	
Viscous friction coefficient c_v (kg/s)	0.1	0.3	1
Take-off speed v^* (m/s)		**15.7**	
Propeller's diameter $d/(2\mathcal{R})$ (m)	0.25	0.5	1
Peak additional ground power $\overline{P}_g$ (kW)	**8**	**31**	**124**
Peak additional on-board power $\overline{P}_{ob}$ (kW)	**2**	**9**	**37**
Additional on-board mass Δm (kg)	**2**	**5**	**20**
Required ground area A_g (m^2)	**132**	**192**	**428**

Table 20.1 Design parameters for the take-off. Bold-faced parameters are the results obtained according to the assumptions and analysis described in Sects. 20.3.1 and 20.3.2. If only one numerical value is given per line, then this value holds for all three aircraft

- The required on-board power is rather small because the aircraft does not need to be accelerated any further.
- On-board propellers and batteries are necessary in any case to power the on-board control systems. The use of slightly larger and more powerful on-board motors does not appear to be critical. Moreover, the on-board propellers can also be used to re-charge the batteries to supply energy to the control system during long periods of power generation.
- Since the whole setup can be built in a such a way that it is rotatable, the take-off is independent of the current prevalent wind direction.

These results indicate that the linear take-off approach will have only a rather small impact on the overall system design because it provides a good tradeoff between on-board and on-ground power. Other launching procedures would require extensive system modifications as we discussed in Ref. [7]. However, there we only considered a rigid aircraft with a fixed wing loading of $w_l = 15\,\text{kg/m}^2$. In Fig. 20.4, we show how the required peak power on the ground and on board depend on the wing loading, for two different wing sizes, namely 1.6 m and 10 m. This is important because it is not clear at this stage of AWE development what the typical wing loading will be and—if they turn out to be high ($\sim 20\,\text{kg/m}^2$)—what the impact will

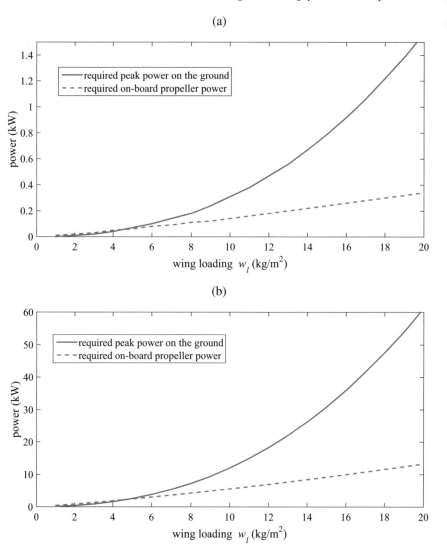

Fig. 20.4 Required peak power for acceleration on the ground (solid line) and propeller power on board (dashed line). (a) small model aircraft like the one used in our experimental setup ($d = 1.6$ m corresponding to 7.7 kW mechanical power with the other parameters given in Table 20.1) as a function of wing loading. (b) larger aircraft ($d = 10$ m with 300 kW mechanical power as shown in Table 20.1) as a function of wing loading

be on the launching procedure. Apparently, the smaller the wing loading, the better for the overall system performance in terms of aircraft maneuverability, operating wind range and ultimately for the power generation capability. On the other hand, at large power levels the structural requirements might call for stronger structures with higher wing loading.

Fig. 20.4 demonstrates that the linear launch approach in combination with small on-board propellers can also cope well with high wing loading. The required power for the on-board propellers increases linearly with the wing loading and represents only about 4% of the mechanically generated peak power for a wing loading of 20kg/m^2, up from about 3% with $w_l = 15 \text{kg/m}^2$ and 2% with $w_l = 8 \text{kg/m}^2$. As expected from Eq. (20.4), the power for acceleration to the take-off speed on the ground increases more than quadratically with the wing loading. However, this additional power does not affect the aircraft design. Moreover, even at high wing loadings of 20kg/m^2, the on-ground peak power is only about 20% of the estimated generated peak mechanical power during crosswind flight. If the main winch is used for acceleration of the slide to take-off speed, high wing loadings do not present a significant disadvantage for the linear launch procedure, either.

So far, we focused our attention on the take-off phase only, and on its impact on the overall system design. In the following section, we present results concerning the full cycle of take-off, flight with low tether tension, and landing. In particular, we consider the modeling and control aspects of this strategy and study it by means of numerical simulations.

20.4 Modeling and Control of Linear Take-Off and Landing Maneuvers

20.4.1 Control Objectives and Problem Formulation

For each phase described in Sect. 20.2.2, there are specific control objectives. During the launch, the ground station shall be able to synchronously accelerate the slide and the main winch to allow the aircraft to take-off with low tether tension. During the flight phase, the on-board control unit shall follow the desired path despite the perturbation of the tether and wind. At the same time, the ground-station control system shall adapt the tether length such that the tension is low but non-zero, to have minimal impact on the aircraft's flight but at the same time avoiding tether entangling on the winch and an excessive line sag. Finally, in the landing phase, the aircraft shall land within the area covered by the rails so that it can engage with the slide again. In Ref. [14], we described a control system able to achieve the above-mentioned goals, and we showed simulation results for an aircraft with small wing loading. We recall here the main aspects of the control strategy, provide simulation results for an aircraft with larger wing loading, and compare them with the ones presented in Ref. [14].

20.4.2 Mathematical Model of the System

We describe the system dynamics with a hybrid model (see Fig. 20.5): a first operating mode accounts for the system's behavior from zero speed to the take-off speed, during which the aircraft and the slide can be considered as a unique rigid body; a second operating mode describes the aircraft motion after take-off, when it is separated from the slide.

We consider an inertial, right-handed reference frame (X, Y, Z) with the origin corresponding to the central point of the rails, which are assumed parallel to the ground, the X–axis aligned with them, and the Z–axis pointing downwards, see Fig. 20.1. We denote a generic vector in the three-dimensional space as $\mathbf{v}$ and we specify the reference frame considered to compute the vector's components with a subscript notation, e.g. $\mathbf{v}_{(XYZ)}$. Each scalar component of the vector will be followed by its axis, i.e. $\mathbf{v}_{(XYZ)} = [v_X, v_Y, v_Z]^\top$. For the sake of brevity, we omit the explicit dependence of the model variables on the continuous time t.

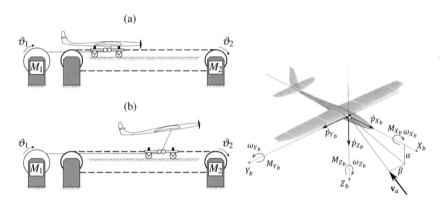

Fig. 20.5 Sketch of the dynamical model. Left: (a) First operating mode, with the aircraft carried by the slide up to take-off speed; (b) second operating mode, with the aircraft detached from the slide. Right: Variables and parameters describing the flight expressed in the body frame of reference. ω_{X_b} is the roll rate, ω_{Y_b} is the pitch rate, and ω_{Z_b} is the yaw rate. α and β are the velocity angles or the angle of attack and the side slip angle, respectively

In the model of the ground station, we denote the motor/generator linked to the winch with M_1 and with M_2 the one connected to the slide drum. $\vartheta_1, \dot{\vartheta}_1, \vartheta_2, \dot{\vartheta}_2$ denote the angular positions and speeds of M_1 and M_2, respectively, while u_{M_1}, u_{M_2} denote the torques applied by the motors. The state and input vectors of the ground-station model are then given by

$$x_{GS} \doteq [\vartheta_1, \dot{\vartheta}_1, \vartheta_2, \dot{\vartheta}_2]^\top,$$
$$u_{GS} \doteq [u_{M_1}, u_{M_2}]^\top. \tag{20.9}$$

The dynamical model of the ground station accounts for inertia and viscous friction of the winch and of the linear motion system and for the external forces exerted by the aircraft (lift, drag and inertia) and by the tether when under tension. The switch between the first and second operating mode occurs when the lift force developed by the aircraft equals its weight, hence lifting it from the slide. From that instant on, the ground station and the aircraft are considered as separate rigid bodies. The full model equations of the ground station are omitted here for brevity; the interested reader is referred to Ref. [14] for the full details.

To model the aircraft's dynamics after detaching from the slide, we consider the body reference frame (X_b, Y_b, Z_b), represented in Fig. 20.5, which is fixed to the plane and whose rotation relative to the inertial frame (X, Y, Z) is defined by the Euler angles ϕ (roll), θ (pitch) and ψ (yaw). Denoting with ω the angular velocity vector of the aircraft (see Fig. 20.5),we have:

$$
\begin{aligned}
\dot{\phi} &= \omega_{X_b} + \left(\omega_{Y_b} \sin(\phi) + \omega_{Z_b} \cos(\theta)\right) \tan(\theta), \\
\dot{\theta} &= \omega_{Y_b} \cos(\phi) - \omega_{Z_b} \sin(\phi), \\
\dot{\psi} &= \tfrac{1}{\cos(\theta)} \left(\omega_{Y_b} \sin(\phi) + \omega_{Z_b} \cos(\phi)\right).
\end{aligned}
\tag{20.10}
$$

We further denote the position of the aircraft relative to the origin of the inertial system (X,Y,Z) with $\mathbf{p}$. The manipulated variables available for control are denoted with u_a (ailerons), u_e (elevator), u_r (rudder), u_f (flaps), and u_m (motor thrust). We then define the state and input vectors of the aircraft model as:

$$
\begin{aligned}
x_g &\doteq \left[p_X, p_Y, p_Z, \dot{p}_{X_b}, \dot{p}_{Y_b}, \dot{p}_{Z_b}, \phi, \theta, \psi, \omega_{X_b}, \omega_{Y_b}, \omega_{Z_b}\right]^\top, \\
u_g &\doteq \left[u_a, u_e, u_r, u_f, u_m\right]^\top,
\end{aligned}
\tag{20.11}
$$

as well as the full system's state and input vectors as:

$$
\begin{aligned}
x &\doteq \begin{bmatrix} x_{GS} \\ x_g \end{bmatrix} \in \mathbb{R}^{16}, \\
u &\doteq \begin{bmatrix} u_{GS} \\ u_g \end{bmatrix} \in \mathbb{R}^{7}.
\end{aligned}
\tag{20.12}
$$

For a given wind vector $\mathbf{v}_w$, the apparent wind speed $\mathbf{v}_a$ is given by:

$$
\mathbf{v}_a = \mathbf{v}_w - \dot{\mathbf{p}}.
\tag{20.13}
$$

We denote the angle of attack with α and the side slip angle of the aircraft with β (see Fig. 20.5). The angles α, β and their time derivatives $\dot{\alpha}$, $\dot{\beta}$ are used to compute the aerodynamic coefficients that, together with $\|\mathbf{v}_a\|_2$ and the control inputs u_g, determine the magnitudes of the aerodynamic force $\mathbf{F}_a$ and moment $\mathbf{M}_a$. The orientations of $\mathbf{F}_a$, and $\mathbf{M}_a$ depend on the aircraft's attitude and on the control inputs as well as on α and β. In addition to the aerodynamic effects, we include the thrust of the propellers, $\mathbf{F}_T$, the aircraft's weight $\mathbf{F}_W$, and the force $\mathbf{F}_t$ and moment $\mathbf{M}_t$ exerted by the tether. The tether force is computed by considering its elasticity, aerodynamic drag and weight, which are functions of its length. The total force and moment applied to the aircraft are then computed as $\mathbf{F} = \mathbf{F}_a + \mathbf{F}_W + \mathbf{F}_t + \mathbf{F}_T$ and $\mathbf{M} = \mathbf{M}_a + \mathbf{M}_t$, respectively. They are, in general, a function of the full system's state x and input u. We make the following main assumptions:

- The earth is assumed to be flat and serves as inertial reference.
- The aircraft is a rigid body with constant mass.
- The flight takes place at very low Mach number; thus the compression effects are neglected.
- The body axis lies in the plane of symmetry of the glider.
- The coupling between the longitudinal and the lateral motion is negligible.

Based on the above definitions and assumptions, we can write the model dynamics as a system of first-order non-linear differential equations:

$$\dot{x} = f(x, u, \mathbf{v}_w) \qquad (20.14)$$

For the sake of space, we omit the full derivation of the forces and of the model equations here (see Refs. [14] and [3] for the full details).

20.4.3 Control Design

We propose a decoupled control approach, where the controller of the ground station (respectively of the aircraft) computes the values of u_{GS} (resp. u_g) according to local information. Thus, there is no active communication between the aircraft and the ground station. Rather, the coordination between the two control systems is realized by exploiting the measurement of the tether tension. We further assume that the two controllers are aware of whether the aircraft is on the slide (first operating mode in Sect. 20.4.2) or not (second operating mode). This information can be easily obtained with contact or proximity sensors installed on both the ground station and the aircraft.

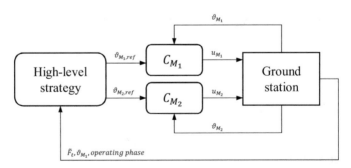

Fig. 20.6 Controller for the ground station. The "operating phase" in the outer feedback path refers to whether the system is into the first or second operating mode; this is a Boolean type of information that can be detected by means of e.g. a proximity switch.

The controller for the ground station is hierarchical (see Fig. 20.6): two low-level position control loops track the reference angular positions for motors M_1 and M_2, issued by a high-level strategy. The low-level controllers C_{M_1}, C_{M_2} are linear, designed using standard loop-shaping techniques [15] since the ground station dynamics are essentially linear as long as the tether force is kept at zero, i.e. when the tether is slightly slack. Hard limits $\bar{u}_{M_1}, \bar{u}_{M_2}$ on the magnitude of the torques u_{M_1}, u_{M_2} that the motors can deliver are accounted for as simple saturations of the input in the control strategy.

On the other hand, the high-level controller changes on the basis of the operating mode. During the take-off, a step reference position equal to the desired take-off travel is issued to the slide controller. During the consequent motion, the slide reaches the take-off speed. At the same time, the reference position for the winch motor is latched to the slide movement. After take-off, the slide motor is stopped, while the winch motor employs a reeling strategy aimed to control the load on the tether, in order not to influence the aircraft's motion significantly while at the same time avoiding too large tether sag and entanglement. The full details of the described control strategy are reported in Ref. [14].

For the aircraft controller, we adopt a hierarchical approach, too, where a low-level Linear Quadratic Regulator (LQR) tracks a reference state for the aircraft velocity and attitude. A high-level controller is used to compute such a reference state in order to control the flight path.

The LQR is designed considering the linearization of the system's model around a steady state $\mathbf{x}_{g,\text{trim}}$ and corresponding input $\mathbf{u}_{g,\text{trim}}$. In the approach used so far, the pair $(\mathbf{x}_{g,\text{trim}}, \mathbf{u}_{g,\text{trim}})$ corresponds to a straight flight, constant altitude motion. The linearized dynamics are computed by neglecting the presence of the tether, which is then an external disturbance from the point of view of the aircraft's controller.

Regarding the high-level controller for the aircraft, we define a sequence of target way points in space, denoted as $[p_{i,X}^w, p_{i,Y}^w, p_{i,Z}^w]^\top, i = 1, \ldots, N$, that are used to compute reference altitude and heading for the low-level LQR. The switching from one to the next way point is based on a proximity condition. The choice of the way points (number and position) is done manually in this study, in order to achieve a roughly rectangular flight pattern. Indeed, their position has to be adapted to the features of the aircraft, like aerodynamic efficiency and wing loading, and in general according to its maneuverability. For example, given the same aerodynamic coefficients and roll angle of the aircraft, the resulting turning radius will be larger with a larger wing loading. As a consequence, the position of the target points has to be adapted to reflect such a change of turning radius. One research direction to improve the approach described here is to select the way points via numerical optimization techniques.

For a given way-point, the high-level strategy issues two reference signals: one to control the altitude of the aircraft, and one to control its heading. The altitude controller computes a reference pitch rate $\omega_{Y_b,\text{ref}}$ on the basis of the measured path angle γ, defined as:

$$\gamma \doteq \alpha - \theta, \tag{20.15}$$

where the angle of attack α is defined as :

$$\alpha \doteq \arctan\left(\frac{\dot{p}_{Z_b}}{\dot{p}_{X_b}}\right), \tag{20.16}$$

A reference path angle γ_{ref} is derived from the current aircraft's altitude and that of the current target way-point:

$$\gamma_{\text{ref}} = \arctan\left(\frac{p_{i,Z}^w - p_Z}{p_{i,X}^w - p_X}\right). \tag{20.17}$$

Then, the reference pitch rate given to the LQR is computed as:

$$\omega_{Y_b,\text{ref}} = -k_\gamma(\gamma_{\text{ref}} - \gamma), \tag{20.18}$$

where $k_\gamma > 0$ is a constant gain chosen by the control designer.

The second reference signal issued by the high-level controller is for the heading of the aircraft. The reference heading needed to reach the current target point is computed as

$$\psi_{\text{ref}} = \arctan\left(\frac{p_{i,Y}^w - p_Y}{p_{i,X}^w - p_X}\right) \tag{20.19}$$

where the four-quadrant arctangent is used. The LQR then tracks such a reference yaw angle. To obtain smooth transitions from one target point to the next, we filter the reference heading signal with a first order low-pass filter. Computing the yaw reference with Eq. (20.19) is sufficient to control the heading during the flight. However, this approach does not consider the alignment of the aircraft with the orientation of the ground station, which is required to land with high accuracy. Hence, in the landing phase another strategy is used within the high-level controller. In particular, assuming without loss of generality that the last target point is the origin of the inertial system, we consider the angle β_y, defined as (see Fig. 20.7):

$$\beta_y = \arctan\left(\frac{p_Y}{p_X}\right) = \beta_t + \psi \tag{20.20}$$

where β_t is the angle between the tether projected on the ground, and the inertial X-axis, as shown in Fig. 20.7. In a way similar to the altitude controller described by Eqs. (20.17) to (20.18), we set a reference yaw rate as:

$$\dot{\psi}_{\text{ref}} = k_\beta(\beta_{y_{\text{ref}}} - \beta_y) \tag{20.21}$$

with k_β being a design parameter. In order to align the aircraft with the rails, we set $\beta_{y,\text{ref}} = 0$ throughout the landing maneuver.

We refer the reader to Ref. [14] for the stability and robustness analysis of this high-level controller, as well as for a brief stability analysis of the closed-loop dynamics when both the high-level and low-level loops are implemented. For an air-

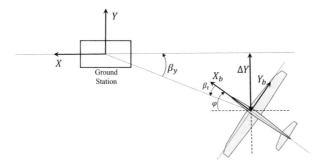

Fig. 20.7 Lateral positioning analysis for high-level controller design

craft with a relatively small wing loading, this approach led to satisfying results. Here, we employ the same approach for an aircraft with larger wing loading. In this case, as we will further comment in Sects. 20.5 and 20.6, it turns out that the increased mass renders the control problem more difficult and can lead to stall behaviors in some circumstances, e.g. in presence of strong turbulence, and to lower landing accuracy. Hence, more sophisticated control approaches like gain scheduling for the low-level and trajectory tracking algorithm for the high level might be required. On the other hand, for moderate or no turbulences, the simple approach with a fixed LQR coupled with the high-level controllers described earlier, is able to achieve the prescribed tasks with good performance also with the larger wing loading.

20.5 Simulation Results

We implemented the model and the control system in Matlab/Simulink. The main parameters of the ground station and of the aircraft are described in Table 20.2. In addition, the values $\rho = 1.2\,\text{kg/m}^3, g = 9.81\,\text{m/s}^2$ were used for the air density and gravity acceleration.

These parameters correspond to an aircraft with the same design as the one considered in Ref. [14], but wing loadings of $3.8\,\text{kg/m}^2$ and $8\,\text{kg/m}^2$. Similarly to what is discussed in Ref. [14], we computed the aerodynamic coefficients using XFLR5 [13]. The numerical values for the lighter aircraft are reported in Ref. [14], while those for the heavier one are omitted for the sake of space.

20.5.1 Take-Off Phase

Examples of simulation results for the take-off phase are shown in Figs. 20.8 and 20.9, for the higher wing loading of 8 kg/m². In Fig. 20.8(a), it can be noted that

Ground station parameters		Aircraft and tether parameters	
Winch radius	0.1 m	Wingspan	1.68 m
Slide drum radius	0.1 m	Aspect Ratio	8.9
Winch m. of inertia	0.08 kg m^2	Wing loading	$\{3.8, 8\}$ kg/m^2
Slide drum m. of inertia	0.01 kg m^2	Mass	$\{1.2; 2.54\}$ kg
Winch visc. fr. coeff.	0.04 kg m^2/s	Propeller power	$\{320; 600\}$ W
Slide drum visc. fr. coeff.	0.01 kg m^2/s	Peak motor thrust	$\{8; 16.6\}$ N
Slide mass	2 kg	Tether Young mod.	5.3×10^9 Pa
Visc. friction coeff. of rails	0.6 kg/s	Tether breaking elong.	0.02
Peak torque M_1	220 N m	Tether drag coeff.	1
Peak torque M_2	22 N m	Tether diameter	0.002 m

Table 20.2 Simulation parameters for aircraft with wing loadings of 3.8 kg/m^2 and 8 kg/m^2

the total travel distance of the slide is equal to about 5 m, and that the aircraft starts the ascend after 2.9 m, i.e. when the take-off speed of 13 m/s has been reached. As shown in Fig. 20.8(b), the slide motor exploits the full rated torque to accelerate and then to brake the slide. The propeller is engaged only after take-off and, after a short transient, it settles to a steady value sufficient to achieve the desired vertical velocity, see Fig. 20.9(a). Fig. 20.9(b) presents the power consumption of the slide motor, winch motor and propellers during the take-off. The simulated peak power values are 3 kW to accelerate the slide, and 0.5 kW for the ascend phase. The former value is in line with the theoretical results of Sect. 20.3 which provide, for the same parameters, 3.1 kW for the ground acceleration. On the other hand, the on-board power predicted by the simulation is larger than the 0.1 kW given by the theoretical results, essentially due to a larger climb rate (twice the one considered

(a)   (b)

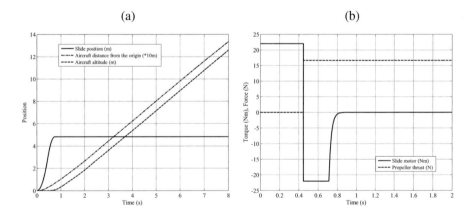

Fig. 20.8 Simulation results for (a) courses of the aircraft height, slide position and aircraft distance from the ground station (divided by 10 for the sake of clarity) and (b) courses of the slide motor torque and of the propeller thrust. Wing loading: 8 kg/m^2

(a)   (b)

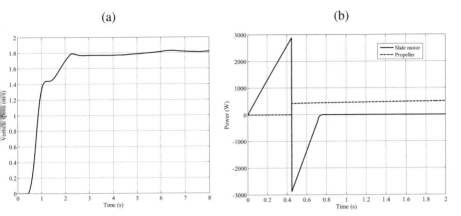

Fig. 20.9 Simulation results for (a) course of the vertical speed of the aircraft and (b) courses of the slide motor's and propeller's power. Wing loading: 8 kg/m^2

in Sect. 20.3), due to the pitch of the aircraft, which has the effect of decreasing the thrust in horizontal direction and adding a braking contribution from the lift force projected onto the x_g–axis, and due to the fact that in the simulation the plane operates at a lift coefficient of 0.6. The on-board power is anyways a reasonably small fraction (approximately 5%) of the system's power.

The results obtained with the lower wing loading, i.e. 3.8 kg/m^2, are qualitatively identical. The aircraft in this case takes off after 1.1 m at a take-off speed of about 9 m/s. The simulated peak power values are 2 kW to accelerate the slide, and 0.2 kW for the ascend phase. The same considerations as those drawn above for the higher wing loading, about the matching between the theoretical results and the simulations, hold also in this case.

20.5.2 Flight and Landing Phases

In Fig. 20.10(a), we present the flight patterns for an aircraft with 3.8 kg/m^2 wing loading (the same as considered in Ref. [14]) and for 8 kg/m^2, with the main parameters reported in Table 20.2. As discussed in Sect. 20.4.3, a larger wing loading leads to a larger flight pattern, with longer climbing and approaching phases. The reasons for this behavior are essentially the larger velocity required by the heavier aircraft (while the climbing velocity remains the same) combined with the increased inertia, while the maximum position of the control surfaces and the maximum roll and pitch angles are roughly the same. In other words, the increased mass requires a larger fraction of the lift force to be exploited to keep the aircraft airborne, leaving a smaller lift contribution available for maneuvering, hence increasing the minimum turning radius. Another approach to reduce the turning radius could be to exploit the tether tension in order to enforce geometrically the pattern curvature, eventually by reeling-in during the turn. Fig. 20.10(b) shows the tether load during the whole cycle. A low tension is kept throughout the flight, notwithstanding the pronounced changes in tether length, also shown in Fig. 20.10(b), which matches closely the distance between the aircraft and the origin. This result indicates that indeed it should be possible to avoid continuous communication between the ground station and the aircraft and still obtain satisfactory results. Our recent experimental results further confirm this aspect [5]. If available, communication could be then added to further improve the performance, e.g. by injecting additional apparent wind speed by reeling-in in case of close-to-stall situations

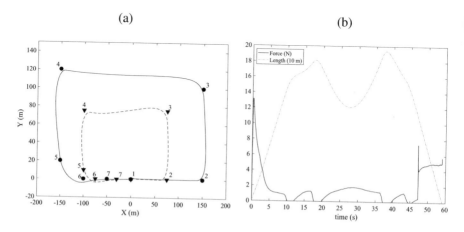

Fig. 20.10 Simulation results: (a) A three-dimensional illustration of the flight path with reference points for the aircraft with $3.8\,\text{kg/m}^2$ wing loading (dashed) and with $8\,\text{kg/m}^2$ wing loading and (b) tether tension (solid) and length (dashed) for the aircraft with $8\,\text{kg/m}^2$ wing loading

Since the high-level controller is based on way points instead of a trajectory tracking approach, and since such way points are different between the two considered aircraft due to their different wing loading, evaluating the tracking performance of the controller during the low-tension flight is of little interest. Instead, comparing the landing positioning performance between the two aircraft and as a function of the wind conditions is of high interest for the sake of this study. To this end, a comparison of the positioning precision achieved by the two aircraft in the landing phase is shown in Fig. 20.11. We consider increasing nominal wind speed (aligned with the rails such that take-off and landing are performed with the aircraft facing the incoming wind) and random wind disturbances with an amplitude equal to 30% of the nominal speed in all three directions. We compute 50 simulated flights for each nominal wind speed and each aircraft. The obtained positioning precisions are shown, in terms of average X and Y positions, $\overline{X}$ and $\overline{Y}$, and standard deviations σ_X and σ_Y, in Table 20.3, together with the statistics of the horizontal landing speed (average $\overline{V}$ and standard deviation σ_V) and of the unsuccessful landings.

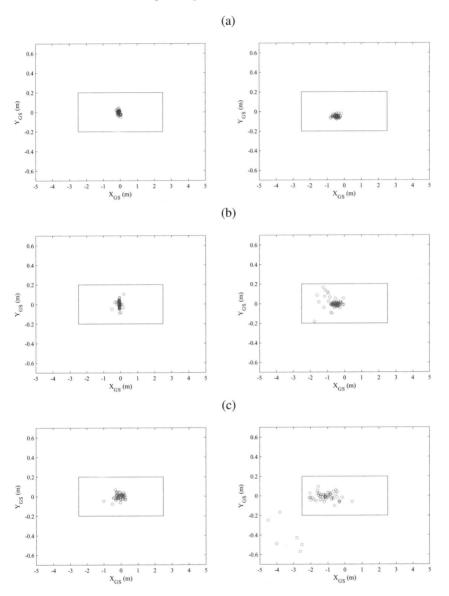

Fig. 20.11 Simulation results. Touch-down position for the aircraft with $3.8\,\text{kg/m}^2$ wing loading (left plots) and for $8\,\text{kg/m}^2$ wing loading (right plots) with nominal wind of (a) 2 m/s, (b) 4 m/s and (c) 6 m/s and uniformly distributed 3D wind disturbances in the range of $\pm 30\%$ of the nominal wind. The rectangle in the plots correspond to the dimensions of the rails (note the different scales of the two axes)

Figures 20.11(a) to (c) show the touch-down points on the ground station with the origin being in the middle of the rails which cover an area of $0.4\,\text{m} \times 5\,\text{m}$. With

Nom. wind	$\overline{X}$ (m)	σ_X (m)	$\overline{Y}$ (m)	σ_Y (m)	$\overline{V}$ (m/s)	σ_V (m/s)	Failures (%)
\multicolumn{8}{c}{Wing loading: 3.8 kg/m^2}
0 m/s	0.18	0.03	−0.002	0.03	16.2	2.3	0
2 m/s	−0.07	0.07	−0.004	0.013	15.0	2.2	0
4 m/s	−0.06	0.1	0.005	0.026	12.9	1.9	0
6 m/s	−0.02	0.16	−0.003	0.034	10.6	1.7	0
\multicolumn{8}{c}{Wing loading: 8 kg/m^2}
0 m/s	−0.46	0.14	−0.05	0.035	17.1	2.5	0
2 m/s	−0.45	0.17	−0.007	0.017	15.0	2.18	0
4 m/s	−0.54	0.54	0.015	0.046	12.5	2.14	4
6 m/s	−1.01	1.65	−0.11	0.31	10.6	2.4	4

Table 20.3 Landing precision and velocities for different wind conditions. For each nominal wind speed, wind gusts with a magnitude of ±30% of the nominal wind are considered. For the case of zero nominal wind speed, wind gusts of ±1 m/s are considered

increasing nominal wind speed, the average touch-down point is pushed backwards and the lateral positioning accuracy improves. This is due to the fact that a higher front wind reduces the aircraft's speed relative to ground, hence giving more time to align with the rails. These effects are expected and indicate a good overall performance of the control system.

For the lighter aircraft, touch-down is always within the area spanned by the rails. For the heavier aircraft, the situation is more critical. In fact, with the strongest nominal wind speed (6 m/s), in 4% of the cases the aircraft stalls during the flight. In 14% of the cases, the controller is not able to land the aircraft on the rails, finally the average accuracy of the successful landings is worse than that obtained with the lighter aircraft. The main reasons for this outcome are similar to what is discussed above, i.e. the larger mass reduces maneuverability and makes it more difficult to counteract the wind disturbances. Moreover, the wider flight trajectory implies a longer tether, with consequently larger weight and drag with respect to the case of the lighter aircraft. Using a more advanced control approach (e.g. gain scheduling) might help improving robustness, as well as having more powerful on-board propellers, a more efficient wing design and larger actuation limits of the control surfaces, or even additional control surfaces.

20.6 Conclusions and Future Developments

We presented an overview of recent results pertaining to the take-off and landing phases of AWE systems based on a rigid aircraft and pumping-power conversion. A theoretical analysis derives the main links among the ground and on-board power, the on-board mass, and the land occupation required to carry out the take-off. A simulation study further shows the system behavior in a full cycle consisting of take-off, low-tension flight and landing again on the rails used for the initial start-up. A decentralized control approach to carry out the cycle is described as well.

Regarding the take-off phase, the results indicate that the additional ground and on-board equipment constitutes a rather small cost fraction of the total system costs, even with large wing loading. At the same time, the required land occupation is reasonably small.

On the other hand, when it comes to the low-tension flight and landing phases, the mass has an important effect on stability and landing accuracy, so that a trade-off between maneuverability, wing loading and on-board power has to be achieved. Better control strategies and ad-hoc wing design and actuators can mitigate the issue, and this will be subject of future research. For example, one could schedule the controller according to the incoming airspeed, employ feed-forward contributions to improve the control response time, or develop additional control strategies which, for example, could monitor the behavior during landing and eventually decide to take-off again before touch-down and then attempt another landing.

In summary, the present study shows that the wing loading is a crucial parameter in the system design not only for power generation, where it affects the cut-in and cut-out wind speeds, the flight trajectory, and ultimately the overall capacity factor, but also for the landing phase. Which condition is the most critical one and hence provides the main design criteria for the whole system is also a subject of future studies. It is well true that similar results, concerning the effects of wing loading on take-off, flight and landing for aircraft, are largely available in the literature of flight dynamics, however without considering the tether. The added value of our study in this respect is to study these aspects also when the tether is present. Another finding of our research is that with a proper force measurement and a large enough torque of the winch motor, the coordination between the winch and the aircraft can be carried out without active communication among the two, since the winch can react fast enough to avoid stalling the plane. This behavior can be further improved by using a damping system on the ground, e.g. by means of a mass-spring-damper system, and by using the measured state of such a system to improve the winch control strategy.

Recent experimental tests carried out at ABB Switzerland, Corporate Research, demonstrated the feasibility of autonomous launch and low-tension flight of a small rigid aircraft. The employed approach and results are presented in Refs. [4, 5].

References

1. Ampyx Power B.V. http://www.ampyxpower.com/. Accessed 6 Feb 2017
2. Bontekoe, E.: How to Launch and Retrieve a Tethered Aircraft. M.Sc.Thesis, Delft University of Technology, 2010. http://resolver.tudelft.nl/uuid:0f79480b-e447-4828-b239-9ec6931bc01f
3. Etkin, B.: Dynamics of Atmospheric Flight. Dover Publication, New York, NY (1972)
4. Fagiano, L., Nguyen-Van, E., Rager, F., Schnez, S., Ohler, C.: A Small-Scale Prototype to Study the Take-Off of Tethered Rigid Aircrafts for Airborne Wind Energy. IEEE/ASME Transactions on Mechatronics **22**(4), 1869–1880 (2017). doi: 10.1109/TMECH.2017.2698405

5. Fagiano, L., Nguyen-Van, E., Rager, F., Schnez, S., Ohler, C.: Autonomous Take-Off and Flight of a Tethered Aircraft for Airborne Wind Energy. IEEE Transactions on Control Systems Technology **26**(1), 151–166 (2018). doi: 10.1109/TCST.2017.2661825
6. Fagiano, L., Milanese, M.: Airborne Wind Energy: an overview. In: Proceedings of the 2012 American Control Conference, pp. 3132–3143, Montréal, QC, Canada, 27–29 June 2012. doi: 10.1109/ACC.2012.6314801
7. Fagiano, L., Schnez, S.: On the Take-off of Airborne Wind Energy Systems Based on Rigid Wings. Renewable Energy **107**, 473–488 (2017). doi: 10.1016/j.renene.2017.02.023
8. Fagiano, L., Schnez, S.: The Take-Off of an Airborne Wind Energy System Based on Rigid Wings. In: Schmehl, R. (ed.). Book of Abstracts of the International Airborne Wind Energy Conference 2015, pp. 94–95, Delft, The Netherlands, 15–16 June 2015. doi: 10.4233/uuid: 7df59b79 - 2c6b - 4e30 - bd58 - 8454f493bb09. Presentation video recording available from: https://collegerama.tudelft.nl/Mediasite/Play/2ebb3eb4871a49b7ad70560644cb3e2c1d
9. Fritz, F.: Application of an Automated Kite System for Ship Propulsion and Power Generation. In: Ahrens, U., Diehl, M., Schmehl, R. (eds.) Airborne Wind Energy, Green Energy and Technology, Chap. 20, pp. 359–372. Springer, Berlin Heidelberg (2013). doi: 10.1007/978-3-642-39965-7_20
10. Gros, S., Zanon, M., Diehl, M.: A relaxation strategy for the optimization of Airborne Wind Energy systems. In: Proceedings of the 2013 European Control Conference (ECC), pp. 1011–1016, Zurich, Switzerland, 17–19 July 2013
11. Kruijff, M., Ruiterkamp, R.: Status and Development Plan of the PowerPlane of Ampyx Power. In: Schmehl, R. (ed.). Book of Abstracts of the International Airborne Wind Energy Conference 2015, pp. 18–21, Delft, The Netherlands, 15–16 June 2015. doi: 10.4233/uuid: 7df59b79 - 2c6b - 4e30 - bd58 - 8454f493bb09. Presentation video recording available from: https://collegerama.tudelft.nl/Mediasite/Play/2e1f967767d541b1b1f2c912e8eff7df1d
12. Loyd, M. L.: Crosswind kite power. Journal of Energy **4**(3), 106–111 (1980). doi: 10.2514/3.48021
13. Meschia, F.: Model analysis with XFLR5. Radio Controlled Soaring Digest **25**(2), 27–51 (2008). http://www.rcsoaringdigest.com/pdfs/RCSD-2008/RCSD-2008-02.pdf
14. Nguyen Van, E., Fagiano, L., Schnez, S.: Autonomous take-off and landing of a tethered aircraft: a simulation study. In: Proceedings of the American Control Conference, pp. 4077–4082, Boston, MA, USA, 6–8 July 2016. doi: 10.1109/ACC.2016.7525562
15. Skogestad, S., Postlethwaite, I.: Multivariable Feedback Control. 2nd ed. Wiley, New York (2005)
16. Vander Lind, D.: Analysis and Flight Test Validation of High Performance Airborne Wind Turbines. In: Ahrens, U., Diehl, M., Schmehl, R. (eds.) Airborne Wind Energy, Green Energy and Technology, Chap. 28, pp. 473–490. Springer, Berlin Heidelberg (2013). doi: 10.1007/978-3-642-39965-7_28
17. Vermillion, C., Glass, B., Rein, A.: Lighter-Than-Air Wind Energy Systems. In: Ahrens, U., Diehl, M., Schmehl, R. (eds.) Airborne Wind Energy, Green Energy and Technology, Chap. 30, pp. 501–514. Springer, Berlin Heidelberg (2013). doi: 10.1007/978-3-642-39965-7_30

Chapter 21
Kite Networks for Harvesting Wind Energy

Roderick Read

Abstract This chapter presents a simple new wind energy concept based on airborne rotary power generation and tensile rotary power transfer to the ground. The inexpensive prototypes use flexible inflatable wings that are arranged on ring kites, similar to how the rotor blades of a wind turbine are arranged on the hub. These autorotating rotary ring kites are stacked and integrated into a tensile structure that transfers the collected rotational power to a ground-based generator. A separate lifting kite provides additional lift to elevate the stack of rotary ring kites. Simulations and prototype testing show that network kite rigging provides the stabilizing benefits of wide tethering to networked individual kites even during fast flight for power generation. Turbulence effects are largely smoothed on individual kites. Stacked rotary ring kites can be integrated into a lattice of interconnected lifting kites, to concurrently run, at close proximity and thus allowing for greater land use efficiency. Solutions for joining the work of multiple ground stations to a single, more efficient generator are discussed. Software for kite network design is discussed. The designs are licensed as open source hardware to encourage engagement.

21.1 Introduction

Wind power as a renewable energy source is desirable. Taller wind turbines can harvest stronger and more persistent winds. However, the upscaling of conventional tower-based concepts has huge material use implications. Searching for innovative system concepts that scale better is thus key for achieving a sustainable and economic electricity generation in the future. Airborne wind energy systems (AWES) can operate at higher altitudes without the need for a tower. The technology can potentially supersede established tower-based wind energy technologies for large-

Roderick Read (✉)
Windswept and Interesting Ltd, 15a Aiginis, Isle of Lewis UK HS2 0PB
e-mail: rod.read@windswept-and-interesting.co.uk

scale energy generation at lower costs, land use and CO_2 output. Lightweight kite systems scale better and can operate at higher altitudes with a smaller ground footprint. In essence, more wind power can be harvested with less material.

AWES are commonly presented as a further development of conventional wind turbines. Figure 21.1 outlines the derivation of the rotary ring kite and tensile torque transfer concept. The tip of a conventional rotor blade is the fastest moving part,

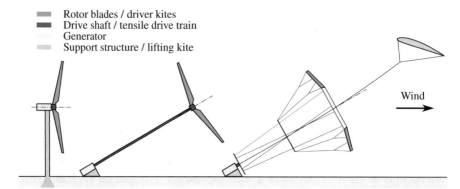

Fig. 21.1 Downwind horizontal axis wind turbine (left), intermediate conceptual step (center) and rotary ring kite and tensile torque transfer concept (right). In the first step, the heavy mast of the HAWT is eliminated by placing the generator on the ground while tilting the drive shaft upwards and extending it to allow unconstrained operation of the now pitched rotor. In the second step, the rigid and heavy rotor blades are replaced by lightweight membrane kites and the inclined drive shaft is replaced by a tensile rotational drive train. An additional lifting kite ensures the inclination of the tensile drive train

sweeping also the largest flow cross section. Although it is the lightest part of the blade it converts most of the wind power. The rigid tower has to support the weight of the rotor and the generator. It is loaded by compression and has to also resist the bending moment that the aerodynamic loading of the rotor generates.

AWES harvest wind energy with fast-flying wings that are connected to the ground by lightweight tethers. The wings use the kinetic energy of the wind and the ground tethering to fly predefined maneuvers and convert some of the wind energy into mechanical or electrical energy. The wing is bridled in such a way that it is inclined with respect to the local relative flow. The generated aerodynamic lift force propels the wing on its flight path and also generates a tensile force in the tether.

The central question is how to convert this aerodynamic force into energy that can be used on the ground and, as a matter of fact, there is a broad variety of different conversion concepts that are currently being developed. Some AWES generate the electricity on the flying wing and transmit it through the conducting tether to the ground. Other AWES use the tether or a tensile structure to transmit mechanical energy to the ground where it is then converted into electricity. This chapter treats only the case of ground-based electricity generation.

Two types of mechanical energy transfer to the ground station can be distinguished. The first one uses cable drums or similar mechanisms on the ground to convert the traction power of kites, defined as product of tether force and reeling velocity, into shaft power, defined as product of torque and angular speed. Most implemented systems operate a single kite [3, 14, 19] or two kites [9] in pumping cycles. A traction kite operated on a linear track has been implemented as a first step towards multi-kite systems that collaboratively generate electricity on a horizontal loop track [1].

The second type uses a tensile structure to directly transfer rotational power from a rotating kite configuration to the ground station. The tensile structure consists of several tethers that are kept separated from each other and tensioned while rotating around a common axis. The working principle of such a concept is illustrated in Fig. 21.1 (right). Similar to the tips of turbine rotor blades, the lightweight driver kites sweep a relatively large flow cross section at high speed. The tensile rotational drive train is optimized to transfer the generated torque at a minimum airborne mass of the structure. An additional lifting kite is used to ensure a stable inclination of the rotary system.

It is unusual in engineering applications to transmit shaft power over long distances and it is even more unusual to do this with a lightweight tensile structure. Using the experimental setup shown in Fig. 21.2 we have successfully demonstrated the working principle of the rotary ring kite concept and the feasibility of tensile torque transfer to the ground. The shaft power available at the ground station can easily be used for continuous electricity generation. The prototype designs have been published under open hardware licensing at [18].

Fig. 21.2 The "daisy stack" developed by Windswept and Interesting Ltd employs a tensile structure to transmit the rotational power of stacked ring kite configurations (30 August 2017)

This chapter describes prototyping, experimentation and design proposals for rotary ring kite configurations. The "daisy stack" illustrated in Fig. 21.2 integrates rotary ring kites into a tensile rotational drive train that provides continuous positive shaft power to the ground station. Much of the experimentation investigated the rotational power transmission and the application of network designs to kite bridle systems. The daisy stack was the first AWES to win the "100 × 3 challenge" announced on [18], which had the goal to fly an AWES at an altitude of 100 foot and generating an average of at least $P_{net} = 100$ W for 100 minutes. Tests in December 2017 with the latest system illustrated in Fig. 21.2 have yielded a net power output of $P_{net} \approx 600$ W [11].

The design process at Windswept and Interesting Ltd (W&I) essentially was open trial and error. Experimental designs evolved from experiences with kites, adventure sports and crafts. Occasionally, trails have been dangerous. The work has only been possible with the help of open online forums and the published work of the AWES community [20]. W&I considered many workable AWES schemes. Of those designs, rotary kite network and lift kite network designs are recommended.

Daisy ring kites can be classified as gyrokites that rely on wind-powered autorotation to develop aerodynamic lift in order to fly. By integrating the ring kites into a tensile drive train the rotational power can be transferred to the ground for conversion into electricity. Notable similarity and inspiration is seen in classic kites, like the spin bol, and in the works of Dave Santos [15], Rudy Hardburg on a "Coaxial Multi-Turbine Generator" [8], Bryan Roberts on a "Flying Electric Generator" [13], Doug Selsam on a "Serpentine Superturbine" [17] and Pierre Benhaïem on "Rotating Reeling" [2] and together with Roland Schmehl in Chap. 22 of this book.

Multiple kites bridled together establish a larger meta-kite. Even when only tied to a single arched load line in crosswind direction such a meta-kite will remain in stable flight. Meta-kites accumulate energy from a large harvesting area and can thus be dangerously powerful!

Kite networks with wide spacing and interconnections constrain the freedom of motion of the individual component kites. Kites, which would otherwise fly independently of one another, can work cooperatively flying in a network formation. Kite networks can also be formed into more complex three-dimensional lattice configurations. Networked kites simplify AWES flight control by using bridle network geometries and aerodynamic effects in combination to constrain the possible flight patterns of individual kites. The simple autogyro prototype has no cyclic pitch control. Without power curve profiling nor even controls this prototype is not optimised for rotary power generation yet. It does however provide a smooth continuous generation from inexpensive kites in a range of workable wind conditions. For larger daisy stacks automated controls including launching and landing systems would be preferred for safe operation. Passive control from a network geometry, force alignment and aeroelasticity effects [4] can be used to control a working kite network AWES.

The rest of this chapter focuses on experimentation results and conceptual designs of six key elements of an AWES farming architecture developed by W&I:

- Rotary "daisy" ring kites

- Power over rotating tethers (PORT)
- Stacked ring kite configurations
- Lifting isotropic network kite (LINK)
- Ground control and generation
- Open source design.

Each of these is presented separately in the following as a section. An AWES combining the six elements is then briefly considered at the end of this chapter.

21.2 Daisy Ring Kites

Rotary ring kites are the central functional elements converting the kinetic energy of the wind into a rotational motion and at the same time providing aerodynamic lift to stay airborne at the operational altitude. In the following we will detail the design principles, motives for the choices of kite components, present options for tuning and control of rotary rings kites and describe experiential results.

21.2.1 Driven Ring Kite Design Principle

As illustrated in Fig. 21.3 a set of asymmetric driver kites is mounted along the perimeter of a ring kite and bridled to fly on a circular path much like the blades of a wind turbine rotate around its hub. This analogy is also illustrated schematically in Fig. 21.1. The propulsive power of the driver kites is transferred via the bridle line system into the rotational drive train, which drives the generator on the ground. The radial expansion forces tension the driver kites in spanwise direction and ensure that the bridle line systems of the driver kites and the ring kite are separated. The

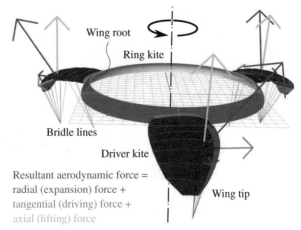

Fig. 21.3 Aerodynamic forces acting on the asymmetric driver kites (blades) mounted along a ring kite (hub). The radial, tangential and axial directions refer to the local reference frame of the entire rotary kite assembly, denoted as "daisy ring kite". For the individual driver kite the radial direction is roughly coinciding with the spanwise direction of the wing while the tangential direction is roughly coinciding with the chordwise direction

resultant axial force of the driver kites constitutes the thrust of the rotary ring kite. In contrast to a horizontal axis wind turbine this thrust force is tilted upwards with the elevation angle and thus has both lift and drag components. Because the presence of a lift component qualifies the rotary assembly as a kite we denote the thrust force in the following as lifting force. As can be seen in Fig. 21.2, the rotary kite assembly is also tethered to a lifting kite, which flies at a higher altitude and provides an additional lifting force which tensions the rotational drive train in axial direction. The tether running along the axis of the drive train up to the lifting kite is denoted in the following as "lifting line".

The tensioning of the drive train by axial and radial forces is essential for maintaining its three-dimensional shape and its capability to transfer rotational power to the ground. The transferable torque increases with the torsion of the drive train up to the point where a beginning constriction starts to impede the capability of torque transfer. The radial aerodynamic force components are caused by the anhedral arc shape of the driver kites and their bank angle [12]. Another radial aerodynamic force component is contributed by the conical shape of the ring kite, which is stabilized by a 3 mm diameter carbon epoxy stiffening rod integrated into the circular leading edge. In newer designs of the system, such as illustrated in Fig. 21.1, the ring kites with conical shape are replaced by rings. An additional radial tensioning is caused by the centrifugal forces acting on the rotating system components.

The axis of rotation of the daisy ring kite is approximately coinciding with the lifting line. The fixed path of the driver kites around the lifting line allows stacking of connected rotary rings to incrementally increase the generated power. The lifting kite is a standard kite used in kite displays, normally of a sled design. The customary use of such a kite is for steady lifting of payload, providing a high degree of stability. The lifting kite can be seen in the top of the right photo shown in Fig. 21.4.

Fig. 21.4 A single daisy ring kite with three driver kites using two separated rotating tethers ("torque ladder") to transmit rotational power to a ground-based generator (18 May 2015)

All of the kites align in downwind direction. Ring kites alone generate only a very low lift. Such kite will rest inflated on the ground, occasionally hopping into the air. However, when supported manually in its operating position a rotary ring kite in autorotation generates a significant lift force. Thus, when in operation at nominal position, ring kites require only a small additional lift contribution from the lifting line. Also, the tension provided by the lifting kite helps to guide the entire stack to align in downwind direction. In flight, the continuous motion of the driver kites is approximately in the plane perpendicular to the lifting line. If this line is at a low elevation angle the driver kites are approximately in cross wind operation.

21.2.2 Driver Kite Choices and Reasoning

The cyclical variation of aerodynamic forces generated by the driver kites were not fully accounted for and exploited in the prototypes. Instead simple beginners "forgiving" steerable two-line parafoil kites were used. Such kites can reliably fly loops with small turning radius and at high speed, despite a strongly asymmetric bridle line control input. They are continuously propulsive over a wide range of angle of attack. The kites were arranged so that they would all loop clockwise when seen from the ground.

Parafoils are most powerful when flying crosswind, in the power zone, towards the center of the wind window [7]. A normal two-line parafoil can be controlled to keep looping in the power zone, until excessive friction between the twisting tethers inhibits control. At this point the kite will spin ever lower until it eventually crashes. Mounting parafoil kites on a rotary ring eliminates the problem of twisting tethers, but creates the new problem of how to maintain a looping flight path at altitude. Luckily, it takes very little extra lift to maintain a looping altitude for a parafoil-driven kite ring. It takes very little vertical force from a lifting kite to raise the trailing edge of a ring kite. With a small lifting kite a ring kite can operate deep in the power zone.

The driver kites shown in Fig. 21.4 had their lower leading edge stiffened with a 3 mm diameter carbon rod, to prevent spanwise collapse. The photos also show a forward overdrive rod supporting the driver kites. This 4 mm diameter carbon rod prevented that driver kites collapsed by flying ahead of the daisy ring kite assembly. The rods were unnecessary for the upper kites of a rotary ring kite stack.

21.2.3 Bridle Layout, Tuning and Control

To tune a kite its bridle line system is adjusted to achieve a specific desired flight behavior. The adjustments include the lengths and attachment points of the bridle lines. In the following we propose options for tuning and flight control specifically developed for the rotary kite system.

Static Tuning The prototypes have relied on a relatively fixed tuning. The center of bridling of the driver kite is shifted towards the attachment on the ring kite. This bridling increases the tangential component of the aerodynamic force, which causes the rotational motion. As illustrated in Fig. 21.5 the sweep angle is fixed when the root is sewn onto a ring and an overdrive rod fixes the relative positioning of the leading edge. An overdrive rod can be seen in Fig. 21.4 and 21.6.

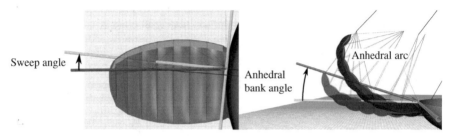

Fig. 21.5 Some tuning parameters for fixed position driver kites

The anhedral bank angle and arc were set by bridling a driver kite to a lower ring kite in the drive train. A spanwise twisting of the wing was achieved by the bridle layout shown in Fig. 21.6. The outer bridle was cascaded onto a tether connecting to the next ring kite towards the ground, slightly forward of the driver kite.

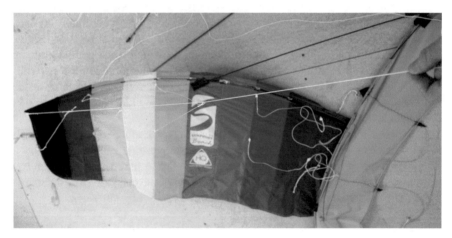

Fig. 21.6 Ring separation distance, the amount of blade twist and forward staggering between ring layers, determines how the static outer bridling cascade has to be tied

Static tuning with two-line driver kites made the manufacturing of prototypes easy. Dynamic tuning, where tethers are bifurcated and connected to match the dynamics of operational force transmission, has improved the performance. Experiments in using reactive and elastic tethering on multi-line kites are being proposed.

Passive-Dynamic Tuning Methods (Without Active Control Systems) Rotary kite nets have a workable range of power output. Beyond this range kites and lines will deform and possibly become damaged. A rotary, autonomously generating AWES can adjust its power throughput to the wind conditions by passively stretching some key lines or surfaces. This can extend the workable wind range of the device and mitigate effects of turbulence. Passive-dynamic tuning complements the use of matching ground generation levels to wind conditions. If either method is flawed or failing, the other will help to balance the operation of the system.

Speed Regulation by Bank Angle Variation With increasing rotational speed the aerodynamic forces at the wing tip of a driver kite increase more than the forces at the wing root. The tensile membrane structure adjusts to this increasing load imbalance by gradual deformation. Conventional symmetric kites experience a similar aero-elastic deformation effect when flying sharp turns [4]. We can use the mechanism to regulate the rotational speed of the daisy ring kites.

Figures 21.3 and 21.5 (right) illustrate how the tip and the root of a driver kite are supported by two separate branches of the bridle line system. As speed increases, the center of the aerodynamic load shifts radially outwards. In response, the anhedral bank angle of the wing decreases and the tip flexes in axial direction with the load. This passive depowering mechanism is used to limit the rotational speed of the rotary ring kite stack. The flexibility of the wing is greatly influenced by the geometry of the bridle line system and its attachment in the stack. Using a rigid leading edge on the driver kites allows the banking angle to increase to a dihedral to spill wind whilst also maintaining span and inflation [7]. The banking angle can also be used for active speed control.

Configurations for collective as well as individual bank angle control are under investigation.

Proposed Speed Regulation by Twist Variation Just as the tips of windsurfing sails twist to spill wind and prevent overloading, the same mechanism can be employed in stiffened versions of the existing model driving blades. Elastic tip response can be set as a function of "mast" stiffness, downhaul tension and panel forming.

Proposed Speed Regulation by Brake and Steering Surf kites use leading edge bridling for support of the inflatable tubular frame, to ensure good aerodynamic performance and to allow for full depower. Three- and four-line single skin kites can also be fully depowered. It is expected that cross bridling these more complex driver kites will allow rear edge bridle lines to automatically tension and depower using the same aeroelastic dynamic response mechanism as bank angle variations.

Ground-Based Cyclic and Collective Line Control This control method has barely been tested on W&I rotary kite nets. We are investigating whether swashplates around the rotation axis can passively or actively set kite attitude from the ground. By completely tilting the ground station ring interface back into the wind a little lift can be induced on ring kites with short tethering. However, kite response on long tethers is always lagging in time and it seems practically unfeasible to use swashplate control of the outer lines to send synchronized control signals to a stack of rotary kites integrated into a long tensile drive train.

A more promising collective control method would be to vary the relative length of the central lifting line with respect to the outer tethers. This signal will propagate up the stack well and it should be simple enough to take steering or power control references for each driver kite from the central line.

Active Ground Station and Wing Tuning Options Active control both from the ground and in the kites themselves may be more appropriate. Controlling the torque at the ground has a crucial impact on the performance of the rotary ring kite stack and can completely stall the assembly of looping driver kites, stopping rotation if needed. AWES companies have used small, powered onboard actuators to adjust the performance of wings. The whole stack could be actively tuned by shortening or lengthening the central lift line with respect to the outer tethering lines.

Lifting Kite Tuning The elevation angle and tension of the lifting line directly influence the performance of the rotary ring kite. A low elevation angle keeps the rotary ring kite deep in the power zone but requires a longer drive train to reach a given altitude. Also, with a low elevation angle the higher rings in a stack will operate to a large part in the wake flow of lower rings. A well-tensioned lifting line provides a good working reference for network rigging and dynamic tuning.

A lifting kite, which, like a weathercock, stays aligned with the downwind direction in all wind speeds is desirable. Designing a single line lifting kite that aligns in downwind direction and is stable for a wide variety of winds is challenging without an active control system. For prototyping, we used a simple three-tether (tripod) configuration to stabilize a "Peter Lynn" single skin lifting kite. The main kite line was supplemented by two lightly loaded steering lines, which were set apart, downwind of the main tether. The steering lines attached to the B line bridle points of the first inside ribs [10]. This tethering configuration does not achieve the desired directional stability without ground-based intervention. Tripod lines provide a fail safe rigging. Their usefulness in breakaway prevention has been accidentally demonstrated.

21.2.4 Experimental Results for a Single Rotary Kite

The rotary ring kites are remarkably stable in flight without any control input. Driver kites adhered very well to the lead, tail and tether guided path of their ring base. The development of the daisy ring kites has been undertaken on a household budget. No reliable performance data was recorded. The only performance record of single ring setup is from a challenge to make enough energy for a cup of tea for my mother. The challenge was completed in approximately 3 hours of flying at a wind speed $v_w \approx$ 5.3 m/s. The generator produced an average power $P_{net} \approx 9.3$ W, which resulted in a net energy $E_{net} \approx 100.8$ kJ. The low power was due mainly to mismatched generator torque demand. The problem was overcome in later tests by stacking ring kites for cumulative torque output and using a multi-tether rotational drive train.

To assess the efficiency of the energy conversion we first determine the available wind power. With a wind velocity $v_w = 5.3$ m/s and an air density $\rho = 1.3$ kg/m^3 the wind power density evaluates to

$$P_w = \frac{1}{2}\rho v_w^3 = 96.8 \text{ W/m}^2. \tag{21.1}$$

Given a driver kite wing span $b = 0.8$ m rotating on a ring radius of $r_h = 0.9$ m the total swept area in the plane of rotation is $A = 6.53$ m^2. Given that the ring is tilted by an angle of attack $\alpha \approx 35°$, the swept area perpendicular to the wind direction becomes $A \cos \alpha = 5.35$ m^2. This leads to a total wind power passing the flow cross section $P_w A \cos \alpha = 518$ W and a total conversion efficiency

$$\eta = \frac{P_{net}}{P_w A \cos \alpha} = 0.018. \tag{21.2}$$

It should be noted that the aerodynamics of the rotary ring kite is in general very similar to the aerodynamics of a yawed horizontal axis wind turbine rotor, which is analyzed in more detail for example in [6, Chap. 3].

Given a driver kite total wing surface area $S = 0.9$ m^2 an alternative reference power can be calculated as $P_w S = 87$ W. The power harvesting factor is evaluated

$$\zeta = \frac{P_{net}}{P_w S} = 0.1, \tag{21.3}$$

which is a quite low value for an AWES [16]. The poor result is however less a consequence of the rotary ring kite design but mainly due to a mismatch of generation equipment used.

For this specific test the rotational power was transferred to the ground by a two-tether rotational drive train, denoted as "torque ladder", and there converted into electricity by a mountain bike crank connected to a Falco emotors 500W Hxm2.0 hub motor. The lowest bike gear had to be used to overcome torque demands of the motor whilst keeping within the workable tensile rotary power transmission parameter range. Later attempts with stacked rotary kite rings and a multi-tether rotational drive train allowed for much greater torque loading. Both drive train concepts are described in more detail in the following section.

21.3 Power Transmission by Rotating Tethers

Transferring rotational power instead of traction power has certain advantages for AWE applications. The concept allows for continuous power output without the need for the phased generation characteristic of reeling on a drum motor / generator. There is no tether abrasion with rotational power transmission because no tether has to run—it just has to fly and be held by abrasion resistant components. It is easy to add multiple rotor blades to a rotary harvesting mechanism. Power transfer over rotating tethers (PORT) relies on keeping the tensioned tethers apart, at sufficient radial distance, as they rotate around the common axis.

Rotational power is generally transferred over tubular drive shafts that can sustain large shear stresses. Using ropes or net tubes initially seems unfeasible. We know that an applied torque leads to twisting and compression of flexible fiber materials. There has not yet been a need for a tensile rope rotational power transmission system. Meaningful rotational power can only be transferred when a constriction of the drive train—a geometric singularity at which the tethers pass through the axis of rotation—can be avoided. Accordingly, the axial and radial tensioning of the drive train by aerodynamic forces is essential for rotational power transfer.

Excessive torsion of the tensile drive train causes lines to overtwist and cross (hockle) if the lines are long enough. Hockled (overtwisted) lines will not transfer torque effectively. In general, longer and closer tethers need more tension to avoid hockling, while short and well-separated tethers can easily transfer torque without much line tension.

21.3.1 Two-Tether Rotational Drive Train

In the most simple configuration of a tensile rotational drive train, two rods are connected at each end by tethers of equal length. One rod is fixed at its center to the axis of a generator and is perpendicular to this axis. When tensioning the tethers by pulling the second rod in axial direction and at the same time turning this rod around the axis, the first rod and the connected generator are also forced to turn. Both rods must maintain a common rotation axis to work efficiently and avoid tangling. The torque is transferred by the tangential components of the tether forces, while the axial components are required for the tensioning of the system.

This unit setup can be extended into a "torque ladder", which is illustrated in Fig. 21.7. The testing has revealed, however, that this double helix "ladder" structure was impractical. It was prone to hyper coiling when line tension in the system

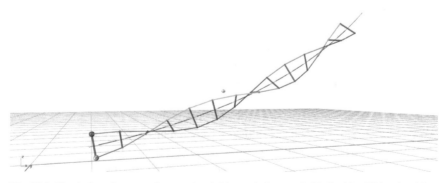

Fig. 21.7 Simulation of a two-tether rotational drive train "torque ladder" with additional guiding lifting line

dropped. Rungs easily got caught inside the tethers. The central guiding lifting line was used to align the rods along a common axis of rotation. Yet, power transfer was jerky given any misalignment.

21.3.2 Multi-Tether Rotational Drive Train

A smoother, more resilient transmission method uses multiple tethers connecting a stack of rings to form a tubular tensile structure, as illustrated in Fig. 21.8. Us-

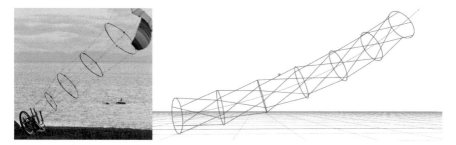

Fig. 21.8 Ring-to-ring transmission of rotational power employs multiple tethers forming a tensile tubular drive train. Power transmission is smooth despite of the misalignment of the crank at the ground to the lift-normal plane in this prototype and model. Also this setup makes use of a central guiding lifting line

ing ring-to-ring rotational power transmission is in fact analogous to torque transfer with inflatable beams. Experiments have shown that the transferable torque increases with the diameter of the inflated beam [5]. Relatively stiff and wide rings, connected at close distance will not lead to hockling of the tethers, even at full propulsion with no axial tension.

The most basic dynamic description of the system assumes that the resultant aerodynamic force and torque contributions of the rotary ring kites will be available at the ground ring. However, this lossless force and torque transmission ignores effects of gravity due to the mass of rings and tethers, aerodynamic line drag and friction in bearings. A full dynamic analysis of rotational power transmission over separated tethers is now being conducted through PhD research by Oliver Tulloch at the University of Strathclyde.

The experimental tests have shown that the ring-to-ring method is well-suited for torque transmission and that it is fail safe and "fail soft". If a component were to break the system continues to run in a diminished condition. Ring-to-ring transmission allowed easier launching of the AWES. The rotary ring kites were evenly pretensioned, inflated and inspected on the ground before being allowed to ascend.

21.4 Stacked Rotary Ring Kite Nets

Integrating rotary ring kites with a multi-tether rotational drive train leads to a systematic modular design. By stacking rotary rings the power output of the system can be incremented in discrete steps. This requires an adjustment of the drive train and lifting kite dimensioning, taking into account the increasing gravitational and aerodynamic drag effects. Exactly how many rotary ring kites of a given size a lifting line with given tension can reliably support is still unknown, however.

21.4.1 Design Considerations and Vision

The extension of the single rotary ring kite illustrated in Fig. 21.3 to a setup of four stacked rings is shown in Fig. 21.9. The drawing indicates how the tethers of driver

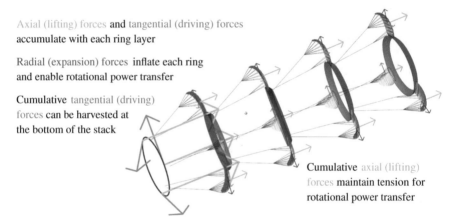

Axial (lifting) forces and tangential (driving) forces accumulate with each ring layer

Radial (expansion) forces inflate each ring and enable rotational power transfer

Cumulative tangential (driving) forces can be harvested at the bottom of the stack

Cumulative axial (lifting) forces maintain tension for rotational power transfer

Fig. 21.9 Integration of rotary ring kites with a multi-tether rotational drive train. Lifting and propulsion forces generated by the rings kites are cumulative through the stack

kites attach to the respectively lower ring in the stack. Because of the networked tethering, the rings rotate all at the same angular velocity, each contributing to the resultant torque of the stack, while using the generated radial expansion forces to tension the ring structure and the generated axial lifting force to tension the entire drive train. The energy harvesting varies along the perimeters of the rotary rings. In general, the upward going kites are facing a stronger apparent wind speed and hence generate a larger line tension. Furthermore, the bottom parts of the rings operate in the wake flow of their upstream neighbors, while the top parts penetrate into the free stream. Because the wake effects decrease with increasing elevation angle of the net the amount of wind energy available for harvesting increases with increasing elevation angle. However, because of the low solidity and large spacing between rotary ring kite layers the impact on the power output seems to be practically unaffected by the elevation angle.

For the investigated prototype each rotary ring kite has its axis of rotation inclined by an angle $\alpha \approx 35°$ with respect to the horizontal wind velocity. A variance in α along the stack can occur due to the sagging of the tensile drive train as a result of gravitational loading, aerodynamic drag and conditions leading to low lifting line tension.

The driver kites of each rotary ring kite are tethered to a upwind ring base, that is lower in the stack. Tethering of a driver kite to a wider ring base generally improves the structural stability of the bridled ram air wing compared to a tethering to a narrower ring base. As can be seen from Fig. 21.5 the bridle attachment angle depends also on the bank angle of the wing. Tethering of the driver kites to a wider ring also increases the generated torque of the rotary ring. Driver kites on wider orbits travel faster and by that make the conversion system more efficient. More rigid and larger driver kites will be suited for such wider orbits. Rings with different properties can be flown to complement each other. Further design details are listed in Fig. 21.10.

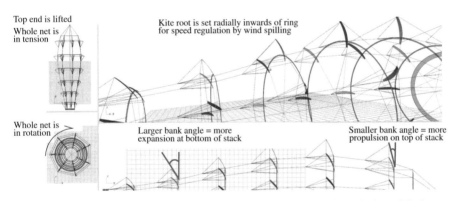

Fig. 21.10 Driver kites at higher ring levels are set flatter, with a smaller bank angle, and fly faster for overall net dynamic

All AWES are affected by the aerodynamic drag of tethers and bridle lines and reducing this important loss factor by design is one of the key goals of current development efforts. In a stack the tethers of the individual kite are very short and therefore the drag loss per driving kite area is greatly reduced. However, line thicknesses should progressively increase with tension and torsion toward the bottom of the stack.

The daisy stack prototype is fail safe and "fail soft". It is fail safe in the way that each component is prevented from breaking away as it links to at least two other components. It is "fail soft" in the way that if something breaks the failing stack has less lift and less power and will eventually bring itself to ground.

We think that the research into rotary kite power networks should be intensified and propose open testing and development to cover the following aspects

- Single skin soft kites for use in hugely scaled arrays

- Rigid kite "blades" with high aerodynamic efficiency
- Short tethered hybrid stiff and flexing tip wings
- Asymmetric parafoils specifically for rotor work
- Active and cyclical control of the angle of attack
- Flying controlled wings outward from the ring surface
- Buoyant or sinking kite configurations for tidal electricity generation
- Optimizing the blade count (solidity) based on ring diameters and blade profile
- Elastic anhedral to dihedral stack models for smoother operation
- Dynamic model of tensile rotary power transmission
- Mixed-function soft and rigid ring layers for generation and transmission needs
- Parametric system optimization

Manual launching and landing the rotary ring kite stack has been mostly easy, only occasionally dangerous in strong winds. Automated launching and landing, particularly of larger systems, would be desirable. The current launch method could easily be mechanized but does not scale well. Concepts for modular attachment of new ring layer parts on live systems have been suggested, but no such device has yet been made. A complete ground handling solution is desirable for high end versions of rotary ring kite nets.

21.4.2 Experimental Results for a Rotary Ring Kite Net

Test data from the "100×3 challenge" [18] is used for the following analysis. This data is based on the 2016 daisy stack prototype using a ground generator adapted from an e-bike. More detailed results based on the latest prototype illustrated in Fig. 21.2, which has peaked at $P = 616$ W so far [11], is due to be published through the PhD research of Oliver Tulloch at the University of Strathclyde.

Three rotary ring kites as described in Sect. 21.2.4 are operated in a stack, amounting in a total swept area in the plane of rotation of $A = 19.6$ m^2 and, with a tilting by an angle of attack $\alpha \approx 35°$, the swept area perpendicular to the wind direction becomes $A \cos \alpha = 16.06$ m^2. With a wind velocity $v_w = 5.5$ m/s and an air density $\rho = 1.3$ kg/m^3 the wind power density evaluates to

$$P_w = \frac{1}{2}\rho v_w^3 = 108.1 \text{ W/m}^2.$$

(21.4)

This leads to a total wind power passing the flow cross section $P_w A \cos \alpha = 1736$ W. Measuring an average power $P_{net} = 111$ W with this setup, the total efficiency is

$$\eta = \frac{P_{net}}{P_w A \cos \alpha} = 0.064.$$

(21.5)

Given a total wing area of $S = 2.7$ m^2 an alternative reference power can be calculated as $P_w S = 291$ W. The power harvesting factor becomes

$$\zeta = \frac{P_{\text{net}}}{P_{\text{w}}S} = 0.38. \qquad (21.6)$$

This value for three rotary rings is significantly larger than the value for the single rotary ring, given in Eq. (21.3), despite the additional losses due to wake effects on downstream rotary rings. This improvement is most likely due to the better match between the torque available from the stack and the torque demands of the generator. In this recorded test the rotational power transmission worked well throughout the tensile drive train. By applying the brake at the ground ring interface it was possible to stop the rotation of the entire stack on demand.

Experimentation and accident showed that leading edge stiffening and the forward overdrive rods are not necessary when flying driver kites on wide diameter stacked rings. This has positive implications for scaling. The driver kites have survived harsh treatment in testing, repeatedly showing how soft blades can handle crashing over the ground when the lifting kite tension drops. The Peter Lynn single skin lifting kite, as used in prototyping, was stabilized with two spread tag lines, anchored downwind of the full stack. Realigning the lifting kite to windward, is unnecessarily time consuming. A more reliable networked lift system with monitoring and control is recommended before risking hitting rigid blades on the ground. A host of improvements have been proposed for future models. Given the power return on the minuscule model cost, we are confident this system can be useful in a range of markets.

Stacked rotary ring kites guided by a lifting line can be arranged in a dense array because rotary drive trains can be operated in parallel both across and down wind. Lifting kite lines can also be conjoined across the entire array to achieve network stability and, therefore, improve land use efficiency. Stacked ring kites have also been suspended from solid structures. It is supposed that stacked ring kites could be set on three-dimensional lattice work to fill the void between mountain gaps whilst generating electricity. Again this would improve land use efficiency. Various floating networks for AWES deployment and channel rope networks for tidal energy generation have been suggested.

21.5 Lifting Isotropic Network Kites

A lattice of interconnected lifting kites can stay airborne in wind from any direction. Because of its network layout and mutual stabilization of the member kites such a lifting meta-kite is generally resilient towards local fluctuations of the wind field. A computational simulation of a lifting isotropic network kite (LINK) exposed to a turbulent wind field is shown in Fig. 21.11.

Simulation and prototype testing has confirmed the stabilizing effect of a wide outer anchoring and wide net tethering of a LINK. A method for steering individual kites by line through network nodes has been demonstrated experimentally. Methods

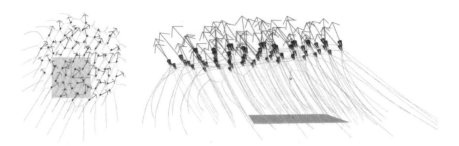

Fig. 21.11 Model representation of a lifting isotropic network kite (LINK). The resultant aerodynamic force vectors from turbulent wind acting on the individual lifting kites are displayed as blue arrows. Line tension varies from high (red) to low (green). The central lifting lines are held apart and aligned to the average downwind direction despite the action of turbulence on the individual kites

to align kite flight with its nodal network normal plane to maintain the deployment of the meta-kite have been suggested and are currently being investigated.

Although networking of kites provides additional stability it would be dangerous to build large power projects without automatic monitoring and control. The system design and performance of a LINK can be improved by controlling the tether length of the individual lifting kites. An example is illustrated in Fig. 21.12. Because the top part of the meta-kite is generally stable due to its exposure to the free stream multiple lift lines can be kept sufficiently apart for safe and dense ring kite farming applications.

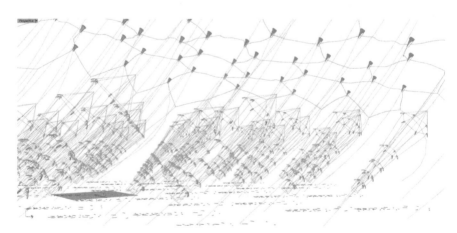

Fig. 21.12 Meta-kite concept to collaboratively farm with large numbers of rotary ring kites

Using a LINK to suspend rotary ring kite stacks would simplify the operation and reduce costs. Algorithms and geometric patterns suitable for stacking LINK layers into a taller three-dimensional energy harvesting lattice are being developed.

Methods have been suggested to extract useful power from coordinated or even harmonic meta-kite motions. Openly proposed ideas for meta-kite swaying, swirling and pumping energy extraction models have been briefly considered, as have coordinated fields of meta-kites working against each other. The control and actuation needed for these designs seems complex and beyond the current work scope.

21.6 Ground Control and Generation for Ring Kite Stacks

For the sake of minimal airborne mass the generation equipment and kite controls, including launching, landing and storage equipment, is placed on the ground. Because the axis of rotation of the rotary ring stack is tilted from the vertical into the wind direction also the ground ring, to which the tensile drive train attaches, needs to be tilted. We have tested generator mounting with both gimbal and following wheel configurations to follow the lifting line axis.

The earlier prototype illustrated in Figs. 21.4 and 21.8 (left) used an e-bike as a ground station, operating the rotating ring kite stack through the crank. In the latest development version shown in Fig. 21.13 a custom-made ground station is used.

Fig. 21.13 Rotary ring kite stack in operation, generating $P_{net} \approx 600$ W, showing the portable tracking ground station, the new ring configuration and the force scale. Andrew Reeve (left) and the author (1 December 2017). See also [11]

The photographic footage depicted in Fig. 21.4 shows that some daisy ring kite prototypes have the coaxial generator-crank assembly not well aligned with the wind direction and kite elevation. That was not too problematic at such a small scale, however, it is clear that better and more controllable alignment will improve system performances.

For an efficient rotational power transfer the track of the rotating tethers on the ground ring interface will have a diameter closely matching the connected airborne ring. This is illustrated by the concept design shown in Fig. 21.14.

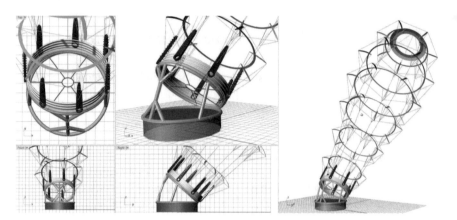

Fig. 21.14 Concept sketch of an all-in-one tracking ground control and generation system

It seems unlikely that a rotary ring kite deployment and recovery system or a cyclical tether tension control will feature on small-scale ring kite stacks soon. However, at larger scales manual handling and intervention with rotary ring kites will not be safe and as consequence automated systems will be required. Solutions are being designed for these utility sector device scenarios.

For rotary ring kite stacks operated in lattice configurations it might be desirable to combine the rotational power of several stacks to jointly drive a generator. We produced simple freewheel collection and field arrangement algorithms, to match network spaced rotary power outputs to a central generator.

21.7 Open Source Design

Windswept and Interesting Ltd has released all of our core design work to date as open source hardware because we believe that it is a better way to start a technology. We are convinced that better work comes from the design integrity of open source hardware. Obtaining funding for open source hardware projects is challenging, but the benefits are obvious. Your right to patent novel components relating to these designs is not affected.

Our company develops three-dimensional AWES models using parametric algorithm design software. Collections of kites and their parameters can be rapidly reconfigured this way. Parametric designs are particularly suited for evolutionary development of design algorithms. The number and variety of parameters, which govern a kite network algorithm, is large. Parametric designs can be automatically evaluated, restructured and optimized with evolutionary iterations of Artificial Intelligence (AI) software. An AI system can evaluate large numbers of combinations of the parameters governing a network kite to derive AWES optimization models, which will otherwise take years to derive experimentally.

AWES, is the kind of complex and valuable design challenge where multiple objective optimization solvers can be applied to great effect. The required tools are openly available. The current work will benefit greatly from a more organized implementation of AI architecture. AWES design should therefore embrace AI.

21.8 Conclusions

Rotary ring kite stacks can work together in networks, harvesting energy continuously and autonomously. The tethering geometry of lifting and rotary kite networks stabilizes the flight of individual kites. A lifting kite can guide working rotary ring kites into suitable operational positions. The rotational power of ring kite stacks can be collected and transferred to the ground by a tensile drive train. Combinations of complementary lifting and rotary kites can be arranged to harvest wind energy in three-dimensional wind farming arrays. Kite and line fatigue has been very low. The performance of an exceptionally inexpensive airborne wind energy prototype improved with upscaling.

The practical prototyping approach left little verifiable data. More accurate measurements are being performed in the frame of a University of Strathclyde study. Improvements in kite performance as well as practical operations such as launching, landing and ground handling routines will soon be tested and published openly. Many areas for performance improvement have been identified. A specially commissioned asymmetric soft wing is currently being discussed.

A more comprehensive and thorough approach to the full working scope of Windswept and Interesting Ltd is being sought. Universities have expressed interest in analyzing the dynamic and optimization challenges posed by the kite methods demonstrated. The potential of using larger diameter rings has been shown. There appears to be large potential for soft, rigid and hybrid kite rotor turbine networks but this potential is mostly unverified.

Business models are being considered for a next iteration daisy kite AWES. Manufacturers have been able to produce kite ring parts remotely. A simple ground generator torsion control based on measurements of line tension and the ring spacing dynamic is likely to be built soon. The open source hardware design methods used are available for anyone to improve. The parametric design software used will be suitable for artificial intelligence development. The methods are very promising. There is scope and reason to vastly increase the work being done on this project.

Acknowledgements The author would like to thank the anonymous donor of a €5000 research sponsorship, which allowed much of this work to go ahead. The ratios of prototype power and efficiency per cost of development are amazing. My friends and family have been incredibly patient with me throughout this work. A host of advice has come from online AWES forums. Thanks also to the peer review group for good advice. Thanks to Roland Schmehl for a fantastic reorganisation of this paper. Thanks mum for the challenge: I hope you enjoyed your cup of tea.

References

1. Ahrens, U., Pieper, B., Töpfer, C.: Combining Kites and Rail Technology into a Traction-Based Airborne Wind Energy Plant. In: Ahrens, U., Diehl, M., Schmehl, R. (eds.) Airborne Wind Energy, Green Energy and Technology, Chap. 25, pp. 437–441. Springer, Berlin Heidelberg (2013). doi: 10.1007/978-3-642-39965-7_25
2. Benhaïem, P.: Rotating Reeling. In: Schmehl, R. (ed.). Book of Abstracts of the International Airborne Wind Energy Conference 2015, p. 100, Delft, The Netherlands, 15–16 June 2015. doi: 10.4233/uuid:7df59b79-2c6b-4e30-bd58-8454f493bb09. Poster available from: http://www.awec2015.com/images/posters/AWEC25_Benhaiem-poster.pdf
3. Bormann, A., Ranneberg, M., Kövesdi, P., Gebhardt, C., Skutnik, S.: Development of a Three-Line Ground-Actuated Airborne Wind Energy Converter. In: Ahrens, U., Diehl, M., Schmehl, R. (eds.) Airborne Wind Energy, Green Energy and Technology, Chap. 24, pp. 427–437. Springer, Berlin Heidelberg (2013). doi: 10.1007/978-3-642-39965-7_24
4. Bosch, A., Schmehl, R., Tiso, P., Rixen, D.: Nonlinear Aeroelasticity, Flight Dynamics and Control of a Flexible Membrane Traction Kite. In: Ahrens, U., Diehl, M., Schmehl, R. (eds.) Airborne Wind Energy, Green Energy and Technology, Chap. 17, pp. 307–323. Springer, Berlin Heidelberg (2013). doi: 10.1007/978-3-642-39965-7_17
5. Breukels, J.: An Engineering Methodology for Kite Design. Ph.D. Thesis, Delft University of Technology, 2011. http://resolver.tudelft.nl/uuid:cdece38a-1f13-47cc-b277-ed64fdda7cdf
6. Burton, T., Jenkins, N., Sharpe, D., Bossanyi, E.: Wind Energy Handbook. 2nd ed. John Wiley & Sons, Ltd, Chichester (2011). doi: 10.1002/9781119992714
7. Dunker, S.: Ram-Air Wing Design Considerations for Airborne Wind Energy. In: Ahrens, U., Diehl, M., Schmehl, R. (eds.) Airborne Wind Energy, Green Energy and Technology, Chap. 31, pp. 517–546. Springer, Berlin Heidelberg (2013). doi: 10.1007/978-3-642-39965-7_31
8. Harburg, R. W.: Coaxial Multi-turbine generator. US Patent 5,040,948, Aug 1991
9. Kite Power Systems. http://www.kitepowersystems.com/. Accessed 10 July 2017
10. Lynn, P.: Pilot Tuning – Twenty pilot kites flying side by side, stable and straight. http://www.peterlynnhimself.com/Pilot_Tuning.php. Accessed 10 Oct 2017
11. Read, R.: 3 stack Daisy flying wind turbine 608W test 4 Dec 2017. https://www.youtube.com/watch?v=x6btemB3hKo (2017). Accessed 15 Dec 2017
12. Rimkus, S., Das, T.: An Application of the Autogyro Theory to Airborne Wind Energy Extraction. Paper DSCC2013-3840. In: Proceedings of the ASME 2013 Dynamic Systems and Control Conference, vol. 3, Palo Alto, CA, USA, 21–23 Oct 2013. doi: 10.1115/DSCC2013-3840
13. Roberts, B. http://altitudeenergy.com.au/. Accessed 30 June 2016
14. Ruiterkamp, R., Sieberling, S.: Description and Preliminary Test Results of a Six Degrees of Freedom Rigid Wing Pumping System. In: Ahrens, U., Diehl, M., Schmehl, R. (eds.) Air-

borne Wind Energy, Green Energy and Technology, Chap. 26, pp. 443–458. Springer, Berlin Heidelberg (2013). doi: 10.1007/978-3-642-39965-7_26

15. Santos, D.: Toward Gigawatt-scale Kite Energy. In: Diehl, M. (ed.). Book of Abstracts of the International Airborne Wind Energy Conference 2011, p. 39, Leuven, Belgium, 24–25 May 2011. http://resolver.tudelft.nl/uuid:0677ccde-8335-40e6-afcf-c56317d52864

16. Schmehl, R., Noom, M., Vlugt, R. van der: Traction Power Generation with Tethered Wings. In: Ahrens, U., Diehl, M., Schmehl, R. (eds.) Airborne Wind Energy, Green Energy and Technology, Chap. 2, pp. 23–45. Springer, Berlin Heidelberg (2013). doi: 10.1007/978-3-642-39965-7_2

17. Selsam, D. S.: Serpentine Wind Turbine. US Patent 6,616,402, June 2001

18. The Airborne Wind Energy Community. http://www.someawe.org. Accessed 3 July 2016

19. Vlugt, R. van der, Peschel, J., Schmehl, R.: Design and Experimental Characterization of a Pumping Kite Power System. In: Ahrens, U., Diehl, M., Schmehl, R. (eds.) Airborne Wind Energy, Green Energy and Technology, Chap. 23, pp. 403–425. Springer, Berlin Heidelberg (2013). doi: 10.1007/978-3-642-39965-7_23

20. Yahoo! Groups: Airborne Wind Energy Forum. https://groups.yahoo.com/neo/groups/AirborneWindEnergy/info. Accessed 8 July 2016

Chapter 22
Airborne Wind Energy Conversion Using a Rotating Reel System

Pierre Benhaïem and Roland Schmehl

Abstract The study proposes a new airborne wind energy system based on the carousel concept. It comprises a rotary ring kite and a ground-based rotating reel conversion system. The moment generated by the ring kite is transferred by several peripheral tethers that connect to winch modules that are mounted on the ground rotor. A generator is coupled to this rotor for direct electricity generation. Because the ring kite is inclined with respect to the ground-rotor the length of the peripheral tethers has to be adjusted continuously during operation. The proposed system is designed to minimize the used land and space. This first study describes the fundamental working principles, results of a small-scale experimental test, a kinematic analysis of steady-state operation of the system and a power transmission analysis. Design choices for the ring kite are discussed, a strategy for launching and landing and methods for passive and active control are described.

22.1 Introduction

The potential of airborne wind energy conversion has been investigated by early explorative research [17, 20, 26] and confirmed by a larger number of recent theoretical and experimental studies [5, 6, 25, 27, 31]. It is however also clear that despite of the advantages of reduced material consumption, access to a larger wind resource and higher yield per installed system, the system-inherent use of a flexible tether requires a comparatively large surface area [9]. This contrasts the general

Pierre Benhaïem (✉)
7 Lotissement des Terres Blanches, 10160 Paisy-Cosdon, France
e-mail: pierre-benhaiem@orange.fr

Roland Schmehl
Delft University of Technology, Faculty of Aerospace Engineering, Kluyverweg 1, 2629 HS Delft, The Netherlands

© Springer Nature Singapore Pte Ltd. 2018
R. Schmehl (ed.), *Airborne Wind Energy*, Green Energy and Technology, https://doi.org/10.1007/978-981-10-1947-0_22

539

motivation for designing an economically competitive wind energy that sweeps the whole frontal airspace, using less land and airspace.

Several concepts have been proposed to maximize the land use efficiency. For single kite systems operating on single ground stations the surface density can be increased by optimizing the spacial arrangement and operation of the systems while accounting for sufficient safety margins to avoid hazardous mechanical or aerodynamic interactions. The next conceptual improvement leads towards systems that operate multiple wings on a single ground stations [15]. For such systems the useful swept area can reach the occupied swept area, however, the technical complexity of such systems also increases significantly. Alternatively, single kite systems operating on single ground stations can be upscaled to increase the land use efficiency [15]. Finally, the complexity of the ground conversion can be increased, for example, using a large rotating structure (carousel) driven by several kites [14] or, alternatively, using carts that are pulled by kites on a round track [1, 2].

The present study proposes a new airborne wind energy system, the Rotating Reel Parotor (RRP), which combines a rotary ring kite with a ground-based rotating reel conversion system [8]. The concept has also been presented at the Airborne Wind Energy Conference 2015 [10]. Other airborne wind energy systems involving rotary kites are the "Gyromill" [23, 25], presented also in Chap. 23 of this book, which is based on onboard electricity generation, and the "Daisy Stack" [24], presented also in Chap. 21, which is transmitting shaft power to the ground, as the present concept. A related technology in the field of aviation is the tethered gyrocopter. In Sect. 22.2 the components of the system and their functions are described while Sect. 22.3 details the fundamental working principles. In Sect. 22.4 a small-scale model is presented and experimental results are discussed. In Sects. 22.6 and 22.5 the kinematics of the system and the torque transmission characteristics are investigated. Section 22.7 elaborates on ongoing and future investigations and Sect. 22.8 presents the conclusions of this study.

22.2 System Design

A conceptual sketch of the ground-based part of the system is illustrated in Fig. 22.1. Similar carousel-type configurations have been proposed for airborne wind energy

Fig. 22.1 The ground-based horizontal ring and its vertical axis of rotation. For direct conversion of the rotational motion a generator is coupled to the ring. The winch modules for the traction tethers are mounted on the ring and are indicated by circles

conversion [1, 14]. To convert the rotational motion of the ring structure directly into electricity it can be coupled to a generator using a gear mechanism. The peripheral traction tethers (not depicted) which drive the rotational motion of the ring are deployed from winch modules that are mounted at equidistant intervals along the ring. Each winch module comprises a cable drum with a connected generator that can also be used in motor mode.

A conceptual sketch of the rotary ring kite, denoted as Parotor, is illustrated in Fig. 22.2. The flying rotor is represented as an actuator ring which defines the swept

Fig. 22.2 The flying rotor is represented as actuator ring which is inclined to the flow by an angle α, its axis of rotation tilted downwind from the vertical axis by the same angle (for simplicity a sideslip angle β_s is not included here)

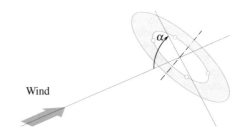

area of the physical rotor. A possible implementation of a small-scale model for test purposes will be discussed in Sect. 22.4.1. The flying rotor has a size that is about the size of the ground rotor and it is inclined with respect to the wind by an angle α. This inclination angle, also denoted as angle of attack, is identical to the angle between the axes of rotation of the ground and flying rotors.

Figure 22.3 shows how the flying rotor is connected to the ground rotor by peripheral tethers. Because the axes of rotation of the two rotors are not aligned the geometric distance between the ground and flying rotor attachment points changes continuously during rotation. As consequence the length of the connecting traction tethers needs to be adjusted continuously. This is the function of the ring-mounted

Fig. 22.3 The assembled Rotating Reel Parotor (RRP) in flight, just before operation. The tether attachment points at the flying rotor are indicated by circles. The radial line from the center of the ground rotor to one of the tether attachment points is an illustration element indicating the phase lag δ of the ground rotor. Before transmitting a torque the phase lag of the ground rotor is zero. The axis of rotation of the ground rotor is always vertical

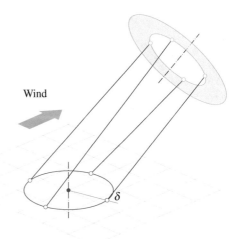

Fig. 22.4 The RRP system
in operation with an angular
speed ω and a phase lag
angle $\delta = 35°$. The arrows at
the winch modules indicate
whether the corresponding
tether is reeled out and energy
is generated (green) or reeled
in and energy is consumed
(red). This definition implies
that the reeling motion is
relative to the winch modules
which move on a circular
path around the center of the
ground rotor

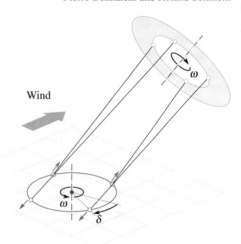

winch modules shown in Figs. 22.1 and 22.3. When the geometric distance between two attachment points of a tether is increasing the corresponding winch module functions as a generator. When the distance is decreasing in the second half of the revolution, the winch is retracting the tether and is consuming energy. Figure 22.4 illustrates the Rotating Reel Parotor in operation. The flying rotor and the ground rotor are co-rotating at identical angular speeds, however, the driven ground rotor lags the flying rotor in phase.

A system of additional suspension lines can be added to support the flying rotor from the center of the ground rotor. Three different options are illustrated in Fig. 22.5, using lines or line segments of constant length. When in tension, all three implementations enforce a constant distance between the centers of the two rotors.

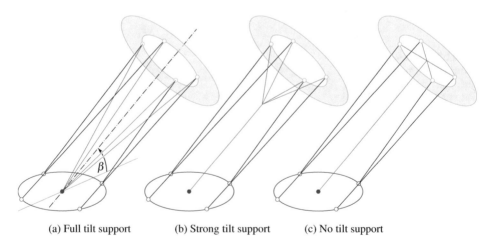

(a) Full tilt support   (b) Strong tilt support   (c) No tilt support

Fig. 22.5 Implementation options for suspension lines (in red) to support the flying rotor

The variant sketched in Fig. 22.5(a) additionally enforces a kinematic coupling between the orientation of the flying rotor, quantified by its angle of attack α and sideslip angle β_s, and the position of the rotor, quantified by the ground elevation angle β and azimuth angle ϕ of the rotor center point. Although this constraint could be a way to stabilize the operation of the system, the additional lines increase the losses due to aerodynamic drag. The bridle-type variant sketched in Fig. 22.5(b) reduces the drag losses and and allows for some tilt motion of the rotor while the central line variant sketched in Fig. 22.5(c) has no additional drag losses and does not impose any constraint on the tilt motion. It should be noted that the suspension lines for the flying rotor can alternatively be attached to an additional lifting kite.

22.3 Working Principles

A general feature of airborne wind energy is the use of flying devices to extract kinetic energy from the wind and to transfer it as either mechanical or electrical energy to the ground, using flexible tethers. Because flexible tethers can only transfer tensile forces an additional mechanism is required on the ground to convert the traction power into shaft power, which can be converted by electrical generators.

22.3.1 Power Transfer and Power Takeoff

The proposed concept employs a set of peripheral tethers to transfer the rotational motion of a flying rotor to a ground rotor. This tensile torque transmission system makes use of the tangential components of the tether forces acting on the ground rotor. The function of the normal force components is to keep the transmission system in tension, which is an obvious prerequisite for the functioning of the system.

It is important to note that the transmission of torque implies torsion of the tether system. As can be seen in Fig. 22.4 the angle of twist, which is identical to the phase lag angle δ of the ground rotor, determines how the tether force is decomposed into tangential and normal components. At small to moderate values of the twist angle, an increasing torsion reduces the angle at which the tethers attach to the ground rotor. This geometric effect increases the tangential components and it allows the tether system to adjust to variations of the torque which can occur, for example, as a result of a fluctuating wind speed. At larger values of the twist angle, for $\delta > 90°$, the effect decreases because the tether system increasingly constricts in a point on the axis of rotation. At $\delta = 180°$ the tether system reaches the singular condition at which all tethers intersect in one point and no practically relevant torque can be transmitted.

The transmission characteristics are also influenced by the distance between the two rotors in relation to their diameter. The further the rotors are apart the smaller the tangential components of the tether forces, the less effective the above mentioned coupling effect between torsion and torque and the lower the torsion stiffness of the tether system. If the rotors are many diameters apart the tether system can not be used effectively for torque transmission.

It can be concluded that on the level of the individual tethers the torsion stiffness of the system is caused by tensile forces, the rotational motion generated by circular traction of the ground rotor. Because the axis of rotation of the flying rotor is tilted downwind the rotational motion requires that the tether lengths are adjusted continuously to the varying geometric distances between the attachment points. As described in Sect. 22.2 this is the function of the winch modules on the ground rotor which compensate the distance variations by reeling the tethers in and out. The two fundamental modes of energy generation are discussed in the following.

22.3.2 Direct Mode of Energy Generation

In this mode the rotational motion of the ground rotor is converted directly into electricity, using one or more generators that are coupled to the rotor by a gear mechanism, as illustrated schematically in Fig. 22.1. The winch modules manage the kinematically induced length variation of the peripheral tethers, as shown in Fig. 22.4. They are controlled in such a way that the tension in the tethers is equal and constant during operation. The modules are electrically interconnected such that the generated and consumed energy is balanced, avoiding the implementation of expensive temporary energy storage. To account for losses in the electrical machines a small amount of electricity is provided by the main generator which is driven directly by the rotor.

By adding suspension lines, as shown in Fig. 22.5, the force level in the system of peripheral tethers is lowered and, as consequence, also the generated and consumed amounts of energy. Because of the reduced losses in the electrical machines the total amount of electrical energy required for the actuation of the tether system is decreased. However, with the addition of suspension lines the tensile torque transmission system becomes more complex and in particular also statically indeterminate (hyperstatic). As consequence this poses additional challenges to the control systems of the winch modules.

22.3.3 Secondary Mode of Energy Generation

In this mode the length variation of the peripheral tethers is converted into electricity, using the winch modules on the ground rotor alternatingly as generators and motors. The suspension lines are essential and are used to selectively reduce the tether tension during reel-in. As consequence, the winch modules consume less energy during reel-in than they generate during reel-out, resulting in a positive net energy of the phase-shifted interconnected modules. The proposed technique is illustrated

Fig. 22.6 The secondary mode of energy generation with two tensioned tethers and two tensioned suspension lines highlighted. Unloaded tensile components are hinted. The two winch modules producing electricity are next to the green arrows, pointing away from the modules, while the two winch modules reeling the tethers in are next to the red arrows, pointing towards the modules. The doted loop is the ground track of the resultant tensile force in the system assuming perfect unloading during reel-in

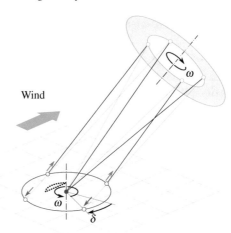

schematically in Fig. 22.6. The two winch modules in reel-out mode operate on tensioned tethers while the two winch modules in reel-in mode operate on untensioned tethers. The shift from tensioned reel-out to untensioned reel-in is managed by the force control of the winch modules. When switching from reel-out to reel-in the set value of the tether force is decreased from its nominal value to a low value. As consequence, the tensile load shifts from the peripheral tether to the corresponding suspension line which inevitably affects the static force balance and geometry of the entire torque transmission system. Accordingly, the set value of the tether force is increased back to the nominal value when switching to reel-out and the tensile load shifts from the suspension line back to the peripheral tether.

Because of the induced rotational asymmetry of the force transmission the resultant force acting on the ground rotor does not pass through a constant point on the ground plane anymore, as it does for the direct mode of energy generation. The resultant tensile force in the transmission system is essentially unsteady and tracks a periodic loop on the ground plane which is shifted sideways towards the half of the ground rotor that moves against the wind. This is indicated as dotted line in Fig. 22.6. The rotational asymmetry affects also the flying rotor which inevitably performs a tumbling motion. In particular the switching of the force transfer, which, in the illustrated example affects two winch modules at the same time, introduces a strong discontinuity in the transmission system. In practice, the switching needs to be replaced by a sufficiently smooth process to avoid a periodic jolting of the entire system.

22.3.4 Discussion

The direct and secondary modes of energy generation differ only in the force control strategy implemented for winch modules. Because of this, the two modes can in principle be blended by the control algorithm. However, because of its rotational asymmetry and unsteadyness it is still an open question whether the secondary mode has any practical relevance.

22.4 Experimental Tests of a Small-Scale Model

A physical model of the proposed RRP system has been designed and built at small scale. Initial tests have been performed to demonstrate the fundamental working principles and to provide an initial assessment of the transmitted torque.

22.4.1 Test Setup

The small-scale model is shown in operation in Fig. 22.7 and the parameters of the test setup are summarized in Table 22.1. The geometric proportions and the elevation angle are roughly the same as for the intermediate-scale system described in Sect. 22.6.5. In place of the winch modules that a larger production system would

Fig. 22.7 Small-scale system built with two spars, a ring, four retractable leashes, a rotating tray, a parachute kite and semi-rigid rotor blades. The system uses four peripheral tethers and several suspension lines. The flying rotor measures 1.3 m from tip to tip

Table 22.1 Design and operational parameters of the small-scale system. Because of the close proximity of the flying rotor to the ground (about 1 m) it was exposed to significant turbulent fluctuations of the wind velocity. The setup uses suspension lines

Parameter name	Symbol	Value	Unit
Average wind speed	$\bar{v}_w$	6.0	m/s
Ground rotor diameter	d_g	0.8	m
Flying rotor inner diameter	d_k	0.6	m
Flying rotor outer diameter	$d_{k,o}$	1.3	m
Number of rotor blades	b	8	
Blade span		0.35	m
Blade root chord		0.12	m
Blade tip chord		0.04	m
Flying rotor swept area	S	1.0	m^2
Lifting kite area		2.0	m^2
Number of peripheral tethers	N	4	
Tether length, minimum	$l_{t,min}$	0.8	m
Tether length, maximum	$l_{t,max}$	1.4	m
Tether length lifting kite		10	m
Elevation angle kite center	β	40	deg

use, this technology demonstrator has off-the-shelf retractable leashes mounted on the ground rotor. As they are equipped with a rotational spring mechanism, these leashes do not produce a constant force but one that is linearly increasing with the deployed tether length. This is an important aspect for the interpretation of the results and the comparison with the analytical calculations and numerical simulations in the following sections. For standalone testing of the rotating reel conversion system the ring kite is replaced by a top ring which is rotated by hand. To assess the torque transmission characteristics the torque imposed on the top ring, τ_k, and the torque arriving at the ground rotor, τ_g, are measured with two torque meters. These tests showed that the torque transmission coefficient is about $\tau_g/\tau_k = 0.5$.

The design challenge of this small-scale test setup was the matching of the torques generated by the ring kite and converted by the described rotating reel conversion system. The baseline design of the ring kite shown in Fig. 22.7 uses eight semi-rigid rotor blades. To operate this kite at wind speeds between 5 to 6 m/s a sled kite was added to provide additional lift. With active conversion system a rotational speed of one revolution per second has been obtained for short times. This relatively high value is due to the small dimensions of the technology demonstrator. The rotor with 8 blades has a high solidity, so a low efficiency compared to the Betz limit [16]. However, the generated torque was appropriate for the tests. A rotor with 16 blades has also been tested and, as expected, produced a higher torque, while achieving lower angular speeds. More complete test data is provided in Sect. 22.4.2.

As concluded in Sect. 22.3.1 the rotating reel conversion system works only if the tethers are not too long compared to the inner diameter of the ring kite. This diameter is indeed approximately equal to the tip height of the system, as shown in Sect. 22.6. Because the wind is generally stronger at higher altitudes [3] the RRP system will have to be quite large. However, the implementation of a motorized ground rotor could be studied for the purpose of increasing the transmitted torque with longer tethers and for applying the second mode of energy generation, as described in Sect. 22.3.3. Such a motorized ground rotor could also be used for launching.

22.4.2 Experimental Results

The objective of the experimental tests has been to demonstrate the fundamental working principles and to quantitatively assess the effectiveness of the energy conversion mechanisms. As none of the elements was optimized the coefficient of the transmitted power cannot be directly deduced. Because the test setup does not include a central generator the achievable direct power takeoff of the ground rotor is assessed by the power that is required to overcome the internal friction torque of the central swiveling tray. Because the test setup uses retractable leashes instead of controlled winch modules, the energy budget related to the tether actuation is assessed on the basis of the stored potential energy of the leashes. The test results for the setup defined in Table 22.1 are summarized in Table 22.2. The limiting values ω_{min}

Parameter name	Symbol	Value	Unit
Angular speed, minimum	ω_{min}	2	rad/s
Angular speed, maximum	ω_{max}	6	rad/s
Angular speed, average	$\overline{\omega}$	3	rad/s
Angular speed, freewheel[a]	$\omega_{\tau=0}$	12	rad/s
Tip speed ratio, minimum	λ_{min}	0.216	
Tip speed ratio, maximum	λ_{max}	0.648	
Tip speed ratio, average	$\overline{\lambda}$	0.324	
Tip speed ratio, freewheel[a]	$\lambda_{\tau=0}$	1.3	
Tether force, minimum	$F_{t,min}$	0.88	N
Tether force, maximum	$F_{t,max}$	1.76	N
Tether reeling power, average	$\overline{P}_{reel}$	1.5	W
Friction torque central swivel	τ_μ	0.225	Nm
Friction power central swivel	P_μ	0.675	W
Flying rotor power, Betz limit	P_{max}	35	W

Table 22.2 Measured properties of the small-scale system

[a] peripheral tethers detached

and ω_{max} describe the range of measured angular speeds of the system, $\overline{\omega}$ a representative average value. The value $\omega_{\tau=0}$ is achieved without conversion system, using only suspension lines. Similarly the values λ_{min} and λ_{max} describe the range of measured tip speed ratios, $\overline{\lambda}$ a representative average and $\lambda_{\tau=0}$ the ratio without conversion system. $F_{t,min}$ and $F_{t,max}$ describe the limiting values of the tether forces that correspond with the tether lengths $l_{t,min}$ and $l_{t,max}$.

Assuming linear elastic behavior, the potential energy stored in the spring mechanism of the leash can be calculated as

$$E = \frac{1}{2} \left(F_{t,\max} + F_{t,\min} \right) \left(l_{t,\max} - l_{t,\min} \right). \tag{22.1}$$

The tether extends from $l_{t,\min}$ to $l_{t,\max}$ during half a revolution of the rotor which is associated with the time period

$$\Delta t = \frac{\pi}{\omega}. \tag{22.2}$$

Considering that two leashes of the system are continuously in reel-out mode we can derive the average equivalent power for these two leashes as

$$\overline{P}_{reel} = 2\frac{E}{\Delta t} = \left(F_{t,\max} + F_{t,\min} \right) \left(l_{t,\max} - l_{t,\min} \right) \frac{\omega}{\pi}. \tag{22.3}$$

Based on the numerical values in Tables 22.1 and 22.2, and using the average value of the angular speed, we can calculate the value of $\overline{P}_{reel}$ specified in Table 22.2. The friction torque τ_μ of the central swivel was measured at the average angular speed and using this value we can calculate the value of the friction power P_μ listed in Table 22.2.

The power values P_μ and $\overline{P}_{reel}$ provide a first insight into the energy budget of the proposed concept. Assuming that the friction in the swivel can be reduced substantially, a power in the order of P_μ would be available for direct continuous conversion into electricity. In contrast to this, the potential energy E quantified by Eq. (22.1) is cyclically progressing through the spring mechanisms of the leashes but in balance for the entire system. This potential is only accessible when using suspension lines to selectively reduce the tether tension during reel-in, however, this was not possible in this simple test setup. As a general conclusion it should be noted that an extrapolation of these values to larger systems is critical if not questionable because of the small scale and the significant measurement uncertainties in this setup.

The efficiency of the flying rotor was not measured, but as it uses numerous semi-rigid blades forming a high-solidity rotor the efficiency is considered to be far below the value of the Betz limit. Defining the wind power density as

$$P_w = \frac{1}{2}\rho v_w^3, \tag{22.4}$$

this limiting power value can be computed as

$$P_{\max} = P_w S \frac{16}{27} \cos^3 \beta, \tag{22.5}$$

where the factor $\cos^3 \beta$ accounts for the misalignment of the flying rotor with respect to the wind [13, p. 98]. By inserting the applicable numerical values we can calculate the value listed in Table 22.2.

The initial tests have shown the potential but also the challenges of the concept. Indeed there have been jolts during rotation of the system and the tests indicated that the turbulent fluctuations of the wind at close proximity to the ground was a possible cause of these jolts. Another contribution is due to the use of retractable leashes with spring mechanisms. The inevitable force variations during rotation induce a tumbling motion of the flying rotor, which becomes stronger with decreasing elevation angle.

Following the initial tests, the effect of parameter and design variations has been studied. Firstly, leashes with lower tensile strength were used. While the baseline design used leashes which generated a force of 1.91 N for 1.30 m of reeled out tether, these generated the same force with 2.20 m of reeled out tether. Secondly, the tensile strength was increased by pairing leashes such that each pair of leashes generated a force if 1.91 N with 0.82 m of reeled out tether. These tests indicated that the tensile strength must be sufficiently high to avoid excessive twist of the tether system and eventually entangling of the tethers. On the other hand if the tensile strength is to high the tether system can not transfer the torque required for a continuous rotation. A larger Rotating Reeling Parotor system of about 5 m diameter would allow harnessing better wind at a height of 5 m.

To address the problem of turbulent wind fluctuations and their effect on the reproducibility of results a leaf blower was used to produce a constant airflow. The center of the ring kite was suspended in space by means of a bar. The modified design and test setup is summarized in Table 22.3. Parameters that are not listed

Table 22.3 Design and operational parameters for the modified design with 16 rotor blades and an increased flow velocity. To increase the tensile strength leashes are arranged in pairs. The setup does not use suspension lines

Parameter name	Symbol	Value	Unit
Number of rotor blades	b	16	
Elevation angle kite center	β	65	deg
Number of peripheral tethers	N	4	
Tether length, minimum	$l_{t,min}$	0.20	m
Tether length, maximum	$l_{t,max}$	0.62	m
Tether force, average	$\overline{F}_t$	0.91	N
Angular speed	ω	5	rad/s
Angular speed, freewheel[a]	$\omega_{\tau=0}$	9	rad/s
Tether reeling power, average	$\overline{P}_{reel}$	1.16	W

[a] peripheral tethers detached and suspension lines added

have not been modified from the baseline design summarized in Table 22.1. The average tether force is calculated as

$$\overline{F}_t = \frac{1}{2}\left(F_{t,max} - F_{t,min}\right) \quad (22.6)$$

In these tests it was possible to operate the RRP system in a steady state rotation without jolts and generating some power. It is envisioned that more thorough results including the torque transmission efficiency as a function of the elevation angle can be achieved using a wind tunnel.

22.5 Kinematics of Steady-State Operation

The revolving system of peripheral tethers has the double function of anchoring the rotary ring kite to the ground and transferring the generated aerodynamic moment to the ground-based conversion system. Uncommon for airborne wind energy systems, the combination of these two functions entails comparatively complex tether kinematics which is governed by strong nonlinear coupling effects. In this section a kinematic model for the steady-state operation of the tensile torque transmission system is derived. This model is used to formulate analytical expressions for the instantaneous tether length and rotor attachment angles which are the starting base for the analysis of the power transmission characteristics in the following section.

22.5.1 Steady-State Operation as an Idealized Condition

The distinguishing feature of the ring kite is that it employs the effect of autorotation to convert kinetic energy from the wind into aerodynamic lift and usable shaft power. To analyze the steady-state flight of this kite the spinning rotor is represented as a non-spinning planar actuator ring. This abstraction, which hides the implementation details of the physical rotor, is shown in Fig. 22.8. The orientation of the actuator ring with respect to the flow is described by the sideslip angle β_s and the angle of attack α. The actuator ring is regarded as a flying object with three translational and two rotational degrees of freedom. The two rotational degrees of freedom of the actuator ring, roll and pitch, tilt the spinning axis of the rotor. The aerodynamic lift

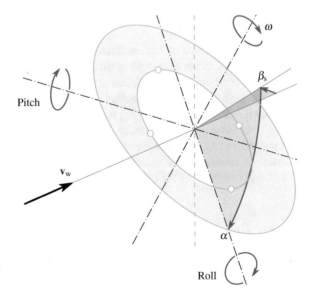

Fig. 22.8 The actuator ring model of the rotary ring kite. The inclination of the ring with respect to the flow is described by two successive rotations. The sideslip angle β_s describes the rotation around the vertical axis while the angle of attack α describes the rotation of the ring around its pitch axis. Roll and pitch axes are attached to the actuator ring and not to the physical rotor. The angular speed ω of the rotor is an operational parameter which, next to the flow angles β_s and α, affects the aerodynamic lift and drag of the ring

L and drag D of the actuator ring are functions of the sideslip angle β_s, the angle of attack α, the angular speed ω, the physical dimensions of the rotor and the wind speed v_w.

The objective of the study is to add a system of actuated peripheral tethers, as outlined in Sect. 22.2, to constrain the degrees of freedom of the ring kite to a steady flight state at a constant position with a constant axis of rotation. However, although the length of the tethers is adjusted continuously to the required geometrical distance, the tether attachment angles at the rotors vary periodically with the rotation angle. Caused by the rotational asymmetry of the tilted tether system, the directional variations of the tether forces lead to transverse resultant forces that induce periodic compensating motions of the flying rotor.

For the purpose of the kinematic analysis these compensating motions are neglected, assuming an idealized condition of steady-state operation in which the ring kite has a constant position with a constant axis of rotation. By prescribing this condition, the length of the individual tethers can be formulated as analytic functions of time and other relevant problem parameters. For the purpose of the analysis it is assumed that all tethers are inflexible and tensioned and can thus be represented as straight lines.

Figure 22.9 shows the configuration of the RRP system with four tethers and without any additional suspension lines. For simplicity we restrict the analysis to the case of steady-state operation of the ring kite with its center point K always in the $x_w z_w$-plane. In this particular case the azimuth angle ϕ vanishes at all times. When using additional suspension lines, as illustrated in Fig. 22.5, the distance l_K of the kite center point from the origin is constant and the axis of rotation of the ring kite has to pass through the origin O which the following kinematic constraints

$$\alpha = 90° - \beta, \tag{22.7}$$
$$\beta_s = 0. \tag{22.8}$$

A and B denote a pair of representative tether attachment points at the flying rotor and the ground rotor, respectively. Because the angular speed ω of both rotors is assumed to be constant the rotation angle is given by ωt, adding a constant phase lag δ for the flying rotor. The tethers are attached on the ground rotor at a distance R_g from the center O, on the flying rotor at a distance R_k from the center K. The tips of the rotor blades are at a distance $R_{k,o}$ from the center K. The distance l_K between the centers of the two rotors is regarded as a parameter that is prescribed either as a distance constraint when using suspension lines, as shown in Fig. 22.5, or by the controlled actuation of the tether system.

22.5.2 Dimensionless Problem Parameters and Reference Frames

From the illustration of the steady-state operation of the system in Fig. 22.9 we can identify $\alpha, \beta_s, \beta, \delta, \omega t, R_g, R_k$ and l_K as the fundamental parameters of the kine-

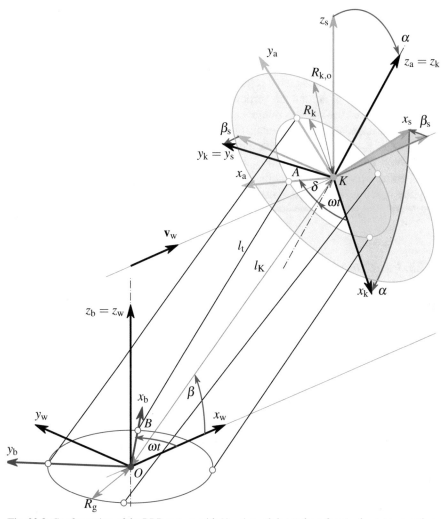

Fig. 22.9 Configuration of the RRP system with $N = 4$ revolving tethers for steady-state operation in the $x_w z_w$-plane ($\phi = 0$). The winch modules and the attachment points on the flying rotor are indicated by circles, A and B denote a representative pair and l_t denotes the length of the connecting tether. The distance of the kite center point K from the origin O is denoted as l_K. The ground rotor lags the flying rotor in phase by and angle δ

matic problem of steady-state operation with the kite center restricted to the $x_w z_w$-plane ($\phi = 0$). The corresponding set of dimensionless parameters are the angles $\alpha, \beta_s, \beta, \delta$ and ωt together with the geometric ratios R_g / R_k and l_K / R_k.

Included in Fig. 22.9 are the right-handed Cartesian reference frames which are used to describe the relative positions on the two rotors. The wind reference frame (x_w, y_w, zw) is considered to be an inertial frame with origin O, its x_w-axis aligned with the wind velocity vector $\mathbf{v}_w$ and its z_w-axis pointing towards zenith. The reference frame (x_b, y_b, z_b) is attached to the ground rotor, with origin at O, its x_b-axis pointing towards the tether attachment point B and rotating with angular velocity ω around the z_w-axis.

The sideslip reference frame (x_s, y_s, z_s) has its origin at the kite center point K and is constructed from the wind reference frame by rotating the x_w- and y_w-axes by the sideslip angle β_s around the vertical axis. The kite reference frame (x_k, y_k, z_k) is constructed from the sideslip reference frame by rotating the x_s- and z_s-axes by the angle of attack α around the y_s-axis. Following a common aeronautical convention, the x_k- and y_k-axes coincide with the roll- and pitch-axes of the actuator ring, respectively. The reference frame (x_a, y_a, z_a) is attached to the flying rotor, with origin at K, its x_a-axis pointing towards the tether attachment point A and rotating with angular speed ω around the z_k-axis, leading the rotation of the ground rotor by an angle δ.

22.5.3 Kinematic Properties

In the following the kinematic relations for the two rotors are derived formulating the positions of points A and B as functions of the geometric and kinematic parameters of the steady-state problem. Point B is fixed to the ground rotor at radius R_g and its coordinates in the wind reference frame can be written as

$$\mathbf{r}_B = \begin{bmatrix} \cos(\omega t) \\ \sin(\omega t) \\ 0 \end{bmatrix} R_g. \tag{22.9}$$

The coordinates of the kite center point K are

$$\mathbf{r}_K = \begin{bmatrix} \cos\beta \\ 0 \\ \sin\beta \end{bmatrix} l_K. \tag{22.10}$$

Point A is fixed to the flying rotor at radius R_k. To determine its coordinates in the wind reference frame we first define the transformation matrices $\mathbf{T}_{ws}$ and $\mathbf{T}_{sk}$ which describe the individual rotations by angles β_s and α, respectively,

$$\mathbf{T}_{ws} = \begin{bmatrix} \cos\beta_s & -\sin\beta_s & 0 \\ \sin\beta_s & \cos\beta_s & 0 \\ 0 & 0 & 1 \end{bmatrix}, \qquad \mathbf{T}_{sk} = \begin{bmatrix} \cos\alpha & 0 & \sin\alpha \\ 0 & 1 & 0 \\ -\sin\alpha & 0 & \cos\alpha \end{bmatrix}.$$

Combining these by multiplication we can derive the matrix $\mathbf{T}_{wk}$ which describes the coordinate transformation from the kite reference frame to the wind reference frame by two successive rotations

$$\mathbf{T}_{wk} = \mathbf{T}_{ws}\mathbf{T}_{sk} = \begin{bmatrix} \cos\beta_s\cos\alpha & -\sin\beta_s & \cos\beta_s\sin\alpha \\ \sin\beta_s\cos\alpha & \cos\beta_s & \sin\beta_s\sin\alpha \\ -\sin\alpha & 0 & \cos\alpha \end{bmatrix}. \tag{22.11}$$

Using this transformation matrix we can formulate the coordinates of point A in the wind reference frame as

$$\mathbf{r}_A = \mathbf{T}_{wk} \begin{bmatrix} \cos(\omega t + \delta) \\ \sin(\omega t + \delta) \\ 0 \end{bmatrix} R_k + \begin{bmatrix} \cos\beta \\ 0 \\ \sin\beta \end{bmatrix} l_K. \tag{22.12}$$

Defining the instantaneous distance vector pointing from point B to point A as

$$\mathbf{r}_A - \mathbf{r}_B = \begin{bmatrix} l_{t,x} \\ l_{t,y} \\ l_{t,z} \end{bmatrix}, \tag{22.13}$$

the coordinates of this vector can be calculated as

$$\mathbf{r}_A - \mathbf{r}_B = \begin{bmatrix} \cos\beta_s\cos\alpha\cos(\omega t + \delta) - \sin\beta_s\sin(\omega t + \delta) \\ \sin\beta_s\cos\alpha\cos(\omega t + \delta) + \cos\beta_s\sin(\omega t + \delta) \\ -\sin\alpha\cos(\omega t + \delta) \end{bmatrix} R_k$$
$$+ \begin{bmatrix} \cos\beta \\ 0 \\ \sin\beta \end{bmatrix} l_K - \begin{bmatrix} \cos(\omega t) \\ \sin(\omega t) \\ 0 \end{bmatrix} R_g, \tag{22.14}$$

and used to determine the geometric distance as

$$l_t = |\mathbf{r}_A - \mathbf{r}_B| = \sqrt{l_{t,x}^2 + l_{t,y}^2 + l_{t,z}^2}. \tag{22.15}$$

Following the convention used in Sect. 22.5.2 the dimensionless tether length is defined as l_t/R_k.

To derive the tether reeling velocity as the rate of change of tether length, $v_t = dl_t/dt$, we apply the general differentiation rule

$$\frac{d}{dt}\sqrt{\mathbf{r}\cdot\mathbf{r}} = \frac{\mathbf{r}}{\sqrt{\mathbf{r}\cdot\mathbf{r}}}\cdot\frac{d\mathbf{r}}{dt}, \tag{22.16}$$

to Eq. (22.15) to get

$$v_t = \frac{1}{l_t}\left(l_{t,x}\frac{dl_{t,x}}{dt} + l_{t,y}\frac{dl_{t,y}}{dt} + l_{t,z}\frac{dl_{t,z}}{dt} \right). \tag{22.17}$$

The individual coordinate derivatives included in the right hand side of this equation are obtained by differentiating Eq. (22.14) as

$$\frac{d}{dt}(\mathbf{r}_A - \mathbf{r}_B) = \begin{bmatrix} -\cos\beta_s\cos\alpha\sin(\omega t + \delta) - \sin\beta_s\cos(\omega t + \delta) \\ -\sin\beta_s\cos\alpha\sin(\omega t + \delta) + \cos\beta_s\cos(\omega t + \delta) \\ \sin\alpha\sin(\omega t + \delta) \end{bmatrix} R_k\omega$$

$$- \begin{bmatrix} -\sin(\omega t) \\ \cos(\omega t) \\ 0 \end{bmatrix} R_g\omega. \quad (22.18)$$

The dimensionless tether reeling velocity is defined as $v_t/(\omega R_k)$.

Next to the tether length l_t and its rate of change v_t a third important derived kinematic property is the angle γ at which the tethers attach to the rotor rings. This angle controls the transfer of torque from the flying rotor to the tether system and further to the ground rotor. Considering the attachment of the tether to the ground rotor and defining the unit vectors pointing along the tether and from the origin to point B as

$$\mathbf{e}_t = \frac{\mathbf{r}_A - \mathbf{r}_B}{l_t}, \quad (22.19)$$

$$\mathbf{e}_x^b = \frac{\mathbf{r}_B}{R_g}, \quad (22.20)$$

the tether attachment angle γ_g can be computed from the z_w-component of the cross product of both vectors as

$$\cos\gamma_g = \mathbf{e}_y^b \cdot \mathbf{e}_t = (\mathbf{e}_z \times \mathbf{e}_x^b) \cdot \mathbf{e}_t = (\mathbf{e}_x^b \times \mathbf{e}_t) \cdot \mathbf{e}_z, \quad (22.21)$$

$$= \frac{1}{R_g l_t}(r_{B,x} l_{t,y} - r_{B,y} l_{t,x}). \quad (22.22)$$

This derivation involves the unit vectors $\mathbf{e}_x^b, \mathbf{e}_y^b$ and $\mathbf{e}_z^b = \mathbf{e}_z$ of the rotating reference frame (x_b, y_b, z_b) and is illustrated in Fig. 22.10.

Fig. 22.10 Definition of the tether attachment angle γ_g for the ground rotor. The cosine of this angle is obtained as orthogonal projection of the tether unit vector $\mathbf{e}_t$ onto the tangential unit vector $\mathbf{e}_y^b$

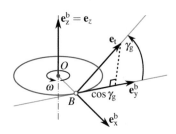

In a similar way, the tether attachment angle γ_k at the flying rotor can be computed from the unit vectors $\mathbf{e}_x^a, \mathbf{e}_y^a$ and $\mathbf{e}_z^a = \mathbf{e}_z^k$ of the rotating reference frame (x_a, y_a, z_a) and the tether unit vector $\mathbf{e}_t$ as

$$\cos \gamma_k = \mathbf{e}_y^a \cdot \mathbf{e}_t = (\mathbf{e}_z^k \times \mathbf{e}_x^a) \cdot \mathbf{e}_t = (\mathbf{e}_x^a \times \mathbf{e}_t) \cdot \mathbf{e}_z^k, \qquad (22.23)$$

$$= \frac{1}{R_k l_t} \left[(r_{A,y} l_{t,z} - r_{A,z} l_{t,y}) \sin \alpha + (r_{A,x} l_{t,y} - r_{A,y} l_{t,x}) \cos \alpha \right]. \qquad (22.24)$$

Physically, Eqs. (22.21) and (22.23) represent the contribution of the tether force to the dimensionless torque in the system. This kinematic expression will be used as a starting point for the analysis of the torque transfer in Sect. 22.6.

The derivations in this section are for a representative pair of tether attachment points. For the other pairs similar relations can be formulated by applying additional phase shifts to the phase angle ωt.

22.5.4 Parametric Case Study

The kinematics of the torque transmission system is fully described by the N distance vectors which connect the flying rotor to the ground rotor and which are given by Eq. (22.14) for a representative pair of tether attachment points. In the following the effect of the angular parameters $\alpha, \beta_s, \beta, \delta$ and ωt on the geometry of a tether system with representative proportions $R_g/R_k = 1$ and $l_K/R_k = 2$ is analyzed.

The variation of the minimum and maximum tether lengths with the elevation angle is quantified in Fig. 22.11(left). At the limiting case of a vertical tether system, $\beta = 90°$, the axes of rotation of both rotors coincide and accordingly the tethers are of constant length $l_{t,min} = l_{t,max}$. For vanishing phase lag angle, $\delta = 0$, the tether

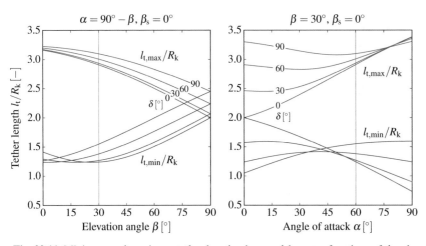

Fig. 22.11 Minimum and maximum tether lengths, $l_{t,min}$ and $l_{t,max}$, as functions of the elevation angle β (left) and angle of attack α (right) for $R_g/R_k = 1$ and $l_K/R_k = 2$. The left diagram illustrates the special case of kinematically coupled angle of elevation and angle of attack, e.g. by means of suspension lines, while the right diagram illustrates the study for a specific constant elevation angle. The vertical lines at 30° and respectively 60° indicate identical conditions in both diagrams

length equals the distance between the two rotors, $l_t = l_K$, and for increasing phase lag also the tether length increases continuously. For decreasing elevation angle the variation of tether length increases. At practically relevant values $30° < \beta < 60°$ the dimensionless length difference $\Delta l_t / R_k$ is roughly between 1.5 and 1.0.

The variation of the minimum and maximum tether lengths with the angle of attack of the flying rotor is quantified in Fig. 22.11(right) for a representative value of the elevation angle, $\beta = 30°$, and a vanishing sideslip angle. At the limiting case of a horizontal flying rotor and vanishing phase lag the tethers are aligned with the axis of rotation and accordingly the tether length is constant. It should be noted that this holds only for the special case of $R_g / R_k = 1$ because for any other value the tethers are generally not aligned with the axis of rotation.

The variation of the tether attachment angles during one full revolution of the system is illustrated in Fig. 22.12. For the interpretation of the diagrams it is im-

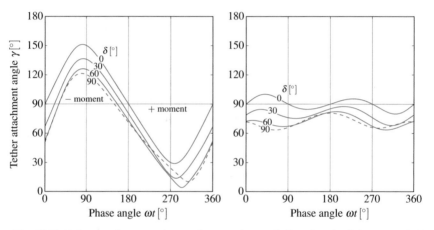

Fig. 22.12 Tether attachment angle γ at the ground rotor (left) and at the flying rotor (right) as functions of the phase angle ωt for $\beta = 30°, \alpha = 60°, \beta_s = 0, R_g / R_k = 1, l_K / R_k = 2$

portant to note that for $\gamma < 90°$ the tensile force in the tether contributes a positive moment, acting in the direction of the rotation, while for $\gamma > 90°$ it contributes a negative moment, acting against the direction of the rotation. At the limiting case $\gamma = 90°$ the moment contribution vanishes (see also Fig. 22.10).

Figure 22.12(left) shows the tether attachment angle γ_g, as defined by Eq. (22.21), for different values of the phase lag angle. It can be seen that the step from $\delta = 0$ to $30°$ results in a consistent and nearly uniform shift of the sine-type curve to lower values. The steps from 30 to 60° and further to 90° follow this trend and increase the asymmetry of the curves with respect to the limiting case $\gamma = 90°$, however, they are also characterized increasingly by nonlinear kinematic effects. The asymmetry with respect to $\gamma = 90°$ directly affects the transfer of torque to the generator because it quantifies the net moment contribution of the corresponding force per revolution of the system. It can be concluded that for the analyzed case a phase lag angle between 60 and 90° results in the best achievable moment contribution. The curve for $\delta = 0$ shows the expected change of sign of the moment contribution at $\omega t = 180$ and 360°, however, the extreme values $\gamma_{max} = 151.1$ and $\gamma_{min} = 28.9$ do not occur at $\omega t = 90$ and 270°, as one might expect, but at $\omega t = 81.6$ and 278.4°. This is a consequence of the geometric asymmetry of the revolving tether system tilted in downwind direction.

Figure 22.12(right) shows the tether attachment angle γ_k, as defined by Eq. (22.23), for different values of the phase lag angle. Compared to the ground rotor attachment angle the variation is substantially smaller, for this particular case almost one magnitude. Furthermore, the frequency of the variation is doubled, for example, the curve for $\delta = 0$ changes the sign of the moment contribution at $\omega = 90, 180, 270$ and 360°. For practically required values of the phase lag angle, as can be seen for $\delta \gtrsim 30°$, the moment contribution is shifted entirely to positive values.

This behavior can be explained by the fact that for the case of kinematically coupled angle of elevation and angle of attack, for which the axis of rotation of the flying rotor passes through the center of the ground rotor, the tether system attaches orthogonally to the flying rotor, which minimizes the kinematically induced variation of the attachment angle of the individual tethers and allows a stable counterbalancing of the aerodynamic moment. On the other hand the tether system attaches to the ground rotor at the elevation angle which causes a fundamental asymmetry of the moment transfer to the rotor and as consequence the tether attachment angle and the moment contribution of the tether force alternate periodically, as illustrated in Fig. 22.12. The torque transfer mechanism will be investigated in more detail in Sect. 22.6.

22.5.5 Conclusions

The objective of this section was to derive a kinematic model for the steady-state operation of the tensile torque transmission system. To achieve this, it was assumed that the system configuration in steady-state operation is known and can be described by the angle of attack α and sideslip angle β_s of the flying rotor, the elevation angle β of the kite center point, the phase lag angle δ of the ground rotor, the distances R_g and R_k of the tether attachment points from the centers of the ground and flying rotors, respectively, and the distance l_K of the kite center point from the origin. For such a prescribed operational state Eq. (22.14) describes the time evolution of the vector connecting the ground and flying rotor attachment points of the tether, Eq. (22.15) of the length of the tether, Eq. (22.17) of the reeling velocity of the tether and Eq. (22.21) of the attachment angle of the tether at the ground rotor.

The noncoaxial arrangement of the rotors and the phase lag distort the geometry of the tether system to an asymmetric state and introduce nonlinear kinematic effects. The parametric case study has shown how these effects intensify with increasing distortion of the tether system. Furthermore, the tether attachment angle was identified as an important kinematic property for the moment transfer. Because the special case of kinematically coupled angle of elevation and angle of attack leads to nearly constant tether attachment geometry at the flying rotor, which is optimal for a stable torque transfer, we will only consider this configuration in the remainder of the chapter.

22.6 Power Transmission in Steady-State Operation

The aerodynamic force and moment of the ring kite are transferred to the ground conversion system by tensile forces only. The particular feature of the system is the power takeoff by two different, intrinsically coupled energy conversion mechanisms. The direct mechanism is based on the resultant moment that the tensile forces exert on the ground rotor, whereas the secondary mechanism is based on the length variation of the tethers. In this section a model for the power transmission characteristics of the tether system is formulated for steady-state operation. This model is used to assess the transmission efficiency as a function of the problem parameters, as well as the relation between transmitted torque and aerodynamic force. The focus of the analysis is on the tether system and not on the ring kite itself. It should be noted that the use of suspension lines is not considered in this analysis.

22.6.1 Energy Equation of the Single Tether

To assess the power transmission by the revolving tether system we first analyze the energy balance of the single tether. For this purpose the tether is cut free at the attachment points, as illustrated in Fig. 22.13. Neglecting the effects of aerodynamic

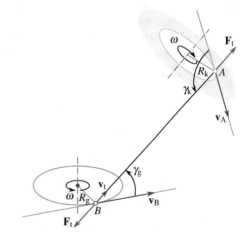

Fig. 22.13 Forces and velocities at the attachment points of a representative tether. The other tethers and their attachment points are not depicted. The attachment points A and B move with circumferential velocities $v_A = \omega R_k$ and $v_B = \omega R_g$. In the depicted situation the tether length is decreasing which requires the winch module at attachment point B to reel the tether in with a velocity v_t

drag and inertial forces and assuming that the tether is straight and inelastic, it can be concluded that the tensile forces at the two attachment points are of equal magnitude and pointing in opposite direction. In reference to Fig. 22.13 the energy equation can be formulated as

$$F_t \cos \gamma_k \omega R_k = F_t \cos \gamma_g \omega R_g + F_t v_t. \tag{22.25}$$

The left hand side represents the power transferred from the flying rotor to the tether by the circular motion of the attachment point A, while the first term on the right hand side represents the power transferred from the tether to the ground rotor by the circular motion of the attachment point B. The third contribution is the mechanical power that is transferred to the winch module that is attached to the rotor at point B. If we define a characteristic power of the tensile torque transmission problem as $F_t \omega R_k$ and divide Eq. (22.25) by this expression we obtain the dimensionless equation

$$\cos \gamma_k = \cos \gamma_g \frac{R_g}{R_k} + \frac{v_t}{\omega R_k}. \tag{22.26}$$

This fundamental equation relates the two tether attachment angles and the dimensionless tether reeling velocity introduced in the context of Eq. (22.18).

The variation of the three dimensionless power contributions is shown in Fig. 22.14. The case of vanishing phase lag is depicted in Fig. 22.14(left) and, as expected, in-

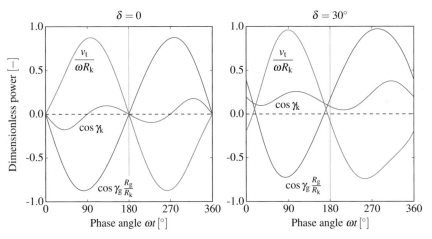

Fig. 22.14 Kinematic modulation of the dimensionless power balance at the tether during one revolution for $R_g/R_k = 1, l_K/R_k = 2, \beta = 30°, \beta_s = 0°$ and $\alpha = 60°$. The dashed line represents the sum of all contributions

dicates that the net power that is transferred from the flying rotor to the tether during one revolution is close to zero. As consequence, the other two power contributions, the shaft power contribution to the ground rotor and the reeling power transferred to the winch module have to cancel out each other. When applying a phase lag angle of $\delta = 30°$ the net power transferred from the flying rotor to the tether is positive, which is indicated by the upwards shift of the corresponding curve. It is obvious from Fig. 22.14 that for this particular case, the input power is balanced by comparatively large variations of the output power contributions. In the real system, the associated losses would be significant, which is a point of concern.

It is important to note that Eq. (22.25) does not provide any information about the actual values of the tensile force and their power contributions but only the relative distribution of these contributions depending on the instantaneous kinematics of the system. To derive the actual values of the tensile forces the equations of motion of the ground and flying rotors have to be considered, which is the topic of the following section.

22.6.2 Quasi-Steady Motion of the Flying Rotor

Because of the relatively low mass of the flying rotor and the tethers the airborne system adjusts rapidly to force imbalances. The resulting quasi-steady motion is governed by the equilibrium of the aerodynamic force distribution, the tether forces and gravitational forces. If we neglect, for simplicity, the effect of gravity, the equilibrium of forces and moments acting on the flying rotor can be formulated as

$$\mathbf{F}_a = -\sum_{i=1}^{N} \mathbf{F}_{t,i}, \tag{22.27}$$

$$\mathbf{M}_a = -\sum_{i=1}^{N} (\mathbf{r}_{A,i} - \mathbf{r}_K) \times \mathbf{F}_{t,i}, \tag{22.28}$$

which is illustrated in Fig. 22.15.

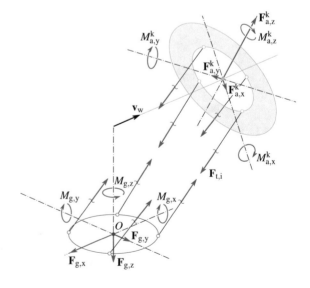

Fig. 22.15 Forces and moments acting on the ground and flying rotors. The resultant aerodynamic force and moment are represented by their components in the kite reference frame. The reaction force and moment acting in the bearing mechanism of the ground rotor are represented by their components in the wind reference frame. $M_{g,z}$ denotes the transmitted moment that is available for conversion into electricity

The resultant aerodynamic force and moment vectors, $\mathbf{F}_a$ and $\mathbf{M}_a$, are represented by their components in the kite reference frame. $\mathbf{F}_{a,z}^k$ is the main force component acting along the rotor axis, while $\mathbf{F}_{a,x}^k$ and $\mathbf{F}_{a,y}^k$ are the two transverse components. Accordingly, $M_{a,z}^k$ is the main moment component acting around the rotor axis, while $M_{a,y}^k$ and $M_{a,x}^k$ are the components around the pitch and roll axes of the kite. The z_k-components of the aerodynamic force and moment are the two key functional elements of the RRP system, responsible for tensioning the tether system and for generating torque that is transferred to the ground to be converted into electricity.

The aerodynamic loading of the flying rotor is balanced by the N tether forces $\mathbf{F}_{t,i}$. The calculation of the individual moment contributions specified by Eq. (22.28) differs from the calculation of the tether attachment angle γ_k, as specified by Eq. (22.23), only by the additional multiplicative factors $F_{t,i}$, the magnitudes of the tether forces.

The difficulty in solving the quasi-steady equilibrium equations for the unknown tether forces $\mathbf{F}_{t,i}$ comes from the fact that except for the design parameters R_g, R_k and $R_{k,o}$, the actuated tether lengths $l_{t,i}$ and the wind velocity v_w all other problem parameters, $\alpha, \beta_s, \beta, \phi, \delta, \omega$ and l_K have to be regarded as degrees of freedom, subject to additional kinematic coupling conditions. This differs from the starting point of the kinematic analysis in Sect. 22.5 where we assumed steady-state operation of the system with known values of these problem parameters.

22.6.3 Approximate Solution of Steady-State Operation

Instead of attempting to solve the problem of quasi-steady motion of the flying rotor exactly, as described by Eqs. (22.27) and (22.28), we derive an approximate solution of the idealized problem of steady-state operation. Following the approach described in Sect. 22.5 we consider only the principal force axis of the system, which is the axis of rotation of the flying rotor. To fulfill the force equilibrium in this axis we assume that the components of the tether forces in this direction are all of equal magnitude, which is formally expressed by the conditions

$$\mathbf{F}_{t,i} \cdot \mathbf{e}_z^k = -\frac{F_{a,z}^k}{N}, \qquad i = 1, \ldots, N. \tag{22.29}$$

Representing the force vectors as $\mathbf{F}_{t,i} = F_{t,i}\mathbf{e}_{t,i}$, where $\mathbf{e}_{t,i}$ represents the unit vector along tether i, the individual force magnitudes can be derived as

$$F_{t,i} = -\frac{F_{a,z}^k}{N\mathbf{e}_{t,i} \cdot \mathbf{e}_z^k}, \qquad i = 1, \ldots, N. \tag{22.30}$$

The tether forces defined by these equations exactly balance the axial aerodynamic force component $F_{a,z}^k$. Furthermore, the resultant roll and pitch moments of the tether forces vanish because the geometric center of the tether attachment points coincides with the kite center K and the moment-contributing force components $\mathbf{F}_{t,i} \cdot \mathbf{e}_z^k$ are all equal. As consequence, the corresponding aerodynamic moment components $M_{a,x}^k$ and $M_{a,y}^k$ vanish and

$$M_a = M_{a,z}^k. \tag{22.31}$$

However, the tether forces defined by Eq. (22.30) induce transverse force components which need to be balanced by the transverse aerodynamic force components $\mathbf{F}_{a,x}^k$ and $\mathbf{F}_{a,y}^k$ and which lead to transverse compensating motions. We can derive the following expressions for the ratios of the transverse aerodynamic force components to the axial force component

$$\frac{F_{a,x}^k}{F_{a,z}^k} = \frac{1}{N} \sum_{i=1}^{N} \frac{\mathbf{e}_{t,i} \cdot \mathbf{e}_x^k}{\mathbf{e}_{t,i} \cdot \mathbf{e}_z^k}, \tag{22.32}$$

$$\frac{F_{a,y}^k}{F_{a,z}^k} = \frac{1}{N} \sum_{i=1}^{N} \frac{\mathbf{e}_{t,i} \cdot \mathbf{e}_y^k}{\mathbf{e}_{t,i} \cdot \mathbf{e}_z^k}. \tag{22.33}$$

The moment components acting around the rotational axes of the flying rotor and the ground rotor can be evaluated as

$$\frac{M_a}{R_k F_{a,z}^k} = \frac{1}{N} \sum_{i=1}^{N} \frac{(\mathbf{e}_{x,i}^a \times \mathbf{e}_{t,i}) \cdot \mathbf{e}_z^k}{\mathbf{e}_{t,i} \cdot \mathbf{e}_z^k} = \frac{1}{N} \sum_{i=1}^{N} \frac{\cos \gamma_{k,i}}{\mathbf{e}_{t,i} \cdot \mathbf{e}_z^k}, \tag{22.34}$$

$$\frac{M_{g,z}}{R_k F_{a,z}^k} = \frac{1}{N} \frac{R_g}{R_k} \sum_{i=1}^{N} \frac{(\mathbf{e}_{x,i}^b \times \mathbf{e}_{t,i}) \cdot \mathbf{e}_z}{\mathbf{e}_{t,i} \cdot \mathbf{e}_z^k} = \frac{1}{N} \frac{R_g}{R_k} \sum_{i=1}^{N} \frac{\cos \gamma_{g,i}}{\mathbf{e}_{t,i} \cdot \mathbf{e}_z^k}, \tag{22.35}$$

using the product $R_k F_{a,z}^k$ as a characteristic moment of the tensile torque transmission problem, for normalization of the moment components.

To compute an approximate solution of the steady-state operation of the flying rotor we regard the transverse aerodynamic force components given by Eqs. (22.32) and (22.33) as perturbations. Based on the formulation of an optimization problem we minimize the perturbations to find the best solution. Starting point of the optimization is a specific configuration defined by the dimensionless parameters β, δ and l_K/R_k. The orientation of the flying rotor with respect to the wind, defined by the flow angles α and β_s, is varied to minimize the perturbations. Because the transverse forces oscillate periodically we use the following objective function

$$f(\alpha, \beta_s) = \left| \max F_{a,x}^k - \min F_{a,x}^k \right| + \left| \max F_{a,x}^k + \min F_{a,x}^k \right|$$
$$+ \left| \max F_{a,y}^k - \min F_{a,y}^k \right| + \left| \max F_{a,y}^k + \min F_{a,y}^k \right| \tag{22.36}$$

applying the min and max operators to to the complete interval $0° \leq \omega t \leq 360°$. The solution of the optimization problem is the combination of flow angles α and β_s that minimizes Eq. (22.36). The solution is approximative because the residual transverse forces are causing compensating motions which are not taken into ac-

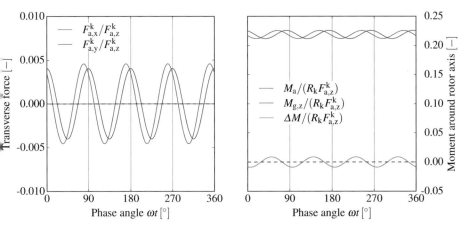

Fig. 22.16 Steady-state operation of the flying rotor with dimensionless transverse force components (left) and moment components around the rotor axes (right) for $N = 4$, $R_g/R_k = 1$, $l_K/R_k = 2$, $\beta = 30°$ and $\delta = 30°$. Initial values for the minimum search are $\alpha_0 = 60°$, $\beta_{s,0} = 0°$ and the solution values are $\alpha = 62.36°$, $\beta_s = 4.38°$

count in the analysis. However, the following results indicate that the effect of the compensating motions is minor and can be neglected.

A representative result is illustrated in Fig. 22.16. The left diagram shows the periodic variations of the transverse force components acting on the flying rotor which are of the order of 1% of the axial force component. The mean values $\overline{F}^k_{a,x}$ and $\overline{F}^k_{a,y}$ vanish. The right diagram shows the periodic variations of the generated aerodynamic moment and the usable moment at the ground rotor, as well as the difference of both curves. It should be noted that the product $R_k F_a$ is only a reference moment used for normalization and does not have any other physical meaning than providing a characteristic order of magnitude value. Compared to the single-tether behavior, as shown in Figs. 22.12 and 22.14, the frequency of the oscillation is increased by a factor of $N = 4$, which is caused by the superposition of phase-shifted data.

It can be recognized that the periodic variations of the moments $M^k_{a,z}$ and $M_{g,z}$ are shifted in phase by the angle $\delta = 30°$. The moment difference $\Delta M = M_a - M_{g,z}$ is associated with the periodic variation of the net mechanical energy processed by the winch modules. For a single tether this relationship is given by Eq. (22.26). For the entire system the normalized moment difference is computed as

$$\frac{\Delta M}{R_k F^k_{a,z}} = \frac{1}{N\omega R_k} \sum_{i=1}^{N} \frac{v_{t,i}}{\mathbf{e}_{t,i} \cdot \mathbf{e}^k_z}. \tag{22.37}$$

It can further be recognized that the mean value of the moment difference for a full revolution of the tether system is zero, which means that the average moments are identical,

$$\overline{M}_a = \overline{M}_{g,z}. \tag{22.38}$$

This essentially means that the transmission efficiency for the ideal system in steady-state operation is, as expected, 100%. For a real system the electrical interconnection of the winch modules will cause conversion losses that will significantly reduce the transmission efficiency. Based on the presented analytic modeling framework these losses as well as all other types of losses (tether aerodynamic drag, bearing friction losses, etc.) can be taken into account in a future study. It is also obvious from the analysis that the number of peripheral tethers affects the frequency of variation the instantaneous kinematic properties and the associated forces and moments but has no effect on the mean values.

In Fig. 22.17 the representative example is expanded to the full range of values of the phase lag angle δ. The left diagram shows the computed values of the flow

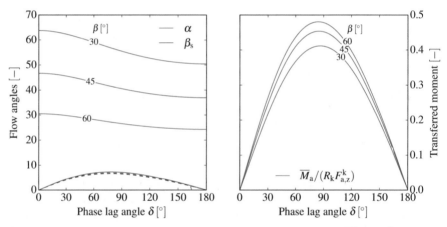

Fig. 22.17 Flow angles α and β_s (left) and dimensionless average moment $\overline{M}_a/(R_k F_{a,z}^k)$ (right) as functions of the phase lag angle δ, for various values of the elevation angle β and for $N = 4$, $R_g/R_k = 1$ and $l_K/R_k = 2$. Initial values for the minimum search are $\alpha_0 = 90° - \beta$ and $\beta_{s,0} = 0°$

angles α and β_s, while the right diagram shows the average moment $\overline{M}_a$ normalized by the reference value $R_k F_{a,z}^k$. It can be recognized that for the limiting values $\delta = 0°$ and $\delta = 180°$ no moment can be transmitted, while the maximum moment $\overline{M}_{a,max}$ can be transmitted for δ_{max}. For this particular example we have $\delta_{max} \lesssim 90°$. This maximum moment increases with increasing elevation angle β.

22.6.4 Requirements for the rotor aerodynamic design

To this point the focus of the analysis has been the transmission of torque from the flying rotor to the ground rotor. From Fig. 22.17(right), or similar diagrams for other combinations of problem parameters, the possible range of the transmittable aerodynamic moment $\overline{M}_a$ can be determined as a function of the aerodynamic force $F_{a,z}^k$. It can further be determined how within this range the moment varies with the phase lag angle δ. From Fig. 22.17(left) we can determine the required orientation of the flying rotor to transmit this moment to the ground while in a steady state of operation. However, these parameters also have a major effect on the aerodynamics of the flying rotor. In fact, the two key functional components of the RRP system, the generation of the aerodynamic moment and the transmission of this moment to the ground rotor, need to be matched properly to achieve steady-state operation. It is the purpose of this section to derive the top-level requirements for the rotor aerodynamic design. The specific implementation of the rotary ring kite is however not within the scope of the present analysis.

To determine the aerodynamic characteristics of the ring kite it is useful to decompose the resultant aerodynamic force $\mathbf{F}_a$ into lift and drag components. By definition the drag force $\mathbf{D}$ is aligned with the apparent wind velocity $\mathbf{v}_a = \mathbf{v}_w - \mathbf{v}_k$, while the lift force $\mathbf{L}$ is perpendicular to the drag component. Assuming that the velocity of the kite $\mathbf{v}_k = d\mathbf{r}_K/dt$ can be neglected during steady-state operation we can use the components of $\mathbf{F}_a$ in the wind reference frame to calculate

$$L = \sqrt{F_{a,y}^2 + F_{a,z}^2}, \tag{22.39}$$

$$D = F_{a,x}. \tag{22.40}$$

Using the transformation matrix $\mathbf{T}_{wk}$ defined by Eq. (22.11) the components of the instantaneous aerodynamic force and its mean value can be calculated as

$$\mathbf{F}_a = \begin{bmatrix} F_{a,x} \\ F_{a,y} \\ F_{a,z} \end{bmatrix} = \mathbf{T}_{wk} \begin{bmatrix} F_{a,x}^k \\ F_{a,y}^k \\ F_{a,z}^k \end{bmatrix} \quad \text{and} \quad \overline{\mathbf{F}}_a = \begin{bmatrix} \cos\beta_s \sin\alpha \\ \sin\beta_s \sin\alpha \\ \cos\alpha \end{bmatrix} F_{a,z}^k, \tag{22.41}$$

because $\overline{F}_{a,x}^k = \overline{F}_{a,y}^k = 0$. Furthermore, the mean values of lift and drag can be calculated as functions of the axial aerodynamic force component and the flow angles

$$L = F_{a,z}^k \sqrt{\sin^2\beta_s \sin^2\alpha + \cos^2\alpha}, \tag{22.42}$$

$$D = F_{a,z}^k \cos\beta_s \sin\alpha, \tag{22.43}$$

which are related by

$$\frac{L}{D} = \frac{\sqrt{\sin^2\beta_s \sin^2\alpha + \cos^2\alpha}}{\cos\beta_s \sin\alpha}. \tag{22.44}$$

Equations (22.42), (22.43) and (22.44) define the required aerodynamic characteristics of the airborne system as functions of the axial aerodynamic force $F_{a,z}^k$, the angle of attack α and the sideslip angle β_s.

The dimensional forces and the moment are generally expressed in terms of dimensionless aerodynamic coefficients

$$L = \frac{1}{2}\rho C_L v_w^2 S, \qquad \text{with} \quad C_L = C_L(\alpha_{\text{eff}}, \lambda), \qquad (22.45)$$

$$D = \frac{1}{2}\rho C_D v_w^2 S \qquad \text{with} \quad C_D = C_D(\alpha_{\text{eff}}, \lambda), \qquad (22.46)$$

$$M_a = \frac{1}{2}\rho C_M v_w^2 R_{k,o} S \qquad \text{with} \quad C_M = C_M(\alpha_{\text{eff}}, \lambda), \qquad (22.47)$$

where $S = \pi(R_{k,o}^2 - R_k^2)$ is the swept rotor area, λ is the tip speed ratio defined by

$$\lambda = \frac{\omega R_{k,o}}{v_w} \qquad (22.48)$$

and α_{eff} is the angle between wind velocity vector $\mathbf{v}_w$ and the rotor disk, defined by

$$\cos\alpha_{\text{eff}} = \mathbf{e}_x^k \cdot \mathbf{e}_x, \qquad (22.49)$$

$$\alpha_{\text{eff}} = \arccos(\cos\beta_s \cos\alpha). \qquad (22.50)$$

The sideslip angle and the angle of attack contribute equally to α_{eff} because of the ring-shaped swept area of the rotor. For a static wing this is not the case and the effects of sideslip angle and angle of attack have to be differentiated. It can be shown that the axial moment coefficient C_M is formally related to the more customary power coefficient C_p [13, p. 45] by the relation

$$C_M = \frac{C_p}{\lambda}. \qquad (22.51)$$

It should also be noted that the induced velocity is not taken into account in the above simplified aerodynamic analysis. An excellent follow-up study in this direction is [13, p. 99–103] which assesses Glauert's momentum theory for a gyrocopter in autorotation.

Aside of the influence of the operational parameters α_{eff} and λ, the aerodynamic coefficients depend also on design parameters, for example, the solidity σ of the rotor. Because the rotor aerodynamic design is out of the scope of the present study the analysis will not be continued at this point. It should be noted though that rotary kites with flexible wings have not been studied scientifically so far.

It has to be assumed that the aerodynamic characteristics required for steady-state operation of the tensile torque transmission system, namely Eqs. (22.42), (22.43) and (22.44), can not necessarily be achieved by a specific design of the ring kite. This problem can be overcome by first designing the ring kite for the required aerodynamic moment and then, in a second step, designing an additional lifting kite which is tethered to the center of the ring kite and which supplements the aerodynamic characteristics of the ring kite to meet the overall requirements for the combined system.

22.6.5 Conceptual Design Example

In this section we outline a conceptual design process based on the developed modeling framework. Starting point is a tensile torque transmission system with a given geometry. We chose the intermediate-scale system defined in Table 22.4. The value

Parameter name	Symbol	Value	Unit
Ground rotor diameter	d_g	50	m
Flying rotor inner diameter	d_k	40	m
Distance between rotors	l_K	100	m
Elevation angle	β	30	deg
Phase lag ground rotor	δ	45	deg
Wind speed	v_w	12	m/s
Angular speed	ω	2.05	rad/s
Nominal power	P	1.4	MW

Table 22.4 Geometric and operational parameters of an intermediate-scale tensile torque transmission system. This configuration is also portrayed in Fig. 22.9

of the phase lag angle is set well below the limiting value for maximum torque transfer, δ_{max} to ensure good control behavior. From Table 22.4 we get

$$\frac{R_g}{R_k} = 1.25 \tag{22.52}$$

$$\frac{l_K}{R_k} = 5. \tag{22.53}$$

In a first step we calculate the aerodynamic moment that is required for transmitting the nominal power P at an angular speed ω of the rotor as

$$M_a = \frac{P}{\omega} = 683\,\text{kNm.} \tag{22.54}$$

We then determine the orientation of the flying rotor, in terms of the flow angles α and β_s, which minimizes the transverse perturbation forces defined by Eqs. (22.32) and (22.33). To compute this best approximation of steady-state operation we minimize the objective function defined by Eq. (22.36). Starting from the initial values $\alpha_0 = 60°$ and $\beta_{s,0} = 0°$ the iterative optimization procedure leads to the values

$$\alpha = 60.08°, \tag{22.55}$$

$$\beta_s = 1.05°, \tag{22.56}$$

which reduce the oscillation amplitudes of $F_{a,x}^k / F_{a,z}^k$ and $F_{a,y}^k / F_{a,z}^k$ to below 0.05%. From Eq. (22.34) we can then calculate

$$\frac{M_a}{R_k F_{a,z}^k} = 0.1365, \tag{22.57}$$

which, using Eq. (22.54) and the value of R_k can be solved for the axial aerodynamic force

$$F_{a,z}^k = 250\,\text{kN}. \tag{22.58}$$

Using Eqs. (22.42), (22.43) and (22.44) we can now compute the lift and drag force as

$$L = 125\,\text{kN}, \tag{22.59}$$
$$D = 217\,\text{kN}, \tag{22.60}$$
$$L/D = 0.576 \tag{22.61}$$

It is important to note that the numerical values given by Eqs. (22.58), (22.59) and (22.60) are not the result of an aerodynamic analysis but instead are required to transmit the aerodynamic moment specified by Eq. (22.54) to the ground rotor while maintaining a steady state of operation of the revolving tether system.

As a next step we analyze the aerodynamic requirements of the airborne subsystem. For conventional wind turbines the tip speed is generally limited by a noise constraint. In [13, p. 339] this tip speed limit is given as 65 m/s. Considering the value of ω listed in Table 22.4 and using a tip speed limit of ≈ 70 m/s we can calculate the outer diameter of the rotor, the swept area of the rotor and from Eq. (22.48) the tip speed ratio as

$$d_{k,o} = 70\,\text{m}, \tag{22.62}$$
$$S = 2592\,\text{m}^2, \tag{22.63}$$
$$\lambda = 5.98. \tag{22.64}$$

Based on these values we can compute the aerodynamic coefficients from Eqs. (22.45), (22.46), (22.47) and (22.51) as

$$C_L = 0.546, \tag{22.65}$$
$$C_D = 0.948, \tag{22.66}$$
$$C_M = 0.0854, \tag{22.67}$$
$$C_p = 0.51. \tag{22.68}$$

The practical design of a ring kite would aim to achieve the required moment coefficient specified by Eq. (22.67) and then, in a second step, supplement its lift and drag forces by tethering an additional lifting kite to the center of the ring kite, as explained in Sect. 22.6.4.

22.6.6 Design Recommendations and Conclusions

The geometric proportions of the tensile torque transmission system have a decisive role. It is evident that the longer the tether system and the smaller the ground ro-

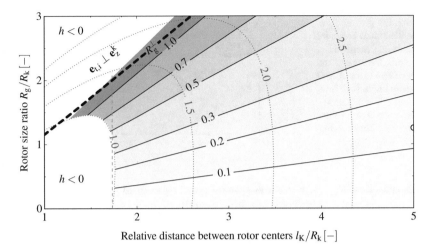

Fig. 22.18 Transferable aerodynamic moment $M_a/(R_k F_{a,z}^k)$ as function of the relative distance l_K/R_k between rotor centers and the rotor size ratio R_g/R_k for $\beta = 30°$, $\delta = 45°$. The colored contour plot and the solid black isolines cover only valid regions with positive ground distance of the tether attachments on the flying rotor ($h > 0$). The dashed line $R_g^\perp$ marks the condition of maximum transferable moment which is also a validity limit for the approximate solution of steady-state operation. The dotted red isolines mark the condition of ground contact for different values of the relative outer size $R_{k,o}/R_k$ of the flying rotor. The limiting isoline $(R_{k,o}/R_k)_{h=0} = 1$ coincides with the border of the contour plot. The dashed line at $l_K/R_k = \cot\beta \approx 1.73$ is the reference for the lower limit, for $\alpha = 90° - \beta$

tor the lower the transferable torque. The fundamental relationship is quantified by Eq. (22.26) which describes the influence of the tether attachment angles and the rotor size ratio. To support this recommendation quantitatively we have computed the transferable aerodynamic moment as a function of the geometric proportions of the tether system. The result of this analysis is illustrated in Fig. 22.18 for a system with representative elevation angle and phase lag angle. The contour plot and the solid isolines show that the transferable moment decreases for increasing distance between the rotors and that it increases with increasing size of the ground rotor. The diagram also includes the condition of ground contact of the flying rotor for different values of its relative outer size $R_{k,o}/R_k$. As shown in Fig. 22.19 this condition $(R_{k,o}/R_k)_{h=0}$ can be derived from the ground distance function

$$\frac{h}{R_k} = \frac{l_K}{R_k}\sin\beta - \frac{R_{k,o}}{R_k}\sin\alpha, \tag{22.69}$$

by setting $h = 0$ and solving for $R_{k,o}/R_k$. For example, if we consider a system with a relative outer size $R_{k,o}/R_k = 2$ only the region to the right of the dotted isoline labeled by the value $(R_{k,o}/R_k)_{h=0} = 2$ is physically feasible because of positive ground distance ($h > 0$). The data point at $l_K/R_k = 5$ and $R_g/R_k = 1.25$ refers to the specific calculation example in Sect. 22.6.5 which results in values $M_a/(R_k F_{a,z}^k) =$

Fig. 22.19 Calculation of the distance h of the flying rotor from the ground. The specific illustrated geometric case has been described in Sect. 22.6.5. Because of the relatively large value of l_K/R_k the sideslip angle is in this case small ($\beta_s < 1°$) and the axis of the flying rotor approximately points to the origin ($\alpha = 90° - \beta$)

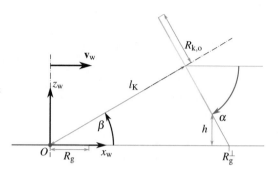

0.1365 and $(R_{k,o}/R_k)_{h=0} = 2.88$. A general conclusion from Fig. 22.18 is that an increasing size of the flying rotor requires generally an increasing distance between the rotors.

The deviation of the limiting isoline $(R_{k,o}/R_k)_{h=0} = 1$ from the dashed reference line at $l_K/R_k = \cot\beta \approx 1.73$ indicates how much the angle of attack α in steady-state operation deviates from the value $\alpha = 90° - \beta = 60°$. For $1.2 < R_g/R_k < 1.8$ the physically feasible region extends to values far below $l_K/R_k = 1.73$. This indicates that for these geometric proportions the flying rotor has an increased ground clearance as consequence of a relatively low angle of attack.

The dashed line in Fig. 22.18 marks the condition where for $\alpha = 90° - \beta$ the flying rotor plane touches the ground rotor. Considering Fig. 22.19 where this point is marked as $R_g^\perp$ the equation of the limiting line can be derived as

$$\frac{R_g^\perp}{R_k} = \frac{1}{\cos\beta}\frac{l_K}{R_k} = 1.155\frac{l_K}{R_k}. \tag{22.70}$$

Above this line the geometric proportions of the system are such that the revolving tethers pass through constellations in which they are momentarily orthogonal to the axis of the flying rotor. In this specific situation the fundamental assumption on which the approximate solution of steady-state operation is based, Eq. (22.30), becomes singular which introduces large artificial forces in the system. However, this anomaly of the theoretical model occurs in a region of the design space that has to be avoided for the sake of operational stability and for this reason the region is excluded as an invalid region ($\mathbf{e}_{t,i} \perp \mathbf{e}_z^k$).

It is obvious from Fig. 22.18 that the size of the flying rotor cannot be much larger than the diameter of the ground rotor if a practically significant torque is to be transferred. This has been confirmed also by the experimental tests. From a rotor aerodynamics point of view, the rotational speed $\omega = 2.05$ rad/s, implies a tip speed ratio of around 6, which is a typical value for conventional wind turbines, and which is achievable with rigid wings. Lower values of the tip speed ratio, for example 4, are achievable with soft wings.

The Parotor should be restudied in all possible variants, including soft and rigid rotors, parachutes with a large opening [24] (with a diameter of 2 to 3 times the wing span), including also some adaptations of centrifugally stiffened rotors [18, 22, 29], and above all C-shaped modular rigid structures with hinges [19] rotor components. Wings or blades should sweep more area and travel faster, like kites making loops [4], but within the rotating structure.

22.7 Current and Future Investigations

The focus of the present study has been the technical feasibility of the proposed RRP system. We have used a small-scale test setup to demonstrate the fundamental working principles and a theoretical model of the revolving tether system to show that steady-state operation is in principle feasible for a specific combination of design and operational parameters. A next important development step will be the design of a rotary ring kite and lifting kite combination with specific aerodynamic characteristics. Of similar importance will be the design of a rotating reel system with efficient energetic balancing of the interconnected winch modules. A possible realization could be mechanical coupling of the winch modules using differential reeling to avoid the additional losses of electrical conversion. With the worked out conceptual and preliminary designs of these two key technology components the assessment of the energy harvesting potential of the RRP system can be further refined.

Next to the overall system design and the conversion performance we will also investigate other important aspects of the technology. A key advantage is vertical take off and landing (VTOL) of the flying rotor with the help of the ground rotor. For this purpose the generator, which is connected to the ground rotor, acts as a motor to power the rotation of the launching Parotor. A possible VTOL configuration of the small-scale model described in Sect. 22.4.1 is shown in Fig. 22.20.

Fig. 22.20 Possible configuration of the small-scale system shown in Fig. 22.7 before vertical launch maneuver

As shown in Fig. 22.9, and also in the conceptual design example discussed in Sect. 22.6.5, the outer diameter of the flying rotor can exceed the diameter of the ground rotor to some degree. It is not a problem in the case of the implementation of a rigid or semi-rigid [11] flying rotor. But in the case of the implementation of a flexible rotor, the diameter of the Parotor should not exceed the diameter of the ground rotor.

If there is no or very low wind the generator of the ground rotor is operated as a motor to keep the flying rotor airborne in a helicopter mode. The winches are also suitable to assure a fast landing in hazardous whether conditions. The envisaged emergency strategy for urgent depower uses a central depower line as illustrated in Fig. 22.5(c). The peripheral tethers are detached from the Parotor which is only kept by the central rope. Thus the Parotor turns around, losing its lift and drag, coming down towards the central station. In case of implementation of the second mode of generation, as described in Sect. 22.3.3, the suspension lines are also detached.

In case of failure of the electrical system and/or in case of rupturing of one or more tethers, the Parotor can be held by the central rope. The Parotor can also be held by the suspension lines (Figs. 22.6 and 22.5) if the second mode of generation is implemented as described in Sect. 22.3.3.

The wind velocity can vary significantly over the swept area of a huge flying rotor. A flexible rotor could employ active deformation of its blades to change their aerodynamic characteristics [12] and to adjust to varying wind conditions.

The RRP system follows the topology model of a single large rotary kite in steady-state rotation, anchored to the ground or sea surface by tethers [7, 30]. Thanks to its uniform motion a huge flying rotor is more easily recognized by other users of the airspace than a farm of smaller units with wings moving in multiple directions. Small wind turbines carrying lights are launched then move along the peripheral tethers. Then they are fixed at a desired height. They provide needed visibility markings.

An implementation of superimposed rotors is also studied. According to some observations [21, 24] there are possible interesting aerodynamic features increasing the efficiency of each rotor from a stack with regard to an identical but single rotor. A phase lag angle of a rotor with the nearby rotor can increase the transmission with relatively longer peripheral tethers. The rotors act then as "ring torque" [24].

22.8 Conclusions

In this chapter we have presented a rotary ring kite which uses a revolving tether system to transfer the generated aerodynamic torque to a ground rotor which is connected to a generator. To analyze this novel concept we have developed a kinematic model of the tether system and a numerical procedure to determine an approximate solution for steady-state rotation. This operational mode is characterized by minimal periodic compensation motions of the flying rotor. To realize this mode the flying rotor has to have specific aerodynamic characteristics and a specific inclination with respect to the flow, also the length of the tethers has to be adjusted continuously with the rotation of the system.

The analysis has further revealed that the power transmission is an interplay between three periodically varying terms of equal magnitude: the power generated by the ring kite, the shaft power available at the ground and the net reeling power of the interconnected winch modules. For an ideal lossless system the net reeling power vanishes over an entire revolution and the transmission efficiency is 100%. In reality, however, energy conversion losses in the winch modules will reduce the transmission efficiency significantly. A possible solution to reduce these losses would be a mechanical interconnection, using electrical machines only to provide a differential reeling power.

We have also analyzed a secondary mode of energy conversion which is based on the selective unloading of the tethers during reel in. This is realized by periodically shifting the tensile load to additional suspension lines, with the theoretical result of a positive net reeling power of the interconnected winch modules. However, because of the inevitable cyclical force imbalance the system will tumble and a steady-state of operation can not be achieved.

Experimental tests with a small-scale model of the Rotating Reel Parotor system have confirmed some of the theoretical findings. In place of electrical machines, which would allow for precise actuation of the tethers, this first physical demonstrator uses winch modules with rotational spring mechanisms. As consequence, the results of this experiment can hardly be used to assess the original concept.

The present study serves as a starting point for future investigations. The planned prototypes increase in logical scaling steps: 5 m rotor diameter and tip height, then 10 m as small-scale models; 25 m, 50 m and 100 m as intermediate-scale models along with a critical assessment of the market opportunities in remote locations; 500 m, 1 km and more as large-scale models harnessing high-altitude winds at utility scale. An essential part of this roadmap is the question about the scalability of the Parotor towards very large dimensions.

The Python source code of the analysis tools developed in the frame of this chapter is available from a public repository [28].

Acknowledgements The authors would like to thank Antonello Cherubini for his help with the mechanical analysis; Antoine Delon, for the geometrical and mathematical representations of the reference axis and kinematics; Ben Lerner for the reorganization of some elements; David Murray for proofreading.

References

1. Ahrens, U.: Wind-operated power generator. US Patent 8,096,763, Jan 2012
2. Ahrens, U., Pieper, B., Töpfer, C.: Combining Kites and Rail Technology into a Traction-Based Airborne Wind Energy Plant. In: Ahrens, U., Diehl, M., Schmehl, R. (eds.) Airborne Wind Energy, Green Energy and Technology, Chap. 25, pp. 437–441. Springer, Berlin Heidelberg (2013). doi: 10.1007/978-3-642-39965-7_25

3. Archer, C. L., Caldeira, K.: Global Assessment of High-Altitude Wind Power. Energies **2**(2), 307–319 (2009). doi: 10.3390/en20200307
4. Argatov, I., Silvennoinen, R.: Asymptotic modeling of unconstrained control of a tethered power kite moving along a given closed-loop spherical trajectory. Journal of Engineering Mathematics **72**(1), 187–203 (2012). doi: 10.1007/s10665-011-9475-3
5. Argatov, I., Rautakorpi, P., Silvennoinen, R.: Estimation of the mechanical energy output of the kite wind generator. Renewable Energy **34**(6), 1525–1532 (2009). doi: 10.1016/j.renene. 2008.11.001
6. Argatov, I., Silvennoinen, R.: Energy conversion efficiency of the pumping kite wind generator. Renewable Energy **35**(5), 1052–1060 (2010). doi: 10.1016/j.renene.2009.09.006
7. Beaujean, J. M. E.: 500MW Wind Turbines. Windtech International, 11 Nov 2011. https: //www.windtech-international.com/content/500mw-wind-turbines Accessed 21 July 2016
8. Benhaïem, P.: Eolienne aéroportée rotative. French Patent 3034473, Oct 2016
9. Benhaïem, P.: Land and Space used. In: Lütsch, G. (ed.). Book of Abstracts of the International Airborne Wind Energy Conference 2013, p. 59, Berlin, Germany, 10–11 Sept 2013. http: //resolver.tudelft.nl/uuid:e3e8aaa4-8ae1-498a-82ce-fdc7a149963f
10. Benhaïem, P.: Rotating Reeling. In: Schmehl, R. (ed.). Book of Abstracts of the International Airborne Wind Energy Conference 2015, p. 100, Delft, The Netherlands, 15–16 June 2015. doi: 10.4233/uuid:7df59b79-2c6b-4e30-bd58-8454f493bb09. Poster available from: http: //www.awec2015.com/images/posters/AWEC25_Benhaiem-poster.pdf
11. Breuer, J. C. M., Luchsinger, R. H.: Inflatable kites using the concept of Tensairity. Aerospace Science and Technology **14**(8), 557–563 (2010). doi: 10.1016/j.ast.2010.04.009
12. Breukels, J., Schmehl, R., Ockels, W.: Aeroelastic Simulation of Flexible Membrane Wings based on Multibody System Dynamics. In: Ahrens, U., Diehl, M., Schmehl, R. (eds.) Airborne Wind Energy, Green Energy and Technology, Chap. 16, pp. 287–305. Springer, Berlin Heidelberg (2013). doi: 10.1007/978-3-642-39965-7_16
13. Burton, T., Jenkins, N., Sharpe, D., Bossanyi, E.: Wind Energy Handbook. 2nd ed. John Wiley & Sons, Ltd, Chichester (2011). doi: 10.1002/9781119992714
14. Canale, M., Fagiano, L., Milanese, M.: Power kites for wind energy generation - fast predictive control of tethered airfoils. IEEE Control Systems Magazine **27**(6), 25–38 (2007). doi: 10. 1109/MCS.2007.909465
15. Diehl, M., Horn, G., Zanon, M.: Multiple Wing Systems – an Alternative to Upscaling? In: Schmehl, R. (ed.). Book of Abstracts of the International Airborne Wind Energy Conference 2015, p. 96, Delft, The Netherlands, 15–16 June 2015. doi: 10.4233/uuid:7df59b79-2c6b-4e30-bd58-8454f493bb09. Presentation video recording available from: https://collegerama. tudelft.nl/Mediasite/Play/1065c6e340d84dc491c15da533ee1a671d
16. Duquette, M. M., Visser, K. D.: Numerical Implications of Solidity and Blade Number on Rotor Performance of Horizontal-Axis Wind Turbines. Journal of Solar Energy Engineering **125**, 425–432 (2003). doi: 10.1115/1.1629751
17. Fletcher, C. A. J., Roberts, B. W.: Electricity generation from jet-stream winds. Journal of Energy **3**(4), 241–249 (1979). doi: 10.2514/3.48003
18. Hodges, T.: Centrifugally Stiffened Rotor. NIA Task Order Number 6528 Final Report, National Institute of Aerospace, 1 June 2015, pp. 57–147. http://ntrs.nasa.gov/archive/nasa/casi. ntrs.nasa.gov/20160001625.pdf
19. Ippolito, M.: Kite wind energy collector. Patent WO2014199407 A1, Dec 2014
20. Loyd, M. L.: Crosswind kite power. Journal of Energy **4**(3), 106–111 (1980). doi: 10.2514/3. 48021
21. Michel, D., Koyama, K., Krebs, M., Johns, M.: Build and test a three kilowatt prototype of a coaxial multi-rotor wind turbine. Independent Assessment Report CEC-500-2007-111, Dec 2007. http://www.energy.ca.gov/2007publications/CEC-500-2007-111/CEC-500-2007-111.PDF
22. Moore, M. D.: Eternal Flight as the Solution for X. Presented at the NIAC 2014 Symposium, Stanford University, Palo Alto, CA, USA, 4–6 Feb 2014. https://www.nasa.gov/sites/default/ files/files/Moore_EternalFlight.pdf

23. Rancourt, D., Bolduc-Teasdale, F., Demers Bouchard, E., Anderson, M. J., Mavris, D. N.: Design space exploration of gyrocopter-type airborne wind turbines. Wind Energy **19**, 895–909 (2016). doi: 10.1002/we.1873
24. Read, R.: Opportunities and Progress in Open AWE Hardware. In: Schmehl, R. (ed.). Book of Abstracts of the International Airborne Wind Energy Conference 2015, pp. 118–120, Delft, The Netherlands, 15–16 June 2015. doi: 10.4233/uuid:7df59b79-2c6b-4e30-bd58-8454f493bb09. Poster available from: http://www.awec2015.com/images/posters/AWEC23_Read-poster.pdf
25. Roberts, B. W., Shepard, D. H., Caldeira, K., Cannon, M. E., Eccles, D. G., Grenier, A. J., Freidin, J. F.: Harnessing High-Altitude Wind Power. IEEE Transactions on Energy Conversion **22**(1), 136–144 (2007). doi: 10.1109/TEC.2006.889603
26. Rye, D. C., Blacker, J., Roberts, B. W.: The Stability of a Tethered Gyromill. AIAA-Paper 81-2569. In: Proceedings of the AIAA 2nd Terrestrial Energy Systems Conference, Colorado Springs, CO, USA, 1–3 Dec 1981. doi: 10.2514/6.1981-2569
27. Schmehl, R.: Large-scale power generation with kites. Journal of the Society of Aerospace Engineering Students VSV Leonardo da Vinci March, 21–22 (2012). http://resolver.tudelft.nl/uuid:84b37454-5790-4708-95ef-5bc2c60be790
28. Schmehl, R.: Parotor. https://github.com/rschmehl/parotor. Accessed 25 Oct 2016
29. Selfridge, J. M., Tao, G.: Centrifugally Stiffened Rotor: A Complete Derivation and Simulation of the Inner Loop Controller. AIAA-Paper 2015-0073. In: Proceedings of the AIAA Guidance, Navigation, and Control Conference (AIAA SciTech), Kissimmee, FL, USA, 5–9 Jan 2015. doi: 10.2514/6.2015-0073
30. Snieckus, D.: Giant airborne 'power station' could blow rivals out of the water. Recharge News, 6 Mar 2012. http://www.rechargenews.com/news/technology/article1295509.ece Accessed 21 July 2016
31. Williams, P., Lansdorp, B., Ruiterkamp, R., Ockels, W.: Modeling, Simulation, and Testing of Surf Kites for Power Generation. AIAA Paper 2008-6693. In: Proceedings of the AIAA Modeling and Simulation Technologies Conference and Exhibit, Honolulu, HI, USA, 18–21 Aug 2008. doi: 10.2514/6.2008-6693

Part IV
Implemented Concepts

Chapter 23
Quad-Rotorcraft to Harness High-Altitude Wind Energy

Bryan W. Roberts

Abstract Wind at higher altitudes is generally stronger and more persistent than near-surface wind. At many locations the atmospheric flows have annual average power densities that by far exceed these of any other renewable energy sources. Capturing this energy potential has been the objective of a pioneering airborne wind energy concept based on a tethered rotorcraft which was invented in Australia in the 1980s. The chapter summarizes early research with a towed generating rotor, wind tunnel tests and a low-altitude atmospheric test vehicle. These tests have confirmed the feasibility of kite-like flight of a craft having twin or quadruple rotors with the rotors simultaneously generating electricity. Using high-altitude wind data statistics for Australia and the USA it is shown that near base-load electrical outputs can be achieved at capacity factors of 70 to 80%. The governing physical relations of the technology are derived from classical helicopter theory leading to the rotor thrusts and the rotors' limits to power generation. The range of useful tip-speed ratios is presented for the complete range of rotor disk incidence angles. This mathematical model is used to describe the low-altitude operation of a small quad-rotorcraft. The model is suitable to predict the performance of a multi-megawatt machine. The final contribution of the chapter is a dynamic analysis of the system to devise a control strategy for the craft's power output, pitch, roll and yaw, using purely blade collective pitch action.

23.1 Introduction

It is proposed that a tethered quad-rotorcraft can harness the enormously powerful winds at higher altitude, thereby generate electricity from these winds [3]. It is well known that two major jet streams exist in each Earth hemisphere at higher altitudes.

Bryan W. Roberts (✉)
Altitude Energy Pty. Ltd., Australia
e-mail: trytables@bigpond.com

© Springer Nature Singapore Pte Ltd. 2018
R. Schmehl (ed.), *Airborne Wind Energy*, Green Energy
and Technology, https://doi.org/10.1007/978-981-10-1947-0_23

These streams are called the sub-tropical jet and the polar-front jet. The former are of particular relevance as they exist in bands approximately 1000 km wide over the Mediterranean, Northern India, China, Southern Japan, North America, Africa, Australia, South America and elsewhere. These streams have enormous energy and persistence compared to near-surface winds. They are formed by sunlight falling on the tropics in combination with the Earth's rotation. The formation of these jet streams can be seen in Fig. 23.1, which is a section through the Earth showing how heated air rises in the tropics (0° latitude) and then moves towards the North and South poles (90° latitude) after it rises to tropopause altitude. Subsequently at

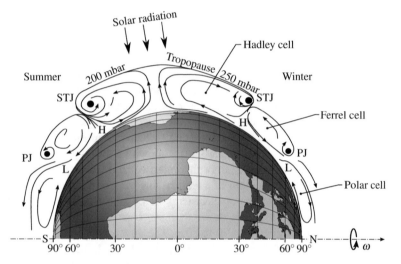

Fig. 23.1 Global atmospheric flow mechanisms and formation of jet streams (STJ: subtropical jet, PJ: polar jet, H: high pressure region, L: low pressure region). Thickness of atmosphere not to scale

increasing latitudes, due to the Coriolis acceleration, the air moves from west to east to form the jet streams.

Furthermore, compared to ground-based turbines operating in low-velocity winds at the bottom of the Earth's boundary layer, these jet stream winds offer a potential annual energy output of about two orders of magnitude greater than that obtainable from ground-based turbines of equivalent rotor area.

The following Sect. 23.2 presents two resource studies comprising wind data for near-surface altitudes up to the tropopause level, where the jets generally reside. Section 23.3 outlines several harvesting systems based on the tethered rotorcraft principle. Section 23.4 describes a wind tunnel analysis of a small-scale model while Sect. 23.5 provides details on the design and testing of an outdoors rotorcraft. In Sects. 23.6 to 23.9 the quad-rotorcraft configuration is analyzed in detail, discussing important aspects of its operation. Section 23.10 describes the stability and control and Sect. 23.11 concludes the chapter. The preliminary content of the present chapter has been presented at the Airborne Wind Energy Conference 2015 [12].

23.2 Upper Wind Data for Australia and USA

The southern and northern sub-tropical jet streams (around 30 to 40° latitude) cross the planet in a W-E direction. The jet stream is invariably present, sometimes bifurcated, with annual average velocities of around 130 km/h. The passage of the jet is observed to meander north and south so that any fixed land or ocean site is swept by the jet. Extensive studies of the wind statistics have been undertaken by Atkinson et al [1] for Australia and O'Doherty and Roberts [10] for the USA. In Australia an annual average wind power density of 19 kW/m^2 is achievable, while in the USA the maximum annual average power density is 17 kW/m^2. It might be argued that the generally higher power densities in the southern hemisphere are due to a colder pole in the south relative to that in the north.

Figure 23.2 shows the isopleths of annual average power density over Australia at an altitude of 250 mbar. It may be seen therein that the power distribution is

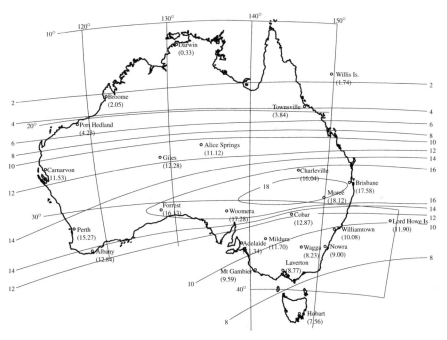

Fig. 23.2 Isopleths of the annual average power density P_w [kW/m^2] over Australia at an altitude of 250 mbar which corresponds to 10 km altitude

spatially well organized because of the lack of high mountains tending to upset the orderly flow of air.

A similar graph for USA is not so well organized [10], possibly due to the presence of the Rocky Mountains. A standard wind energy technique is to represent the cumulative probability distribution $F(v_w)$ of wind speed v_w by a Weibull model

$$F(v_w) = 1 - \exp\left[-\left(\frac{v_w}{v_{w,0}}\right)^n\right], \quad \text{for} \quad v_w > 0, \qquad (23.1)$$

where $v_{w,0}$ and n are two constants chosen to give a good fit to the observed data.

Figure 23.3 shows the cumulative probability distribution for Albany, NY, USA. It is typical of the US states in that general area. Because special "Weibull paper" is

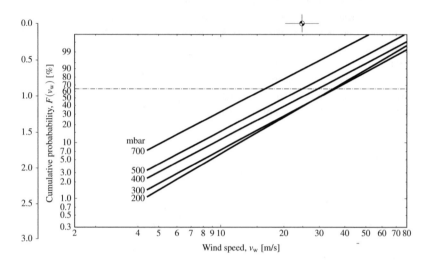

Fig. 23.3 Annual probability distribution of velocity in Albany, NY, USA [10]

used the cumulative probability distributions of the wind speed are straight lines. A sample use of the diagram is as follows. At an altitude of 300 mbar wind speeds of 10 m/s or lower will occur for approximately 7% of the time, namely 613 hours per annum. This period per annum below the so called threshold velocity of 10 m/s is made-up by the number of down-times per annum multiplied by the average down-time in each event. An average down-time is about 24 hours. Furthermore, it is shown in references [1] and [10], the latter providing details for some 50 sites across the USA, that the winds are generally stronger in winter than in summer. The data given in Fig. 23.3 is for one of the best sites in the USA and it is almost identical, although a little less optimal, than the best site in Australia, around Moree in the state of North South Wales.

The average annual power densities quoted above are the highest power densities on Earth for any large-scale renewable resource. These power densities vastly exceed that of solar radiation. The latter is generally around 0.25 kW/m² at the surface depending on latitude. Furthermore, high-altitude wind exceeds the power density of any other renewable energy resource found on Earth. They exceed the resource of near surface winds, that of ocean currents, tidal and the geothermal resources.

Hoffert, Caldeira et al. [7] estimate that the total thermal power consumption by human civilization is 10 TW, which is about 1% of the total amount of power dissipated in the planet's wind system. Most of this planetary wind energy is concentrated in the jet stream system, so that a maximum energy extraction of around 1% would not have any adverse impact on the Earth's environment or climate.

This high-altitude wind resource is just a few kilometers above the surface of the planet, where the energy is needed. Because of its vast power and persistent nature it would, if captured, be a very attractive and inexhaustible power supply.

23.3 Various Capture Systems

One of the earliest suggestions for high-altitude capture was that made by Manalis [9] in 1976. Various systems have been examined since. These range from tethered balloons, tethered fixed-winged craft, tethered kites in simple or crosswind flight, climbing and descending devices and rotorcraft. The preferred option here is a tethered rotorcraft, a variant of the gyroplane principle, where conventional rotors operating at a significant disk incidence generate power in the on-coming wind, while simultaneously producing sufficient lift to keep the system aloft.

In engineering design, if some component performs two important functions simultaneously, then this component should be featured. In the current rotorcraft concept the rotors provide lift while simultaneously generating power. It is this important dual function that is the centerpiece of the current work.

Roberts and Blackler [17] confirmed the power generating characteristics of a rotor at incidence to the wind by mounting a simple, flap-articulate rotor above a test vehicle as seen in Fig. 23.4. By moving the test vehicle through still air they

Fig. 23.4 Test setup for rotor at incident angle

were able to obtain a confirmation of the power generation principle along with an estimate of the accuracy of the theoretical predictions.

The two-bladed ($b = 2$) rotor had a linear twist angle of $\theta_1 = +8°$, while the collective pitch angle was pre-set to $\theta_0 = -8°$ using specially machined blade grips. The rotor's torque output was measured, while the rotor thrust was not recorded because if the torque output was shown to agree with the extended theory of Gessow and Crim [4], then it is highly likely the thrust coefficient would agree since it was derived from a common theory. This avoided the added complication of constructing a thrust measuring and recording system. The results of these experiments are shown in Fig. 23.5. Therein the measured rotor-shaft torque coefficient C_Q is somewhat

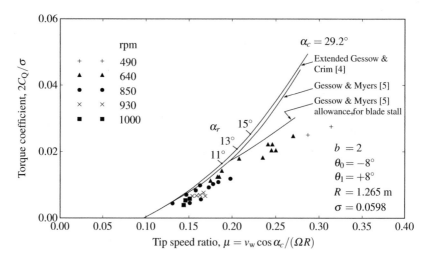

Fig. 23.5 Torque parameter versus tip speed ratio (test date at Schofields: 28 November 1979)

less than the calculated values using accepted profile drag values. Note that the coefficient C_Q used in Fig. 23.5 is in helicopter terminology, namely based on the rotor tip speed ΩR, not the wind speed v_w. Symbols R and Ω denote the radius and angular velocity of the rotor. The parameter $2C_Q/\sigma$ is a conventional helicopter parameter, where σ is the rotor solidity defined as the total blade area divided by the swept area of the rotor. The rotor's control axis angle is $\alpha_c = 29.2°$ and three specific values $\alpha_r = 11, 13$ and $15°$ are the angle of attack values on the retreating blade at a standard reference location on the blade's span as defined in the helicopter texts such as Gessow and Myers [5]. This technique is used in helicopter work to set limits on rotor operation without having excessive retreating blade stall. A limit of $13°$ has been used throughout the current work.

However, the results agree well when allowance is made for increased aerodynamic drag at the test Reynolds number. The conventional method of allowing for increased profile drag due to retreating blade stall is also demonstrated by the test results. The small angle theory of Gessow and Myers [5] can be seen to correspond with the extended theory of Gessow and Crim [4].

Encouraged by the results of Fig. 23.5 it was decided to construct a wind-tunnel model of a twin, side-by-side rotorcraft to explore the handling in a conversion of the craft from helicopter to generate mode and vice versa, while in-flight.

23.4 Wind Tunnel Model

A wind-tunnel model was constructed as described in detail in reference [17]. The model had twin, contra-rotating rotors, with $R = 0.335$ m and $\sigma = 0.0462$ and is shown in Fig. 23.6. The airflow in the side view is from right to left and the twin

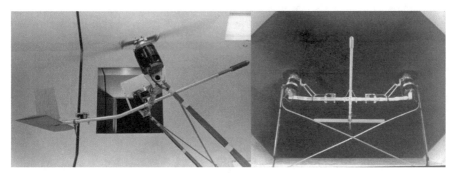

Fig. 23.6 Wind tunnel model in side view (left) and front view (right) in generating mode

rotors are driven by two separate, permanent-magnet DC motor/generators.

Because of a slight mismatch in the manually controlled rotational speed the closely spaced tips tended to interact and induce vibration. Therefore a thin vertical, edge-on partition was used to aerodynamically isolate the rotor tips. A horizontal tailplane was used for longitudinal stability and control, while the tethering arrangement was a twin-bar frame hinged under the rotors and at the tunnel floor. This frame was cross-braced to effectively eliminate any yaw or roll freedom in the model. In this way with the frame pivoted at the tunnel floor and at the craft, it was possible to investigate the pitch performance of the system in a hover mode with low tunnel flow. Then as the tunnel speed was increased the craft was converted to a generation mode, all on both rotors. The tailplane incidence and the collective pitch on each rotor were manually controlled by a standard radio-control servo link. The servos can be clearly identified in Fig. 23.6. No cyclic pitch action was provided.

In summary, it is envisaged that a craft having twin, and sometime later having quadruple or more rotors, can generate electrical power at altitude with the rotors inclined at an adjustable angle to the on-coming wind. In general the rotor disks operate at an angle of about $\alpha_c = 40°$. The wind then acts on the inclined rotors producing lift, gyroplane-style, while simultaneously driving the rotors to generate electricity, windmill-style. The electricity so generated is conducted down the tether to a ground station.

It is also important to consider that the craft can also function as an elementary powered helicopter with electrical energy supplied from the ground, with the generators then functioning as motors. The craft can then ascend, descend or maintain altitude during any short wind lull aloft. A ground winch, which could reel the tether, would be used to retrieve the craft in an emergency. Obviously a single conducting tether would be preferable.

23.5 Atmospheric Craft

An atmospheric test vehicle was next designed and constructed. A picture of the craft close to auto-rotation is shown in Fig. 23.7. A full report on the design and

Fig. 23.7 The Gyromilll Mk2 was equipped with twin, single bladed, counterweighted rotors of solidity 2.2% and a diameter of 3.65 m. The total mass of the craft was 29 kg. Again no cyclic pitch capability was employed. Thus in hover and wind it was found necessary to use three parallel tethers to maintain craft attitude. These three tethers were used in all tests, but the craft was difficult to control particularly in low winds. Two tethers were attached near the rotor thrust lines while the third tether was attached well forward on the forward pointing boom

preliminary performance of this craft is given in reference [14]. The craft generated power at about 50 feet altitude for a short period, but in hindsight it was very difficult to control without the use of a then unobtainable modern gyro-stabilizing avionics. Next it was decided to use a four or more rotor system in order to have active attitude control by employing differential collective pitch action on at least four rotors without the use of cyclic action. The avoidance of cyclic action should greatly enhance the fatigue life, while making each rotor's control system almost identical to that of well-proven ground-based wind turbines.

23.6 Quad-Rotor System

This arrangement consists of four identical rotors in mutual counter-rotation. They can be mounted in a suitable airframe which is tethered by a single tether in the powerful and persistent winds aloft. The strength element of the electro-mechanical cable can be of the Kevlar family. This element is wound together with insulated aluminum, or possibly copper, conductors. For high-altitude operation a high-voltage, direct current (HVDC) system is preferred with transmission voltages of about 15 kV, or more. This amounts to about 3 volts per meter of operating altitude. It is acknowledged that conductor insulation could be an issue, but current advice is that these voltages are achievable in a twin-conductor, DC system. This high voltage is necessary if the tether weight is not to be excessive compared to the weight of the craft. This important weight issue will be discussed further below. The rated output per unit from these high-altitude, multiple-rotor systems is envisaged to be in the 3 to 30 MW range, making them useful for commercial electricity production. These generators at high altitude would avoid community concerns about the visual and noise intrusions usually associated with conventional ground-based wind turbines. Also there is a lesser of a bird-strike problem. Nevertheless, they would need to be placed in restricted airspace to avoid intrusion by other aircraft and to be located away from populated areas. While an array of these generators at altitude would be similar to conventional wind farms, in most instances the craft can be located much closer to the demand load centers than that of ground-based wind farms.

When operating as an electric generator the quadruple, or more rotors, are inclined at an adjustable and controllable angle to the on-coming wind. In general the rotors have disk incidence angles up to no more than $\alpha_c = 50°$. The disk incidence is reduced in increasing wind conditions so as to hold the power output at its rated value without exceeding the safe tether load. This can be achieved while maintaining altitude and by varying the rotor's rotational speed, all without introducing excessive retreating blade incidences.

It can be shown that with all factors considered, the capacity (availability or generating) factor of these craft is far higher than that obtainable from the very best ground-based wind turbines. Reference [18] quotes capacity factors for high-altitude rotorcraft at between 71 and 90% for a number of US sites. Typical capacity factors for ground-based turbines are only about 30%. Therefore, it can be concluded that high-altitude craft can be classed as base-load generators, if the above capacity factors were to be demonstrated.

Figure 23.8 shows a quad-rotorcraft with four identical rotors arranged with a forward pair of rotors ahead of a rearward pair. The rotors are in mutual counter-

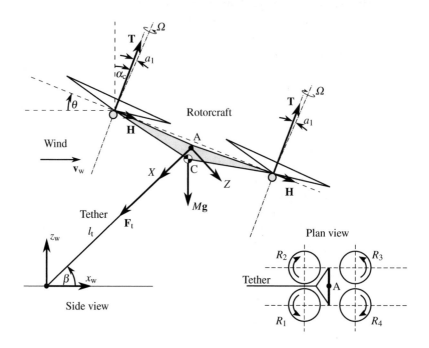

Fig. 23.8 Configuration of a typical quad-rotorcraft in side and plan view for the case of no cyclic pitch ($\theta = \alpha_c$), showing the wind reference frame (x_w, z_w) and the tether reference frame (X, Z)

rotation [16]. Thus each rotor rotates in an opposite direction to that of its two adjacent neighbours. With this particular arrangement the craft's pitch, roll and yaw can be controlled by the application of collective pitch changes to the rotors. No cyclic pitch action is necessary. This will help reduce construction costs and maintenance expenses. Variation of collective pitch thus changes the thrust developed by each rotor in a format described below using gyro-derived error signal data [13].

- Total craft thrust (i.e. craft altitude and power output) is controlled by collective pitch action on all rotors applied simultaneously by equal amounts.
- Roll control is by differential collective action between the port and starboard pair of rotors by an equal amount.
- Pitch control is by differential collective pitch action between the forward and rearward pair of rotors by an equal amount.
- Yaw control is through differential torque reaction. This by the application of differential collective pitch changes on pairs of opposite rotors by equal amounts.

It should be noted that there is a yaw control reversal at low wind speeds, so it is recommended that the differential collective action described above be used for hover and low wind speeds only. At wind speeds above the yaw reversal it is recommended a vertical stabilizer, namely a fin and rudder, be designed with sufficient control authority to enable the yaw control system described above to be disabled at higher wind speeds.

Tethered craft at high altitude have a further inherent advantage over ground-based wind turbines. This is their ability to reduce the effects of gust induced loads and torques. This is due to the flexibility of the tether cable which does not exist in the rigidly mounted ground-based equivalent. This flexibility arises from cable elasticity and from the change in cable shape under gust conditions. This inherent flexibility results in a very significant alleviation in the gust loads and torques applied to rotors, gearboxes etc. This alleviation is estimated to be more than an order of magnitude reduction. Further work is required on this matter.

23.7 Equilibrium Flight Performance of a Quad-Rotorcraft

In this section we will examine the equilibrium performance of a typical quad-rotorcraft in generating flight in reference to the configuration illustrated in Fig. 23.8. Various forms of rotor arrangements are conceivable, however, for simplicity of the analysis, a rectangular layout in plan view is assumed.

The rotorcraft is exposed to a steady wind of velocity v_w. The nose-up angle of the craft is denoted by θ, which is identical to the control axis angle α_c of the rotors because no cyclic pitch is used. The rotor's flapping angle a_1 is shown as the angle between the normal to the tip-path plane and the control axis. The total rotor thrust component along the control axis is T and normal to this axis is the component force H. To account for the aerodynamic drag of the fuselage the additional force D_{fus} is added. Because at equilibrium flight conditions the rotorcraft is not moving the drag of the fuselage is aligned with the wind velocity.

A single, straight tether of length l_t is attached at point A to the craft on its plane of longitudinal symmetry. This attachment point is the origin of the tether reference frame which has its X- and Z-axes aligned with and normal to the tether. The gravitational force Mg acts on the craft's center of mass which is denoted as point C. The points A and C may be coincident, but this is not a necessary requirement. The tether is assumed for simplicity to be massless, inextensible and with an infinitesimal diameter, at the present stage of this analysis. It is represented by a straight line. These simplifying assumptions for the tether are reasonable provided its length is around 300 m or less. However, for higher altitudes the analysis has been extended to include tether mass and the wind loads. The tether force F_t is aligned with the tether because of the straight-line assumption and the fact that the flexible tether can only transfer a tensile force.

A number of equilibrium studies have been made by Ho [6], Roberts [15] and Jabbarzadeh [8] for the twin and quad-rotor arrangements described above. These used the classical rotor theory of Gessow and Crim [4], which is applicable to the current high disk incidence angles at high inflow conditions. It was assumed for simplicity in all these studies that all four rotors were in identical operation and each are in isolated air flows. Thus, aerodynamic interference between rotors was not taken into account, nor any aerodynamic interaction with the fuselage. The resulting equilibrium force polygon for the quad-rotor system is shown in Fig. 23.9.

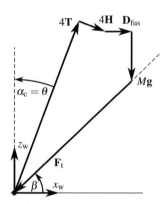

Fig. 23.9 Equilibrium force polygon for the quad-rotorcraft involving the total rotor thrust 4**T** and total H-force 4**H**, the tether force **F**$_t$, the craft gravitational force M**g** and the aerodynamic drag force on the fuselage **D**$_{fus}$

The power coefficient C_p and aerodynamic lift coefficient C_L of an individual rotor are shown in Figs. 23.10 and 23.11, respectively. It should be noted that C_p is normalized using the wind velocity, as in wind turbine practice, instead of normalized using the rotor tip speed, as in helicopter practice. Note that a positive sign of

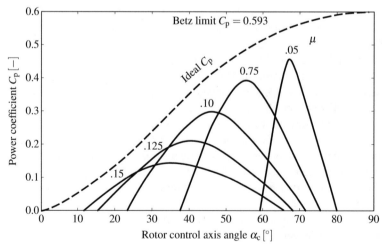

Fig. 23.10 Power coefficient C_p of a single rotor as function of the rotor control axis angle α_c, for various values of the tip speed ratio μ

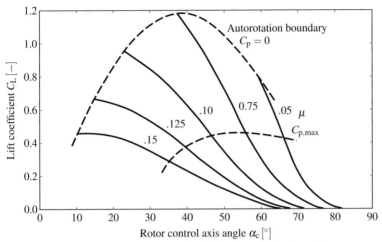

Fig. 23.11 Aerodynamic lift coefficient C_L of a single rotor as function of the rotor control axis angle α_c, for various values of the tip speed ratio μ

C_p implies a power output, this being opposite to that used in helicopter theory. The lift coefficient is normalized using the wind velocity. The tip speed ratio used is as defined in helicopter theory. Both figures were calculated by Jabbarzadeh [8] using a rotor solidity $\sigma = 0.05$, linear twist of the blades $\theta_1 = +8°$, Lock number $\gamma = 10$, tip loss factor $B = 0.97$ and operating all rotors at a retreating blade incidence limit, as normally defined, with a value 13°.

The dashed curve in Fig. 23.10 represents the ideal, maximum power output, that could be obtained for zero profile drag of the rotor blades. Hence the bell-shaped curves derived from Gessow and Crim [4], incorporating profile drag effects, will always lie below the ideal, dashed curve. The curves are terminated at $C_p = 0$ which represents autorotation conditions, where no power is being developed or supplied. The favored autorotation condition, to be discussed below, is one of the left-hand, abscissa crossings of the bell-shaped curves. Only the left-hand, zero crossings will be considered in what follows.

In Fig. 23.11 the C_L-curves all terminate on the uppermost dashed curve. At this limit the rotors are in autorotation, with the values of lift coefficient at their individual maxima and with the power output at zero. It can be seen that the maximum lift coefficient occurs at $\alpha_c \approx 40°$ and $\mu \approx 0.075$. In addition, it is important to note the following. The maxima of the bell-shaped curves shown in Fig. 23.10 are shown as a locus-line drawn as the lowermost dotted curve of Fig. 23.11. In other words, equilibrium operations are best performed anywhere between the two dotted curves shown on Fig. 23.11. The closer operations are made to the lower dotted curve the greater will be the power output, but the craft's available thrust will reduce as operations approach this lower dotted curve. Operations are possible below the abscissa of Fig. 23.10, but power must be supplied to the system to develop sustainable lift.

23.8 Best Autorotation Conditions

Autorotation relates physically to the flight condition where the system is on the point of collapse due to insufficient wind speed being available to support the craft and its tether without any input of power to the rotors. The left-hand side crossing of the bell-shaped curves with the ordinate axis in Fig. 23.10 implies that no power is being produced and all the wind's kinetic energy is being used to generate lift. A left-hand cutting at a lesser control axis angle than on the right-hand crossing is preferred, because this condition is more favorable from a tether viewpoint. This means that a lower nose-up attitude of the craft gives a tether closer to the vertical with a resulting lower tether length. The question then arises as to which of the left-hand crossings is most favorable to give the lowest wind speed to keep the system aloft in the critical autorotation condition? This minimum wind speed can be determined by considering vertical force equilibrium in Fig. 23.9. It can be shown that the ratio of the craft's weight disk loading to the free stream dynamic pressure,

$$\frac{\frac{Mg}{4\pi R^2}}{\frac{1}{2}\rho v_{\rm w}^2} = C_{\rm L}[1 - \tan(\alpha_{\rm c} + a_1)\tan\beta], \qquad (23.2)$$

is the important relevant parameter, with $\alpha_{\rm c}$ denoting the rotor control axis angle and a_1 denoting the rotor backward tilt angle. The left-hand side of this equation has to be organized to be at its maximum value in order to achieve the minimum wind speed for a given craft weight disk loading.

The variation of the weight disk loading to dynamic pressure ratio is shown in Fig. 23.12. A reasonable value for β may be, say, 35°. For this particular case, the

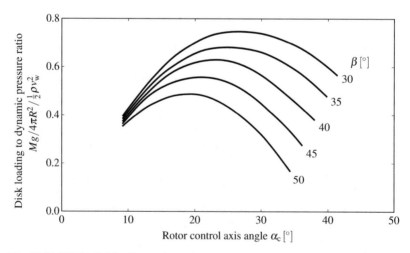

Fig. 23.12 Weight disk loading to dynamic pressure ratio as a function of rotor control axis angle $\alpha_{\rm c}$, for various values of the elevation angle β

best autorotation conditions can be read-off Fig. 23.12 to give a control axis angle of 25° at a corresponding tip speed ratio of 0.093. An extremely important conclusion can now be drawn from Fig. 23.12.

If a high-altitude craft, with 5% solidity, were to have a straight, massless tether arranged at an angle of, say, 35° to the horizontal, then for the craft to stay aloft the craft's weight disk loading to dynamic pressure ratio, as read from Fig. 23.10, cannot exceed 0.69. If we wish to fly it in autorotation at 15,000 feet in, say, a 10 m/s wind, then the weight disk loading must be at or less than 0.69×38.5 Pa, or 0.553 lb/ft^2. In other words, the craft's weight disk loading must be low by rotorcraft standards, and it must not exceed 0.553 lb/ft^2 to achieve an autorotation speed of 10 m/s at the nominated 35° cable angle.

In making the above statement it is realized that the tether has been assumed to be weightless. Of course, a change in the autorotation speed quoted above will be in proportion to the square root of any change in the weight disk loading, all other factors remaining the same. It should be noted that the disk loading is based on the weight Mg of the craft. Because the tether force acting on the craft is much larger than the gravitational force, it means that the craft's disk loading based on weight should be much less than that used on untethered rotorcraft.

23.9 Consideration of Tether Weight

We will now formally introduce tether weight to calculate the operating envelope for an example craft. Wind loads on the tether and the craft's fuselage drag will be here neglected for simplicity. However Roberts [11] has available full computer codes which do include these two effects. Put simply, it is considered here to be more explicit if we consider the cable mass without the imprecision of the actual tether's wind profile, along with the uncertainty of the craft's drag coefficient.

A preliminary system analysis, to be used for a demonstration of the analysis technique, can now be developed assuming that the tether is of uniform mass per unit length. Thus the tether forms a catenary attached at the point A shown in Fig. 23.8. Thus from point A in the craft, the tether drapes down to an anchor point on the ground.

Consider, for the demonstration, a quad-rotorcraft with rotors of radius $R = 12.35$ m (≈ 80 ft) with a solidity of $\sigma = 0.05$. In this example we use a NACA 0012 blade section with the conventional blade and rotor parameters. Four basic modes of operation can be defined:

Mode A: Rated power output in any wind above the rated speed.
Mode B: Rated power at rated wind speed.
Mode C: Part power output in light winds.

Mode D: Autorotation at the minimum sustainable wind speed.

Next, assume the craft is configured to give a rated power output of 3.1 MW at an altitude of 15,000 feet. A tether weight of 460 kg/km has been assumed, using Kevlar as the tensile member and incorporating twin, insulated aluminum conductors. A combined electrical efficiency for the generator and tether transmission has been taken as 90%. The result would be a tether about 15 mm in diameter with the Kevlar stressed to an adequate and safe level.

A central aspect of the design would be operation in mode B. Here the craft is best at a nose-up attitude of 47° at a wind speed of 25.8 m/s. It then develops the rated power of 3.1 MW. This produces a tether tension of 300 kN. At any wind speed greater than 25.8 m/s, such as in mode A, the system should not exceed its peak rating both electrically and structurally. Thus the craft maximum power output and its maximum tether tension have been frozen at the above values, never to be exceeded.

The craft weight has been estimated to be 3135 kg. This gives the weight disk loading for the vehicle of 0.333 lb/ft^2 (c.f. this value with the 0.553 lb/ft^2 statement in the previous section). In this example craft, the rated output power loading is 150 W/ft^2.

The system characteristics for the different modes of operation are shown in Table 23.1. Mode C is shown for part-power operation in a wind of 18.3 m/s, while

| | Operating Mode | | | |
Description	A	B	C	D
Electrical power output, MW	3.13	3.13	1.26	0
Altitude of craft, km	4.57	4.57	4.57	4.57
Incidence of rotors, deg	27.4	47.0	49.4	26.0
Wind speed, m/s	36.6	25.8	18.3	10.2
Total craft mass, kg	3135	3135	3135	3135
Mass of tether, kg	2592	3846	4402	4620
Maximum tether tension, kN	300	300	232	61
h/y at or above wind speed	570	2280	4330	6950

Table 23.1 System characteristics for different modes of operation

mode D is autorotation at the lowest sustainable wind of 10.2 m/s. The hours of operation at or above the wind speeds for each mode have been extracted from the Weibull charts for Albany in the USA [10], or for Moree in Australia [8] at an altitude of 15,000 feet. These two sites are almost identical in their wind probability data.

The above results can be used to construct a power-duration curve for the system. This gives an annual capacity factor of about 50% with a total annual energy output of 13.3 GWh. The percentage of time per annum that the craft would need to be landed is 20.3%.

Finally, it is interesting to calculate side elevations of the craft and its tether in the modes A through D. These are given in Fig. 23.13, again assuming no aerodynamic loading on the tether, but solely taking into account the effect of gravity.

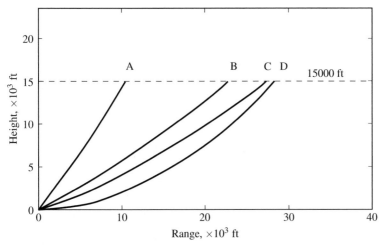

Fig. 23.13 Side elevations of the craft and its tether in the modes A through D

23.10 Quad-Rotorcraft Stability and Control

It is well-known that rotorcraft are inherently unstable if left uncontrolled. A control strategy is therefore essential for the operation of the system, stabilizing pitch, roll and yaw, with pitch being the important variable.

We will discuss longitudinal and lateral stability assuming for simplicity that the tether is straight, massless and inextensible, as shown in Fig. 23.8. This assumption is reasonable, provided the tethering cable is relative light and short in length, namely about 100 m. If the tether is straight and inextensible, the craft's longitudinal motions will be solely pitch and heave, with the point A in Fig. 23.8 moving only in tangential direction (Z-axis), perpendicular to the tether. Motion in the radial diraction (X-axis) is not possible. Lateral motion (Y-axis) can also be examined under the same assumptions. On the above basis the stability and control can be simply examined without the introduction of tether dynamics. Of course the tether could be included, but considerably more complication would be required.

Ho [6], Strudwicke [20] and Roberts [15] have studied the longitudinal stability of a twin-rotorcraft, while Roberts [15] has extensively studied both the longitudinal and lateral stability of the quad-rotor system. For the quad-rotor system, under the assumptions given above, the coupled pitch θ and z-motions from an equilibrium position of the rotorcraft are coupled and unstable, when the craft's controls are fixed. However, the system can be easily stabilized using a proportional and damped (PD) controller using differential collective pitch action on the rotors, as described in Sect. 23.6. This form of controller uses as its input only the error-signal between the actual and desired pitch angles of the craft. This error-signal is used to action the four rotors' collective pitch angle, differentially from their equilibrium positions, by an equal positive or negative amount of magnitude $\Delta\theta_0$.

The equations of motion for small perturbations of the rotorcraft from its equilibrium configuration, namely its position, pitch angle (nose-up angle) θ and rotor collective pitch θ_0, can be studied by firstly setting-up the relevant rotor force derivatives for disturbed flight from the equilibrium conditions. Begin by calculating the force derivatives X_u, Z_u, X_w, Z_w, X_q, Z_q, X_θ and Z_θ in the wind reference frame following standard texts, such as Bramwell [2]. It should be noted that Bramwell's formulation uses x- and z-axes in opposite direction to the x_w- and z_w-axes of the wind reference frame shown in Figs. 23.8 and 23.9. In addition, the control force derivatives, $T_{\theta,0}$ and $H_{\theta,0}$, can be computed.

The next step in the calculation schedule is to transform the rotor force derivatives to the tether reference frame, using the equations of Seckel [19, p. 463]. For the purpose of this perturbation analysis the tether reference frame is spacially locked to its equilibrium state. With reference to Fig. 23.8 this transformation from wind to tether reference frame is a counter-clockwise rotation with an angle $180° + \beta$. This transforms all the force derivatives into perturbations associated with displacements and motions measured in the tether reference frame (X, Z). The reason for this transformation, as mentioned above, is that any x-perturbation is eliminated, if the tether is straight and inextensible. Therefore, the current analysis reduces from a system having three degrees of freedom to one of only two degrees, namely simply studied with z- and θ-perturbations.

The next step in the rotor calculations is to sum the derivatives for all of the four rotors to calculate the forces in total on the vehicle. Also in addition, calculate the moment derivatives knowing the physical size and configuration of the rotors in the craft's airframe. Care should be taken to express all moment derivatives as moments about the craft's center of mass. Referring to Fig. 23.8 the center of mass C is generally not coincident with the attachment point A. This latter point is important when compiling the rotorcraft's equations of motion for small perturbations from the equilibrium position. The equations of motion can now be written as

$$\mathbf{M}\begin{bmatrix} \ddot{z} \\ \ddot{\theta} \end{bmatrix} + \mathbf{D}\begin{bmatrix} \dot{z} \\ \dot{\theta} \end{bmatrix} + \mathbf{K}\begin{bmatrix} z \\ \theta \end{bmatrix} = \begin{bmatrix} \delta \\ \eta \end{bmatrix}\Delta\theta_0, \tag{23.3}$$

$$\theta_{0;1,2} = \bar{\theta}_{0;1,2} + \Delta\theta_0, \tag{23.4}$$

$$\theta_{0;3,4} = \bar{\theta}_{0;3,4} - \Delta\theta_0, \tag{23.5}$$

where $\mathbf{M}$ represents the mass matrix, $\mathbf{D}$ the damping matrix and $\mathbf{K}$ the stiffness matrix. This matrix equation represents a double-output, single-input control system. Such a system is often called a "follower system", where in the present case the displacement z simply follows the value of θ. The right-hand side column matrix shows the force and moment control derivatives, where the craft's control derivatives are the coefficients δ and η. Now because we have chosen the rotor arrangement as in Fig. 23.8 and having incorporated the differential collective pitch action, as defined in Eqs. (23.4) and (23.5), it follows that $\delta = 0$. However, the η-term, being the control moment, has a finite value, which in effect controls the craft's pitch attitude.

The matrices on the left-hand side of Eq. (23.3) can be derived using the appropriate derivatives involving the craft's position, pitch angle, velocity, pitch rate and the acceleration of point A. The **D**-matrix is determined using the above rotor derivatives, the **M**-matrix by using the mass and moment of inertia of the craft noting that the point A may not be coincident with C, while the **K**-matrix is essentially determined by the tether tension and the perturbation of point A in the Z-direction.

In Eqs. (23.4) and (23.5) a collective pitch change, $\Delta\theta_0$, is applied equally to the R_1 and R_2 rotors, while an opposite change of the same magnitude is applied to rotors R_3 and R_4. After considerable work it has been found that for a craft of almost any size, it can be stabilized by the application in a PD controller with a proportional gain of about 0.1 to 0.2° of collective pitch change per degree of error in the craft's pitch angle. This value of controller gain is strongly dependent on the distance between the fore and aft rotor mounting in the fuselage. A damping term in the controller could also be useful.

A similar philosophy for the control of the roll and yaw should lead to a favorable outcome. However, the control of yaw is subject to the yaw reversal effect discussed in Sect. 23.6. Differential collective pitch for yaw control, in or near hover, has the above mentioned control reversal. To counter this effect a vertical stabilizer (vertical fin) and rudder should be used for yaw control when generating power in windy conditions. In the latter condition yaw control by differential collective would be disabled.

23.11 Conclusions

It has been shown from atmospheric data that that the wind speed and wind power increases with increasing altitude, up to the tropopause level. In order to harness this enormous energy a quad-rotorcraft has been proposed and analyzed.

Graphs are shown in Figs. 23.10 and 23.11 for the power and lift coefficients as functions of control axis angle, parametrized by the tip speed ratio, for a rotor solidity of 5%. Other solidities would give similarly shaped graphs. In Fig. 23.11 it can be seen that realistic operations can occur anywhere between the two dashed curves therein. Fig. 23.12 shows for various tether elevation angles how different weight disk loadings to dynamic pressure ratios are necessary in order to maintain operations at the limiting autorotation condition.

For demonstration purposes the above theory has been applied to a sample craft operating at an altitude of 15,000 feet. This altitude has been chosen simply because it is well established that an electro-mechanical tether to this altitude is feasible. These altitudes have been used twenty four hours a day, seven days a week, for border protection duties in the USA for some years. In this situation the tether is attached to and restrains the tethered balloon. Next a 3.15 MW quad-rotor system at 15,000 feet has been chosen to demonstrate that the technology is feasible, but this example is not proposed as the optimal altitude for any construction. This craft has been shown to give a generating capacity factor of 50%. It is suggested here that operations at somewhat higher altitudes, namely 20,000 to 25,000 feet could give significant power outputs at capacity factors of between 70 and 80%.

The chapter concludes by examining the stability and control of a quad-rotorcraft. This theory is applicable at any trim rotor incidence and rotor tip speed ratio. It is shown that differential collective pitch action on the rotors can control the rotorcraft in pitch, roll and yaw. However, the yaw control theory confirms that a control inversion occurs early in the craft's operating range. To avoid yaw difficulties, differential collective pitch action is proposed only for low wind speeds. At higher wind speeds a conventional vertical stabilizer and rudder is proposed, with the collective pitch action disabled.

Editors note After the compilation of this chapter a very interesting contemporary analysis of the gyrocopter-type airborne wind energy system has been published by Rancourt et al. [11].

References

1. Atkinson, J. D. et al.: The Use of Australian Upper Wind Data in the Design of an Electrical Generating Platform. Charles Kolling Research Laboratory Technical Note TN D-17, 1–19 (1979)
2. Bramwell, A. R. S.: Helicopter Dynamics. Edward Arnold (Publishers) Ltd., London, UK (1976)
3. Fletcher, C. A. J., Roberts, B. W.: Electricity generation from jet-stream winds. Journal of Energy 3(4), 241–249 (1979). doi: 10.2514/3.48003
4. Gessow, A., Crim, A. D.: An extension of lifting rotor theory to cover operation at large angles of attack and high inflow conditions. Technical Report NACA TN-2665, Langley Aeronautical Laboratory, Langley Field, VA, US, Apr 1952. http://naca.central.cranfield.ac.uk/reports/1952/naca-tn-2665.pdf
5. Gessow, A., Myers Jr., G. C.: Aerodynamics of the Helicopter. Macmillan Co., New York, NY (1952)
6. Ho, R. H. S.: Lateral Stability and Control of a Flying Wind Generator. M. E. (Res) Thesis. M.Sc.Thesis, University of Sydney, Nov 1992. http://hdl.handle.net/2123/2609
7. Hoffert, M. I., Caldeira, K., Jain, A. K. et al.: Energy Implications of Future Stabilization of Atmospheric CO_2 Content. Nature **395**, 881–884 (1998). doi: 10.1038/27638
8. Jabbarzadeh Khoei, A.: Optimum Twist for Windmill Operation of a Tethered Helicopter. M. E. Studies Thesis. M.Sc.Thesis, University of Sydney, Aug 1993. http://hdl.handle.net/2123/2608
9. Manalis, M. S.: Airborne Windmills and Communication Aerostats. Journal of Aircraft **13**(7), 543–544 (1976). doi: 10.2514/3.58686
10. O'Doherty, R. J., Roberts, B. W.: The Application of U.S. Upper Wind Data in One Design of Tethered Wind Energy Systems. SERI/TR-211-1400, Solar Energy Research Institute, Golden, CO, USA, Feb 1982. doi: 10.2172/5390948

11. Rancourt, D., Bolduc-Teasdale, F., Demers Bouchard, E., Anderson, M. J., Mavris, D. N.: Design space exploration of gyrocopter-type airborne wind turbines. Wind Energy **19**, 895–909 (2016). doi: 10.1002/we.1873
12. Roberts, B. W.: Quad-Rotorcraft to Harness High Altitude Wind Energy. In: Schmehl, R. (ed.). Book of Abstracts of the International Airborne Wind Energy Conference 2015, pp. 84–85, Delft, The Netherlands, 15–16 June 2015. doi: 10.4233/uuid:7df59b79-2c6b-4e30-bd58-8454f493bb09. Presentation video recording available from: https://collegerama.tudelft.nl/Mediasite/Play/102d7cb3437542acbf4078bac1e853eb1d
13. Roberts, B. W.: Control System for A Windmill Kite. Australian Patent 2009238195, Apr 2009
14. Roberts, B. W.: Design and Preliminary Performance of the Gyromill Mk2. End of grant report 380, Department of Resources and Energy, Canberra, Australia, Oct 1984. http://nla.gov.au/nla.cat-vn2242481
15. Roberts, B. W.: Private papers
16. Roberts, B. W.: Windmill Kite. US Patent 6,781,254, Aug 2004
17. Roberts, B. W., Blackler, J.: Various Systems for Generation of Electricity Using Upper Atmospheric Winds. In: Proceedings of the 2nd Wind Energy Innovation Systems Conference, pp. 67–80, Solar Energy Research Institute, Colorado Springs, CO, USA, 3–5 Dec 1980
18. Roberts, B. W., Shepard, D. H., Caldeira, K., Cannon, M. E., Eccles, D. G., Grenier, A. J., Freidin, J. F.: Harnessing High-Altitude Wind Power. IEEE Transactions on Energy Conversion **22**(1), 136–144 (2007). doi: 10.1109/TEC.2006.889603
19. Sechel, E.: Stability and Control of Airplanes and Helicopters. Academic Press, New York (1964)
20. Strudwicke, C. D.: A Control System for a Power Generating Tethered Rotorcraft. M. E. (Res) Thesis. M.Sc.Thesis, University of Sydney, Nov 1995. http://hdl.handle.net/2123/4993

Chapter 24
Pumping Cycle Kite Power with Twings

Rolf Luchsinger, Damian Aregger, Florian Bezard, Dino Costa, Cédric Galliot, Flavio Gohl, Jannis Heilmann, Henrik Hesse, Corey Houle, Tony A. Wood and Roy S. Smith

Abstract Pumping cycle kite power has attracted considerable interest over the last years with several start-ups and research teams investigating the technology. While all these groups produce electrical power with a ground-based generator in a cyclic process, there is no consent about the shape, structure and control of the flying object. In particular the launching and landing strategy has not been settled yet. TwingTec has followed a pragmatic approach focusing on the flying part of the system. The spin-off from Empa and FHNW has developed over the last years in close collaboration with leading research institutes from Switzerland the twing, an acronym for tethered wing. The guiding principle behind the design of the twing was to combine the light weight property of a kite with the aerodynamic properties of a glider plane. Launching and landing was solved by integrating rotors into the structure allowing the twing to hover. Launching, transition into crosswind, autonomous power production, transition into hover and landing has been demonstrated with the current small-scale test system.

Rolf H. Luchsinger (✉) · Dino Costa · Cédric Galliot · Flavio Gohl · Corey Houle
TwingTec AG, Überlandstrasse 129, 8600 Dübendorf, Switzerland
e-mail: rolf.luchsinger@twingtec.ch

Rolf H. Luchsinger · Florian Bezard · Dino Costa · Cédric Galliot · Flavio Gohl
Empa, Center for Synergetic Structures, Überlandstrasse 129, 8600 Dübendorf, Switzerland

Damian Aregger · Corey Houle
FHNW, Institute of Aerosol and Sensor Technology, Klosterzelgstrasse 2, 5210 Windisch, Switzerland

Henrik Hesse · Tony A. Wood · Roy S. Smith
ETH Zurich, Automatic Control Laboratory, Physikstrasse 3, 8092 Zurich, Switzerland

© Springer Nature Singapore Pte Ltd. 2018
R. Schmehl (ed.), *Airborne Wind Energy*, Green Energy
and Technology, https://doi.org/10.1007/978-981-10-1947-0_24

24.1 Introduction

Looking for efficient ways to convert renewable energy sources into electrical power becomes more and more important. Airborne wind energy is a new approach to harness the power of the wind. Among the various concepts, pumping cycle kite power has attracted considerable interest over the last years. Several start-ups and research teams investigate this technology, particularly in Europe. The basic concept of pumping cycle kite power is well understood and theoretical and experimental investigations have revealed the potential of this technology [2, 5, 14, 16]. However, there are still some key elements of the technology where there is so far no consent among the different teams on how to solve them. In particular, the design of the kite and the launching and landing concept are pursued in very different ways.

Several teams operate with flexible tube and foil kites. These kites are controlled either by the ground station through a multiline configuration [3, 17] or by means of a control pod below the kite [11, 20]. The advantages of soft kites are their minimal weight, their stable flight behavior and that they can be manufactured at low cost building on experiences of the surf kite industry. Furthermore, they are comparatively crash resistant. The disadvantages are their poor depower behavior which is key for an efficient pumping cycle system [16] and the very limited life-time of the fabrics involved. Overall, soft kites are interesting models to investigate basics concepts and control algorithms, but at the presently achieved state of material durability are considered inferior for commercial systems which have to run continuously for many years in order to be economical.

The other end of the application spectrum is defined by rigid gliders with all the control surfaces of an airplane. In such a setting, the main control authority is shifted from the ground station to the wing [18]. Rigid wings can certainly meet the demands for the aerodynamic performance and durability, but with more than $9\,\mathrm{kg/m^2}$ [18] their weight per wing surface area limits their performance at low wind conditions. More severe, it is up to now not clear how launching, landing and relaunching can be accomplished fully autonomously without human interaction.

TwingTec is convinced that the ideal wing for pumping cycle kite power is a synergetic combination of the light weight property of the surf kite with the aerodynamic and structural properties of the glider.

Initial work was focused on increasing the stiffness of tube kites [4]. Detailed simulations with our tool KiteSim 2.0 revealed that precise pitch control is instrumental to fly efficient pumping cycles [12]. This brought us to a plane like configuration, where pitch is with an on-board activated elevator controlled and eventually to our twing design. In all these steps we have ensured that the weight of the twing is kept as low as possible.

With respect to launching and landing, we are convinced that only active systems can fulfill the requirements of a commercial pumping cycle kite power system. By investigating a number of different approaches, we came to the conclusion that a system where rotors are integrated into the twing is the best option to fulfill all the necessary requirements. With such a tricopter design, the twing hovers during launching and landing, ensuring full control authority during these critical phases.

The transitions into and out of the pumping cycles are done at high elevations enabling sufficient time and space for these processes. Finally, motor thrust during hover can be augmented with the aerodynamic forces of the wind resulting in increased stability of the launching and landing maneuver.

This paper gives an overview of our small-scale TwingPower test system which has been built up over the last years in close collaboration between TwingTec, Empa, FHNW and ETH. In Sect. 24.2 we will describe the developed mobile platform composed of a ground station, the aerodynamic lifting devices and the control system. In Sect. 24.3 we describe the various test procedures, such as hardware in the loop testing, tow testing and flight tests with a full system setup. In Sect. 24.4 we explore the market and possible applications before we draw the conclusions in Sect. 24.5. The preliminary content of the present chapter has been presented at the Airborne Wind Energy Conference 2015 [15].

24.2 Mobile Test Platform

24.2.1 Ground Station

A dedicated ground station for testing twings and advanced control algorithms was designed and constructed at FHNW during 2013. The design and components used were largely based on the previous ground station, which was developed during the SwissKitePower project, but with only two tethers instead of three. With a lightweight, aluminum frame, the ground station can easily be lifted and secured onto the back of a truck, allowing for easy transportation for testing. Images of the ground station during a field test are shown in Fig. 24.1.

Fig. 24.1 Mobile ground station on the back of a truck (left) and in detailed view (right)

In order to maximize flexibility, two independent winch drives were used, each with a 15kW servo motor, connected to a drum via a coupling and gearbox. Each winch has its own level-winding mechanism, also driven by an independent servo drive, allowing for easy adjustment of the spooling pitch, depending on the size of the tethers used. A schematic of the main electrical and automation components is shown in Fig. 24.2.

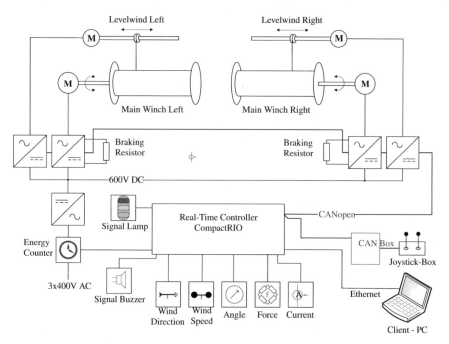

Fig. 24.2 Schematic view of the main electrical and automation components of the ground station

All four drives are powered from a common 600 V DC bus, which is supplied from a 3-phase, 400 V AC connection via a passive rectifier. This can be fed from a standard, 16 A/32 A grid connection, or from a fuel-based generator. During the power phase, the energy produced by the motors acting as generators is dissipated over a set of brake resistors, connected to brake choppers, integrated in the main drives. Each brake resistor is rated for a continuous output of 10 kW. As shown in the figure above, the automation software is implemented in NI Labview on a CompactRIO (cRIO) real-time controller. A number of sensor signals are read directly into the cRIO via its expandable I/O hardware, while other inputs and outputs are sent over a CANopen bus. All sensor inputs are acquired at 100 Hz by the main control loop, but an integrated FPGA allows for certain signals to be acquired and pre-filtered at 50 kHz. The basic control functionality that is implemented on the cRIO is divided into reeling and steering control, as shown in Fig. 24.3.

Control of the winches is divided into two main tasks: reeling and steering control. The functionality of the two winches is separated in order to simplify the control tasks, with a master winch (left) and a slave winch (right). Reeling control refers to the control of the speed and torque of the master winch (left), while steering control adjusts the speed of the slave winch (right), in order to maintain a constant

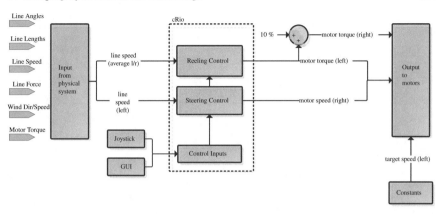

Fig. 24.3 Basic control functionality implemented on the cRIO

tether offset. In order to maintain a minimum tether force, and wing loading, the master winch is always supplied with a reel-in speed command and its torque limit is adjusted depending on the reel-out speed achieved. A simple mapping is used to generate a force set point from the reel-out speed, which is converted into a feed-forward torque command supplied to the master winch drive. An additional feedback controller using the measured line force adds or subtracts addition torque from the output to achieve better control over the line force, accounting for friction, inertia and errors in the drive torque estimate.

24.2.2 Twings

The development of twings is a key know-how which has been built up over the last years. First twings were flown in 2012, as illustrated in Fig. 24.4, making us the first team to demonstrate that two-line control works well with rigid wings. From these early versions onward, pitch was controlled by means of an activated onboard elevator. The design was subsequently improved. The airfoil was changed to a high lift airfoil and the load per wing area was increased. In parallel the in-house manufacturing process has been modified. The current twing design relies solely on materials proven in the aviation industry which is important to estimate the life-time of the wing structure. In order to minimize the production efforts a rectangular wing shape was chosen. Coupled numerical simulations including cost analysis with KiteSim 2.0 [12] have revealed, that high aspect ratios beyond 10 do not constitute a major benefit, as tether drag becomes dominant. Currently, aspect ratios in the range of 5 to 8 are employed. Using a single tail-boom instead of two has simplified the design of the tail and improved its overall stiffness. With better designed connections and by integrating the electronics on the tail-boom, the assembling time of the twing

Fig. 24.4 Early prototype of a twing (2012)

on the field for flight tests has been also significantly reduced. Twings with wing surface areas ranging from $1.3\,\text{m}^2$ to $6\,\text{m}^2$ have been built so far.

Launching of these early versions of twings was accomplished by towing up the wing. A dedicated launch rig was built, which made physical interactions of the test team with the wing obsolete, a key element in order to increase safety of this procedure. For landing, the wing was brought to the edge of the wind window, where it was gliding towards the ground. This procedure allowed us to achieve a lot of flight hours with twings, to optimize the flight pattern and the autonomous control system. Nevertheless, it was clear from the very beginning that such a launching and landing concept is not feasible for a product, which has to operate fully autonomously. Launching needs a lot of space in particular on a spot with changing wind directions as the wing needs to be launched mainly against the wind. The landing procedure has proven to be critical in particular in gusty winds.

Passive launching and landing in the ambient wind is also not feasible for rigid wings due to the increased weight per area compared to soft kites. High winds needed for launching contradicts the promise of kite power that stronger winds at higher altitude can be accessed while there are low winds at the ground. A stable and reliable launching and landing concept which works under all wind conditions is a key element for pumping cycle kite power. Based on detailed simulations [12] it appeared that the most controllable way to launch and land the twing was to integrate rotors onto the structure for this purpose. This allows us to increase the overall robustness of the launching and landing process. The drawback of this approach is that the overall weight of the twing is increased. However, since the twing structure was

Fig. 24.5 Twing with integrated launching and landing concept in hover mode (2015)

designed from the very beginning for minimal weight, we are able to handle this, as successful demonstration of hover during launching, transition into crosswind flight, autonomous pumping cycles and transition back to hover mode for landing reveals. A tricopter concept was chosen as it allows for an easy integration of the rotors into the structural design of the twing, as shown in Fig. 24.5. During hovering, the rotors are controlled with an on-board controller based on signals from an IMU.

24.2.3 Control

In this section we outline the development of a steering controller to achieve autonomous pumping operation of twings following the control strategies developed in [9] and [21] for the traction and retraction phases, respectively. The resulting control approach uses only ground-based line angle and length measurements to compute a steering input applied at the ground station in form of differential line length. A crucial feedback variable for the developed control approach is the twing velocity vector orientation, γ, commonly referred to as velocity angle [9]. It represents the kite heading and is defined as the angle between the projection of the twing velocity vector onto the tangent plane at the twing position and the local North pointing to the zenith. Regardless of the complexity of the controlled system—twings or soft kites—the definition of the velocity vector orientation describes the wing state in one scalar and is well suited as a feedback variable. Since the twing position and velocity vector orientation cannot be measured directly, we estimate the feedback variables from line angle measurements at the ground station based on a Kalman filtering approach using a kinematic model [10]. Details on definition and estimation of the velocity vector orientation for ground-based systems are given in Chap. 17.

To control twings during traction and retraction phases we use the hierarchical control scheme of [9] which consists of a cascaded control architecture, as shown in Fig. 24.6. Based on the estimated twing position and wind direction at flight al-

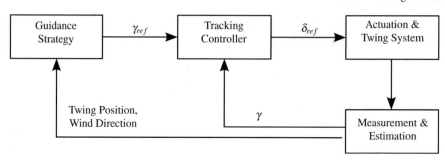

Fig. 24.6 Pumping cycle scheme overview for steering controller

titude, a high-level controller computes a reference velocity vector orientation, γ_{ref}, which is tracked using a proportional controller. The resulting steering input, γ_{ref}, is commanded at the ground station as differential line length to actuate the twing. As a first approximation we measure the wind direction at the ground station but the developed control approach is significantly improved using estimates of the instantaneous wind conditions at flight altitude. The latter can be estimated from the shape of the figure-eight trajectories during the traction phase or by maximizing the average power over one figure-eight following the approach in [22]. This optimization approach indirectly aligns the target points with the wind window.

Because of the cascade in the described control approach, the different phases of the pumping cycle can be implemented in the guidance strategy, as illustrated in Fig. 24.7. During the traction phase a target switching strategy, as developed in [9], is used to achieve figure-eight trajectories. Given the estimate of the twing position and two user-defined target points, a reference velocity vector orientation, γ_{ref}, is computed and low-pass filtered using a Butterworth filter. Switching azimuth angles have been defined to alternate between the target points leading to figure-eight uploop paths in the power zone unreeling the tethers under high traction forces. Note that the steering control in this work is decoupled from the torque-based reeling controller implemented on the GS.

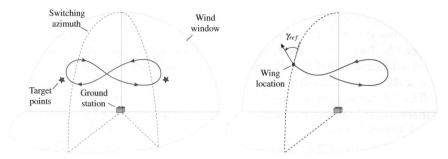

Fig. 24.7 Guidance strategy for steering controller: traction phase (left) and retraction phase (right)

Once the twing has reached a defined maximum line length, the guidance strategy switches to the retraction phase which is also illustrated in Fig. 24.7. During retractions the twing is actively depowered using an elevator and guided towards the side of the wind window where the tethers are reeled in under low tension. The elevator can be used to optimize the power output and line tension during traction and retraction phases. In this work however we followed a decoupled approach where a constant elevator setting is set for each phase and the line tension is controlled through the torque-based reeling controller. As the wing velocity in the tangent plane converges to zero during the retraction, we use a regularized version of the velocity vector orientation, introduced in [21], during this maneuver. This control strategy was found to be more robust against line sag compared to another approach in [21] which purely controls the twing elevation and magnitude of the azimuth angle during retractions.

Unlike conventional soft kites, which tend to slowly adapt to changes in aerodynamics, twings behave like glider aircraft which can deteriorate the controllability during the retraction phase and lead to excessive oscillations during this system critical phase. We therefore initiate the retraction phase in the power zone to reduce delay in the estimation of the twing velocity vector orientation. At the end of the retraction phase the twing is guided directly to the power zone to ensure sufficient line tension before the traction phase is initiated.

The resulting control approach is able to achieve autonomous pumping cycles with few tuning parameters. Experimental results in Figs. 24.8 and 24.9 demonstrate the implementation of the pumping cycle controller for a $3\,\text{m}^2$ twing operating between 100 and 150 m line length. At 346 s the guidance strategy switches to retraction, as shown in Fig. 24.9, and the twing is depowered and stabilized at a reference of 0.35 rad using the regularized version of the velocity vector orientation. At 354 s the tether has been recoiled to 100 m and the guidance strategy switches from retraction to traction phase repeating the pumping cycle.

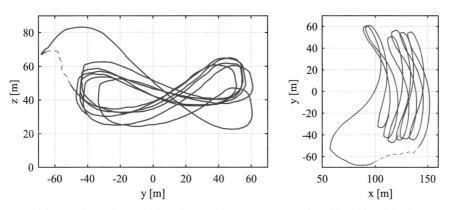

Fig. 24.8 Pumping cycle results from line angle measurements projected in the inertial reference frame with its origin at the ground station. Traction phase with figure-eight paths in blue solid and retraction phase in orange dashed

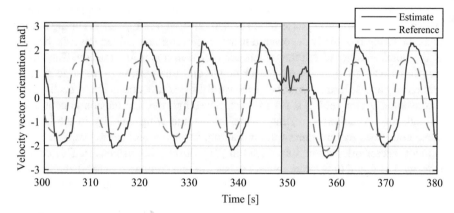

Fig. 24.9 Time history of velocity vector orientation tracking over one pumping cycle shown in Fig. 24.8. During retraction (346–354 s) the regularized version of velocity vector orientation is shown

To ensure extended autonomous operation we further model the steering dynamics of the twing as a delayed dynamical system [1]. The model information, which is identified online from measured data, is explicitly utilized to schedule the steering gain of the proportional tracking controller to adapt to varying operating conditions, e.g. due to varying wind conditions or uncoiling of tethers. The adaptive model-based control approach in [1] can further incorporate information about system delay, due to actuation, estimation and tether dynamics, to improve the tracking performance during the traction phase. This is particularly relevant to ground-based airborne wind energy systems operating at high altitudes.

24.3 Test

A crucial part of the system development is the validation of analytical results and simulation data. Often there exists a gap between theory and real life test data because of unsteady wind conditions or hardware constraints for example. At present, classic design and testing strategies for airborne wind energy system development include analytical calculations, modeling, simulation, wind tunnel testing, tow testing or field testing. In the following sections a more detailed insight of different test strategies is given.

24.3.1 Hardware in the Loop Test (HIL)

Field testing the entire system requires comparably high resources because of transport and setup overhead. Test spots might be not accessible the whole year round and wind conditions are changing all the time. Therefore as much system or subsystem testing should happen prior to a field test in order to maximize the field testing outcome. A possible solution is the integration of real hardware subsystems into simulation, called hardware in the loop (HIL) testing. In a HIL simulation one splits the system into two parts, one subpart is referred to as the system under test (SUT). This part can be a controller or an actuator that exists in reality. The other subpart is referred to as the plant simulation. This part is a mathematical model of a dynamic system that interacts with the SUT.

A ground generation based airborne wind energy system consists in general of three main components: The ground station, the tether and the kite. For the particular experimental HIL setup a testing method including two ground stations was developed. The first ground station GS1 was built in the course of the SwisskitePower Project [19], whereas the second ground station GS2 was commissioned by TwingTec AG for the current project. The two ground stations are directly connected by means of two tethers. This number originates from TwingTec's two line kite setup. Because of the tether connection between the two ground stations a physical force interaction is possible. In addition, TwingTec's simulation tool KiteSim 2.0 [12] has been employed.

In the context of the HIL terminology GS2 acts as the SUT whereas GS1 acts as a kite emulator. Steering inputs to GS2 result in certain tether displacements. These tether movements are fed to the simulator, which in turns calculates the resulting tether forces F_{sim} at every simulation time step, see Fig. 24.10. These calcu-

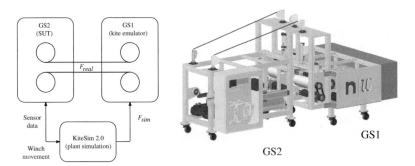

Fig. 24.10 Schematic of the HIL testing setup (left). GS2 acting as the SUT is connected via two tethers (red and blue) to GS1 emulating the flying kite. Both ground stations communicate with a plant simulation (kite and tether model). CAD rendering of the HIL setup (right)

lated forces are then translated into physical tether pulling forces F_{real} by GS1. The physical line forces finally translate into line movements on GS2, which closes the hardware loop.

Performing tests with the concept described above allows operating the ground station weather independent and 24 hours per day. Not only will there be the true ground station dynamics fed into the simulator but also all the real ground station interfaces are in operation as when testing in reality with the full test setup. On the other hand the kite and tether model do not represent the fully realistic flight hardware and show differences to the real world dynamics. Figure 24.11 shows different quantities of a HIL test performed in April 2015. The simulated $1.5\,\mathrm{m}^2$

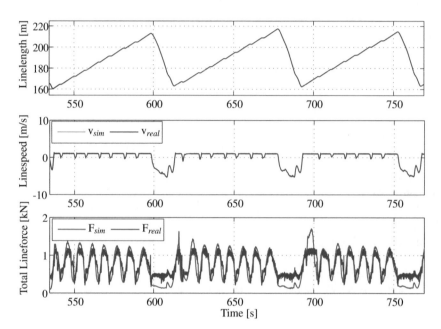

Fig. 24.11 Samples of simulated line length (top), simulated and real line speed (middle - both curves match very close therefore only one shape is visible) and simulated and real line force (bottom) from a fully autonomous pumping HIL test

rigid wing was flying virtual pumping cycles fully autonomously.

At the moment with the HIL setup we carry out pre-field test verifications as for example new ground station software implementations or autopilot controller strategies. In the future a ground station performance optimization such as for example electric drive train efficiency, winching scheme, force control optimization, up to long run software and hardware stability verifications are planned.

24.3.2 Tow Test

An important tool we have developed for the validation of the performance of the twings are tow-tests, as illustrated in Fig. 24.12. In such tests, the twings are pulled by a vehicle over a long track in order to generate apparent wind, ideally on a windless day. This way we can control the apparent wind conditions in a range from 0 to 30 m/s. Measuring the lift as a function of apparent airspeed and angle of attack gives valuable insight into the aerodynamic performance of the twings. Furthermore, the structural performance can be validated under high loads. We have even flown twings on tethers in such a setup. This allowed us to improve and validate e.g. the stability of our launching and landing concept, in particular by gradually moving towards stronger wind conditions.

Fig. 24.12 Tow test on the former military airfield in Dübendorf, Switzerland, with a 6 m² twing

24.3.3 Full System Flight Test

While component testing with a HIL setup or tow-tests help to speed up the development process, the ultimate tests are what we call full system flight tests. In such tests, the reliability and performance of the system can be investigated and improved in the real operational environment. TwingTec has obtained permission of BAZL (the Swiss equivalent of the FAA) for five different test sites in different parts of Switzerland, where we are allowed to fly depending on the site between 150 m and 300 m above ground. Since 2014 we also started to do flight tests in Valkenburg, the Netherlands, in cooperation with TU Delft. As described in Sect. 24.1, our ground station can be easily transported on the back of a truck making us flexible in terms of choosing the site with the optimal wind and weather conditions. A view of a full system flight test in the Western part of Switzerland is shown in Fig. 24.13.

Fig. 24.13 Full system flight test

A set of data of a twing flying autonomous pumping cycles is given in Fig. 24.14. The data was collected during the autonomous flight of a $3\,\text{m}^2$ twing over a period of approximately 1 hour, a period of 300 s of which is shown below, representing 9 pumping cycles. The pumping cycles can be clearly seen by the line speed values, the second trace in the plot, with positive speeds during traction and negative speeds during retraction. Also important are the high forces during reel-out, typically oscillating between 1 and 2 kN and the low forces during reel-in, approximately 200 N. The result is an average mechanical power production of 2.4 kW over the 9 pumping cycles, or $0.8\,\text{kW}/\text{m}^2$ when normalized by the area of the Twing. Peak power goes up to 20 kW. This represents a significant advance over the power production measured during our former SwissKitePower project [19] based on soft kites, which was approximately $0.1\,\text{kW}/\text{m}^2$ at comparable wind speed, demonstrating the twing's superior aerodynamic efficiency even on this small scale.

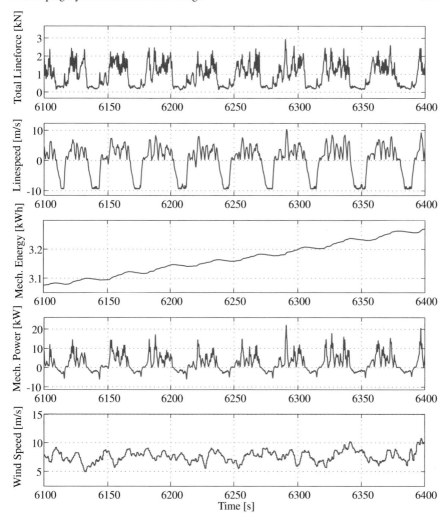

Fig. 24.14 Measured data over 300 s of a 1 h flight with a 3 m² twing: Total line force [N], line speed [m/s], net mechanical energy produced [kWh], wind speed at 5 m above the ground [m/s] as a function of time [s]

24.4 Market and Applications

In a petition letter from the Airborne Wind Energy Community to the European Commission from 2011, it was indicated that "AWE-based power plants have the potential to generate electric power at a price below 0.03€/kWh" [7, 8]. Although this seemed like a significant reduction in the cost of energy produced from wind at the time, recent advances and market forces have driven the cost of energy production from conventional horizontal axis wind turbines down significantly over the past 5 years and in the best wind sites LCOEs almost as low as 0.03€/kWh are achievable [13]. For most companies engaged in the development of airborne wind energy systems, releasing a first product that can compete with utility-scale wind turbines is very difficult, so niche markets are needed where small systems can gain a foothold, provide value for real customers and can eventually be scaled up to larger sizes to compete in mainstream markets. TwingTec aims to enter the market with a system in the range of 100kW, the TT100 as shown in Fig. 24.15.

Fig. 24.15 Rendering of the TT100, a mobile wind energy system dedicated to off-grid markets

One such niche market is off-grid, where electricity is generated using diesel fuel, resulting in high costs and environmental impacts. Burning 1 liter of diesel produces about 3KWh of electricity, with a diesel cost of $1.5/L, which is typical in remote areas due to the costs to transport and store the fuel, this results in an electricity cost of approximately $0.50/kWh. One such area with such high energy costs is Wainwright, a small, remote community in Alaska with approximately 500 inhabitants. In 2011 they commissioned a study to install 5, 100kW wind turbines and found it would have cost $11,000/kW installed, about 5-10 times higher than typical costs for an on-grid wind turbine project. The turbines themselves only represent about 32% of the total project cost, with 53% and 15% representing the construction and integration, respectively [6]. The deployment of TT100s into such an application would have the advantage that the construction costs, representing transportation and installation of the systems are significantly lower than for the wind turbines. This advantage is directly enabled by the inherent mobility of the technology, with the entire ground station being integrated into a standard 20 foot shipping container. A further benefit of this mobility is that the systems can be removed if the power is no longer needed, or if the customer cannot continue to pay their bills. This will allow for TT100s to be leased or rented to customers who prefer to reduce their upfront capital expenditures or to keep depreciating assets off their books. The flexibility and mobility of the technology will allow for mobile wind farms, or TwingFarms, to be deployed, as shown in the rendering in Fig. 24.16, even if the power demands are temporary or the future development of demand is uncertain. Such a scenario can be of particular interest for the mining industry.

Fig. 24.16 Rendering of a TwingFarm

24.5 Conclusions

The technology of pumping cycle kite power has seen recent advances. Through the development of the twing with an integrated launching and landing concept we are able to demonstrate the key technical challenges: launching independent of the strength of the wind, autonomous power production with pumping cycles and landing independent of the strength of the wind. Major elements of this progress are the strong collaboration with leading research institutes in Switzerland over the last years, the optimization of the test processes including HIL tests and tow tests and a very detailed numerical simulation tool. As a next step the system will be further optimized on the current test scale before we move into the pilot phase with up-scaled versions.

References

1. A., W. T., Hesse, H., Zgraggen, A. U., Smith, R. S.: Model-based Identification and Control of the Velocity Vector Orientation for Autonomous Kites. In: Proceedings of the 2015 American Control Conference, Chicago, IL, USA, 1–3 July 2015. doi: 10.1109/ACC.2015.7171088
2. Argatov, I., Rautakorpi, P., Silvennoinen, R.: Estimation of the mechanical energy output of the kite wind generator. Renewable Energy **34**(6), 1525–1532 (2009). doi: 10.1016/j.renene.2008.11.001
3. Bormann, A., Ranneberg, M., Kövesdi, P., Gebhardt, C., Skutnik, S.: Development of a Three-Line Ground-Actuated Airborne Wind Energy Converter. In: Ahrens, U., Diehl, M., Schmehl, R. (eds.) Airborne Wind Energy, Green Energy and Technology, Chap. 24, pp. 427–437. Springer, Berlin Heidelberg (2013). doi: 10.1007/978-3-642-39965-7_24
4. Breuer, J. C. M., Luchsinger, R. H.: Inflatable kites using the concept of Tensairity. Aerospace Science and Technology **14**(8), 557–563 (2010). doi: 10.1016/j.ast.2010.04.009
5. Canale, M., Fagiano, L., Milanese, M.: High Altitude Wind Energy Generation Using Controlled Power Kites. IEEE Transactions on Control Systems Technology **18**(2), 279–293 (2010). doi: 10.1109/TCST.2009.2017933
6. D., V.: Wainwright Wind-Diesel Hybrid Feasibility Study, V3 Energy, LLC, Eagle River, Alaska, 2011. http://www.v3energy.com/wp-content/uploads/2010/08/v3-energy-wainwright-feasibility-study.pdf
7. Diehl, M., Ockels, W. et al.: On the Development of Airborne Wind Energy in Europe. Letter to the Members of the European Parliament and European Commissioners. Leuven, Belgium, 2 Dec 2011. http://homes.esat.kuleuven.be/~highwind/wp-content/uploads/2012/02/Letter_petition-no-sig-final_version.pdf Accessed 1 Oct 2016
8. European Parliament Committee on Petitions: Notice to Members on Petition 1326/2011. PE498.097 v01-00, 24 Oct 2012. http://www.europarl.europa.eu/meetdocs/2009_2014/documents/peti/cm/917/917158/917158en.pdf
9. Fagiano, L., Zgraggen, A. U., Morari, M., Khammash, M.: Automatic crosswind flight of tethered wings for airborne wind energy:modeling, control design and experimental results. IEEE Transactions on Control System Technology **22**(4), 1433–1447 (2014). doi: 10.1109/TCST.2013.2279592
10. Fagiano, L., Huynh, K., Bamieh, B., Khammash, M.: On sensor fusion for airborne wind energy systems. IEEE Transactions on Control Systems Technology **22**(3), 930–943 (2014). doi: 10.1109/TCST.2013.2269865
11. Fritz, F.: Application of an Automated Kite System for Ship Propulsion and Power Generation. In: Ahrens, U., Diehl, M., Schmehl, R. (eds.) Airborne Wind Energy, Green Energy and Technology, Chap. 20, pp. 359–372. Springer, Berlin Heidelberg (2013). doi: 10.1007/978-3-642-39965-7_20
12. Gohl, F., Luchsinger, R. H.: Simulation Based Wing Design for Kite Power. In: Ahrens, U., Diehl, M., Schmehl, R. (eds.) Airborne Wind Energy, Green Energy and Technology, Chap. 18, pp. 325–338. Springer, Berlin Heidelberg (2013). doi: 10.1007/978-3-642-39965-7_18
13. Lazard: Levelized Cost of Energy Analysis - Version 8.0. https://www.lazard.com/media/1777/levelized_cost_of_energy_-_version_80.pdf (2014). Accessed 19 May 2016
14. Loyd, M. L.: Crosswind kite power. Journal of Energy **4**(3), 106–111 (1980). doi: 10.2514/3.48021
15. Luchsinger, R. H. et al.: Closing the Gap: Pumping Cycle Kite Power with Twings. In: Schmehl, R. (ed.). Book of Abstracts of the International Airborne Wind Energy Conference 2015, pp. 26–28, Delft, The Netherlands, 15–16 June 2015. doi: 10.4233/uuid:7df59b79-2c6b-4e30-bd58-8454f493bb09. Presentation video recording available from: https://collegerama.tudelft.nl/Mediasite/Play/646b794e7ac54320ba48ba9f41b41f811d

16. Luchsinger, R. H.: Pumping Cycle Kite Power. In: Ahrens, U., Diehl, M., Schmehl, R. (eds.) Airborne Wind Energy, Green Energy and Technology, Chap. 3, pp. 47–64. Springer, Berlin Heidelberg (2013). doi: 10.1007/978-3-642-39965-7_3
17. Milanese, M., Taddei, F., Milanese, S.: Design and Testing of a 60 kW Yo-Yo Airborne Wind Energy Generator. In: Ahrens, U., Diehl, M., Schmehl, R. (eds.) Airborne Wind Energy, Green Energy and Technology, Chap. 21, pp. 373–386. Springer, Berlin Heidelberg (2013). doi: 10.1007/978-3-642-39965-7_21
18. Ruiterkamp, R., Sieberling, S.: Description and Preliminary Test Results of a Six Degrees of Freedom Rigid Wing Pumping System. In: Ahrens, U., Diehl, M., Schmehl, R. (eds.) Airborne Wind Energy, Green Energy and Technology, Chap. 26, pp. 443–458. Springer, Berlin Heidelberg (2013). doi: 10.1007/978-3-642-39965-7_26
19. SwissKitePower. http://www.swisskitepower.ch/. Accessed 10 July 2012
20. Vlugt, R. van der, Peschel, J., Schmehl, R.: Design and Experimental Characterization of a Pumping Kite Power System. In: Ahrens, U., Diehl, M., Schmehl, R. (eds.) Airborne Wind Energy, Green Energy and Technology, Chap. 23, pp. 403–425. Springer, Berlin Heidelberg (2013). doi: 10.1007/978-3-642-39965-7_23
21. Zgraggen, A. U., Fagiano, L., Morari, M.: Automatic Retraction and Full-Cycle Operation for a Class of Airborne Wind Energy Generators. IEEE Transactions on Control Systems Technology 24(2), 594–608 (2015). doi: 10.1109/TCST.2015.2452230
22. Zgraggen, A. U., Fagiano, L., Morari, M.: Real-Time Optimization and Adaptation of the Crosswind Flight of Tethered Wings for Airborne Wind Energy. IEEE Transactions on Control Systems Technology 23(2), 434–448 (2015). doi: 10.1109/TCST.2014.2332537

Chapter 25
Fast Power Curve and Yield Estimation of Pumping Airborne Wind Energy Systems

Maximilian Ranneberg, David Wölfle, Alexander Bormann, Peter Rohde, Florian Breipohl and Ilona Bastigkeit

Abstract Besides other aspects such as safety, capital expenditures, lifetime and maintenance of a wind energy converter, the power curve is the defining performance characteristic in order to derive its economic viability. Power curves for horizontal axis wind turbines have been studied, validated and optimized for decades. This study tackles the power curve estimation and optimization of airborne wind energy converters, in particular systems that use the so-called pumping, or Yo-Yo principle. A fast but detailed model of the pumping airborne wind energy system is used to calculate a family of power curves at different fixed altitudes. Based on these power curves a yield estimation method is presented which also considers power losses due to ice accretion, insufficient conditions for take-off and low visibility situations. Furthermore estimated yield values are presented for an example location.

25.1 Introduction

Without towers, the ecological impact of airborne wind energy converters (AWEC) can be reduced to a fraction compared to conventional wind turbines (WEC). High-altitude operation offers the potential to reach significantly higher onshore wind speeds. With variable operating altitudes and hence wind speeds, the machines could operate near nominal power without increased fatigue.

Maximilian Ranneberg (✉) · Alexander Bormann · Florian Breipohl
EnerKíte, Fichtenhof 5, 14532 Kleinmachnow, Germany
e-mail: m.ranneberg@enerkite.com

David Wölfle
EWC Weather Consult GmbH, Schönfeldstraße 8, 76131 Karlsruhe, Germany

Ilona Bastigkeit · Peter Rohde
Fraunhofer IWES, Am Seedeich 45, 27572 Bremerhaven, Germany

© Springer Nature Singapore Pte Ltd. 2018
R. Schmehl (ed.), *Airborne Wind Energy*, Green Energy
and Technology, https://doi.org/10.1007/978-981-10-1947-0_25

623

EnerKíte is developing and operating airborne wind energy converters according to the reverse Yo-Yo concept. Yo-Yo AWE systems generate electricity on the ground and are operated in two phases. Initially, during the traction phase, the wing flies crosswind, unfurling the lines with optimal force and speed. Later, during the retraction phase, the wing returns to the starting point as fast and smooth as possible with minimal energy expenditure. The ground station generates electrical power from the torque of the unfurling lines whilst steering the wing using differential drum drives.

Since 2012, EnerKíte is operating the EK30, a mobile 30 kW research and development platform in Brandenburg, Germany [4, 23]. Several other teams are also operating Yo-Yo prototypes. Ampyx Power is using small airplanes with full onboard actuation and a single tether winch. At TU Delft, a 20 kW prototype with a single winch and a soft-kite wing has been in operation since 2010, for the purpose of various academic research from controller design [16] to system modeling [8]. Their system is controlled using an airborne actuating unit, which is controlled from the ground by wireless transmission. SkySails is using a similar airborne-controlled system, but with significantly larger kites [6].

Until now, few detailed descriptions and models of power curves of Yo-Yo systems have been presented. In [5] the power curve of a given system is estimated by using model-predictive control with a point-mass system and evaluating the power output over a set of points. In [8] a model is used to establish the basic characteristics of a ground station. The model uses an approximate formula for the reel-out phase and a set of discretization points over time for the reel-in phase. The aerodynamic characteristics of the model lead to an acceleration and increase in elevation angle during the reel-in phase, and hence a set of points is used to evaluate the total increase in elevation. In [9], simulation results with a point-mass and four-point-mass approach are compared, with good agreement, to measurements with the prototype. The SkySails prototype has been used to measure power output and forces [7]. For the evaluation of the economic feasibility, power curves are the most important aspect of a wind energy generator.

Here, a model is presented that allows the rapid calculation of power curves for specific AWEC designs at different altitudes. The model is described in detail in Sect. 25.2 and compared to the state of the art in fast airborne wind energy models. This model is then used to calculate several distinct power curves for a 100 kW system, assuming operation at fixed average altitudes in Sect. 25.3 as well as optimized operation at variable altitudes assuming a logarithmic wind profile in Sect. 25.4. These power curves are used to estimate the yield at a specific site in Sect. 25.4 by using simulated wind data and including yield losses due to visibility, ice-accretion and minimal take-off conditions. The preliminary content of the present chapter has been presented at the Airborne Wind Energy Conference 2015 [24].

25.2 Model Description

25.2.1 State of the Art

Several fast simulation models that may be used for power evaluation for airborne wind energy systems have been published to date. In one of the first publications on Yo-Yo airborne wind energy converters by Loyd [19], a simple formula for estimating power output can be deduced and is given by

$$P = \frac{2}{27}\rho A \frac{c_L^3}{c_D^2} w^3. \tag{25.1}$$

This holds under the assumption of exact crosswind conditions at zero elevation angle without mass and high lift-to-drag ratios. In [8] the formula is adapted for different elevation angles and uses an efficiency factor to estimate the loss due to non-ideal crosswind conditions during the traction phase. This model is also compared to point-mass models, which are the next step in terms of model complexity. Using a point-mass model for the wing neglects all inertia effects due to rotational movements but can include the basic inertial effects and aerodynamics quite well. Such point-mass models have been used extensively in the literature, either to evaluate control strategies [7] or to solve optimal control problems [14]. Some point-mass models have been used to estimate power curves. In [8] the power curve of a 20 kW system is calculated. In [5] such a model is used for control design, which was subsequently evaluated at different wind speeds.

However, dynamic point mass models still need to be controlled. Either by casting an optimal control problem, or by directly designing and simulating a feedback controller. Both approaches are time-consuming, in terms of computational time as well as in terms of time spent on control development. Furthermore, they cannot be easily automated. Another approach is to prescribe the trajectory and calculate the resulting forces with the assumption of a quasi-steady state. That is, neglecting necessary accelerations between states of the system and assuming a force equilibrium. In [1] the mechanical power output is estimated under different shapes of the prescribed trajectories. A similar approach is used in [21]. However, both approaches are using approximations of the true force equilibrium by assuming very high glide ratios and low masses. Additionally, the reel-out speed was set using ad-hoc approaches and not optimized.

In this work, a trajectory is prescribed as well. Instead of using approximations to enable fast equilibrium calculations and assume tether velocities, the true equilibrium is calculated at every point. The complete trajectory is discretized and used in an optimization problem to calculate the optimal tether velocity, ground station torque and roll angle. All relevant machine design limitations, such as force and power constraints, are taken into account.

25.2.2 Traction Phase

In the traction phase the trajectory of the figure-eight is prescribed, as illustrated in Fig. 25.1. This path is discretized using N points. At every discretization point, a set

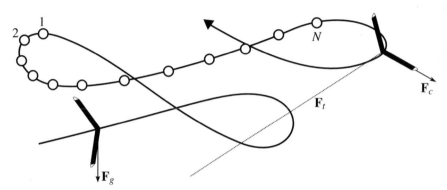

Fig. 25.1 N discretized points and non-aerodynamic forces during the traction phase. The operational altitude is defined in the center of the figure-eight

of forces is calculated and must be brought into an equilibrium. The aerodynamic forces are the same used in [23] and defined as follows. Let A be the aerodynamic area of the kite and ρ the density of the air. Airspeed is given by $\mathbf{v}_a = \mathbf{w} - \mathbf{v}$, with the wind vector $\mathbf{w}$, and the kite velocity $\mathbf{v}$, both given in Cartesian coordinates. The drag force is always parallel to the airspeed vector and calculated with the drag coefficient c_D as

$$\mathbf{F}_D = \frac{1}{2}\rho A c_D |\mathbf{v}_a| \mathbf{v}_a. \tag{25.2}$$

The lift force is perpendicular to this drag force. The notion of rolling around the principal axis, like an airplane, is used to model the effect of steering inputs. Assuming negligible side-slip, the roll axis of the kite aligns with the airspeed. An initial lift vector points perpendicular to the airspeed but parallel to the tether. This can be interpreted as a kite without line length differences with respect to the main tether line. The control input then results in a lift vector $\mathbf{Z}$ by a rotation of this vector around the airspeed axis. With the lift coefficient c_L the force is then given by

$$\mathbf{F}_L = \frac{1}{2}\rho A c_L |\mathbf{v}_a|^2 \mathbf{Z}(\psi), \tag{25.3}$$

where ψ is the roll angle.

The tether drag force acting on the kite is integrated over every point s between the ground station at length 0 and the kite at length L, given by

$$\mathbf{F}_{Ds} = \frac{1}{2}\rho d c_{Ds} \int_0^L \left| \mathbf{w} - \frac{s}{L}\mathbf{v} \right| \left(\mathbf{w} - \frac{s}{L}\mathbf{v} \right) ds, \tag{25.4}$$

where d is the effective diameter and c_{Ds} is the drag coefficient of the lines. The gravitational force $\mathbf{F}_g = -mg\hat{\mathbf{z}}$ with the system mass m comprising tether and wing mass. The centrifugal force $\mathbf{F}_c = m_w v_c^2 / R\,\hat{\mathbf{c}}$ with the wing mass m_w, the curve radius R and the tangential speed v_c. This force points perpendicular to the path and outwards w.r.t. the curve into the direction $\hat{\mathbf{c}}$. The ground station applies a torque to the winch and thus a force $\mathbf{F}_t$ onto the tether.

These forces are cast in the standard Cartesian coordinate system. During the traction phase, the tether will be reeled out from its minimum cyclic length to its maximum cyclic length. Over that time, the mass and drag can differ significantly. To include this effect on the power curve, the points describing a diagonal are evaluated over a set of tether lengths and the points describing the curve are evaluated at the mean tether length.

25.2.3 Retraction Phase

During the retraction phase, the kite moves quickly towards the ground station until the lower cyclic tether length is reached. This maneuver is simply described in the $x-z$ plane and roll angles and subsequent movements out of this plane are neglected. As in the traction phase, an equilibrium of the forces acting on the wing is needed. To understand the physics behind the retraction phase, in the following a simplified analytical calculation of this maneuver is presented. The airspeed of the wing at an elevation angle of θ, that is the angle between the ground and the straight tether, is given by

$$\mathbf{v}_a = (w + \cos\theta v_t, \sin\theta v_t). \qquad (25.5)$$

Now the direction of the aerodynamic forces is known: The drag force $\mathbf{F}_D = \rho/2A c_D |\mathbf{v}_a|\mathbf{v}_a$ is parallel to the airspeed and the lift force $\mathbf{F}_L = \rho/2A c_L |\mathbf{v}_a| R_{90}\mathbf{v}_a$ is perpendicular. R_{90} describes a counter-clockwise rotation of $90°$. Additional forces are the force acting from the ground station on the tether and the gravitational force acting on the mass of the tether and wing.

For the analytical calculations it is assumed that the gravitational effects and tether drag during retraction are negligible and the following holds: Let ϕ be the angle between the airspeed $\mathbf{v}_a$ and the tether direction $\hat{\mathbf{t}} = (\cos\theta, \sin\theta)$. For a resulting aerodynamic force parallel to the tether, and hence the possibility of an equilibrium, the force perpendicular to the tether needs to vanish

$$\sin\phi\,|\mathbf{F}_D| - \cos\phi\,|\mathbf{F}_L| = 0. \qquad (25.6)$$

This results in a constraint on the lift-to-drag ratio of the wing

$$\tan\phi = \frac{c_L}{c_D}. \qquad (25.7)$$

The angle ϕ can be replaced using the wind speed w, elevation angle θ and tether velocity v_t. The definition of ϕ leads to $|\mathbf{v}_a \times \hat{\mathbf{t}}| = \sin\phi |\mathbf{v}_a|$ and $\mathbf{v}_a \cdot \hat{\mathbf{t}} = \cos\phi |\mathbf{v}_a|$.

From the definition of $\mathbf{v}_a$ and the tether direction $\hat{\mathbf{t}}$ the terms can be calculated to be

$$|\mathbf{v}_a \times \hat{\mathbf{t}}| = (w + \cos\theta v_t)\sin\theta - (\sin\theta v_t)\cos\theta = \sin\theta w \qquad (25.8)$$

$$\mathbf{v}_a \cdot \hat{\mathbf{t}} = (w + \cos\theta v_t)\cos\theta + (\sin\theta v_t)\sin\theta = \cos\theta w + v_t. \qquad (25.9)$$

This leads to

$$\frac{c_L}{c_D} = \frac{\sin\theta}{\cos\theta + v_t/w}. \qquad (25.10)$$

This enforces very low lift-to-drag ratios for reasonable reel-in speeds, that is between 0 and 1 for $v_t \geq w$.

However, with gravitational force the constraint on the maximum allowed lift-to-drag ratio is relaxed depending on the mass per area. In the numerical evaluations in the following sections, gravity is always incorporated. In [8] several discretization points are used in the non-equilibrium retraction phase, but assuming equilibrium these states only differ slightly due to different tether lengths. For the complete cycle, the retraction phase is therefore discretized using a single evaluation point, with the tether length set to the mean length of the complete cycle. The variables for an equilibrium are the tether force F and the tether speed v_t.

The aerodynamic coefficients C_L and C_D differ from the parameters used in the traction phase, but are not part of the optimization process and instead chosen to be design parameters.

25.2.4 Optimization

The work over the complete cycle is given by the work done during the traction phase W_∞, and the work necessary during the retraction phase $W_\nearrow$. To calculate the mean power output over the complete cycle, the work is divided by the time spent in each phase, t_∞ and $t_\nearrow$, and an additional time loss t_{Lost}. This time loss is necessary since the stationary retraction phase is not instantly acquired in real application, but instead the wing is changed in aerodynamic state and the drums must change the direction of rotation and accelerate.

The problem is then discretized at $N+1$ points with 2 degrees of freedom (tether speed and tether force) for the retraction point and 4 degrees of freedom (tether speed, tether force, roll angle, wing speed) for every traction point. This leads to the following optimization problem with the set of all variables $\mathbf{x}$:

$$\underset{x}{\text{maximize}} \qquad P = \max_x \frac{W_\infty + W_\nearrow}{t_\infty + t_\nearrow + t_{Lost}} \qquad (25.11)$$

$$\underset{i=1,\dots,N+1}{\text{subject to}} \qquad \sum \mathbf{F}(i) = 0, \qquad (25.12)$$

$$F_{min} \leq F_t(i) \leq F_{t,max}, \qquad (25.13)$$

$$P_{min} \leq F_t(i)v_t(i) \leq P_{max}. \qquad (25.14)$$

The problem is solved using sequential quadratic programming with Lagrange multipliers for the equality constraints. For the force inequalities a barrier method using logarithmic barrier functions is employed. The power inequality is handled by an active set strategy, using the set I to describe the points where the constraints are active:

1. Set I to an empty set.
2. Optimize with power constraints $F_t(I)v_t(I) = P_{max}$.
3. Using current optimum with current active set I, find active constraints by choosing I_+ where $F_t(I)v_t(I) \geq P_{max}$.
4. If $I_+ = I$, return. Otherwise, set $I = I_+$ and go to 2.

25.2.5 Component Efficiencies

There are additional losses in the power output due to the system component efficiencies, which are simplified to be constant values regardless of the actual tether velocity or force. Electric machines, gears and tethers all have associated efficiencies, which define what percentage of power is additionally converted into heat. In the generation phase, this leads to a reduction by $P_\infty = \xi_{electric}\xi_{gears}\xi_{tether}F_t v_t$. In the retraction phase more than the necessary mechanical power must be applied, $P_\swarrow = \xi_{electric}^{-1}\xi_{gears}^{-1}\xi_{tether}^{-1}F_t v_t$. If a storage system is used, the efficiency has to be calculated differently. Only the work that has to be stored and released in the storage system will be reduced by its efficiency factor, which is only applied during the release of the stored energy. The retraction phase needs to be powered through the storage system and hence $W_{\swarrow,\xi} = W_\swarrow \xi_{storage}^{-1}$. In addition, the mean power output needs to be supplied by the storage system during the retraction phase. The total work that is supplied over the whole time $t = t_\infty + t_\swarrow + t_{Lost}$ is thus reduced by the efficiency within that period. The real mean supplied power P_{mean} from the cycle power P without storage system efficiency (but including all other efficiencies and the storage system efficiency for the retraction energy above) is then given implicitly by

$$P_{mean} = \frac{Pt - (1 - \xi_{storage})P_{mean}(t_\swarrow + t_{Lost})}{t} \tag{25.15}$$

$$= \frac{t}{t + (1 - \xi_{storage})(t_\swarrow + t_{Lost})}P. \tag{25.16}$$

This scenario is only concerned with the contribution of the storage system that is used to maintain a homogeneous power supply during the cyclic operation.

Simple powertrain or direct drive configurations with a single motor usually suffer from the two equally important and different operating points of Yo-Yo AWECs. For a high powertrain efficiency EnerKíte proposes a combination of efficient drives which are operating near their best points both in the reel-out and reel-in phase [17].

25.3 Power Curves

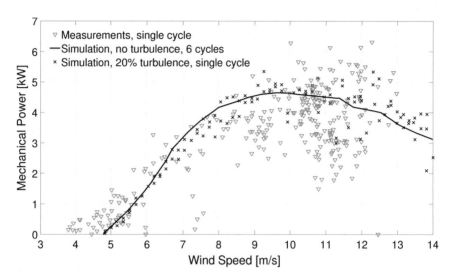

Fig. 25.2 Comparison between detailed simulation of the EK30 with the old ram-air wings and measurements of single cycles obtained during continuous operation over three days in 2013. The wind speed is given at the mean operational altitude of the kite, while the simulations were run assuming the same wind speed over all altitudes

25.3.1 Comparison with Detailed Simulations and Measurements

For the EK30 research and development platform, detailed models exist that incorporate realistic aerodynamics, accurate bridle and tether dynamics along with models of the actual machinery. In the Summer of 2013, LiDAR measurements were available during the continuous operation of the EK30 with ram-air wings over the course of three days. These wind measurements were obtained by the Fraunhofer IWES within the project OnKites [11] and were previously published in [4, 23]. A comparison between the detailed simulation using ram-air wings and measured power outputs is shown in Fig. 25.2. The EK30 is currently equipped with relatively low torque and, using ram-air wing, is able to produce roughly 5 kW peak cyclic power output. Every measurement is the mean power over a single complete cycle. The measurements agree with the simulated power curve, and the deviations are in the expected range. In the low wind speed range, the turbulence is significantly larger than 20%, and this turbulence leads to a net increase in power due to the cubic influence of the wind speed. This increase in wind power is captured by the system

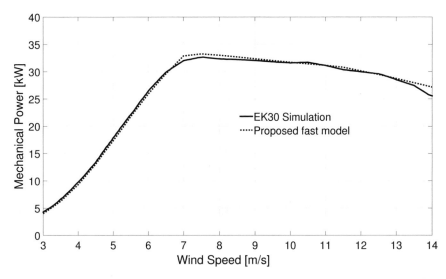

Fig. 25.3 Comparison between detailed simulations of the EK30 and a fast model of the EK30, assuming the same wind speed over all altitudes. Only mechanical power is compared

and results in power measurements above the low turbulence simulation. In the high wind speed range, the generation phase is relatively short. Only one or two figure-eight maneuvers are flown during generation. Turbulence has a significant effect on the power output, as gusts decide whether the system can fly an additional crosswind swipe - or must abort the generation phase sooner than expected.

Even though the power curves of the semi-rigid wings have not been measured in the field, they have been analyzed with simulations. To estimate the validity of the results generated using the fast model, the fast results for the EK30 parameters are compared to power curves obtained using the detailed simulation with a directly retractable wing. In Fig. 25.3 the simulation of a semi-rigid wing at the EK30 platform, optimized for power output at constant wind speed across altitudes and with a set maximal tether length of 400 m, is compared to the fast simulation. In addition to the new wings, the simulations were run with the assumption of increased available torque at the same power. The simulation parameters for the fast model were chosen as close as possible to the values used in the detailed simulations. However, the aerodynamic forces in the detailed simulation depend on the side-slip, rotational rates and angle of attack. Hence these variables vary in time. A non-physical parameter in the fast model is the additional time loss due to the two changes between phases. This parameter has been chosen to be 6 s which lead to the close agreement in Fig. 25.3.

25.3.2 Power Curves of a 100 kW System

Different operating altitudes and different elevation angles lead to different tether lengths. These tether lengths can change the efficiency and operating conditions dramatically, as long tethers may double the weight and increase the drag and hence the system performance significantly. At every site, the wind conditions and the variation in wind speed across the altitudes is different. In Fig. 25.4, a family of power curves for a system operating at different altitudes is shown. The simulations are based on a wing design similar to the one used for the evaluation in the previous section, but all design parameters (generator power, wing area, nominal force) are scaled and chosen to a 100 kW EnerKíte machine with a nominal wind speed of 7.5 m/s at an altitude of 200 m.

A few notes on this design are in order. EnerKíte systems, or more generally all tethered airborne wind energy systems, can change their operating altitude. And they have another significant advantage compared to conventional wind turbines: For WEC, low nominal design wind speeds result in large blades and high torques. AWECs also need higher tether forces and larger wings for lower design speeds. But for WEC the tower must carry the large blades and withstand the big torques. A tethered system carries itself and the tether only needs to be strong enough to withstand the high tensions. This creates the opportunity to design a wind energy system for low wind speeds without the significant structural penalty that occurs for WEC.

For each wind speed and altitude, the mean elevation angle and other operational parameters were optimized and the mean operating altitude was kept constant. The yield at a specific site, where detailed wind speed measurements over several altitudes are available, can be estimated with such a family of power curves. It is also assumed that a control system is in place that ensures operation always at the optimal altitude during every time step over the observed period.

25.4 Yield Estimation

If wind speeds at several altitudes are either given by measurements or by simulations as a set of discrete values for a range of time steps, it is then possible to calculate the ideal power production P_{ideal} by applying a set of altitude dependent power curves, as shown in Fig. 25.4, to the wind speed profile. It is furthermore assumed that the AWECs is controlled in such a way that the yield is maximized. Hence the yield estimation considers the optimal operational height for every investigated time step. In following section a set of additional yield losses is derived, and yield estimates are presented for a sample site.

But first a simplified approach to incorporate site conditions is presented, which assumes a logarithmic wind profile. This leads to roughness-length specific power curves, one of which will be called reference curve due to the connection to the reference yield.

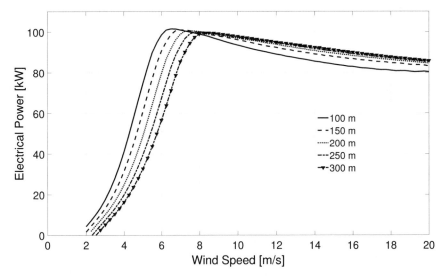

Fig. 25.4 Family of power curves of a 100 kW EnerKíte system (design for nominal power at 7.5 m/s wind speed and at 200 m altitude) over different altitudes. The wind speed given is the wind speed at the operational altitude of the system. Higher altitude results in longer tethers and/or increased elevation angles and hence less efficient systems

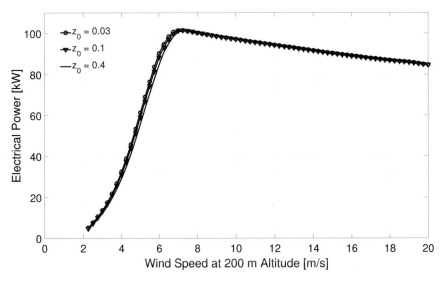

Fig. 25.5 Power curve of a 100 kW EnerKíte system at roughness lengths of 0.4 to 0.03 over the wind speed at 200 m altitude. The machine is operating at various altitudes, see Fig. 25.4, optimizing for power output

25.4.1 Reference Power Curve

Usually the time series of wind speeds are not readily available at different altitudes at the site of interest. Then, a logarithmic wind profile can be assumed in order to derive an estimate of the power output. A logarithmic profile is commonly used to estimate the wind speed at the operating altitude $w(z)$ from a base wind speed at some altitude $w(z_{ref})$ using a terrain-roughness parameter z_0 which is typically between 0.0001 ("Class 0": Offshore) and 0.4 ("Class 3": Landscape with a good amount of trees and buildings):

$$w(a) = w(a_0) \frac{\ln (z/z_0)}{\ln (z_{ref}/z_0)}. \tag{25.17}$$

In Fig. 25.5 a power curve for the system operating at a roughness lengths of 0.4, 0.1 and 0.03 is shown. At every wind speed at the reference altitude of 200 m, the operating altitude is optimized for power output. These power curves can be used to calculate the yield at a site which is only defined by a roughness parameter and a mean wind speed at a reference altitude. Note the modest differences due to roughness length.

In the following chapters a specific site in Brandenburg, Germany is analyzed. The DTU Windatlas [15] estimates a mean wind speed of 7.72 m/s at an altitude of 200 m at this site (March 2016). Using the reference power curves in Fig. 25.5 this leads to a yearly yield of 618 MWh, 613 MWh and 604 MWh for the roughness lengths $z_0 = 0.03, z_0 = 0.1$ and $z_0 = 0.4$, respectively.

In fact, due to the relatively low influence of roughness on the power curve, initial yield estimates may be calculated regardless the roughness length for the presented design. As a reference curve the resulting power curve for $z_0 = 0.1$ and a reference altitude of 200 m is chosen. The influence of slight changes in surface roughness are about 1%.

However, this evaluation should only be used for preliminary investigations of a site, as the logarithmic profile is usually only assumed to be valid up to 100 m [25]. Additionally, the described process assumes that at every point in time, the wind speed distribution can be described by a logarithmic profile and an optimal altitude at that time can be chosen. The logarithmic wind profile is valid only for the mean values over a significant period of time, for example months. Not included is these yield calculations is a varying wind speed distribution across the operating altitudes by assuming a constant Rayleigh distribution.

25.4.2 Reference Yield and Capacity Factor

In order to allow the comparison of AWEC between each other and with today's wind turbines a reference yield estimation in accordance to the German Renewables Energies Act (EEG) is proposed. A comparison of the presented design with two conventional wind turbines is shown in Table 25.1. The reference yield is calculated for a period of five years by assuming an idealized wind described by a Rayleigh distribution, a logarithmic wind profile with a roughness length of 0.1 and a mean wind speed of 5.5 m/s at 30m altitude [13]. The calculated yield hereby is deducted by the availability of the system, which for today's wind turbines is 98% or higher. For an airborne wind energy system with respect to its novelty 95% is proposed. The above derived reference power curve leads to a reference yield of roughly 3000 MWh or an annual yield of 600 MWh. With regard to a 100% renewable world and system integration into grid and off grid applications, besides the levelized cost of electricity the capacity factor becomes crucial for integration, security of supply and the base load capability of the energy system. The capacity factor CF is derived from the annual yield divided by nominal power. For the presented system under reference conditions a capacity factor of CF = 68% is calculated.

	EnerKíte 100 kW System	Fuhrländer FL100	Siemens SWT-2.3 113
Rated Power [kW]	100	100	2300
Rated Wind Speed [m/s]	7.5	13.0	11.5
Hub Height [m]	80-300	35	92.5-122.5
Capacity Factor [%]	68	24.3	41.7-44.3

Table 25.1 Comparison between the presented airborne wind energy system with optimized altitude operation under reference conditions with conventional wind turbines. Rated wind speed is defined at hub-height for conventional wind turbines and defined at 200 m for the EnerKíte system. The data for the wind tubines was taken from the website of the FGW e.V. [10]

The comparison in Table 25.1 shows that both the operational altitude and the chosen rated wind speed have a strong impact on the capacity factor or the availability and security of the electricity supply. A more realistic approach to site specific yields uses advanced wind models and simulation, which is presented in the following section.

25.4.3 Yield Deductions

The above mentioned availability of 95% may cover both unpredictable interruptions and deductions resulting from the expected need to land at harsh weather conditions. A more realistic estimation of power production, P_{real}, may be computed by lowering P_{ideal} for the following reasons:

- A certain minimum wind speed w_{start} at a corresponding altitude z_{start} may be required to allow the AWEC's take-off.

- Interruption of operation may be necessary if the visibility $vis(t)$, as defined by e.g. Met Office [20], falls below a certain minimum value vis_{min}, depending on the implemented system for obstruction marking and collision avoidance.
- In case of air temperatures $T_{air}(t)$ below the freezing point T_{freeze}, operation of the AWEC may be stopped to allow de-icing of the system, which is assumed to consume a certain amount of deicing power P_{de-ice}. Furthermore it is considered that the deicing operation will take only a certain fraction p_{de-ice} of the total operation time. The time left for standard power production follows hence as
$p_{ice-oper} = 1 - p_{de-ice}$
- Lightning strike or threat may require an interruption of operation which would reduce power production by a factor $p_{lightning}(t)$. However, as the probability of lightning striking an AWEC has not been investigated before, it is not possible to quantify the the actual yield losses, which are therefore neglected in following discussions.

Going on, the influence of the above losses on P_{real} can be formulated as a function of reduction factors p:

$$P_{real}(t) = (P_{ideal}(t) - P_{de-ice}p_{de-ice}p_{ice-oper})p_{start}(t)p_{ice}(t)p_{vis}(t)p_{lightning}(t) \tag{25.18}$$

whereby the p(t) values can be computed by:

$$p_{start}(t) = \begin{cases} 0, & w(z_{start},t) < w_{start} \\ 1, & w(z_{start},t) \geq w_{start} \end{cases} \tag{25.19}$$

$$p_{vis}(t) = \begin{cases} 0, & vis(t) < vis_{min} \\ 1, & vis(t)) \geq vis_{min} \end{cases} \tag{25.20}$$

$$p_{ice}(t) = \begin{cases} p_{ice-oper}, & t_{air}(t) < t_{freeze} \\ 1, & t_{air}(t) \geq t_{freeze} \end{cases} \tag{25.21}$$

The energy yield within a certain time range $t_0,....,t_1$ is then given by

$$E_{real} = \int_{t_0}^{t_1} P_{real}(t)dt. \tag{25.22}$$

25.4.4 Site, Model Validation and Yield Results

Based on the methodology introduced before, a yield estimation has been carried out for one location in Germany. Information about the site, along with the parameters used and the estimated yield results, are summarized in Table 25.2. The wind speed profiles used for yield estimation have been produced by COSMO-DE, a numerical weather predication system. COSMO-DE is a non-hydrostatic, compressible limited-area model, with a spatial resolution of ca. 2.7km, operated by Deutscher

Wetterdienst DWD [2]. The wind speed values have been taken from a grid point located very close to the investigated location and from the six lowermost altitude layers. Furthermore the wind speed values in hourly resolution are based on the forecast hours 0, 1 and 2 of each model run of COSMO-DE in 2013 and 2014. The COSMO-DE wind speed data has been compared with a LiDAR measurement taken on site for a range of two months. For the yield estimation and also the comparison with measured data it has been necessary to interpolate and extrapolate intermediate wind speed values from the given six altitudes. Cubic splines have been used for interpolation while extrapolation to altitudes higher then 258 m have been computed by a routine which applies the so called power law [22] to the two uppermost available wind speed values.

Site	Sommersberg (Brandenburg, Germany)
Coordinates COSMO-DE	53.180447° N, 12.207835° E
Coordinates LiDAR	53.179646° N, 12.189038° E
Available Altitudes [m]	258.21, 183.93, 122.32, 73.03, 35.72, 10.0
Visibility and Temperatures	DWD Station Goldberg, 53.36° N, 12.06° E
Timeframe	Jan 2013 - Dec 2014
Model	COSMO-DE
Yield Parameters	
Freezing Temperature	0° C
Starting Condition	$w \geq 2$ m/s at an altitude of 25 m
p_{de-ice}	0.025
Necessary Power for De-icing	2 kW
Yield and Losses	kWh
E_{ideal}	685,567
E_{real}	647,809

Table 25.2 Yield estimates at the site. Visibility was excluded here. If a minimal visibility of 4 km would be included as a necessary condition, an additional yield loss of 91,512 kWh (13%) would occur. This is an unacceptable loss in yield, and appropriate methods to allow operation during low visibility are thus an important topic for future developments

Three years of research in the field, during the execution of the project OnKites, have shown that a comprehensive assessment of the potential of AWE systems is extremely difficult. One reason consists in the fact that there is only limited data available of the meteorological conditions in heights over 150 m. Therefore, several measurement campaigns were carried out in the framework of the project by Fraunhofer IWES. To achieve this aim, the WindCube V2 LiDAR system developed by Leosphere was utilized [18]. The WindCube V2 is a pulsed LiDAR system that can measure wind velocity up to 150 m altitude with good data availability. Results show that in the case of sunny and warm days, the wind is very turbulent with similar wind speeds at all altitudes. This phenomenon was observed during all measurement campaigns and can be seen for example in [4]. It is important to re-mark that the described effect appears only in the time series with a resolution that captures the variation over the day. When considering annual time series, where the wind data are daily averaged, the phenomenon remains hidden because of the large averaging range.

Because AWE systems have to be kept airborne all the time, it is very important to determine the wind resources above 150 m with one minute sampling intervals. The new project OnKites II (2014-2016) deals with more accurate investigations of specific questions which have remained open in the first project [3, 12]. During two ongoing measurement campaigns planned by Fraunhofer IWES, the scanning LiDAR system Galion G4000 of the company Sgurr Energy is used. The scanning LiDAR system measures the wind speed up to 1100 m height during six months at each site which will allow a better estimation of the wind resources at these heights. An advantage of a scanning LiDAR system is to operate in an individual scanning geometry, for example an arc scan. The investigation of the data of an arc scan will provide information about the horizontal wind field which is interesting for the trajectory of the AWE systems—for their power curves as well as for yield estimation.

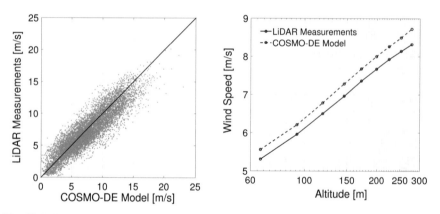

Fig. 25.6 Comparison between COSMO-DE model data with LiDAR measurements taken from 1 September 2015 until 31 October 2015. Left: Scatter plot of all measurements. Right: Mean velocities over altitudes

The comparison between the model and the measurements can be found in Fig. 25.6. It is thereby evident that model and measured data correlate generally well. There is a systematic bias of around 4.2%. Since only measurements over two months were obtained, it could not be validated if the bias has a general or a seasonal character. Additionally, there is a distance of 1250 m between the LiDAR measure-ment setup and COSMO-DE evaluation point. Hence, the wind speeds used for the yield estimation have not been corrected. If the wind speed would be corrected, the yield would be reduced by 1.7%. No other systematic error is observed.

25.5 Summary and Outlook

A model was presented to estimate the power curves of Yo-Yo AWE systems at different operating altitudes. The presented fast approach only takes seconds for the computation of a full set of power curves, similar to other approaches where the dynamics are replaced by approximate formulas, but can predict the output with significantly more confidence and compares well with detailed dynamic simulations. A key aspect of the model is the inclusion of the force and power constraints and their effect on the trajectories and on the power output.

With these families of power curves it is possible to estimate the yield, based on a set of wind speed profiles. In addition, a set of power curves under the assumption of a logarithmic wind profile with different roughness lengths was presented. Interestingly, the variation in shape and in the resulting yield for the presented case is quite modest. While a logarithmic wind profile is not considered a valid model for reliable site assessment, it enables a simple comparison between AWE systems and designs, as well as preliminary investigations into potential sites of interest.

For a specific site, a yield estimation has been carried out using detailed wind speed profiles taken from COSMO-DE, a numerical weather predication system for a time range of two years. The modeled wind speed profiles have been compared favorably with LiDAR measurements. Even though a systematic bias of 4.2% was found between the COSMO-DE data and the LiDAR measurements, a correction of the model data would only result in a yield reduction of 1.7%. This modest reduction is due to the relatively low nominal design wind speed, and is one of the strengths of airborne wind energy converters. The yield analysis included additional yield losses such as low visibility, insufficient conditions for take-off, and ice accretion. The influence of lightning was neglected, and the issue of low visibility was shown to have a great influence to the yield. This points to important directions in research and development.

The model has been compared with good agreement to more detailed simulations of the EK30 research platform. However, there are currently no measurements of power curves with semi-rigid wings available. In the near future, the presented calculations need to be validated against actual measurements.

A possible addition to the presented method is the inclusion of several non-equilibrium states during retraction instead of a single equilibrium state. This may allow to drop the rather arbitrary additional time loss during the complete cycle. Not shown here is the operating altitude during operation. For the low wind speed design the altitude chosen in the logarithmic power curves was quite low, increasing only to decrease the load at higher wind speeds. How the operational altitude is chosen at different sites, and how this changes with different AWEC designs and wind profile assumptions, is an important topic that will be addressed in the future.

To the best knowledge of the authors, this is the first published in-depth yield estimation for airborne wind energy systems that includes the specific peculiarities of airborne wind energy: A system that operates at variable altitudes, the necessity for wind data over several altitudes and the validation of such data using LiDAR measurements.

Acknowledgements The wind data was collected by Fraunhofer IWES Northwest within the projects OnKites and OnKites II (Studies of the potential of flight wind turbines, Phase I and II). OnKites (finished 2013, FKZ 0325394) and OnKites II (2014-2016, FKZ 0325394A) are funded by the German Federal Ministry for Economic Affairs and Energy (BMWi) on the basis of a decision by the German Bundestag and project management Projektträger Jülich.

References

1. Argatov, I., Rautakorpi, P., Silvennoinen, R.: Estimation of the mechanical energy output of the kite wind generator. Renewable Energy **34**(6), 1525–1532 (2009). doi: 10.1016/j.renene. 2008.11.001
2. Baldauf, M., Förstner, J., Klink, S., Reinhardt, T., Schraff, C., Seifert, A., Stephan, K.: Kurze Beschreibung des Lokal-Modells Kürzestfrist COSMO-DE (LMK) und seiner Datenbanken auf dem Datenserver des DWD, Deutscher Wetterdienst, Geschäftsbereich Forschung und Entwicklung, Offenbach, Germany, 13 June 2014. http://www.imk-tro.kit.edu/download/LMK_082006.pdf Accessed 12 May 2016
3. Bastigkeit, I., Rohde, P., Wolken-Möhlmann, G., Gambier, A.: Study on wind resources at mid-altitude. In: Schmehl, R. (ed.). Book of Abstracts of the International Airborne Wind Energy Conference 2015, p. 83, Delft, The Netherlands, 15–16 June 2015. doi: 10.4233/uuid:7df59b79-2c6b-4e30-bd58-8454f493bb09. Presentation slides available from: http://awec2015.eu/images/presentations/AWEC63_bastigkeit-presentation.pdf
4. Bormann, A., Ranneberg, M., Kövesdi, P., Gebhardt, C., Skutnik, S.: Development of a Three-Line Ground-Actuated Airborne Wind Energy Converter. In: Ahrens, U., Diehl, M., Schmehl, R. (eds.) Airborne Wind Energy, Green Energy and Technology, Chap. 24, pp. 427–437. Springer, Berlin Heidelberg (2013). doi: 10.1007/978-3-642-39965-7_24
5. Canale, M., Fagiano, L., Milanese, M.: High Altitude Wind Energy Generation Using Controlled Power Kites. IEEE Transactions on Control Systems Technology **18**(2), 279–293 (2010). doi: 10.1109/TCST.2009.2017933
6. Erhard, M., Strauch, H.: Theory and Experimental Validation of a Simple Comprehensible Model of Tethered Kite Dynamics Used for Controller Design. In: Ahrens, U., Diehl, M., Schmehl, R. (eds.) Airborne Wind Energy, Green Energy and Technology, Chap. 8, pp. 141–165. Springer, Berlin Heidelberg (2013). doi: 10.1007/978-3-642-39965-7_8
7. Erhard, M., Strauch, H.: Flight control of tethered kites in autonomous pumping cycles for airborne wind energy. Control Engineering Practice **40**, 13–26 (2015). doi: 10.1016/j.conengprac.2015.03.001
8. Fechner, U., Schmehl, R.: Model-Based Efficiency Analysis of Wind Power Conversion by a Pumping Kite Power System. In: Ahrens, U., Diehl, M., Schmehl, R. (eds.) Airborne Wind Energy, Green Energy and Technology, Chap. 14, pp. 249–269. Springer, Berlin Heidelberg (2013). doi: 10.1007/978-3-642-39965-7_14
9. Fechner, U., Vlugt, R. van der, Schreuder, E., Schmehl, R.: Dynamic Model of a Pumping Kite Power System. Renewable Energy (2015). doi: 10.1016/j.renene.2015.04.028. arXiv:1406.6218 [cs.SY]
10. Fördergesellschaft Windenergie und andere Erneuerbare Energien (FGW e.V.) http://www.wind-fgw.de. Accessed 12 May 2016
11. Gambier, A.: Projekt OnKites : Untersuchung zu den Potentialen von Flugwindenergieanlagen (FWEA). Final Project Report, Fraunhofer Institute for Wind Energy and Energy System Technology IWES, Bremerhaven, Germany, 2014. 155 pp. doi: 10.2314/GBV:81573428X
12. Gambier, A., Bastigkeit, I., Nippold, E.: Projekt OnKites II : Untersuchung zu den Potentialen von Flugwindenergieanlagen (FWEA) Phase II. Final Project Report, Fraunhofer Institute for Wind Energy and Energy System Technology IWES, Bremerhaven, Germany, June 2017.

105 pp. https://www.tib.eu/de/suchen/id/TIBKAT%3A1002309476/Projekt-OnKites-II-Untersuchung-zu-den-Potentialen/

13. German Federal Ministry for Economic Affairs and Energy (BMWi): Gesetz für den Ausbau erneuerbarer Energien. https://www.gesetze-im-internet.de/eeg_2014/anlage_2.html (2014). Accessed 12 May 2016

14. Houska, B., Diehl, M.: Optimal control for power generating kites. In: Proceedings of the 9th European Control Conference, pp. 3560–3567, Kos, Greece, 2–5 July 2007

15. International Renewable Energy Agency (IRENA): DTU Global Wind Atlas. http://irena.masdar.ac.ae/?map=103. Accessed 12 May 2016

16. Jehle, C., Schmehl, R.: Applied Tracking Control for Kite Power Systems. AIAA Journal of Guidance, Control, and Dynamics 37(4), 1211–1222 (2014). doi: 10.2514/1.62380

17. Kövesdi, P., Dreier, J.-E.: Drive train and method for drives having widely spaced operating points. German Patent WO/2015/032491, 2015

18. Leosphere: Windcube V2 LiDAR system. http://www.leosphere.com/products/vertical-profiling/windcube-v2-site-assessment-lidar (2016). Accessed 12 May 2016

19. Loyd, M. L.: Crosswind kite power. Journal of Energy 4(3), 106–111 (1980). doi: 10.2514/3.48021

20. Met Office: Observer's Handbook. 4th ed. OH Met.O.1028 2000 (Reprint). Met Office (2000). https://digital.nmla.metoffice.gov.uk/archive/sdb:collection%7C0d531225-ff0a-44c9-8c58-a9df90dc9038/

21. Noom, M. N.: Theoretical Analysis of Mechanical Power Generation by Pumping Cycle Kite Power Systems. M.Sc.Thesis, Delft University of Technology, 2013. http://repository.tudelft.nl/view/ir/uuid:1c1a3e90-11e6-4fe7-8808-8c6a1227dadb/

22. Panofsky, H. A., Dutton, J. A.: Atmospheric turbulence: models and methods for engineers and scientists. Wiley, New York (1984)

23. Ranneberg, M.: Sensor Setups for State and Wind Estimation for Airborne Wind Energy Converters. (2013). arXiv:1309.1029 [cs.SY]

24. Ranneberg, M., Bormann, A.: Estimation, Optimisation and Validation of Power Curves for Airborne Wind Energy. In: Schmehl, R. (ed.). Book of Abstracts of the International Airborne Wind Energy Conference 2015, pp. 50–51, Delft, The Netherlands, 15–16 June 2015. doi: 10.4233/uuid:7df59b79-2c6b-4e30-bd58-8454f493bb09. Presentation video recording available from: https://collegerama.tudelft.nl/Mediasite/Play/a5fcf164487546849230da29a8a81f421d

25. Stull, R. B.: An introduction to boundary layer meteorology, vol. 13. Atmospheric and Oceanographic Sciences Library. Springer Netherlands (1988). doi: 10.1007/978-94-009-3027-8

Chapter 26
A Roadmap Towards Airborne Wind Energy in the Utility Sector

Michiel Kruijff and Richard Ruiterkamp

Abstract The development path of the Ampyx Power airborne wind energy system is described. It is intended for the utility sector and large-scale grid connection. The technology generates energy by flying a tethered glider-aircraft attached to a ground-based generator following a crosswind pattern as the tether unwinds under high tension, and rewinds under near-zero tension. The benefits, drawbacks and decision rationales of major design choices are discussed: crosswind operation, rigid aircraft concept, ground-based generator. The development plan is shared and an indication is given how we defined our performance targets by prototype tests and extrapolations based on validated dynamic simulation. The development plan is to first build a system aimed to demonstrate safety and autonomy. Next, the first commercial system shall minimize Levelized Cost of Energy (maximizing the customer's return on investment). A larger system then maximizes productivity (maximizing the customer's net profit). Offshore operation is targeted. Safety levels are continuously improved to enable co-use of the land under the tethered aircraft.

26.1 Introduction

Ampyx Power develops a novel airborne wind energy system (AWES) which will eventually allow sustainable production of power at lower costs than fossil-fueled alternatives. The availability of such technology will likely trigger a paradigm shift in the electricity sector. The AWES converts wind power into mechanical power by having an autopilot-controlled glider aircraft creating pull on a tether by flying repetitive crosswind patterns at an altitude of 200 to 450 m, as described in [10] and illustrated in Fig. 26.1 (left). Conversion to electrical power happens in a ground-based generator from which the tether is extracted. Once the tether has

Michiel Kruijff (✉) · Richard Ruiterkamp
Ampyx Power B.V., Lulofsstraat 55 Unit 13, 2521 AL The Hague, The Netherlands
e-mail: michiel@ampyxpower.com

© Springer Nature Singapore Pte Ltd. 2018
R. Schmehl (ed.), *Airborne Wind Energy*, Green Energy
and Technology, https://doi.org/10.1007/978-981-10-1947-0_26

Fig. 26.1 Comparison of Ampyx Power's 2 MW AWES with typical 2 MW wind turbine in flight (left) and in storage (right)

been extracted to full length, the glider aircraft is controlled to glide back to the pattern starting point, during which phase the tether is retracted. During this reel-in phase, tether tension is minimal and power consumption is only a fraction of the power produced during the reel-out phase. Automatic land and launch cycles are made possible through a platform-based solution. The chapter will address Ampyx Power's ambitions with this concept and how we intend to achieve them, in terms of our development plan as well as our sizing rationale.

Ampyx Power targets the utility market with its AWES, therefore our concept firstly will have to be able to complement or compete with conventional wind turbine plants. Following an introduction into the key architectural features of our concept, a comparison against the conventional wind turbine performance is provided. Ampyx Power recognizes that one of the key additional challenges of developing AWE technology is securing the required levels of safety and reliability for the fully autonomous operation in a large range of weather conditions. The means in which we intend to meet these are described. With the Ampyx Power mindset and ambition clarified, we describe our current status and development plan, with a focus respectively on the aircraft/power generator combination and the launch and land platform. The second part of this chapter describes the sizing methodology that we have developed, which is based on a tool chain including an aero-structural model, a performance model and a cost model.

In Sect. 26.2 we present the overall development plan of our company. In Sect. 26.3 we detail this plan by outlining key architectural choices, design processes and certification, explaining the development status and the pursued approach as well as our strategy for launching and landing. In Sect. 26.4 important considerations about the sizing for the commercial system are discussed, such as the aerodynamic and structural models, the performance model and the trade-offs and finally the cost model and the prediction of the achievable LCoE. Conclusions are discussed in Sect. 26.5. The preliminary content of the present chapter has been presented at the Airborne Wind Energy Conference 2015 [8].

26.2 The Ampyx Power AWES Development Plan

The main architectural design choices that define the Ampyx Power AWES are:

- crosswind flight rather than static,
- a rigid wing rather than a flexible kite,
- ground-based generator rather than on-board,
- utility-scale power generation rather than off-grid.

Table 26.1 highlights the benefits and drawbacks for each of these choices. The aircraft makes use of a platform for its landing rather than e.g. a runway or some dynamic capturing system (such as a rotating arm). Even if we do not consider this a fundamental constraint for the AWES concept, we believe the platform landing has fundamental advantages, further detailed in a dedicated section below. We have rather firmly settled on a single tether solution rather than a double tether.

Architectural choice	Ampyx Power considerations
Crosswind	Crosswind systems much more efficiently convert wind power than static systems [9].
Rigid wing	Compared to a flexible system (kite). The reel-in phase is more efficient due to an aircraft's natural and fast glide dynamics vs. drag behavior of kite. This greatly improves cycle efficiency[a]. Rigid wing dynamic control has less degrees of freedom due to less wing flexing. Rigid wing control can be reliably performed also following tether rupture. A rigid wing can be designed for a high lift coefficient compared to a kite. Hence the same power can be generated with a more compact solution—though likely more heavy/costly. Composite structures can be designed to be damage tolerant and should meet a 20-year lifetime requirement. Lifetime is then orders of magnitude longer than that of flexible kites that typically survive only for hundreds of hours of operation [1, 4]. Regular replacement of a kite is costly and creates downtime.
Ground-based generator	An airborne generator would have to be custom designed and mass optimized, so likely costlier than off-the-shelf ground-based equipment. It puts high value in the air, at higher risk of loss in case of a crash. For a ground-based generator the tether does not need to carry (significant) current and can be thinner (less drag losses) and simpler (lower cost). However, due to the reel-in reel-out cycles, it will wear faster. The pumping winch is more complex than one for on-board power generation. Also, efficiency is lost since during reel-in no energy is generated. Furthermore, the on-board generator can double as oversized propulsion and allow for relatively simple spot landing techniques [12].
Utility scale	It is Ampyx Power's belief that a significant commercial potential lies in providing large amounts of energy at low cost. We also think that off-grid solutions may well be viable, however we think that such a solution requires its own dedicated design effort not necessarily applicable to a utility-scale solution. It would thus be for Ampyx Power more of a distraction on the path to a utility-scale system than a stepping stone.

[a] We define cycle efficiency as ratio of energy that is actually generated over reel-out/reel-in cycle and energy the system would generate when only reeling out over the same period of time.

Table 26.1 Key design choices for the Ampyx Power AWES

We realize that a double tether provides some safety through redundancy but the increased wear and replacement of tether material that comes with it would drive up the energy price. At the same time, the doubled tether drag would contribute to additional power generation losses, as shown in Fig. 26.11. A double tether would also allow (part of) control of the wing dynamics to be done from the ground, so there could be placed on-board less costly and less heavy avionics and actuators. Some avionics there would still be required such that the aircraft can be autonomously controlled and safely landed following tether rupture. However, the control and systems on-ground would be more complex and control would be less direct, so likely less precise.

Other design choices we have made for the current prototype are considered subject to trade-off and may vary from concept to concept, such as:

- the lemniscate pattern (figure of eight) vs. ellipse,
- the single-point tether-aircraft connection without bridle,
- the rather conventional aircraft layout and typical aircraft controls,
- on-board power generation for the avionics rather than provision of power through the cable.

26.3 Ampyx Power AWES Versus Conventional Wind Turbines

26.3.1 Key Architectural Choices

In the classical wind turbine concept only a small fraction of the structure (namely the blade tips) generate the majority of the power output. In contrast (as with all AWE crosswind concepts) the full wing span of the AWES aircraft is exposed to the high speed of the air flow that is obtained by its crosswind trajectory [9, Fig. 1]). Therefore, every part of the wing structure generates the associated high level of lift. A wing of size less than a single turbine blade generates the same power as the whole wind turbine. The material usage is thus highly effective.

Furthermore, there is almost no torque on the foundation, so that can be less massive than for conventional wind turbines. This is particularly valuable for offshore applications, where the generator can be placed on a spar rather than a pole, as illustrated in Fig. 26.2. No data exists yet on wake losses for large rigid-wing AWE farms. Two factors suggest these losses will be quite limited in comparison to conventional turbines. The smaller 'blade' area will disturb less air (though more intensely) and this disturbance is distributed through the flight pattern over a much larger vertical spacing.

With synchronized systems positioned closely together, our commercial design is expected to eventually achieve a park level power density in the range of 10 to 25 MW/km^2 depending on the system sizing. For prudence reasons (obtaining flight hour statistics), initial parks will operate at lower density. A set-up of an initial commercial park of AWE systems should have a facility density of slightly better than

Fig. 26.2 Ampyx Power offshore AWES impression

$1/L^2$, L being the maximum tether length. During power generation, the aircraft trajectories do not interfere with each other, and some margin can be appreciated, as depicted in Fig. 26.3. During launching and landing maneuvers that exit the footprint, some additional space would need to be created. For this, surrounding pattern movements could be narrowed, shifted and/or temporarily paused (loitering). Later, a packing of about $2/L^2$ to $3/L^2$ should be achievable, requiring synchronization of the patterns to within half the pattern width and allowing some overlap of the tether footprints. Even higher densities can be achieved using a cylindrical rather than conical volume constraint for the 3D flight path [5].

Fig. 26.3 Tether footprints for a given wind direction in a plant set-up. Left: a packing density of $1.2/L^2$, without pattern footprint overlap. L being the maximum tether length Right: a packing density of $\sim 2.7/L^2$, with flight overlap allowed over the bottom half of the downwind tether as well as synchronization required within half a pattern width (half a footprint shown)

26.3.2 Processes and Certification

Ampyx Power has opted to engage with authorities to identify suitable civil aviation standards and recommended design processes, including airworthiness, design organization and operational aspects, with the goal to achieve the required reliability and safety for commercial operation as well as certification (Table 26.2 and Chap. 29 in this book). We believe these design processes are enabling and have a number of proven benefits. They result in more traceability in case of issues and increase transparency of the project. Hence they greatly aid the convergence of design and provide better learning and improvement possibilities. This should result in better planning capability and lower end-to-end development cost. A certified design and production has furthermore the obvious benefit that the design can be produced in large numbers with only functional testing required for each delivered item rather than full verification.

Adhering to such standards can be challenging and requires some investment. In order to make it possible for a small company like Ampyx Power to provide full design traceability, support configuration management and process-based engineering, we believe a single place for our data and a single interface for the whole team is the way forward. Therefore, we have commissioned the Polish-based company Xignum to develop a tool tailored to the needs of AWE certification, cov-

Permitting element	Description
E.Y013-01	Current European Aviation Safety Agency (EASA) policy for Unmanned Aerial Systems (UAS), deemed applicable as our system is regarded a tethered UAS. It prescribes the tailoring of a suitable Certification Specification and an appropriate systems safety analysis.
CS-22	Certification Specification for sailplanes (gliders) which we have used as a starting point for tailoring, since our aircraft, once off the tether (the most risky situation) behaves as an unmanned glider with limited range and kinetic energy.
SC-RPAS 1309-01 Issue 2	Safety standard for Very Light Aircraft (VLA) classes of Remotely Piloted Aircraft Systems (RPAS) that we selected as starting point.
ED-79A	Accepted means of compliance for aircraft design processes (planning, requirements derivation, validation, verification, safety analysis, configuration management, product assurance, certification)
DO-178C	Accepted means of compliance for software design process. We will tailor this with the help of selected software standards of the European Cooperation for Space Standardization (ECSS).
AS9100	Aerospace supplier organizational quality system, which we use as a guideline.
(A)DOA	(Alternative to) Design Organization Approval: organizational processes, mainly product assurance and certification that should be in place when a commercial product is designed. Not required for experimental/development planes.
Operational requirements	The National Aviation Authority (NAA) requires on the operational side Safety management system, operations manual, pilot training

Table 26.2 Overview of some permitting standards being implemented by Ampyx Power

ering all required project and process data, including e.g. workflows, versioning and data interrelations. Some common problems and the approach by Xignum are shown in Table 26.3. Its specifications are based on experience with space projects

Team issue during development	Xignum approach
Why are we doing this exactly again?	Customer requirements
Am I doing the right thing?	Requirements
Am I doing the thing right?	Validation & Verification workflows
How reliable is this document?	Document approval workflow
Where is the most recent version of this document?	Document access & version control
Did not someone think of this before?	Decision (Trade-Off)
We really better take action now and not forget to think of this next time.	Risk, Lesson Learned work flows
Why again did we choose this option?	Decision (Trade-Off)
I am not quite sure about this but don't want to wait for the input to arrive. How can I continue my work?	Decision (Assumption) workflow, Task (To Be Confirmed / To Be Determined).
Where is this piece of scratched hardware?	Inventory management, Assembly
What software version and parameter settings did we use in that test flight again?	Software version control (GIT link), test logs
This is never going to work, in my opinion. But if you insist on going ahead...	Risk
I am not sure we agree on which thing should do what exactly.	Design tree, function definitions
I need to manufacture the exact same piece of hardware as 3 months ago	Manufacturing Reference
What changed since that meeting where we all agreed?	Baseline definition, history log
There is a problem with this document, hardware that should be handled properly	Review Item Disposition (RID), Task (approval workflow), Non-conformance Report (NCR)
How do we know whether it is safe enough?	Safety analysis (Functional Hazard analysis, Fault Tree diagram, Failure Mode and Effect Analysis)
What is the next step and who should do it?	Workflow Which inventory items exactly where in that test and where is the test data? Assembly, Validation & Verification Plan

Table 26.3 Some common problems encountered in complex development projects relieved through the Xignum tool

[7], the aerospace industry standard AS9100 [11], the European Cooperation for Space Standardization (ECSS), as well as the aeronautical design guideline ED-79A (Table 26.2). Note that this solution is available to the AWE community through Xignum.

26.3.3 Ampyx Power AWES Development Status and Approach

Ampyx Power currently operates 2 prototypes in a test field in The Netherlands, for which it has obtained type registration and an exemption (license to operate) from the national authorities based on a safety analysis, implementation of a safety system, pilot training and operations manual. Representative footage is shown in Fig. 26.4.

The aircraft are designated AP-2A1 and AP-2A2, and can produce about 20 kW net power. These 5.5 m prototypes serve to demonstrate the principle of a fully automatic operation (power generation → land → launch → power generation), as well as to raise the technology readiness level. AP-2 has a single autopilot. Its safety is based on a mitigation (of autopilot failure) through redundant remote control piloting. AP-2A1 is equipped with propulsion (for landing and take-off from a compact platform). AP-2A1 has an on-board power generator (small turbine installed on top of the fuselage) to power the actuators and avionics and thus provide the capability

Fig. 26.4 AP-2A aircraft, 90 kW generator and control center

for unlimited flight. Autonomous flight without intervention has been demonstrated for flights of over 2 hours. Net power production has been demonstrated [10], and matched with simulation data.

The next steps in Ampyx Power's AWES development are the certifiable pre-commercial prototype AP-3 (200 kW) and the to-be certified commercial version AP-4 (2 MW). These systems shall be operational in the coming few years. Table 26.4 sketches out some details for this development plan.

The AP-3 is to demonstrate full autonomy, design for reliability and safety, as well as predictability of performance and of cost. During its operation we aim to ev-idence high cycle efficiency, and low maneuvering and drag losses. It shall demon-strate operation at high g-load. It also serves as learning platform to meet the chal-lenges of site development, grid connection, maintainability and 24/7 operations.

The AP-3 design shall be such that no single failure is to lead to loss of life or (if in any way possible) to loss of the aircraft. The probability of a combination of failure conditions leading to loss of life shall be extremely remote. Since the necessary avionics for full autonomy under such constraints are rather complex, they will be implemented in steps during the AP-3 verification phase.

AP-3 is to meet reliability requirements using a triple-redundant autopilot. How-ever, as the individual autopilots are based on the same hardware and software, a common cause failure cannot be prevented (or in safety wording, independence

	AP-2 Proof-of- concept	AP-3 Commercial prototype	AP-4A Utility-scale prototype	AP-4B/C Utility-scale commercial	AP-5(TBD) Utility-scale commercial
Development	2011–2015	2014–2018	2018–2020	TBD	TBD
Wingspan [m]	5.5	12	35	35	50
Mass [kg]	35	400	3500	3000	5000
Windturb. eq.[MW]	0.01	0.2	2.0	2.0	4.0
Density [MW/km^2]	-	0.5	5	$10 - 15$	$15 - 25$
Co-use	Limited	Limited	Limited	Yes	Yes
Offshore	No	No	Yes	Yes	Yes
Optimized for	Breadboarding	Autonomy & safety demo	LCoE	LCoE	Plant output
Technology (mostly)	Custom	COTS	COTS Aircraft Custom ground segment	COTS Aircraft Custom ground segment	Custom
Safety mitigation	RC Pilot	Tether	Tether	Independent back-up autopilot	Independent back-up autopilot
Certification target / policy	Exemption (NAA)	Permit (EASA E.Y0.13)	Permit (EASA E.Y0.13)	Certified (EASA E.Y0.13)	Certified (EASA CoO)

Table 26.4 Ampyx Power AWES generations AP2 (in operation), AP3 (under development), AP4 (commercial version), technological approach and current estimates of performance. COTS: Com-mercial Off-The-Shelf. TBD: To Be Determined. CoO: Concept of Operations

cannot be claimed). Hence mitigations need to be in place. Firstly, the general public will not be allowed underneath the operating aircraft. The tether is then to secure that, upon autopilot failure, a "safe" crash would result within known radius. Such a crash would still be a costly event. Thus, during early verification, as for AP-2, a pilot will be available to take over control when necessary. In later flights, an independently developed emergency landing software will be added, allowing us to fly more comfortably without pilot oversight.

AP-3 size is limited to a 12 m wing span as deemed practical for a developmental aircraft. The wing aspect ratio is selected for high power output per kilogram of wing mass (Figs. 26.5 and 26.12). With a mass of about 300 kg, the AP-3 is designed to sustain 42 kN operational tether tension.

We have initiated the certification trajectory with EASA for AP-3, as a stepping stone to AP-4. It is foreseen though that only a small number of units of AP-3 will fly, and this will be under a permit as a developmental aircraft.

It is noted that uptime and reliability can be analyzed and improved by strict design rigor but have eventually to be demonstrated in the field and will undoubtedly need to improve over time as flight hours and operational experience accumulates. An extended flight campaign of AP-3 may be used for this, alternatively, this objective can be met by the early AP-4 models.

The AP-4 is scaled with the target to operate at a minimal Levelized Cost of Energy (LCoE). For commercial viability this should be well below that of conven-

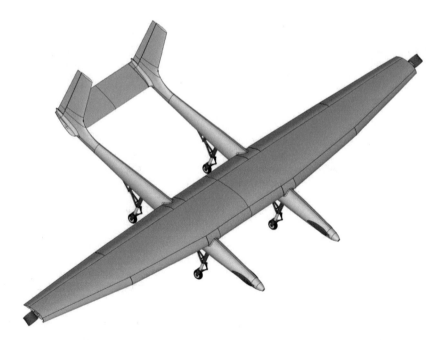

Fig. 26.5 AP-3 conceptual sketch (Autumn 2015)

tional wind energy, and our cost model predicts this can indeed be achieved (see section below for more details). AP-4 shall be ready for offshore deployment. The generator is to be optimized to limit conversion losses.

Three versions of AP-4 are foreseen over time (4A,B and C): to increase safety, and to reduce manufacturing and maintenance cost.

AP-4A will still be an experimental system with its power output already at final commercial levels (2 MW). It is a much larger system than AP-3, but shares the same avionics, control software and autonomy.

AP-4B will be the first certified system, and is to be deployed on a large scale. Its aircraft will feature design independence added to the redundancy in the avionics, which would make it possible, eventually, to allow the general public underneath the operating hardware. AP-4B is otherwise identical to AP-4A. Cost is lowered due to better deals with suppliers (economy of scale).

AP-4C would be the cost optimized version of AP-4. Based on flight and maintenance experience, we can slowly steer away from initially conservative values for component quality, safety margins and material grades. Furthermore, we would start developing our own production lines for selected components, systems and materials, in order to further benefit from our economy of scale and reduce system costs. Its LCoE shall beat the price of the traditional fossil fuel alternatives.

AP-5 can be roughly expected to be double the size of AP-4 and should optimize for the customer's net benefit, given availability of a limited plant area or number of systems. Its sizing is still immature. It depends on the projected energy price vs. cost, the packing density that can be achieved (scaling of tether length with wing size), scaling of wing mass with size and possibly, for the case of large offshore farms, scaling of wake losses.

26.3.4 Launching and Landing

The Launch and Land (L&L) solution is an immensely challenging part of the AWE system, that we have managed to cover using a combination of conventional technologies, adapted to our purpose. The requirement for a compact L&L system derives from the need to perform launch and land:

- in rough terrain and, eventually, at sea;
- for any wind direction;
- without disconnecting the tether;

An extensive conceptual trade-off has been performed. A conventional field landing for an AWES aircraft would require about 130×100 m runway field to cover all wind directions and wind levels. For commercial operation, that is not a practical option. A rotor-type vertical lift solution such as employed in [12] is considered not viable for Ampyx Power's ground-based generator concept at utility scale. A solution where the aircraft is gently captured by some type of controlled interceptor whilst still in the air is considered technologically challenging and not sufficiently robust.

A platform solution for a compact landing, rotated into the wind, has a number of benefits, when compared to the main alternatives. The L&L platform hardly scales with aircraft size. So the larger the aircraft, the smaller, relatively, will be the platform. Our requirement, derived from the cost model, is that the platform shall be shorter than the wingspan of our commercial model, the AP-4.

Platform launch and land can be achieved using conventional technology, such as a catapult and an arresting line (as we are applying for AP-2) or a net. Alternatively, we can use the tether to accelerate (as in a standard winch launch) and even to decelerate the aircraft—it is already attached. A complication is that the L&L shall be fully autonomous, and not require human intervention, even for post-landing guidance towards the launch position. This makes the use of e.g. a conventional net impractical for nominal landing. Furthermore, the aircraft shall be kept restraint until the next launch, such that gusts do not levitate the aircraft off the platform and cause damage.

The platform concept allows to use alternative implementations as introduced above for the mitigation of single failures, such as tether release, failure of propulsion or failure of one of the platform systems itself. Our approach is that any functional failure that can happen that has a major consequence (risk to damage the aircraft) is mitigated by a secondary solution.

Expressed in functional terms of the ED-79A design process and safety approach, the solution adopted by Ampyx Power can be described schematically as in Fig. 26.6. An impression of the resulting design is provided in Fig. 26.7. A detailed description of functionality will be subject of a future publication.

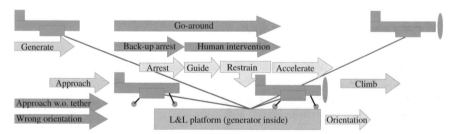

Fig. 26.6 Autonomous Launch and Land concept based on safety considerations. Mission phases Generate, Approach and Climb connect to the Land and Launch phases, which are covered by the required functions Arrest, Guide (to restrain position), Restrain and Accelerate. Furthermore, some form of Orientation with the wind is required to deal with any wind direction. Some indicative failures/mitigations to be covered are indicated as well. Whereas the overall functionality and safety is thus defined, each function in it can be implemented by one of several existing technologies depending on the scaling of the system (in our case AP-2, AP-3, AP-4 etc.)

Fig. 26.7 Impression of the AP-4 (2 MW) in comparison to a conventional 2 MW wind turbine with aircraft in operation (left) and in storage configuration on the platform (right)

26.4 Sizing of a Commercial System

An integrated set of sizing tools has been developed by Ampyx Power for a preliminary sensitivity analysis to understand the primary cost factors and perform first order sizing of the AP-3 and AP-4 designs. Figure 26.8 shows the main elements and parameters involved. The primary loop is that, given the necessary constraints, tether tension and wing planform are determined to achieve optimal LCoE (AP-4), or, employing also a secondary loop, to achieve optimal energy production given a

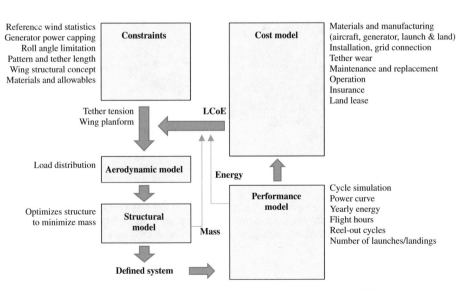

Fig. 26.8 Ampyx Power toolbox and methodology for system sizing to minimal LCoE

certain mass or size limitation (AP-3). Within each cycle, tether tension and wing planform are fed into an aerodynamic and a structural model. This defines a system for which energy output and LCoE are estimated.

26.4.1 Aerodynamic and Structural Model

The structural optimization model is centered around the optimization of wing mass for a given cable tension, from which, with simpler models, the total aircraft mass is estimated. It has a parametric description of the wing structural cross-section (spars, spar caps, skin), its dimensions (aspect ratio, tapering etc.) and its mass breakdown (composites, honeycomb, glue etc.). Based on a non-linear lifting-line aerodynamic model the loads and stresses are computed and parameters are adjusted to achieve optimal mass within given boundary constraints (allowable for torsional stiffness, wing tip bending, stress/strain levels for fatigue, buckling, breaking), as illustrated in Fig. 26.9. The model has been validated for a number of data points using hand calculations as well as Computational Fluid Dynamics (CFD) and Finite Element Methods (FEM). To complete the mass estimate, the avionics mass is added as a constant. The fuselage is estimated as a fraction of the wing mass. Paint, glue, cabling, actuators and equipment masses are estimated based on interpolation of properties of components available on the market.

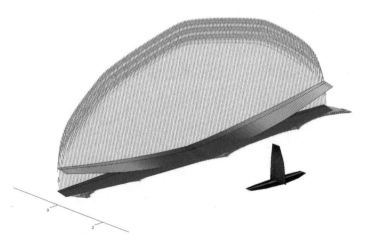

Fig. 26.9 Ampyx Power sizing tool computes for wing and tail surfaces the distribution of aero loads (shown as vectors) and then the resulting deflection of the structure (shown in color coding), based on a reference layout (spars, caps, skin), material properties and wall/skin thickness

26.4.2 Performance Model and Trade-Offs

The performance predictions of the Ampyx Power AWES are based on a number of analyses:

- Simulation in 6-DOF using detailed modeling of aircraft flight characteristics and controller, as visualized in Fig. 26.10. A flexible tether model is included [10].
- Prototype flights using AP-1 and AP-2, as shown in Fig. 26.4, matching closely the results of step 1 [10].
- Fast simulation using three independently developed cross-validated point mass tools [3]. This simulation is part of the integrated sizing toolset.

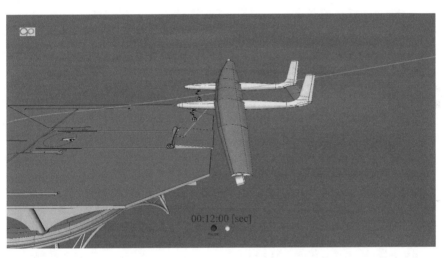

Fig. 26.10 Ampyx Power 6-DOF simulation including environmental and sensor disturbance models for Monte Carlo analysis. Here a tethered landing onto the platform under a gusty 8 Beaufort (17.2 to 20.7 m/s) wind conditions is shown

The resulting power curves from these analyses typically look like the curves shown in Fig. 26.11. Three regimes can be distinguished:

1. Cubic law regime. This is the regime where Loyd's estimation works well, at least as a trend [9]. The power rises with the cube of the wind speed.
2. Linear regime. The maximum cable tension has been reached, and the cable is operated at maximum tension. The reel-rate increases with the wind speed to limit the tension.
3. Flat regime. The maximum reel rate and generator power has been achieved. E.g. through angle of attack control the power is limited. In fact, in this regime, power production will generally decrease somewhat with increasing wind speed, e.g. due to larger margins to be taken for gust loads.

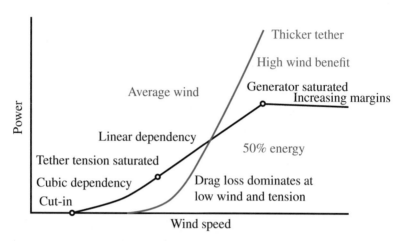

Fig. 26.11 Typical power curves as function for a typical optimal tether and a thicker tether. A tether designed for high tension has large drag losses at medium wind speeds. The typical transition points for an optimal tether selection are indicated. The tether is operated at maximum tension above approx. average wind speed, the power generator is sized (and power curve capped) to the power output achieved at approx. the 50% yearly energy wind level

The lesson learned from this analysis is the impact that the drag of a realistic tether has on performance. With a thin tether, the maximum tension is reached already at low wind speed, and the power gain with higher wind speed becomes linear through most of the regime. It would seem more favorable to have a thicker tether, such that the (seemingly beneficial) cubic growth of power with wind speed is stretched deep into the high-wind regime. However, such a tether has a large drag and at low wind speeds, it is not used to its potential (not fully loaded), and unnecessary tether drag is suffered. The result is that the power curves, though cubic, are so flat initially that the benefit of the thick tether only becomes apparent at very high wind speeds, which in turn are exceedingly rare and contribute little to the yearly energy production [3].

When the power curves are multiplied by the wind statistics distribution (e.g. a Weibull), the optimal tether thickness for a given aircraft design and wind statistics can become clear. It seems that approximately, the optimal tether thickness would lead to transition to the linear regime around the average wind speed [3, Fig. 11].

Primary factors that have a significant impact on power production have been analyzed to be:

- Reel-in speed: increase from 20 to 30 m/s raises power by 10%
- Capping of maximum power to limits of generator: relatively small effect: limiting the maximum power to 50% of the peak level at 20 m/s capping has only 5% effect on yearly power produced. The cost model indicates that the optimum generator is sized to reach full power at about 13 to 15 m/s, a value similar to conventional turbines. Techniques to cap the power include angle of attack / flap angle control, pattern elevation control and pattern width.

- System efficiencies: with respect to the mechanical reel-out power in ideal conditions (point mass), an overall 50 to 60% efficiency is considered a fair target to account for control and power conversion losses.
- Lift of wing profile: can be improved by multi-element wings, flow control etc. A coefficient of $C_L = 2.4$ for the entire vehicle, or higher, seems feasible.
- Roll angle: to fly the steep curves of the pattern, a system with large mass or short tether needs either a large roll angle, a longer tether or a wider pattern. A large roll angle between tether and aircraft body causes unfavorable loading of the tether (power loss), yet a shorter tether means less drag.
- Drag of tether: can be improved by 30% with e.g. latex coating, and by having a thinner cable although it would have to be replaced more often [6], experiences with Dyneema®material in [2]. The tether length is to be minimized, and is determined by the (optimal) roll angle (initial length) and the high wind reel-out (final length).

Figure 26.12 illustrates how our toolset can be used to optimize for a given mass and find the design that provides the best cycle power output. In fact, this is what

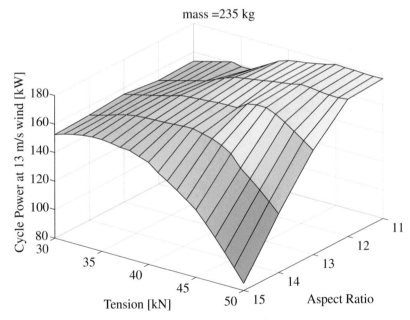

mass =235 kg

Fig. 26.12 Optimization of cable tension and wing aspect ratio towards maximum power output for a 235 kg aircraft

we have done for AP-3. For AP-3, being primarily an autonomy demonstrator, size is the primary constraint, not LCoE.

26.4.3 Cost Model and LCoE

In case of our commercial system AP-4, we are optimizing for minimal LCoE. Our model for LCoE includes a break-down into the known subsystems of the full facility and balance of plant parametric dependencies of cost on system sizing (esp. power output, cable length and cable tension, based on proxies from the wind turbine and aeronautical industry), as well as operational cost and modeled cable wear. Each input parameter has a range that is based on commercial and aviation cost levels. Uncertainty margin on the absolute results for this model and our input parameters is about 0.5 ct/kWh. Some typical results for an AP-4 analysis are provided in Fig. 26.13. Results of our model are roughly confirmed by Chap. 30 in this book.

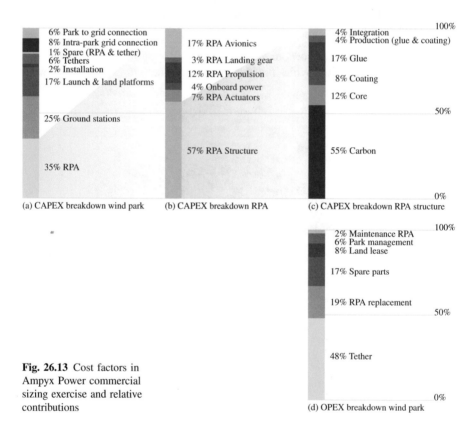

Fig. 26.13 Cost factors in Ampyx Power commercial sizing exercise and relative contributions

From a sensitivity analysis on LCoE, based on variation of all discussed design parameters, there appears to be an optimal wing area for minimal LCoE, as shown in Fig. 26.14. Best net profit for a given plant area may be obtained for wing areas roughly twice that size.

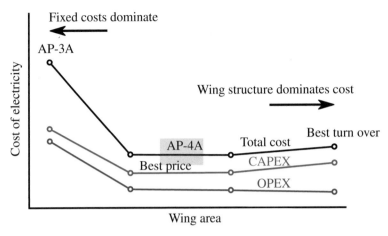

Fig. 26.14 Trends of Ampyx Power commercial sizing exercise and AP-4A target

26.5 Conclusions

The Ampyx Power AWES generates energy by flying a tethered rigid glider-aircraft attached to a ground-based generator. It follows a crosswind pattern as the tether unwinds under high tension and the aircraft spirals away from the generator. The tether rewinds under near-zero tension while the aircraft glides back to the generator. Ampyx Power targets the utility market, including offshore, and therefore requires high power output and high levels of autonomy, safety and reliability. This objective led us to a rotating-platform type horizontal Launch & Land solution. Ampyx Power currently operates 2 prototypes (AP-2A1 and AP-2A2, about 20 kW net power production demonstrated) in a test field in The Netherlands. The AP-3 with 200 kW is currently under development for which we have implemented and customized aeronautical design processes, airworthiness and safety standards. The commercial product AP-4 has been sized for minimal LCoE, resulting in a 2 MW class system. It shall operate at a LCoE well below that of conventional wind energy. The final offshore solution should feature a high power density per square kilometer surface area. The price estimate and system sizing are based on dynamic simulations over a range of wind speeds for typical sites, coupled to an aero/structural model and a cost model including capital and operational aspects. The dynamic simulations have been cross-validated as well as matched against AP-2 flight data. A step-wise development of avionics redundancy has been described, that should eventually allow co-use of the terrain overflown by the tethered aircraft. The presented approach has been developed for a rigid wing utility-scale AWE system, but may be more generally applicable within the wider AWE community.

Acknowledgements The financial support of the European Commission through the project AMPYXAP3 (H2020-SMEINST-666793) is gratefully acknowledged.

References

1. Bormann, A., Ranneberg, M., Kövesdi, P., Gebhardt, C., Skutnik, S.: Development of a Three-Line Ground-Actuated Airborne Wind Energy Converter. In: Ahrens, U., Diehl, M., Schmehl, R. (eds.) Airborne Wind Energy, Green Energy and Technology, Chap. 24, pp. 427–437. Springer, Berlin Heidelberg (2013). doi: 10.1007/978-3-642-39965-7_24
2. Bosman, R., Reid, V., Vlasblom, M., Smeets, P.: Airborne Wind Energy Tethers with High-Modulus Polyethylene Fibers. In: Ahrens, U., Diehl, M., Schmehl, R. (eds.) Airborne Wind Energy, Green Energy and Technology, Chap. 33, pp. 563–585. Springer, Berlin Heidelberg (2013). doi: 10.1007/978-3-642-39965-7_33
3. Breukelman, P., Kruijff, M., Fujii, H. A., Maruyama, Y.: A new wind-power generation method employed with high altitude wind. In: Proceedings of Grand Renewable Energy 2014 International Conference, Tokyo, Japan, 27 July–1 Aug 2014
4. Dunker, S.: Ram-Air Wing Design Considerations for Airborne Wind Energy. In: Ahrens, U., Diehl, M., Schmehl, R. (eds.) Airborne Wind Energy, Green Energy and Technology, Chap. 31, pp. 517–546. Springer, Berlin Heidelberg (2013). doi: 10.1007/978-3-642-39965-7_31
5. Goldstein, L.: Density of Individual Airborne Wind Energy Systems in AWES Farms. http://www.awelabs.com/wp-content/uploads/AWES_Farm_Density.pdf (2014). Accessed 19 May 2016
6. Jung, T. P.: Wind Tunnel Study of Drag of Various Rope Designs. AIAA Paper 2009-3608. In: Proceedings of the 27th AIAA Applied Aerodynamics Conference, San Antonio, TX, USA, 22–25 June 2009. doi: 10.2514/6.2009-3608
7. Kruijff, M.: Tethers in Space, A propellantless propulsion in-orbit demonstration. Ph.D. Thesis, Delft University of Technology, 2011. http://resolver.tudelft.nl/uuid:9d437e58-82c0-4af1-935f-69ba5573c7a2
8. Kruijff, M., Ruiterkamp, R.: Status and Development Plan of the PowerPlane of Ampyx Power. In: Schmehl, R. (ed.). Book of Abstracts of the International Airborne Wind Energy Conference 2015, pp. 18–21, Delft, The Netherlands, 15–16 June 2015. doi: 10.4233/uuid: 7df59b79-2c6b-4e30-bd58-8454f493bb09. Presentation video recording available from: https://collegerama.tudelft.nl/Mediasite/Play/2e1f967767d541b1b1f2c912e8eff7df1d
9. Loyd, M. L.: Crosswind kite power. Journal of Energy **4**(3), 106–111 (1980). doi: 10.2514/3.48021
10. Ruiterkamp, R., Sieberling, S.: Description and Preliminary Test Results of a Six Degrees of Freedom Rigid Wing Pumping System. In: Ahrens, U., Diehl, M., Schmehl, R. (eds.) Airborne Wind Energy, Green Energy and Technology, Chap. 26, pp. 443–458. Springer, Berlin Heidelberg (2013). doi: 10.1007/978-3-642-39965-7_26
11. Society of Automotive Engineers: Quality Management Systems – Requirements for Aviation, Space and Defense Organizations, AS9100. http://standards.sae.org/as9100d/
12. Vander Lind, D.: Analysis and Flight Test Validation of High Performance Airborne Wind Turbines. In: Ahrens, U., Diehl, M., Schmehl, R. (eds.) Airborne Wind Energy, Green Energy and Technology, Chap. 28, pp. 473–490. Springer, Berlin Heidelberg (2013). doi: 10.1007/978-3-642-39965-7_28

Part V
Technology Deployment

Chapter 27
Niche Strategies to Introduce Kite-Based Airborne Wind Energy

Linda M. Kamp, J. Roland Ortt and Matthew F. A. Doe

Abstract Kite-based airborne wind energy systems are new high-tech systems that provide sustainable wind energy. Instead of using a wind turbine, these systems use a kite to generate energy. Commercializing such new high-tech systems is a risky strategy, the failure rate is high. This chapter identifies barriers that block large-scale diffusion of kite-based airborne wind energy systems and specific niche strategies to deal with these barriers. The results are based upon literature research and interviews with six academic and industry experts active in the field of airborne wind energy. We identified the most important barriers to large-scale implementation of airborne wind energy. We show how particular barriers, such as the lack of knowledge of the technology and the lack of support and investment opportunities, interact and together block large-scale production and diffusion. The second result is that several niche strategies can be identified to tackle the barriers in this field. The "geographic niche strategy", the "demo, experiment and develop niche strategy" and the "educate niche strategy" are identified as good strategies to introduce the kite-based systems. The chapter ends with a discussion of these niche strategies and how they relate to previous research into introduction of sustainable energy technologies.

27.1 Introduction

This chapter focuses on airborne wind energy systems and explores how specific niche strategies can be selected for introducing these systems by analysing barriers to their large-scale implementation.

Airborne wind energy (AWE) is a cluster of technologies with the ability to extract wind power by using airborne elements. How the wind energy is converted into

Linda M. Kamp (✉) · J. Roland Ortt · Matthew F. A. Doe
Delft University of Technology, Faculty of Technology, Policy and Management, Jaffalaan 5, 2600 GA Delft, The Netherlands
e-mail: l.m.kamp@tudelft.nl

© Springer Nature Singapore Pte Ltd. 2018
R. Schmehl (ed.), *Airborne Wind Energy*, Green Energy and Technology, https://doi.org/10.1007/978-981-10-1947-0_27

(predominantly) electrical energy is what differentiates the technologies in the cluster, as different mechanisms are applied to lift the systems into the air and convert wind energy into electrical energy [3]. In this chapter we focus on one particular AWE configuration: kite-based traction power systems.

A serious problem with radically new high-tech systems in general is that it takes long before large-scale diffusion starts [27]. That is particularly true for sustainable energy provision systems such as biomass gasifiers [34, 40], wind turbines [21] or solar PV [19, 39]. An explanation for the time span between invention and large-scale diffusion can be found by looking at barriers. Many barriers have to be faced before large-scale diffusion is possible. A way to deal with these barriers is to introduce the product in a small part of the market first—a niche market. The term niche market refers to a relatively small group of customers with specific wants and demands regarding a product [11, 38].

This chapter has two goals. Firstly, it investigates the types of barriers that exist for the introduction of an innovative kite-based airborne wind energy system and the relative importance of these barriers. Secondly, it explores how these barriers can be dealt with by means of specific niche strategies that either break away or circumvent these barriers.

In recent years, a number of papers, such as [9, 15, 29] have been published that used a strategic niche management approach to investigate the introduction of sustainable energy systems. The current chapter is the first to apply such an approach to the case of airborne wind energy systems. It is also the first to investigate specific niche strategies that companies can use. Within the research field of airborne wind energy systems, most publications so far have focused on technical aspects. Some publications have taken another viewpoint and investigate issues such as economic aspects [12], and patent analyses [33]. However, in this field no research has been published yet on barriers to large-scale implementation of these technologies and on strategies to deal with these barriers.

The remainder of this chapter is organized as follows. Section 27.2 describes the kite-based airborne wind energy systems. Section 27.3 describes theoretical notions derived from earlier work on barriers and niche strategies. Section 27.4 covers the research methodology. In Sect. 27.5 are the research findings with regard to barriers and strategies. Section 27.6 presents the conclusion, discussion and recommendations. The preliminary content of the present chapter has been presented at the Airborne Wind Energy Conference 2015 [30].

27.2 Practice: Airborne Wind Energy Systems

The idea of using airborne devices, mainly kites, goes back many centuries. Yet, it was not until 1827 that the first book on the topic was published by George Pocock. In this book Pocock describes his successful experiments with carriages driven by kites. After that came the "Golden Age of Kites" (1860—1915) in which kites developed technically to a high level before being pushed out by the rising aviation industry. But it took many decades, dominated by fossil fuels, before the idea of airborne wind energy was reinvigorated after the oil crises of the 1970s. In 1975, space pioneer Hermann Oberth published a book on airborne wind energy, "Das Drachenkraftwerk" [26] and Payne and McCutchen patented airborne wind power concepts. In 1979, Bryan Roberts conducted demonstration experiments of "flying electricity generators" in Australia. In 1980 Miles Loyd published his chapter "Crosswind Kite Power", in which the foundations for quantitative analysis of airborne wind power systems was laid [23].

In 1997, the late Dutch astronaut and university professor Wubbo Ockels patented the Laddermill and started a research group at Delft University of Technology [3, 10]. In 2001 in Germany SkySails developed the first commercial kite system for ship traction. The company Makani Power was founded in 2006 with substantial funds from Google. That same year, Windlift in the US and NTS in Germany were founded, while the KiteGen project realized a pumping kite system [10]. Following a High Altitude Wind Power Conference in Chico, California, in 2009, the first international Airborne Wind Energy Conference (AWEC) was held 2010 in Stanford, California, and from then on six annual international conferences have taken place, including the 2015 event. In 2013 Makani Power was fully acquired by Google and the first book on "Airborne Wind Energy" was published by Springer Verlag, in part cataloging the history and development of airborne wind energy. Currently, the technologies occupy a niche in a fossil-fuel driven landscape but the number of research institutes involved in the development of AWE systems has grown enormously [3].

Airborne wind energy systems require exceptionally strong yet lightweight system designs. Currently, different airborne systems are proposed. Some systems use conventional turbines to generate electricity, either suspended in the air by a helium-filled structure or on a crosswind flying wing. There are also system configurations and technologies that use drag sails to harness wind power or systems that use the auto-gyro effect for both lift and power generation.

27.3 Theory: Barriers and Niche Strategies

Rogers [35] describes a model in which the diffusion of a product follows a smooth S-shaped pattern. This model is based upon two assumptions: (1) that a new product is directly introduced into the large market and (2) that the product remains essentially invariant over the life cycle. However, evidence shows that often products are not introduced into the large market directly because in this market there are barriers for market introduction. A way to deal with these barriers is to first introduce the new product in a small market—a niche market—using a niche strategy. This may also involve developing the product further through incremental and radical innovations, as also found in [1]. Several types of niches exist. We define a "strategic niche" as a niche that emerges prior to industrial production and large-scale diffusion of a new high-tech product in a mainstream application.

There are many examples of new innovations that were introduced in a strategic niche before large-scale market introduction [28, 32]. A typical example is the use of solar PV in satellites. This strategic niche appeared prior to the use of solar PV for generating electricity for households. Strategic niches appear when one or more factors that are needed for large-scale diffusion of a new high-tech product are missing. Here we define a high-tech product using three elements: a high-tech product is an artifact with a certain functionality, based on technological principles and consisting of a number of main components. Using this definition, we define an AWE system as a high-tech product. An overview of the factors needed for large-scale diffusion is presented in Table 27.1. This overview, taken from [31], is based on literature research in [6, 13, 16, 22, 24]. These sources investigate factors that have to be present in the market and in the wider social context in order to make development and large-scale diffusion of innovations possible. All of the resulting factors were ordered and combined into twelve categories, presented in Table 27.1.

After analyzing these factors, specific niche strategies can be derived. In practice that is not as straightforward as it seems. If, for example, the factor 'availability of customers' (factor 5 in Table 27.1) is not present in the system, then this is a barrier which seriously hampers large-scale diffusion. However, the mere existence of this barrier does not reveal what type of niche strategy can be adopted. More knowledge of the market and the context is required to derive possible niche strategies. The twelve categories of factors have different roles. The absence of some factors can directly block large-scale diffusion (such as the lack of customers) whereas the absence of other factors serve as a cause of that barrier. Customers can be lacking, for example, because they miss the knowledge required to understand and use a product (factors 7 and 8 in Table 27.1) or they can be lacking because these customers cannot afford the product (factor 11 in Table 27.1). In these cases, completely different niche strategies should be considered. In the first situation, a niche strategy should be aimed at educating customers. In the second situation, a niche strategy can aim at a simple and cheaper version of the product or a niche strategy can supply the product to a wealthy top customer segment. In this way, the twelve factors in Table 27.1 can be divided in six core factors and six causes, as shown in Fig. 27.1.

Factors 1-6 in Table 27.1 and Fig. 27.1 have a direct effect on the large-scale diffusion of the high-tech product whereas factors 7-12 have a more indirect effect, because they influence one or more of the factors 1-6.

Figure 27.1 is built up in two layers. Factors 1-6 (middle part of Fig. 27.1), referred to as core factors, represent the core technological and market system required for large-scale diffusion. Some of these core factors refer to technical components and subsystems such as the product itself, the production system and complementary products and services. Some other factors refer to availability of actors such as customers or the availability of support and investors. The institutional aspects refer to the laws, rules, norms and values used to guide processes such as production, supply, adoption and use. Each of these core factors need to be in place in order to enable large-scale diffusion to occur. The second layer of factors (left part of Fig. 27.1), referred to as influencing factors, contains contextual factors that explain why problems in the core system emerge. Two of these influencing factors relate

Factors	Description
1. Availability of a new high-tech product	The product needs to have a good price/performance ratio compared to competitive products in the perception of customers before large-scale diffusion is possible. If (one or more components of) the product is/are not available, large-scale diffusion is not (yet) possible.
2. Availability of a production system	Availability of a system to produce the technology is required for large-scale diffusion. In some cases a product can be created in small numbers but if industrial production technologies are not yet available, then large-scale diffusion is not possible.
3. Availability of complementary products and services	Complementary products and services refer to products and services required for the production, distribution, adoption and use. The unavailability of such products and services means that large-scale diffusion is not (yet) possible.
4. Availability of support & investments	The availability of local support and partners that facilitate the investment of suppliers, customers and others in the technology
5. Availability of customers	The availability of customers means that a market application for the product is identified, that customer segments for these applications exist and that the customers are knowledgeable about the product and its use and are willing and able to pay for adoption. If applications are unknown or if customer groups do not exist, are not able to obtain the product or are unaware of the benefits of the product, large-scale diffusion is blocked.
6. Availability of supporting institutional aspects (laws, rules and standards)	The regulatory and institutional environment refers to the laws and regulations that indicate how actors (on the supply and demand side of the market) deal with new product. These laws and regulations can either stimulate the diffusion of radically new high-tech products (such as subsidy that stimulates the use of sustainable energy) or completely block it (such as laws prohibiting something).
7. Availability of knowledge of technology	The knowledge of the technology refers to the knowledge required to develop, produce, replicate and control the technological principles in a product. In many cases a lack of knowledge blocks large-scale diffusion.
8. Availability of knowledge of application	Knowledge of the application can refer to knowing potential applications. If a technological principle is demonstrated but there is no clue about its practical application, large-scale diffusion is impossible. A lack of knowledge of the application can also refer to customers that do not know how to use a new product in a particular application. In that case large-scale diffusion is not possible either.
9. Availability of relevant natural resources and labour	Natural resources and labour are required to produce and use a new high-tech product. These resources and labor can be required for the production system, for complementary products and services or for the product itself. In many cases a lack of resources and labor block large-scale diffusion.
10. Availability of supporting socio-cultural aspects	Socio-cultural aspects refer to the norms and values in a particular culture. These aspects might be less formalized than the laws and rules in the institutional aspects but their effect can completely block large-scale diffusion.
11. Availability of supporting macro-economic aspects	Macro-economic aspects refer to the national or global economic situation. For example, a recession can stifle the diffusion of a new high-tech product.
12. Availability of a positive vision and image	If the main actors and the broader public hold a negative perception regarding the technology and its potential, this can block large-scale diffusion.

Table 27.1 Actors and factors necessary for large-scale diffusion [31]

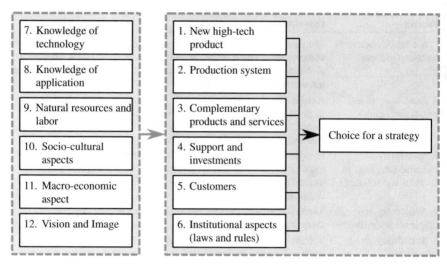

Fig. 27.1 Factors important for the development and large-scale diffusion of new high-tech products and hence for the choice of niche strategies [31]

to knowledge, i.e., knowledge of the technology and knowledge of the application. One factor represents the natural resources and labor. Two other factors refer to the socio-cultural aspects and the macro-economic aspects that drive or hamper the core system. Finally, the vision and image aspects were added to the model because they were found to be important in particular for AWE systems.

After distinguishing the barriers, we formulated specific niche strategies in three steps (see also [31]). First, we identified 21 logical combinations of a core factor and an influencing factor that together can form a barrier for large-scale market introduction. For example, lack of customers because of lack of knowledge of the application (example 1), or lack of investments because of macro-economic aspects (example 2). Second, we argued what would be needed to overcome or circumvent each of these 21 barriers. For example, educating customers can overcome a lack of knowledge (example 1 above) and subsidizing can overcome a lack of spending power among potential customers (example 2 above). This systematic search for strategies in each of these situations resulted in a list of ten strategies that can be applied to remove or circumvent at least one of the 21 identified barriers. Third, we found 50 historical cases of high-tech products in which these strategies were applied in practice to deal with specific barriers. More information on these cases can be found in [31]. Table 27.2 shows the ten niche strategies, as also published in [31].

Description	Description of the specific niche strategy
1. Demo, experiment and develop niche strategy	A niche strategy can be adopted to demonstrate the product in public in a controlled way so the limited quality of performance is not a problem. As part of the strategy experimenting with the product is important to develop the product further.
2. Top niche strategy	A niche strategy can be adopted where specially made products can be made to order, in small numbers, for a specific top-end niche of the market. A skimming strategy can be adopted in which the top niche of customers is supplied first with a special product.
3. Subsidized niche strategy	A niche strategy can be adopted where the product is subsidized if its use by a particular segment of users is considered as societally relevant or important.
4. Redesign niche strategy	A niche strategy can be adopted where the product is introduced in a simpler version that can be produced with the existing knowledge, less use of resources and therefore for a lower price
5. Dedicated system or stand-alone niche strategy	A niche strategy can be adopted where the product is used in stand-alone mode or a dedicated system of complementary products and services is designed (e.g., a local network when an infrastructure is not available on a wider scale).
6. Hybridization or adaptor niche strategy	A niche strategy can be adopted by which the new product is used in combination with the old product and thereby all existing complementary products and services can be re-used. Or an adaptor/convertor is provided to make the product compatible with existing complementary products and services.
7. Educate niche strategy	A niche strategy can be adopted aimed at transferring the knowledge to suppliers and customers.
8. Geographic niche strategy	A niche strategy that can be adopted to introduce the new product into the market in another geographic area where the conditions are more favorable.
9. Lead user niche strategy	A niche strategy can be adopted finding innovators or lead users. These users can co-develop the product because they are willing to experiment with the product.
10. Explore multiple markets niche strategy	A niche strategy can be adopted in which multiple customer applications can be explored. Visibility of the first applications can stimulate explorative use in new applications.

Table 27.2 Specific niche strategies and the conditions in which they can be considered [31]

27.4 Research Methodology for the AWE Analysis

Our research methodology to find out the main barriers for large-scale diffusion of kite-based high-altitude wind energy and the strategies to deal with these barriers can be divided into two parts. The first part consisted of literature research into barriers and suggested strategies for market introduction of AWE systems. The sources used in the literature research are [3, 8, 17, 20, 25]. The second part consisted of six interviews with AWE experts from different countries, two of which from academia and four from business (founders or managing directors from AWE companies).

The interview method is described in more detail below. For a still more detailed description of the interview method see [29].

Assessing the market for radically new high-tech systems is very difficult. Even experts can experience difficulties [37, 41]. In general they tend to be prone to bias and inconsistency, both of which can damage their accuracy [18, 37], they tend to place too much trust in their own predictions [4, 7], they only consider a very limited set of alternative strategies [14] and they tend to choose strategies intuitively rather than systematically [2]. Several expert methods have been developed (see e.g. [5, 36]). The Delphi technique, for example, can lead to a consensus between a range of experts on a specific topic.

The goal of the method used in this chapter is to assess the market situation, to indicate the most important barriers for large-scale diffusion of AWE systems, and to select niche strategies to deal with these barriers. Rather than consensus, we aim to seek consistency in each individual expert's evaluations. Therefore, we did not use the Delphi approach but, instead, decided to interview experts separately. In order to make the outcomes of each interview as consistent as possible and therefore as reliable as possible, we designed four steps in each interview, as described below. As input for the interviews we used the pre-specified list of theoretical barriers (describing the factors required for large-scale diffusion and their causes) as presented in Fig. 27.1 and the pre-specified list of theoretical niche strategies as presented in Table 27.2.

In the first step we asked the experts to indicate their experience and expertise regarding technological and market aspects of AWE systems. The information was used to describe the general market situation and to indicate the case specific knowledge of the experts. This knowledge serves as a proxy or indicator of the validity of subsequent expert evaluations in the interview. In the second step we addressed the market situation for AWE systems further and in particular discussed the barriers for large-scale diffusion and their causes. In this step we started asking for barriers (open question) to find out what the experts thought. The answers also served as a check whether our pre-specified list of barriers was complete. We then proceeded by showing the experts the pre-specified list of barriers and requested them to select the most important ones (closed question). Also, we asked them to reflect upon a list of barriers for diffusion of high-altitude wind energy that we had extracted from written sources [3, 8, 17, 20, 25] (closed question).

Finally we discussed possible discrepancies between the answers to the open and closed questions. In the third step we addressed possible niche strategies in a similar process: an open question about possible niche strategies, a closed question to rate our pre-specified list of theoretical niche strategies and suggested strategies for high-altitude wind energy, and a discussion to address possible discrepancies between the answers to the open and closed questions. In the fourth step we asked the experts about the linkages between the most important barriers on the one hand and the selected niche strategies on the other hand. After the interviews, we checked the interview outcomes with our findings from the literature on barriers and strategies for market introduction of AWE systems. Using all these steps and checks we ensured that the outcomes were as consistent and as reliable as possible.

27.5 Findings from the AWE Analysis

This section presents the barriers (combinations of core and influencing factors) blocking large-scale diffusion of AWE systems and the main strategies to deal with these barriers.

27.5.1 Findings regarding the barriers and their relative importance

Table 27.3 provides an overview of the barriers blocking large-scale diffusion of AWE systems. We list eight of these barriers, starting with the most important one. Each barrier is formulated in terms of a core factor and an influencing factor. For example, the first line in Table 27.3 indicates that a lack of knowledge regarding the AWE technology has a negative effect on support and investment, which represents a serious barrier to large-scale diffusion.

The results in Table 27.3 clearly indicate that a lack of knowledge of technology is one of the most important influencing factors. This factor is present five times in the eight most important influencing factors. Some of the remaining influencing factors have a clear link with the knowledge of technology. The vision & image has a negative effect because of the uncertainty regarding reliability, operation and safety and that is also related to a lack of knowledge of the technology. The same applies to the uncertainty about economic performance.

27.5.2 Findings regarding the strategies and their relative importance

The strategy ranking provides an overview of the niche strategies to introduce kite-based AWE systems, which can be used to tackle certain barriers. In Table 27.4 we list the three main niche strategies, starting with the most important one.

The table indicates that the most important strategy is the geographic niche strategy. It is interesting to find out that this most important niche strategy has no obvious link to the predominant barrier of a lack of technological knowledge. However, the geographic niche strategy is an obvious choice for high-altitude wind energy. Firstly, rules and regulations as well as investment climate for AWE systems vary widely across geographic regions. As indicated by our interviewees, this is the most important reason to opt for the geographic niche strategy. Other reasons can be that the average wind speed varies per region and that AWE systems require a large area to lift and use them safely, which limits the regions where they can be used.

The other two niche strategies are more logically linked to the predominant barrier of a lack of technological knowledge of the AWE systems. This current lack of knowledge of the technology makes a "demo, experiment and develop" niche strategy a logical choice. If this knowledge is mastered but has not yet diffused among stakeholders then an "educate niche" strategy is a logical strategy.

27.6 Conclusion and Discussion

Based on literature research and a structured interview method we have detected the most important barriers that block large-scale diffusion for kite-based airborne wind energy systems (AWE systems) and we selected three niche strategies for introducing these systems.

A central problem for the AWE systems is a lack of knowledge of the technology. This indicates that the principle is still experimental. For this problem a so-called "demo, experiment and develop" niche strategy is suggested by our results. Also during the early stages of the innovation process of quite a number of other sustainable energy technologies such as wind turbines or PV the "demo, experiment and

No.	Barriers in terms of influencing factors and core factors	Description
1.	Knowledge of Technology—Support & Investment	Lack of proof of concept and performance of the technology inhibit investment and support for further development of kite-based AWE systems.
2.	Vision & Image—Support & Investment	Uncertainty regarding reliability, operation and safety of kite-based AWE systems amongst the general public and investors has a negative effect on the support for and investment in kite-based AWE systems.
3.	Macro-economic aspects—Customers	Competition of other (renewable) energy systems inhibits market access for kite-based AWE system customers.
4.	Knowledge of Technology—Institutional aspects	Lack of knowledge and experience inhibits access and regulation of airspace for kite-based AWE systems.
5.	Knowledge of Technology—Customers	Lack of experience and data regarding safety and reliability inhibits customers of kite-based AWE systems.
6.	Knowledge of Technology—Technological development	Technical challenges regarding control systems and materials inhibit a marketable AWE system.
7.	Macro-economic aspects—Support	Investment Uncertainty about the economic performance of kite-based AWE systems undermines investment and support.
8.	Knowledge of Technology—Product processes	Experience and knowledge of the manufacturing and the supply chain of kite-based AWE systems is minimal.

Table 27.3 Ranking of barriers (in terms of core factors and influencing factors) for AWE systems

develop" niche strategy has been used [21, 39]. Especially in the Netherlands this strategy was the predominant one. However, it should be noted that sources such as [21] and [39] show that too much focus on this strategy alone can slow down the innovation process because of the risk of too much focus on R&D aspects and not involving the demand side of the market enough.

Another central problem for the AWE systems is the lack of knowledge of these systems that relevant stakeholders have. The lack of knowledge explains the poor image of these systems. This image, in turn, has an impact on available investment funds. For this problem a so-called "educate niche strategy" is suggested by our results.

The previous two problems are related to the early and experimental stage of AWE systems. In due course, we expect these problems to be solved. The third and most important strategy that we found, the geographic niche strategy, at first sight seems unrelated to the selected barriers. This strategy is required because of the large differences in rules and regulations between different countries. Another reason for this strategy is that AWE systems require a significant amount of space both on land and in the air to be used safely. This requirement calls for a selection of regions that fulfill this requirement, and that implies a geographic niche strategy. In the early stages of wind turbine development in the early 1980s, this strategy was also pursued by, among others, Dutch and Danish wind turbine manufacturers that entered the market in California because more space was available there and regulations and subsidies were more favorable [21]. As [21] shows, when pursuing this strategy it is important to build up strong relationships with other local stakeholders such as the demand side of the market and local policy makers since this improves the knowledge flows between stakeholders and therefore the innovation process. Pursuing the suggested strategies will remove the barriers and therefore turn them into opportunities for accessing new (niche) markets.

No.	Strategy	Description
1.	Geographic niche strategy	A niche strategy can be adopted to introduce AWE systems in the market in another geographic area where the conditions are more favorable and there are less barriers.
2/3.	Demo, experiment and develop niche strategy	A niche strategy can be adopted to demonstrate AWE systems in public in a controlled way. As part of the strategy experimenting with the product it is important to develop the product further, for example in a research environment.
2/3.	Educate niche strategy	A niche strategy can be adopted aimed at transferring the knowledge of AWE to consumers, suppliers, policy makers and other relevant actors.

Table 27.4 Strategy ranking for AWE systems

References

1. Abernathy, W. J., Utterback, J. M.: Patterns of innovation in technology. Technology Review **80**(7), 40–47 (1978)
2. Agor, W. H.: The logic of intuition: How top executives make important decisions. Organizational Dynamics **14**(3), 5–18 (1986). doi: 10.1016/0090-2616(86)90028-8
3. Ahrens, U., Diehl, M., Schmehl, R. (eds.): Airborne Wind Energy. Green Energy and Technology. Springer, Berlin Heidelberg (2013). doi: 10.1007/978-3-642-39965-7
4. Arkes, H. R.: Overconfidence in judgmental forecasting. In: Armstrong, J. S. (ed.) Principles of Forecasting: A Handbook for Researchers and Practitioners, pp. 495–515. Kluwer Academic Publishers, Dordrecht (2001). doi: 10.1007/978-0-306-47630-3_22
5. Armstrong, J. S. (ed.): Principles of Forecasting: A Handbook for Researchers and Practitioners. Springer Science+Business Media, New York (2001). doi: 10.1007/978-0-306-47630-3
6. Bergek, A., Jacobsson, S., Carlsson, B., Lindmark, S., Rickne, A.: Analyzing the functional dynamics of technological innovation systems: A scheme of analysis. Research Policy **37**(3), 407–429 (2008). doi: 10.1016/j.respol.2007.12.003
7. Brenner, L. A., Koehler, D. J., Liberman, V., Tversky, A.: Overconfidence in Probability and Frequency Judgments: A Critical Examination. Organizational Behavior and Human Decision Processes **65**(3), 212–219 (1996). doi: 10.1006/obhd.1996.0021
8. Bronstein, M. G.: Harnessing rivers of wind: A technology and policy assessment of high altitude wind power in the U.S. Technological Forecasting and Social Change **78**(4), 736–746 (2011). doi: 10.1016/j.techfore.2010.10.005
9. Caniëls, M. C. J., Romijn, H. A.: Actor networks in Strategic Niche Management: Insights from social network theory. Futures **40**(7), 613–629 (2008). doi: 10.1016/j.futures.2007.12.005
10. Cherubini, A., Papini, A., Vertechy, R., Fontana, M.: Airborne Wind Energy Systems: A review of the technologies. Renewable and Sustainable Energy Reviews **51**, 1461–1476 (2015). doi: 10.1016/j.rser.2015.07.053
11. Dalgic, T., Leeuw, M.: Niche Marketing Revisited: Concept, Applications and Some European Cases. European Journal of Marketing **28**(4), 39–55 (1994). doi: 10.1108/03090569410061178
12. De Lellis, M., Mendonça, A. K., Saraiva, R., Trofino, A., Lezana, Á.: Electric power generation in wind farms with pumping kites: An economical analysis. Renewable Energy **86**, 163–172 (2016). doi: 10.1016/j.renene.2015.08.002
13. Edquist, C.: Design of innovation policy through diagnostic analysis: identification of systemic problems (or failures). Industrial and Corporate Change **20**(6), 1725–1753 (2011). doi: 10.1093/icc/dtr060
14. Ehrlinger, J., Eibach, R. P.: Focalism and the Failure to Foresee Unintended Consequences. Basic and Applied Social Psychology **33**(1), 59–68 (2011). doi: 10.1080/01973533.2010.539955
15. Eijck, J. van, Romijn, H.: Prospects for Jatropha biofuels in Tanzania: An analysis with Strategic Niche Management. Energy Policy **36**(1), 311–325 (2008). doi: 10.1016/j.enpol.2007.09.016
16. Geels, F. W.: From sectoral systems of innovation to socio-technical systems. Research Policy **33**(6–7), 897–920 (2004). doi: 10.1016/j.respol.2004.01.015
17. GL Garrad Hassan: Market Status Report High Altitude Wind Energy, now merged with DNV GL, Aug 2011
18. Hogarth, R. M.: Judgment and Choice: The Psychology of Decision. 2nd ed. John Wiley, New York (1987)
19. Huijben, J. C. C. M., Verbong, G. P. J.: Breakthrough without subsidies? PV business model experiments in the Netherlands. Energy Policy **56**, 362–370 (2013). doi: 10.1016/j.enpol.2012.12.073
20. Inman, M. (ed.): Energy High in the Sky: Expert Perspectives on Airborne Wind Energy Systems, Near Zero, 9 Sept 2012. http://www.nearzero.org/reports/airbornewind/pdf

21. Kamp, L. M., Smits, R. E. H. M., Andriesse, C. D.: Notions on learning applied to wind turbine development in the Netherlands and Denmark. Energy Policy **32**(14), 1625–1637 (2004). doi: 10.1016/S0301-4215(03)00134-4
22. Kemp, R., Schot, J., Hoogma, R.: Regime shifts to sustainability through processes of niche formation: The approach of strategic niche management. Technology Analysis & Strategic Management **10**(2), 175–198 (1998). doi: 10.1080/09537329808524310
23. Loyd, M. L.: Crosswind kite power. Journal of Energy **4**(3), 106–111 (1980). doi: 10.2514/3. 48021
24. Malerba, F.: Sectoral systems of innovation and production. Research Policy **31**(2), 247–264 (2002). doi: 10.1016/S0048-7333(01)00139-1
25. Megahed, N. A.: Landscape and Visual Impact Assessment: Perspectives and Issues with Flying Wind Technologies. International Journal of Innovative Research in Science, Engineering and Technology **3**, 11525–11534 (2014). http://www.ijirset.com/upload/2014/april/88_Landscape.pdf
26. Oberth, H.: Das Drachenkraftwerk. Uni Verlag, Dr. Roth-Oberth, Feucht, Germany (1977)
27. Ortt, J. R.: Understanding the Pre-diffusion Phases. In: Tidd, J. (ed.) Gaining Momentum Managing the Diffusion of Innovations, pp. 47–80. Imperial College Press, London (2010)
28. Ortt, J. R., Delgoshaie, N.: Why does it take so long before the diffusion of new high-tech products takes off? In: Proceedings of 17th International Conference on Management of Technology, International Association for Management of Technology (IAMOT), Dubai, United Arab Emirates, 6–10 Apr 2008
29. Ortt, J. R., Kamp, L. M., Doe, M. F.: Niche strategy selection to introduce radically new systems – The case of kite-based airborne wind energy. In: Proceedings of the 21st IEEE International Conference on Engineering, Technology and Innovation/ International Technology Management Conference (ICE/ITMC), Belfast, Northern Ireland, 22–24 June 2015. doi: 10.1109/ICE.2015.7438672
30. Ortt, J. R., Kamp, L. M., Doe, M. F. A.: How to Introduce Kite-Based Airborne Wind Energy Systems – The Selection of Niche Strategies to Overcome Barriers to Adoption. In: Schmehl, R. (ed.). Book of Abstracts of the International Airborne Wind Energy Conference 2015, pp. 64–65, Delft, The Netherlands, 15–16 June 2015. doi: 10.4233/uuid:7df59b79-2c6b-4e30-bd58-8454f493bb09. Presentation video recording available from: https://collegerama. tudelft.nl/Mediasite/Play/d923a66b221b4117872189e7a73d29e71d
31. Ortt, J. R., Langley, D. J., Pals, N.: Ten Niche Strategies To Commercialize New High-Tech Products. In: Proceedings of the International Conference on Engineering, Technology and Innovation (ICE) & IEEE International Technology Management Conference, The Hague, The Netherlands, 24–26 June 2013. doi: 10.1109/ITMC.2013.7352687
32. Ortt, J. R., Suprapto, M.: The role of strategic niches in creating large-scale applications for high-tech products. In: Proceedings of 20th International Conference of the International Association for Management of Technology, Miami Beach, Florida, USA, 10–14 Apr 2011
33. Phan, K., Daim, T.: Forecasting the Maturity of Alternate Wind Turbine Technologies Through Patent Analysis. In: Daim, T., Oliver, T., Kim, J. (eds.) Research and Technology Management in the Electricity Industry: Methods, Tools and Case Studies, Chap. 8, pp. 189–211. Springer, London (2013). doi: 10.1007/978-1-4471-5097-8_8
34. Raven, R. P. J. M.: Strategic Niche Management for Biomass – A comparative study on the experimental introduction of bioenergy technologies in the Netherlands and Denmark. Ph.D. Thesis, TU Eindhoven, 2005. http://alexandria.tue.nl/extra2/200511821.pdf
35. Rogers, E. M.: Diffusion of Innovation. 1st ed. Free Press of Glencoe, New York (1962)
36. Rowe, G., Wright, G.: Expert opinions in forecasting: the role of the Delphi technique. In: Armstrong, J. S. (ed.) Principles of Forecasting: A Handbook for Researchers and Practitioners, pp. 125–146. Springer Science+Business Media, New York (2001). doi: 10.1007/978-0-306-47630-3_7
37. Schnaars, S. P.: Megamistakes; Forecasting and the Myth of Rapid Technological Change. 29th ed. Free Press, New York, London (1989)

38. Shani, D., Chalasani, S.: Exploiting Niches Using Relationship Marketing. Journal of Consumer Marketing **9**(3), 33–42 (1992). doi: 10.1108/07363769210035215
39. Vasseur, V., Kamp, L. M., Negro, S. O.: A comparative analysis of Photovoltaic Technological Innovation Systems including international dimensions: the cases of Japan and The Netherlands. Journal of Cleaner Production **48**, 200–210 (2013). doi: 10.1016/j.jclepro.2013.01.017
40. Verbong, G. P. J., Christiaens, W., Raven, R. P. J. M., Balkema, A.: Strategic Niche Management in an unstable regime: Biomass gasification in India. Environmental Science & Policy **13**, 272–281 (2010). doi: 10.1016/j.envsci.2010.01.004
41. Wheeler, D. R., Shelley, C. J.: Toward More Realistic Forecasts for High-Technology Products. Journal of Business & Industrial Marketing **2**(3), 55–63 (1987). doi: 10.1108/eb006036

Chapter 28
Ecological Impact of Airborne Wind Energy Technology: Current State of Knowledge and Future Research Agenda

Leo Bruinzeel, Erik Klop, Allix Brenninkmeijer and Jaap Bosch

Abstract In this first review on the subject we describe the ecological impact of airborne wind energy technologies in general, with a particular focus on the rigid wing system developed by Ampyx Power. The chapter outlines a framework consisting of disturbance, ecological sensitivity, impact and legal aspects. We conclude that between 2–13 birds will collide annually with the autonomous aircraft alone. A challenging aspect is to estimate the mortality caused by the tether. Based on data from studies on power lines we find that a tether, that is one kilometer long and active all year round, will cause approximately 11 bird victims per year. For a tethered aircraft active only during the day and only with sufficiently strong wind an estimate of 5–15 bird fatalities per year will be realistic. This estimate is comparable to the number of fatalities found at average wind turbines. These figures can be ten times higher or lower depending on the bird activity at the specific deployment site. We provided a model for the mortality based on the specific characteristics of a bird species. A challenging future task will be the validation of this model considering that evidence suggest that birds can survive an encounter with the tether.

28.1 Introduction

Airborne wind energy (AWE) is an emerging renewable energy technology which accesses wind resources that are at higher altitudes than these in reach of conventional wind turbines. Ampyx Power is developing a novel airborne wind energy sys-

Leo Bruinzeel (✉) · Erik Klop · Allix Brenninkmeijer
Altenburg & Wymenga ecological consultants B.V., Suderwei 2, 9269 ZR, Feanwâlden, The Netherlands
e-mail: l.bruinzeel@altwym.nl

Jaap Bosch (✉)
Ampyx Power B.V., Lulofsstraat 55 Unit 13, 2521 AL The Hague, The Netherlands
e-mail: jaap@ampyxpower.com

© Springer Nature Singapore Pte Ltd. 2018
R. Schmehl (ed.), *Airborne Wind Energy*, Green Energy
and Technology, https://doi.org/10.1007/978-981-10-1947-0_28

679

tem (AWES), which uses a tethered autonomous aircraft for the conversion of wind energy into electricity. In the coming years Ampyx Power is planning to optimize the current prototype for commercial exploitation. With this in mind, it is important to anticipate on national and international laws and legislation. Many countries have environmental legislation in place that may be relevant for this type of projects. This chapter aims to provide a framework for the assessment of the ecological impact of the innovative technology. The impact of tethered aircraft on landscapes is also important, but not included in this review. This chapter does not provide a full assessment but rather a guideline on how to identify and assess the relevant ecological aspects and possible impacts in order to comply with relevant environmental legislation.

28.2 Ampyx Power Airborne Wind Energy System Concept

The Ampyx Power AWES converts the kinetic energy of wind into mechanical energy by having an autopilot-controlled glider aircraft creating pull on a tether by flying repetitive crosswind patterns at an altitude of 200–450 meters (Fig. 28.1). A ground-based generator converts this energy during reeling out into electric power. Once the tether has been unrolled, the glider aircraft is controlled to descent to lower altitudes, reeling in the tether with a minimal tether tension and a power consumption that is only a fraction of the power produced during the reel-out phase.

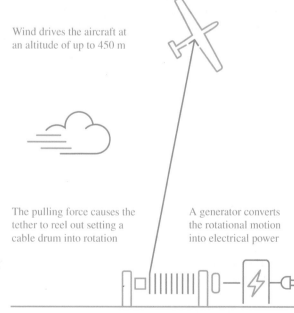

Wind drives the aircraft at an altitude of up to 450 m

The pulling force causes the tether to reel out setting a cable drum into rotation

A generator converts the rotational motion into electrical power

Fig. 28.1 Schematic representation of the Ampyx Power AWES. A more detailed outline of the technology is provided in [31–33]

The generator and electronic back end will be of similar dimensions as used in the nacelle of conventional wind turbines. A more detailed description of the Ampyx Power AWES and the envisioned roadmap to large-scale deployment is presented in Chap. 26 of this book.

28.3 Environmental Impact Assessment

An environmental impact assessment (EIA) describes all the impacts the project may have on its surroundings. Effects on flora and fauna are usually described under the header "ecological impact". In this review we focus only on the ecological impact. In general, impact assessments consist of four components. The relation between these components is illustrated in Fig. 28.2.

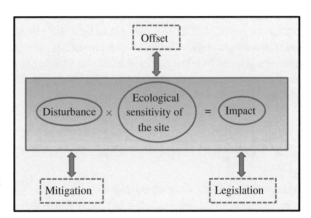

Fig. 28.2 Schematic representation of the four different components of the impact assessment framework

The first component is the disturbance resulting from the whole project. Disturbance is here broadly defined as all stimuli associated with the project that may cause a reaction among natural systems. This describes along which routes the device is fundamentally interacting with the environment and will result in a general overview, summarizing the theoretical pathways how the device interacts or may interact with the environment. This needs to be exhaustive. At a later stage some factors may be described as being not relevant. However, ignoring these factors beforehand would lead to an incomplete environmental impact assessment.

The second step is to investigate the ecological values and sensitivity of the site. The ecological sensitivity is defined by the species, communities and habitats at the site, their occurrence in a wider context (at regional, national and international scale), their behavior and their susceptibility to disturbance. The second step should be focused on natural values occurring at the site, but also on natural values in the vicinity. Some species have no ecological relation with the site, but can use the airspace above the site to commute or migrate.

The third step is the impact assessment of the combined result of disturbance and the ecological sensitivity of the site. This is not a physical equation, but rather an interaction term, describing that the impact is related to these two factors. A reduction of the total impact can be achieved by reducing the disturbance caused by the device or choosing a site with less ecological sensitivity, or both. As the disturbance caused by the tethered aircraft is caused by mechanistic properties of the device (speed, noise etc.), we conclude that in general the ecological impact will vary, to a large extent, as a result of the sensitivity of sites.

In the fourth and last step, the impact on ecological values is mirrored against the legal protection of flora, fauna and natural sites in general. The ecological impact may comply with the legal situation, but in some cases a conflict may arise between one or more specific ecological values being at stake due to the device. In that case there are in general two routes to follow: mitigation and offset. Mitigation is adjusting the project as such that the disturbance part of the equation will be reduced to acceptable levels. One can adjust the project in space, time, magnitude or nature. For instance another location or operation outside specific time periods (e.g the breeding season of birds). Offset is a method to increase the ecological value of the site as such that the ecological impact will become less. If mitigation does not suffice, one can—in general as a last resort—offset the ecological damage by realizing a bonus for the environment. Offset is in general a challenge, especially when the ecological cost and benefits are measured in different units.

28.4 Ecological Impact of AWE Projects: Current State

Many AWE initiatives have started and all have in common that they utilize the wind resources that are available in higher strata above the ground, therefore concepts and approaches developed for one device can be applicable for others. Currently there is not much information available concerning the environmental impact of AWE projects. A search in the Web of Science database resulted in 180 publications for the combined keywords "Airborne Wind Energy" but none of the studies dealt with environmental effects (Table 28.1). In comparison a total of over 36,000 studies are related to conventional wind energy. Specific internet sites are devoted to the mechanical and technical side of AWE projects, but not to the ecological or environmental impacts. Furthermore, directions, suggestions or otherwise helpful information that might guide us towards finding information on the ecological impacts are currently lacking. We can thus conclude that not much is known about the ecological and environmental impacts of the novel field of AWE. A more detailed up-to-date literature analysis has been presented in [40], however, the same conclusion can be drawn from this. To our knowledge we present the first review addressing the ecological aspects of an AWE project. The most important ecological effects to be expected are mortality of birds and bats and disturbance of mammals, birds and bats (and possibly "barrier" effects, blocking migration or commuting routes).

Keywords (Web of Science)	Publications	Relevant for this review
Airborne wind energy	180	0
Airborne wind energy + ecology	1	0
Airborne wind energy + impact	22	0
Airborne wind energy + birds	3	0
Airborne wind energy + environment	15	0
Wind energy	36,656	–
Wind energy + ecology	200	–
Wind energy + impact	4,500	–
Wind energy + birds	492	–
Wind energy + environment	3,043	–

Table 28.1 Web of Science results (October 2015) for search entries "Airborne Wind Energy" in combination with keywords (ecology, impact, birds & environment). As a comparison similar results are given for conventional wind energy (_= not assessed)

Since AWE projects use new technologies, the disturbance cannot directly be deduced from other studies. However, we can break down the technology into functional components and the disturbance effects of these components can be, directly or indirectly, estimated from literature sources on related subjects, so-called proxies.

28.5 Proxies for the Airborne Wind Energy Systems

In Table 28.2 we identified the components of the Ampyx Power AWES that are responsible for the main ecological effects, mortality and disturbance, and for which no direct estimates are available. These components are related to similar components or devices that can be used as reference or proxy for the disturbance caused by the AWES. In this section we discuss the various proxies.

AWES component	Similar component (proxy)
Gliding aircraft	Glider aircraft (airfield), small aircraft (ultralights)
Repetitive flight pattern aircraft	Glider aircraft (airfield)
Tether (stable vertical object)	Television towers, wind turbines, other towers/vertical objects, power lines/horizontal objects
Tether (high-altitude moving object)	Wind turbines, kite surfers, kites

Table 28.2 The AWES broken down in functional components and the similar component (proxy) of which disturbance information may be available that might act as a model

28.5.1 Gliders and Motorized Aircraft

The disturbance associated with the tethered aircraft during the retraction phase is comparable to that of regular glider aircraft of similar dimensions. However, glider aircraft in general undertake relatively short flights, while the Ampyx Power aircraft performs near-continuous flights. We could not trace information on the ecological impacts of specific glider airfields, as a frequently used airfield might be used as proxy for a continuous flying gliders at one location. The disturbances of glider aircraft are not thoroughly investigated [36]. In general, evidence from the literature suggests that the impact is less compared to that of motorized aircraft [36]. Observations on silent aircraft such as balloons and Zeppelins in Switzerland showed that waterbirds (ducks, geese, herons, cormorants and gulls) took off on average up to a balloon height of 300 m or less. Above a balloon height of 300 m usually no response was visible among the birds. This height threshold to take-off was higher (500 m) for geese. However, incidentally a Zeppelin at 500 m evoked a response in all waterbird species, resulting in birds flying around for 10–15 minutes [12].

Motorized aircraft cause visual and audible disturbance. Most impact studies designed to measure the effect of aircraft on animals are investigating the combined effect [23]. In order to understand only the visual aspects of disturbance (glider aircraft hardly produce any noise), it is important to separate the influence of these factors. Several lines of evidence suggest that the disturbance of motorized aircraft is mainly caused by audible disturbance (noise) and not by visual disturbance. Studies performed in mountainous terrain showed that the effect of low-flying jet aircraft was attributable to the noise [28, 49]. Animals displayed a similar response to artificial jet noise from speakers (lacking the visual cues) compared to the combined effect of real jets. Birds in Australia responded strongly to only aircraft noise generated by speakers, mimicking aircraft at various altitudes [11]. Another study measured a difference in the response of geese to aircraft depending on the amount of noise they produced, which was also apparent with relatively small aircraft [48]. However hang gliders, not producing any sound, can induce a strong flight (escape) response in mammals such as Chamoix, Red deer and Alpine Ibex [41, 45].

In Germany, research has been undertaken on the disturbance effects of paragliders on launch sites throughout the country [9]. The density of breeding birds around take-off positions was studied and results were compared with nearby (undisturbed) control locations. There were no lower breeding bird densities observed around launch locations. However, the results were mainly based on small passerines, for larger species—present in relatively lower densities—the sample size was too limited to test for an effect, but sufficed to conclude that there was no major effect. There is in general no information available concerning the disturbance effect of glider aircraft flying during the night. Disturbance caused by leisure sports, such as kite surfing, is not well investigated. Studies conducted in the Grevelingen [46] and Wolderwijd [22], both in The Netherlands, and a general investigation [30] all have in common that they investigate the effect of an airborne moving object in relation with a rapidly-moving person near ground level. Therefore, these effects are not discernible from effects caused by jet ski's, power boats or other fast-approaching humans at water level. To estimate the impact of AWES these studies are not of much fundamental use.

28.5.2 Stationary and Moving Vertical Objects

Stationary vertical objects such as television towers and communication towers are well investigated in the USA [18, 19, 37–39] and can be used as a proxy for the tether. They consist of a monopile, often with accompanying guy wires, and are equipped with FAA obstruction lights. There is a significant correlation between the annual bird mortality and the height of the tower, with tall towers resulting in more casualties among nocturnal migrants and height explains 84% of the variation in number of fatalities [37]. However, the relation describing the avian casualties as a function of the height of the tower, is not suited to model the relationship between the height of the aircraft/tether and the number of bird fatalities, for two reasons: the guy wires and the lights. Guy wires at communication towers are responsible for approximately 85% of the victims [19] and the tethered aircraft is lacking additional guy wires (in fact the tether can be regarded as a single guy wire). For communication towers the total volume of airspace occupied by guy wires increases nonlinearly with the height of the tower, providing a mechanistic explanation for the relationship [37]. Communication towers are supplied with red lights that cause birds to circle around the towers (and to collide with the tower and guy wires), especially in clouded weather [37, 38]. Furthermore, television towers are mainly causing casualties at night among nocturnal migrants. The Ampyx Power aircraft will initially only be used during daylight hours.

Moving vertical objects, like rotating wind turbines might act as a proxy for the tether. The ecological impacts of wind farms are well known and many published studies are available. The main ecological impacts are disturbance during the construction activities, and mortality (bird and bats) during the operational phase because of collisions with the rotor blades. In addition, wind farms may cause fragmentation of habitat, avoidance of the area and may be a barrier in migrating routes of birds and bats. The species composition of turbine fatalities reflects that of the species community present in the area [50]. In general, wind farms are responsible for 7 bird victims (median value) per turbine, per year (Table 28.3).

28.5.3 Stationary Horizontal Objects

Stationary horizontal objects such as power lines have an impact both night and day and are well investigated [6, 7]. A few studies have calculated fatality rates per km power line per time unit (Table 28.4). A median value of approximately 0.3 fatalities per km per day is reported. Casualties range between 0.02 to 1.89 collisions per km per day (a factor 100 difference). This variation is to a large extent

Windpark	Time period	T [months]	N [turbine^{-1} year^{-1}]	Source
Vansycle, USA	1999	12	0.6	[15]
Näsudden, Sweden	–	–	0.7	[21]
Altamont, USA	1988–2000	106	0.9	[44]
Buffalo Ridge, USA	1994–1999	92	1.0	[15]
Nieuwkapelle Diksmuide, Belgium	2005–2006	12	1.0	[16]
Blyth Harbour, UK	–	–	1.3	[21]
Obersdorf, Austria	–	–	1.5	[21]
Foote Creek Rim, USA	1998/1999	12	1.8	[15]
Woolnorth Tasmania, Australia	2002/2003	14	1.9	[21]
Gent, Belgium	2004	12	2.8	[16]
Simonsberger Koog, Germany	–	–	>2.2	[21]
San Gorgino, USA	–	–	2.3	[15]
Friedrich-Wilhelm-L.-K., Germany	–	–	>2.6	[21]
Steinberg-Prinzendorf, Austria	–	–	3	[21]
Delfzijl-Zuid, The Netherlands	2006–2011	60	2.2–6.8	[10]
Kreekraksluizen, The Netherlands	–	–	3.3	[42]
Nine Canyon Wind Project, USA	2002/2003	12	3.6	[21]
Alaiz-Echague, Pyrenees, Spain	2000/2001	12	>3.6	[34]
Kluizendok Gent, Belgium	2005–2007	24	6.6	[16]
Urk, The Netherlands	1987–1989	24	7.3–18.3	[51]
Breklumer Koog, Germany	–	–	>7.5	[21]
Eemshaven, The Netherlands	2009–2014	60	7–33	[24]
Guerinda, Pyrenees, Spain	2000/2001	12	>8.5	[34]
Bremerhaven-Fischereihafen, Germany	–	–	9	[21]
Schelle, Belgium	2002–2004	36	11.3	[16]
Fehmarn, German Baltic Sea	2009	12	13	[8]
Prellenkirchen, Austria	–	–	13.9	[21]
Oosterbierum, The Netherlands	1986–1991	60	18.3–36.5	[50]
Almere, The Netherlands	2004	3	20	[29]
Zeebrugge, Belgium	2001–2007	84	21.4	[16]
Salajones, Pyrenees, Spain	2000/2001	12	>21.7	[34]
Izco-Aibar, Pyrenees, Spain	2000/2001	12	>22.6	[34]
Brugge (Boudewijnkanaal), Belgium	2001–2006	72	23.7	[16]
Wieringerwerf, The Netherlands	2004	3	27	[29]
Middenmeer, The Netherlands	2004	3	39	[29]
Kleine Pathoekewg. Brugge, Belgium	2005–2006	24	42.3	[16]
Solano County, USA	–	–	54	[15]
El Perdon, Pyrenees, Spain	2000/2001	12	>63.3	[34]
Minimum estimate			0.6	
Maximum estimate			>63.3	
Median			**7**	
Range			0.6–63.3	

Table 28.3 Overview of the (corrected) number of bird casualties N per turbine per year in wind farms in The Netherlands, Europe, Australia and USA, investigated over a period of T months

caused by the natural variation in bird numbers at sites. In order to compare power lines (consisting of a few parallel lines) with the tether (a single line) we divided all estimates by 10 (assuming a power line consists of on average 10 parallel lines and mortality is linearly related to the number of lines). In those situations we arrive at a median value of 0.03 fatalities per km per day per single line (range 0.002–0.189). We propose to use these figures to model the number of collisions victims by the tether.

The studies are mainly conducted in the low countries (Netherlands, parts of Germany) and are worldwide located in the temperate zone and in the center of the East Atlantic Flyway a migratory highway for birds migrating between northern Eurasia and Africa [1]. These sites are generally characterized by high densities of birds year round. Based on these arguments we believe that the estimates are reliable representatives for many sites on the globe. In general, the number of birds on many sites will in fact be lower, however finding sites with higher numbers of birds in space and time will be restricted to areas with very condensed bird migration (see [1] for locations). Fatalities can be the result of electrocution and collision, although in many modern power (high voltage) lines the conductors are spaced far apart, making electrocution unlikely. Birds may also be affected due to fragmentation of their habitat. The number of collision fatalities varies widely and is dependent on many factors, including location, spatial configuration of the lines, numbers of birds and flight movements in the area, season, the surrounding terrain, and the visibility of the lines. The latter factor is influenced by weather conditions and time of day, but also the presence of markers [4, 20].

Location	N [km^{-1} day^{-1}]	N [km^{-1} day^{-1} line^{-1}]	Study
Four areas in Mid-Germany	0.02	0.002	[5]
Eemshaven	0.18	0.02	[26]
"Moors", Netherlands	0.23	0.02	[27]
Polder Mastenbroek	0.26	0.03	[43]
Westerbroekstermadepolder	0.32	0.03	[47]
North Dakota, USA	0.34	0.03	[17]
17 areas in The Netherlands	0.36	0.04	[47]
Mid-Germany	0.43	0.04	[5]
"Meadows" in The Netherlands	0.44	0.04	[27]
Muiden	0.51	0.05	[43]
Eemshaven	0.36–1.11	0.04–0.11	[25]
Locations with high bird density	1.89	0.19	[27]
Minimum estimate	0.02	0.002	
Maximum estimate	1.89	0.19	
Median	0.3	0.03	
Range	0.02–1.89	0.002–0.189	

Table 28.4 Bird casualties N caused by power lines and by location, assuming that power lines consist of 10 separate parallel lines

28.6 Animal Collision Rate

To model the animal collision rate of the tethered aircraft we distinguish between the aircraft and the tether.

28.6.1 Mortality by the Tether

In general, the speed of the tether is much higher than the speed of the bird. We assume that birds cannot anticipate an approaching tether. For a bird which is slowly approaching the impact zone the probability P_t of a lethal encounter with the tether is proportional to the characteristic dimension l_b of the bird

$$P_t \propto l_b. \tag{28.1}$$

The larger the bird, the larger the contact area and the larger the risk of collision, all else being equal. However exposure time to the approaching tether is also dependent on the flight speed v_b of the bird. The faster the bird flies, the less time it spends in de the danger zone, hence we can formulate

$$P_t \propto \frac{1}{v_b}. \tag{28.2}$$

In addition, birds can display specific behavior close to the contact zone, that may increase the chance of a lethal encounter. This can be a range of behaviors that all result in more time spent in the contact zone compared to birds that just fly in a straight line, for instance soaring birds that are circling in the vicinity

$$P_t \propto \text{behavior bird}. \tag{28.3}$$

A small bird approaching the tether with a high speed has a low risk of getting hit by the tether because it passes rapidly through the contact zone and its small dimensions make a hit less probable. Other birds traveling slow or being large or displaying flight behavior deviating from a straight line are exposed to a higher risk of getting hit. Based on this simple model we can estimate collision risks for various bird species. Power lines are in our opinion the best proxy to estimate bird mortality of the tether in general. The relative collision risk, i.e. the risk for a bird to get hit assuming that it is present in the area, is independent of altitude or speed of the tether. Instead, it is solemnly determined by the size of birds or bats, their speed and their flight behavior.

  The number of casualties per bird species per site can be calculated by first estimating the total number of casualties in general as a function whether the site is categorized as an area with moderate, low or high bird activity (Table 28.5). Subsequently, the estimated total number of casualties should be broken down to species level. This is a function of the species present in the area, the numbers and their

Bird activity	N [km^{-1} year^{-1}]	Source (see Table 28.4)
Low	1.095	10× lower
Moderate	10.95	Median value
High	109.5	10× higher

Table 28.5 Estimated bird casualties N per year for a tether of 1 km length for year-round 24/7-operation at sites with varying degrees of bird activity

flight behavior. Furthermore it is dependent of species-specific traits, such as bird size and bird speed that determines the relative collision risk (see Sect. 28.6.3).

28.6.2 Mortality by the Aircraft

The rate of animal collisions with general aviation aircraft in the US is 1.20 collisions per 100,000 movements, which are departures and arrivals [14]. Birds were involved in 97% of the collisions, terrestrial mammals in only 2.2% and bats in only 0.7%. One aircraft movement consists of various flight phases (park, taxi, take-off run, climb, en route, descent, approach and landing roll). The "en route" and flight phases do not pose a real threat since these take place at heights above 1500 ft (approximately 500 m) where collisions are generally rare [14]. We can relate the movements of the tethered aircraft to movements by aviation aircraft, assuming that the aircraft operates mostly in air layers where collisions occur. Therefore, we need to translate the continuous flight movements of the tethered aircraft into standard aircraft (airport) movements.

If we assume that one aircraft movement (either take-off run plus climb or approach and landing roll) takes between 60 and 180 seconds, we arrive for a fully operational tethered aircraft (active 24 hours, 365 days a year) at between one million movements per year (assuming one general aviation movement takes 30 seconds) and 175,000 movements per year, assuming that one aviation movement takes 180 seconds (Table 28.6). Under these assumptions the tethered aircraft will cause on average between 2–13 bird casualties per year. The number of bats that get hit are negligible. This is under assumption that the characteristics of a tethered aircraft is comparable with average general aircraft. However, the Ampyx Power aircraft is considerable smaller and flies slower, which attributes to a lower collision risk, but are also generally silent, which may attribute to a higher collision risk. Furthermore, the aircraft will not be active daily and year-round.

Assumption movement duration [s]	Corresponding AWES movements per year (24 hours × 365 days)	Animal collisions N [year^{-1}]		
		Total	Birds	Bats
30	1,051,200	12.6	12.2	0.3
60	525,600	6.4	6.2	0.1
120	262,800	3.2	3.1	0.1
180	175,200	2.2	2.1	<0.1

Table 28.6 Total animal collisions per year broken down into bird and bat collisions for an Ampyx Power AWES for year-round 24/7 operation. This was calculated on the basis of four different assumptions (30–180 s) for the conversion of one airport/aircraft movement into time

28.6.3 Species-Specific Relative Collision Risk

Continuing the derivation outlined in Sect. 28.6.1, we combine Eqs. (28.1) and (28.2) and use the airspeed v_e and wing span b, listed in Table 28.8 in the appendix of this chapter for a range of European bird species, to formulate the relative collision risk

$$P_\dagger \propto \frac{b}{v_e}. \tag{28.4}$$

The airspeed v_e is the flight speed of the bird relative to the air. It is this kinematic property which characterizes the flight of a bird in a wind field because it is related to the aerodynamic properties of the animal. However, the risk of collision with a stationary, ground-attached object depends on the flight speed v_b of the bird measured with respect to the ground. The velocity vectors are related by $\mathbf{v}_e = \mathbf{v}_b - \mathbf{v}_w = -\mathbf{v}_a$, where v_a is the apparent wind speed experienced by the bird. The airspeed was derived in [3] from the flight speed v_b of the birds, measured by tracking radar, and corrected for the wind speed v_w at the location of the bird, measured by radar tracking balloons at the flight altitude of the birds. We use in our analysis the airspeed of the bird instead of the flight speed assuming that in average the birds fly in all directions and also that the wind direction varies such that the effect of the background wind speed v_w is statistically canceled out.

The last entry of Table 28.8 indicates an average airspeed of $v_e = 14.1$ m/s and average wing span of $b = 0.91$ m for an average bird (the standard bird in this study), leading to a ratio

$$c = \frac{b}{v_e} = 0.065. \tag{28.5}$$

On the basis of this reference value we define the relative standardized collision risk

$$P_\dagger \propto \frac{b}{v_e c}, \tag{28.6}$$

which allows a species-specific calculation of the relative standardized collision risk using the tabulated wing span and airspeed data. This probability quantifies the risk for a bird of a certain species to collide with the tether (given presence of the species in the area) compared to the risk of a standard bird. This will allow to identify species with relative high or low collision risks. Based on Eqs. (28.4), (28.5) and (28.6) we have calculated the relative collision risk and the relative standardized collision risk and appended this as two extra columns in Table 28.8 to the original data of [3].

There are two routes providing a relative collision risk for a species that is absent in the list. The easiest method is to search for a comparable species in the list, based on size and ecological group, and use these values as estimate for the unlisted bird species. Or, alternatively, find an estimate for the wing span of the unlisted species and divide this by the airspeed v_e of the bird determined on the basis of body mass m_b by the allometric relation [2]

$$v_e = 15.9 \, m_b^{0.13} \qquad (28.7)$$

The data listed in Table 28.8 implies the assumption that birds are flying in a straight line. In reality, however, birds are not always flying in a straight line. Especially local birds, that breed or forage in the vicinity show deviating flight patterns. During fieldwork one has to get an impression of the number and species present in the area and the regular flights they undertake. Ultimately, one needs to get an idea how much extra time an individual is spending in the collision risk zone, compared to an individual of the same species flying in a straight line.

28.7 Ecological Impact of Tethered Aircraft

In this section we outline the potential disturbance caused by tethered aircraft. The potential impact is derived from impact studies on comparable systems. In compliance with environmental impact assessment (EIA) standards we distinguish three phases: construction phase, utilization phase and decommissioning phase. The activities in the decommissioning phase (and maintenance during the utilization phase) are similar in nature to the construction phase, so not separately treated here. The main effects in the construction phase are: presence of man, noise, vibrations, emissions and physical activity (ground works, construction works). The main effects in the utilization phase are: movement (landing, launching of motorized aircraft), continuous movement (unpowered aircraft noise, movement and glitter/glare) and mortality due to collisions with the aircraft/tether, noise, vibrations, electromagnetism and glitter/glare (Table 28.7).

The main response variables are habitats (quality and quantity) and species (survival, breeding and behavior). In compliance with EIA standards the physical effects of the project are first described. These are broken down into components that are the fundamental routes through which the project interacts with the environment.

Table 28.7 Effect indicator for plants, habitats and small animals (including fish, insects) (0 No effects, 1 Very small / negligible effects, 2 Moderate small effects, 3 Medium effects, 4 Large effects)

A) Plants, habitats, small animals	Quality	Quantity
movement (disturbance)	0	0
movement (mortality)	0	0
noise	1	0
vibrations	0	0
emissions	1	1
electromagnetism	0	0
glitter and glare	0	0
B) Birds	Survival	Breeding	Behavior
movement (disturbance)	1	2	1
movement (mortality)	3	1	2
noise	0	2	2
vibrations	0	0	0
emissions	0	0	0
electromagnetism	0	0	0
glitter and glare	0	0	0
C) Bats	Survival	Breeding	Behavior
movement (disturbance)	1	1	2
movement (mortality)	1	0	0
noise	0	0	0
vibrations	0	0	0
emissions	0	0	0
electromagnetism	0	0	0
glitter and glare	0	0	0
D) Ground-based birds, mammals	Survival	Breeding	Behavior
movement (disturbance)	0	2	1
movement (mortality)	0	0	0
noise	0	2	1
vibrations	0	0	0
emissions	0	0	0
electromagnetism	0	0	0
glitter and glare	0	0	0

For different ecological groups we summarized the magnitude of these effects on their ecology. As ecological groups we distinguished between A) plants, habitats and small animals (usually, but not always, species with limited protection on sites and limited mobility), B) birds and C) bats (both groups of flying animals with large home ranges and usually well protected species) and D) ground-based birds and mammals (Table 28.7). The listed effects should be interpreted as a general and relative susceptibility to the stimulus of an average species within the relevant ecological group. Noise, vibrations, electromagnetism, physical activity, emissions and glitter/glare are factors that need to be investigated, but beforehand it is expected that in general these will not pose a realistic threat for wildlife. Animals can be affected in their survival, their breeding output and in their behavior. In EIAs studying wind farms, there is usually a fourth factor involved related to connectivity or related to the device acting as a barrier in the migration route. The barrier effect is not an affected trait of the birds, but rather the outcome of a change in behavior. Birds for instance, decide to avoid an area with a tethered aircraft. In fact the barrier effect is occurring when animals avoid the dangerous object, which in part is a desired effect. In some cases this may result in longer migration routes (although the increase in energy expenditure is generally small) or in extreme examples may lead to habitat loss because certain areas may no longer be used by animals because it is no longer profitable. This can happen, for instance, when foraging and breeding locations are spaced apart.

28.7.1 Disturbance

In general, the AWE study sites or testing sites will be located on regular farmland or other privately owned land where normal human activities take place. In general, we assume AWE locations to be at intermediate remote locations, at sites where there is space for the project, but always in proximity to electricity grid connection points. The activities during building and utilization of the tethered aircraft will be of similar magnitude. In general, there will be no or very limited additional disturbance related to human activity, i.e. the movement of humans, on the ground and this will mainly affect larger and skittish animals. Larger animals tend to be generally more susceptible to human disturbance. The susceptibility is strongly species-specific. For the impact one can rely on other EIAs related to infrastructure. In general, we judge the impact of movement by humans and machines on wildlife as negligible.

The disturbance associated with the tethered aircraft is comparable to that of regular glider aircraft of similar dimensions. During operation at daytime, the tethered aircraft will not to cause much disturbance, because it is not generating considerable noise and flies at higher altitudes in a repetitive and predictive manner. Most local animals will get used to the aircraft in due time, because they can learn that it is not harmful (just as animals may get used to regular traffic). There is no information available concerning the disturbance effect of nocturnal flying glider aircraft. The effect of a glider aircraft above 300 m will usually not evoke a reaction in breeding or staging birds on the grounds and the disturbance caused by a silent unmotorized aircraft is expected to be very small [12, 36]. Foraging bats will detect the flying aircraft by echolocation and are thus able to avoid collisions. We propose to use an impact (disturbance) distance of 300 m for both birds and mammals [12, 36].

The disturbance caused by motorized aircraft is mainly due to noise and a lesser extent to visual cues. During take-off and landing, the tethered aircraft is generating some noise, at that stage its disturbance may be similar to that of micro light aviation [35]. The noise of the engine is probably small and it is likely that the impact contour of the noise (dB contour) is below the disturbance contour of visual cues. Therefore, one can probably use the disturbance contour of 300 m for the aircraft to model the disturbance, irrespective whether the engine is on or off.

28.7.2 Mortality

An object that is easily identified as harmful by an animal will result in an appropriate reaction (avoidance) and will result in a certain space that will be permanently or temporarily unavailable for the animal. In this case disturbance is translated in (permanent or temporary) habitat loss. An animal that is not responding to the object with avoidance behavior might lethally (or non-lethally see Fig. 28.3) collide with the moving object.

Fig. 28.3 Photo sequence capturing the rare situation of a bird, as part of a group of domestic pigeons (Columba livia domestica), impacting a tensioned tether, with the tether indenting and the bird recovering and continuing its flight. We can conclude that 1) birds can survive a tether impact and 2) the impact is clearly visible and may be used for monitoring. Depicted is the 20 kW kite power system of Delft University of Technology using a commercially available Genetrix Hydra LEI tube kite of 14 m^2 surface area, steered by a suspended remote-controlled control unit and a tether of 4 mm diameter made of Dyneema®. The photo was taken on 28 June 2011 at Valkenburg airfield, The Netherlands, by Max Dereta

Increasing the visibility of for instance the tether might decrease the collision risk, but may increase habitat loss. So in a sense disturbance and mortality are two sides of the same coin. As described previously, the number of collisions with bats and birds will strongly depend on the location and should also be interpreted in a suitable spatial context. The estimated number of casualties per bird species should be interpreted in a local ecological context. In the EU it is nowadays common practice to scale the mortality in relation to the normal or background mortality of a species. If the additional mortality is less than 1% of the natural or background mortality of a species, the project can be regarded as having for certain no impact on the species. This does not imply that a mortality level that exceeds this level has significant or important consequences for the focal species, but this needs to be investigated.

28.8 Conclusion and Discussion

The studies needed in the construction phase of an airborne wind energy system (AWES) are not deviating from a regular environmental impact assessment (EIA) that is aimed at construction at a specific site, the activities and impacts are similar. The utilization phase is the phase with most uncertainties. The most decisive factor will be the mortality caused by the moving aircraft and tether. This is (relatively) ranked as having a medium impact. All other factors are ranked lower, having less ecological impact. The first theoretical predictions reveal that the number of casualties will be relatively low. The mortality caused by the aircraft can be derived from aviation statistics and under specific worst-case assumptions we conclude that between 2–13 birds will collide annually with the glider aircraft. For the moving tether we conclude that data derived from studies on power lines offer the best basis. Based on these studies we conclude that a tether will cause—for an average site— approximately 11 bird fatalities per year. The total number of fatalities are therefore expected to be between 13–24 per year. This is the prediction for a tethered aircraft active all year round, 24 hours a day, with a tether of 1 km long and located at a site with moderate bird activity. For a tethered aircraft active only at daylight hours and only during days with sufficient wind force and with a shorter tether these figures are considerable less, an estimate of 5–15 bird fatalities per year will be realistic for this situation. The conclusion at this stage is that the number of bird fatalities predicted for a tethered aircraft is comparable with the range of fatalities registered at conventional wind farms (0.6–63 fatalities per year, median value 7, see Table 28.3). A future challenge will be to collect empirical data and to compare these with data collected for wind farms under similar conditions. Ultimately, one needs to scale the energy production with the ecological footprint to make a sound comparison. In order to estimate species-specific mortality we provide a simple model based on species-specific bird traits (size and speed) and presence and flight behavior in the area.

Validation of these calculations will be difficult. The common way of validation is to meticulously search the area for casualties and, after correction for several factors such as predation risk and detection probability, estimate the casualties for a year-round situation. Given the large area over which the tethered aircraft is active and the predicted low numbers of casualties, this will not be feasible. Instead impact-triggered cameras (aimed at the aircraft and at part of the tether) could be installed on the aircraft to monitor collisions. Although anecdotic, the rare photograph shown in Fig. 28.3 illustrates that these techniques are feasible, since the tether is clearly showing an aberrant shape after the impact with a bird. Furthermore it proves that not all collisions with a tether are fatal, so our calculations are worst-case models, which is common in environmental impact assessments. A next step would be to compare the ecological footprint (expressed as the magnitude of the footprint scaled in relation to the generated energy) of AWE projects in comparison with, for instance, conventional wind farms.

Another gap in the current knowledge and fuel for the research agenda is the environmental impact when multiple AWE installations are built at one site. Wind-farms offer here a sound comparison, and there is sufficient information available on the effects of different constellations and of single wind turbines versus wind turbines installed in groups to conduct this assessment.

Acknowledgements We would like to thank the boards of Ampyx Power and Altenburg & Wymenga, in particular Joris Latour, to allocate time for us to write this chapter. We would like to thank Roland Schmehl and Prabu Sai Manoj M of Delft University of Technology for assistance during the writing stage. Roland Schmehl kindly provided the rare photograph that was taken by Max Dereta. The financial support of the European Commission through the project AMPYXAP3 (H2020-SMEINST-666793) is gratefully acknowledged.

Appendix

Table 28.8 Airspeed v_e, body mass m_b, wing span b (maximum wing tip to wing tip distance), relative collision risk b/v_e and relative standardized collision risk $P = b/(v_e c)$, with $c = 0.065$. Birds are arranged in taxonomical groups g, with values 1: swans, geese & ducks, 2 Flamingo, pigeons, swifts, 3: divers, cormorants, pelican, herons, storks & crane, 4: falcons, crows, songbirds and 5: hawks, eagles, osprey & bee-eater. Except for the collisions risks, the data originates from [3], supplemented by data from [13], which is marked by •

Species	g	v_e	m_b	b	b/v_e	P	Species	g	v_e	m_b	b	b/v_e	P
		[m/s]	[kg]	[m]	[s]	[-]			[m/s]	[kg]	[m]	[s]	[-]
Cygnus olor	1	16.2	10.597	2.3	0.142	2.20	Corvus corone	5	13.5	0.566	0.91	0.067	1.04
Cygnus columbianus	1	18.5	6.637	1.98	0.107	1.66	Corvus corax	5	14.3	1.149	1.21	0.085	1.31
Cygnus cygnus	1	17.3	8.689	2.29	0.132	2.05	Sturnus vulgaris	5	16.2	0.083	0.38	0.023	0.36
Anser fabalis	1	17.3	3.035	1.62	0.094	1.45	Sturnus vulgaris •	5	12.4				
Anser albifrons	1	16.1	2.582	1.41	0.088	1.36	Fringilla coelebs	5	12.8	0.022	0.26	0.020	0.31
Anser anser	1	17.1	3.326	1.55	0.091	1.40	Fringilla coelebs •	5	12.8				
Branta canadensis	1	16.7	3.628	1.69	0.101	1.57	Fringilla montifringilla	5	15.0	0.024	0.27	0.018	0.28
Branta leucopsis	1	17.0	1.705	1.08	0.064	0.98	Limosa lapponica	3	18.3	0.318	0.73	0.040	0.62
Branta bernicla	1	17.7	1.306	1.01	0.057	0.88	Numenius phaeopus	3	16.3	0.383	1.07	0.066	1.02

Species	g	v_e [m/s]	m_b [kg]	b [m]	b/v_e [s]	P [-]
Tadorna tadorna	1	15.4	1.193			
Anas penelope	1	20.6	0.783	0.82	0.040	0.62
Anas crecca	1	19.7	0.348	0.59	0.030	0.46
Anas platyrhynchos	1	18.5	1.082	0.88	0.048	0.74
Anas acuta	1	20.6	1.024	0.90	0.044	0.68
Aythya ferina	1	23.6	0.823	0.77	0.033	0.51
Aythya fuligula	1	21.1	0.694	0.71	0.034	0.52
Aythya marila	1	21.3	0.931	0.82	0.038	0.60
Somateria mollissima	1	17.9	2.015	0.98	0.055	0.85
Somateria spectabilis	1	16.0	1.591	0.93	0.058	0.90
Polysticta stelleri	1	21.9	0.805			
Clangula hyemalis	1	22.0	0.874	0.71	0.032	0.50
Melanitta nigra	1	22.1	0.990	0.85	0.038	0.60
Melanitta fusca	1	20.1	1.743	0.97	0.048	0.75
Bucephala clangula	1	20.3	0.901	0.70	0.034	0.53
Mergus serrator	1	20.0	1.004	0.87	0.044	0.67
Mergus merganser	1	19.7	1.489	0.93	0.047	0.73
Phoenicopterus ruber •	2	15.2	3.053	1.53	0.101	1.56
Columba oenas	2	15.8	0.295	0.75	0.047	0.74
Columba palumbus	2	16.3	0.490	0.75	0.046	0.71
Columba palumbus •	2	17.6				
Apus apus	2	9.7	0.038	0.40	0.041	0.64
Apus apus •	2	10.6				
Apus pallidus •	2	10.5	0.042	0.44	0.042	0.65
Apus melba •	2	12.6	0.078	0.57	0.045	0.70
Haematopus ostralegus	3	13.0	0.523	0.82	0.063	0.98
Charadrius hiaticula	3	19.5	0.064	0.41	0.021	0.33
Pluvialis dominica	3	13.7	0.145			
Pluvialis squatarola	3	17.9	0.219	0.62	0.035	0.54
Vanellus vanellus	3	12.8	0.219	0.75	0.059	0.91
Vanellus vanellus •	3	11.9				0.00
Calidris canutus	3	20.1	0.128	0.50	0.025	0.39
Calidris alpina	3	15.3	0.054	0.36	0.024	0.36
Philomachus pugnax	3	17.4	0.114	0.55	0.032	0.49
Philomachus pugnax •	3	13.6				
Gallinago gallinago	3	17.1	0.132	0.52	0.030	0.47
Falco vespertinus •	5	12.8	0.165	0.72	0.056	0.87
Falco subbuteo •	5	11.3	0.238	0.74	0.065	1.01
Falco eleonorae •	5	12.8	0.387	0.95	0.074	1.15
Falco peregrinus •	5	12.1	0.789	1.02	0.084	1.31
Lullula arborea •	5	9.8	0.027	0.29	0.030	0.46
Alauda arvensis	5	15.1	0.039	0.35	0.023	0.36
Alauda arvensis •	5	12.7				
Riparia riparia	5	14.3	0.015	0.27	0.019	0.29
Riparia riparia •	5	11.3				
Hirundo rupestris •	5	9.9	0.019	0.32	0.032	0.50
Hirundo rustica	5	10.0	0.016	0.32	0.032	0.50
Hirundo rustica •	5	11.3				
Delichon urbica	5	9.7	0.015	0.29	0.030	0.46
Delichon urbica •	5	11.0				
Anthus trivialis	5	12.7	0.022	0.27	0.021	0.33
Anthus trivialis •	5	12.0				
Anthus pratensis •	5	10.5	0.018	0.26	0.025	0.38
Motacilla flava •	5	12.7	0.018	0.26	0.020	0.32
Motacilla alba	5	14.1	0.021	0.26	0.018	0.29
Motacilla alba •	5	13.0				
Prunella modularis •	5	12.2	0.020	0.21	0.017	0.27
Oenanthe oenanthe •	5	12.8	0.023	0.28	0.022	0.34
Turdus pilaris	5	13.0	0.105	0.42	0.032	0.50
Turdus pilaris •	5	12.4				
Turdus philomelos	5	11.0	0.068	0.36	0.033	0.51
Turdus philomelos •	5	11.7				
Turdus iliacus	5	13.8	0.061	0.36	0.026	0.40
Turdus viscivorus	5	11.9	0.114	0.44	0.037	0.57
Numenius arquata	3	16.3	0.794	0.97	0.060	0.92
Tringa nebularia	3	12.3	0.174	0.61	0.050	0.77
Tringa glareola	3	9.6	0.066	0.40	0.042	0.65
Arenaria interpres	3	14.9	0.111	0.47	0.032	0.49
Phalaropus lobatus	3	13.1	0.033	0.34	0.026	0.40
Phalaropus fulicarius	3	12.4	0.054	0.42	0.034	0.52
Stercorarius pomarinus	3	15.2	0.688	1.18	0.078	1.20
Stercorarius parasiticus	3	13.8	0.438	1.06	0.077	1.19
Stercorarius longicaudus	3	13.6	0.297	1.00	0.074	1.14
Larus minutus	3	11.5	0.118			
Larus ridibundus	3	11.9	0.283	0.97	0.082	1.26
Larus canus	3	13.4	0.411	1.11	0.083	1.28
Larus fuscus	3	13.1	0.719	1.34	0.102	1.58
Larus fuscus •	3	11.9				
Larus argentatus	3	12.8	1.142	1.34	0.105	1.62
Larus glaucoides	3	15.9	0.819			
Larus hyperboreus	3	13.4	1.445			
Larus marinus	3	13.7	1.669	1.67	0.122	1.89
Rissa tridactyla	3	13.1	0.408	0.96	0.073	1.14
Sterna caspia	3	12.1	0.655			
Sterna paradisaea	3	10.9	0.110	0.80	0.073	1.14
Chlidonias leucopterus •	3	12.0	0.054	0.65	0.054	0.84
Gavia stellata	4	18.6	1.505	1.04	0.056	0.87
Gavia arctica	4	19.3	2.543	1.20	0.062	0.96
Gavia adamsii	4	18.7	5.500			
Phalacrocorax carbo	4	15.2	2.227	1.40	0.092	1.43
Pelecanus onocrotalus •	4	15.6	8.504	2.91	0.187	2.89
Botaurus stellaris	4	8.8	1.133	1.26	0.143	2.22
Nycticorax nycticorax •	4	11.2	0.763	1.06	0.095	1.47
Ardeola ralloides •	4	11.7	0.287	0.86	0.074	1.14
Egretta alba •	4	10.2	0.888	1.44	0.141	2.19
Ardea cinerea	4	12.5	1.439	1.73	0.138	2.14
Ardea cinerea •	4	11.2				
Ardea purpurea •	4	10.8	0.906	1.35	0.125	1.94
Ciconia nigra •	4	16.0	3.000	1.50	0.094	1.45
Ciconia ciconia •	4	16.0	3.432	1.91	0.119	1.85
Plegadis falcinellus •	4	12.6	0.566	0.89	0.071	1.09
Platalea leucorodia •	4	14.1	1.857	1.30	0.092	1.43
Porzana porzana •	4	13.9	0.078	0.38	0.027	0.42
Grus grus	4	15.0	5.614	2.22	0.148	2.29
Grus grus •	4	13.6				
Falco naumanni •	5	11.3	0.151	0.65	0.058	0.89
Falco tinnunculus	5	10.1	0.203	0.73	0.072	1.12
Fringilla montifringilla •	5	11.6				
Carduelis chloris •	5	12.2	0.028	0.25	0.020	0.32
Carduelis carduelis •	5	12.8	0.016	0.24	0.019	0.29
Carduelis spinus	5	14.5	0.014	0.21	0.014	0.22
Carduelis spinus •	5	12.4				
Carduelis cannabina	5	14.8	0.015	0.24	0.016	0.25
Pyrrhula pyrrhula	5	13.4	0.022	0.27	0.020	0.31
Pernis apivorus	5	12.5	0.778	1.26	0.101	1.56
Pernis apivorus •	6	10.1				
Milvus migrans •	6	11.7	0.815	1.52	0.130	2.01
Milvus milvus	6	12.0	1.012	1.66	0.138	2.14
Haliaeetus albicilla	6	13.6	4.967	2.18	0.160	2.48
Neophron percnopterus •	6	12.6	2.062	1.65	0.131	2.03
Circus aeruginosus	6	11.2	0.653	1.16	0.104	1.60
Circus aeruginosus •	6	10.1				
Circus cyaneus	6	9.1	0.433	1.10	0.121	1.87
Circus macrourus •	6	9.6	0.420	1.09	0.114	1.76
Circus pygargus •	6	8.4	0.291	1.09	0.130	2.01
Accipiter nisus	6	11.3	0.277	0.67	0.059	0.92
Accipiter nisus •	6	10.0				
Accipiter brevipes •	6	11.1	0.195	0.70	0.063	0.98

Species	g	v_e [m/s]	m_b [kg]	b [m]	b/v_e [s]	P [-]	Species	g	v_e [m/s]	m_b [kg]	b [m]	b/v_e [s]	P [-]
Turdus viscivorus •	5	12.4					Buteo buteo	6	11.6	0.885	1.24	0.107	1.66
Parus ater	5	10.6	0.009	0.18	0.017	0.26	Buteo buteo •	6	13.3				
Parus major •	5	13.6	0.019	0.23	0.017	0.26	Buteo lagopus	6	10.5	0.943	1.35	0.129	1.99
Garrulus glandarius	5	6.7	0.162	0.54	0.081	1.25	Aquila pomarina •	6	11.7	1.391	1.47	0.126	1.95
Garrulus glandarius •	5	12.9					Aquila nipalensis •	6	7.7	2.900	2.03	0.264	4.08
Nucifraga caryocatactes	5	13.4	0.173	0.58	0.043	0.67	Aquila chrysaetos	6	11.9	4.069	2.03	0.171	2.64
Corvus monedula	5	12.5	0.245	0.65	0.052	0.81	Hieraaetus pennatus •	6	11.3	0.828	1.11	0.098	1.52
Corvus monedula •	5	14.7					Pandion haliaetus	6	13.3	1.578	1.60	0.120	1.86
Corvus frugilegus	5	11.5	0.488	0.93	0.081	1.25	Pandion haliaetus •	6	11.4				
Corvus frugilegus •	5	13.0					Merops apiaster •	6	12.2	0.057	0.47	0.039	0.60
							Mean (species)		14.1	1.065	0.91	0.065	1.00

References

1. Alerstam, T.: Bird Migration. Cambridge University Press, Cambridge (1990)
2. Alerstam, T., Rosén, M., Bäckman, J., Ericson, P. G. P., Hellgren, O.: Flight Speeds among Bird Species: Allometric and Phylogenetic Effects. PLOS Biology 5(8), 1–7 (2007). doi: 10.1371/journal.pbio.0050197
3. Alerstam, T., Rosén, M., Bäckman, J., Ericson, P. G. P., Hellgren, O.: Protocol S1. Supplementary List of Flight Speeds and Biometry of Bird Species. PLOS Biology 5(8), 1–7 (2007). doi: 10.1371/journal.pbio.0050197.sd001
4. Barrientos, R., Alonso, J. C., Ponce, C., Palacín, C.: Meta-Analysis of the Effectiveness of Marked Wire in Reducing Avian Collisions with Power Lines. Conservation Biology 25(5), 893–903 (2011). doi: 10.1111/j.1523-1739.2011.01699.x
5. Bernshausen, F. von, Strein, M., Sawitzky, H.: Vogelverhalten an Hochspannungsfreileitungen – Auswirkungnen von elektrischen Freileitungen auf Vögel in durchschnittlich strukturierten Kulturlandschaften. Vogel und Umwelt 9. Sonderheft: 59–92 (1997)
6. Bevanger, K.: Estimating bird mortality caused by collision and electrocution with power lines; a review of methodology. In: Ferrer, M., Janss, G. F. E. (eds.) Birds and Power Lines: collision, electrocution, and breeding, pp. 29–56. Quercus, Madrid, Spain (1999)
7. Bevanger, K.: Biological and conservation aspects of bird mortality caused by electricity power lines: a review. Biological Conservation 86(1), 67–76 (1998). doi: 10.1016/S0006-3207(97)00176-6
8. BioConsult SH GmbH, ARSU GmbH: Zum Einfluss von Windenergieanlagen auf den Vogelzug auf der Insel Fehmarn – Gutachterliche Stellungnahme auf Basis der Literatur und eigener Untersuchungen im Frühjahr und Herbst 2009. Im Auftrag der Fehmarn Netz GmbH & Co. OHG, Husum & Oldenburg, 2010. http://www.naturschutzstandards-erneuerbarer-energien.de/images/literatur/2010_bioconsult_vogelzug%20fehmarn.pdf
9. Brendel, U.: Der Einfluss von Hängegleitern und Gleitseglern auf die Avifauna. Ornithologische Bewertung von Startplatzbereichen auf ausgewählten Fluggeländen in repräsentativen Lebensraumtypen, Zukunft Biosphäre Gmbh, Bischoffswiesen, Germany, 2003. https://natursportinfo.bfn.de/index.php?id=25036
10. Brenninkmeijer, A., Weyde, C. van der: Monitoring aanvaringsslachtoffers Windpark Delfzijl-Zuid 2006–2011. Eindrapportage vijf jaar monitoring. A&W rapport 1656, Altenburg & Wymenga ecologisch onderzoek BV, Feanwâlden, The Netherlands, 2011. http://www.altwym.nl/uploads/file/533_1427453204.pdf
11. Brown, A. L.: Measuring the effect of aircraft noise on sea birds. Environment International 16(4), 587–592 (1990). doi: 10.1016/0160-4120(90)90029-6

12. Bruderer, B., Komenda-Zehnder, S.: Einfluss des Flugverkehrs auf die Avifauna – Schluss-bericht mit Empfehlungen. Schriftenreihe Umwelt 376, Bundesamt für Umwelt, Wald und Landschaft, Bern, 2005. http://www.batsandwind.org/pdf/postconpatbatfatal.pdf

13. Bruderer, B., Boldt, A.: Flight characteristics of birds: I. radar measurements of speeds. International Journal of Avian Science 143(2), 178–204 (2001). doi: 10.1111/j.1474-919X.2001.tb04475.x

14. Dolbeer, R. A., Wright, S. E., Weller, J. R., Begier, M. J.: Wildlife Strikes to Civil Aircraft in the United States 1990–2013. Federal Aviation Administration National Wildlife Strike Database Serial Report 20, U.S. Departments of Transportation and Agriculture, Federal Aviation Authority, Washington, DC, USA, July 2014. https://wildlife.faa.gov/downloads/Wildlife-Strike-Report-1990-2013-USDA-FAA.pdf

15. Erickson, W. P., Johnson, G. D., Strickland, D. M., Young, J., Sernka, K. J., Good, R. E.: Avian collision with wind turbines: a summary of existing studies and comparisons to other sources of avian collision mortality in the United States, National Wind Coordinating Committee, Washington, DC, USA, Aug 2001. doi: 10.2172/822418

16. Everaert, J.: Effecten van windturbines op de fauna in Vlaanderen. Onderzoeksresultaten, discussie en aanbevelingen.(effects of wind turbines on fauna in Flanders. Study results, discussion and recommendations). INBO.R.2008.44, Instituut voor Natuur- en Bosonderzoek, Brussel, Belgium, 2008. http://ebl.vlaanderen.be/publications/documents/24633

17. Faanes, C. A.: Bird behavior and mortality in relation to power lines in prairie habitats. Fish and Wildlife Technical Report 7, U.S. Fish and Wildlife Service, Washington, D.C., 1987. https://pubs.er.usgs.gov/publication/2000102

18. Gehring, J., Kerlinger, P., Manville, A. M.: Communication towers, lights, and birds: successful methods of reducing the frequency of avian collisions. Ecological Applications 19(2), 505–514 (2009). doi: 10.1890/07-1708.1

19. Gehring, J., Kerlinger, P., Manville, A. M.: The role of tower height and guy wires on avian collisions with communication towers. The Journal of Wildlife Management 75(4), 848–855 (2011). doi: 10.1002/jwmg.99

20. Hartman, J. C., Gyimesi, A., Prinsen, H. A. M.: Veldonderzoek naar draadslachtoffers en vliegbewegingen bij een gemarkeerde 150 kV hoogspanningslijn. Rapport nr. 10-082, Bureau Waardenburg BV, Culemborg, The Netherlands, 8 Nov 2010. https://buwa.nl/fileadmin/buwa_upload/Bureau_Waardenburg_rapporten/09-355_Effectiviteit_vogelflappen_onderzoek_Bureau_Waardenburg.pdf

21. Hötker, H.: Auswirkungen des 'Repowering' von Windkraftanlagen auf Vögel und Fledermäuse, Michael-Otto-Institut im NABU-Forschungs- und Bildungszentrum für Feuchtgebiete und Vogelschutz, Bergenhusen, Germany, Oct 2006. https://bergenhusen.nabu.de/imperia/md/images/bergenhusen/windkraft_endbericht.pdf

22. Jansen, M.: Monitoring Kitesurfzone Wolderwijd. Final Report, May 2011

23. Kempf, N., Hüppop, O.: Auswirkungen von Fluglärm auf Wildtiere: ein kommentierter Überblick. Journal für Ornithologie 137(1), 101–113 (1996). doi: 10.1007/BF01651502

24. Klop, E., Brenninkmeijer, A.: Monitoring aanvaringsslachtoffers Windpark Eemshaven 2009–2014. Eindrapportage vijf jaar monitoring. A&W-rapport 1975, Altenburg & Wymenga ecologisch onderzoek BV, Feanwâlden, The Netherlands, 2014. http://www.altwym.nl/uploads/file/534_1427453591.pdf

25. Klop, E., Brenninkmeijer, A.: Vervolgmonitoring vogelslachtoffers hoogspanningslijnen Eemshaven. Jaarrapportage 2013–2014. A&W-rapport 2062, Altenburg & Wymenga ecologisch onderzoek BV, Feanwâlden, The Netherlands, 2014

26. Koolstra, B. J. H.: Habitattoets 380 kV-station Oude Schip en -hoogspanningsleiding. Passende Beoordeling Natuurbeschermingswet en quick scan Flora- en faunawet, Arcadis Nederland BV, Den Bosch, The Netherlands, 2008

27. Koops, F. B. J.: Draadslachtoffers in Nederland en effecten van markering, KEMA Nederland, Arnhem, 1987

28. Krausman, P. R., Wallace, M. C., Hayes, C. L., De Young, D. W.: Effects of Jet Aircraft on Mountain Sheep. The Journal of Wildlife Management **62**(4), 1246–1254 (1998). doi: 10. 2307/3801988

29. Krijgsveld, K. L., Akershoek, K., Schenk, F., Dijk, F., Dirksen, S.: Collision Risk of Birds with Modern Large Wind Turbines. Ardea **97**(3), 357–366 (2009). doi: 10.5253/078.097.0311

30. Krijgsveld, K. L., Smits, R. R., Winden, J. van der: Verstoringsgevoeligheid van vogels: update literatuurstudie naar de reacties van vogels op recreatie. Final Report 08-173, Bureau Waardenburg BV, Culemborg, The Netherlands, 23 Dec 2008. www.vliz.be/imisdocs/publications/211691.pdf

31. Kruijff, M.: The Technology of Airborne Wind Energy – Part I: Launch & Land. https://www.ampyxpower.com/2017/04/1002 (2017). Accessed 10 Oct 2017

32. Kruijff, M.: The Technology of Airborne Wind Energy – Part II: the Drone. https://www.ampyxpower.com/2017/04/the-technology-of-airborne-wind-energy-part-ii-the-drone (2017). Accessed 10 Oct 2017

33. Kruijff, M.: The Technology of Airborne Wind Energy – Part III: Safe Power. https://www.ampyxpower.com/2017/05/the-technology-of-airborne-wind-energy-part-iii-safe-power (2017). Accessed 10 Oct 2017

34. Lekuona, J. M.: Uso del espacio por la avifauna y control de la mortalidad de aves murcielagos en los parques eolicos de navarra durante un ciclo annual, Direccion General de Medio Ambiente, Departemento de medio Ambiente, Ordenacion del Territorio y Viviends, Gobierno de Nevarra, Pamplona, Spain, Apr 2001. http://www.gurelur.org/p/es/proyectos/energia-eolica/i-Descargables/estudio-eolica.pdf

35. Lensink, R.: Verstorende effecten van MLA Venlo in relatie tot groene wet- en regelgeving: notitie op hoofdzaken. Report 09-028, Bureau Waardenburg BV, Culemborg, The Netherlands, 15 Mar 2009

36. Lensink, R., Aarts, B. G. W., Anema, L. S.: Bestaand gebruik kleine luchtvaart en beheerplannen Natura 2000: Naar een uniforme en transparante behandeling van dit onderwerp in alle beheerplannen. Report 10-180, Bureau Waardenburg BV, Culemborg, The Netherlands, 7 Feb 2011. http://www.natura2000.nl/files/10-431-eindrapport_7feb_rle_klein.pdf

37. Longcore, T., Rich, C., Mineau, P., MacDonald, B., Bert, D. G., Sullivan, L. M., Mutrie, E., Gauthreaux, S. A., Avery, M. L., Crawford, R. L., Manville, A. M., Travis, E. R., Drake, D.: An Estimate of Avian Mortality at Communication Towers in the United States and Canada. PLOS ONE **7**(4). edited by Krkosek, M., 1–17 (2012). doi: 10.1371/journal.pone.0034025

38. Longcore, T., Rich, C., Mineau, P., MacDonald, B., Bert, D. G., Sullivan, L. M., Mutrie, E., Gauthreaux, S. A., Avery, M. L., Crawford, R. L., Manville, A. M., Travis, E. R., Drake, D.: Avian mortality at communication towers in the United States and Canada: which species, how many, and where? Biological Conservation **158**, 410–419 (2013). doi: 10.1016/j.biocon.2012.09.019

39. Longcore, T., Rich, C., Gauthreaux, S. A.: Height, guy wires, and steady burning lights increase hazard of communication towers to nocturnal migrants: a review and meta-analysis. The Auk: Ornithological Advances **125**(2), 485–492 (2008). doi: 10.1525/auk.2008.06253

40. Mendonça, A. K. d. S., Vaz, C. R., Lezana, Á. G. R., Anacleto, C. A., Paladini, E. P.: Comparing Patent and Scientific Literature in Airborne Wind Energy. Sustainability **9**(6), 915 (2017). doi: 10.3390/su9060915

41. Mosler-Berger, C.: Störungen von Wildtieren: Umfrageergebnisse und Literaturauswertung, Bundesamt für Umwelt, Wald und Landschaft (BUWAL), Bern, Switzerland, 1994

42. Musters, C. J. M., Noordervliet, M. A. W., Ter Keurs, W. J.: Bird casualties caused by a wind energy project in an estuary. Bird Study **43**(1), 124–127 (1996). doi: 10.1080/00063659609461003

43. Renssen, T. A.: Vogels onder hoogspanning : een studie betreffende de invloed van hoogspanningslijnen op vogelsterfte. Reeks Natuur en Millieu 10 (1977)

44. Smallwood, K. S., Thelander, C.: Bird Mortality in the Altamont Pass Wind Resource Area, California. Journal of Wildlife Management **72**(1), 215–223 (2008). doi: 10.2193/2007-032

45. Szemkus, B., Ingold, P.: Behaviour of Alpine ibex (Capra ibex ibex) under the influence of paragliders and other air traffic. Zeitschrift für Säugetierkunde **72**, 84–89 (1998). http://biostor. org/reference/biostor/183706
46. Verbeek, R. G., Krijgsveld, K. L.: Kitesurfen in de Delta en verstoring van vogels en zeehonden: onderbouwing van locaties waar kitesurfen via het Beheerplan kan worden toegestaan. Report 12-143, Bureau Waardenburg BV, Culemborg, 2013. http://www.kitesurfeur.be/wp-content/uploads/2013/11/12-414-kitesurflocaties-Delta-20130124-verkleind.pdf
47. Vlas, M. J. de, Butter, M.: Draadslachtoffers in de Westerbroekstermadepolder, Schatting van het aantal dode vogels als gevolg van een hoogspanningslijn in een natuurgebied. 61, Rijksuniversiteit Groningen, 2003. https://www.rug.nl/research/portal/files/14483269/rap61.pdf
48. Ward, D. H., Stehn, R. A., Erickson, W. P., Derksen, D. V.: Response of Fall-Staging Brant and Canada Geese to Aircraft Overflights in Southwestern Alaska. The Journal of Wildlife Management **63**(1), 373–381 (1999). doi: 10.2307/3802522
49. Weisenberger, M. E., Krausman, P. R., Wallace, M. C., De Young, D. W., Maughan, O. E.: Effects of Simulated Jet Aircraft Noise on Heart Rate and Behavior of Desert Ungulates. The Journal of Wildlife Management **60**(1), 52 (1996). doi: 10.2307/3802039
50. Winkelman, J. E.: De invloed van de Sep-proefwindcentrale te Oosterbierum (Fr.) op vogels (The impact of the Sep Wind Park near Oosterbierum (Fr.), The Netherlands, on birds). RIN-Rapport 92/2–5, DLO-Instituut voor Bos- en Natuuronderzoek, Arnhem, The Netherlands, 1992. http://edepot.wur.nl/383034
51. Winkelman, J. E.: Vogels en het windpark nabij Urk (NOP): aanvaringsslachtoffers en verstoring van pleisterende eenden, ganzen en zwanen. RIN-Rapport 89/15, Rijksinstituut voor Natuurbeheer, Arnhem, The Netherlands, 1989. http://edepot.wur.nl/387439

Chapter 29
Current and Expected Airspace Regulations for Airborne Wind Energy Systems

Volkan Salma, Richard Ruiterkamp, Michiel Kruijff, M. M. (René) van Paassen and Roland Schmehl

Abstract Safety is a major factor in the permitting process for airborne wind energy systems. To successfully commercialize the technologies, safety and reliability have to be ensured by the design methodology and have to meet accepted standards. Current prototypes operate with special temporary permits, usually issued by local aviation authorities and based on ad-hoc assessments of safety. Neither at national nor at international level there is yet a common view on regulation. In this chapter, we investigate the role of airborne wind energy systems in the airspace and possible aviation-related risks. Within this scope, current operation permit details for several prototypes are presented. Even though these prototypes operate with local permits, the commercial end-products are expected to fully comply with international airspace regulations. We share the insights obtained by Ampyx Power as one of the early movers in this area. Current and expected international airspace regulations are reviewed that can be used to find a starting point to evidence the safety of airborne wind energy systems. In our view, certification is not an unnecessary burden but provides both a prudent and a necessary approach to large-scale commercial deployment near populated areas.

29.1 Introduction

Due to the emerging interest in airborne wind energy (AWE), a considerable number of prototype installations is approaching the stage of commercial development. As

Volkan Salma · Richard Ruiterkamp · Michiel Kruijff
Ampyx Power B.V., Lulofsstraat 55 Unit 13, 2521 AL The Hague, The Netherlands

Volkan Salma (✉) · M. M.(René) van Paassen · Roland Schmehl
Delft University of Technology, Faculty of Aerospace Engineering, Kluyverweg 1, 2629 HS Delft, The Netherlands
e-mail: volkan.salma@esa.int

© Springer Nature Singapore Pte Ltd. 2018
R. Schmehl (ed.), *Airborne Wind Energy*, Green Energy and Technology, https://doi.org/10.1007/978-981-10-1947-0_29

consequence, operational safety and system reliability are becoming crucially important aspects and it is evident that a certification framework addressing safety and reliability of airborne wind energy systems will be required for a successful market introduction and broad public acceptance.

Compared to conventional wind turbines, AWE systems operate at higher altitudes and for most concepts this operation is not stationary. Because of their substantially larger operational envelope the interaction with the aviation system is potentially stronger. For these reasons, AWE systems introduce risks to third parties in the air and objects on the ground. Thus, besides addressing the safety issues for wind turbines, such as the risk of lightning or fire within the equipment, additional considerations are required for managing the aviation-related risks.

The main system components are one or more flying devices, one or more tethers and the energy conversion system, which can be part of the flying device or part of a ground station. The flight control system can be either part of the flying device, a separate airborne device or part of the ground station. A thorough classification of implemented prototypes is provided in [2]. Although existing standards can be partially applied to some of the components, such as the low voltage directive (LVD) 2006/95/EC for electrical installation, and machine directives 2006/42/EC or IEC 61400 for wind turbines, there are no standards for the tether and the flying devices. This study investigates the applicability of existing rules and standards whose objective is to manage the aviation-related risks. In essence these standards define the acceptable risks to other airspace users or to people and property on the ground, often denominated as third-party risk. The aim of this chapter is to provide an overview of the current situation of AWE applications from the aviation perspective and to outline a permitting and certification approach for different types of AWE systems.

Ampyx Power is one of the early movers in this area and is currently pursuing with the European Aviation Safety Agency (EASA) the certification of an utility-scale, grid-connected rigid glider [33–35, 44]. It is the aim of this chapter to position the experience of Ampyx Power in the broader context of commercial-scale AWE systems of any design. We limit ourselves though to systems whose operation requires permitting by aviation authorities and takes place near populated areas and/or critical infrastructure. In other words, we consider only deployment scenarios which create an actual safety risk. We assume that the commercial operations for which a permit is sought take place initially over land restricted to qualified personnel. Our focus is on the European regulatory framework and on those AWE systems for which the current unmanned aerial vehicle (UAV) regulations seem most appropriate as a starting point. We merely provide the basic context for other cases.

At the Airborne Wind Energy Conference 2015, Glass mentioned the unique challenge of airborne wind turbine (AWT) certification because of the combined elements of wind turbine together with aircraft and the additional tether considerations [20]. In addition, he suggested a unified framework for the certification of AWTs. This framework starts with reviewing the existing standards in related sectors, including wind turbine standards and aviation standards. Then, identification of the AWT operation regime is required to see what must be addressed in the standards. Afterwards, a conservative gap analysis has to be performed for identifying the areas that are not adequately covered by the standards. Lastly, new requirements have to be developed to fill the gaps. Glass recommends collaboration with the standards developing organizations through the entire standard making process.

At the same conference, Ruiterkamp provided an overview of existing and expected rules and the standards for ensuring safe operation of AWE applications [44]. He further described possible risks introduced by AWE systems and supplemented his study with the expected legislation for a rigid wing concept.

Langley investigated AWE systems from a legal perspective [36]. In this study, he introduces the environmental impacts of AWE systems and the current legal landscape. An early mover in the US, Makani Power, which was acquired in 2013 by Google and is currently one of the "moonshot" projects of the Alphabet subsidiary X, has published a detailed document about the operation of an AWE system [21], responding to a "Notification for Airborne Wind Energy Systems (AWES)" issued by the Federal Aviation Authority (FAA) [17].

The chapter is structured as follows. Section 29.2 describes the commonly used terms in the study such as "regulation", "certification" and "flying permit". Section 29.3 provides an overview of the flying permit status of current prototypes to grasp the variety of architectures currently considered. The information has been collected by means of a survey and reveals that each architecture faces its own specific safety challenges requiring a tailoring of the mitigation measures. In Sect. 29.4, we start to explore the perspective for a large-scale deployment of such prototypes, and for this, the place of the AWE applications in the airspace is studied. Possible interference between AWE systems and current aviation activities is described. In Sect. 29.5 we introduce the civil aviation authorities which would most likely be important in the regulation making process. We then highlight in Sect. 29.6 three possible starting points to obtain operation permits for AWE prototypes. The first one is unmanned aerial vehicle (UAV) registration, the second one is air navigation obstacle registration, the third one is tethered gas balloon registration. Concerning UAV regulations, the current certification framework and future expectations are described, highlighting also different views of aviation authorities on tethered aircraft. Concerning air traffic obstacle regulation, ICAO rules for air traffic obstacles which might be applicable to AWE applications are referenced. Lastly, yet importantly, we assess in Sect. 29.7 different permitting and certification paths for AWE systems in the light of current regulations and future projections.

29.2 Concepts of Regulation, Certification and Flying Permit

As a starting point we describe how regulation, certification, permitting and standard relate to each other for the European context. The relevant overarching European laws are the Regulation (EC) No. 216/2008 [12], which distributes the responsibilities between EASA and the national aviation authorities (NAA), defines the mechanism of certification and lists the high-level airworthiness requirements, as well as Commission Regulation (EU) No. 748/2012 [10], which implements Part 21, the globally agreed requirements for certification in aviation. The Certification Specification (CS) and Special Conditions (SC) are type-specific soft regulations for airworthiness (incl. safety through the respective articles 1309), and suitable starting points for tailoring. Certification is done with respect to a certification basis agreed between applicant and aviation authority: a selection and tailoring of the appropriate CS/SC and definition of an acceptable means of compliance, using e.g. ARP/ED standards.[1] A Permit to Fly is given by the NAA for small, experimental or developmental systems. The necessary airworthiness and safety evidence shall be approved by a certification body or a qualified entity (this may be as part of a certification trajectory, but does not have to be). Key question addressed here is: can the system be flown safely? The NAA in addition considers local constraints and operational safety.

29.3 Current Operation Permit Status of AWE Systems

Airborne wind energy is currently in the development and testing phase. In this phase, companies and research groups conduct their tests with special permissions. Most of these permissions are issued by local civil aviation authorities. AWE application examples from different high-level architectures are shown in Table 29.1. Comprehensive information for each architecture and up-to-date implementation details for practically demonstrated AWE systems can be found in [2].

To understand the current status and extent of these exemptions, a survey was conducted in the context of the International Airborne Wind Energy Conference 2015 [45, p. 9]. Companies and research groups around the globe were invited to provide the technical specifications of their prototypes and information on the flight permit. The analysis of this data shows that current prototypes have a small airborne

	Ground generator, single tether	Ground generator, multiple tether	Onboard generator, single tether
Flexible Wing	TU Delft [47] Politecnico di Torino [13] SkySails Power [19] Kite Power Systems [30]	Kitenergy [38]	
Rigid Wing	Ampyx Power [46] Kitemill [31] eWind Solutions [32]	TwingTec [37] EnerKite [1]	Makani Power [52] Windlift [54]
Other	Omnidea [43]		Altaeros Energies [53]

Table 29.1 Selection of current AWE applications and architectures. System concepts with onboard generator (as a primary means for electricity generation) and multiple tethers are not known to the authors

---

[1] The acronym ARP stands for Aerospace Recommended Practices and the acronym ED stands for EUROCAE (European Organisation for Civil Aviation Equipment) Document.

Organization	Prototype category	Size^a	Tether#	Weight (kg)
TU Delft	Flexible wing / generator on ground	25 m^2	1	20
Kontra Engineering	Flexible wing / generator on ground	2.5 m^2	2	0.5
Kitemill	Rigid wing / generator on ground	3.7 m	1	4.5
Windswept and		~2 m^2 driving		
Interesting Ltd	Flexible wing / generator on ground	~3 m^2 lifting	Many	1.6
FlygenKite	Flexible wing / generator on ground	2 m	Many	0.2
	Flexible wing / airborne generation			
Kite Power Systems	Flexible wing / generator on ground	7 m	1	45
Kite Power Systems	Flexible wing / generator on ground	up to 40 m		450
EnerKite	Semi-rigid wing / generator on ground	11 m	3	20
Ampyx Power	Rigid wing / generator on ground	5.5 m	1	35
Federal University	Flexible wing / no electricity gen.	3 m^2	1	2
of Santa Catarina	(flight control purposes only)			
kPower	Rigid wing / generator on ground	1–300 m^2	Many^b	0.5–100
	Flexible wing / generator on ground			
	Rigid wing / airborne generation			
	Flexible wing / airborne generation			
Altaeros Energies	Lighter than air / airborne generation	N/A	3	N/A
TwingTec	Rigid wing / generator on ground	3 m^2	2	15

^{a} m^2 for projected wing area, m for wing span
^{b} 3D lattices form for topological stability

Table 29.2 Reported AWE prototypes in the certification survey

mass, occupy only a small volume of the airspace and generally have human pilots in the loop or supervising the system. They are operating in a selected safe area to mitigate the risks to third parties. It is expected that the final commercial products will be significantly larger with higher airborne mass, will occupy larger volumes of the airspace and will ultimately have to comply with international airspace regulations. Conference participants were asked to fill out a web-based survey. Among the responses from 26 different organizations, 15 different AWE prototypes are reported. Table 29.2 shows the main properties of the reported prototypes. According to the responses, 10 out of 15 prototypes are formally registered with a civil certification authority. Three systems are registered as an air navigation obstacle, 6 systems are registered as unmanned glider or tethered kite. The remaining system holds an environmental permit (Dutch: "omgevingsvergunning") from the responsible local municipality. A selection of collected flying permit data is provided in Table 29.3. Results show that there is currently no consensus among the certification authorities. For technically similar concepts, some aviation authorities require personnel training, while others do not impose this requirement. Some prototypes need licensed personnel to operate. While most of the prototypes are allowed to operate at night, some can operate only during daylight hours.

Organization	Operation permit type	Issuing authority	Validity country code	Validity (Location)	Permitted altitude (m)	(P)ilot/op. Required (T)raining required (N)ight flight permitted Full (A)utonomy permitted	Other notes
TU Delft	Kite power system	ILT	NL	Valkenburg Airfield	500	N	–
Kitemill	Air traffic obstacle	CAA	NO	Lista	520	P – T – A – N	–
Ampyx Power	Unmanned glider	NAA	NL	Kraggenburg	300	P – T – A	*[a]
kPower	*[b]	FAA	US	*[c]	609	P – T – N	–
		FAA	US	*[d]	5486		NOTAM[f] req.
		FAA	US	*[e]	>10000		NOTAM[f] req.
QConcepts	*[g]	*[h]	NL	Doetinchem	300	P – A	–
Altaeros Energies	Air traffic obstacle	FAA	US	Confidential	240	Confidential	–
TwingTec	Tethered kite	BAZL	CH	Chasseral, Diegenstal, Silvaplana	150, 300[i]	A	*[j]

[a] 5000 meters of visibility required, off-cable flight below 450 m, not above people, 150 m horizontal distance from people, traffic and buildings, visual line-of-sight (VLOS)
[b] Legacy kite rules (FARs, part 101)
[c] Any place where legacy kites can be operated
[d] Warm Springs FAA UAS Test Range
[e] The Tillamook FAA UAS Test Range
[f] Notice to Airman
[g] Environmental permit (Dutch: "omgevingsvergunning")
[h] Local municipal
[i] Depending on location
[j] max 20 m², max 25kg

Table 29.3 Selection of flight permit data

29.4 AWE Systems in the Airspace

Aviation authorities divide the airspace into segments. These segments are called classes and labeled with the letters A through G. Each class has its own rules. For example, in Class A, all operations must be conducted under instrument flight rules (IFR) and air traffic control (ATC) clearance is required for flights. Even though most countries adhere to ICAOs standard rules for classes, individual nations can adapt the rules for their own needs. Current AWE system prototypes operate in Class G airspace, which is normally near to the ground. Figure 29.1 shows the airspace separation and Class G airspace. Class G is typically up to 1200 feet above ground level (AGL). However, Class G can be limited to 700 feet AGL if there is an airport close by, which requires Class B airspace in its vicinity as shown in Fig. 29.1. Class G is known as uncontrolled air space. There is no specific aircraft equipment or pilot specifications to enter Class G. Moreover, no ATC communication is required to fly in Class G. Although Class G is uncontrolled, civil aviation rules are still valid. There are visibility and cloud clearance requirements for flights in Class G, and most flights operate under visual flight rules, meaning that separa-

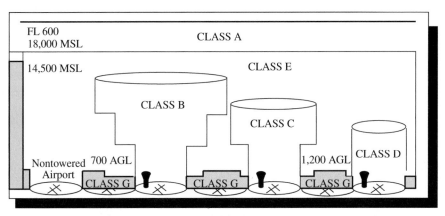

Fig. 29.1 Airspace separation and Class G airspace by FAA [14]

tion is based on the "see and avoid" principle. Since class G airspace is open to all users, interference between AWE systems and aircraft is possible.

In addition to interference risk, there are other aviation related risks posed by AWE systems. For example, uncontrolled crash (while the tether is attached or not) or uncontrolled departure from the designated flight area (with the tether partly attached or also detached) are the aviation risks which have to be managed.

29.5 Relevant Aviation Certification Bodies

This section discusses the regulatory bodies that provide rules for safe aviation and civil airspace. There are national aviation organizations as well as international aviation organizations that strive to harmonize aviation rules, in order to facilitate international air travel. These organizations all could have a role in the AWE relevant rule making process.

29.5.1 International Civil Aviation Organization (ICAO)

The ICAO was founded in 1944 upon the signing of the Convention on International Civil Aviation, commonly known as Chicago Convention. Since 1947, the organization works with the Convention's 191 Member States and with global aviation organizations as a specialized agency of the United Nations (UN). ICAO develops International Standards and Recommended Practices (SARPs) which are used by member states as a framework for their aviation law making processes.

29.5.2 Federal Aviation Authority (FAA)

The FAA is the civil aviation agency of United States Department of Transportation. The agency makes Federal Aviation Regulations (FARs) and puts them into practice to ensure the safety of civil aviation within the United States. The FAA is authorized to certify a civil aircraft for international use.

29.5.3 European Aviation Safety Agency (EASA)

The EASA was established in 2002 by the European Commission (EC) to ensure the safety of civil aviation operations. The agency advises the EC and member states of the European Union (EU) regarding new legislation. EASA is a second agency, next to the FAA, authorized to certify civil aircraft for international use.

29.5.4 National Aviation Authority (NAA)

The national regulatory body which is responsible for aviation is denoted as NAA or civil aviation authority (CAA). These authorities make national legislation in compliance with ICAO SARPs.

29.5.5 Joint Authorities for Rulemaking on Unmanned Systems (JARUS)

The JARUS is a group that consists of experts from national aviation authorities or regional aviation safety organizations which aims to define the certification requirements for UAVs to safely integrate them to the current aviation system. JARUS defines its objective for UAVs as follows [29]:

> ...to provide guidance material aiming to facilitate each authority to write their own requirements and to avoid duplicate efforts.

Working groups in JARUS publish recommended certification specifications for interested parties such as ICAO, EASA and NAAs.

29.6 Regulations for Airborne Wind Energy Systems

At the time of this study, there is no directly applicable regulation for AWE technologies. However, regulations for UAVs, air traffic obstacles or unmanned balloons are available as a starting point for tailoring to the specifics of a selected AWE architecture. In this section, current regulations for unmanned aerial vehicles from different regulatory bodies, air traffic obstacle regulations and tethered gas balloon regulations are summarized. Similar-looking AWE systems can be categorized differently depending on the modes of operation and the inherent safety measures and the proper starting point should be selected accordingly, together with the responsible aviation authority. Note that our focus is mostly on the developing UAV regulation since we expect that most tethered aircraft that will have the ability to (aerodynamically) leave their restricted safe area as a result of a single tether failure will be considered unmanned aircraft. Hence, for those systems, the UAV regulation seems the most appropriate starting point.

29.6.1 Regulations for the Unmanned Aerial Vehicle Category

Unmanned aerial vehicles were first used in the military sector. The technology then evolved also for civil applications and nowadays there are already many commercial products on the market, such as UAVs for high-quality aerial photography or 3D mapping. However, the increasing interest in UAVs has also led to a rise in safety concerns. As a consequence, national aviation agencies and international aviation organizations have directed their attention to developing certification processes, regulations and standards for UAVs including those related to airworthiness. One of the main challenging factors for UAV regulation is the wide variety of systems in the UAV domain. For instance, the UAV concept includes devices from micro UAVs which are extremely lightweight (e.g. 16 grams [3]) to High Altitude Long Endurance (HALE) class UAVs up to 14 tons [42]. Consequently, there is no consensus on a classification method which is able to cover this broad range yet. Several different classification approaches have been proposed for UAVs, such as classification according to aircraft weight, avionics complexity level, aircraft configuration (number and type of engines, etc.), aircraft speed, operation purpose (e.g., aerial work), operation airspace (segregated, non-segregated), overflown area, kinetic energy, operational failure consequence, and operation altitude.

The first publicly accepted standardization agreement, the STANAG 4671 [41] compiled by the North Atlantic Treaty Organization (NATO), was an important step forward in UAV registration, even though it is limited to military UAVs. The standard is based on EASA's CS-23 [6] civil airworthiness code. In addition to CS-23, STANAG 4671 includes subparts which are specific to UAVs such as ground control station and datalink. The standard provides a broad range of requirements for flight, aircraft structure, design, construction, power plant, equipment, command and control and the control station. However, the standard only addresses fixed-wing UAVs with a weight between 150 and 20,000 kg. As a result a considerable number of UAV types are not covered by the standard, among which the designs that are not structurally similar to conventional aircraft. With the following STANAG 4703 [40], the

NATO Standardization Agency (NSA) defined the airworthiness requirements also for lighter military UAVs whose take-off weight does not exceed 150 kg.

At the time of writing this chapter, required rules for integrating UAVs to civil airspace are still subject to change and different certification proposals from different certification authorities exist. In addition, it is known that a limited number of UAV applications are certified for civil operations by FAA and EASA with a case-by-case risk evaluation and only for specific operations. Depending on the definition of UAV in the upcoming regulations by different aviation authorities, some of the AWE applications may fall into the UAV category. In the following we will explore the possibilities in the light of current regulations, known regulatory views and the published regulatory proposals.

For AWE applications falling in the UAV category an airworthiness certificate would be sought for commercial operation. Currently, two types of airworthiness certificates are common for manned aviation. In contrast to the standard airworthiness certificate, the restricted airworthiness certificate has operational limitations such as restrictions on maneuvers, speed, activities undertaken or where the flights may be conducted. According to first drafts of UAV certification method proposals, a similar type scheme (standard and restricted airworthiness) will be used for UAVs. Considering that current AWE applications have very specific characteristics, such as being tethered to a ground station or operating in a specific area, it can be expected that the restricted type certificate will apply.

29.6.1.1 ICAO Regulations for Unmanned Aerial Vehicles

On 7 March 2012, ICAO adopted Amendment 6 to the International Standards and Recommended Practices, Aircraft Nationality and Registration Marks, which is identical to Annex 7 to the Convention on International Civil Aviation (also known as Chicago Convention). This revision included UAVs as remotely piloted aircraft (RPA), defining an RPA as "an unmanned aircraft which is piloted from a remote pilot station" [23]. At the same time Amendment 43 to Annex 2 "Rules of the Air" to the Chicago Convention was adopted. This amendment stipulates that an RPA shall be operated in such a manner as to minimize hazards to persons, property or other aircraft. Amendment 43 is the first regulation by ICAO that introduces the operation of remotely piloted aircraft systems (RPAS) in the Chicago Convention.

The current regulation [24] requires a certification of all types of aircraft that intend to fly in controlled and uncontrolled airspace, even though the certification framework for UAVs is not clear in Chicago Convention yet. In March 2011, ICAO published Circular 328 specifically addressing "Unmanned Aircraft Systems (UAS)" [28]. The aim of this circular is to establish a basis by properly defining the new technology, clarifying the differences between unmanned and manned aircraft. In March 2015, Circular 328 was superseded by the "Manual on Remotely Piloted Aircraft Systems (Doc 10019)" [27]. The following excerpts from this document are deemed representative of ICAO's current perspective on UAVs [27, Chap. 1, Sect. 6]

1.6.3 These hazards relate to all RPAS operations irrespective of the purpose of the operation. Therefore, the recommendations in this manual, unless specified otherwise, apply equally to commercial air transport and general aviation, including aerial work, operations conducted by RPAS.
1.6.4 In order for RPAS to be widely accepted, they will have to be integrated into the existing aviation system without negatively affecting manned aviation (e.g. safety or capacity reduction). If this cannot be achieved (e.g. due to intrinsic limitations of RPAS design), the RPA may be accommodated by being restricted to specific conditions or areas (e.g. visual line-of-sight (VLOS), segregated airspace or away from heavily populated areas).

and further [27, Chap. 2, Sect. 2]

2.2.7 Categorization of RPA may be useful for the purpose of a proportionate application of safety risk management, certification, operational and licensing requirements. RPA may be categorized according to criteria such as: maximum take-off mass (MTOM), kinetic energy, various performance criteria, type/area of operations, capabilities. Work is underway in many forums to develop a categorization scheme.

Autonomous unmanned aircraft and their operations, including unmanned free balloons or other types of aircraft which cannot be managed on a real-time basis during flight, is not in the scope of the Doc 10019. At the time of writing, there are no rules for AWE applications or tethered aircraft in the ICAO regulations.

29.6.1.2 EASA Regulations for Unmanned Aerial Vehicles

EC-2008 is the European Union's law that converts the ICAO SARPs to the EU structure, describing the responsibilities of EASA and NAAs [12]. Annex II of EC-2008 defines the exceptional cases which are outside EASA's area of responsibility. For example, the following cases do not lie within the responsibility of EASA[2]:

(b) aircraft, specifically **designed or modified for research**, experimental or **scientific purposes**, and likely **to be produced in very limited numbers**...
...(I) unmanned aircraft with an operational mass of **no more than 150 kg**

NAAs of member states are responsible for the regulation of these cases. Apart from the above mentioned exception cases, EASA makes the common European rules for UAV certification.

One of the important steps in civil UAV airworthiness certification is the interim Policy Statement EASA E.Y013-01 [4], which is still in use and aims at protecting people and property on the ground but not the UAV itself. The policy provides a kinetic energy-based classification method and a systematic certification guideline which suggests tailoring of fixed manned aircraft certification regulations. According to the tailoring principle, class determination has to be done as a first step using the kinetic energy evaluation method, which is defined in the regulation. Then, a tailoring process is required, adjusting an already existing certification specification for a conventional aircraft, which is in the same kinetic energy class with the new

---

[2] In this and the following quotations the emphasis is added by the authors

type that is intended to be certified. During this process, each requirement of the existing certification specification has to be reviewed and its applicability for new type has to be evaluated. Depending on the new type, special conditions may be added. This conditions may provide a starting point for the future applicants. It is further stated in the policy [4, Paragraph 21A.17]

> At an applicant's request, the Agency may accept USAR version 3, STANAG 4671, or later updates, as the reference airworthiness code used in setting the type certification basis

It should be noted that Ampyx Power and EASA have come to the conclusion that the tethered aircraft of Ampyx Power resembles more an unmanned glider than the typical tactical UAV that STANAG 4671 is templating. Therefore, the company has chosen to tailor CS-22 for its airworthiness baseline. These examples show EASA's willingness to accept the most suitable pre-existing airworthiness certification standard as a starting point for the tailoring process. The EASA E.Y013-01 has been amended regarding system safety to cover the class of very light aircraft (VLA) by Special Condition SC-RPAS.1309 [8], leaning on CS23.1309 [6]. This amendment was also adopted by Ampyx Power as a starting point for system safety.

The EASA E.Y013-01 provides guidance for restricted type certificates, as well as for standard type certificates for UAVs. However, it is not aimed at regulating public operations such as UAVs that are used by the military, police or firefighting department. Regarding mass criteria, EASA advises the NAAs of the member states to develop their own regulations for the UAVs which are lighter than 150 kg. As a consequence of this rule, current laws for light UAVs in the European countries are not harmonized and some of the countries do not yet have regulations.

EASA publishes the drafts of amendments on ICAO regulations as Notice of Proposed Amendment (NPA) in order to collect the comments of member states. In September 2014, the agency published the NPA-2014-09 with the first mention of operations of tethered aircraft [9]. In this notice, EASA identifies the tethering of the aircraft as a recognized mode of operation for remotely piloted aircraft

> **TAXONOMY OF OPERATIONS**
> RPA typical flight pattern may comprise a wide range of scenarios, which could be categorized in the following types of operations:
> (a) Very low level (VLL) operations below the minimum heights prescribed for normal IFR or VFR operations: for instance below 500 ft ($\approx$ 150 m) above ground level (AGL); they comprise:
> (1) **operations of tethered aircraft**;
> (2) Visual line of sight (VLOS) within a range from the remote pilot, in which the remote pilot maintains direct unaided visual contact with the RPA and which is not greater than 500 meters;
> (3) Extended visual line of sight (E-VLOS) where the remote pilot is supported by one or more observers and in which the remote crew maintains direct unaided visual contact with the RPA;
> (4) Beyond VLOS (B-VLOS) where neither the remote pilot nor the observer maintain direct unaided visual contact with the RPA.
> (b) **Operations of tethered aircraft**, above the minimum height in (a); ...

This statement can bring rigid wing AWE systems under Amendment 43 to Annex II of the Chicago Convention. Annex II covers other aspects related to RPAS besides their integration in airspace, namely the principles that RPAS shall be airworthy, the remote pilots licensed and the RPAS operator certified. However, specific ICAO standards and recommended practices—the SARPs—for the airworthiness and operation of RPAS as well as for licensing of the remote pilot have not been developed yet.

In addition to the EASA E.Y013-01 and NPA 2014-09, EASA has recently published a "Concept of Operations for Drones" [7]. This new proposal starts from the application rather than the aircraft used, applying a risk-based classification and regulation scheme for UAV operation. With this new scheme, EASA aims to cover a broad range of types and operations of UAVs, applying the three categories "Open", "Specific" and "Certified". Operations in the "Open" category would not require any certification as long as they operate in a defined boundary, for example not close to aerodromes, not in populated areas, being very small. The boundary conditions are not defined in the proposal but it is mentioned that conditions for the "Open" category are expected to be clarified in a collaboration with member states and industry. The "Specific" category is for UAVs whose conditions will not fit the "Open" category. These will require a risk assessment process specific to the planned operations. Depending on the output of the risk assessment process they might be certified case by case with specific limitations adapted to the operations. Permitting for the "Specific" category would be delegated to the NAAs. If the risk assessment shows that the UAV introduces a very high risk then the "Certified" category would be applicable. This requires multiple certificates similar to those for the manned aviation system, such as pilot licenses, approvals for design and manufacturer organizations. In addition to the certificates which are currently in use for manned aviation industry, the "Specific" category may also require new additional certifications that are specific to UAV operations, such as command and control link certification.

The operation-specific, case-by-case safety assessment method for the "Specific" category provides a mechanism to cover unconventional machines flying in civil airspace. If these machines have sufficient risk mitigation factors, such as being connected to the ground or being operated away from populated areas, an operation specific certificate could be sought. Current AWE applications would fall most likely into the "Specific" category, whereas utility-scale commercial systems would fall into the "Certified" category.

29.6.1.3 FAA Regulations for Unmanned Aerial Vehicles

The Title 14 of the Code of Federal Regulation [49] regulates the aeronautics and space operations conducted within the boundaries of USA. According to the current version [49, Part 91, Sect. 2031]

...every civil aircraft that operates in the US must have a valid airworthiness certificate.

Currently, unmanned aircraft systems can be certified by the FAA to operate in the national airspace (NAS) with a special airworthiness certificate in the experimental category [49, Part 21, Sect. 191]. However, FAA is regarding the aircraft as a part of a system, which includes command and control link, ground control systems and ground crew and accordingly, the entire system has to be certified. Nevertheless, the subsystems which do not exist in conventional aircraft, such as command and control links, ground control systems or sense and avoid systems, do not have any regulations yet. As a result, general use of commercial UAVs for civil use is highly restricted in US airspace at present.

The Title 14 of the Code of Federal Regulation (14 CFR) classifies the operation purpose of UAVs at a very high level [22]. In this classification, the first category is "Civil use", which refers to operation by a company or individual. The second category is "Public use", which includes the operations for scientific research and governmental purposes such as military operations. The last category is recreational use of model aircraft which is covered by FAA Advisory Circular 91-57 [16]. Currently, UAVs which are used for public operations require a Certificate of Waiver or Authorization (COA) from the FAA that permits public agencies and organizations to operate in a particular airspace. There are many COAs in use today by the several organizations, such as the Departments of Agriculture (USDA), Commerce (DOC), Defense (DOD), Energy (DOE), Homeland Security (DHS), Interior (DOI), Justice (DOJ) as well as NASA, State Universities and lastly State/Local Law Enforcement [51]. UAVs in the "Civil use" category can only operate with a special airworthiness certificate in the experimental category with limits on the operation to not create any risk for other airspace users or for people on the ground [49].

In February 2012, the United States Congress enacted the Federal Aviation Administration Reauthorization Legislation, which seeks to provide a framework for integrating UAVs safely into American airspace [48]. Following this action, the Next Generation Air Transportation System (NextGen) partner agencies, which are the Department of Transportation (DOT), DOD, DOC and DHS as well as NASA and FAA, started to work together to develop the Unmanned Aircraft Systems (UAS) Comprehensive Plan [50]. This report defines the interagency goals, objectives and approach to integrating UAS into the national airspace. Following the release of this report, FAA published a UAS roadmap [15] which includes a timeline for tasks required for integration of UAVs into the current aviation system. In accordance with this roadmap, FAA together with NexGen agencies established test sites for UAV research and development and studied new UAV-specific technologies such as detect-and-avoid systems.

While the FAA works on new regulations, the interim policy "Special Rules for Certain Unmanned Aircraft Systems" [48] has been enacted in 2012. Briefly, the Sect. 333 law authorizes the Secretary of Transportation to give a permit to civil operations of UAVs after an evaluation.

Regarding AWE applications, there is a discrepancy between EASA and FAA. On the one hand EASA recognizes the tethered aircraft as unmanned aircraft, on the other hand FAA clearly excludes the tethered aircraft from unmanned aircraft category [18, Appendix A];

41. Unmanned Aircraft (UA). A device used or intended to be used for flight in the air that has no onboard pilot. This device excludes missiles, weapons, or exploding warheads, but includes all classes of aircraft, helicopters, airships, and powered-lift aircraft without an onboard pilot. UA do not include traditional balloons (see 14 CFR part 101), rockets, tethered aircraft and un-powered gliders

In December 2011, the FAA had issued a "Notification for Airborne Wind Energy Systems" [17], according to which each deployment of an AWE system needs to be assessed on a case-by-case basis, accounting for the surrounding aviation environment to ensure aviation safety. Makani Power submitted a detailed response to this notification in February 2012 [21].

29.6.2 Regulations for Air Traffic Obstacle Category

Air navigation obstacles can be an impediment to civil air traffic. Some of the AWE companies registered their current AWE prototypes as air navigation obstacle (see Table 29.2). The aim of such a registration is to inform the aviation system to prevent incidents. For example, masts and wind turbines have to be registered as air traffic obstacles. This information is visualized in aviation charts and it is taken into account during flight route planning or emergency situations. If we consider the typical operation altitudes of AWE systems, obstacle registration might be sought in the future. ICAO defines "obstacle" in the Chicago Convention, Annex 4 [25] as follows

All fixed (whether temporary or permanent) and mobile objects, or parts thereof, that:
a) are located on an area intended for the surface movement of aircraft; or
b) extend above a defined surface intended to protect aircraft in flight; or
c) stand outside those defined surfaces and that have been assessed as being a hazard to air navigation.

According to this definition, air traffic obstacles can be mobile as many AWE systems are.

The Chicago Convention, Annex 14 [26] is about aerodromes and it includes the definition of the surrounding zones. Obstacle limitation surfaces are zones which have to be free of obstacles to permit regular civil use of the airspace. However, many AWE applications will potentially operate outside of these zones, about which the ICAO recommends to the civil aviation authorities the following in Annex 14 [26]

4.3 Objects outside the obstacle limitation surfaces
    4.3.1 Recommendation.— Arrangements should be made to enable the appropriate authority to be consulted concerning proposed construction beyond the limits of the obstacle limitation surfaces that extend above a height established by that authority, in order to permit an aeronautical study of the effect of such construction on the operation of aeroplanes.
    4.3.2 Recommendation.— In areas beyond the limits of the obstacle limitation surfaces, at least those objects which extend to a **height of 150 m or more above ground elevation** should be regarded as obstacles, unless a special aeronautical study indicates that they do not constitute a hazard to aeroplanes.

According to Annex 14, obstacles have to be conspicuous to air vehicles. Its Chap. 6 on "Visual aids for denoting obstacles" describes the required marking and lighting scheme for different types of obstacles. Regarding marking methods for increasing the visibility the following is recommended

> 6.1.2.2 Recommendation –Other objects outside the obstacle limitation surfaces should be marked and/or lighted if an aeronautical study indicates that the object could constitute a hazard to aircraft (this includes objects adjacent to visual routes e.g. waterway, highway).

Similarly, Article 6.2.2 defines marking requirements for mobile objects and Article 6.2.3 defines lighting requirements for objects with a height exceeding 150 m above ground. Article 6.2.4 addresses wind turbines separately, which is important because it defines the required marking for a wind farm setup. A similar or hybrid approach might be sought for future AWE farms.

29.6.3 Regulations for Tethered Gas Balloons Category

For static AWE systems that resemble the system developed by Altaeros Energies [53] a more applicable basis is the EASA certification specification for tethered gas balloons, CS-31TGB [5]. The lack of a complex control system which is required for the crosswind AWE systems, in conjunction with the self-stabilizing nature of a tethered lighter-than-air gas balloon will probably be sufficient to make CS-31TGB applicable.

We note here that the certification specification provides two more important inputs for the generic safety requirements and certification basis of AWE systems:

1. CS 31TGB.25 where the required tether safety factor of 3.5 is given.
2. AMC 31TGB.53(a) where it is stated that acceptable means of compliance to CS 31TGB.25(a) can be shown by a certificate of compliance to the Machinery Directive 2006/42/EC [11]. This means that a winch system can be certified to the Machinery Directive 2006/42/EC and thereby show compliance with an airspace certification specification. For AWE systems that use a winch as part of the ground station this can be important to limit the certification efforts for non-flying parts.

29.7 Discussion

Since no unified legal framework for AWE systems exists to the present day, the categories mentioned above are just starting points for a discussion with the authorities. They are a reference from which deviations can be defined systematically on a case-by-case basis. Nevertheless, we can derive some generally valid considerations.

AWE systems introduce potential hazards for other airspace users and people or critical infrastructure on the ground. These inherent risks have to be mitigated to successfully commercialize AWE technologies. It should be noted that this risk mitigation is not only sensible for saving lives, but also, from a commercial perspective, to reduce the costs resulting from accidents and crashes. It may well be a property of AWE that the commercial requirement for reliability is even more stringent than that coming from aviation regulations.[3]

If we define "normal operation" of the AWE system as the expected continuous operation within a limited airspace, with limited altitude and horizontal boundaries, we have to account for potential situations in which the AWE system interacts with the current civil aviation system. To prevent such undesirable interaction, regardless of the type of AWE system, some form of airspace segregation has to be arranged.

Furthermore, independent of the selected regulatory starting point, as UAV, obstacle or otherwise, and independent of the degree of permitting or certification sought, it will be fundamental that any risk of one or multiple fatalities as a result of a single functional failure is mitigated. The aviation approach to safe systems design is based on the presumptions that

- any single function can fail, so it must be assumed the tether can rupture, and
- any single failure with potential catastrophic consequence shall be demonstrably mitigated.[4]

---

[3] Consider, as an example, a fully autonomous utility-scale system that has a design lifetime of 20 years and is in operation 5000 hours per year. Suppose that the airborne element replacement cost represents 10% of the levelized cost of energy (LCOE). As a complex system, the airborne element may have 100 failure conditions that would lead to loss of the aircraft ("hazardous"). If any of those failure conditions occurs during the design lifetime, the energy cost would be driven up by 10%, say 0.5 eurocent per kWh, which is more than significant and will negatively affect the commercial viability. It is commonly argued that the probability of a failure condition that might lead to death of someone from the general public ("catastrophic failure") must be at least 10 times less than a hazardous failure, leading to a required probability level per catastrophic failure condition of 10^{-8} per flight hour (pfh), which is once every $5000 \times 20 \times 100 \times 10$ flight hours. This number is two orders of magnitude more stringent than the 10^{-6} pfh requirement of Special Condition SC-RPAS.1309 [8] regarding UAVs or Certification Specification CS-23.1309 [6, Paragraph 23.1309] regarding general aviation.

To make the argument more vivid, one can also turn it around. For general aviation, a catastrophic incident is accepted every 10,000 flight hours. Yet, this number of flight hours is reached every other year by a single utility-scale AWE system and every week for a park of 100 systems. This is clearly something the general public would not accept. Note that utility-scale AWE cannot be installed too far away from the population, since they are supposed to provide the population with electricity, and long-distance cabling cost is forbiddingly expensive, so part of the solution has to come from additional design for safety. Still, even with the 10^{-8} pfh reliability level calculated above, in a park of 100 systems, nearly every 2 months an aircraft would be expected to crash within the park, which hardly seems economically viable. So, a further reduction of the number of hazardous failure conditions and/or a further improvement in reliability, and accordingly in design rigor, may be recommendable for this example.

What sets utility-scale AWE systems apart from general aviation aircraft and typical RPAS is the number of flight hours and the complexity, which determines the number of failure conditions. The challenge is that AWE systems are in this regard more in the direction of commercial airliners, albeit not quite as critical or complex, and an intermediate reliability approach and design rigor is to be pursued.

[4] The certification requirement for catastrophic failure probability applies to accidental death of someone from the general public during commercial operation. This is not to be confused with

Thus, assuming that the commercial AWE system is operated near a populated area—the consumer of the generated electricity—the risk of uncontrolled flight outside of the designated safe zone shall be mitigated in case of tether rupture or intentional release of the aircraft. Having a controlled flight following a mechanical disconnection is one possible option to mitigate such an event. Having a second, structurally independent tether is another option. Or one could otherwise demonstrate that the detached kite is not able to reach people or critical infrastructure on the ground.

It should be noted that if one aims to operate a kite with significant kinetic energy directly above people, the tether solution alone cannot act as sufficient mitigation, for example in case of a faulty flight controller that would lead to a crash onto the populated area. It shall then be shown that there are independent means of overcoming a single failure of any flight control function.

Factors that will affect the authorities' assessment of the overall risk posed by the system furthermore include the kinetic energy, the availability of onboard propulsion, which determines the flight range, and the complexity, including autonomy, with which AWE aims to enter new territory.

Ampyx Power interprets the above review in such a way, that single-tether AWE systems that can still (aerodynamically) reach populated areas after tether failure or release are likely to be considered to be UAVs. Therefore, the certification approach for UAVs seems to be a suitable starting point and the level of certification will depend on the risk factor which the system presents [7]. A different approach, such as obstacle registration, may arguably be followed, for example for kites that are steered from the ground above a restricted area using two structurally independent tethers.

In any case, certification of design and operation to some defined standard will, in our view, be a necessity for commercial deployment. Apart from the expected positive impact that the introduction of rigorous processes will have on system reliability and maintenance, design certification enables the concept of similarity as evidence for quality and safety. This is a proven way to cost-effectively deploy the large numbers of complex systems that the AWE industry aspires to. This means that also production and maintenance aspects shall be standardized. These further certifications are outside the scope of this study. It should be noted, that we only

---

examples that may come to mind, such as the unfortunate recent SpaceshipTwo incident [39] that illustrate the higher level of acceptance for accidents during development affecting flight crew only. Secondly, the Certification Specification CS-23.1309 [6, Paragraph 23.1309] for mitigation of catastrophic failures applies to the functions of aviation systems, such as avionics, complex mechanisms, not to structures. For structures, it is recognized that redundancy could make the aircraft too heavy. The accepted approach there is to include the proper design safety factor and design for damage tolerance, for example, due to fatigue following barely visible tooling damage, hail, bird strike.

We argue that the tether is more than a structural element, but a functional part of a complex mechanism. It is used to control and restrict the dynamics of the airborne element, it is subject to wear during reeling, its integrity is affected by weather, subject to salt spray, dirt and lightning, it is subject to complex loading dynamics, such as jerks, shocks etc. At the same time, the tether is designed for minimal drag so the design safety factor may be limited. Hence we have to assume its incidental failure as part of a safety analysis.

considered so far the aviation-related risks and the regulation aspects of the AWE systems from an aviation perspective.

AWE systems are complex systems which consist of many components. There are additional regulations requirements, such as electric machinery regulations, grid connection regulations, noise emission regulations, environmental regulations and lighting regulations for the subcomponents which should be taken into consideration. It is noted here that those system elements and operations certified by an aviation authority are generally not required to comply also to machine standards, but these standards may be supporting guidance for the design or verification.

29.8 Conclusions

AWE systems have to be regulated for a successful commercial introduction and broad public acceptance. Ultimately, AWE systems are expected to be larger and heavier than current prototypes. They are expected to operate in Class G airspace where interaction with other airspace users is possible. In addition, AWE systems introduce risks to the people on ground. Therefore, it is expected that commercial AWE systems will have to comply with international airspace regulations.

The regulation framework for AWE systems is not yet mature. Current prototypes operate with special permits. These operation permits are issued by local aviation authorities and there is little commonality among the permits. Registration of the prototype as an air traffic obstacle or unmanned aerial vehicle (UAV) is the main approach followed by AWE companies and academic research groups. Classifying the AWE systems as UAV is a controversially discussed topic: on the one hand, current EASA view recognizes the tethered unmanned aircraft as UAV, on the other hand FAA excludes the tethered aircraft from the UAV category.

Each AWE system category has its own operation characteristics. The path for flight permitting and/or product certification goes through hazard analysis and mitigation independently from the category into which the system falls.

A regulation set which is specific to AWE systems will be built up over time, based on the specifically negotiated cases of first movers. As long as such a regulation is not in place, the most appropriate existing certification specifications and standards will have to be selected with authorities and tailored as necessary.

Lastly, yet importantly, AWE developers should accept the shared responsibility to avoid any incidents involving other airspace users, people on the ground or critical infrastructure. Such an incident, if no proper prevention or mitigation approach was in place, could well put the entire AWE industry under the most stringent aviation rules, which would jeopardize its commercial viability and eventual success.

Acknowledgements The financial support of the European Commission through the projects AMPYXAP3 (H2020-SMEINST-666793) and AWESCO (H2020-ITN-642682) is gratefully acknowledged.

References

1. Bormann, A., Ranneberg, M., Kövesdi, P., Gebhardt, C., Skutnik, S.: Development of a Three-Line Ground-Actuated Airborne Wind Energy Converter. In: Ahrens, U., Diehl, M., Schmehl, R. (eds.) Airborne Wind Energy, Green Energy and Technology, Chap. 24, pp. 427–437. Springer, Berlin Heidelberg (2013). doi: 10.1007/978-3-642-39965-7_24
2. Cherubini, A., Papini, A., Vertechy, R., Fontana, M.: Airborne Wind Energy Systems: A review of the technologies. Renewable and Sustainable Energy Reviews **51**, 1461–1476 (2015). doi: 10.1016/j.rser.2015.07.053
3. Croon, G. C. H. E. de, Groen, M. A., De Wagter, C., Remes, B., Ruijsink, R., Oudheusden, B. W. van: Design, aerodynamics and autonomy of the DelFly. Bioinspiration & Biomimetics **7**(2), 025003 (2012). doi: 10.1088/1748-3182/7/2/025003
4. European Aviation Safety Agency: Airworthiness Certification of Unmanned Aircraft Systems (UAS), Policy Statement EASA E.Y013-01, 25 Aug 2009. https://www.easa.europa.eu/system/files/dfu/E.Y013-01_%20UAS_%20Policy.pdf
5. European Aviation Safety Agency: Certification Specifications and Acceptable Means of Compliance for Tethered Gas Balloons, EASA CS-31TGB, 1 July 2013. https://www.easa.europa.eu/system/files/dfu/Annex%20to%20ED%20Decision%202013-011-R.pdf
6. European Aviation Safety Agency: Certification Specifications for Normal, Utility, Aerobatic, and Commuter Category Aeroplanes, EASA CS-23, 14 Nov 2003. https://www.easa.europa.eu/system/files/dfu/decision_ED_2003_14_RM.pdf
7. European Aviation Safety Agency: Concept of Operations for Drones. https://www.easa.europa.eu/system/files/dfu/204696_EASA_concept_drone_brochure_web.pdf. Accessed 9 May 2016
8. European Aviation Safety Agency: Equipment, Systems and Installations in Small Remotely Piloted Unmanned Systems (RPAS), EASA SC-RPAS.1309-01, July 2015. https://www.easa.europa.eu/system/files/dfu/SC-RPAS.1309-01_Iss01-public%20consultation.pdf
9. European Aviation Safety Agency: Transposition of Amendment 43 to Annex 2 to the Chicago Convention on remotely piloted aircraft systems (RPAS) into common rules of the air, EASA NPA 2014-09, 3 Apr 2014. https://www.easa.europa.eu/system/files/dfu/NPA%202014-09.pdf
10. European Commission: Commission Regulation (EU) No 748/2012 of 3 Aug 2012 laying down implementing rules for the airworthiness and environmental certification of aircraft and related products, parts and appliances, as well as for the certification of design and production organisations, 3 Aug 2012. http://eur-lex.europa.eu/eli/reg/2012/748/oj
11. European Parliament and Council of the European Union: Directive 2006/42/EC of the European Parliament and of the Council of 17 May 2006 on machinery, and amending Directive 95/16/EC, 17 May 2006. http://eur-lex.europa.eu/eli/dir/2006/42/oj
12. European Parliament and Council of the European Union: Regulation (EC) No 216/2008 of the European Parliament and of the Council of 20 Feb 2008 on common rules in the field of civil aviation and establishing a European Aviation Safety Agency, and repealing Council Directive 91/670/EEC, Regulation (EC) No 1592/2002 and Directive 2004/36/EC, 20 Feb 2008. http://eur-lex.europa.eu/eli/reg/2008/216/2013-01-29
13. Fagiano, L., Milanese, M., Piga, D.: High-altitude wind power generation. IEEE Transactions on Energy Conversion **25**(1), 168–180 (2010). doi: 10.1109/TEC.2009.2032582
14. Federal Aviation Administration: Aeronautical Information Manual. Official Guide to Basic Flight Information and ATC Procedures. (2015). http://www.faa.gov/air_traffic/publications/media/AIM.pdf
15. Federal Aviation Administration: Integration of Civil Unmanned Aircraft Systems (UAS) in the National Airspace System (NAS) Roadmap, 1st ed., 7 Nov 2013. https://www.faa.gov/uas/media/uas_roadmap_2013.pdf

16. Federal Aviation Administration: Model Aircraft Operating Standards, FAA Advisory Circular 91-57, 1 June 1981. https://www.faa.gov/documentLibrary/media/Advisory_Circular/91-57.pdf
17. Federal Aviation Administration: Notification for Airborne Wind Energy Systems (AWES), FAA-2011-1279, Dec 2011. https://www.gpo.gov/fdsys/pkg/FR-2011-12-07/pdf/2011-31430.pdf
18. Federal Aviation Administration: Unmanned Aircraft Systems (UAS) Operational Approval, FAA N 8900.227, 30 July 2013. https://www.faa.gov/documentLibrary/media/Notice/N_8900.227.pdf
19. Fritz, F.: Application of an Automated Kite System for Ship Propulsion and Power Generation. In: Ahrens, U., Diehl, M., Schmehl, R. (eds.) Airborne Wind Energy, Green Energy and Technology, Chap. 20, pp. 359–372. Springer, Berlin Heidelberg (2013). doi: 10.1007/978-3-642-39965-7_20
20. Glass, B.: A Review of Wind Standards as they Apply to Airborne Wind Turbines. In: Schmehl, R. (ed.). Book of Abstracts of the International Airborne Wind Energy Conference 2015, pp. 80–81, Delft, The Netherlands, 15–16 June 2015. doi: 10.4233/uuid:7df59b79-2c6b-4e30-bd58-8454f493bb09. Presentation video recording available from: https://collegerama.tudelft.nl/Mediasite/Play/90b60bc1e2bf44759ddc1b18185383791d
21. Hardham, C.: Response to the Federal Aviation Authority. Docket No.: FAA-2011-1279; Notice No. 11-07; Notification for Airborne Wind Energy Systems (AWES), Makani Power, 7 Feb 2012. https://www.regulations.gov/#!documentDetail;D=FAA-2011-1279-0014
22. Hayhurst, K. J., Maddalon, J. M., Morris, A. T., Neogi, N., Verstynen, H. A.: A Review of Current and Prospective Factors for Classification of Civil Unmanned Aircraft Systems. NASA TM-2014-218511, NASA Langley Research Center, Aug 2014. https://shemesh.larc.nasa.gov/people/jmm/NASA-TM-2014-218511.pdf
23. International Civil Aviation Organization: Adoption of Amendment 6 to Annex 7, ICAO State Letter AN 3/1-12/9, 4 Apr 2012. https://www.icao.int/Meetings/UAS/Documents/Adoption%20of%20Amendment%206%20to%20Annex%207.pdf
24. International Civil Aviation Organization: International Standards and Recommended Practices. Annex 2 – Rules of the Air, 10th ed., July 2005. https://www.icao.int/Meetings/anconf12/Document%20Archive/an02_cons%5B1%5D.pdf
25. International Civil Aviation Organization: International Standards and Recommended Practices. Annex 4 – Aeronautical Charts, 11th ed., July 2009
26. International Civil Aviation Organization: International Standards and Recommended Practices. Annex 14, Vol. 1 – Aerodrome Design and Operations, 6th ed., July 2013
27. International Civil Aviation Organization: Manual on Remotely Piloted Aircraft Systems (RPAS), ICAO 10019, Mar 2015
28. International Civil Aviation Organization: Unmanned Aircraft Systems (UAS), ICAO Circular 328-AN/190, Apr 2012. https://www.icao.int/Meetings/UAS/Documents/Circular%20328_en.pdf
29. Joint Authorities for Rulemaking on Unmanned Systems. http://jarus-rpas.org/ (2017). Accessed 1 Oct 2017
30. Kite Power Systems Ltd. http://www.kitepowersystems.com/. Accessed 4 Oct 2017
31. Kitemill AS. http://www.kitemill.no/. Accessed 16 July 2015
32. Kronborg, B., Schaefer, D.: eWind Solutions Company Overview and Major Design Choices. In: Schmehl, R. (ed.). Book of Abstracts of the International Airborne Wind Energy Conference 2015, pp. 32–33, Delft, The Netherlands, 15–16 June 2015. doi: 10.4233/uuid: 7df59b79-2c6b-4e30-bd58-8454f493bb09. Presentation video recording available from: https://collegerama.tudelft.nl/Mediasite/Play/748f1290e610439dab221365c521bdfd1d
33. Kruijff, M.: The Technology of Airborne Wind Energy – Part I: Launch & Land. https://www.ampyxpower.com/2017/04/1002 (2017). Accessed 10 Oct 2017
34. Kruijff, M.: The Technology of Airborne Wind Energy – Part II: the Drone. https://www.ampyxpower.com/2017/04/the-technology-of-airborne-wind-energy-part-ii-the-drone (2017). Accessed 10 Oct 2017

35. Kruijff, M.: The Technology of Airborne Wind Energy – Part III: Safe Power. https://www. ampyxpower.com/2017/05/the-technology-of-airborne-wind-energy-part-iii-safe-power (2017). Accessed 10 Oct 2017
36. Langley, W. R.: Go, Fly a Kite: The Promises (and Perils) of Airborne Wind-Energy Systems. Texas Law Review **94**, 425–450 (2015). http://www.texaslrev.com
37. Luchsinger, R. H. et al.: Closing the Gap: Pumping Cycle Kite Power with Twings. In: Schmehl, R. (ed.). Book of Abstracts of the International Airborne Wind Energy Conference 2015, pp. 26–28, Delft, The Netherlands, 15–16 June 2015. doi: 10.4233/uuid:7df59b79-2c6b-4e30-bd58-8454f493bb09. Presentation video recording available from: https://collegerama.tudelft.nl/Mediasite/Play/646b794e7ac54320ba48ba9f41b41f811d
38. Milanese, M., Taddei, F., Milanese, S.: Design and Testing of a 60 kW Yo-Yo Airborne Wind Energy Generator. In: Ahrens, U., Diehl, M., Schmehl, R. (eds.) Airborne Wind Energy, Green Energy and Technology, Chap. 21, pp. 373–386. Springer, Berlin Heidelberg (2013). doi: 10.1007/978-3-642-39965-7_21
39. National Transportation Safety Board: In-Flight Breakup During Test Flight Scaled Composites SpaceShipTwo, N339SS, Near Koehn Dry Lake, California October 31, 2014. NTSB/AAR-15/02, Washington, DC, USA, 28 July 2015. https://www.ntsb.gov/investigations/AccidentReports/Reports/AAR1502.pdf
40. North Atlantic Treaty Organization: Light Unmanned Aircraft Systems Airworthiness Requirements, NATO STANAG 4703 draft, 1st ed., Sept 2014
41. North Atlantic Treaty Organization: UAV Systems Airworthiness Requirements (USAR) for North Atlantic Treaty Organization (NATO) Military UAV Systems, NATO STANAG 4671 draft, 1st ed., Mar 2007
42. Northrop Grumman: RQ-4 Block 40 Global Hawk. http://www.northropgrumman.com/Capabilities/GlobalHawk/Documents/Datasheet_GH_Block_40.pdf. Accessed 16 July 2015
43. Pardal, T., Silva, P.: Analysis of Experimental Data of a Hybrid System Exploiting the Magnus Effect for Energy from High Altitude Wind. In: Schmehl, R. (ed.). Book of Abstracts of the International Airborne Wind Energy Conference 2015, pp. 30–31, Delft, The Netherlands, 15–16 June 2015. doi: 10.4233/uuid:7df59b79-2c6b-4e30-bd58-8454f493bb09. Presentation video recording available from: https://collegerama.tudelft.nl/Mediasite/Play/e51a679525fe491990de3a55a912f79d1d
44. Ruiterkamp, R., Salma, V., Kruijff, M.: Update on Certification and Regulations of Airborne Wind Energy Systems – The European Case for Rigid Wings. In: Schmehl, R. (ed.). Book of Abstracts of the International Airborne Wind Energy Conference 2015, pp. 78–79, Delft, The Netherlands, 15–16 June 2015. doi: 10.4233/uuid:7df59b79-2c6b-4e30-bd58-8454f493bb09. Presentation video recording available from: https://collegerama.tudelft.nl/Mediasite/Play/c8a9806aea024394a36cc35f9d6e98a81d
45. Schmehl, R. (ed.): Book of Abstracts of the International Airborne Wind Energy Conference 2015. Delft University of Technology, Delft, The Netherlands (2015). doi: 10.4233/uuid:7df59b79-2c6b-4e30-bd58-8454f493bb09
46. Sieberling, S., Ruiterkamp, R.: The PowerPlane an Airborne Wind Energy System. AIAA Paper 2011-6909. In: Proceedings of the 11th AIAA Aviation Technology, Integration, and Operations (ATIO) Conference, Virginia Beach, VA, USA, 20–22 Sept 2011. doi: 10.2514/6.2011-6909
47. Terink, E. J., Breukels, J., Schmehl, R., Ockels, W. J.: Flight Dynamics and Stability of a Tethered Inflatable Kiteplane. AIAA Journal of Aircraft **48**(2), 503–513 (2011). doi: 10.2514/1.C031108
48. United States Congress: FAA Modernization and Reform Act of 2012. 112th Congress (2011–2012), House Resolution 658, Became Public Law No 112-95, Feb 2012. http://www.gpo.gov/fdsys/pkg/BILLS-112hr658enr/pdf/BILLS-112hr658enr.pdf
49. United States Government: Title 14 Code of Federal Regulations – Aeronautics and Space, http://www.ecfr.gov/cgi-bin/text-idx?tpl=/ecfrbrowse/Title14/14tab%5C_02.tpl Accessed 29 May 2016

50. United States Next Generation Air Transportation System Joint Planning & Development Office: Unmanned Aircraft Systems (UAS) Comprehensive Plan, Washington, DC, USA, Sept 2013. http://purl.fdlp.gov/GPO/gpo42116
51. University of Washington Technology Law and Public Policy Clinic: Domestic Drones – Technical and Policy Issues. Clinic Policy Report, University of Washington, School of Law, 2013, pp. 1–20. https://www.law.washington.edu/clinics/technology/reports/droneslawandpolicy.pdf
52. Vander Lind, D.: Developing a 600 kW Airborne Wind Turbine. In: Schmehl, R. (ed.). Book of abstracts of the International Airborne Wind Energy Conference 2015, pp. 14–17, Delft, The Netherlands, 15–16 June 2015. doi: 10.4233/uuid:7df59b79-2c6b-4e30-bd58-8454f493bb09. Presentation video recording available from: https://collegerama.tudelft.nl/Mediasite/Play/639f1661d28e483cb75a9a8bdedce6f11d
53. Vermillion, C., Glass, B., Rein, A.: Lighter-Than-Air Wind Energy Systems. In: Ahrens, U., Diehl, M., Schmehl, R. (eds.) Airborne Wind Energy, Green Energy and Technology, Chap. 30, pp. 501–514. Springer, Berlin Heidelberg (2013). doi: 10.1007/978-3-642-39965-7_30
54. Windlift, Inc. http://www.windlift.com. Accessed 16 July 2015

Chapter 30
Life Cycle Assessment of Electricity Production from Airborne Wind Energy

Stefan Wilhelm

Abstract Renewable energies are superior to conventional electricity generating technologies in most environmental categories but are not completely free of environmental burdens. Especially when large-scale deployment is the goal, the effects of renewable energy use can have significant effects. As of now, there is no profound evaluation of the ecological aspects of airborne wind energy systems in the literature. By applying the life cycle assessment approach, this study investigates the global warming potential and cumulative energy demand associated with the production of 1 kWh of electricity from an AWE plant. In addition, the greatest global warming contributors and the energy payback time are evaluated and compared to conventional wind energy. For that purpose, energy and material flows of all life cycle processes, from exploitation of raw materials, manufacturing, assembly, transportation, installation, operation and maintenance to decommissioning and disposal, are analyzed. The study is based on a fictitious 1.8 MW airborne wind energy system including all required components up to connection to the electricity grid. As an example case, a generalized fixed wing aircraft with a ground-based generator is considered. Then, this system is compared to a conventional wind turbine of a similar power rating. This information can support system developers in an eco-friendlier system design and decision-makers in economy, public and politics to evaluate their support of this technology.

30.1 Introduction

The emerging airborne wind energy (AWE) technology is a promising contributor to the solution for meeting one of the world's greatest problems: that of global energy supply. Rising energy demands worldwide are expected, whereas availability

Stefan Wilhelm (✉)
ZIM network HWN 500, c/o VIP Innovation GmbH, Am Oberhafen 12, 13597 Berlin, Germany
e-mail: stefan.wilhelm@hwn500.de

© Springer Nature Singapore Pte Ltd. 2018
R. Schmehl (ed.), *Airborne Wind Energy*, Green Energy
and Technology, https://doi.org/10.1007/978-981-10-1947-0_30

of today's main energy source, fossil fuels, is depleting and is raising concerns about fossil fuels' potential to negatively affect the earth's climate.

Governments are in a position to set goals and the scope of the solutions by which to handle these issues. The German Bundestag (lower house of federal parliament) stipulated an energy concept in 2010 that comprises (1) the reduction of greenhouse gas (GHG) emission by at least 40% from 1990 to 2020, (2) a decrease in primary energy use by 20% in the same time span, (3) an increase in energy productivity of 2.1% per year related to final energy use, and (4) a share of renewable sources of electricity production of 35% by 2020 with a 15% increase every 10 years, reaching 80% by 2050. In 2015, the actual share of renewables in electricity generation in Germany was 30.0% [6], requiring intensive efforts in the coming years [13].

Renewable energy sources help achieve all of these goals and are therefore the focal point of many researchers' efforts in the energy field GHG emissions and the share of renewables are closely connected, since around half the emissions are caused by the electricity industry [14]. By using renewables for energy production, the use of fossil primary energy is reduced tremendously - but not entirely.

Wind power for example uses an abundant resource, and harnessing it causes practically no carbon dioxide emissions during operation. However, during its life cycle from manufacturing to disposal there is a quantifiable amount of GHG emitted. Related issues are material and energy use for manufacturing, rare metal and aluminum consumption, toxicity of lacquers, bird and bat death, and waste blade handling. Current limited availability of sites on land for wind turbine installations leads to construction offshore, where civil engineering efforts are higher and environment conditions are harsher. Considerable amounts of aluminum, zinc and other metals might be released from protective galvanic anodes [17].

Airborne wind energy is expected to pose an additional renewable technology that might overcome some of the problems with wind energy within a few years [42, 43]. Driven by the 1970's energy crisis, AWE was scientifically investigated [27], but only recently has AWE research gained significant ground with the availability of high performance and lightweight tether material, computational power and advanced control technologies [4, 28]. SkySails GmbH was founded in 2001, developing kite-based ship propulsion systems [35]. In 2006, Makani Power was founded in the US, developing fast flying airfoils to generate electricity with small turbines mounted on the wing [29]. Ampyx Power was founded in 2008 using a ground-based electricity generation system that it is currently the largest enterprise solely focusing on AWE in Europe [5]. More than 50 organizations in industry and academia are involved in research and development in the AWE field today. Accessing better and unused wind resources with considerably less material requirements, AWE appears beneficial from both an economic and ecological perspective.

The remarks above highlight that although renewables present an environmentally superior alternative in many aspects, they do have effects on the environment that should be assessed. With respect to the proceeding maturity and closer large-scale deployment of AWE technology, more sophisticated analyses in this area should be conducted. The life cycle assessment tool was chosen to achieve these goals because it allows holistic accounting of a certain environmental indicator through the entire life cycle of a product, from raw material production through manufacturing, installation, operation and maintenance, to decommission and disposal.

The life cycle assessment tool is first outlined in Sect. 30.2 and subsequently applied on an AWE plant, as described in Sect. 30.3, starting with the definition of the goal and scope of the assessment study. This includes the definition of the characteristics and boundaries of the studied system, the impact categories, and limitations. In Sect. 30.4 the life cycle inventory analysis, the data collection for all life cycle phases and a reference case is described. With that data, calculations for the life cycle impact assessment are executed, the results of which are presented in Sect. 30.5 for the AWE plant, analyzed in a sensitivity study and compared to other electricity generating technologies. Finally, the life cycle interpretation and conclusions are drawn in Sect. 30.6.

30.2 Life Cycle Assessment

The effects of technological activities on the environment are abundant. Acidification potential, metal depletion, eutrophication potential, global warming potential, human toxicity: these are just a few of the many impact categories that have been defined for life cycle assessments. The tool can be defined as a "compilation and evaluation of inputs, outputs and the potential environmental impacts of a product system throughout its life" [22]. That means that natural resources like energy and materials that are taken from the environment, as well as emissions to air, soil and water, are recorded over the entire life cycle and analyzed with respect to their effect on the environment. This enables a holistic analysis of selected impact categories throughout a product life from cradle to grave. As illustrated in Fig. 30.1 for an AWE power plant, this starts with the analysis of the extraction of natural resources for raw material and energy supply. The manufacturing phase requires further resources and thus, causes emissions. Transportation of all materials is taken into consideration, as well as (sub-) products and waste streams. Further exchanges occur during the long years of operation, including maintenance and replacement parts. Finally, end-of-life routes of the materials are studied. After decommissioning they can be disposed of in a landfill, energetically recovered via incineration plants, or they can take a route by which they are at least partially fed back to the raw material stream of its own lifecycle or that of a different product by recycling or even reuse.

However, since categories other than those selected are not considered, the tool does not allow for a general evaluation of environmental superiority. Neither can it serve for finally deciding over the sustainability of different products, since only one of the three pillars of sustainability—which consists of society, economy and environment—is part of a LCA.

Several sources, such as the ISO standards 14040 [22] and 14044 [23] or Guinée et al. [16] suggest a procedure for LCA in four interdependend stages:

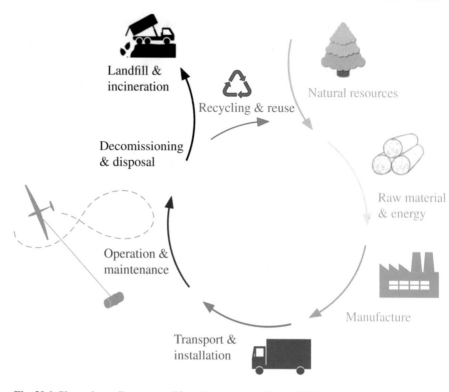

Fig. 30.1 Phases in cradle-to-grave life cycle assessment for an AWE power plant

- *Goal and scope definition* Explanation of the study frame including intentions, functional unit, system boundaries, impact categories, data quality, etc.
- *Inventory analysis (LCI)* Constellation of inputs and outputs of energy and material
- *Impact assessment (LCIA)* Selection of impact categories, category indicators and characterization model, as well as classification (assignment of the LCI results to the selected impact categories) and characterization (calculation of category indicator results)
- *Interpretation* Identification of significant issues, evaluation of completeness, sensitivity and consistency, as well as conclusions and limitations

Several software tools exist to support the implementation of an LCA. Umberto NXT LCA v7.0 [20] was used in this analysis to model the AWE system and calculate the material and energy flows as well as the LCIA results. A component tree with respective manufacturing and service processes, energy and material inputs and outputs and other necessary data was created and edited. The impact categories may be chosen and the factors to connect the streams to the categories are considered within the software.

LCA of wind power In conventional power plants, production and combustion of fossil fuels has a 10 times higher impact on GWP than construction of the infrastructure [25]. For renewable energies this "combustion" factor is missing, and construction of the plant therefore becomes a dominant factor. According to a study of a 5 MW offshore wind power plant, the rotor causes 20% of the categories' resulting GHG emissions, and the tower and nacelle 40% each; this may serve as a rule of thumb estimation [40]. Results, especially the absolute values, can vary significantly for the same technology and the same impact category. They are influenced by the system boundary and level of detail of the model, rated power and actual energy production, recycling scenario or whether the installation is offshore or onshore.

Table 30.1 compiles data for global warming potential (GWP) and cumulative energy demand (CED) adopted or calculated from different literature sources. The rated power of the turbines is between 1.5–3 MW. The GWP ranges from 5–45 gCO_2-eq/kWh and the CED ranges from 54–648 kJ-eq/kWh. The wind speed that is underlying the calculations is indicated, too. The results for the 3 MW turbine [8] at 7 and 8 m/s show how relevant the wind speed can be. Maintaining all other parameters, a wind speed reduction of 1 m/s yields 23% higher resulting CED values.

Source	Turbine power [MW]	Wind speed [m/s]	GWP [gCO₂/kWh]	CED [kJ/kWh]	Remarks
Guezuraga [15]	2.0	7.4 (at hub, ≈ 104 m)	9.7	118	Includes two gear replacements
	1.8	6 (at hub, ≈ 104 m)	8.8	116	Gearless nacelle
Kaltschmitt [24]	2.5	4.5 (at 50 m)	30	(128)	12.5 MW plant
	2.5	5.5 (at 50 m)	18	(77)	CED only from fossils
	2.5	6.5 (at 50 m)	13	(56)	
	1.5	4.5 (at 50 m)	31	(121)	7.5 MW plant
	1.5	5.5 (at 50 m)	18	(74)	CED only from fossils
	1.5	6.5 (at 50 m)	13	(54)	
Martinez [32]		n.d.	6.6	105	$CF = 22.8\%$
Marheineke [31]	1.0	5.5 (at 10 m)	21	(288)	$CF = 27\%$; backup and primary energy from wind excluded by recalculation
	1.0	6.5 (at 10 m)	45	(648)	
	1.0	4.5 (at 10 m)	15	(216)	
	1.5	5.5 (at 10 m)	19	(288)	
Siemens [33]	2.3	8.5 (at hub, ≈ 100 m)	5	76.6	46 MW plant
Enercon [11]	2.3	(medium)	8.7	100	
Vestas [8]	3.0	8 (at hub, ≈ 84 m)	7	120	$CF = 43\%$
	3.0	7 (at hub, ≈ 84 m)	8.61	148	100 MW plant
Vestas [41]	1.65	7.38 (at hub)	7.05	108	$CF = 40.8\%$, 300 MW plant

Table 30.1 Selected literature values for global warming potential and cumulative energy demand of wind turbines (CF denotes the capacity factor)

The differences in wind speed must be considered when comparing these values. It is remarkable that values for GWP stated by companies (Enercon, Vestas, Siemens) rank below the averages presented here, where capacity factors are positively estimated and plant sizes are rather high.

30.3 Goal and scope

The life cycle assessment was conducted in the context of the author's M.Sc. thesis [42] and preliminary results had been presented at the Airborne Wind Energy Conference 2015 [43]. The present chapter is based on this work.

30.3.1 Goal definition

The goals of this study are:

1. Determination of the environmental burden of electricity generation with AWE systems by means of contribution to global warming and consumption of primary energy resources.
2. Identification of the main contributors to the environmental burdens of AWE and potential for savings.
3. Determination of the time that the plant needs to be operated to recover the energy invested over the life cycle.
4. Assessment whether developing this new technology would lower global warming potential of electricity supply.

The results of this study might be used by the developing companies to assess their own systems for improvement of its carbon footprint, in certification process and marketing. The information could also support decision-makers in policy decisions (national and international), financing for strategic and planning purposes, and also the public might be interested in these particular aspects of the technology.

30.3.2 Function and functional unit

The functional unit denotes the quantified primary function that a product system fulfills and allows comparability between different product systems. All energy and material flows are assigned to the functional unit. As the main function of the investigated system is the generation of electricity, the functional unit in this study is defined as *1 kWh electricity that is delivered to the grid by an airborne wind energy plant as defined and operating under low wind conditions (IEC III)*. Wind classification is specified in DS/EN 61400-1:2005 for wind turbines, which differentiates low, medium and high wind class designations. Supplement E details wind classes further. IEC III conditions were chosen since it is a very conservative estimate. Compared to better wind sites, more facilities or more material are required to yield the same power. At the same time, it is a realistic scenario for early application of AWE since most sites with good wind conditions are not available and because, most likely, AWE plants will not be installed at the best wind sites until maturity of the technology is proven.

30.3.3 Investigated AWE plant

The investigated product system in this study is a fictitious AWE plant that generates electricity in utility scale, which is of a size comparable to other generation technologies. The AWE farm delivers 327.6 MW to the grid and comprises 182 facilities of 1.8 MW-rated (cycle) power each. This energy is expected at a wind speed of 7.4 m/s at "hub height", which is the average operation altitude in this case. The plant has a lifetime of 20 years. These facilities are not built nor are being developed in this exact design by any institution at the time of this writing. Against the background of the developments in the industry, the design choices are considered a possible dominating design in the future, or to be a conservative choice for future systems with respect to the study results. Further characteristics can be found in Table 30.2. The facilities are ground-based lift power generation systems of yo-yo type, cf. Chap. 26. Each of the 182 AWE facilities consist of five distinct component systems, as illustrated in Fig. 30.2.

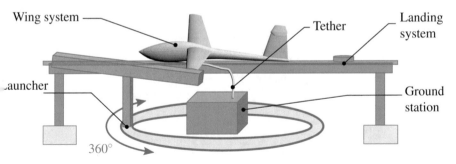

Fig. 30.2 Schematic drawing of the investigated AWE facility indicating the five distinct component systems

It was decided to separate the terms ground station, launch system and landing system, since the first is mainly state of the art technology and the others can vary considerably depending on design choices by the AWE developer. The foundation is part of the ground station and the landing system. Additionally, the facility has a share of the balance-of-station components:

- Internal cabling
- Energy storage (buffer)
- Hangar
- Control tower
- Transformer station
- External cabling (connection of the wind farm to grid)

30.3.4 System boundary

Following a cradle-to-grave approach, the product system and its boundaries for this analysis are illustrated in Fig. 30.3. Elementary input and output flows for the system are calculated from energy and material flows of the processes in the life cycle phases raw material acquisition, manufacturing, installation, operation and maintenance, and decommissioning and disposal.

Raw Materials and Manufacturing are merged into one phase for practical reasons. For some components, better data can be found when these two phases are considered combined. It is expected that raw materials and manufacturing are the major contributors to the selected impact categories. It is shown with LCAs that this is the case for conventional wind energy, and the case is assumed to be similar for AWE, cf. [8, 24, 34, 37]. Raw material acquisition includes the extraction and processing of resources from the natural environment for energy or material demand. Manufacturing includes production of main components of the AWE plant as

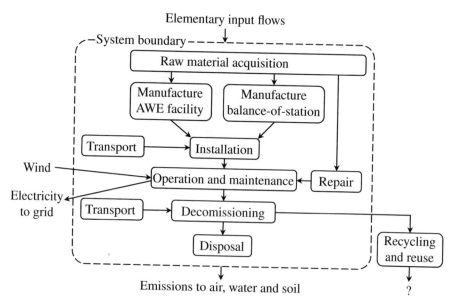

Fig. 30.3 Schematic representation of the product system "AWE plant" and system boundaries

well as intermediate materials and subcomponents; all of these together make up the product system. Production facilities are neglected in this study since the production of one AWE plant is assumed to be only an insignificant fraction of the lifetime production from capital equipment.

Installation includes transportation from a manufacturer to the installation site by a specific type of transport for a given distance. Installation also includes excavations for cabling on site. Possible road construction and other erecting efforts appear insignificant and are neglected.

Operation, maintenance and replacement includes energy for launches and steering of the wing systems, daily transportation of staff to the site, and petroleum consumption over the plant lifetime. Plant renovations and equipment replacement (including transportation of replaced equipment) over the plant's lifetime are included and assigned to this phase. Finally, electrical losses are considered within the plant, but losses in the grid on the way to the consumer are not included.

Decommissioning and disposal are modeled with the "cut-off rule". All material is assumed to go to landfilling for which only transportation of waste is considered. Over the long run, degrading of waste has an impact on waste and soil but as of now this is not easily possible to model with current LCA tools [41]. In addition, the effects of such degradation are considered to be insignificant in this case, since most decommissioned material is inert. Recycling and reuse are not included in the analysis. This is expected to be a conservative approach since recycled or reused material or energy might be credited to the product system. Energetic recovery from incineration of plastics or replacement of virgin steel by recycled steel, for example, could lower the calculated overall environmental burden significantly.

30.3.5 Data collection

The procedure for data collection starts with the creation of a component tree for the AWE plant. Data for material type and mass is defined for all items on the list along with information about the production process of each and associated emissions with respect to the selected impact categories.

For the standard components that are used, supplier data of specific sites is preferred, either from measurements or calculations. This is the case, for example, for a power transformer, generator and cabling. As it is unclear how AWE systems will be designed in the future, the gathered AWE-specific data is of a relatively general type. Data for AWE-specific components come from literature values of similar applications, expert consultations, or other types of sound estimations. Average production values, standard processes for intermediate works and typical characteristics are assumed. If not indicated, the respective data for material type and manufacturing are taken from the ecoinvent database [10].

30.3.6 Impact categories

To reflect the stated problems of energy industry (climate change and resource depletion) and, at the same time, to limit efforts for data collection, the following two impact categories have been chosen: 1) global warming potential over a 100 year perspective (GWP100a, in short GWP) and 2) the cumulative energy demand (CED) of primary energy resources.

Global warming, or climate change, is a consequence of increased radiative forcing on earth, enhanced by the ability of several gases in the atmosphere to reflect a part of the heat radiation to the earth. The intensity of contributing to this effect varies from gas to gas. When carbon dioxide is defined with factor 1, methane, as an example, has the factor 21, and sulphur hexafluoride 23,900 [21]. The concentration of those gases in the atmosphere is decisive for the magnitude of the global warming effect. This effect is implemented herein according to the CML2001 method [16].

Cumulative energy demand represents the "energy intensity" of a product. It can serve to assess and compare the demand for primary energy throughout the life cycle of a good or service to achieve a certain function. As of now, there is no standardized implementation of this method. For this study, the approach, as described in [19], is used as a basis but the dispersal into eight source categories (three for non-renewables, five for renewables) are summarized in one value in this study.

Another interesting number is the energy payback time EPT. It can be computed from the indicator result CED, the annual and lifetime energy production of the plant (AEP_{plant} and $EP_{\mathrm{plant,lt}}$) as

$$EPT = \frac{CED}{AEP_{\mathrm{plant}}} EP_{\mathrm{plant,lt}}, \qquad (30.1)$$

which is equivalent to the multiplication of CED and plant lifetime.

30.3.7 Key assumptions

The LCA tool can only support an environmental assessment of this technology, but it is not a decision maker on the overall superiority of a product since not all environmental aspects are covered in an LCA. This is even more relevant for statements about sustainability since the other pillars of sustainability - the social and economic aspects - not included in the assessment.

The relative nature of the LCA approach and the specific uncertainties should be considered when interpreting the results. Reliability of numbers, obtained with the LCA tool in comparative analyses, increase with similarity system boundaries, functional unit and data quality in both, reference case and analyzed case. The results depend strongly on design choices, estimated system performance, procedure of data collection and the system boundaries, including farm size and recycling scenario. One important assumption is the disposal by all material to landfilling. The resulting environmental impacts could be significantly lower if recycling of metals (possible to between 90% and 96%) and energetic recovery would be included and credited to the AWE plant. Only typical production averages for use of recycled material were assumed.

Long-term operation of actual AWE-systems in a real-world environment still has to be proven; Lifetime expectations and data sources of AWE-specific components, as well as performance estimations all contain significant levels of uncertainty. It may be possible that capacity factors, actual power output, full load hours, replacement frequency and other relevant parameters will be different in a commercial product and should be updated from this analysis accordingly.

30.3.8 Reference case

Since there is no case for LCA of AWE in literature, a conventional wind power plant with similar parameters was modeled in a similar approach for validation of the AWE model. That way, the results AWE model are known in relation to conventional wind power, which can be compared to literature. Characteristic parameters for both models are listed in Table 30.2. The Vestas V82 serves as a reference in this case [41].

Parameter	HAWT	AWES	unit
Rated facility power	1,650	1,800	kW
Installed generating power	1,650	2,500	kW
Capacity factor	40.8	41.07	%
Number of facilities	182	182	units
Rated plant power	300	328	MW
Farm efficiency	95	95	%
AEP (plant)	1,020	1,118	MWh
Plant lifetime prod.	20,393	22,356	MWh
Plant lifetime	20	20	yrs
Distance to grid	50	50	km
Distance between facilities	408	400	m
Average wind speed at respective hub height	7.38	7.4	m/s

Table 30.2 Characteristic facility and plant parameters of a towered and an airborne wind energy system

30.4 Life Cycle Inventory analysis

30.4.1 Data collection

Manufacturing The wing system is automatically steerable and has the purpose to generate aerodynamic lift. It comprises an unmanned glider-like carbon fiber reinforced polymer structure and an aramid core (modeled with nylon 6-6), equipped with actuators for steering, ram-air-turbine and battery for onboard electricity supply and auxiliary propulsion as well as several sensors and other components. The wing system's total weight is 2,500 kg.

The tethering is lightweight and transmits the aerodynamic forces as tensile forces to the ground. It consists of a coated ultra-high molecular weight polyethylene rope. One tether set has a weight of 495 kg.

The ground station handles the tether and converts mechanical power to electrical power. It consists of a winch (5,300 kg), gearbox, generator, converter (3,300 kg), and steel structure with foundation (6,700 kg). To achieve 1.8 MW cycle power, the generator has a rated power of 2.5 MW with a weight of 12,700 kg. At 21,600 kg, the gearbox is almost twice as heavy as the generator. The materials are mainly different kinds of steel, cast iron, concrete and copper.

The landing system allows safe retrieval of the wing system from any direction. It consists of a steel structure with wooden deck and a foundation (28,900 kg). Additional equipment for deceleration of the wing system is mounted on the deck (2,300 kg).

The launcher should launch the system independently of the wind speed and direction. It consists of a pneumatic catapult (5,400 kg, mainly steel) that can rotate 360 deg on a rail track (101,600 kg gravel and steel).

The investigated plant consumes a total of 249 tons of material per facility over the lifecycle, whereas 230 tons are for the facility manufacture, replacements and maintenance and the rest for its share in balance-of-station. The material of the defined product system is mainly gravel (32%), metals (42%), plywood (7%) and plastics (5%). Carbon fiber of the wings accounts for less than 1% of total material weight.

The data on materials and production processes are mainly taken from the ecoinvent database as described in the respective reports for metal processing [19], metals [7], plastics [18] and transport [36]. Production averages are considered for many component, which include also typical contents of recycled materials. Especially for metals in the AWE plant, the actual recyclability would likely be much higher than considered in this approach. In some cases specific data were chosen when a) data for standard components was available as for generator [1], converter [2], transformer [3] and cabling [38] or b) the database is outdated or does not cover a similar material or process as for CFRP (average of values from [9] and [39]) or the tether ([26] for fiber, own modelling for processing and [30] for coating).

Transport and installation For transportation efforts to the site, the mass of loads, distance, and type of transportation are considered. The transport effort is defined as product of the load and the distance the load traveled and has the unit tkm. It is modeled with respective ecoinvent activities with totals for the plant of:

- 2,181,687 tkm: market for transport, freight, lorry 3.5-7.5 metric tons
- 11,674,111 tkm: market for transport, freight, lorry 7.5-16 metric tons
- 181,864 tkm: transport, freight, lorry >32 metric tons
- 1,772,134 tkm: market for transport, freight train
- 1,636,106 tkm: market for transport, freight, sea, transoceanic ship

Since the entire installation of the AWE facility takes place at ground level, no lifting efforts are considered for this study. Only the laying of 182×400 m internal and 50,000 m external cabling is estimated with 100,400 MJ of diesel for excavating a bed with a cross section of $0.48\,\mathrm{m}^2$.

Operation, maintenance and replacement Launch energy is required every two days in the worst case scenario according to expectations of an AWE developer company. Using data from a potential catapult launcher supplier, the energy demand accumulates to 415,200 kWh over 20 years for 182 plants.

O&M over 20 years requires a total distance of 521,400 km traveled by a car weighing 2 tons.

Oil changes are required for maintenance of the gearbox, generator and motors, winch, and launcher. The resin molded transformer does not require oil changes. It is assumed that 500 kg oil must be replaced after 45,000 h which corresponds to 3 oil changes per facility over the 20-year lifetime, resulting in a total of 273,000 kg oil.

For replacement, the manufacturing of components according to their specific lifetime, as well as transportation to the site, are included. Per facility, the following parts are replaced over the plant's lifetime:

- Wing system (1×)
- Gearbox (1×)
- Plywood (landing deck, 8,600 kg)
- Li-Po Batteries (345 kg)
- Tether (2,050 kg)
- Converter (1×)
- Lubricating oil (1,500 kg)

Decommissioning and disposal As described in system boundaries, it was conservatively assumed that all material is landfilled, whereas no relevant emissions occur since the material is mostly inert (i.e. gravel and metals). Only dismantling and transport to the landfill are considered.

Dismantling of reinforced concrete is considered with an energy consumption of 264,800 MJ. For excavation of cabling bed, as for cable laying, 100,400 MJ are required. All other components are assumed to be loaded on trucks without significant effort or environmental impact.

Transportation from site to landfill is modeled for a transport distance of 400 km, which is assumed to also account for the unloaded distance traveled to the site. The transport effort is modeled using:

- 363,727 tkm: transport, freight, lorry >32 metric tons, EURO5
- 14,552,970 tkm: market for transport, freight, lorry 7.5-16 metric tons, EURO5
- 3,080,744 tkm: market for transport, freight, lorry 3.5-7.5 metric tons, EURO5

30.4.2 Conventional turbine

Components, weights and materials used for the LCA of the conventional wind power plant with the V82 turbines as described in [41] are also examined using data from supplier sources and, where possible, in the same way as described for the AWE plant. More modern wind turbines might have improved in some aspects since this supplier statement. However, the data is still considered as "beneficial" for the representation of the conventional wind power plant since its parts are widely state of the art, whereas processes for AWE are partly far from that and expected future improvements would introduce uncertainty into such a dataset.

30.5 Results and discussion

30.5.1 Indicator results

Cumulative energy demand The CED for electricity production with the AWE plant is 75.22 kJ-eq/kWh, which correlates to 2.1% of the lifetime energy production needed for manufacturing, operation and disposal.

The energy payback time can be calculated by simplifying Eq. 30.3, using the simulation results of the CED and the plant lifetime of 20 years as

$$EPT = CED \frac{1\,kWh}{3.6\,MJ} lifetime_{plant} \tag{30.2}$$

$$= 7.522 \times 10^{-2} \frac{1\,kWh}{3.6\,MJ} 20 \times 12\,months = 5.01\,months \tag{30.3}$$

This result means that the cumulative energy spent over the entire life cycle of the AWE plant is recovered within 5 months of operation.

Similarly, the energy yield ratio EYR, which indicates the ratio between energy that is produced by the plant over its lifetime and energy that is invested, can be calculated $EYR = 47.86$. It means that around 48 times more energy is produced by the AWE plant than it required for manufacturing, operation and disposal.

Global warming potential The overall indicator result in the impact category global warming potential over a 100 year perspective is 5.611 gCO_2-eq/kWh of electricity production with the AWE plant. The distribution over the life cycle phases and the main contributor groups are represented in Fig. 30.4. The GHG are caused

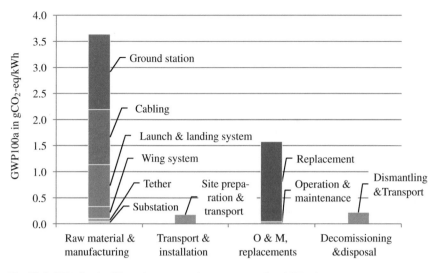

Fig. 30.4 GWP for the life cycle stages and components of an AWE plant

by 65% in raw material and manufacturing phase, 3% in installation and transport, 28% in operation, maintenance and replacements and 4% in end-of-life processes.

Besides the container term "replacement" (27%), main contributions come from the ground station (26%), cabling (19%) and launch and landing systems (14%); these will be analyzed in more detail. The contribution of the initially installed tether is minor, as well as contributions of the substation, installation, operation and maintenance, and disposal. The hangar and control tower cannot be displayed due to their extremely low contribution.

The gearbox (51%) and generator (29%) combined make up over 80% of the ground station in both impact categories and 21% in the overall result, not including replacements.

The gross contribution of cabling is mostly attributed to external cabling (94%).

In the chosen design, half of the GHG caused by the launch and landing systems come from landing deck or, specifically, 39% comes from plywood alone. The launcher and yaw system account for 21% and 18%, respectively.

The emissions from replacements are dominated again by the gearbox with 48%. The second biggest contributor here is tether material with 17%, which corresponds with an overall impact of 4.5%. Its initial share in mass increased over the plant lifetime from 0.2% to 1.5% due to replacements. The wing and plywood that are replaced once account for 15% and 10%, respectively, of all replacements.

When analyzing the results for different components, one cannot find a general correlation between mass and environmental impact. Fig. 30.5 shows selected components and their contribution to mass, CED, and GWP for one facility at installation. The three parameters are mostly strongly imbalanced. Tether, wing system and cabling have a comparatively low contribution to the consumed materials but high contributions to the selected impact categories. The tether contributes with 0.2% to

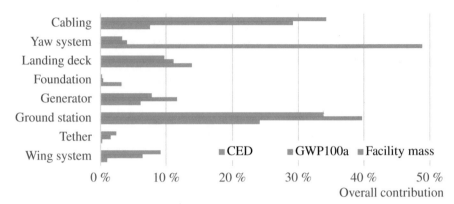

Fig. 30.5 Contribution of selected components to mass, CED and GWP of the facility as installed

the mass of a facility at installation but contributes with 6.1% and 8.7%, respectively, to GWP and CED. This is mainly due to the energy intensive process from granulate to fiber on the one hand and the low density of polymers on the other hand. In contrast, the foundation and yaw system in the chosen design have very low impacts compared to their mass. The yaw system consists mainly of gravel which is used as a resource that is little processed, and thus has a high mass but no strong impacts on environment. It contributes to 40.8% of the mass but only 2.6% to GWP and 2% to CED. The other big material components of the yaw system combined, concrete and steel, have a more balanced relationship between those parameters and reduce the discrepancy compared to gravel alone.

To better understand this finding, several materials are listed in Table 30.3 with their share in mass, GWP and CED, over the life cycle of the plant. The numbers show that, compared to their mass, impacts from gravel, plywood and reinforced concrete are low, but for the tether the behavior is contrary.

Table 30.3 Share in mass, GWP and CED over the plant life cycle for selected materials

Material	Mass	GWP	CED
Gravel	31.7%	0.2%	0.2%
Plywood	7.0%	2.0%	2.0%
Tether	1.0%	5.5%	8.1%
Reinforced concrete	13.1%	3.7%	0.3%

30.5.2 Sensitivity Analysis

In a sensitivity study several parameters are varied to show their respective effects on the overall result. Deviations from the baseline scenario, which is indicated in brackets, are analyzed for:

- Plant power output (327 MW)
- Frequency of wing launches (0.5 per day)
- Distance to the grid (50 km)
- Replacement of tether (lifetime lower part 2 years, upper part 10 years)
- Replacement of gearbox (all gearboxes replaced)

Figure 30.6 gives an overview over the sensitivity in GWP with respect to the baseline scenario. The impact of the tether as a single component was of special interest in this study. If the average tether lifetime is increased significantly, the overall result is not significantly affected. Reducing lifetime of its two sections from 2 and 10 years to 8 months and 2 years, respectively, GWP increases by 13%. It should be noted that the material today is not yet optimized specifically for AWE applications since comparable applications are not common. The lifetime expectation in the baseline scenario is considered achievable and a realistic value. Not only from an environmental, but also from an economic perspective, there is motivation to yield these values.

Plant power output could change due to several reasons, including: higher or lower actual wind speed at the site than predicted, downtimes due to legal requirements, or technical availability. This assessment is equivalent to changes in capacity factor. Only the power output is changed and not the design of the plant. If output changes +/- 20%, GWP changes by -17% and +25% respectively. The behavior of these numbers for a plant of smaller power rating strongly depends on the external

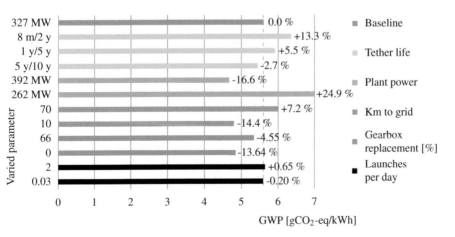

Fig. 30.6 Summary of baseline scenario and selected values from the sensitivity study

cabling. Whereas most significant parameters, such as internal cabling, energy storage, and transformer station, reduce nearly linearly with the number of facilities, external cabling might not. It is possible to reduce its cross section with a smaller transmitted power to a certain extent without significant transmission losses. In the reference study, a reduction from 182 to 30 facilities caused an increase in CED of only 8%, which can be expected to be similar for this study due to its similar product system [41].

Impacts are around 15% lower when distance to grid is lowered from 50 to 10 km. A reduction of cable length is considered for this analysis, while diameter and other parameters remain unchanged.

The overall impacts decrease linearly by approximately 13% when reducing the share of replaced gearboxes from all to none. Linear interpolation is possible for other shares. For this analysis, the production of gearboxes in the O&M life cycle phase, as well as transportation to the site and to disposal, are considered.

Energy for wing launches is irrelevant for the whole range of reasonable launch frequencies, from 0.03 to 2 per day. Only energy for launches is considered for this analysis. Effects like reduced energy yield, component lifetime and other parameters remain unchanged.

30.5.3 Comparison to other electricity generation technologies

An interesting comparison is to compare the mass of the analyzed AWE plants with a conventional wind energy plant. Table 30.4 lists the masses of comparative component groups of both technologies at installation and over the lifetime, including replacements. Considering the plant lifetime, the wing mass is about 10% of the blade mass, tether mass is around 2% of the tower mass and the ground station requires about 15% more material than the nacelle. The launch and landing system is only present in the AWE plant but the heavy foundation of the HAWT plant is already included in other AWE plant components. The balance of station is the same

HAWT	Mass [t]		Mass [t]		AWES
	at installation	in lifetime	at installation	in lifetime	
Rotor	42.2	2.2	2.1	4.5	Wing system
Nacelle	51	68	50.2	76.6	Ground station
Tower	136	136	0.5	2.5	Tethering
	(0)	(0)	138	147	Launch/Landing
Foundation	832	823	(0)	(0)	
Cabling	15.7	15.7	15.7	15.7	Cabling
Substation	0.8	0.8	0.8	0.8	Substation
Total	1,078	1,095	208	249	Total

Table 30.4 Masses of component systems of comparable conventional and airborne wind energy plants at installation and in a plant lifetime (data for horizontal axis wind turbine from [41])

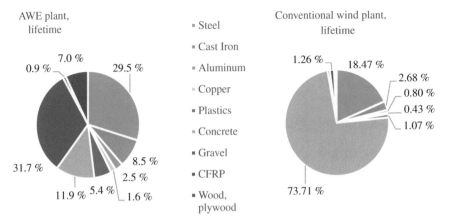

AWE plant, lifetime

- Steel
- Cast Iron
- Aluminum
- Copper
- Plastics
- Concrete
- Gravel
- CFRP
- Wood, plywood

Conventional wind plant, lifetime

Fig. 30.7 Composition of consumed materials of the AWE plant (left) and conventional wind energy plant (right) over the lifetime

for both plants. In total the AWE plant consumes 249 tons, or approximately 20%, of the material mass of the conventional wind energy plant.

The composition of these masses by the different material groups are presented in Fig. 30.7. Whereas the HAWT plant consists of around 75% of concrete and 25% of metals, the AWE plant consumption is only 12% concrete but over 40% metals, over 30% gravel and around 13% plywood, plastics and carbon fiber reinforced polymer (CFRP).

The category indicator results of the conventional wind energy plant model are 11.5 gCO$_2$-eq/kWh in GWP and a CED of 142.6 kJ-eq/kWh. The CED corresponds to an energy payback time of 9.5 months. The conventional wind power plant model is within the range found in literature [24, 31, 32] and with the values found in Table 30.1. Thus, the studied AWE plant consumes only 22.7% of the mass, causes 48.8% of the GWP and has only 54.9% of the CED of a conventional wind energy plant. If a different kind of yaw system were chosen, weight difference would be even much greater.

Figure 30.8 shows the results for those and further technologies. The comparison to other electricity generation technologies is of more limited reliability. The values for wind in the 1-3 MW range, lignite, and the electricity mix are taken from the respective ecoinvent models as they are in the database. The definitions of product systems and system boundaries are likely to be significantly different. However, it allows for a rough evaluation and better understanding of the results of AWE in relation to current technology. The GWP from the AWE plant compared to an average HAWT turbine of 1-3 MW size is approx. 3.7 times less. However, all considered wind power technologies are marginal compared to the electricity mix and even more compared to lignite, which had a 25.6% share in electricity production in Germany in 2014. GWP and CED of the AWE plant are less than 1% (0.87% and 0.74%, respectively) of that of the German electricity mix, which causes, on average, 644.2

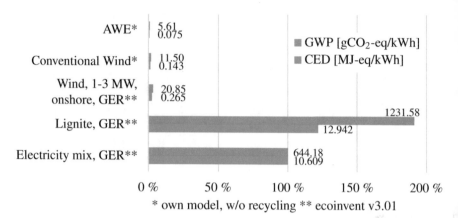

Fig. 30.8 Composition of consumed materials of the AWE plant (left) and conventional wind energy plant (right) over the lifetime

gCO_2-eq/kWh and requires 10,609 kJ/kWh [12]. It should be noted that there is a wide range of European grid supply. According to the dataset, the electricity mix in Norway is associated with 31 gCO_2-eq/kWh and 1040 in Poland [12].

30.6 Life cycle interpretation and conclusions

The study analyzes the global warming potential and cumulative energy demand associated with the generation of 1 kWh electricity with an airborne wind energy plant and provides the first numerical results for such an analysis. Using a 100-year perspective, the impact categories cumulative energy demand and global warming potential are selected in order to reflect two of the greatest challenges of energy industry: energy resource depletion and climate change.

The indicator result for CED is 75.22 kJ-eq/kWh. It can be derived that 2.1% of the lifetime energy production is needed for all life cycle phases, (I) raw material acquisition and manufacturing, (II) transport and installation, (III) operation, maintenance and replacements and (IV) end-of-life. Thus, the AWE plant generates 48 times more energy than it requires. The energy payback time is five months.

The indicator result for GWP100a is 5.611 gCO_2-eq/kWh. The emissions are mainly caused in phase (I) and (III) with 65% and 28%, respectively, whereas both are mainly due to manufacturing of the components for the first installation and replacements.

Over the entire life cycle, manufacturing of the generator and gearbox combined account for a considerable share of more than one third of the GWP, including replacements but not transport. This arises mainly from the provision of the necessary metals and can be improved by using secondary metals, for example, in structural parts. The external cabling accounts for 18% of the GHG emissions but cannot be significantly influenced or improved as long as distance to grid is constant. The specific landing deck in this study accounts for 10% over the lifetime, whereas more than half of it is caused by the plywood platform and its replacement. Especially material consumption can be much lower, when choosing a different launch and landing system. The study results suggest, even though a higher consumption of a certain material leads to higher environmental impacts, massive material consumption alone does not necessarily lead to smaller environmental impacts, but instead the choice of material type should be evaluated. Another savings potential that is AWE-specific rests in the tether, which accounts for 5.5% of the life cycle impacts. If the assumed lifetimes cannot be reached, the emissions are significantly higher. The sensitivity study shows that the actual power output of the plant or capacity factor has great influence on the results. Accordingly, performance improvements are a key for lowering indicator results. Frequency of launch and landing has no significant impact on the results. For most component systems, composition of GWP and CED are well correlated.

All results are rather overestimated due to conservative assumptions. However, the impacts from AWE-electricity are comparatively low. AWE seems to be a preferable option in terms of global warming and primary energy requirements for electricity production. Compared to a conventional wind energy plant that was modeled in comparable size and procedure, the AWE plant needed a 50% larger generator and gearbox. Nevertheless, it achieved a two times lower energy payback time of five months (compared to 9.5 months). The studied AWE plant consumes only 23% of the mass, causes 49% of the GWP, and consumes only 55% of the CED compared to a conventional wind turbine. Both wind power technologies, however, cause less than 1% of those impacts in total compared to German electricity mix.

The study allows a fostering of the understanding of the environmental implications associated with the use of airborne wind energy. Developers might use the outcomes of the study for consideration in an environmentally optimized system design at early stage. Decision-makers in industry, economics and politics might use the results for their evaluation whether to engage in this technology.

Acknowledgements The author would like to thank VIP Innovation GmbH and its network HWN500 for providing time to condense the original M.Sc. thesis into the present book chapter. Furthermore, the provision of data by Ampyx Power and the Endowed Professorship of Technical Textiles and Textile Mechanical Components at TU Chemnitz are highly appreciated. Special thanks for proof reading and valuable feedback go to the author's co-students Andrew C. Toth, Fahim Sadat and Adam Beck and to the unknown peer reviewers of this chapter.

References

1. ABB Automation: AC machine type AMG in the 500–5000 kVA power range. Environmental Product Declaration FREPD_001 rev. A, 2003. Retrieved from http://www.abb.com/abblibrary/DownloadCenter

2. ABB Oy: Drive^{IT} Low Voltage AC Drive ACS800 frequency converter, 630 kW power. Environmental Product Declaration 3AFE64726536 rev. B, 2003. Retrieved from http://www.abb. com/abblibrary/DownloadCenter

3. ABB Transformers AB: Power transformer TrafoStar 500 MVA. Environmental Product Declaration SEEPD_TPT_TrafoStar0001_1 rev. A, 2003. Retrieved from http://www.abb.com/ abblibrary/DownloadCenter

4. Ahrens, U., Diehl, M., Schmehl, R. (eds.): Preface. Airborne Wind Energy. Green Energy and Technology. Springer, Berlin Heidelberg (2013). doi: 10.1007/978-3-642-39965-7

5. Ampyx Power B.V. http://www.ampyxpower.com/. Accessed 6 Feb 2017

6. Arbeitsgemeinschaft Energiebilanzen e.V.: Energy Consumption in Germany in 2015. Report, Berlin, Germany, Mar 2015. http://www.ag-energiebilanzen.de/index.php?article_id=29& fileName=ageb_jahresbericht2015_20160523_engl.pdf

7. Classen, M., Althaus, H.-J., Blaser, S., Tuchschmid, M., Jungbluth, N., Doka, G., Faist Emmenegger, M., Scharnhorst, W.: Life Cycle Inventories of Metals. ecoinvent report No. 10, v2.1, Swiss Centre for Life Cycle Inventories, Dübendorf, Switzerland, 2009

8. D'Souza, N., Gbegbaje-Das, E., Shonfield, P.: Life Cycle Assessment Of Electricity Production from a Vestas V112 Turbine Wind Plant. Final Report, PE North West Europe ApS, 31 Jan 2011. https://www.vestas.com/~/media/vestas/about/sustainability/pdfs/lca_v112_study_report_2011.pdf

9. Duflou, J. R., Deng, Y., Van Acker, K., Dewulf, W.: Do fiber-reinforced polymer composites provide environmentally benign alternatives? A life-cycle-assessment-based study. MRS Bulletin 37(04), 374–382 (2012). doi: 10.1557/mrs.2012.33

10. ecoinvent Association: ecoinvent 3 database. http://www.ecoinvent.org (2013). Accessed 5 Feb 2015

11. Enercon GmbH: LCA of ENERCON Wind Energy Converter E-82 E2, Aurich, Germany, 2011

12. Frischknecht, R., Tuchschmid, M., Faist Emmenegger, M., Bauer, C., Dones, R.: Strommix und Stromnetz. In: Dones, R. (ed.) Sachbilanzen von Energiesystemen: Grundlagen für den ökologischen Vergleich von Energiesystemen und den Einbezug von Energiesystemen in Ökobilanzen für die Schweiz. ecoinvent report No. 6 Teil XVI, v2.0. Paul Scherrer Institut Villigen, Swiss Centre for Life Cycle Inventories, Dübendorf, Switzerland (2007). retrieved from http://www.ecoinvent.org

13. German Federal Ministry for Economic Affairs and Energy (BMWi): Energie in Deutschland – Trends und Hintergründe zur Energieversorgung. Brochure, Feb 2013. http://www.bmwi.de/DE/Mediathek/publikationen,did=251954.html

14. German Federal Ministry for the Environment, Nature Conservation, Building and Nuclear Safety (BMUB): Climate Protection in Figures – Facts, Trends and Incentives for German Climate Policy. Report, June 2014. http://www.bmub.bund.de/N52751-1

15. Guezuraga, B., Zauner, R., Pölz, W.: Life cycle assessment of two different 2 MW class wind turbines. Renewable Energy 37(1), 37–44 (2012). doi: 10.1016/j.renene.2011.05.008

16. Guinée, J. (ed.): Handbook on Life Cycle Assessment. Operational Guide to the ISO Standards. Eco-Efficiency in Industry and Science, vol. 7. Springer Netherlands (2002). doi: 10. 1007/0-306-48055-7

17. Heins, O., Krebs, T., Baumann, M., Binder, G.: Korrosionsschutz von Offshore-Windenergieanlagen. In: Proceedings of the HTG-Kongress 2011, Würzburg, Germany, 9 Sept 2011. http: //www.htg-online.de/Tagungsberichte.475.0.html

18. Hischier, R.: Life Cycle Inventories of Packaging and Graphical Papers. ecoinvent report No. 11, v2.0, EMPA St. Gallen, Swiss Centre for Life Cycle Inventories, Dübendorf, Switzerland, 2007. retrieved from http://www.ecoinvent.org

19. Hischier, R., Weidema, B., Althaus, H.-J., Bauer, C., Doka, G., Dones, R., Frischknecht, R., Hellweg, S., Humbert, S., Jungbluth, N., Köllner, T., Loerincik, Y., Margni, M., Nemecek, T.: Implementation of Life Cycle Impact Assessment Methods. ecoinvent report No. 3, v2.2, Swiss Centre for Life Cycle Inventories, Dübendorf, Switzerland, 2010. https://db.ecoinvent. org/reports/03_LCIA-Implementation-v2.2.pdf

20. ifu Hamburg GmbH: Umberto NXT LCA. http://www.umberto.de/en/versions/umberto-nxt-lca/. Accessed 9 Mar 2015
21. Intergovernmental Panel on Climate Change (IPCC): IPCC Second Assessment: Climate Chance 1995, 1995. https://www.ipcc.ch/pdf/climate-changes-1995/ipcc-2nd-assessment/2nd-assessment-en.pdf
22. International Organizational for Standardization: Environmental management – Life cycle assessment – Principles and framework, ISO Standard 14040:2006
23. International Organizational for Standardization: Environmental management – Life cycle assessment – Requirements and guidelines, ISO Standard 14044:2006
24. Kaltschmitt, M., Streicher, W., Wiese, A. (eds.): Erneuerbare Energien. Springer, Berlin Heidelberg (2006). doi: 10.1007/3-540-28205-X
25. Khan, F. I., Hawboldt, K., Iqbal, M. T.: Life Cycle Analysis of wind-fuel cell integrated system. Renewable Energy **30**(2), 157–177 (2005). doi: 10.1016/j.renene.2004.05.009
26. Louwers, D., Steeman, R., Meulman, J. H.: Sustainability: market trends, case study on carbon footprint for a vest made with Dyneema® and proposed waste solution. In: Proceedings of the Personal Armor Systems Symposium PASS, Cambridge, UK, 8 Sept 2014. Retrieved from http://www.dsm.com
27. Loyd, M. L.: Crosswind kite power. Journal of Energy **4**(3), 106–111 (1980). doi: 10.2514/3.48021
28. Loyd, M. L.: Foreword. In: Ahrens, U., Diehl, M., Schmehl, R. (eds.) Airborne Wind Energy, Green Energy and Technology. Springer, Berlin Heidelberg (2013). doi: 10.1007/978-3-642-39965-7
29. Makani Power Inc. http://www.makanipower.com. Accessed 4 July 2013
30. Mammitzsch, J.: Expert consultation, TU Chemnitz, Germany, 2015. http://www.innozug.de
31. Marheineke, T.: Life cycle assessment of fossil, nuclear and renewable electricity generation techniques. Ph.D. Thesis, University of Stuttgart, 2002. doi: 10.18419/opus-1576
32. Martínez, E., Sanz, F., Pellegrini, S., Jiménez, E., Blanco, J.: Life cycle assessment of a multi-megawatt wind turbine. Renewable Energy **34**(3), 667–673 (2009). doi: 10.1016/j.renene.2008.05.020
33. Siemens AG: A clean energy solution – from cradle to grave. Onshore wind power plant employing SWT-2.3-108. Environmental Production Declaration, 2015. Retrieved from http://www.energy.siemens.com
34. Singh, A., Pant, D., Olsen, S. I. (eds.): Life Cycle Assessment of Renewable Energy Sources. Green Energy and Technology. Springer London, London (2013). doi: 10.1007/978-1-4471-5364-1
35. SkySails GmbH: SkySails Antrieb für Frachtschiffe. http://www.skysails.info/deutsch/skysails-marine/skysails-antrieb-fuer-frachtschiffe/bedienung/ (2014). Accessed 30 May 2016
36. Spielmann, M., Dones, R., Bauer, C., Tuchschmid, M.: Transport Services. ecoinvent report No. 14, v2.0, Swiss Centre for Life Cycle Inventories, Dübendorf, Switzerland, 2007. https://db.ecoinvent.org/reports/14_transport.pdf
37. Stein, C.: Lebenszyklusanalyse einer Offshore-Windstromerzeugung. M.Sc.Thesis, University of Technology Hamburg, 2010
38. Südkabel GmbH: XLPE Power Cable Systems for High and Extra-High Voltages. Brochure, Mannheim, Germany, 2004. http://www.suedkabel.de/cms/upload/pdf/hoch_und_hoechstspannungskabelsysteme.pdf
39. Suzuki, T., Odai, T., Hukui, R., Takahashi, J.: LCA of Passenger Vehicles Lightened by Recyclable Carbon Fiber Reinforced Plastics. In: Proceedings of the International Conference on Life Cycle Assessment 2005, pp. 1–3, San Jose, Costa Rica, 25–28 Apr 2015. http://j-t.o.oo7.jp/publications/050425.pdf
40. Tryfonidou, R.: Energetische Analyse eines Offshore-Windparks unter Berücksichtigung der Netzintegration. Ph.D. Thesis, Ruhr University Bochum, 2006. http://nbn-resolving.de/urn:nbn:de:hbz:294-18549

41. Vestas Wind Systems A/S: Life cycle assessment of electricity produced from onshore sited wind power plants based on Vestas V82-1.65 MW turbines, Randers, Denmark, 29 Dec 2006. https://www.vestas.com/~/media/vestas/about/sustainability/pdfs/lca%20v82165%20mw% 20onshore2007.pdf

42. Wilhelm, S.: Life Cycle Assessment of Electricity Production from an Airborne Wind Energy System. M.Sc.Thesis, University of Technology Hamburg, 2015. doi: 10.15480/882.1302

43. Wilhelm, S.: Life Cycle Assessment of Electricity Production from an Airborne Wind Energy System. In: Schmehl, R. (ed.). Book of Abstracts of the International Airborne Wind Energy Conference 2015, p. 75, Delft, The Netherlands, 15–16 June 2015. doi: 10.4233/uuid: 7df59b79 - 2c6b - 4e30 - bd58 - 8454f493bb09. Presentation video recording available from: https://collegerama.tudelft.nl/Mediasite/Play/a64fec03f3d245599272db23155c9d381d

Author Index

© Springer Nature Singapore Pte Ltd. 2018
R. Schmehl, *Airborne Wind Energy*, Green Energy and
Technology, https://doi.org/10.1007/978-981-10-1947-0

Printed in the United States
By Bookmasters